EXAMINATION GUIDE
HOLT INTRODUCTORY ALGEBRA 2

Holt Introductory Algebra 1 and *Holt Introductory Algebra 2* represent a comprehensive program in first-year algebra that is designed to be studied over a two-year period. *Holt Introductory Algebra 2* can be used by students who complete at least the minimum course described on page M–54 of the *Teacher's Edition* of *Holt Introductory Algebra 1*.

The following is a list of key features of *Holt Introductory Algebra 2*. A description of each feature appears on the page indicated.

SKILLS LESSONS

Pivotal Exercises (p. M-2)

Classroom Exercises (p. M-2)

Written Exercises (p. M-2)

Mixed Practice (p. M-2)

Applications Exercises (p. M-2)

PROBLEM SOLVING AND APPLICATIONS LESSONS

These lessons are integrated throughout the book. That is, each *Problem Solving and Applications* lesson immediately follows the related skills lesson. (p. M-3)

THINKING SKILLS

In the Examples that appear in *Problem Solving and Applications* lessons, the focus is on thinking. (p. M-3) *Focus on Reasoning* topics, *non-routine* problems, and *critical thinking* exercises are aimed primarily at improving higher-order thinking skills. (p. M-4)

REAL-WORLD APPLICATIONS

Each chapter contains special topic applications that relate to real-world situations. These applications, like the *Problem Solving and Applications* lessons, present problem-solving strategies, develop decision-making skills, and introduce students to new content areas. (p. M-5)

REVIEW–MAINTENANCE–TESTING

Mid-Chapter Reviews (p. M-6)

Review Capsules (p. M-6)

Chapter Summary (p. M-6)

Chapter Reviews (p. M-6)

Cumulative Reviews (p. M-6)

USING MANIPULATIVES

Pages 539–569 contain fifteen manipulative lessons. Each lesson is referenced to the related textbook lesson that introduces a key algebraic concept. A set of manipulatives designed for the teacher's use in demonstrating the manipulative lessons on the overhead projector is also available.

Holt, Rinehart and Winston

Harcourt Brace Jovanovich

Austin • Orlando • San Diego • Chicago • Dallas • Toronto

HOLT
INTRODUCTORY
ALGEBRA 2
Annotated Teacher's Edition

Russell F. Jacobs

Holt, Rinehart and Winston

Harcourt Brace Jovanovich HBJ

Austin • Orlando • San Diego • Chicago • Dallas • Toronto

CONTENTS

Requests for permission to make copies of any part of the work should be mailed to:
Permissions Department, Holt, Rinehart and Winston, Inc., 8th Floor, Orlando, Florida 32887.

Material from earlier edition: Copyright © 1988, 1982 by Harcourt Brace Jovanovich, Inc.

Printed in the United States of America

ISBN 0-03-076986-8

2 3 4 5 6 7 8 9 041 98 97 96 95 94 93 92

DESCRIPTION OF THE PROGRAM

Textbook

Pages M-2 through M-6 describe, both verbally and pictorially, the features of the student textbook.

Using Manipulatives

These 15 lessons (see pages 539–569) provide a resource for teachers who wish to help students bridge the gap from the concrete to the abstract by modeling algebraic concepts.

Teacher's ResourceBank™

The *Teacher's ResourceBank*™ contains blackline masters, organized by chapter, that provide material for additional practice, review, and testing. The Teacher's ResourceBank includes:

- Review of Basic Algebra Skills
- Practice Worksheets
- Problem-Solving Exploration Lessons
- Quizzes
- Chapter Tests, Forms A and B
- Cumulative Tests
- Teaching Aids

Instructional Transparencies

The *Instructional Transparencies* package provides 57 colorful transparencies for teachers who use an overhead projector in the classroom. Teacher's Notes on the blackline-master version of the transparency offer suggestions for using the transparency and include objectives, a correlation to the related textbook lesson, and answers to problems on the transparency.

Annotated Teacher's Edition

The numerous aids that are provided in the margin of the annotated textbook pages are described on pages M-10 and M-11. The textbook page, which is reduced approximately ten percent, contains the objectives, commentary for the teacher, and answers to the exercises. Answers that cannot be annotated are listed under *Additional Answers* in the margin.

Teacher's Manual

The bound-in *Teacher's Manual* consists of three components: chapter overviews with section-by-section commentary, warm-up exercises for each lesson (see page M-11 for a description), and suggested assignment-guide charts for each chapter as well as a master timetable for the school year. (See page M-17.) An overview of the program is presented on pages M-12 through M-15.

The Lesson

Important ideas are highlighted.

Pivotal exercises involve the student in the learning.

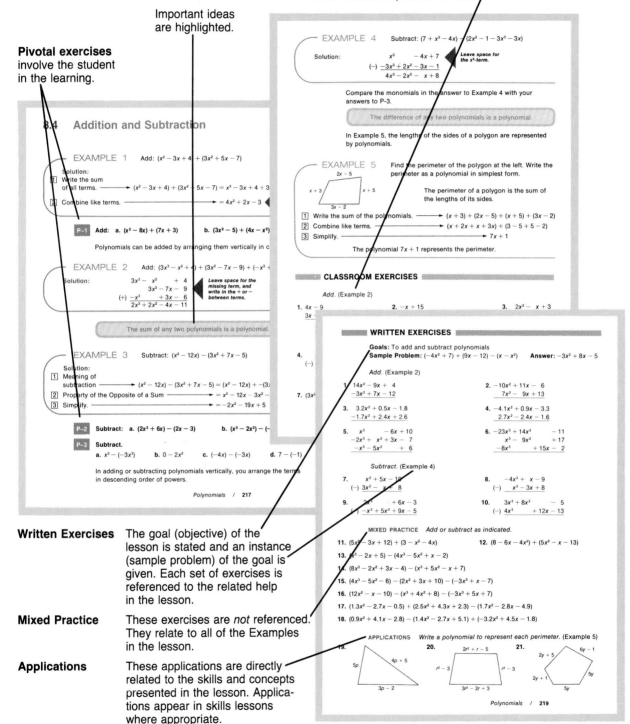

Classroom Exercises
Each set of exercises is referenced to the related help in the lesson.

EXAMPLE 4   Subtract: $(7 + x^3 - 4x) - (2x^2 - 1 - 3x^3 - 3x)$

Solution: x^3 $- 4x + 7$ *Leave space for the x^2-term.*
       $(-) \underline{-3x^3 + 2x^2 - 3x - 1}$
       $4x^3 - 2x^2 - x + 8$

Compare the monomials in the answer to Example 4 with your answers to P-3.

The difference of any two polynomials is a polynomial.

In Example 5, the lengths of the sides of a polygon are represented by polynomials.

EXAMPLE 5   Find the perimeter of the polygon at the left. Write the perimeter as a polynomial in simplest form.

$2x - 5$
$x + 3$ $x + 5$
$3x - 2$

The perimeter of a polygon is the sum of the lengths of its sides.

1. Write the sum of the polynomials. → $(x + 3) + (2x - 5) + (x + 5) + (3x - 2)$
2. Combine like terms. → $(x + 2x + x + 3x) + (3 - 5 + 5 - 2)$
3. Simplify. → $7x + 1$

The polynomial $7x + 1$ represents the perimeter.

3.4   Addition and Subtraction

EXAMPLE 1   Add: $(x^2 - 3x + 4) + (3x^2 + 5x - 7)$

Solution:
1. Write the sum of all terms. → $(x^2 - 3x + 4) + (3x^2 + 5x - 7) = x^2 - 3x + 4 + 3$
2. Combine like terms. → $= 4x^2 + 2x - 3$

P-1   Add: a. $(x^2 - 8x) + (7x + 3)$ b. $(3x^2 - 5) + (4x - x^2)$

Polynomials can be added by arranging them vertically in c

EXAMPLE 2   Add: $(3x^3 - x^2 + 4) + (3x^2 - 7x - 9) + (-x^3 + $

Solution: $3x^3 - x^2 + 4$ *Leave space for the missing term, and write in the + or − between terms.*
       $3x^2 - 7x - 9$
       $(+) \underline{-x^3 + 3x - 6}$
       $2x^3 + 2x^2 - 4x - 11$

The sum of any two polynomials is a polynomial.

EXAMPLE 3   Subtract: $(x^2 - 12x) - (3x^2 + 7x - 5)$

Solution:
1. Meaning of subtraction → $(x^2 - 12x) - (3x^2 + 7x - 5) = (x^2 - 12x) + -(3$
2. Property of the Opposite of a Sum → $= x^2 - 12x - 3x^2 - $
3. Simplify. → $= -2x^2 - 19x + 5$

P-2   Subtract: a. $(2x^2 + 6x) - (2x - 3)$ b. $(x^3 - 2x^2) - ($

P-3   Subtract.
   a. $x^3 - (-3x^3)$ b. $0 - 2x^2$ c. $(-4x) - (-3x)$ d. $7 - (-1)$

In adding or subtracting polynomials vertically, you arrange the terms in descending order of powers.

Polynomials / 217

CLASSROOM EXERCISES

Add. (Example 2)

1. $4x - 9$ 2. $-x + 15$ 3. $2x^2 - x + 3$
   $3x$

4.
   $(-)$

7. $(3x^2$

WRITTEN EXERCISES

Goals: To add and subtract polynomials
Sample Problem: $(-4x^2 + 7) + (9x - 12) - (x - x^2)$ **Answer:** $-3x^2 + 8x - 5$

Add. (Example 2)

1. $14x^2 - 9x + 4$ 2. $-10x^2 + 11x - 6$
   $\underline{-3x^2 + 7x - 12}$ $\underline{7x^2 - 9x + 13}$

3. $3.2x^2 + 0.5x - 1.8$ 4. $-4.1x^2 + 0.9x - 3.3$
   $\underline{-1.7x^2 + 2.4x + 2.6}$ $\underline{2.7x^2 - 2.4x - 1.6}$

5. $x^3 - 6x + 10$ 6. $-23x^3 + 14x^2 - 11$
   $-2x^3 + x^2 + 3x - 7$ $x^3 - 9x^2 + 17$
   $\underline{-x^3 - 5x^2 + 6}$ $\underline{-8x^3 + 15x - 2}$

Subtract. (Example 4)

7. $x^2 + 5x - 10$ 8. $-4x^2 + x - 9$
   $(-) \underline{3x^2 - x + 8}$ $(-) \underline{x^2 - 3x + 8}$

9. $3x^3 + 6x - 3$ 10. $3x^3 + 8x^2 - 5$
   $(-) \underline{-x^3 + 5x^2 + 9x - 5}$ $(-) \underline{4x^3 + 12x - 13}$

MIXED PRACTICE   *Add or subtract as indicated.*

11. $(5x^2 - 3x + 12) + (3 - x^2 - 4x)$ 12. $(8 - 6x - 4x^2) + (5x^2 - x - 13)$
13. $(x^3 - 2x + 5) - (4x^3 - 5x^2 + x - 2)$
14. $(8x^3 - 2x^2 + 3x - 4) - (x^3 + 5x^2 - x + 7)$
15. $(4x^3 - 5x^2 - 6) - (2x^2 + 3x + 10) - (-3x^3 + x - 7)$
16. $(12x^2 - x - 10) - (x^3 + 4x^2 + 8) - (-3x^3 + 5x + 7)$
17. $(1.3x^2 - 2.7x - 0.5) + (2.5x^2 + 4.3x + 2.3) - (1.7x^2 - 2.8x - 4.9)$
18. $(0.9x^2 + 4.1x - 2.8) - (1.4x^2 - 2.7x + 5.1) + (-3.2x^2 + 4.5x - 1.8)$

APPLICATIONS   *Write a polynomial to represent each perimeter.* (Example 5)

19.   20.   21.
$5p$ $4p + 5$ $2r^2 + r - 5$ $2y + 5$ $6y - 1$
          $r^2 - 3$ $r^2 - 3$ $2y + 1$ $5y$
$3p - 2$ $3r^2 - 2r + 3$ $5y$

Polynomials / 219

Written Exercises   The goal (objective) of the lesson is stated and an instance (sample problem) of the goal is given. Each set of exercises is referenced to the related help in the lesson.

Mixed Practice   These exercises are *not* referenced. They relate to all of the Examples in the lesson.

Applications   These applications are directly related to the skills and concepts presented in the lesson. Applications appear in skills lessons where appropriate.

Problem Solving and Applications Lesson

Word Rule/Formula
Formulas are introduced first with
words, then with symbols.

Strategies
The approach to problem solving used in
the program is an application of George
Polya's four-step method: **Understand,
Plan, Solve,** and **Check** (Look back).

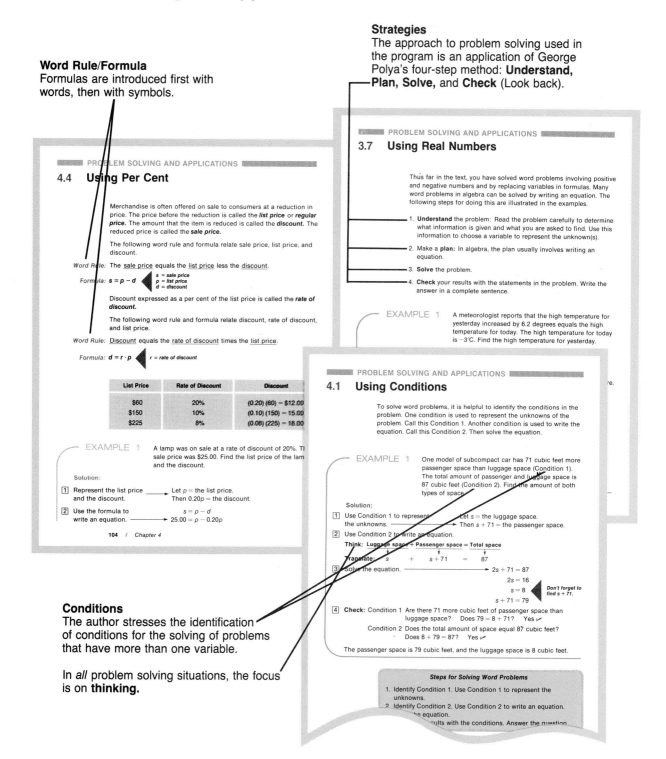

4.4 Using Per Cent

Merchandise is often offered on sale to consumers at a reduction in
price. The price before the reduction is called the **list price** or **regular
price.** The amount that the item is reduced is called the **discount.** The
reduced price is called the **sale price.**

The following word rule and formula relate sale price, list price, and
discount.

Word Rule: The <u>sale price</u> equals the <u>list price</u> less the <u>discount.</u>

Formula: $s = p - d$
$\quad s$ = sale price
$\quad p$ = list price
$\quad d$ = discount

Discount expressed as a per cent of the list price is called the **rate of
discount.**

The following word rule and formula relate discount, rate of discount,
and list price.

Word Rule: <u>Discount</u> equals the <u>rate of discount</u> times the <u>list price.</u>

Formula: $d = r \cdot p$ $\quad r$ = rate of discount

List Price	Rate of Discount	Discount
$60	20%	(0.20) (60) = $12.00
$150	10%	(0.10) (150) = 15.00
$225	8%	(0.08) (225) = 18.00

EXAMPLE 1 A lamp was on sale at a rate of discount of 20%. The
sale price was $25.00. Find the list price of the lamp
and the discount.

Solution:

1. Represent the list price → Let p = the list price.
   and the discount. Then $0.20p$ = the discount.

2. Use the formula to $s = p - d$
   write an equation. → $25.00 = p - 0.20p$

104 / Chapter 4

3.7 Using Real Numbers

Thus far in the text, you have solved word problems involving positive
and negative numbers and by replacing variables in formulas. Many
word problems in algebra can be solved by writing an equation. The
following steps for doing this are illustrated in the examples.

1. **Understand** the problem: Read the problem carefully to determine
   what information is given and what you are asked to find. Use this
   information to choose a variable to represent the unknown(s).

2. Make a **plan:** In algebra, the plan usually involves writing an
   equation.

3. **Solve** the problem.

4. **Check** your results with the statements in the problem. Write the
   answer in a complete sentence.

EXAMPLE 1 A meteorologist reports that the high temperature for
yesterday increased by 6.2 degrees equals the high
temperature for today. The high temperature for today
is $-3°C$. Find the high temperature for yesterday.

4.1 Using Conditions

To solve word problems, it is helpful to identify the conditions in the
problem. One condition is used to represent the unknowns of the
problem. Call this Condition 1. Another condition is used to write the
equation. Call this Condition 2. Then solve the equation.

EXAMPLE 1 One model of subcompact car has 71 cubic feet more
passenger space than luggage space (Condition 1).
The total amount of passenger and luggage space is
87 cubic feet (Condition 2). Find the amount of both
types of space.

Solution:

1. Use Condition 1 to represent Let s = the luggage space.
   the unknowns. → Then $s + 71$ = the passenger space.

2. Use Condition 2 to write an equation.

   Think: Luggage space + Passenger space = Total space

   Translate: $s + s + 71 = 87$

3. Solve the equation. → $2s + 71 = 87$
   $2s = 16$
   $s = 8$ *Don't forget to
   $s + 71 = 79$ find s + 71.*

4. Check: Condition 1 Are there 71 more cubic feet of passenger space than
   luggage space? Does $79 = 8 + 71$? Yes ✓
   Condition 2 Does the total amount of space equal 87 cubic feet?
   Does $8 + 79 = 87$? Yes ✓

The passenger space is 79 cubic feet, and the luggage space is 8 cubic feet.

Steps for Solving Word Problems

1. Identify Condition 1. Use Condition 1 to represent the
   unknowns.
2. Identify Condition 2. Use Condition 2 to write an equation.
   ... the equation.
   ... sults with the conditions. Answer the question

Conditions
The author stresses the identification
of conditions for the solving of problems
that have more than one variable.

In *all* problem solving situations, the focus
is on **thinking.**

Thinking Skills

Focus on Reasoning
Each *Focus On Reasoning* topic is designed to develop each student's ability to identify efficient strategies to solve problems.

In this lesson, the student is shown how to identify the **pattern** in a sequence of geometric figures.

In this lesson, the Example shows how to make a **comparison** of two quantities by using **logical reasoning.**

Focus on Reasoning: Geometric Patterns

Sometimes a sequence is a succession of geometric figures. Look for what changes and what remains the same. This will help you to identify the pattern.

EXAMPLE   What is the next figure?

Think: The figures are in pairs.
The pairs have the same shape.
The second figure of each pair is upside down.

Next figure:

EXERCISES

Identify the pattern. Then draw the next figure in each sequence.

1.
2.
3.
4.
5.
6.
7.
8.
9.
10.
11.
12.

Focus on Reasoning / 167

Focus on Reasoning: Comparisons

Many problems involving the **comparison of two quantities** can be solved by logical reasoning. Little or no computation with paper and pencil may be necessary.

Refer to these instructions for the Examples and Exercises.

Each problem consists of two quantities, one in Column I and one in Column II. Compare the two quantities.

* Write **A** if the quantity in Column I is greater.
* Write **B** if the quantity in Column II is greater.
* Write **C** if the two quantities are equal.
* Write **D** if there is not enough information to determine how the two quantities are related.

EXAMPLE 1	Column I	Column II
	$1800 \times 9 \times 32$	$18 \times 32 \times 900$

Solution: Since 32 appears in both products, compare (1800×9) and (18×900).
Since $(1800 \times 9) = 18 \times 100 \times 9$ and $(18 \times 900) = 18 \times 9 \times 100$, the products are equal.
Answer: **C**

EXERCISES

	Column I	Column II
1.	$5 \times 600 \times 3$	$500 \times 7 \times 3$
2.	$0.25 \times 800 \times 9$	$9 \times 25 \times 8$
3.	6% of 1010	1% of 606
4.	$\dfrac{(19)(7)(6)}{(32)(4)}$	$\dfrac{(2)(10)(42)}{(16)(8)}$
5.	$\dfrac{8}{8 - 7.9}$	$\dfrac{8}{8 - 7.99}$
6.	$(4)(6)(8)(2^5)$	$(2)(6)(16)(5^2)$

426 / Focus on Reasoning

Non-Routine Problems
These problems require the application of logic and deductive reasoning.

NON-ROUTINE PROBLEM

24. The figure at the right shows three different views of a cube.
   a. What shape is shown on that side of the cube which is opposite the side showing a circle?
   b. What shape is shown on that side of the cube which is opposite the side showing a triangle?

286 / Chapter 10

Critical Thinking
These problems are designed to develop and improve higher order thinking skills.

MORE CHALLENGING EXERCISES

21. If $\dfrac{1}{3c} - \dfrac{1}{6d} = 0$, how does the value of c compare with the value of d?

22. If $x^4 - \dfrac{1}{x} = \dfrac{x^2 - 1}{x}$, what is the value of a?

23. Write a rational expression that, when subtracted from $\dfrac{3x - 7}{2x + 3}$, gives a difference of $\dfrac{x - 5}{2x + 3}$.

24. Write a rational expression that, when added to $\dfrac{3}{x + 4}$, gives a sum of $\dfrac{7x + 5}{x^2 + 4x}$.

Sums and Differences of Rational Expressions / 305

M-4   *Teacher's Manual*

Real-World Applications

Each chapter contains special topic applications that relate to real-world situations.

This lesson demonstrates the use of **models** for developing the meaning of factoring.

Topics from **geometry** and **statistics** are integrated throughout the program.

Using a Geometric Model — Factoring

An algebraic method for solving a quadratic equation by **completing the square** was shown on pages 494–495. The following method of completing the square uses a **geometric model**. This method works for finding positive solutions only to quadratic equations of the form $ax^2 + bx + c = 0$ where $a > 0$, $b > 0$, and $c < 0$.

EXAMPLE: Solve $x^2 + 4x - 12 = 0$ by completing the square.

SOLUTION:

1. Write the equation in the form $x^2 + bx = -c$. ⟶ $x^2 + 4x = 12$

2. Represent x^2 as a square with sides of length x.

   Area: x^2

3. Represent $4x$ as two rectangles, each with an area of $\frac{1}{2}(4x)$, or $2x$.

   Position the rectangles as shown.

   Area of the L–shaped figure: $2x + x^2 + 2x$, or $x^2 + 4x$.

   From step 1, $x^2 + 4x = 12$.
   Thus, the area of the L–shaped figure is 12.

   Area: $x^2 + 4x$, or 12

4. Co...
   reg...

   Wi...
   rec...

   Wi...

   L–...

5. Sin...
   the...

   Th...

6. Ch...

Making Tables — Surface Area/Volume

An important goal in industrial production is to keep the amount of raw materials to a minimum. For example, a food processor wants containers that provide the desired **capacity** but with dimensions that require the least amount of material (**surface area**). The formula below gives S, the surface area of a cylindrical container in square centimeters.

$$S = \frac{(3.14r^3 + c)2}{r}$$

c = capacity in milliliters
r = radius in centimeters

EXAMPLE: Find, to the nearest whole number, the value of r that will require a minimum amount of material for a cylindrical can with a capacity of 354 milliliters.

SOLUTION: $S = \frac{(3.14r^3 + 354)2}{r}$

Use a calculator to complete the table shown at the right.

By calculator: When $r = 3.0$, $S = (3^3 \times 3.14 + 354) \times 2 \div 3$.

r	3.0	3.5	4.0	4.5
S	?	?	?	?

Solve also for $r = 3.5$, 4.0, and 4.5.

```
3 × = = × 3 . 1 4
+ 3 5 4 × 2 ÷ 3 =        292.52
```

In the table, the smallest value for S is 277.5. The corresponding value of r is about 4 centimeters.

r	3.0	3.5	4.0	4.5
S	292.5	279.2	277.5	281.5

EXERCISES

1. In the Example, show that 3.8 is the value of r to the nearest *tenth*. (HINT: Find the values of S for $r = 3.7$, 3.8, and 3.9)

2. In the Example, the formula for the height of the can is $h = \frac{c}{3.14r^2}$. Find, using $r = 3.8$, the value of h that requires a minimum amount of material.

112 / Chapter 4

The use of the **calculator** is integrated throughout the textbook as a problem-solving tool.

Using Statistics — Misleading Graphs

Statistical data is often presented in bar graphs. Graphs are an efficient, easily–read way to display information. However, graphs can be **misleading** in the way in which they display data. Special attention should be paid to the horizontal and vertical scales used in the graphs.

In the graph at the right, are the profits really going up as indicated by the rising line?

PROFITS

Dec. Nov. Oct. Sept. Aug. July June May

70   50   30   10
Thousands of Dollars

EXERCISES

The two bar graphs below show the number of stereos sold by two electronics stores for each of the past five years.

Woofer Stereo
Stereo Sales, 1983–1987
(Hundreds of Stereos)

Tweeter Electronics
Stereo Sales, 1983–1987
(Hundreds of Stereos)

1. Which company appears to have sold the greater number of stereos each year?

2. What is the value of each vertical unit on the graph for Woofer Stereo? for Tweeter Electronics?

3. How many stereos did Woofer Stereo sell in 1983? in 1984? in 1987?

4. How many stereos did Tweeter Electronics sell in 1983? in 1984? in 1987?

5. Did Tweeter Electronics sell more stereos than Woofer Stereo in any year from 1983 to 1987?

6. The graphs are the same size and are placed side by side. Do you think this encourages the consumer to compare the height of the bars without examining the scales? Explain.

7. Suppose the graphs were not side by side. Do you think Tweeter Electronics would still appear to have sold more stereos? Explain.

Sums and Differences of Rational Expressions / 317

Review

Mid-Chapter Review
This review appears at the approximate mid-point of each chapter. Each set of exercises is referenced to the related lesson.

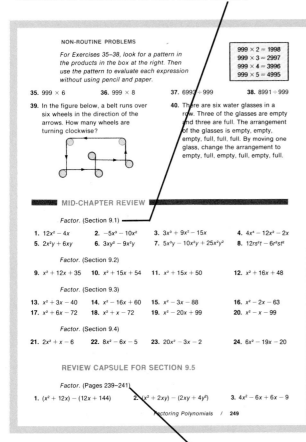

NON-ROUTINE PROBLEMS

For Exercises 35–38, look for a pattern in the products in the box at the right. Then use the pattern to evaluate each expression without using pencil and paper.

$$999 \times 2 = 1998$$
$$999 \times 3 = 2997$$
$$999 \times 4 = 3996$$
$$999 \times 5 = 4995$$

35. 999×6 **36.** 999×8 **37.** $6993 \div 999$ **38.** $8991 \div 999$

39. In the figure below, a belt runs over six wheels in the direction of the arrows. How many wheels are turning clockwise?

40. There are six water glasses in a row. Three of the glasses are empty and three are full. The arrangement of the glasses is empty, empty, empty, full, full, full. By moving one glass, change the arrangement to empty, full, empty, full, empty, full.

MID-CHAPTER REVIEW

Factor. (Section 9.1)

1. $12x^2 - 4x$ **2.** $-5x^3 - 10x^2$ **3.** $3x^3 + 9x^2 - 15x$ **4.** $4x^4 - 12x^2 - 2x$
5. $2x^2y + 6xy$ **6.** $3xy^2 - 9x^2y$ **7.** $5x^3y - 10x^2y + 25x^2y^2$ **8.** $12rs^2t - 6r^2st^2$

Factor. (Section 9.2)

9. $x^2 + 12x + 35$ **10.** $x^2 + 15x + 54$ **11.** $x^2 + 15x + 50$ **12.** $x^2 + 16x + 48$

Factor. (Section 9.3)

13. $x^2 + 3x - 40$ **14.** $x^2 - 16x + 60$ **15.** $x^2 - 3x - 88$ **16.** $x^2 - 2x - 63$
17. $x^2 + 6x - 72$ **18.** $x^2 + x - 72$ **19.** $x^2 - 20x + 99$ **20.** $x^2 - x - 99$

Factor. (Section 9.4)

21. $2x^2 + x - 6$ **22.** $8x^2 - 6x - 5$ **23.** $20x^2 - 3x - 2$ **24.** $6x^2 - 19x - 20$

REVIEW CAPSULE FOR SECTION 9.5

Factor. (Pages 239–241)

1. $(x^2 + 12x) - (12x + 144)$ **2.** $(x^2 + 2xy) - (2xy + 4y^2)$ **3.** $4x^2 - 6x + 6x - 9$

Factoring Polynomials / **249**

Chapter Summary
Important terms and important ideas are listed at the end of each chapter.

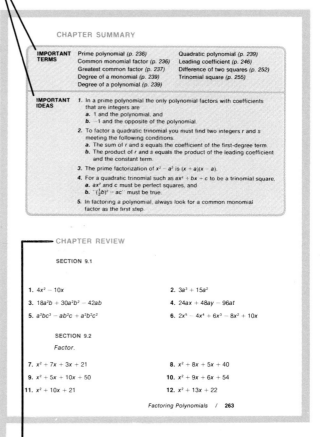

CHAPTER SUMMARY

IMPORTANT TERMS

Prime polynomial *(p. 236)*
Common monomial factor *(p. 236)*
Greatest common factor *(p. 237)*
Degree of a monomial *(p. 239)*
Degree of a polynomial *(p. 239)*

Quadratic polynomial *(p. 239)*
Leading coefficient *(p. 246)*
Difference of two squares *(p. 252)*
Trinomial square *(p. 255)*

IMPORTANT IDEAS

1. In a prime polynomial the only polynomial factors with coefficients that are integers are
   a. 1 and the polynomial, and
   b. −1 and the opposite of the polynomial.
2. To factor a quadratic trinomial you must find two integers r and s meeting the following conditions.
   a. The sum of r and s equals the coefficient of the first-degree term.
   b. The product of r and s equals the product of the leading coefficient and the constant term.
3. The prime factorization of $x^2 - a^2$ is $(x + a)(x - a)$.
4. For a quadratic trinomial such as $ax^2 + bx + c$ to be a trinomial square,
   a. ax^2 and c must be perfect squares, and
   b. $(\frac{1}{2}b)^2 = ac$ must be true.
5. In factoring a polynomial, always look for a common monomial factor as the first step.

CHAPTER REVIEW

SECTION 9.1

1. $4x^2 - 10x$ **2.** $3a^3 + 15a^2$
3. $18a^2b + 30a^2b^2 - 42ab$ **4.** $24ax + 48ay - 96at$
5. $a^2bc^3 - ab^2c + a^2b^2c^2$ **6.** $2x^5 - 4x^4 + 6x^3 - 8x^2 + 10x$

SECTION 9.2
Factor.

7. $x^2 + 7x + 3x + 21$ **8.** $x^2 + 8x + 5x + 40$
9. $x^2 + 5x + 10x + 50$ **10.** $x^2 + 9x + 6x + 54$
11. $x^2 + 10x + 21$ **12.** $x^2 + 13x + 22$

Factoring Polynomials / **263**

Review Capsule
Each *Review Capsule* reviews prior-taught skills and concepts that will be needed in the lesson that immediately follows. The exercises are referenced to the related pages.

Chapter Review
Each *Chapter Review* prepares the student for the formal chapter test. Two forms of each chapter test are included in the **Teacher's ResourceBank**.

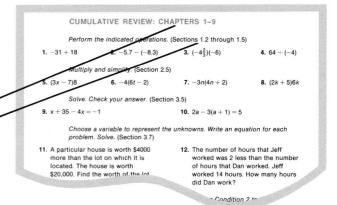

CUMULATIVE REVIEW: CHAPTERS 1–9

Perform the indicated operations. (Sections 1.2 through 1.5)

1. $-31 + 18$ **2.** $-5.7 - (-8.3)$ **3.** $(-4\frac{2}{3})(-6)$ **4.** $64 \div (-4)$

Multiply and simplify. (Section 2.5)

5. $(3x - 7)8$ **6.** $-4(6t - 2)$ **7.** $-3n(4n + 2)$ **8.** $(2k + 5)6k$

Solve. Check your answer. (Section 3.5)

9. $x + 35 - 4x = -1$ **10.** $2a - 3(a + 1) = 5$

Choose a variable to represent the unknowns. Write an equation for each problem. Solve. (Section 3.7)

11. A particular house is worth $4000 more than the lot on which it is located. The house is worth $20,000. Find the worth of the lot.

12. The number of hours that Jeff worked was 2 less than the number of hours that Dan worked. Jeff worked 14 hours. How many hours did Dan work?

Condition 2 to

Cumulative Review
This review appears after Chapters 3, 6, 9, 12, 15, and 19. Each set of exercises is referenced to the related section.

Tests

Chapter Tests
Two forms of each chapter test, Form A and Form B, are included.

NOTE: Each test item is coded to the related section (lesson) in the student textbook.

Name _____ Date _____ Score _____

CHAPTER 6   POWERS   FORM A

Write each product as one power.

(6.1)   1. $5^6 \cdot 5^2$ 1. _____

     2. $3^2 \cdot 3^4 \cdot 3^5$ 2. _____

     3. $a^7 \cdot a \cdot a^3$ 3. _____

Write each quotient as one power. No divisor equals 0.

(6.2)   4. $\dfrac{7^6}{7^2}$

     5. $(3.2)^{14} \div (3.2)^5$

     6. $\dfrac{x^{11}}{x^3}$

Solve each equation.

(6.1)   7. $a^3 \cdot a^4 = 5^7$

     8. $16^c \cdot 16^7 = 16^{11}$

(6.2)   9. $\dfrac{3^{15}}{3^a} = 3^6$

    10. $\dfrac{10^{7y}}{10^{3y}} = 10^{12}$

Write as one power. Simplify

Name _____ Date _____ Score _____

CHAPTER 6   POWERS   FORM B

Write each product as one power.

  $4^3 \cdot 4^4$ 1. _____

  $2^2 \cdot 2^6 \cdot 2^3$ 2. _____

  $m^2 \cdot m \cdot m^6$ 3. _____

Write each quotient as one power. No divisor equals 0.

  $\dfrac{10^7}{10^4}$ 4. _____

  $(4.6)^{10} \div (4.6)^3$ 5. _____

        6. _____

        7. _____

        8. _____

        9. _____

       10. _____

       11. _____

       12. _____

Name _____ Date _____ Score _____

CUMULATIVE TEST: CHAPTERS 1-3

Choose the correct answer. Write the letter of your choice in the answer column.

(1.1)   1. Simplify: $-|-6.3|$

     a. -6.3 b. 6.3 c. $|-6.3|$ d. $|6.3|$ 1. _____

(1.2)   2. Add: $(-17.5) + 12.4$

     a. -29.9 b. 29.9 c. -5.1 d. 5.1 2. _____

     3. Add: $-32 + 45$

     a. -79 b. 79 c. -13 d. 13 3. _____

(1.3)   4. Subtract: $28 - 34$

     a. -6 b. 6 c. 62 d. -62 4. _____

     5. Subtract: $-1.5 - 3.6$

     a. -2.1 b. -5.1 c. 5.1 d. 2.1 5. _____

(1.4)   6. Multiply: $(-4)(36)$

     a. -154 b. 144 c. -144 d. 154 6. _____

     7. Multiply: $-\dfrac{4}{5}(-45)$

     a. -36 b. 36 c. 55 d. -55 7. _____

(1.5)   8. Divide: $-12.5 \div (0.5)$

     a. -25 b. -6.25 d. 25 8. _____

Cumulative Tests
Each *Cumulative Test* is presented in a **multiple-choice format.** The six *Cumulative Tests* cover Chapters 1-3, 4-6, 7-9, 10-12, 13-15, and 16-19.

Practice Worksheets

The practice worksheets (see the *Teacher's ResourceBank* on page M-9) provide additional exercises and examples for each section (lesson) in Chapters 1–19.

NOTE: Each page is clearly referenced to the related section by title and pages.

Instructional Transparencies

This package contains 57 overhead transparencies. A blackline-master version of each transparency is also included for those who wish to have students work at their desks as a transparency is shown on the overhead projector.

NOTE: Each transparency is clearly referenced to the related section by title and pages.

2.6 COMBINING LIKE TERMS   (Pages 50-51)

Simplify.

Example: $-3a^2 + 4a + 5a^2 + 2a = 2a^2 + 6a$

1. $2x + 13x$ _____

2. $5b^2 - 10b^2$ _____

3. $-3a + 4 + 8a$ _____

4. $6x^2 - 3y + 2x^2 + y$ _____

5. $17m - 2n + 3m + n$ _____

6. $15b^2 - 6a^2 - 4b^2$ _____

7. $-r + 9r - 2r$ _____

8. $12p^2 - 3p^2 + 10p^2$ _____

9. $\frac{1}{2}y + 2x + \frac{5}{2}y$ _____

10. $0.5d + 2.1b - 0.3d$ _____

11. $\frac{2}{3}a^2 - \frac{1}{5}b + \frac{1}{3}a^2 + \frac{4}{5}b$ _____

12. $3\frac{1}{2}c - 4d - 1\frac{1}{2}c + d$ _____

13. $16m^2 - 2m + 3m^2 + 8 + 9m$ _____

14. $ab^2 + 3ab - 5ab^2 - 6a + 10ab$ _____

15. $-3q + 16q^2 + q - 5q^3 - 10q^2 + 7q$ _____

16. $1.3n - 0.8p + 6 + 0.7n - 0.2p - 10$ _____

17. $\frac{1}{3}b + \frac{1}{4}c - \frac{2}{3}b - \frac{1}{2}c - b$ _____

18. $14a^2b - 2ab^2 + 3a^2b - ab + 7ab^2$ _____

19. $-a - b + 3a - 10b + 4c - 11b - 2c$ _____

20. $-10x^2 - 3x + 2 + 4x^2 - 3 - 5x$ _____

21. $2y - 3y^2 - 7y + 9y^2 - 11y$ _____

22. $6b^2 - 5 - 7a^2 - 7b^2 + 1 - b^2$ _____

W-13

---

19 The Rule of Pythagoras

HOLT INTRODUCTORY ALGEBRA 2
Page 198

FIGURE	COMPUTATION
1.	$c^2 = a^2 + b^2$
	$c^2 = $ _____
	$c^2 = $ _____
	$c^2 = $ _____
	$c = $ _____
	$c = $ _____
2.	$c^2 = a^2 + b^2$
	_____
	_____
	_____ $= b^2$
	_____ $= b^2$
	_____ $= b$
	_____ $= b$

Teaching suggestions for each transparency can be found on the back of the corresponding blackline master.

Teacher's ResourceBank

The *Teacher's ResourceBank* contains the following resources which will be especially helpful in accommodating students of varying interests and ability levels.

1. Review of Basic Algebra Skills
2. Practice Worksheets
3. Problem-Solving Explorations
4. Teaching Aids
5. Quizzes

6. Chapter Tests, Form A
7. Chapter Tests, Form B
8. Cumulative Tests
9. Answer Sections

Quizzes
Two *Quizzes* are provided for each chapter. Each *Quiz* is referenced to the lessons being tested.

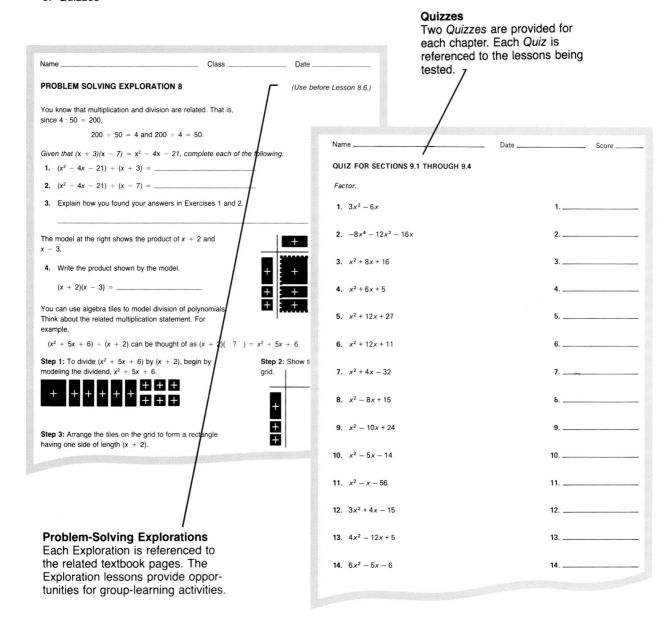

Name _____ Class _____ Date _____

PROBLEM SOLVING EXPLORATION 8   *(Use before Lesson 8.6.)*

You know that multiplication and division are related. That is, since $4 \cdot 50 = 200$,

$$200 \div 50 = 4 \text{ and } 200 \div 4 = 50.$$

Given that $(x + 3)(x - 7) = x^2 - 4x - 21$, *complete each of the following.*

1. $(x^2 - 4x - 21) \div (x + 3) =$ _____

2. $(x^2 - 4x - 21) \div (x - 7) =$ _____

3. Explain how you found your answers in Exercises 1 and 2.

_____

The model at the right shows the product of $x + 2$ and $x - 3$.

4. Write the product shown by the model.

   $(x + 2)(x - 3) =$ _____

You can use algebra tiles to model division of polynomials. Think about the related multiplication statement. For example,

$(x^2 + 5x + 6) \div (x + 2)$ can be thought of as $(x + 2)(\quad ? \quad) = x^2 + 5x + 6$.

Step 1: To divide $(x^2 + 5x + 6)$ by $(x + 2)$, begin by modeling the dividend, $x^2 + 5x + 6$.

Step 2: Show th grid.

Step 3: Arrange the tiles on the grid to form a rectangle having one side of length $(x + 2)$.

Name _____ Date _____ Score _____

QUIZ FOR SECTIONS 9.1 THROUGH 9.4

Factor.

1. $3x^2 - 6x$ 1. _____

2. $-8x^4 - 12x^3 - 16x$ 2. _____

3. $x^2 + 8x + 16$ 3. _____

4. $x^2 + 6x + 5$ 4. _____

5. $x^2 + 12x + 27$ 5. _____

6. $x^2 + 12x + 11$ 6. _____

7. $x^2 + 4x - 32$ 7. _____

8. $x^2 - 8x + 15$ 8. _____

9. $x^2 - 10x + 24$ 9. _____

10. $x^2 - 5x - 14$ 10. _____

11. $x^2 - x - 56$ 11. _____

12. $3x^2 + 4x - 15$ 12. _____

13. $4x^2 - 12x + 5$ 13. _____

14. $6x^2 - 5x - 6$ 14. _____

Problem-Solving Explorations
Each Exploration is referenced to the related textbook pages. The Exploration lessons provide opportunities for group-learning activities.

Annotated Teacher's Edition

Pre-Lesson Activities
Two types are included:

1. Warm-Up Exercises
2. Maintenance

The *Maintenance* exercises help to keep alive prior-taught skills and concepts that are not related to the lesson.

Lesson Resources
The supplementary material that is available for the lesson is listed.

Common Error
This feature states an error commonly made by students, gives a specific instance *(Example)* and a *Prescription* for correcting the error.

OBJECTIVES: To solve and check an equation with the variable on both sides
To solve and check an equation with the variable on both sides but requiring use of the Distributive Property at the first step

Teaching Suggestions p. M-24

Lesson Resources
Warm-Up: p. M-25
Maintenance: See below
Manipulative Activity 7
Visuals 5 and 12
Skills Practice: p. IV-23

Maintenance
1. Multiply: $(-1)(-106)(\frac{1}{2})$
   Ans: 53 (Section 1.4)
2. Simplify: $3x - (y - 2)$
   Ans: $3x - y + 2$
   (Section 2.4)
3. Simplify: $-20x^2 + 17x^2$
   Ans: $-3x^2$ (Section 2.6)
4. Solve and check:
   $-4y = 12.8$
   Ans: $y = -3.2$
   (Section 3.2)
5. The perimeter of a bumper sticker is 30 inches. The width is 3 inches. Find the length.
   Ans: 12 inches
   (Section 3.4)

Additional Examples
Example 1
Solve.
1. $5x + 3 = 3x - 5$
   Ans: $x = -4$
2. $45 + 8y = y + 17$
   Ans: $y = -4$
3. $k + 2 = 6k - 27$
   Ans: $k = -5$

Example 2
Solve.
1. $-5h + 1 = 15 - 12h$
   Ans: $h = 2$
2. $2 - 7a = -2a - 13$
   Ans: $a = 3$
3. $3.9x = -24 - 2.1x$
   Ans: $x = -4$

4.3 Variable on Both Sides

When the variable appears on both sides of an equation, the first step is to eliminate the v...

Solve: $5x$

Solution:

☐ Subtract $3x$ from each side. ⟶ $5x$

☐ Subtract 12 from each side. ⟶ $2x$

☐ Divide each side by 2.

Check:

$5x + 12 = 3x - 6$	
$5(-9) + 12$	$3(-9) - 6$
$-45 + 12$	$-27 - 6$
-33	-33

Solve.

a. $11t + 16 = 4t + 2$
   $t = -2$

Solve: 1

Solution:

☐ Add $7r$ to each side. ⟶ 1

☐ Subtract 42 from each side. ⟶

☐ Divide each side by 4.

Solve.
$p = -14$
a. $3 - 2p = 17 - p$

100 / Chapter 4

100

In equations that contain parentheses, use the Distributive Property first. Then eliminate the variable from one side of the equation.

$$\text{Solve: } 5 + 2x = 2(x - 3) + x$$

Solution:

☐ Use the Distributive Property. ⟶ $5 + 2x = 2(x - 3) + x$

☐ Combine like terms. ⟶ $5 + 2x = 2x - 6 + x$

$5 + 2x = 3x - 6$

☐ Subtract $2x$ from each side. ⟶ $5 + 2x - 2x = 3x - 6 - 2x$

$5 = x - 6$

☐ Add 6 to each side. ⟶ $5 + 6 = x - 6 + 6$

$11 = x$

Check:

$5 + 2x = 2(x - 3) + x$	
$5 + 2(11)$	$2(11 - 3) + 11$
$5 + 22$	$2(8) + 11$
27	27

Solve.

a. $9 + 4y = 3(y + 4) - y$ b. $2 + 10g = 6(2g + 3) + 2g$
   $y = 1\frac{1}{2}$ $g = -4$

CLASSROOM EXERCISES

Write the two equations that are formed in eliminating the variable from each side of the following equations. (Step 1, Examples 1-2)

1. $3x - 2 = 2x + 5$ 2. $3 - 4y = 7 - 5y$ 3. $8 + a = -3a + 6$
4. $10 + 5g = -8 + 3g$ 5. $-4d - 6 = 12 - d$ 6. $10b - 7 = 9 - 4b$

For each equation, write an equivalent equation that has the variable on one side. (Steps 1 and 2, Examples 1-2)

EXAMPLE: $3x - 7 = 2x + 8$ **ANSWER:** $x - 7 = 8$ or $-7 = -x + 8$

7. $4p + 2 = 6 - p$ 8. $5 - 2r = 3r - 2$ 9. $-3z - 8 = z - 5$
10. $3m + 2 = -2m - 8$ 11. $10.8 - 3.4j = 9.2 + 7.1j$ 12. $8.2 - h = 3h - 3.4$

Problem Solving: One Variable / 101

Additional Examples
Example 3
Solve.
1. $3(2 - y) + 5y = 3y + 8$
   Ans: $y = -2$
2. $-4(3t - 20) = 70 - 10t$
   Ans: $t = 5$
3. $6y - 17 = \frac{1}{2}(8y + 2) - y$
   Ans: $y = 6$

Common Error
When solving equations with the variable on both sides, students incorrectly combine like terms.

Example
$9x + 4 = 3x + 20$
$12x + 4 = 20$
$12x = 24$
$x = 2$

Prescription
Remind students that terms can be added only if they are on the same side of the equation. Insist that students check the solution in the original equation.

Additional Answers
Classroom Exercises
1. $x - 2 = 5$; $-2 = -x + 5$
2. $3 = 7 - y$; $3 + y = 7$
3. $8 = -4a + 6$; $8 + 4a = 6$
4. $10 + 2g = -8$;
   $10 = -8 - 2g$
5. $-3d - 6 = 12$;
   $-6 = 12 + 3d$
6. $14b - 7 = 9$;
   $-7 = 9 - 14b$
7. $5p + 2 = 6$ or $2 = 6 - 5p$
8. $5 - 5r = -2$ or
   $5 = 5r - 2$
9. $-4z - 8 = 5$ or
   $-8 = 4z + 5$
10. $5m + 2 = -8$ or
    $2 = -5m - 8$
11. $10.8 - 10.5j = 9.2$ or
    $10.8 = 9.2 + 10.5j$
12. $8.2 - 4h = -3.4$ or
    $8.2 = 4h - 3.4$

101

Additional Examples
Each example in the textbook is supplemented by *Additional Examples* that can be used by the teacher in the following ways:

1. As an alternate to the textbook Example.
2. As exercises to be used after presenting the textbook Example.

Additional Answers
Answers that cannot be annotated on the textbook page are listed in the margin.

Annotated Teacher's Edition

Warm-Up Exercises
A set of *Warm-Up Exercises* is included for each section (lesson). These pre-lesson activities review prior-taught skills and concepts that are used in the section. The *Warm-Up Exercises* can be found on pages M-18 through M-55 in the *Teacher's Manual*. The answers are shown in **bold-faced type.**

Assignment Guide
The *Written Exercises* are accompanied by a three-level *Assignment Guide*. A *Suggested Timetable of Assignments* chart for each chapter (see pages M-56 through M-62) and a *Master Timetable* (see page M-17) are also provided.

WARM-UP EXERCISES

Section 14.1
Pages 378–380

Find the value of y for each given value of x.

1. $2x - y = 3$; $x = 1$ **−1**	2. $-x + 4y = -2$; $x = 6$ **1**	
$; x = 4$ **$-\frac{11}{2}$**	4. $5x - 3y = -2$; $x = 2$ **4**	
$-5; x = -4$ **$-\frac{3}{4}$**	6. $-5x - 2y = 12$; $x = -3$ **$\frac{3}{2}$**	

intercepts for the graphs of the equations below.

$= 0$ **4; 2**	2. $3x + y + 6 = 0$ **−2; −6**
$= 0$ **−1; 3**	4. $x + y + 6 = 0$ **−6; −6**
2 **−2; 2**	6. $x + y = -1$ **−1; −1**
$= 0$ **2; 2**	8. $x - 2y + 4 = 0$ **−4; 2**

ression. Write the result in the x to complete the magic square.

2. $4x - x$	
4. $-7x + 4x$	
$x - 7$	6. $2x - 3 + 3x + 3$
$)$	8. $3x - 4x$

1	2	3
2x	3x	−2x
4 −3x	**5** x	**6** 5x
7 4x	**8** −x	**9** 0

ression.

$2x - 2y)$ **3x**	2. $-2x + 4y + (2x + 3y)$ **7y**
$(-4x - 2y)$ **−5y**	4. $3x + 3y + (6x - 3y)$ **9x**
$+ (8x - 4y)$ **3y**	6. $3x - 2y + (-3x + 12y)$ **10y**

ng value.

y	Equations	x	y
−4	2. $x + y =$ ___?___	−14	−3 −11
−6	4. $x - y =$ ___?___	−17	−12 5
−3 ?	6. $x - y = 10$		11 1 ?
−10 ?	8. $x - y = -12$	−20 ?	−8
−3	10. $x - y = 0$		−9 −9 ?

Assignment Guide
Basic
Day 1 p. 102: 1–15 odd
Day 2 p. 102: 17–35 odd
Average
Day 1 p. 102: p. 102: 1–23 odd
Day 2 p. 102: 25–37 odd
Above Average p. 102:
1–37 odd, 38

Additional Answers
6. $d = -0.8$
14. $x = -72$
18. $d = -7$
20. $g = -2\frac{5}{8}$
34. $x = -16$

WRITTEN EXERCISES

Goal: To solve equations that have a variable on both sides
Sample Problem: Solve: $5 + 4x = 3x - 10 - x$
Answer: $x = -7\frac{1}{2}$

Solve. Check each answer. (Example 1)

1. $3x - 5 = 2x + 12$ $x = 17$
2. $5x - 3 = 4x + 9$ $x = 12$
3. $12a - 9 = 10a + 17$ $a = 13$
4. $8b - 14 = 6b + 28$ $b = 21$
5. $1.6c - 7.5 = 14.1 + 3.4c$ $c = -12$
6. $0.5d - 3 - d = -1.8 + d$
7. $3\frac{1}{2}k + 17 = 2\frac{3}{4}k + 12$ $k = -6\frac{2}{3}$
8. $\frac{1}{20} + \frac{1}{5}s - 11 = \frac{3}{8}s - 11$ $s = \frac{1}{8}$

(Example 2)

9. $12 - x = 26 - 3x$ $x = 7$
10. $20 - 2y = 56 - 5y$ $y = 12$
11. $2 - 3z = z + 22$ $z = -5$
12. $5 - 2m = -22 - 4m$ $m = -13\frac{1}{2}$
13. $2\frac{1}{4} - 1\frac{5}{9}p - 19 = \frac{5}{9}p + 38 - 2\frac{1}{3}$ $p = -21\frac{1}{4}$
14. $9 - 5\frac{1}{4}k = 27 - 2\frac{1}{4}k - 2\frac{3}{4}k$
15. $2t + 1.2 - 5t = 2.8 - t$ $t = -0.8$
16. $3.2 + z = -6.7 - 1.2z$ $z = -4.5$

(Example 3)

17. $3(4 - x) - 22 = x + 46$ $x = -14$
18. $-4(d - 2) - 14 = 36 + 2d$
19. $14 - 3y = -3(4 + 2y) - 7$ $y = -11$
20. $-6g - 8 = 12 - 4(3g + 9)$
21. $0.2(n + 1) = 0.6(n - 1) - 0.7$ $n = 3.75$
22. $3(y + 0.5) = 0.5y + 1.5$ $y = 0$
23. $\frac{4}{5}y - \frac{1}{2}(y - 1) = \frac{3}{4}(y + 1)$ $y = -\frac{5}{9}$
24. $\frac{1}{3}(z + 1) - 2z = 6z + \frac{1}{2}$ $z = 0$

MIXED PRACTICE

25. $10 - 5y = y + 61$ $y = -8\frac{1}{2}$
26. $14 - 3b = -16 - 7b$ $b = -7\frac{1}{2}$
27. $6 - 15q = 11q - 46$ $q = 2$
28. $21 + m = -19 - 7m$ $m = -5$
29. $3(2 - k) - 7 = 4(2k + 8)$ $k = -3$
30. $5z + 2(z - 1) = 3(z + 1)$ $z = 1\frac{1}{4}$
31. $-10 - 18d = 26 - 12d$ $d = -6$
32. $-4f - 19 = 11 + f$ $f = -7\frac{1}{5}$
33. $4.4 - 2.1d - 5.7d$ 17.2 $d = 6$
34. $2.3x - 31.5 = 14.9 + 5.2x$
35. $8p - 2(2p - 1) = 6(p + 2)$ $p = -5$
36. $5(2r + 3) - 4r = 20r + 1$ $r = 1$

MORE CHALLENGING EXERCISES

For Exercises 37–38, a represents an even integer and b represents an odd integer. Solve for x. Identify x as an even integer or an odd integer.

37. $2a + 2x = x + b$ $x = b - 2a$; x is odd
38. $4x - a = 3\frac{1}{2}x - 3b$ $x = 2(-3b + a)$; x is even

102 / Chapter 4

Problem–Solving Skills
Solving a multi-step problem (Ex. 37–38)
Using logical reasoning (Ex. 37–38)

Critical Thinking
See Exercises 37 and 38.

102

Critical Thinking
The exercises that relate to developing higher-level thinking skills are identified for the teacher.

Problem-Solving Skills
This item identifies the problem-solving skills that are used to solve the indicated problems.

OVERVIEW: Holt Introductory Algebra 2

Philosophy

Holt Introductory Algebra 2 is the second course in a two-year program of elementary algebra. This program was developed for students who, by nature of their mathematical backgrounds and maturity, can expect to learn algebra more effectively by having it presented at a slower pace. *Holt Introductory Algebra 1* and *Holt Introductory Algebra 2* have been carefully written and organized to give attention to the varying remediation needs of the target students.

Such students often are weak in mathematical skills, yet are capable of a degree of abstract reasoning. They often have difficulty in reading a standard first-year algebra textbook. They tend to forget skills and concepts quickly and require constant review and reinforcement. On the whole, students such as these may be described as having low-average ability in mathematics or as having demonstrated low-average achievement. They have never enjoyed much success in a mathematics class.

Classroom experience with hundreds of thousands of students has shown that devoting more time to the usual one-year course gives students a chance to mature mathematically. It is this maturing that enables students to learn more mathematics than they would in the usual one-year course. The success that *Holt Introductory Algebra* students experience, perhaps for the first time in mathematics, in turn improves their motivation and interest. *Holt Introductory Algebra* is, indeed, an effective tool for helping students develop their mathematical power.

Pedagogy

Among the students in an *Introductory Algebra* class, there will probably be a wide range and diversity of student ability, interest, academic achievement, and maturity. This diversity alone challenges the classroom methods of the most experienced teacher. Because a method or procedure that is effective for one teacher in a given classroom situation may not be the choice of another, the *Holt Introductory Algebra* program allows for a variety of pedagogical approaches.

Regardless of the instructional style used, students and teachers will find several features of each section that are helpful to the successful learning of algebra. The examples are clear-cut and diverse enough to cover the subtleties of the new concepts and skills being developed. The **pivotal exercises** can serve as a means of keeping students involved and attentive to the presentation, as well as to provide check points of understanding. The **Classroom Exercises,** which are referenced to the examples and pivotal exercises, provide a good vehicle for directed study. The **Review Capsules** preceeding the lesson provide a further aid to give the teacher assurance that special skills needed in the next lesson have been sharpened. The **Written Exercises** provide a basis for homework assignments. These exercises are referenced to the examples in the lesson which provide model solutions for students to look at again if they run into difficulty.

An important feature in the *Teacher's Edition* helps the teacher better accommodate the learning levels of his or her students. This feature is the **Assignment Guide** which contains suggested assignments and overall content to be covered by students at three different instructional levels.

Problem Solving

Probably no one disputes the importance of including instruction on problem solving in various subjects of the school curriculum, although it probably gets more attention in mathematics because of the logical nature of the discipline. *Holt Introductory Algebra* provides problem solving experiences for students in many ways. There are an abundance and variety of problem solving applications throughout the book. Many of these take the form of "word" problems—but with some important distinctions from the classical word problems of the past. First, great diversity is evident in these problems even within a given lesson. Although the same concept may be covered by a given set of word problems, it is not just a matter of replicating the solution of a model problem. Students do have to bring some creativity and critical thinking to each problem. The other aspect of these word problems that is different is that many of them do represent realistic applications. They are not just abstractions which have little interest to students. They are vibrant and current, and students are motivated by them. The plan provided for solving problems in *Holt Introductory Algebra* is modeled after the steps recommended by the late Professor George Polya who wrote the classic book on problem solving titled *How To Solve It*.

Polya's Steps in Problem Solving
1. Understanding the problem
2. Devising a plan
3. Carrying out the plan
4. Looking back

Besides applied word problems, other problem-solving opportunities are provided in *Holt Introductory Algebra*. These appear in various forms in the **Special Topics** of each chapter and the **Focus on Reasoning** topics in Chapters 3, 6, 9, 12, 15, and 19. These special features are described separately in this Teacher's Manual Overview. Also, within each chapter there is a special feature called **Non-Routine Problems.** These non-routine problems are probably the best kind of activity to help students acquire problem-solving skills in the general sense. Students must bring to bear all the problem-solving skills and resources available to them in order to solve non-routine problems. The only model for the solution of a particular problem is what the student might develop. The list below includes some of the most common problem-solving strategies that are used in this text.

Problem-Solving Strategies

Choosing the operation
Completing a table
Determining whether a problem has
    too much or too little information
Drawing a diagram
Identifying a pattern
Interpreting information
Making a comparison
Making a graph
Making a list
Making a table
Reading a diagram
Reading a graph
Reading a table

Solving a multi-step problem
Solving a simpler problem
Translating from English to an
    algebraic expression
Using a diagram
Using a formula
Using estimation
Using guess and check
Using logical reasoning
Using patterns
Working backwards
Writing an equation
Writing and solving an equation

These problem-solving strategies are identified in appropriate places throughout the book.

Logical Reasoning

The nature of algebra suggests that logical reasoning is necessary if students are to understand the conclusions and generalizations that are developed as the course unfolds. Elementary high school algebra is a quasi-postulational system. Concepts are developed from such assumptions as undefined terms, definitions, and number properties. **Deductive methods** are informally used in order to arrive at conclusions in the form of rules, properties, and other generalizations. **Inductive methods** are also used in some cases to arrive at generalizations. In summary, students do develop a sense about logical reasoning as they read through a lesson or as they proceed step by step through a problem—even as simple as evaluating a formula.

Besides this day-to-day exposure to logical reasoning, students have an opportunity to learn more in the **Focus on Reasoning** special topics. These appear in Chapters 3, 6, 9, 12, 15, and 19. Many of the methods suggested in these *Focus on Reasoning* topics are used in college entrance exams. Practice with this type of reasoning exercise will help students prepare for some of the special techniques required on these exams.

Critical Thinking

Questions and exercises that challenge the critical thinking of students appear in various places in the book. Some of the pivotal exercises that appear at key ponts in the lessons require critical thinking on the part of students. However, this is not the main purpose of these exercises. The exercises in the special topics that appear in each chapter are of two types. One type is essentially numerical and, with them, the concepts presented in the special topic are applied. The other type involves critical thinking. As one example, the student may be asked to observe how the value of one variable in a formula is affected by a certain change in the value of another variable. From time to time, the *Written Exercises* in the lessons contain **More Challenging Exercises.** Some of these are just more difficult, but of the same type as the other Written Exercises. However, some of these *More Challenging Exercises* are designed as critical thinking exercises. These come in a variety of forms.

Cooperative Learning

A mode of instruction that has gained in popularity in recent years is the use of cooperative learning groups. This approach seems to be most efficient with about four students to a group. The small group gives each student an opportunity to speak, ask questions, and discuss problems with other students. Many students feel more comfortable interacting with others in a small group than in a large class.

Several types of activities in *Holt Introductory Algebra* are appropriate for cooperative learning groups. The *Classroom Exercises* in most of the lessons are certainly appropriate. Students in a small group can help each other and also check each other's work to make sure it is correctly done. Success in the *Classroom Exercises* will almost certainly assure success in doing the *Written Exercises*.

Another activity in the book that is appropriate for cooperative learning is the solving of *non-routine problems*. Often, an individual student is completely stymied by such problems. The student can gain useful insight into solving problems of this kind by interacting with other students in a small group. The sharing of ideas, hunches, and varying interpretations is the reason a group endeavor may be more successful than an individual one for certain problems. The activity that seems to be most appropriate for cooperative learning is the solving of word problems. In particular, individual students often cannot do the first step of representing the unknowns in a problem. By working in a small group, these students will be encouraged and helped by others in the group who can, perhaps, do the work successfully on their own initiative.

Manipulative Activities

We know from learning theory and from the work of Piaget and Dienes that all students move in their understanding from an ability to grasp the concrete through an understanding of the pictorial to a perception of the abstract expression of a principle. In fact, many students, particularly those characterized as "at risk," still need to learn first at the concrete level. They best grasp the concepts of algebra if they first have the opportunity to manipulate concrete objects such as algebra tiles (see Manipulative Lessons 1–15 on pages 539–569 in *Holt Introductory Algebra 2*).

The *Using Manipulatives* lessons in *Holt Introductory Algebra 2* guide students in using manipulatives and models to discover and develop the principles of algebra. For example, Manipulative Lesson 11 (see pages 560–561) uses algebra tiles to help students understand the procedure for multiplying binomials.

Skills

With the emphasis on problem solving and critical thinking, teachers may tend to lose sight of the importance of skills development. Students can't progress far in the study of mathematics if they do not have competency in certain skills including computation, operations with positive and negative rational numbers, factoring, solving equations and inequalities, graphing, and so on. *Holt Introductory Algebra* provides many features that are devoted to skill development. First, there is an abundance of both *Classroom* and *Written Exercises* for practice. There are *Mid-Chapter Review* exercises which reinforce the skills and concepts of the first half of the chapter. Then the *Chapter Review* exercises review and reinforce the skills and concepts of the entire chapter. Besides, there is additional opportunity for review and maintenance of skills in the *Cumulative Review* exercises which appear at the end of Chapters 3, 6, 9, 12, 15, and 19.

ADMINISTRATIVE CONCERNS

Overview

One of the first questions that arises in offering first-year algebra as a course to be studied over a period of two years is related to the matter of credit. Should students be given credit for two years of high school mathematics, or should they be given credit for just one year, since they have not progressed beyond the usual content of the first-year course? Another question is how to label this credit on students' permanent records; that is, what names should be given to these courses? Some school officials, when considering the implementation of a two-year algebra program, are concerned that *Holt Introductory Algebra 2* may be confused with *Second Year Algebra* or *Algebra Two*.

Solutions

Years of experience by personnel in many school districts across the country with the *Introductory Algebra* courses have provided answers to these administrative problems. Almost without exception, schools <u>do</u> give students credit toward graduation for two years of mathematics upon successful completion of both years of the *Introductory Algebra* courses. The first course is usually offered in the ninth grade and the second course in the tenth grade. The rationale for awarding two years of credit is based upon what typically happened to this type of student prior to the advent of this program. Many of these students traditionally enrolled in a general mathematics course in the ninth grade and the regular one-year algebra course in the tenth grade. Those students who completed the two years of work successfully received two years of credit.

Course Titles
Most schools have chosen to adopt new course titles when they have inaugurated the two-year program in first-year algebra. The most popular course titles have been *Introductory Algebra 1* and *Introductory Algebra 2*. Some school personnel have chosen to call the first year of the course *Pre-Algebra* and the second year of the course *Algebra One*. Most school personnel do not want to label the first year of the course General Mathematics. Students take pride in being enrolled in a course that is called <u>algebra</u> and this not only affects their attitude towards mathematics but also challenges them to achieve a higher level of understanding and competency.

LESSON PLAN GUIDE

Overview
The following pages in this *Teacher's Manual,* along with the annotated pages and the copy in the margins of the annotated pages, contain all the components for effective lesson planning. The lesson objective, references to supplemental materials, exercises for maintaining previously-taught skills, additional examples, attention to correcting common errors, and tri-level assignment guides are included on the annotated pages. (See pages M-10 and M-11.)

Pages M-18 through M-55 include an *Overview* for each chapter, brief commentary for each section (lesson), and *Warm-Up Exercises* for each section. To assist you in planning, a *Master Timetable* is provided on page M-17. In addition, a *Suggested Timetable of Assignments* for each chapter is included (pages M-56—M-62).

The *Master Timetable* and the *Suggested Timetables of Assignments* are structured for three levels.

Basic
This represents the minimum course—Chapters 1-15 less sections 11.7, 15.4, and 15.5. The *Skills Practice Book* is intended primarily for these students.

Average
This represents the "average" course—Chapters 1-16. The *Suggested Timetable of Assignments* for each of these chapters provides one day for selected extension topics—*Special Topics, Non-Routine Problems,* and *Focus on Reasoning.* Certain portions of the *Skills Practice Book* will also benefit many of these students.

Above Average
This represents the maximum course—Chapters 1-19. Students at this level should be able to handle all the extension topics (described in the *Average* above) plus the *More Challenging Exericses.*

MASTER TIMETABLE

Overview

The following *Master Timetable* for three levels of ability is coordinated with the *Suggested Timetables of Assignments* on pages M-56 through M-62 and the *Assignment Guide* that appears in the margin of the annotated pages. Note that the total number of days for each chapter includes additional days for review and testing for each level as well as extension topics for the Average and Above Average levels.

The *Master Timetable*, the *Suggested Timetables of Assignments*, and the *Assignment Guide* should be considered as guidelines only.

MASTER TIMETABLE

Chapter	Basic Sections	Basic Days	Average Sections	Average Days	Above Average Sections	Above Average Days
1	All	12	All	10	All	9
2	All	11	All	9	All	8
3	All	11	All	11	All	9
4	All	11	All	10	All	7
5	All	12	All	11	All	9
6	All	9	All	9	All	8
7	All	15	All	13	All	10
8	All	11	All	11	All	9
9	All	12	All	12	All	10
10	All	10	All	10	All	9
11	11.1–11.6	9	All	10	All	9
12	All	9	All	9	All	8
13	All	10	All	10	All	9
14	All	9	All	8	All	7
15	15.1–15.3 15.6–15.7	13	All	12	All	9
16	Omit	0	All	9	All	7
17	Omit	0	Omit	0	All	10
18	Omit	0	Omit	0	All	10
19	Omit	0	Omit	0	All	7
Cumulative Reviews		6		6		6
Total Days		**170**		**170**		**170**

CHAPTER 1: OPERATIONS ON NUMBERS

OVERVIEW

The first chapter is devoted to a review of important concepts that were introduced in *Holt Introductory Algebra 1*. The material is presented at a more sophisticated level than when introduced earlier. Manipulative Lessons 1 and 2 on pages 540–543 can be used to introduce Lessons 1.2 and 1.3 respectively.

SECTION-BY-SECTION COMMENTARY

Section 1.1 (pages 2–3)
The purpose of this introductory section is to define absolute value in terms of the opposite of a number.

Section 1.2 (pages 4–7)
This section reviews addition of positive and negative real numbers as integers, common fractions, or decimal fractions. It is not necessary to require the students to parrot the formal rules. However, they should have a working knowledge of them.

Section 1.3 (pages 8–10)
Subtracting a number is defined as adding its opposite. This is the best approach to teaching subtraction of real numbers because students do not have to learn two sets of rules.

Section 1.4 (pages 11–14)
The multiplication rules are developed by the use of patterns of products of numbers. This method seems to be convincing to most students.

Section 1.5 (pages 16–18)
In this section division is defined in terms of multiplication. This is a good opportunity to emphasize that zero cannot be a divisor. This concept is an easy one for students to forget.

Section 1.6 (pages 19–20)
The rules for the order of operations in an expression are reviewed.

Section 1.7 (pages 21–23)
The important terms *base, exponent,* and *power* are reviewed in this section. Then emphasis is given to practicing the evaluation of expressions containing powers.

WARM-UP EXERCISES

Section 1.1
Pages 2–3

Write the opposite.

1. 25 km east **25 km west**
2. $10 bank withdrawal **$10 bank deposit**
3. 7 flights up **7 flights down**
4. 7 m below sea level **7 m above sea level**
5. $250 profit **$250 loss**
6. 5° rise in temperature **5° fall in temperature**

Section 1.2
Pages 4–7

Write a positive or negative number for each arrow.

Examples: a. Ans.: $\longmapsto$ 5 b. Ans.: $\longleftarrow$ −3

1. $\longleftarrow$ −6
2. $\longmapsto$ 4
3. $\longleftarrow$ −8
4. $\longleftarrow$ −7
5. $\longmapsto$ 8
6. $\longmapsto$ 2

Add and write the answer in the magic square. The sums in the rows, columns, and diagonals are the same.

1	2	3
−3	13	−7
4	**5**	**6**
−3	1	5
7	**8**	**9**
9	−11	5

1. $4 + (-7)$
2. $(-7) + 20$
3. $(-4) + (-3)$
4. $(-12) + (-5) + 14$
5. $1.8 + (-1.3) + 0.5$
6. $3\frac{1}{2} + 7\frac{3}{4} + (-6\frac{1}{4})$
7. $(-23) + 44 + (-12)$
8. $(-1.7) + (-5.8) + (-3.5)$
9. $19 + (-17) + 23 + (-20)$
10. What is the magic number?   **3**

Give the answer to each addition problem and then express the addition problem as a multiplication problem.

1. $3 + 3 = ?$ **6; 2(3)**
2. $15 + 15 = ?$ **30; 2(15)**
3. $(-4) + (-4) = ?$ **−8; 2(−4)**
4. $(-200) + (-200) = ?$ **−400; 2(−200)**
5. $7 + 7 + 7 = ?$ **21; 3(7)**
6. $4 + 4 + 4 = ?$ **12; 3(4)**

Find the missing factor in each exercise.

1. $(-7)(?) = -28$ **4**
2. $(-9)(?) = 72$ **−8**
3. $(12)(?) = -108$ **−9**
4. $(-6)(?) = 33$ **−5$\frac{1}{2}$**
5. $(?)(-2.5) = -25$ **10**
6. $(?)(-\frac{3}{4}) = 12$ **−16**
7. $(?)(-7) = -28$ **4**
8. $(?)(-8) = 120$ **−15**
9. $(-1.3)(?) = 0$ **0**

Find both answers in each exercise. Perform the operation inside parentheses first.

1. $2 + (3 \cdot 5) = ?$ **17**
  $(2 + 3) \cdot 5 = ?$ **25**
2. $(-8 + 2) \cdot (-4) = ?$ **24**
  $(-8) + (2 \cdot -4) = ?$ **−16**
3. $(-40 \div 8) \cdot (2\frac{1}{2}) = ?$ **−12$\frac{1}{2}$**
  $(-40) \div (8 \cdot 2\frac{1}{2}) = ?$ **−2**
4. $(8 - 6) + 4 = ?$ **6**
  $8 - (6 + 4) = ?$ **−2**

Write the answer of each exercise in the magic square. The sums in the rows, columns, and diagonals are the same.

1	2	3
2	−29	−3
4	**5**	**6**
−15	−10	−5
7	**8**	**9**
−17	9	−22

1. $(-3)(3) + 11$
2. $(-2)(-2)(-2)(-2)(-2) + 3$
3. $(-2) + (-1)(-1)(-1)$
4. $(-3)(-3)(-3) + (2)(2)(2) + (2)(2)$
5. $\dfrac{(-5)(-5)(-5) + (5)(5)}{10}$
6. $(-4) + (-1)(-1)(-1)(-1)(-1)$
7. $(2)(5) + (-3)(-3)(-3)$
8. $(-\frac{1}{3})(-\frac{1}{3})(3)(3)(3)(3)$
9. $(-\frac{5}{2})(-\frac{5}{2})(-\frac{5}{2}) + (-6\frac{3}{8})$
10. What is the magic number for this square?   **−30**

CHAPTER 2: REAL NUMBERS

OVERVIEW

The chapter opens with a review of these subsets of the set of real numbers: counting numbers, whole numbers, integers, rational numbers, and irrational numbers. The Addition and Multiplication Properties of Real Numbers are presented as are special properties of real numbers and the Distributive Property. In Section 2.6 the properties of real numbers are applied to combining like terms. Manipulative Lesson 3 on pages 544–545 can be used to introduce Lesson 2.6.

SECTION-BY-SECTION COMMENTARY

Section 2.1 (pages 30–34)
Starting with the set of counting numbers, you can show students how to define larger number sets. The set of whole numbers consists of the set of counting numbers and 0. The set of integers consists of the set of whole numbers and their opposites. The set of rational numbers consists of quotients of integers in which no divisor equals zero. Finally, the set of real numbers is composed of the sets of rational and irrational numbers.

Section 2.2 (pages 35–38)
Students tend to accept the Commutative Property of Addition as an obvious fact. You can show them that subtraction, for example, is not a commutative operation.

Section 2.3 (page 39–41)
It would be appropriate at this time to point out the similarities between the Addition and Multiplication Properties. Both include Commutative and Associative Properties. The Multiplication Property of One serves the same purpose as the Addition Property of Zero. The Multiplication Property of Reciprocals is like the Addition Property of Opposites.

Section 2.4 (pages 43–46)
Several concepts that often cause difficulty for students are stated as special properties in this section. Perhaps the most troublesome concept is covered by the special property which shows that $a - (b + c)$ and $a - b - c$ are equivalent expressions.

Section 2.5 (pages 47–49)
Factoring is introduced in this section. It is explained in the context of using the Distributive Property to express sums as products. This topic is covered more thoroughly in Chapter 9.

Section 2.6 (pages 50–51)
Addition and subtraction of like terms is covered in this section. This procedure is often called *combining like terms*.

WARM-UP EXERCISES

Section 2.1
Pages 30–34

Cross out the numeral that does not represent the first number shown.

1. $2; \frac{6}{3}, \frac{2}{2}, \frac{8}{4}, \frac{10}{5}$

2. $8; \frac{16}{2}, \frac{24}{3}, \frac{-16}{-2}, \frac{-8}{1}$

3. $\frac{1}{2}; 0.5, \frac{-2}{-4}, \frac{-12}{24}, \frac{3}{6}$

4. $1\frac{1}{2}; \frac{3}{2}, 1.5, 1\frac{2}{10}, 1\frac{5}{10}$

5. $2\frac{3}{4}; \frac{11}{4}, \frac{12}{4}, 2.75, \frac{22}{8}$

6. $-5; \frac{-10}{2}, \frac{-10}{-2}, \frac{5}{-1}, \frac{-5}{1}$

7. $-13; \frac{-13}{1}, \frac{-26}{2}, \frac{13}{-1}, \frac{26}{-2}$

8. $-\frac{3}{8}; \frac{6}{-16}, \frac{3}{-8}, \frac{-6}{16}, \frac{-6}{16}$

9. $-1\frac{1}{4}; \frac{-5}{4}, -1.25, \frac{-5}{4}, -1\frac{2}{8}$

10. $-3\frac{2}{3}; \frac{-11}{3}, \frac{11}{-3}, \frac{22}{-6}, -3.6$

Section 2.2
Pages 35–38

*Evaluate both expressions **a** and **b** in each exercise.*

1. a. $(-5.2) + 3.7$ **−1.5**
   b. $3.7 + (-5.2)$ **−1.5**

2. a. $-1.008 + 0$ **−1.008**
   b. $0 + (-1.008)$ **−1.008**

3. a. $14.3 + (-14.3)$ **0**
   b. $(-14.3) + 14.3$ **0**

4. a. $(-23 + 17) + 9$ **3**
   b. $-23 + (17 + 9)$ **3**

5. a. $(-3\frac{5}{16} + 5\frac{3}{4}) + 1\frac{1}{4}$ $\mathbf{3\frac{11}{16}}$
   b. $-3\frac{5}{16} + (5\frac{3}{4} + 1\frac{1}{4})$ $\mathbf{3\frac{11}{16}}$

6. a. $-1.8 + (2.6 + 5.4)$ **6.2**
   b. $(-1.8 + 2.6) + 5.4$ **6.2**

Section 2.3
Pages 39–41

Evaluate each expression.

1. $(-5\frac{3}{4}) + 4\frac{1}{8} + (-2\frac{1}{4})$ $\mathbf{-3\frac{7}{8}}$
2. $(-6.7 + 1.3) + (0.7 + 6.7)$ **2**
3. $(-356) + (-139) + (-144) + (-361)$ **−1000**
4. $1\frac{7}{8} + (-2\frac{3}{4}) + 4\frac{5}{12} + 3\frac{1}{8} + (-2\frac{1}{4})$ $\mathbf{4\frac{5}{12}}$
5. $(-0.004) + 5.09 + (-0.006) + 2.01 + (-0.09)$ **7.0**

Section 2.4
Pages 43–46

*Evaluate both expressions **a** and **b** in each exercise.*

1. a $-(2.7 + (-5.6))$ **2.9**
   b. $(-2.7) - (-5.6)$ **2.9**

2. a. $(-(-5))(-(-6))$ **30**
   b. $(-5)(-6)$ **30**

3. a. $12.8 - (2.6 + 5.4)$ **4.8**
   b. $12.8 - 2.6 - 5.4$ **4.8**

4. a. $(-12) - 9$ **−21**
   b. $-1(12 + 9)$ **−21**

5. a. $(-1\frac{3}{4}) + (-5\frac{1}{4})$ **−7**
   b. $-(1\frac{3}{4} + 5\frac{1}{4})$ **−7**

6. a. $-(-14.3 + 8.4)$ **5.9**
   b. $-1(-14.3 + 8.4)$ **5.9**

Section 2.5
Pages 47–49

Write an equivalent expression in each exercise.

1. $-(x - 13)$ **−x + 13**
2. $-1(-x)$ **x**
3. $-1(x + 9)$ **−x − 9**
4. $-1(x + y + z)$ **−x − y − z**
5. $(-x)(y)$ **−xy**
6. $(-x)(-y)$ **xy**
7. $8 - (x + 4)$ **8 − x − 4**
8. $x - (4 - y)$ **x − 4 + y**

Section 2.6
Pages 50–51

*Match each of Exercises 1–5 with one of items **A-J** at the right below.*

1. A factor of $6x^2 - 3x$ **E**
2. $2x(2x - 1)$ **I**
3. A factor of $10x^2 - 2$ **H**
4. $(-x + 2)(-4x)$ **C**
5. A factor of $36x^2$ and $48y^2$ **A**

A. 12
B. $12x$
C. $4x^2 - 8x$
D. $2x^2 - 2x$
E. x

F. $4x - 8$
G. $5x^2 - 2$
H. $5x^2 - 1$
I. $4x^2 - 2x$
J. $3x^2$

CHAPTER 3:   EQUATIONS AND PROBLEM SOLVING

OVERVIEW

In this chapter, solutions of equations using one operation are studied first. This is expanded to solutions using two or more operations. Then problem solving based on formulas is introduced. More complicated equations involving like terms are then covered. The last two sections of the chapter introduce the important basic elements of problem solving by equations. One section is devoted to translating words to algebraic symbols. Then the standard Polya model for problem solving is introduced. Manipulative Lessons 4, 5, 6, and 7 on pages 546–553 can be used to introduce Lessons 3.1, 3.2, 3.3, and 3.5 respectively.

SECTION-BY-SECTION COMMENTARY

Section 3.1 (pages 58–60)
The concept of equivalent equations is introduced in this section. This is needed in order to present the Addition and Subtraction Properties of Equations. Students should be required to give detailed steps in their work until they show a good knowledge of equation solving. Remind students that the check is for the purpose of discovering any computational errors.

Section 3.2 (pages 61–63)
The Multiplication and Division Properties of Equations are presented in this section. Students should still be required to give detailed steps in solutions of equations.

Section 3.3 (pages 64–66)
The equations of this section contain more than one operation. Two or more properties of equations need to be used to solve each equation.

Section 3.4 (pages 67–69)
The formula relating distance, rate, and time, and the formula for the perimeter of a rectangle are applied in this section. Known values are substituted in a formula and then it is solved by applying the properties of equations.

Section 3.5 (pages 71–73)
Some of the equations in this section require that you combine like terms as the first step. Others require that, first, you apply the Distributive Property before combining like terms.

Section 3.6 (pages 74–77)
It is important that students have confidence in accurately translating word expressions to algebraic expressions and word sentences to equations. Many of the most common word expressions for the operations are reviewed. Then the examples and exercises provide practice in translating from words to algebraic symbols.

Section 3.7 (pages 78–81)
In this section, the classical problem-solving steps as conceived by the late Professor George Polya are listed. Students are well advised to follow this plan for solving problems by the application of equations.

WARM-UP EXERCISES

Section 3.1
Pages 58–60

Identify which equations are true when the variable is replaced by -3.

1. $n + 5 = 2$ **True**
$(-9) - (-11) = 5 + n$ **True**
$n + 5 = 2 - 5$ **False**

2. $8.6 + r = 2.2 + 3.4$ **True**
$8.6 + r - 8.6 = 5.6 - 8.6$ **True**
$r = -3$ **True**

Section 3.2
Pages 61–63

Solve each equation.

1. $7.2 = t + 5.9$ **$t = 1.3$**   **2.** $m + 4.9 = 2.7$ **$m = -2.2$**

3. $3.9 + w = 6.6$ **$w = 2.7$**   **4.** $p - 5.8 = -3.8$ **$p = 2$**

5. $-3.1 = n - 3.7$ **$n = 0.6$**   **6.** $r - 5.3 = -6.1$ **$r = -0.8$**

Section 3.3
Pages 64–66

Solve each equation.

1. $3r = -21$ **$r = -7$**   **2.** $\frac{r}{-4} = 5$ **$r = -20$**

3. $\frac{x}{2} = -12$ **$x = -24$**   **4.** $-\frac{1}{2}y = -9$ **$y = 18$**

5. $-p = -4.2$ **$p = 4.2$**   **6.** $-t = 3.1$ **$t = -3.1$**

Section 3.4
Pages 67–69

Refer to the formula $p = \ell + \ell + w + w$
and the rectangle shown here.
Values of two variables are given.
Find the value of the remaining variable.

(rectangle with sides labeled w and ℓ)

4. $\ell = 15$ feet
  $w = 12$ feet
  $p = 54$ feet

5. $p = 26$ cm
  $\ell = 8$ cm
  $w = 5$ cm

6. $p = 32$ inches
  $w = 5\frac{1}{2}$ inches
  $\ell = 10\frac{1}{2}$ inches

Section 3.5
Pages 71–73

Solve each equation.

1. $12x - 3 = 8$ **$x = \frac{11}{12}$**   **2.** $-3x + 5 = 6$ **$x = -\frac{1}{3}$**

3. $7 = 5 + 3x$ **$x = \frac{2}{3}$**   **4.** $6x - 5 = -4$ **$x = \frac{1}{6}$**

5. $-3 + 12x = 2$ **$x = \frac{5}{12}$**   **6.** $3x - 10 = -8$ **$x = \frac{2}{3}$**

Section 3.6
Pages 74–77

Write each of the following expressions in words. Use the words
"an unknown number" for the variable. Answers will vary.

1. $m + 12$ **An unknown number plus 12**

2. $r - 3\frac{1}{2}$ **An unknown number minus $3\frac{1}{2}$**

3. $8t$ **8 times an unknown number**

4. $24 \div k$ **24 divided by an unknown number**

5. $5n - 27$ **5 times an unknown number minus 27**

Section 3.7
Pages 78–81

Write an equation for each word sentence.

1. The temperature, t, plus 6° equals -3°F. **$t + 6 = -3$**

2. The amount of a loan, a, divided by 12 is \$150. **$\frac{a}{12} = 150$**

3. Four times the number of kilometers, k, is 480. **$4k = 480$**

4. A company's profit, p, minus \$3000 equals $-\$1300$. **$p - 3000 = -1300$**

5. Three times the cost, c, plus \$50 equals \$170. **$3c + 50 = 170$**

6. Twice the number of hours, h, decreased by 3 equals 13. **$2h - 3 = 13$**

CHAPTER 4:   PROBLEM SOLVING: ONE VARIABLE

OVERVIEW

This chapter continues the approach of Chapter 3. This involves extending students' equation-solving skills and using these equations in solving other kinds of word problems. The chapter opens with the focus on identifying the conditions of a problem and using these conditions to represent the unknown and to form an equation describing the problem. This section is followed by the use of formulas to solve perimeter problems. Then students refine their skills on solving equations with the variable on both sides. The last two sections of the chapter cover percent problems and special distance problems. Manipulative Lesson 8 on pages 554–555 can be used to introduce Lesson 4.3.

SECTION-BY-SECTION COMMENTARY

Section 4.1 (pages 90–94)
This section contains a wide variety of word problems that involve two conditions. The use of these conditions in solving the problem and in checking the solution is highlighted.

Section 4.2 (pages 95–99)
Problems involving the perimeter formulas for rectangles and triangles are covered in this section. The use of the two conditions of each problem in solving and checking is continued.

Section 4.3 (pages 100–102)
In this section the equations to be solved have the variable on both sides of the equal sign. Solutions by both the Addition and Subtraction Properties of Equations are shown. Examples 2 and 3 illustrate that the variable can end up on either side of the equality symbol when the solution is obtained.

Section 4.4 (pages 104–107)
Per cent problems based on the formulas $d = rp$ are studied in this section. Also included are some miscellaneous per cent problems involving sports statistics.

Section 4.5 (pages 108–111)
Problems involving two types of rate formulas are presented in this section. One is the familiar distance/rate/time formula $d = rt$. The other involves cruising range, automobile mileage, and fuel capacity.

WARM-UP EXERCISES

Section 4.1
Pages 90–94

Find the unknown numbers in Exercises 1-4.

1. The sum of two numbers is 3. Their difference is also 3. What are the numbers?   **3; 0**

2. The sum of two number is 13. Their product is 36. What are the numbers?   **9; 4**

3. One number is twice another number. Their sum is 12. What are the numbers?   **8; 4**

4. The difference of two numbers is 4. Their product is 5. What are the numbers?   **5; 1**

5. A cork and a bottle cost $1.10. The bottle costs $1 more than the cork. How much does each cost?   **Bottle: $1.05; cork: $0.05**

Section 4.2
Pages 95–99

Write the simplest expression for each perimeter in the magic square.
The sums in the rows, columns, and diagonals are the same.

1. Triangle with sides of length $2x + 3$, $x + 2$, and $x + 2$
2. Rectangle: length = $3x + 2$; width = $x + 2$
3. Rectangle: length = $\frac{5}{2}x + 4$; width = $2x + 2$
4. Triangle with sides of length $3x + 5$, $4x + 5$, and $5x + 4$
5. Triangle with sides of length $2x + 3$, $2x + 3$, and $3x + 3$
6. Square with each side of length $\frac{1}{2}x + 1$
7. Rectangle: length = $2x + 1$; width = $\frac{1}{2}x + 2$
8. Square with each side of length $\frac{3}{2}x + \frac{5}{2}$
9. Triangle with sides of length $2x + 3$, $5x + 1$, and $3x + 7$
10. What is the magic expression for this square? **$21x + 27$**

1	2	3
$4x + 7$	$8x + 8$	$9x + 12$
4	**5**	**6**
$12x + 14$	$7x + 9$	$2x + 4$
7	**8**	**9**
$5x + 6$	$6x + 10$	$10x + 11$

Section 4.3
Pages 100–102

Write the equation that is formed as the first step in solving each equation.

1. $4x - x + 3 = 3$ **$3x + 3 = 3$**
2. $-2x + 3x + 1 = 4$ **$x + 1 = 4$**
3. $2x - 3 + 3x = 7$ **$5x - 3 = 7$**
4. $3x + 1 - 2x = 15$ **$x + 1 = 15$**
5. $2(3x - 1) = 4$ **$6x - 2 = 4$**
6. $-3(1 - 2x) = 9$ **$-3 + 6x = 9$**
7. $2(4x + 1) - 2 = 8$ **$8x + 2 - 2 = 8$**
8. $-2(x + 1) - 3 = 5$ **$-2x - 2 - 3 = 5$**

Section 4.4
Pages 104–107

Solve each equation.

1. $0.05x = 15$ **300**
2. $0.8t = 12$ **15**
3. $0.04n = 24$ **600**
4. $0.07w = 87.5$ **1250**
5. $0.3r = 39.24$ **130.8**
6. $0.002y = 12.48$ **6240**

Section 4.5
Pages 108–111

Solve each problem.

1. How far will a plane fly in 3 hours at an average speed of 600 kilometers per hour?   **1800 km**

2. How long will it take a runner to go 1500 meters at an average speed of 4 meters per second?   **375 seconds**

3. What average speed is needed for a trip of 225 miles in 5 hours?   **45 miles per hour**

4. How many gallons of fuel will be needed for a car to travel 228 miles at a rate of 24 miles per gallon?   **9.5 gal**

5. What is the average mileage for a car that travels 600 miles and uses 20 gallons of fuel?   **30 miles per gallon**

CHAPTER 5:   INEQUALITIES

OVERVIEW

This chapter presents the various properties of order and shows their application to finding solution sets of inequalities. Since solving inequalities is so closely related to solving equations, it will give students further review of the necessary skills. The Addition, Subtraction, Multiplication, and Division Properties for Inequalities are the main working tools for this chapter. The work proceeds from inequalities with the variable on one side and involving one operation to inequalities with the variable on both sides and involving two or more operations.

SECTION-BY-SECTION COMMENTARY

Section 5.1 (pages 116–119)
Graphs of various inequalities including $<$, $>$, $\leq$, and $\geq$ are covered in this section. In this text, "etc." is used with graphs only when the solution set is an infinite subset of {integers}.

Section 5.2 (pages 120–122)
In this section the Comparison Property, the Order Property of Opposites, and the Transitive Property of Order are studied. The Order Property of Opposites is useful in solving simple inequalities of the form $-x < a$.

Section 5.3 (pages 123–126)
The Addition and Subtraction Properties for Inequalities are introduced in this section. Two inequalities are used to show that the Addition Property is reasonable. The concept of equivalent inequalities is used in stating both properties and again in finding the solution sets in the Examples.

Section 5.4 (pages 127–128)
The first example involves like terms. The second example has a variable on both sides, and the third example requires use of the Distributive Property. Then two important kinds of inequalities are considered. One has an empty solution set and the other has the entire domain as its solution set.

Section 5.5 (pages 129–131)
The Multiplication and Division Properties of Inequalities are developed in this section and are used in solving simple inequalities.

Section 5.6 (pages 132–134)
This section provides additional practice in solving inequalities with several steps. Since the methods are similar to those used in solving equations, students will also be refining those skills at the same time.

Section 5.7 (pages 135–137)
This section applies the solution of inequalities to problem solving. Two conditions of the problem are used in the same way that they were used for solving problems with equations. Condition 1 is used to represent the unknown, and Condition 2 is used to write an inequality.

WARM-UP EXERCISES

Section 5.1
Pages 116–119

Name the greater number in each pair.

1. $1\frac{1}{2}$, $1\frac{1}{4}$ $\mathbf{1\frac{1}{2}}$
2. 5.2, 5.25   **5.25**
3. -3, -4 $\mathbf{-3}$
4. -10, -1 $\mathbf{-1}$
5. -6.2, -6.1 $\mathbf{-6.1}$
6. -0.25, -0.2 $\mathbf{-0.2}$
7. 0, 1   **1**
8. 0, -1 **0**
9. 2.3, -2.4 **2.3**

Section 5.2
Pages 120–122

Write an inequality or an equation to describe each graph. Use the variable x.

1.
$x \leq 3$

2.
$x < 3$

3.
$x = 3$

4.
$x > 3$

5.
$x \geq 3$

6.
$x \leq -3$

Section 5.3
Pages 123–126

Replace the ? with >, <, or = to make a true statement.

1. $0.5 \ ? \ \frac{-5}{-10}$ **Ans.:** $=$

2. $10 \ ? \ 12$ **Ans.:** $<$

3. $6.5 \ ? \ 6$ **Ans.:** $>$

4. $\frac{1}{4} \ ? \ \frac{1}{2}$ **Ans.:** $<$

5. $-2 \ ? \ 2$ **Ans.:** $<$

6. $1 \ ? \ 0$ **Ans.:** $>$

7. If $-x < -2$, then $x \ ? \ 2$. **Ans.:** $>$

8. If $-x > 3$, then $x \ ? \ -3$. **Ans.:** $<$

Section 5.4
Pages 127–128

Write the number that must be added to each side of the given inequality to get the variable by itself.

1. $x + 2 < 5$ **−2**

2. $x - 4 > 3$ **4**

3. $6 < x + 1$ **−1**

4. $3 > -1 + x$ **1**

5. $4 + x > -2$ **−4**

6. $-3 < x + 2$ **−2**

Section 5.5
Pages 129–131

Write the equivalent form of each inequality that has the variable by itself on the left.

1. $x - 4 < -2$ **$x < 2$**

2. $-x > 3$ **$x < -3$**

3. $3x - 2x > -3$ **$x > -3$**

4. $-2x + 3x > -4$ **$x > -4$**

5. $-x > 4$ **$x < -4$**

6. $x - 2x > 1$ **$x < -1$**

7. $-x > -5$ **$x < 5$**

8. $2x < x - 2$ **$x < -2$**

Section 5.6
Pages 132–134

Write Yes or No to tell whether x > 2 is the solution set for each inequality.

1. $x + 3 > 1$ **No**

2. $x - 3 > 1$ **No**

3. $x + 1 > 3$ **Yes**

4. $-x > -2$ **No**

5. $-x < -2$ **Yes**

6. $5x - 6x < -2$ **Yes**

7. $-\frac{1}{2}x > -1$ **No**

8. $-\frac{1}{2}x < -1$ **Yes**

9. $-4x < -8$ **Yes**

Section 5.7
Pages 135–137

Write the inequality for each word sentence. Use n for the variable.

1. A number increased by fourteen is greater than twenty-seven.
$n + 14 > 27$

2. Four times a number minus fifteen is less than thirty-three.
$4n - 15 < 33$

3. The sum of a number and two times the number is less than forty-two. **$n + 2n < 42$**

4. Two more than three times a number is greater than seventeen.
$3n + 2 > 17$

CHAPTER 6:   POWERS

OVERVIEW

This chapter covers various properties of powers including products, quotients, powers of products and quotients, and powers of powers. In the examples and exercises, variables as well as numbers are used for the various bases and exponents that occur. In this course, exponents are restricted to integers, but powers involving both negative and zero exponents are used in this chapter and elsewhere in the text.

SECTION-BY-SECTION COMMENTARY

Section 6.1 (pages 144–146)
The Product Property of Powers is introduced inductively by use of several examples. Although these examples involve positive integral exponents, the property is stated for zero or negative exponents as well.

Section 6.2 (pages 147–150)
Two important concepts are covered in this section—the Quotient Property of Powers and the Property of a Zero Exponent. Both properties are developed in an inductive fashion.

Section 6.3 (pages 151–154)
The examples and exercises apply the Product and Quotient Properties of Powers to both positive and negative exponents.

Section 6.4 (pages 156–158)
In this section student learn to apply the properties that involve raising a product or a quotient to a power. Students should learn to distinguish between expressions such as $(5y)^2$ and $(5y^2)$.

Section 6.5 (pages 159–162)
The inductive method is used in this section to introduce the Property of the Power of a Power. Care should be taken with $(x^a)(x^b)$ and $(x^a)^b$, as these expressions are easily confused.

Section 6.6 (pages 163–165)
Two types of conversions are covered based on the definition of scientific notation. Scientific notation is given and the student is asked to convert it to decimal form. Then the ordinary decimal form is given and the student is asked to convert it to scientific notation.

WARM-UP EXERCISES

Section 6.1
Pages 144–146

Complete the table by replacing each ? with a numeral.

	x	$x \cdot x$		$x \cdot x \cdot x$		$x \cdot x \cdot x \cdot x \cdot x$	
1.	2	$?$	**4**	2^3		$2^?$	2^5
2.	3	3^2		$3^?$	3^3	$3^?$	3^5
3.	4	$?^2$	**4**	$4^?$	4^3	$4^?$	4^5
4.	5	$5^?$	5^2	$5^?$	5^3	$5^?$	5^5

Section 6.2
Pages 147–150

Solve each equation and place the answer in the correct square. The sums of the vertical and horizontal answers are the same. Find that sum. **24**

1. $2^3 \cdot 2^5 = 2^x$
2. $(-2)^3(-2)^4 = (-2)^x$
3. $5^x \cdot 5^2 = 5^7$
4. $(1.2)^3(1.2)^x = (1.2)^4$
5. $x^5 \cdot 3^7 = 3^{12}$
6. $(1.5)^2(1.5)^5 = (1.5)^x$
7. $(\pi)^3(\pi)^5 = (\pi)^x$
8. $(3.2)^x(3.2)^6 = (3.2)^8$
9. $4x^2 = 4^3$

1. **8**					
2. **7**					
3. **5**	4. **1**	5. **3**		6. **7**	7. **8**
8. **2**					
9. **4**					

Section 6.3
Pages 151–154

Write *True* or *False* for each sentence.

1. $\dfrac{5^3}{5^2} = 5^5$ **False**
2. $\dfrac{x^7}{x^3} = x^{10}$ **False**
3. $\dfrac{10^3}{10^2} = 10$ **True**
4. $\dfrac{10^{12}}{10^{10}} = 100$ **True**
5. $\dfrac{3^x}{3^1} = 3^{(x+1)}$ **False**
6. $\dfrac{4^8}{4^x} = 4^{(8-x)}$ **True**

Section 6.4
Pages 156–158

Complete the puzzle by writing each product or quotient in simplest form in the correct square. The product of the terms in each row equals 1.

1. $x^9 \cdot x^{-2}$
2. $x^{-2} \cdot x^{-5}$
3. $x^{-6} \cdot x^3$
4. $x^5 \cdot x^{-2}$
5. $(5^2)(5^4)(5^{-6})$
6. $p^2 \cdot p^{-5}$
7. $\dfrac{p^0}{p^{-3}}$
8. $\dfrac{r^{-3}}{r^{-5}}$
9. $\dfrac{r^{-8}}{r^{-6}}$
10. $\dfrac{2^8 \cdot 2^{-6}}{2^2}$

1.	2.	
x^7	x^{-7}	
3. x^{-3}	4. x^3	5. **1**
6. p^{-3}	7. p^3	
8. r^2	9. r^{-2}	10. **1**

Section 6.5
Pages 159–162

Simplify.

1. $(2x)^3$ $8x^3$
2. $(-2x)^3$ $-8x^3$
3. $(2x)^4$ $16x^4$
4. $(-2x)^4$ $16x^4$
5. $(-2xy)^3$ $-8x^3y^3$
6. $-(2x)^4$ $-16x^4$
7. $\left(\dfrac{2}{x}\right)^3$ $\dfrac{8}{x^3}$
8. $\left(\dfrac{x}{-2}\right)^3$ $\dfrac{-x^3}{8}$
9. $\left(\dfrac{-2}{-x}\right)^3$ $\dfrac{8}{x^3}$

Section 6.6
Pages 163–165

Complete this cross-number puzzle. Simplify each expression and place one digit of the answer in each square.

Across
1. $(2^2)^5$
2. $(2^2)^2$
4. $(2^3)^2$
6. $5(2^{-3})^{-1}$
7. $10(2^{-2})^{-4}$

Down
1. $((-2)^2)^2$
2. $(5 \cdot 2^3)^2$
3. $3(2^5)^2$
4. $(-2^3)^2$
5. $(2^6)^2$

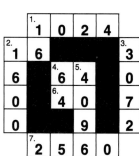

1. **1**	**0**	**2**	**4**
2. **1** **6**			3. **3**
6		4. **6** 5. **4**	**0**
0		6. **4** **0**	**7**
0			**9**	**2**
7. **2**	**5**	**6**	**0**

CHAPTER 7: ROOTS

OVERVIEW

This chapter includes simplification of numbers expressed in radical form and the various operations with radicals. Emphasis is given to radicals that represent *square* roots, but cube and fourth roots are introduced in Section 7.3. The opening sections concentrate on square roots of counting numbers as basic concepts of radicals are presented. Radicals involving variables and algebraic expressions are introduced in Section 7.4. The chapter concludes with quotients of radicals including rationalizing denominators and computing approximations.

SECTION-BY-SECTION COMMENTARY

Section 7.1 (pages 174–176)
Students need a clear understanding of the meaning of a square root. They often think of <u>the</u> square root of a number—forgetting the negative root.

Section 7.2 (pages 177–179)
Students usually have no problem in understanding the general Product Property of Radicals. However, they sometimes get confused with a special product such as $\sqrt{2} \cdot \sqrt{2}$.

Section 7.3 (pages 180–182)
Square root radicals are simplified by expressing the prime factorization of the radicands and applying the Product Property of Radicals. This method is then extended to cube roots and fourth roots.

Section 7.4 (pages 183–186)
Radicands expressed as variables or algebraic expressions are introduced in this section.

Section 7.5 (pages 187–189)
The Distributive Property should be used as the basis for explaining how to simplify the sum of two "like" radicals. Students will quickly see the analogy between "combining like terms" and "combining like radicals."

Section 7.6 (pages 190–192)
The Quotient Property of Radicals is stated. Examples 1 and 2 illustrate two ways in which the Quotient Property is used. Rationalizing the denominator is covered in Examples 3 and 4.

Section 7.7 (pages 193–197)
This section is devoted to finding approximations of square roots by means of a simple table. Some special applications are covered in Examples 3 and 4.

Section 7.8 (pages 198–201)
The Rule of Pythagoras (Pythagorean Theorem) is introduced in this section. Example 1 requires the use of the converse of the Rule of Pythagoras: If the sum of the squares of the lengths of two sides of a triangle equals the square of the length of the longest side, the triangle is a right triangle. Examples 2 and 3 illustrate how to find the length of one side when the lengths of the other two sides are known.

WARM-UP EXERCISES

Section 7.1
Pages 174–176

Replace each ? with a real number to make a true statement. If no real number will make a true statement, write <u>None</u>.

1. $3(?) = 9$ **3**
2. $-3(?) = 9$ **−3**
3. $(?)^2 = 16$ **4, −4**
4. $(?)5 = 25$ **5**
5. $(?)(-5) = 25$ **−5**
6. $(?)^2 = -36$ **None**
7. $(7)^? = 49$ **2**
8. $(7)^? = -49$ **None**
9. $(-7)^? = -49$ **None**
10. Which Exercises 1–9 have *two* answers?   **Exercise 3**

Section 7.2
Pages 177–179

Complete this number puzzle. Write the numbers of the prime factorization of each number. Write one number in each square increasing in size from left to right.

	Across				Down		
1.	10	**6.**	77	**1.**	20	**4.**	245
3.	100	**7.**	35	**2.**	25	**7.**	65
5.	30	**8.**	91	**3.**	66		

(crossword puzzle grid)

1. 2 | 2. 5
3. 2 | 2 | 5 | 4. 5
5. 2 | 3 | 5 | ■ | 6. 7 | 11
11 | ■ | 7. 5 | 7
8. 7 | 13

Section 7.3
Pages 180–182

Simplify.

1. $\sqrt{11} \cdot \sqrt{11}$ **11**
2. $\sqrt{5\frac{1}{2}} \cdot \sqrt{5\frac{1}{2}}$ **$5\frac{1}{2}$**
3. $\sqrt{25} \cdot \sqrt{49}$ **35**

4. $\sqrt{100} \cdot \sqrt{121}$ **110**
5. $\sqrt{2} \cdot \sqrt{32}$ **8**
6. $(-\sqrt{3})(\sqrt{7})$ **$-\sqrt{21}$**

7. $(-\sqrt{3})(-\sqrt{2})$ **$\sqrt{6}$**
8. $\sqrt{\frac{1}{4}} \cdot \sqrt{36}$ **3**
9. $\sqrt{\frac{1}{2}} \cdot \sqrt{72}$ **6**

Section 7.4
Pages 183–186

Simplify.

1. $\sqrt{8}$ **$2\sqrt{2}$**
2. $\sqrt{18}$ **$3\sqrt{2}$**
3. $\sqrt{50}$ **$5\sqrt{2}$**

4. $\sqrt{98}$ **$7\sqrt{2}$**
5. $\sqrt{12}$ **$2\sqrt{3}$**
7. $\sqrt{27}$ **$3\sqrt{3}$**

7. $\sqrt{75}$ **$5\sqrt{3}$**
8. $\sqrt{20}$ **$2\sqrt{5}$**
9. $\sqrt{45}$ **$3\sqrt{5}$**

Section 7.5
Pages 187–189

Simplify. The domain is {nonnegative real numbers}.

1. $\sqrt{25x^2}$ **$5x$**
2. $\sqrt{x^2y^2}$ **xy**
3. $\sqrt{a^4}$ **a^2**

4. $\sqrt{49b}$ **$7\sqrt{b}$**
5. $\sqrt{8p}$ **$2\sqrt{2p}$**
6. $\sqrt{20r^2s}$ **$2r\sqrt{5s}$**

Section 7.6
Pages 190–192

Add and subtract as indicated. Simplify where necessary.

1. $5\sqrt{2} + 2\sqrt{2}$ **$7\sqrt{2}$**   **2.** $9\sqrt{3} - 2\sqrt{3}$ **$7\sqrt{3}$**   **3.** $\sqrt{13} + \sqrt{13}$ **$2\sqrt{13}$**

4. $\sqrt{11} - \sqrt{11}$ **0**   **5.** $2\sqrt{x} + 3\sqrt{x}$ **$5\sqrt{x}$**   **6.** $-\sqrt{t} - 2\sqrt{t}$ **$-3\sqrt{t}$**

7. $2\sqrt{18} + \sqrt{2}$ **$7\sqrt{2}$**   **8.** $\sqrt{12} - \sqrt{3}$ **$\sqrt{3}$**   **9.** $\sqrt{12n} + 5\sqrt{3n}$ **$7\sqrt{3n}$**

Section 7.7
Pages 193–197

Write each radical in simplest form.

1. $\sqrt{242}$ **$11\sqrt{2}$**   **2.** $\sqrt{180}$ **$6\sqrt{5}$**   **3.** $\sqrt{1100}$ **$10\sqrt{11}$**   **4.** $\sqrt{150}$ **$5\sqrt{6}$**

5. $\sqrt{507}$ **$13\sqrt{3}$**   **6.** $\sqrt{\frac{3}{16}}$ **$\frac{\sqrt{3}}{4}$**   **7.** $\sqrt{\frac{1}{3}}$ **$\frac{\sqrt{3}}{3}$**   **8.** $\sqrt{\frac{3}{7}}$ **$\frac{\sqrt{21}}{7}$**

Section 7.8
Pages 198–201

Evaluate each expression.

1. $3^2 + 4^2$ **25**   **2.** $8^2 + 5^2$ **89**   **3.** $10^2 + 12^2$ **244**

4. $6^2 - 3^2$ **27**   **5.** $15^2 - 7^2$ **176**   **6.** $20^2 - 13^2$ **231**

CHAPTER 8:   POLYNOMIALS

OVERVIEW

This chapter covers the operations with polynomials. The first two sections are on monomials. Section 8.3 is concerned with simplifying and evaluating polynomials. The operations with polynomials are developed in the remaining sections with the two final sections covering division. Manipulatives 9, 10, and 11 on pages 556–561 can be used to introduce Lessons 8.3, 8.4, and 8.5 respectively.

SECTION-BY-SECTION COMMENTARY

Section 8.1 (pages 208–209)
It is important to emphasize that an expression such as $\frac{5}{x}$ is not a monomial even though it has a value when x is replaced by a nonzero number. When this expression is written in a monomial form, it becomes $5x^{-1}$. A negative exponent is not acceptable in a monomial.

Section 8.2 (pages 210–212)
The Product Property of Powers and the Commutative and Associative Properties of Multiplication provide a basis for multiplying monomials. Likewise, the Quotient Property of Powers and the Product Rule for Fractions explain the procedure for dividing monomials.

Section 8.3 (pages 213–216)
Polynomials are defined in this section. The examples show how to simplify expressions and write them in best polynomial form.

Section 8.4 (pages 217–219)
Both horizontal and vertical addition and subtraction are covered. Subtraction of polynomials is based on the meaning of subtraction of real numbers.

Section 8.5 (pages 220–223)
Both horizontal and vertical multiplication are covered. The FOIL method for multiplying two binomials is shown. Students are also given the general method of multiplying two polynomials.

Section 8.6 (pages 224–226)
Division of polynomials serves an indirect purpose. Many of the basic concepts and skills of algebra are reviewed when using this algorithm. In this section, only exercises with zero remainders occur.

Section 8.7 (pages 227–230)
Division involving non-zero remainders is included in this section.

WARM-UP EXERCISES

Section 8.1
Pages 208–209

Simplify. The answers form a magic square.

1. $10 - 11$
2. $-18 + 12$
3. $(-1)(-1)$
4. $-10 + 3 + 7$
5. $(0.5)(-4)$
6. $16 \div (-4)$
7. $5(-1)$
8. $(-\frac{1}{7})(-14)$
9. $\frac{-18}{6}$
10. What is the magic number -6

Section 8.2
Pages 210–212

Simplify.

1. $(7^x)(7^y)$ $\mathbf{7^{x+y}}$
2. $(2^{-5})(2^8)$ $\mathbf{2^3}$
3. $(3^{-6})(3^2)$ $\mathbf{\dfrac{1}{3^4}}$

4. $(9^{-2})(9^{-1})$ $\mathbf{\dfrac{1}{9^3}}$
5. $x^3 \cdot x^5$ $\mathbf{x^8}$
6. $\dfrac{a^x}{a^y}$ $\mathbf{a^{x-y}}$

7. $\dfrac{3^6}{3^5}$ $\mathbf{3}$
8. $\dfrac{5^4}{5^8}$ $\mathbf{\dfrac{1}{5^4}}$
9. $\dfrac{2^{-3}}{2^{-5}}$ $\mathbf{2^2}$

Section 8.3
Pages 213–216

Write Yes or No to show whether each expression or number is a monomial over {rational numbers}.

1. $5x$ **Yes**
2. $-2x^3$ **Yes**
3. $4x^0$ **Yes**

4. $\dfrac{-1}{x^3}$ **No**
5. $\dfrac{1}{x^2}$ **No**
6. 2 **Yes**

7. $-\frac{1}{2}x^3$ **Yes**
8. $\sqrt{2}\,x^4$ **No**
9. $x + 3$ **No**

Section 8.4
Pages 217–219

Simplify and arrange in decreasing order of powers of x.

1. $2x^2 + x - x^2 + 7$ $\mathbf{x^2 + x + 7}$
2. $x^3 + 7 - x^3 + x$ $\mathbf{x + 7}$
3. $5x^3 + x^2 - 4x^3 - x^2$ $\mathbf{x^3}$
4. $5x^2 - 2 - 4x^2 + 2x^0$ $\mathbf{x^2}$
5. $-3x^3 + 2x^4 - 4$ $\mathbf{2x^4 - 3x^3 - 4}$
6. $5x^0 - 1$ $\mathbf{4}$

Section 8.5
Pages 220–223

Complete this cross-polynomial puzzle. Place each term and operation symbol in a separate square.

Across

1. $x(-x)$
3. $2(2x - 3)$
5. $(x + 4)(x - 1)$
6. $(-6x)(x^4)$

Down

2. $(x + 1)(x + 2)$
3. $2(2x + 2)$
4. $(-3x^2)(2x^3)$
6. $(-4x^3)(2x^3)$

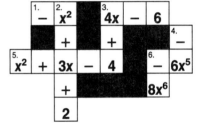

Section 8.6
Pages 224–226

Simplify.

1. $3x(x + 1)$ $\mathbf{3x^2 + 3x}$
2. $2x^2(x - 2)$ $\mathbf{2x^3 - 4x^2}$
3. $-x^2(x - 1)$ $\mathbf{-x^3 + x^2}$
4. $(4x^2 - 5) - (x^2 + 3)$ $\mathbf{3x^2 - 8}$
5. $\dfrac{x^4}{x^2}$ $\mathbf{x^2}$
6. $\dfrac{9x^3}{x}$ $\mathbf{9x^2}$
7. $\dfrac{64x^4}{8x}$ $\mathbf{8x^3}$
8. $\dfrac{56x^4}{8x^2}$ $\mathbf{7x^2}$

Section 8.7
Pages 227–230

Divide.

1. $5\overline{)625}$ **125**
2. $10\overline{)110}$ **11**
3. $3\overline{)8400}$ **2800**

4. $11\overline{)11022}$ **1002**
5. $6\overline{)328}$ $\mathbf{54\frac{2}{3}}$
6. $9\overline{)461}$ $\mathbf{51\frac{2}{9}}$

CHAPTER 9: FACTORING POLYNOMIALS

OVERVIEW

This chapter expands on the definition of factoring given in Section 2.5. At that point, factoring was defined as the process of expressing a sum as a product. The concepts of *prime polynomial* and *greatest common factor* are introduced in order to define factoring of polynomials. The first six sections cover common monomial factoring, factoring the difference of two squares, factoring quadratic trinomials, and factoring trinomial squares. Section 9.7 combines more than one type of factoring in one polynomial. Manipulatives 12, 13, and 14 on pages 562–567 can be used to introduce Lessons 9.3, 9.5, and 9.6 respectively.

SECTION-BY-SECTION COMMENTARY

Section 9.1 (pages 236–238)
The concept of prime polynomial is introduced in this section. This is important because factoring of polynomials involves finding their prime factors. The concept of greatest common factor is also introduced.

Section 9.2 (pages 239–241)
This section expresses factorable polynomials like $x^2 + bx + c$ in the form $x^2 + rx + sx + rs$. This leads to the factoring of quadratic trinomials like $ax^2 + bx + c$ in later sections.

Section 9.3 (pages 242–245)
Students are taught to write a factorable quadratic trinomial of the form $x^2 + bx + c$ as $x^2 + rx + sx + rs$ and then proceed as in Section 9.2.

Section 9.4 (pages 246–249)
An example is used to show that the factoring of a general quadratic trinomial such as $ax^2 + bx + c$ is the reverse of multiplying two binomials. The first step is to find two integers r and s such that $r + s = b$ and $r \cdot s = ac$ in order to write the trinomial as $ax^2 + rx + sx + c$.

Section 9.5 (pages 252–254)
Development of a rule for factoring the difference of two squares is accomplished by first considering the nature of the general product of two binomials. One represents the difference of two numbers and the other represents the sum of the same numbers.

Section 9.6 (pages 255–257)
In this section the conditions necessary for a trinomial of the form $ax^2 + bx + c$ to be a trinomial square are presented. Then factoring of trinomial squares is demonstrated as a special method.

Section 9.7 (pages 258–260)
Students of low-average ability often experience difficulty in applying one or more of the methods learned in the preceding sections. They must first identify the method of factoring to use. Several steps are often involved with factoring a general polynomial, and a student may "lose" some of the factors in the process. Emphasis should be given to the importance of applying common monomial factoring first.

WARM-UP EXERCISES

Section 9.1
Pages 236–238

Multiply.

1. $x(2x - 5)$ $2x^2 - 5x$
2. $x(x + 4)$ $x^2 + 4x$
3. $2(2x + 1)$ $4x + 2$
4. $2(x^2 + 1)$ $2x^2 + 2$
5. $2x(2x^2 + 3)$ $4x^3 + 6x$
6. $5(x - 2)$ $5x - 10$
7. $3x^2(x^2 - 3x)$ $3x^4 - 9x^3$
8. $x^2(3x + 2)$ $3x^3 + 2x^2$

Section 9.2
Pages 239–241

Complete the table below. Write two integers, r and s, that have the given sum and product. The answer to Exercise 1 is shown.

	Sum	Product	r	s			Sum	Product	r	s
1.	3	2	1	2	**2.**	4	3	1	3
3.	5	4	1	4	**4.**	5	6	2	3
5.	6	8	2	4	**6.**	7	10	2	5

Section 9.3
Pages 242–245

Factor.

1. $x^2 + 4x + 3$ $(x + 1)(x + 3)$ **2.** $x^2 + 3x + 2$ $(x + 1)(x + 2)$
3. $x^2 + 8x + 12$ $(x + 2)(x + 6)$ **4.** $x^2 + 7x + 6$ $(x + 1)(x + 6)$
5. $x^2 + 7x + 12$ $(x + 3)(x + 4)$ **6.** $x^2 + 6x + 8$ $(x + 2)(x + 4)$

Section 9.4
Pages 246–249

Complete the table below. Write two integers, r and s, that have the given sum and product.

	Sum	Product	r	s			Sum	Product	r	s
1.	1	−2	−1	2	**2.**	−1	−6	−3	2
3.	1	−6	−2	3	**4.**	1	−12	−3	4
5.	−1	−12	−4	3	**6.**	−3	−10	−5	2

Section 9.5
Pages 252–254

Multiply.

1. $(x + 2)(x + 3)$ $x^2 + 5x + 6$ **2.** $(x + 1)(x + 2)$ $x^2 + 3x + 2$
3. $(x + 3)(x - 2)$ $x^2 + x - 6$ **4.** $(x + 2)(x - 3)$ $x^2 - x - 6$
5. $(x + 2)(x - 2)$ $x^2 - 4$ **6.** $(x + 3)(x - 3)$ $x^2 - 9$

Section 9.6
Pages 255–257

Place a check under each binomial that is a factor of the polynomial in the left column.

	x − 1	x + 1	x − 2	x + 2
1. $x^2 - 1$	✓	✓		
2. $x^2 + 3x + 2$		✓		✓
3. $x^2 - x - 2$		✓	✓	
4. $x^2 + x - 2$	✓			✓
5. $x^2 - 4$			✓	✓

Section 9.7
Pages 258–260

Factor.

1. $4x + 4$ $4(x + 1)$ **2.** $18x^2 + 3x$ $3x(6x + 1)$
3. $x^2 + 5x + 6$ $(x + 2)(x + 3)$ **4.** $x^2 + 7x + 10$ $(x + 2)(x + 5)$
5. $x^2 + x - 12$ $(x - 3)(x + 4)$ **6.** $x^2 - 6x + 5$ $(x - 5)(x - 1)$
7. $2x^2 + 3x + 1$ $(2x + 1)(x + 1)$ **8.** $x^2 - 49$ $(x - 7)(x + 7)$

CHAPTER 10: PRODUCTS AND QUOTIENTS OF RATIONAL EXPRESSIONS

OVERVIEW

The chapter begins with the definition of a rational expression. The opening section also includes evaluating rational expressions and examining restrictions on the domain of the variable(s). The work extends to simplifying (reducing) rational expressions in preparation for the operations of multiplication and division. Note that addition and subtraction of rational expressions follow in Chapter 11. All of the work with rational expressions includes polynominals that are relatively easy to factor. Factoring of polynomials seems to be a difficult skill for low-average ability students to develop and maintain.

SECTION-BY-SECTION COMMENTARY

Section 10.1 (pages 268–270)
In this section *rational expression* is defined. The polynomials used to form rational expressions will be polynomials over {rational numbers}. This means the coefficients of the terms are rational numbers. The examples cover evaluating rational expressions and finding restrictions on the domain of the variable.

Section 10.2 (pages 271–273)
The work of simplifying rational expressions in this section is like reducing fractions of arithmetic. The procedures are similar except that the numerators and denominators are polynomials rather than numbers.

Section 10.3 (pages 274–276)
The examples and exercises of this section involve only rational expressions with numerators and denominators that are monomials. Example 1 is shown in detail. The other examples show how several steps may be combined in order to work problems of this type more expeditiously.

Section 10.4 (pages 277–280)
In this section rational expressions with binomial numerators and denominators are used in the examples and exercises. The work is restricted to types that involve no factoring.

Section 10.5 (pages 281–283)
A brief review of three types of factoring is included in this section. These include quadratic trinomials, difference of two squares, and common monomial factoring. It is usually desirable to have students write results with numerators and denominators left in factored form.

Section 10.6 (pages 284–286)
The Quotient Rule for rational expressions shows that division of rational expressions is similar to division with fractions of arithmetic. In this lesson students learn to divide rational expressions without factoring getting in the way.

Section 10.7 (pages 287–289)
In this lesson the Quotient Rule is applied to problems in which factoring of polynomials is also an essential step. Each quotient must first be expressed as a product, after which the techniques of Section 10.5 can be used. Expressing a quotient as a complex fraction is also introduced as a method of simplifying a quotient.

WARM-UP EXERCISES

Section 10.1
Pages 268–270

Find the value or values of x that will make each expression equal to zero.

1. $x - 2$ **2**
2. $x + 6$ **−6**
3. $4x$ **0**
4. $-8x$ **0**
5. $5x - 3$ $\frac{3}{5}$
6. $3x + 12$ **−4**
7. $x^2 - 16$ **4, −4**
8. $(x - 2)(x + 9)$ **2, −9**
9. $(x + 4)(x + 3)$ **−4, −3**

Section 10.2
Pages 271–273

Complete each of Exercises 1–6 with one of items **A-L** at the right.

1. $50x^3 = 10x(?)$ **F**
2. $3x + 15 = 3(?)$ **K**
3. $7x - 28 = 7(?)$ **G**
4. $x^2 + 10x = x(?)$ **A**
5. $4x^2 - 28x = 4x(?)$ **C**
6. $9x - 9y = 9(?)$ **B**

A. $x + 10$
B. $x - y$
C. $x - 7$
D. $3a + b$
E. $x + 7$
F. $5x^2$

G. $x - 4$
H. $a + 3b$
I. $x + y$
J. $5x$
K. $x + 5$
L. $x - 28$

Section 10.3
Pages 274–276

Identify a name for 1 to use in simplifying each rational expression. (No divisor equals 0.)

1. $\dfrac{3}{9}$ $\dfrac{3}{3}$

2. $\dfrac{8x}{12x}$ $\dfrac{4x}{4x}$

3. $\dfrac{x(x - 1)}{x(x + 1)}$ $\dfrac{x}{x}$

4. $\dfrac{2x(x + 3)}{4x(x - 3)}$ $\dfrac{2x}{2x}$

5. $\dfrac{x^2 - 5x}{x^2 + x}$ $\dfrac{x}{x}$

6. $\dfrac{x - 1}{x^2 - 1}$ $\dfrac{x - 1}{x - 1}$

Section 10.4
Pages 277–280

Write *True* or *False* for each sentence.

1. $(-1(-4) = 4$ **True**
2. $(1)(-6) = 6$ **False**
3. $-5(-2 + 3) = 10 - 15$ **True**
4. $-1(2 + 3) = -2 - 3$ **True**
5. $-1(-x) = -x$ **False**
6. $-1(x + 3) = -x - 3$ **True**

Section 10.5
Pages 281–283

Simplify.

1. $\dfrac{2x}{3x}$ $\dfrac{2}{3}$

2. $\dfrac{4x}{4(x + 1)}$ $\dfrac{x}{x + 1}$

3. $\dfrac{3(x - 1)}{x(x - 1)}$ $\dfrac{3}{x}$

4. $\dfrac{x(x - 3)}{2x(x + 3)}$ $\dfrac{(x - 3)}{2(x + 3)}$

5. $\dfrac{(x + 5)(x - 4)}{x(x - 4)(x - 4)}$ $\dfrac{(x + 5)}{x(x - 4)}$

Section 10.6
Pages 284–286

Divide. Simplify where necessary.

1. $\dfrac{1}{5} \div \dfrac{1}{3}$ $\dfrac{3}{5}$

2. $\dfrac{1}{4} \div \dfrac{2}{3}$ $\dfrac{3}{8}$

3. $\dfrac{1}{6} \div \dfrac{1}{2}$ $\dfrac{1}{3}$

4. $\dfrac{2}{5} \div \dfrac{2}{3}$ $\dfrac{3}{5}$

5. $\dfrac{4}{3} \div \dfrac{2}{9}$ 6

6. $\dfrac{8}{9} \div \dfrac{2}{15}$ $\dfrac{20}{3}$

7. $\dfrac{5}{6} \div \dfrac{25}{24}$ $\dfrac{4}{5}$

8. $\dfrac{7}{10} \div \dfrac{3}{5}$ $\dfrac{7}{6}$

Section 10.7
Pages 287–289

Write each quotient as an equivalent product.

1. $\dfrac{4}{5} \div \dfrac{2}{3}$ $\dfrac{4}{5} \cdot \dfrac{3}{2}$

2. $\dfrac{5}{x} \div (x + 1)$ $\dfrac{5}{x} \cdot \dfrac{1}{x + 1}$

3. $\dfrac{x}{x + 1} \div x$ $\dfrac{x}{x + 1} \cdot \dfrac{1}{x}$

4. $\dfrac{x}{(x - 1)} \div \dfrac{x}{x + 1}$ $\dfrac{x}{x - 1} \cdot \dfrac{x + 1}{x}$

5. $\dfrac{x - 1}{x} \div \dfrac{x}{x - 1}$ $\dfrac{x - 1}{x} \cdot \dfrac{x - 1}{x}$

6. $(x + 1) \div \dfrac{3x}{x - 1}$ $\dfrac{x + 1}{1} \cdot \dfrac{x - 1}{3x}$

CHAPTER 11:   SUMS AND DIFFERENCES OF RATIONAL EXPRESSIONS

OVERVIEW

Adding and subtracting with rational expressions may be a difficult skill for some students to develop. Many steps are involved and students must know many previously studied concepts very well in order to perform the steps correctly. As in the preceding chapter, the polynomial factoring is kept relatively simple in the exercises. The main outcome to be expected of students is a demonstrated understanding of the process for adding and subtracting with rational expressions.

SECTION-BY-SECTION COMMENTARY

Section 11.1 (pages 294–296)
The fundamental rules for the sum of two rational expressions and the difference of two rational expressions are stated in this section. These parallel the same rules for fractions of arithmetic. In this section the like denominators are restricted to monomials and binomials, and the numerators are monomials.

Section 11.2 (pages 297–299)
This section is an extension of Section 11.1. Some of the examples and exercises may need to be simplified after the application of the Sum or Difference Rule. Also, many of the exercises have binomial numerators as well as denominators.

Section 11.3 (pages 300–302)
The goal in this section is for students to learn to find the least common multiple of two or more polynomials. A review of the meaning of the least common multiple of two or more counting numbers is included. For purposes of the work in this chapter it is desirable to express the least common multiple of two or more polynomials in factored form.

Section 11.4 (pages 303–306)
As the title suggests, the work of this lesson will be restricted to expressions having monomial denominators. Numerators are also restricted to monomials. This approach should allow students to develop some proficiency in the process before the manipulative work gets complicated.

Section 11.5 (pages 307–309)
All the examples and exercises of this section require no factoring of polynomial denominators. In each case the least common multiple is the product of the two denominators. A few of the exercises involve binomial numerators.

Section 11.6 (pages 310–313)
This section includes addition and subtraction with rational expressions of the most general type. Students should be encouraged to arrange their work in a form that will help them proceed toward a solution in a step-by-step manner. Factoring of denominators is required.

Section 11.7 (pages 314–316)
It is probably good to have students note the restriction on the domain of the variable in an equation before they transform the equation in any way. This will avoid the problems of students forgetting to check for so-called "extraneous" roots.

WARM-UP EXERCISES

Section 11.1
Pages 294–296

Add or subtract as indicated. Simplify where necessary.

1. $\frac{2}{5} + \frac{1}{5}$ $\frac{3}{5}$

2. $\frac{3}{8} - \frac{2}{8}$ $\frac{1}{8}$

3. $\frac{5}{7} - \frac{6}{7}$ $-\frac{1}{7}$

4. $\left(-\frac{3}{5}\right) + \left(-\frac{1}{5}\right)$ $-\frac{4}{5}$

5. $\left(-\frac{1}{6}\right) + \left(-\frac{2}{6}\right)$ $-\frac{1}{2}$

6. $\frac{2}{7} + \frac{5}{7}$ 1

7. $\left(-\frac{3}{4}\right) + \left(-\frac{1}{4}\right)$ -1

8. $\frac{2}{9} - \left(-\frac{1}{9}\right)$ $\frac{1}{3}$

9. $\left(-\frac{1}{4}\right) - \left(-\frac{3}{4}\right)$ $\frac{1}{2}$

Section 11.2
Pages 297–299

Add or subtract as indicated. Simplify where necessary.

1. $\dfrac{3}{5} + \dfrac{1}{5}$ **$\dfrac{4}{5}$**

2. $\dfrac{3}{9} - \dfrac{4}{9}$ **$-\dfrac{1}{9}$**

3. $\dfrac{6}{x} + \dfrac{3}{x}$ **$\dfrac{9}{x}$**

4. $\dfrac{5}{x} - \dfrac{8}{x}$ **$-\dfrac{3}{x}$**

5. $\dfrac{1}{5x} + \dfrac{3}{5x}$ **$\dfrac{4}{5x}$**

6. $\dfrac{2}{7x} - \dfrac{3}{7x}$ **$-\dfrac{1}{7x}$**

7. $\dfrac{4}{x+1} + \dfrac{1}{x+1}$ **$\dfrac{5}{x+1}$**

8. $\dfrac{x}{x+2} + \dfrac{2x}{x+2}$ **$\dfrac{3x}{x+2}$**

Section 11.3
Pages 300–302

Add or subtract as indicated.

1. $(x^2 + 1) + (2x^2 + 3)$ **$3x^2 + 4$**

2. $(-x - 1) - (x + 1)$ **$-2x - 2$**

3. $(x^3 - 1) + (x^2 + 1)$ **$x^3 + x^2$**

4. $(4x + 3) - (4x - 1)$ **4**

5. $(x + 1) - (x - 1)$ **2**

6. $(x^3 + 1) + (x^2 - 1)$ **$x^3 + x^2$**

Section 11.4
Pages 303–306

Find each LCM.

1. $3; 11$ **33**

2. $r; s$ **rs**

3. $x; 5x$ **$5x$**

4. $6a; 2a$ **$6a$**

5. $3x; 12y$ **$12xy$**

6. $a^2b; ab^2$ **a^2b^2**

7. $rs; rs^2$ **rs^2**

8. $2d; cd$ **$2cd$**

Section 11.5
Pages 307–309

Write the missing name for 1.

1. $\dfrac{3}{5}\left(\dfrac{?}{?}\right) = \dfrac{6}{10}$ **$\dfrac{2}{2}$**

2. $\dfrac{2}{x}\left(\dfrac{?}{?}\right) = \dfrac{4}{2x}$ **$\dfrac{2}{2}$**

3. $\dfrac{x}{3}\left(\dfrac{?}{?}\right) = \dfrac{x^2}{3x}$ **$\dfrac{x}{x}$**

4. $\dfrac{5}{2a}\left(\dfrac{?}{?}\right) = \dfrac{5b}{2ab}$ **$\dfrac{b}{b}$**

5. $\dfrac{3b}{2a}\left(\dfrac{?}{?}\right) = \dfrac{9ab}{6a^2}$ **$\dfrac{3a}{3a}$**

6. $\dfrac{5x}{xy}\left(\dfrac{?}{?}\right) = \dfrac{10xy}{2xy^2}$ **$\dfrac{2y}{2y}$**

Section 11.6
Pages 310–313

Write the simplest name for each numerator.

1. $\dfrac{(x + 1) + x}{x + 1}$ **$2x + 1$**

2. $\dfrac{(3x + 5) - 2x}{x}$ **$x + 5$**

3. $\dfrac{(5x - 5) - x}{x + 3}$ **$4x - 5$**

4. $\dfrac{3 + (2x - 5)}{x - 4}$ **$2x - 2$**

5. $\dfrac{(x + 5) - (x + 3)}{x + 3}$ **2**

6. $\dfrac{(5x + 1) - (x + 1)}{x - 2}$ **$4x$**

Section 11.7
Pages 314–316

Solve each equation.

1. $4x = 24$ **$x = 6$**

2. $3x = 33$ **$x = 11$**

3. $4x + 2 = 6$ **$x = 1$**

4. $2x - 1 = 7$ **$x = 4$**

5. $3x - 5 = 2x$ **$x = 5$**

6. $6x + 10 = 11x$ **$x = 2$**

7. $3x + 4 = 2x + 7$ **$x = 3$**

8. $7x + 35 = 3x + 3$ **$x = -8$**

9. $6x + 18 = 5x + 15$ **$x = -3$**

CHAPTER 12:   RELATIONS AND FUNCTIONS

OVERVIEW

This chapter involves graphing in a coordinate plane. Graphing will be utilized as a vehicle for presenting a number of mathematical topics including relations, functions, linear and quadratic functions, systems of linear equations and inequalities, and solution sets of quadratic equations.

SECTION-BY-SECTION COMMENTARY

Section 12.1 (pages 322–324)
This section develops student skills in locating points that are graphs of ordered pairs of numbers.

Section 12.2 (pages 325–327)
This section develops the concept of one-to-one correspondence between the set of points of a coordinate plane and ordered pairs of real numbers.

Section 12.3 (pages 328–331)
In this section the concept of relation is defined, and three ways of describing a relation are covered. Emphasis should be given to the fact that the elements of a relation are the *ordered pairs* of numbers and not the single numbers that make up the pairs.

Section 12.4 (pages 333–336)
The concepts of domain and range are defined as well as the important concept of a function. The vertical line test of a function is discussed. This is a good device for students to remember.

Section 12.5 (pages 337–339)
The purpose of this section is to show students how to make graphs of functions when the domain is given and the rule is given by an algebraic formula. A linear function is graphed in Example 1, and a quadratic function is graphed in Example 2.

Section 12.6 (pages 340–342)
Example 1 of this section involves interpretation of the graph of a function. Example 2 involves making predictions from a graph by extrapolating data.

WARM-UP EXERCISES

Section 12.1
Pages 322–324

Start at the shaded square in the center each time and name the square reached after making the move or moves described.

A	B	C	D	E
F	G	H	I	J
K	L	■	M	N
O	P	Q	R	S
T	U	V	W	X

1. 1 right   **M**
2. 2 left   **K**
3. 2 up   **C**
4. 1 down   **Q**
5. 2 right, 1 up   **J**
6. 2 right, 2 down   **X**
7. 1 left, 1 up   **G**
8. 1 left, 2 down   **U**
9. 2 left, 2 down   **T**
10. 2 right, 2 up   **E**

Section 12.2
Pages 325–327

Write the ordered pair that corresponds to each point.

1. A **(−3, 2)**   **2.** B **(1, 1)**

3. C **(1, −1)**   **4.** D **(−2, −1)**

5. E **(−2, 2)**   **6.** F **(−3, −2)**

7. G **(2, −2)**   **8.** H **(−1, −2)**

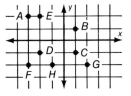

Section 12.3
Pages 328–331

Graph the point that corresponds to each ordered pair in the same coordinate plane. Label each point.

1. (0, 0)   **2.** (4, 0)

3. (1, 0.5)   **4.** (1, −0.5)

5. (0, −2)   **6.** (0, 2)

7. (−3, 1.75)   **8.** (−3, −1.75)

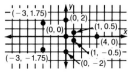

Section 12.4
Pages 333–336

Complete the Magic Square. Find the y value corresponding to the given x value for the relation T described by the equation $y = x + 3$.

1. (5, ___)   **2.** (−9, ___)   **3.** (1, ___)

4. (−5, ___)   **5.** (−1, ___)   **6.** (3, ___)

7. (−3, ___)   **8.** (7, ___)   **9.** (−7, ___)

10. What is the magic sum for this square? **6**

1. 8	2. −6	3. 4
4. −2	5. 2	6. 6
7. 0	8. 10	9. −4

Section 12.5
Pages 337–339

From the graph below, write the coordinates of another point which has the same x value as the given point.

1. (−1, 1) **(−1, −1)**   **2.** (−1.25, −1.25) **(−1.25, 1.25)**

3. (−2, 2) **(−2, −2)**   **4.** (−2.5, −2.5) **(−2.5, 2.5)**

5. (−3, 3) **(−3, −3)**   **6.** (−3.5, −3.5) **(−3.5, 3.5)**

7. (−4, 4) **(−4, −4)**   **8.** (−4.75, −4.75) **(−4.75, 4.75)**

9. Is this the graph of a function? **No**

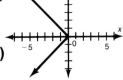

Section 12.6
Pages 340–342

Using {real numbers} as the domain, complete the tables. Then draw a graph of the functions.

1. $y = 2x - 1$ **2.** $y = 0 \cdot x - 2$

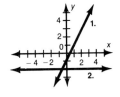

x	−2	−1	0	1	2
y	−5	−3	−1	1	3

x	−2	−1	0	1	2
y	−2	−2	−2	−2	−2

CHAPTER 13:   LINEAR FUNCTIONS

OVERVIEW

The concept of function was introduced and defined in Chapter 12. This chapter treats the linear function in a somewhat more detailed fashion. Such concepts as slope, parallel lines, and intercepts are covered. Some applications of the linear function are shown in Section 13.7.

SECTION-BY-SECTION COMMENTARY

Section 13.1 (pages 350–352)
The linear function is defined in this section, and the standard formula of a linear function $y = mx + b$, is developed.

Section 13.2 (pages 353–355)
The y value of the point of intersection of a line and the y axis is defined to be the y intercept.

Section 13.3 (pages 356–358)
In this section, *slope* is defined as the value of m in the standard formula of a linear function, $y = mx + b$.

Section 13.4 (pages 359–363)
The slope formula based on the coordinates of two points of a line is developed.

Section 13.5 (pages 364–366)
The main idea of this section is that two lines in a coordinate plane are parallel if they have the same slope.

Section 13.6 (pages 367–369)
The x intercept is defined in this section. A technique for graphing a linear function using only the x and y intercepts is demonstrated.

Section 13.7 (pages 370–372)
In this section, some practical applications of linear functions are presented. In particular, emphasis is given to using graphs to estimate values of linear functions.

WARM-UP EXERCISES

Section 13.1
Pages 350–352

Complete each table below. Then graph the ordered pairs, and draw a line containing the points.

1. $y = x + 1$

x	−2	−1	0	1	2
y	−1	0	1	2	3

2. $y = x^2$

x	−2	−1	0	1	2
y	4	1	0	1	4

1.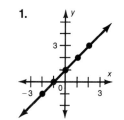

2.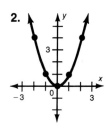

3. Which of these graphs appears to be a straight line?   **The first**

Section 13.2
Pages 353–355

Compare each rule with $y = mx + b$. Then write the values of m and b.

1. $y = 2x + 1$; $m = \underline{\ 2\ }$ $b = \underline{\ 1\ }$ **2.** $y = -3x - 5$; $m = \underline{-3}$ $b = \underline{-5}$

3. $y = 2 - x$; $m = \underline{-1}$ $b = \underline{\ 2\ }$ **4.** $y = 10 - 5x$; $m = \underline{-5}$ $b = \underline{10}$

5. $y = \frac{1}{2}x - \frac{1}{2}$; $m = \underline{\frac{1}{2}}$ $b = \underline{-\frac{1}{2}}$ **6.** $y = \frac{1}{4} - \frac{1}{4}x$; $m = \underline{-\frac{1}{4}}$ $b = \underline{\frac{1}{4}}$

Section 13.3
Pages 356–358

Which function in each pair would have the "steeper" graph?
NOTE: The correct answer is underlined.

1. $\underline{y = 5x}$ **2.** $y = 2x$ **3.** $\underline{y = -3x}$ **4.** $y = 4x$
  $y = x$ $\underline{y = 4x}$ $y = -x$ $\underline{y = 5x}$

5. $y = -6x$ **6.** $y = 5x + 3$ **7.** $\underline{y = -2x - 1}$ **8.** $y = x + 1$
  $\underline{y = -10x}$ $\underline{y = 8x + 3}$ $y = -x - 1$ $\underline{y = -3x + 1}$

Section 13.4
Pages 359–363

Write the slope of the graph of each linear function in the correct square.
The answers form a magic square.

1. $y = 2x + 1$ **2.** $y = 3 - 8x$ **3.** $y = 6x$

4. $y = 4x + 2$ **5.** $y = 2$ **6.** $y = 3 - 4x$

7. $y = -6x$ **8.** $y = 8x$ **9.** $y = -2x + 2$

1. 2	2. −8	3. 6
4. 4	5. 0	6. −4
7. −6	8. 8	9. −2

Section 13.5
Pages 364–366

Write the slope of the line that contains the two given points.

1. $(3, 3)$; $(1, 1)$ **1** **2.** $(-2, 1)$; $(0, 0)$ $-\frac{1}{2}$ **3.** $(0, 0)$; $(2, 1)$ $\frac{1}{2}$

4. $(-2, 2)$; $(-1, 4)$ **2** **5.** $(-1, 4)$; $(0, 2)$ -2 **6.** $(0, 1)$; $(1, 0)$ -1

Section 13.6
Pages 367–369

Write <u>Yes</u> or <u>No</u> to tell whether the graphs of the given functions are parallel.

1. $y = 2x + 1$ **2.** $y = \frac{1}{2}x + 1$ **3.** $y = 1 - 3x$
  $y = 3 - 2x$ **No** $y = 0.5x - 1$ **Yes** $y = -3x - 5$ **Yes**

4. $y = 2$ **5.** $y = -1$ **6.** $y = -2x - 2$
  $y = x$ **No** $y = -x$ **No** $y = 3 - 2x$ **Yes**

Section 13.7
Pages 370–372

Write each x intercept in the correct square. The sums
of the horizontal and vertical answers are equal.

1. $y = x - 2$ **2.** $y = x + 3$ **3.** $y = 2x + 4$

4. $y = 3x - 3$ **5.** $y = 2x - 6$ **6.** $y = 4x + 4$

7. $y = 4x - 2$ **8.** $y = x + \frac{1}{2}$ **9.** $y = -3x$

10. What is the sum of the horizontal and vertical answers? $1\frac{1}{2}$

1. 2			
2. −3			
3. −2	4. 1	5. 3	6. −1	7. $1\frac{1}{2}$
		8. $-\frac{1}{2}$		
		9. 0		

CHAPTER 14:   SYSTEMS OF SENTENCES

OVERVIEW

This chapter covers a variety of methods for finding solution sets of systems of linear equations. Included are the graphical, addition, and multiplication/addition methods.

SECTION-BY-SECTION COMMENTARY

Section 14.1 (pages 378–380)
In this section, attention is directed to the graphs of the solution sets of linear equations in two variables.

Section 14.2 (pages 381–384)
In this section, the graphical method of finding solution sets is presented. One of the examples covers the case in which the solution set is empty.

Section 14.3 (pages 385–387)
The addition method is used in this section. All of the examples and exercises have been designed so that a variable is eliminated by adding the two equations.

Section 14.4 (pages 388–390)
In this section, both multiplication and addition must be applied. The first example requires that only one equation be multiplied by a number. The second example requires that both equations be multiplied by a number.

Section 14.5 (pages 391–393)
The addition method and the multiplication/addition method are applied to solving word problems.

WARM-UP EXERCISES

Section 14.1
Pages 378–380

Find the value of y for each given value of x.

1. $2x - y = 3; x = 1$ **−1**

2. $-x + 4y = -2; x = 6$ **1**

3. $3x + 2y = 1; x = 4$ $-\frac{11}{2}$

4. $5x - 3y = -2; x = 2$ **4**

5. $2x - 4y = -5; x = -4$ $-\frac{3}{4}$

6. $-5x - 2y = 12; x = -3$ $\frac{3}{2}$

Section 14.2
Pages 381–384

Write the x and y intercepts for the graphs of the equations below.

1. $x + 2y - 4 = 0$ **4; 2**

2. $3x + y + 6 = 0$ **−2; −6**

3. $3x - y + 3 = 0$ **−1; 3**

4. $x + y + 6 = 0$ **−6; −6**

5. $x - y = -2$ **−2; 2**

6. $x + y = -1$ **−1; −1**

7. $x + y - 2 = 0$ **2; 2**

8. $x - 2y + 4 = 0$ **−4; 2**

Section 14.3
Pages 385–387

Simplify each expression. Write the result in the corresponding box to complete the magic square.

1. $5x - 3x$

2. $4x - x$

3. $-x + (-x)$

4. $-7x + 4x$

5. $4x + 7 - 3x - 7$

6. $2x - 3 + 3x + 3$

7. $12x + (-8x)$

8. $3x - 4x$

9. $4x - 4x$

1	2	3
2x	**3x**	**−2x**
4	5	6
−3x	**x**	**5x**
7	8	9
4x	**−x**	**0**

Section 14.4
Pages 388–390

Simplify each expression.

1. $x + 2y + (2x - 2y)$ **3x**

2. $-2x + 4y + (2x + 3y)$ **7y**

3. $4x - 3y + (-4x - 2y)$ **−5y**

4. $3x + 3y + (6x - 3y)$ **9x**

5. $-5x + 4y + (8x - 4y)$ **3x**

6. $3x - 2y + (-3x + 12y)$ **10y**

Section 14.5
Pages 391–393

Write each missing value.

Equations		x	y
1. $x + y =$ __?__	**3**	7	−4
3. $x - y =$ __?__	**1**	−5	−6
5. $x + y = 10$		13	−3 **?**
7. $x + y = -12$		−2	−10 **?**
9. $x + y = 0$		3 **?**	−3

Equations		x	y
2. $x + y =$ __?__ −14		−3	−11
4. $x - y =$ __?__ −17		−12	5
6. $x - y = 10$		11	1 **?**
8. $x - y = -12$		−20 **?**	−8
10. $x - y = 0$		−9	−9 **?**

CHAPTER 15: MORE ON SYSTEMS OF SENTENCES

OVERVIEW

This chapter opens with a presentation of the substitution method of solving a system of two linear equations in two variables. Then the work of using systems of equations to solve problems that were introduced in Chapter 14 is now extended. The classic digit problems and mixture problems are covered. The introduction of the determinant method for solving linear systems is a unique feature of the chapter. Then this method and others are applied to money problems and motion problems.

SECTION-BY-SECTION COMMENTARY

Section 15.1 (pages 398–400)
The substitution method is developed in this section. It is, perhaps, the most useful method for working with applications of systems of equations. Example 3 in the text illustrates an inconsistent system of equations. The graph of such a system is two parallel lines.

Section 15.2 (pages 401–404)
The two-variable representation of two-digit numbers is explained. The examples are problem solving applications based on two-digit numbers. Identification and proper use of the two conditions in these problems are emphasized.

Section 15.3 (pages 405–409)
Solving mixture problems is one of the most challenging tasks of an elementary algebra course. In this presentation students are given many helpful suggestions for making their work successful. These include identifying the conditions of the problem, using a table to organize the relationships of the problem, and help in translating the words of the problem to form the equations.

Section 15.4 (pages 412–414)
The purpose of this section is to teach students what a 2×2 determinant is and how to find its value.

Section 15.5 (pages 415–417)
The method of determinants for solving systems of equations in two variables is presented.

Section 15.6 (pages 418–421)
The problems in this section involve money or costs of items. The number of items, the cost per item, and the total cost are quantities either known or to be solved for in each problem. The use of a table to show the relationships of each problem is recommended.

Section 15.7 (pages 422–425)
The motion problems of this section are based on the formula $d = rt$, which relates distance, speed, and time. Again, emphasis is given to using tables to show the relationship of quantities in each problem.

WARM-UP EXERCISES

Section 15.1
Pages 398–400

Write the equivalent equation that has x on one side.

1. $x - 3y = 2$ $x = 2 + 3y$
2. $x + 5y = 3$ $x = 3 - 5y$
3. $x - 2y = 7$ $x = 7 + 2y$
4. $12 = 10y - x$ $x = 10y - 12$

Section 15.2
Pages 401–404

Find the value of (a) $10x + 4$ and (b) xy in each exercise.

1. $x = 5; y = 1$ **a. 54 b. 5**
2. $x = 3; y = 4$ **a. 34 b. 12**
3. $x = 6; y = 0$ **a. 64 b. 0**
4. $x = 7; y = 2$ **a. 74 b. 14**

Section 15.3
Pages 405–409

Write a decimal for each percent.

1. 16% **0.16**   **2.** 3% **0.03**   **3.** 20% **0.2**   **4.** 12% **0.12**

Solve.

5. 15% of 360 is what number? **54**   **6.** 40% of 120 is what number? **48**

7. What number is 8% of 200? **16**   **8.** What number is 20% of 150? **30**

Section 15.4
Pages 412–414

Evaluate each expression.

1. $2(3) - 3(1)$ **3**   **2.** $3(-1) - 2(4)$ **−11**   **3.** $1(4) - (-3)(2)$ **10**

4. $4(-2) - 2(5)$ **−18** **5.** $(-5)(3) - 4(-1)$ **−11** **6.** $(-1)(1) - 2(6)$ **−13**

Section 15.5
Pages 415–417

Complete the magic square by computing the value of each determinant. Write each answer in the correct square.

1. $\begin{vmatrix} 3 & 2 \\ 1 & 1 \end{vmatrix}$ **2.** $\begin{vmatrix} 3 & 2 \\ 1 & 0 \end{vmatrix}$ **3.** $\begin{vmatrix} -3 & 2 \\ 1 & -1 \end{vmatrix}$

4. $\begin{vmatrix} 3 & 2 \\ 0 & 0 \end{vmatrix}$ **5.** $\begin{vmatrix} 0 & 0 \\ 1 & 1 \end{vmatrix}$ **6.** $\begin{vmatrix} 4 & 2 \\ 6 & 3 \end{vmatrix}$

7. $\begin{vmatrix} 3 & -2 \\ 1 & -1 \end{vmatrix}$ **8.** $\begin{vmatrix} 2 & 0 \\ 1 & 1 \end{vmatrix}$ **9.** $\begin{vmatrix} -3 & -2 \\ 1 & 1 \end{vmatrix}$

1.	2.	3.
1	**−2**	**1**
4.	5.	6.
0	**0**	**0**
7.	8.	9.
−1	**2**	**−1**

Section 15.6
Pages 418–421

Write <u>Yes</u> or <u>No</u> to tell whether each given set is the solution set of the system.

1. $x + y = 5$ $\{(4,1)\}$ **2.** $2x - y = 6$ $\{(1,-1)\}$
  $x - y = 3$ **Yes** $x + y = 0$ **No**

3. $x + y = 29$ $\{(20,9)\}$ **4.** $x - y = -6$ $\{(2,10)\}$
  $x - y = 7$ **No** $x + y = 12$ **No**

5. $y = 2x + 4$ $\{(2,3)\}$ **6.** $2x + 2y = 20$ $\{(6,4)\}$
  $x + y = 5$ **No** $x - y = 2$ **Yes**

Section 15.7
Pages 422–425

Multiply.

1. $3(x + 1)$ **3x + 3**   **2.** $5(y - 2)$ **5y − 10**   **3.** $2(x + y)$ **2x + 2y**

4. $8(a - b)$ **8a − 8b**   **5.** $12(-m + 4)$ **6.** $(x - 12)15$ **15x − 180**
                                      −12m + 48

CHAPTER 16:   SYSTEMS OF INEQUALITIES

OVERVIEW

The work with systems of inequalities is a natural combination of the material studied in Chapters 5 and 14. The approach used in Chapter 16 provides students with an opportunity to learn more about the concepts of relation and the linear function. Furthermore, the manipulative algebra required will help students strengthen their understanding and improve their accuracy in fundamental skills. Finally, the emphasis given to graphs should provide students with increased knowledge and skill in graphical interpretation. This chapter and those that follow are considered optional for Basic level classes.

SECTION-BY-SECTION COMMENTARY

Section 16.1 (pages 434–436)
Linear inequalities in two variables are introduced in this section. Students will have to be reminded that multiplying each side of an inequality by a negative number reverses the order.

Section 16.2 (pages 437–439)
Students should understand that the graph of each linear function separates the coordinate plane into three regions. This concept will help students in making graphs of linear inequalities.

Section 16.3 (pages 440–444)
The graphing of an inequality of the form $x + y \le a$ is introduced by considering the graph of the union of two sets. Then students can better appreciate that $x + y \le a$ means $x + y < a$ or $x + y = a$.

Section 16.4 (pages 445–447)
In this section, a system of two sentences in two variables is studied as an example of a compound <u>and</u> sentence. The graph of such a system is the intersection of the graphs of the separate sentences of the system.

Section 16.5 (pages 448–450)
Graphing of systems of linear sentences is extended to sentences in two variables using $\le$ and $\ge$.

WARM-UP EXERCISES

Section 16.1
Pages 434–436

Solve.

1. $x + 1 > -3$ **$x > -4$**   2. $x - 3 \le -5$ **$x \le -2$**   3. $-x < 2$ **$x > -2$**
4. $2x + 6 \le 8$ **$x \le 1$**   5. $11 > 3x + 2$ **$3 > x$**   6. $4x \ge 2x + 8$ **$x \ge 4$**
7. $3x \le x - 2$ **$x \le -1$**   8. $-3x \ge -9$ **$x \le 3$**   9. $\frac{1}{2}x \le 5$ **$x \le 10$**

Section 16.2
Pages 437–439

Try this tic-tac-toe puzzle. Write <u>Yes</u> and <u>No</u> in the correct box to show whether the inequality is in slope-intercept form.

1. $2y > x$ 2. $y \le -2x + 1$ 3. $y > 3x - 5$
4. $y \ge 1$ 5. $y \ge 1 + 3x$ 6. $y < -3$
7. $5x \le y + 1$ 8. $y > -5x$ 9. $x < -y - 1$
10. What is the winning word? **No**

1.	2.	3.
No	**Yes**	**Yes**
4.	5.	6.
Yes	**No**	**Yes**
7.	8.	9.
No	**Yes**	**No**

Section 16.3
Pages 440–444

Write <u>Yes</u> or <u>No</u> to tell whether (0,0) is in each solution set.

1. $y < 2x - 1$ **No**   2. $y > 4x - 3$ **Yes**   3. $y > 3x$ **No**
4. $y < -\frac{1}{4}x$ **No**   5. $y > 1$ **No**   6. $y < 4$ **Yes**
7. $-x > -3$ **Yes**   8. $x < 5$ **Yes**   9. $2x + 3y > 1$ **No**

Section 16.4
Pages 445–447

Graph the linear inequality $y < -x + 1$ and write <u>Yes</u> or <u>No</u> to tell whether the given points are in the solution set.

1. $(0,0)$ **Yes**   2. $(0,2)$ **No**
3. $(0,-4)$ **Yes**   4. $(6,0)$ **No**
5. $(-3,0)$ **Yes**   6. $(1,1)$ **No**
7. $(-1,1)$ **Yes**   8. $(-1,-1)$ **Yes**
9. $(3,-1)$ **No**   10. $(3,0)$ **No**

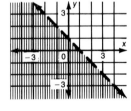

Section 16.5
Pages 448–450

Write <u>Yes</u> or <u>No</u> to tell whether each given ordered pair is in the solution set of the inequality.

1. $y \ge x + 1$; $(2,3)$ **Yes**   2. $y \le 2x - 1$; $(2,3)$ **Yes**
3. $y > x + 1$: $(2,3)$ **No**   4. $y < 3x - 2$; $(-2,-4)$ **No**
5. $y \le x - 2$; $(-2,-4)$ **Yes**   6. $y > 3x + 2$; $(-2,-4)$ **No**
7. $y \le 2x + 3$; $(-1,3)$ **No**   8. $y \le x - 3$; $(0,0)$ **No**

CHAPTER 17:   QUADRATIC FUNCTIONS

OVERVIEW

The quadratic function is studied in this chapter with emphasis on graphing and graphical interpretation. The graph of a parabola was introduced in Section 12.5. At that time it was presented simply as a function to be graphed. The standard form of the quadratic polynomial and the effects of the values of a, h, and k on the graph are presented. The highest and lowest points of the graph of a parabola are discussed in connection with the study of the vertex and the axis of the parabola. This chapter and the one that follows are considered optional for Basic and Average level classes.

SECTION-BY-SECTION COMMENTARY

Section 17.1 (pages 456–457)
In this section, quadratic trinomials are used in forming rules for quadratic functions. The general form is $y = ax^2 + bx + c$.

Section 17.2 (pages 458–460)
The examples of this section involve the graphing of quadratic functions. In Example 2, a method is shown for determining the coordinates of the turning point.

Section 17.3 (pages 461–464)
Several quadratic functions whose rules are of the form $y = ax^2$ are graphed to show the effect of the coefficient a on the graphs. Example 3 shows the graph of a parabola that is not a function.

Section 17.4 (pages 465–468)
The main difficulty that students will have in this section is to identify the correct value for h. For example, there is a natural tendency for students to think that the graph of $y = (x - 2)^2$ is 2 units to the left of the graph of $y = x^2$ rather than to its right.

Section 17.5 (pages 469–472)
In this section, functions defined by rules of the forms $y = ax^2$, $y = a(x - h)^2$, and $y = a(x - h)^2 + k$ are compared by showing their graphs in the same coordinate plane.

Section 17.6 (pages 473–474)
The emphasis in this section is on changing a quadratic polynomial to the standard form, $a(x - h)^2 + k$.

Section 17.7 (pages 475–477)
In this section, students learn how to identify the coordinates of the turning point of a parabola from the standard form of the quadratic function $y = a(x - h)^2 + k$.

WARM-UP EXERCISES

Section 17.1
Pages 456–457

Evaluate each polynomial.

1. $x^2 + 1$ when $x = 2$ **5**
2. $x^2 - 3$ when $x = 4$ **13**
3. $x^2 - 2x + 1$ when $x = 1$ **0**
4. $-x^2 + 3x + 2$ when $x = -1$ **−2**
5. $x^2 - 4$ when $x = -3$ **5**
6. $-2x^2 - 3x + 2$ when $x = 0$ **2**
7. $x^2 + 2x - 5$ when $x = 2$ **3**
8. $x^2 - 3$ when $x = -5$ **22**

Section 17.2
Pages 458–460

Complete each table.

1. $y = x^2 - 3$

x	-2	-1	0	1	2
y	1	-2	-3	-2	1

2. $y = -x^2 + 2$

x	-2	-1	0	1	2
y	-2	1	2	1	-2

3. $y = \frac{1}{2}x^2 - x$

x	-2	-1	0	1	2
y	4	$1\frac{1}{2}$	0	$-\frac{1}{2}$	0

Section 17.3
Pages 461–464

Complete the table, graph the ordered pairs, and draw a parabola containing the points.

$y = x^2 - x$

x	-3	-2	-1	0	$\frac{1}{2}$	1	2	3	4
y	12	6	2	0	$-\frac{1}{4}$	0	2	6	12

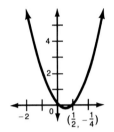

$(\frac{1}{2}, -\frac{1}{4})$

Section 17.4
Pages 465–468

Indicate if the graph of each function opens upward or downward.

1. $y = 2x^2$ **Upward**   **2.** $y = x^2$ **Upward**   **3.** $y = -3x^2$ **Downward**
4. $y = -x^2$ **Down**   **5.** $y = \frac{1}{4}x^2$ **Upward**   **6.** $y = 2x^2 + x + 2$ **Up**
7. $y = -3x^2 + x + 1$ **Downward**   **8.** $y = -x^2 + 2x - 1$ **Downward**

Section 17.5
Pages 469–472

For each exercise, write the number of units and the direction the graph of $y = -x^2$ can be moved to obtain the graph of the given function.

1. $y = -(x - 1)^2$ **1 unit right**   **2.** $y = -(x + 1)^2$ **1 unit left**
3. $y = -(x - \frac{1}{2})^2$ **$\frac{1}{2}$ unit right**   **4.** $y = -(x + \frac{1}{2})^2$ **$\frac{1}{2}$ unit left**
5. $y = -(x - 2)^2$ **2 units right**   **6.** $y = -(x + 2)^2$ **2 units left**

Section 17.6
Pages 473–474

Factor.

1. $x^2 - 6x + 9$ **$(x - 3)^2$**   **2.** $x^2 + 2x + 1$ **$(x + 1)^2$**
3. $x^2 - 4x + 4$ **$(x - 2)^2$**   **4.** $x^2 - 10x + 25$ **$(x - 5)^2$**
5. $x^2 + 14x + 49$ **$(x + 7)^2$**   **6.** $x^2 - 12x + 36$ **$(x - 6)^2$**

Section 17.7
Pages 475–477

Complete this number puzzle by writing the number that must be added (or subtracted) to change the given polynomial to standard form. The sums of the answers along the diagonals are equal. The answer to Exercise 1 is given.

1. $x^2 + 4x + 4 - 4 = (x + 2)^2 - 4$
2. $x^2 - 4x$ **3.** $x^2 + 6x$
4. $x^2 - 6x$ **5.** $x^2 + x$
6. $x^2 + 8x$ **7.** $x^2 - 8x$
8. $x^2 + 10x$ **9.** $x^2 - 10x$
10. What is the sum of the answers along each diagonal? **$54\frac{1}{4}$**

1. **4**		2. **4**
3. **9**		4. **9**
	5. **$\frac{1}{4}$**	
6. **16**		7. **16**
8. **25**		9. **25**

CHAPTER 18: QUADRATIC EQUATIONS

OVERVIEW

The first section covers the graphical method, and the second section covers the factoring method. Then in the third section, special quadratic equations of the form $(x - a)^2 = k$ are solved by inspection. This section is preparatory material for the method of completing the square that is presented in Section 18.4. The quadratic formula is then presented in Section 18.5. The next section covers the quadratic discriminant and the chapter concludes with applications of quadratic equations. This chapter is considered optional for Basic and Average level classes.

SECTION-BY-SECTION COMMENTARY

Section 18.1 (pages 484–486)
A method is shown for estimating the solutions of a quadratic equation by estimating the zeros of a quadratic function from its graph.

Section 18.2 (pages 487–489)
The factoring method of solving quadratic equations is covered in this section. Only factoring with integral coefficients is used here.

Section 18.3 (pages 490–493)
Special equations of the form $(x + r)^2 = s$ are solved by a rule. Then identification of trinomial squares is reviewed. However, emphasis is given to factoring over the set of rational numbers.

Section 18.4 (pages 494–496)
The work of the section leads to the steps for solving the general quadratic $ax^2 + bx + c = 0$ by the method of completing the square.

Section 18.5 (pages 497–499)
If the material of Sections 18.3 and 18.4 has been developed carefully, students should be able to follow the steps of the development of the quadratic formula. However, the important goal is that students learn to apply the quadratic formula.

Section 18.6 (pages 500–502)
The various types of solutions of quadratic equations are discussed based on the value of the discriminant of each. You may need several examples to develop the concept that a perfect square discriminant guarantees rational solutions.

Section 18.7 (pages 503–506)
This section includes two kinds of problems that require quadratic equations for their solution: areas of rectangles and squares and products of integers.

WARM-UP EXERCISES

Section 18.1
Pages 484–486

Match each equation in Exercises 1–6 with an equivalent equation from items A-J.

1. $x^2 = 3x + 1$ **D**

2. $x^2 = -2x + 5$ **G**

3. $x^2 = 14$ **A**

4. $x^2 + 3x = -2$ **J**

5. $6x - x^2 = -10$ **H**

6. $x - 3x^2 = 5$ **C**

A. $x^2 - 14 = 0$

B. $x^2 - 6x + 10 = 0$

C. $-3x^2 + x - 5 = 0$

D. $x^2 - 3x - 1 = 0$

E. $x^2 - 2x - 5 = 0$

F. $3x^2 + x - 5 = 0$

G. $x^2 + 2x - 5 = 0$

H. $-x^2 + 6x + 10 = 0$

I. $x^2 + 14 = 0$

J. $x^2 + 3x + 2 = 0$

Section 18.2
Pages 487–489

Factor.

1. $x^2 - 2x$ **x(x − 2)**   2. $3x^2 + 6x$ **3x(x + 2)**
3. $x^2 - 9$ **(x + 3)(x − 3)**   4. $x^2 + 4x - 5$ **(x + 5)(x − 1)**
5. $x^2 + 8x + 16$ **(x + 4)²**   6. $x^2 + 7x + 6$ **(x + 1)(x + 6)**

Section 18.3
Pages 490–493

Solve each equation.   **x = 8**

1. $x - 3 = 6$ **x = 9**   2. $x - 3 = -6$ **x = −3**   3. $x + 1 = 9$
4. $x + 1 = -9$ **x = −10**   5. $x - 2 = 1$ **x = 3**   6. $x - 2 = -1$
                                                              x = 1

Section 18.4
Pages 494–496

Try this number puzzle. Write the numeral missing from each trinomial square.
Write one digit in each square of the puzzle.

Down

1. $x^2 + 16x + \underline{\ ?\ }$
2. $x^2 + 12x + \underline{\ ?\ }$
4. $x^2 + 22x + \underline{\ ?\ }$
5. $x^2 - 6x + \underline{\ ?\ }$
6. $x^2 + 30x + \underline{\ ?\ }$

Across

3. $x^2 - 24x + \underline{\ ?\ }$
4. $x^2 - 26x + \underline{\ ?\ }$
6. $x^2 + 10x + \underline{\ ?\ }$
7. $x^2 - 18x + \underline{\ ?\ }$
8. $x^2 + 32x + \underline{\ ?\ }$

Section 18.5
Pages 497–499

Write each quadratic equation in the form $ax^2 + bx + c = 0$.

                                                              3x² + 2x + 1 = 0

1. $x^2 - 2x = 5$ **x² − 2x − 5 = 0**   2. $2x + 1 = -3x^2$
3. $x^2 = 6x$ **x² − 6x = 0**   4. $x^2 = 2x + 15$ **x² − 2x − 15 = 0**

Section 18.6
Pages 500–502

Compare each equation with $ax^2 + bx + c = 0$. *Then write the values of*
a, b, and c for each equation.

Equation	a	b	c	Equation	a	b	c
1. $3x^2 + 3x - 2 = 0$	**3**	**3**	**−2**	2. $2x^2 - 5x + 1 = 0$	**2**	**−5**	**1**
3. $-x^2 + 4x - 2 = 0$	**−1**	**4**	**−2**	4. $x^2 - 3x + 1 = 0$	**1**	**−3**	**1**
5. $x^2 + 1 = 0$	**1**	**0**	**1**	6. $x^2 + 2x = 0$	**1**	**2**	**0**

Section 18.7
Pages 503–506

Write an algebraic expression for the missing value.

1. The length of a rectangle is 4 units greater than the width.
 x = width $\underline{\ ?\ }$ = length   **x + 4**
2. The length of a rectangle is 3 times the width.
 x = width $\underline{\ ?\ }$ = length   **3x**
3. One integer is 3 more than another integer.
 x = smaller integer $\underline{\ ?\ }$ = greater integer   **x + 3**
4. One number is 5 less than another number.
 x = greater number $\underline{\ ?\ }$ = smaller number   **x − 5**

CHAPTER 19:   PROBABILITY

OVERVIEW

This chapter presents some of the basic concepts of probability in an informal way. These include the fundamental counting principle, sample spaces, and the definition of probability, independent events, dependent events, and experimental probability. No attempt is made to cover any topic in depth nor to include examples and exercises that are difficult. Manipulative 15 on pages 568–569 can be used to introduce Lesson 19.2.

SECTION-BY-SECTION COMMENTARY

Section 19.1 (pages 512–514)
The Fundamental Counting Principle is developed in this section. It is one of the most important ideas that provide a basis for probability theory. It provides a way of determining a total number of combinations of things without actually making a list of the combinations.

Section 19.2 (pages 515–519)
Many basic terms of probability theory are introduced in this section including *experiment, outcome, sample space, event, probability, favorable outcomes,* and *possible outcomes.* Underlying the entire discussion of probability is the undefined concept of equal likelihood of occurrence of events. This is very important and is sometimes assumed without actually being stated.

Section 19.3 (pages 520–523)
The terms independent events and mutually exclusive events arc discussed in this section. A rule for the probability of both of two independent events is given. Also, a rule for the probability that either of two mutually exclusive events will occur is stated. It should be noted that two mutually exclusive events are not necessarily independent.

Section 19.4 (pages 525–528)
Dependent events are discussed in this section. Emphasis is given to the probability that two dependent events will both occur. This also happens to be the product of the separate probabilities of the two events as with independent events.

Section 19.5 (pages 529–531)
The content of this section concerns the concept of *experimental probability* as opposed to "mathematical" or "theoretical" probability which has been the basis of the three previous sections. There are many subleties about experimental probability that cannot be dealt with in this limited space. Involved is the theory of population sampling, margin or error, and so forth. It should be emphasized that predictions made on the basis of experimental results are not 100 percent reliable.

WARM-UP EXERCISES

Section 19.1
Pages 512–514

Multiply.

1. $2 \times 3 \times 4$ **24**	**2.** $3 \times 4 \times 5$ **60**	**3.** $4 \times 2 \times 2$ **16**
4. $10 \times 6 \times 4$ **240**	**5.** $4 \times 5 \times 2$ **40**	**6.** $6 \times 8 \times 5$ **240**
7. $4 \times 7 \times 8 \times 2$ **448**	**8.** $3 \times 5 \times 6 \times 9$ **810**	**9.** $2 \times 4 \times 7 \times 5$ **280**

Section 19.2
Pages 515–519

Refer to the set of numbers below to complete the table.
The first row in the table is done for you.

{1, 2, 3, 4, 5, 6, 7, 8, 9}

		List of Numbers	How Many?
	Even numbers	2, 4, 6, 8	4
1.	Odd numbers	**1, 3, 5, 7, 9**	**5**
2.	Factors of 8	**1, 2, 4, 8**	**4**
3.	Multiples of 3	**3, 6, 9**	**3**
4.	Multiples of 5	**5**	**1**
5.	Numbers less than 5	**1, 2, 3, 4**	**4**

Section 19.3
Pages 520–523

The arrow of the spinner shown is equally likely to stop on any of the eight regions. The arrow is spun once. Find each probability.

1. Stopping on 2 $\frac{1}{8}$

2. Stopping on an odd number $\frac{1}{2}$

3. Stopping on a whole number greater than 4 $\frac{1}{2}$

4. Stopping on a prime number $\frac{1}{2}$

5. Stopping on a whole number less than 6 $\frac{5}{8}$

Section 19.4
Pages 525–528

Multiply. Write the answers in lowest terms.

1. $\frac{3}{5} \cdot \frac{1}{2}$ $\frac{3}{10}$

2. $\frac{1}{2} \cdot \frac{2}{9}$ $\frac{1}{9}$

3. $\frac{2}{3} \cdot \frac{3}{5}$ $\frac{2}{5}$

4. $\frac{1}{5} \cdot \frac{5}{9}$ $\frac{1}{9}$

5. $\frac{1}{2} \cdot \frac{4}{9}$ $\frac{2}{9}$

6. $\frac{1}{4} \cdot \frac{5}{19}$ $\frac{5}{76}$

Section 19.5
Pages 529–531

A bag contains 6 blue marbles, 5 red marbles, 1 white marble, and 4 green marbles. One marble is drawn without looking. Find each probability.

1. Drawing a blue marble $\frac{3}{8}$

2. Drawing a red marble $\frac{5}{16}$

3. Not drawing a white marble $\frac{15}{16}$

4. Not a drawing a blue marble $\frac{5}{8}$

5. Drawing a green marble $\frac{1}{4}$

6. Drawing a yellow marble **0**

7. Drawing a blue marble or a green marble $\frac{5}{8}$

8. Drawing a red marble <u>or</u> a white marble $\frac{3}{8}$

SUGGESTED TIMETABLES OF ASSIGNMENTS

CHAPTER 1:   OPERATIONS ON NUMBERS

Section	Page(s)	Basic	Average	Above Average
1.1	3	Odds 1–33	Odds 1–33	Odds 1–33
1.2	7	Odds 1–31	Odds 1–17, 31 } 1 day	All 19–32 } 1 day
1.3	9–10	Day 1: Odds 1–43 Day 2: Evens 2–44	Odds 1–27, 41, 43	All 29–40, 41, 43
1.4	13–14	Odds 1–41	Odds 1–45	Odds 1–41, All 43–47
1.5	18	Odds 1–41	Odds 1–41	Odds 1–41
1.6	20	Odds 1–35	Odds 1–37	Odds 1–37, 38
1.7	22–23	Day 1: Odds 1–29 Day 2: Odds 31–67	Odds 1–67	Odds 1–67
Review and Testing		3 days	3 days	2 days
Extension Topics		—	1 day	1 day
Total Days		**12**	**10**	**9**

CHAPTER 2:   REAL NUMBERS

Section	Page(s)	Basic	Average	Above Average
2.1	33–34	Day 1: Odds 1–49 Day 2: Evens 2–48	Odds 1–49	Odds 1–49
2.2	37–38	Odds 1–41	Odds 1–29, 43 } 1 day	All 31–44 } 1 day
2.3	41	Odds 1–23	Odds 1–23	Odds 7–23, All 25–28
2.4	45–46	Odds 1–45	Odds 1–47	Odds 1–45, 46, 47
2.5	49	Day 1: All 1–16 Day 2: Odds 17–39	Odds 1–47	Odds 1–39, All 41–48
2.6	51	Odds 1–35	Odds 1–35	Odds 1–35
Review and Testing		3 days	3 days	2 days
Extension Topics		—	1 day	1 day
Total Days		**11**	**9**	**8**

CHAPTER 3:   EQUATIONS AND PROBLEM SOLVING

Section	Page(s)	Basic	Average	Above Average
3.1	60	Odds 1–41	Odds 1–41	Odds 1–41
3.2	63	Odds 1–39	Odds 1–43	Odds 1–39, All 41–44
3.3	66	Day 1: Odds 1–21 Day 2: Odds 23–41	Odds 1–43	Odds 1–41, 42, 43

3.4	68–69	All 1–6	All 1–6, 7, 9	All 1–9
3.5	73	Odds 1–29	Odds 1–31	Odds 1–31, 32
3.6	76–77	Odds 1–27	Odds 1–27	Odds 1–19 ⎫
3.7	80–81	Odds 1–11	All 1–11	All 1–11 ⎬ 1 day
Review and Testing		3 days	3 days	2 days
Extension Topics		—	1 day	1 day
Total Days		**11**	**11**	**9**

CHAPTER 4:   PROBLEM SOLVING: ONE VARIABLE

Section	Page(s)	Basic	Average	Above Average
4.1	92–94	Day 1: Odds 1–15 Day 2: Evens 2–16	Odds 1–19	Odds 1–15, All 17–20 ⎫ 1 day
4.2	98–99	Day 1: Odds 1–13 Day 2: Evens 2–14	Odds 1–15	All 9–16 ⎭
4.3	102	Day 1: Odds 1–15 Day 2: Odds 17–35	Day 1: Odds 1–23 Day 2: Odds 25–37	Odds 1–37, 38
4.4	107	All 1–7	All 1–7	All 1–7
4.5	111	All 1–6	All 1–6	All 1–6
Review and Testing		3 days	3 days	2 days
Extension Topics		—	1 day	1 day
Total Days		**11**	**10**	**7**

CHAPTER 5:   INEQUALITIES

Section	Page(s)	Basic	Average	Above Average
5.1	118–119	All 1–18	All 1–18, Odds 19–25	Odds 1–17, All 19–26 ⎫ 1 day
5.2	122	Odds 1–31	Odds 1–31	Odds 9–31 ⎭
5.3	125–126	Odds 1–21	Odds 1–33	Odds 1–21, All 23–24
5.4	128	Day 1: Odds 1–11 Day 2: All 13–21	Odds 1–21	Odds 1–21
5.5	131	Odds 1–19	Odds 1–21	Odds 1–21
5.6	134	Day 1: All 1–11 Day 2: All 12–22	Odds 1–21	Odds 1–21
5.7	137	Odds 1–7	All 1–8	All 1–8
Review and Testing		3 days	3 days	2 days
Extension Topics		—	1 day	1 day
Total Days		**12**	**11**	**9**

CHAPTER 6:   POWERS

Section	Page(s)	Basic	Average	Above Average
6.1	146	Odds 1–35	3, 6, 9, · · · , 36	3, 6, 9, · · · , 36
6.2	149–150	Odds 1–45	3, 6, 9, · · · , 45, 47 } 1 day	3, 6, 9, · · · , 45, 47 } 1 day
6.3	153–154	Odds 1–41	Odds 1–45	Odds 1–41, All 43–46
6.4	158	Odds 1–29	Odds 1–29	Odds 1–29
6.5	162	Odds 1–29	Odds 1–31	Odds 1–31, 32
6.6	165	Odds 1–31	Odds 1–31	Odds 1–31
Review and Testing		3 days	3 days	2 days
Extension Topics		—	1 day	1 day
Total Days		**9**	**9**	**8**

CHAPTER 7:   ROOTS

Section	Page(s)	Basic	Average	Above Average
7.1	176	Odds 1–41	Odds 1–41	3, 6, 9, · · · , 39 } 1 day
7.2	179	Odds 1–29	Odds 1–39	3, 6, 9, · · · , 39
7.3	182	Day 1: Odds 1–21 Day 2: Odds 23–41	Odds 1–47	Odds 1–41, All 42–48
7.4	185	Day 1: Odds 1–19 Day 2: Odds 21–35	Odds 1–47	Odds 1–35, All 37–48
7.5	189	Day 1: Odds 1–21 Day 2: Odds 23–41	Day 1: Odds 1–21 Day 2: Odds 23–41	Odds 1–41
7.6	192	Day 1: Odds 1–19 Day 2: Odds 21–35	Odds 1–35	Odds 1–35
7.7	197	Odds 1–25	Odds 1–25	Odds 1–25
7.8	200–201	Odds 1–19	Odds 1–19	Odds 1–19
Review and Testing		3 days	3 days	2 days
Extension Topics		—	1 day	1 day
Total Days		**15**	**13**	**10**

CHAPTER 8:   POLYNOMIALS

Section	Page(s)	Basic	Average	Above Average
8.1	209	Odds 1–33	Odds 1–33	3, 6, 9, · · · , 33
8.2	212	Odds 1–27	Odds 1–27, 28	3, 6, 9, · · · , 27, 28 } 1 day
8.3	215	Odds 1–29	Odds 1–33	Odds 1–29, All 31–34

8.4	219	Odds 1–21	Odds 1–21	Odds 1–21
8.5	222–223	Odds 1–31	Odds 1–31	Odds 1–31
8.6	226	Odds 1–23	Odds 1–25	Odds 1–25
8.7	230	Day 1: Odds 1–17 Day 2: Evens 2–18	Odds 1–27	Odds 1–37
Review and Testing		3 days	3 days	2 days
Extension Topics		—	1 day	1 day
Total Days		**11**	**11**	**9**

CHAPTER 9:   FACTORING POLYNOMIALS

Section	Page(s)	Basic	Average	Above Average
9.1	238	Odds 1–23	Odds 1–23	Odds 1–23
9.2	241	Odds 1–21	Odds 1–25	Odds 1–21, All 23–25
9.3	245	Day 1: Odds 1–29 Day 2: Odds 31–59	3, 6, 9, · · · , 30, Odds 31–59	Odds 1–59
9.4	248–249	Day 1: Odds 1–17 Day 2: Odds 19–33	Day 1: Odds 1–17 Day 2: Odds 19–39	Odds 1–33, All 35–40
9.5	254	Odds 1–35	Odds 1–47	Odds 1–47
9.6	257	Odds 1–41	Odds 1–41	Odds 1–41
9.7	260	1, 5, 9, · · · , 33	3, 6, 9, · · · , 39	3, 6, 9, · · · , 36, All 37–40
Review and Testing		3 days	3 days	2 days
Extension Topics		—	1 day	1 day
Total Days		**12**	**12**	**10**

CHAPTER 10:   PRODUCTS AND QUOTIENTS OF RATIONAL EXPRESSIONS

Section	Page(s)	Basic	Average	Above Average
10.1	270	Odds 1–31	3, 6, 9, · · · , 30, 32 ⎫ 1 day	Odds 1–31 ⎫ 1 day
10.2	273	Odds 1–31	3, 6, 9, · · · , 30, 32 ⎭	3, 6, 9, · · · , 30, 32 ⎭
10.3	276	Odds 1–23	Odds 1–25	Odds 1–25, 26
10.4	279–280	Odds 1–23	Odds 1–25	Odds 1–23, 24, 25
10.5	283	Odds 1–11	Odds 1–19	Odds 1–11, All 13–20
10.6	286	Odds 1–19	Odds 1–23, 24	Odds 1–19, All 21–24
10.7	289	1, 5, 9, 13, 17	Odds 1–21	Odds 1–17, All 19–22
Review and Testing		3 days	3 days	2 days
Extension Topics		—	1 day	1 day
Total Days		**10**	**10**	**9**

CHAPTER 11:   SUMS AND DIFFERENCES OF RATIONAL EXPRESSIONS

Section	Page(s)	Basic	Average	Above Average
11.1	296	Odds 1–25	3, 6, 9, · · · , 24 } 1 day	3, 6, 9, · · · , 24 } 1 day
11.2	299	Odds 1–19	3, 6, 9, · · · , 24	Odds 1–19, All 21–24
11.3	302	Odds 1–19	Odds 1–21	Odds 1–21, 22
11.4	305	Odds 1–19	Odds 1–23	Odds 1–19, All 21–23
11.5	309	Odds 1–15	Odds 1–19	Odds 1–15, All 17–19
11.6	312–313	Odds 1–17	Odds 1–17	Odds 1–17
11.7	316	Omit	Odds 1–15, 17	Odds 1–15, 17
Review and Testing		3 days	3 days	2 days
Extension Topics		—	1 day	1 day
Total Days		**9**	**10**	**9**

CHAPTER 12:   RELATIONS AND FUNCTIONS

Section	Page(s)	Basic	Average	Above Average
12.1	324	All 1–24	Odds 1–23 } 1 day	Odds 1–23 } 1 day
12.2	327	All 1–22	Odds 1–21	Odds 1–21
12.3	330–331	Odds 1–27	Odds 1–31	Odds 1–27, All 29–31
12.4	335–336	Odds 1–25	Odds 1–25	Odds 1–25
12.5	339	Odds 1–11	Odds 1–15, 16	Odds 1–11, All 13–16
12.6	342	All 1–12	All 1–12	All 1–12
Review and Testing		3 days	3 days	2 days
Extension Topics		—	1 day	1 day
Total Days		**9**	**9**	**8**

CHAPTER 13:   LINEAR FUNCTIONS

Section	Page(s)	Basic	Average	Above Average
13.1	352	Odds 1–23	Odds 1–25	Odds 1–25
13.2	355	Odds 1–21	3, 6, 9, · · · , 21 } 1 day	3, 6, 9, · · · , 21 } 1 day
13.3	358	Odds 1–29	3, 6, 9, · · · , 30	3, 6, 9, · · · , 30
13.4	361–362	Odds 1–27	Odds 1–37	Odds 1–27, All 29–37
13.5	366	Odds 1–13	Odds 1–15	Odds 1–15, 16
13.6	369	Odds 1–23	Odds 1–23	Odds 1–23
13.7	372	Odds 1–17	All 1–18	All 1–18
Review and Testing		3 days	3 days	2 days
Extension Topics		—	1 day	1 day
Total Days		**10**	**10**	**9**

CHAPTER 14:   SYSTEMS OF SENTENCES

Section	Page(s)	Basic	Average	Above Average
14.1	380	Odds 1–17	1, 7, Odds 9–17	1, 7, Odds 9–17
14.2	384	Odds 1–17	3, 6, 9, · · · , 18, 19, 22 } 1 day	3, 6, 9, · · · , 18, All 19–22 } 1 day
14.3	386–387	Odds 1–17	Odds 1–17	Odds 1–17
14.4	390	Day 1: Odds 1–17 Day 2: Evens 2–18	Odds 1–21	Odds 1–17, All 19–24
14.5	393	Odds 1–11	Odds 1–11	Odds 1–11
Review and Testing		3 days	3 days	2 days
Extension Topics		—	1 day	1 day
Total Days		**9**	**8**	**7**

CHAPTER 15:   MORE ON SYSTEMS OF SENTENCES

Section	Page(s)	Basic	Average	Above Average
15.1	400	Day 1: Odds 1–11 Day 2: Evens 2–12	Odds 1–11	Odds 1–11
15.2	404	Day 1: Odds 1–9 Day 2: Evens 2–10	Odds 1–9	Odds 1–9
15.3	407–409	Day 1: All 1–4 Day 2: All 5–8	Day 1: All 1–4, 9 Day 2: All 5–8, 11	All 1–12
15.4	414	Omit	Odds 1–21	Odds 1–21 } 1 day
15.5	417	Omit	Odds 1–11	Odds 1–11 } 1 day
15.6	421	Day 1: All 1–4 Day 2: All 5–8	All 1–8	All 1–8
15.7	424–425	Day 1: All 1–3 Day 2: All 4–7	All 1–7	All 1–7
Review and Testing		3 days	3 days	2 days
Extension Topics		—	1 day	1 day
Total Days		**13**	**12**	**9**

CHAPTER 16:   SYSTEMS OF INEQUALITIES

Section	Page(s)	Basic	Average	Above Average
16.1	436	Omit	Odds 1–27	Odds 1–27
16.2	439	Omit	Odds 1–17	3, 6, 9, · · · , 18
16.3	443–444	Omit	Odds 1–27	3, 6, 9, · · · , 24, All 25–27 } 1 day
16.4	447	Omit	Odds 1–11	Odds 1–11
16.5	450	Omit	Odds 1–19	Odds 1–19
Review and Testing		—	3 days	2 days
Extension Topics		—	1 day	1 day
Total Days		**—**	**9**	**7**

CHAPTER 17: QUADRATIC FUNCTIONS

Section	Page(s)	Basic	Average	Above Average
17.1	457	Omit	Omit	Odds 1–19
17.2	460	Omit	Omit	Odds 1–13
17.3	463–464	Omit	Omit	Odds 1–13
17.4	467–468	Omit	Omit	Odds 1–17, All 19–21
17.5	472	Omit	Omit	Odds 1–17
17.6	474	Omit	Omit	Odds 1–19
17.7	477	Omit	Omit	Odds 1–21
Review and Testing	—	—	—	2 days
Extension Topics	—	—	—	1 day
Total Days	—	—	—	**10**

CHAPTER 18: QUADRATIC EQUATIONS

Section	Page(s)	Basic	Average	Above Average
18.1	486	Omit	Omit	Odds 1–17
18.2	489	Omit	Omit	Odds 1–35
18.3	492–493	Omit	Omit	Odds 1–29, All 30–39
18.4	496	Omit	Omit	Odds 1–27
18.5	499	Omit	Omit	Odds 1–23
18.6	502	Omit	Omit	Odds 1–23
18.7	506	Omit	Omit	Odds 1–7
Review and Testing	—	—	—	2 days
Extension Topics	—	—	—	1 day
Total Days	—	—	—	**10**

CHAPTER 19: PROBABILITY

Section	Page(s)	Basic	Average	Above Average
19.1	514	Omit	Omit	All 1–8
19.2	518–519	Omit	Omit	Odds 1–19, All 21–28 } 1 day
19.3	522	Omit	Omit	All 1–14
19.4	527–528	Omit	Omit	All 1–17
19.5	531	Omit	Omit	All 1–14
Review and Testing	—	—	—	2 days
Extension Topics	—	—	—	1 day
Total Days	—	—	—	**7**

HOLT
INTRODUCTORY
ALGEBRA 2

Russell F. Jacobs

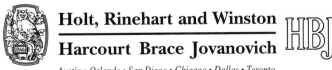

Holt, Rinehart and Winston

Harcourt Brace Jovanovich HBJ

Austin • Orlando • San Diego • Chicago • Dallas • Toronto

ABOUT THE AUTHOR

RUSSELL F. JACOBS

*Formerly Mathematics Supervisor for
the Phoenix Union High School System
Phoenix, Arizona*

EDITORIAL ADVISORS

Alan Margolis
*Mathematics Teacher
Alderdice High School
Pittsburgh, Pennsylvania*

Elaine Sledge
*Mathematics Teacher
Warren Central High School
Bowling Green, Kentucky*

Tim Stephens
*Mathematics Teacher
Canyon High School
Anaheim, California*

Janet Trafican
*Mathematics Teacher
Perry High School
Pittsburgh, Pennsylvania*

Richard Wyllie
*Chairperson, Department of Mathematics
Downers Grove South High School
Downers Grove, Illinois*

ISBN 0-03-076984-1

Contents

Operations on Numbers

The history of computers began thousands of years ago with the *abacus.* Blaize Pascal (1623–1662) built the first mechanical adding machine, the *Pascaline.* Charles Babbage (1791–1871) tried to develop the *analytical engine* to perform arithmetic calculations.

Background

The center photo above shows a model of Charles Babbage's analytical engine.

The photo at the top right shows Pascal's Pascaline which used wheels and gears to add and subtract numbers.

The photo at the bottom right is an example of an abacus.

OBJECTIVES: To write the simplest name of the opposite of a number
To simplify a numeral involving the absolute value symbol
To identify which of two numbers has the greater absolute value

1.1 Opposites and Absolute Value

Note on this number line that 3 and −3 (negative 3) are the same distance from 0 <u>and</u> are in opposite directions from 0. Thus, 3 and −3 are **opposites.**

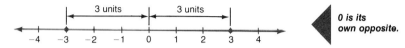

0 is its own opposite.

The symbol for the opposite of a number is the dash, −.

Table 1	Number	Opposite	Simplest Form of the Opposite
	−5	−(−5)	5
	2	−2	−2
	0	−0	0

> The opposite of a negative number is positive. $-(-7)=7$
> The opposite of a positive number is negative. $-7=-7$
> The opposite of 0 is 0. $-0=0$

To indicate distance, but <u>not</u> direction, from zero, you use <u>absolute value.</u> Since 3 and −3 are the same distance from 0, they have the same **absolute value.**

The absolute value of 3 is 3.   The absolute value of −3 is 3.

The symbol | | is used to represent absolute value.

Table 2	Absolute Value	Meaning	Value
	|−5|	Distance of −5 from 0	5
	|2|	Distance of 2 from 0	2
	|0|	Distance of 0 from 0	0
	−|4|	Opposite of the distance of 4 from 0	−4

Remind students that −x does not necessarily represent a negative number.

> The absolute value of a positive number equals the number. The absolute value of a negative number equals the opposite of the number. The absolute value of 0 equals 0.
>
> $|7|=7$
> $|-7|=7$
> $|0|=0$
>
> ▲ $|x| = x$ if $x \geq 0$; $|x| = -x$ if $x < 0$.

EXAMPLE

Which number has the greater absolute value?

a. -15 or 9 **b.** -8 or 13

Solutions:

a. $|-15| = 15;$ $|9| = 9$ **Answer:** -15

b. $|-8| = 8;$ $|13| = 13$ **Answer:** 13

Additional Examples
Which number has the greater absolute value?
1. -19 or 26
   Ans: 26
2. -11 or -21
   Ans: -21
3. 6.03 or -6.3
   Ans: -6.3

Common Error
Students always write the opposite of the number when finding absolute value.

Example
$|-5| = 50$ $|5| = -5$

Prescription
Emphasize that the absolute value of a number is the distance of the number from zero. Show students that the absolute value of any real number except zero is a <u>positive</u> real number.

CLASSROOM EXERCISES

Write the opposite of each number in simplest form. (Table 1)

1. -25 25 2. 0 0 3. -0.001 0.0001 4. 0.001 -0.001 5. $-\frac{2}{3}$ $\frac{2}{3}$
6. $-4\frac{1}{2}$ $4\frac{1}{2}$ 7. $8\frac{7}{8}$ $-8\frac{7}{8}$ 8. $-(-1\frac{2}{3})$ $-1\frac{2}{3}$ 9. $-(-8.5)$ -8.5 10. $-(-12)$ -12

Simplify. (Table 2)

11. $|9|$ 9 12. $|-13|$ 13 13. $|-0|$ 0 14. $|-2.38|$ 2.38 15. $|\frac{4}{5}|$ $\frac{4}{5}$
16. $|-3\frac{5}{8}|$ $3\frac{5}{8}$ 17. $|0.053|$ 0.053 18. $-|-47|$ -47 19. $|-(-\frac{1}{4})|$ $\frac{1}{4}$ 20. $|-100|$ 100
21. $-|0.01|$ -0.01 22. $-|8.7|$ -8.7 23. $-|-2.5|$ -2.5 24. $-|-\frac{1}{2}|$ $-\frac{1}{2}$ 25. $-|-2\frac{1}{2}|$ $-2\frac{1}{2}$

Which number has the greater absolute value? (Example)

26. 42 or -37 42 27. 12 or -3 12 28. 7 or -2 7 29. -5 or 4 -5 30. -3 or -8 -8

WRITTEN EXERCISES

Goal: To write the absolute value of a number
Sample Problems: $|8|;$ $|-7|$ **Answers:** $8;$ 7

Write the opposite of each number in simplest form. (Table 1)

1. 23 -23 2. -41 41 3. -0.12 0.12 4. 0.03 -0.03 5. 0 0
6. -1 1 7. $\frac{3}{5}$ $-\frac{3}{5}$ 8. $1\frac{3}{8}$ $-1\frac{3}{8}$ 9. $-5\frac{6}{7}$ $5\frac{6}{7}$ 10. $-\frac{2}{7}$ $\frac{2}{7}$

Simplify. (Table 2)

11. $|-18|$ 18 12. $|-35|$ 35 13. $-|227|$ -227 14. $-|508|$ -508 15. $|\frac{7}{8}|$ $\frac{7}{8}$
16. $|\frac{2}{3}|$ $\frac{2}{3}$ 17. $|-5\frac{1}{8}|$ $5\frac{1}{8}$ 18. $|-3\frac{3}{4}|$ $3\frac{3}{4}$ 19. $|0.083|$ 0.083 20. $|0.006|$ 0.006
21. $|-2.73|$ 2.73 22. $-|-5.18|$ -5.18 23. $|-(-23)|$ 23 24. $|-(-72)|$ 72 25. $-|-6.8|$ -6.8

Which number has the greater absolute value? (Example)

26. -23 or 21 -23 27. 12 or 7 12 28. -23 or 6 -23 29. 23 or -6 23
30. -8 or -7 -8 31. -13 or -8 -13 32. 7 or -3 7 33. 13 or -5 13

Assignment Guide
Basic p. 3: 1–33 odd
Average p. 3: 1–33 odd
Above Average: p. 3: 1–33 odd

Operations on Numbers / **3**

Lesson Resources
Warm–Up: p. M–18
Maintenance: See below.
Practice Worksheet 1.2
Manipulative 1, p. 540
Teaching Aids 1, 2
Transparencies 50, 51, 52

Maintenance
1. Subtract: $\frac{4}{5} - \frac{1}{3}$
   Ans: $\frac{7}{15}$
2. Add: $0.63 + 2.8 + 4.31$
   Ans: 7.74
3. Locate 2.5 on a number line.
   Ans:

4. Change $2\frac{1}{2}$ to a decimal.
   Ans: 2.5
5. David can type 40 words per minute. How many words can he type in half an hour?
   Ans: 1200 words

OBJECTIVES: To add two negative numbers

1.2 Addition
To add a positive and a negative number
To add 0 and any number

You can use this rule for adding on the number line.

Remind students that addends are the numbers being added.

> **Rule for Adding on the Number Line**
>
> 1. Graph the first addend.
> 2. From this point draw an arrow for the second addend. If this addend is negative, draw the arrow to the left. If this addend is positive, draw the arrow to the right.
> 3. Read the coordinate of the point where the arrow ends.

You may need to explain that the length of the arrow equals the magnitude of the number.

Table

Problems	Solution	
1. $-2 + (-7)$		$-2 + (-7) = -9$
2. $-7 + 9$ NOTE: $\lvert 9 \rvert > \lvert -7 \rvert$		$-7 + 9 = 2$
3. $-8 + 5$ NOTE: $\lvert -8 \rvert > \lvert 5 \rvert$		$-8 + 5 = -3$
4. $-6 + 6$ NOTE: $\lvert -6 \rvert - \lvert 6 \rvert$		$-6 + 6 = 0$
5. $4 + (-4)$ NOTE: $\lvert 4 \rvert = \lvert -4 \rvert$		$4 + (-4) = 0$

Problem 1 above shows that the <u>sum of two negative numbers is a negative number</u>. The following rule tells how to find the sum of two negative numbers without using the number line.

> **Rule for Adding Two Negative Numbers**
>
> 1. Add the absolute values.
> 2. Write the opposite of the result.

4 / Chapter 1

EXAMPLE 1   Add: $-6 + (-3)$

Solution: [1] Add the absolute values. ──────────→ $|-6| + |-3| = 6 + 3 = 9$
          [2] Write the opposite of the result. ──────────→ $-6 + (-3) = -9$

P–1   Add.

a. $-2 + (-7)$ -9 b. $-1 + -5$ -6 c. $-4.7 + 0$ ◀ -4.7 **The sum of any number n and 0 is n.**

Problems 2–5 in the table show that <u>the sum of a positive number and a negative number may be a positive number, a negative number, or zero</u>. The following rules tell how to use absolute value to find the the sum of a positive and a negative number without using the number line.

> **Rule for Adding a Negative Number and a Positive Number**
>
> 1. Subtract the lesser absolute value from the greater absolute value.
>
> 2. The answer is positive if the positive number has the greater absolute value. $8 + (-3) = 5$
>
>   The answer is negative if the negative number has the greater absolute value. $5 + (-7) = -2$
>
>   The answer is zero if the two absolute values are equal. $9 + (-9) = 0$

EXAMPLE 2   Add: $-9 + 15$ ◀ $|15| > |-9|$

Solution:
[1] Subtract the lesser absolute value from the greater. ──────────→ $|15| - |-9| = 15 - 9$
                                                                              $= 6$
[2] The answer is positive because   In this case, the difference of the
    the positive number has the   absolute values is the answer.
    greater absolute value. ──────────→ $-9 + 15 = 6$

P–2   Add.

a. $-7 + 9$ 2 b. $5\frac{3}{4} + (-3\frac{1}{4})$ $2\frac{1}{2}$ c. $-23.1 + (25.7)$ 2.6

Operations on Numbers / 5

EXAMPLE 3   A liquefied gas had a temperature of −13°C. It was slowly heated until its temperature increased 8°C. What was its temperature after being heated?

Solution: $-13 + 8 = \underline{\ ?\ }$

$\boxed{1}$ Subtract the lesser absolute value from the greater. $\longrightarrow |-13| - |8| = 13 - 8$

$= 5$ In this case, the opposite of the difference of the absolute values is the answer.

$\boxed{2}$ The answer is negative because the negative number has the greater absolute value. $\longrightarrow -13 + 8 = -5$

The temperature of the gas was −5°C after being heated.

P–3 **Add.**

a. $-11 + 8 - 3$ b. $-6\frac{2}{3} + 5\frac{1}{3} - 1\frac{1}{3}$ c. $-8.5 + 1.1 - 7.4$

Since the absolute values of opposites are equal, the sum of opposites is 0.

$-3 + 3 = 0$
$1.8 + (-1.8) = 0$

CLASSROOM EXERCISES

*Match each of Exercises 1–4 with one of the graphs **a–d**. (Table)*

1. $1 + (-6)$ b   2. $-7 + 11$ c   3. $4 + (-7)$ d   4. $-5 + 6$ a

a.

b.

c.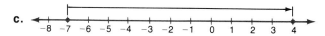

d.

Add. (Example 1)

5. $-28 + (-12)$ −40   6. $-19 + (-13)$ −32   7. $-3\frac{1}{2} + (-2\frac{1}{2})$ −6

8. $-6\frac{1}{5} + (-3\frac{4}{5})$ −10   9. $-0.5 + (-0.2)$ −0.7   10. $-4.3 + (-1.5)$ −5.8

6 / Chapter 1

(Example 2)

11. $28 + (-12)$ 16

12. $-16 + 32$ 16

13. $-2.4 + 5.8$ 3.4

14. $0.28 + (-0.13)$ 0.15

15. $-\frac{2}{5} + \frac{3}{5}$ $\frac{1}{5}$

16. $-\frac{1}{3} + \frac{2}{3}$ $\frac{1}{3}$

(Example 3)

17. $-36 + 30$ −6

18. $52 + (-61)$ −9

19. $22.6 + (-25.9)$ −3.3

20. $-16.8 + 9.3$ −7.5

21. $-12\frac{5}{8} + 5\frac{1}{8}$ $-7\frac{1}{2}$

22. $3\frac{1}{4} + (-6\frac{3}{4})$ $-3\frac{1}{2}$

23. $-56 + 56$ 0

24. $38\frac{1}{9} + (-38\frac{1}{9})$ 0

25. $2.18 + (-2.18)$ 0

WRITTEN EXERCISES

Goal: To add positive and negative numbers
Sample Problems: a. $(-49) + (52)$ **b.** $(-32) + (-12)$
Answers: a. 3   **b.** −44

Add. (Example 1)

1. $-23 + (-19)$ −42

2. $-38 + (-17)$ −55

3. $-47 + (-59)$ −106

4. $-66 + (-48)$ −114

5. $0 + (-0.08)$ −0.08

6. $-17.8 + (-9.5)$ −27.3

(Example 2)

7. $26 + (-18)$ 8

8. $-24 + 32$ 8

9. $14 + (-10)$ 4

10. $12 + (-8)$ 4

11. $-0.12 + 0.25$ 0.13

12. $-2.3 + 8.6$ 6.3

(Example 3)

13. $-39 + 14$ −25

14. $-24 + 15$ −9

15. $23 + (-55)$ −32

16. $25 + (-44)$ −19

17. $-28.7 + 19.8$ −8.9

18. $-43.3 + 29.7$ −13.6

MIXED PRACTICE

19. $-32 + (-21)$ −53

20. $-20 + (-38)$ −58

21. $-31 + 47$ 16

22. $-24 + 14$ −10

23. $-29.6 + (-15.9)$ −45.5

24. $-0.38 + (-0.45)$ −0.83

25. $4.67 + (-4.67)$ 0

26. $81.3 + (-81.3)$ 0

27. $-42 + 42$ 0

28. $-26 + 26$ 0

29. $-\frac{2}{3} + \frac{5}{3}$ 1

30. $-\frac{9}{13} + \frac{7}{13}$ $-\frac{2}{13}$

APPLICATIONS (Example 3)

31. The thermometer reading at 6:00 A.M. was −4°C. By noon, it had risen 9°C. Find the thermometer reading at noon. 5°C

32. On Monday, the change in the price of a stock was $-2\frac{1}{4}$ points. On Tuesday, the change was $-3\frac{3}{4}$ points. Find the total change in price for the two days. −6 points

Assignment Guide
Basic p. 7: 1–31 odd
Average p. 7: 1–17 odd, 31; p. 9: 1–27 odd, 41, 43
Above Average: p. 7: 19–32; p. 10: 29–40, 41, 43

Problem–Solving Skills
Choosing the operation
(Ex. 31–32)

Operations on Numbers / **7**

8

REVIEW CAPSULE FOR SECTION 1.3

*Add or subtract as indicated. Compare the answers in **a** and **b**.*

1. **a.** $8 - 4$ **b.** $8 + (-4)$ 2. **a.** $12 - 8$ **b.** $12 + (-8)$ 3. **a.** $16 - 9$ **b.** $16 + (-9)$

4. **a.** $21 - 11$ **b.** $21 + (-11)$ 5. **a.** $33 - 12$ **b.** $33 + (-12)$ 6. **a.** $1.5 - 1.5$ **b.** $1.5 + (-1.5)$

OBJECTIVES: To subtract a positive number from a positive number, from a negative number, or from 0
To subtract a negative number from a positive number, from a negative number, or from 0

1.3 Subtraction

You can think of subtraction as meaning "add the opposite."

Subtraction	Addition	Answer
$5 - 2$	$5 + (-2)$	3
$3 - 8$	$3 + (-8)$	-5
$7 - (-3)$	$7 + 3$	?
$-2.7 - (-3.5)$	$-2.7 + 3.5$	?
$0 - 37$	$0 + (-37)$	?
$-10.8 - (-10.8)$	$-10.8 + 10.8$	?

P–1 **Find the missing answers in the table above.** $10, 0.8, -37, 0$

To subtract b *from* a, add the opposite of b to a.

To subtract a number, add its opposite.	$7 - 5 = 7 + (-5)$
$a - b = a + (-b)$	$-8 - (-3) = -8 + 3$

EXAMPLE 1   Subtract: $12 - 25$

Solution: 1 Write a sum for the difference. ⟶ $12 - 25 = 12 + (-25)$
2 Add. ⟶ $= -13$

EXAMPLE 2   Subtract: $-13 - (-9)$

Solution: 1 Write a sum for the difference. ⟶ $-13 - (-9) = -13 + 9$
2 Add. ⟶ $= -4$

P–2 **Subtract.**

a. $30 - 12$ 18 **b.** $12 - (-8)$ 20 **c.** $-18 - 7$ -25 **d.** $-5.6 - (-9.8)$

4.2

8 / *Chapter 1*

EXAMPLE 3   The thermometer reading at noon on a January day was −12°C. The wind chill temperature was −17°C. How much lower was the wind chill temperature than the thermometer reading?

Solution: −12 − (−17) = __?__

1. Write a sum for the difference. ⟶ −12 − (−17) = −12 + 17

2. Add. ⟶ = 5

The wind chill temperature was 5°C lower than the thermometer reading.

Additional Example
Example 3
The high temperature for a city on a winter day was 6°C. The low temperature for the same day was −2°C. How much lower was the low temperature than the high temperature?
Ans: 8°C.

Common Error
When writing a sum for a difference, students change the subtraction sign to an addition sign but do not change the sign of the subtrahend.

Example
13 − (−19) = 13 + (−19)
= −6

Prescription
Give students some simple whole number subtraction problems such as 12 − 5. Then ask if changing the problem to 12 + 5 will give the correct answer. Review Step 1 of Examples 1 and 2 and show students that they must change the operation sign and use the opposite of the number they are subtracting.

CLASSROOM EXERCISES

Write a sum for each difference. (Step 1, Examples 1 and 2)

1. 20 − 15 20 + (−15)   **2.** 6 − 11 6 + (−11)   **3.** 0 − 12 0 + (−12)   **4.** 18 − (−5)

5. (−12) − 6 (−12) + (−6) **6.** 18 − (−7) 18 + 7   **7.** −20 − (−18) −20 + 18   **8.** −13 − (−19)

9. 5 − (−23) 5 + 23   **10.** 7 − 12 7 + (−12)   **11.** −3.3 − 6.2 −3.3 + (−6.2) **12.** −2.6 − 3.4

Subtract. (Example 1)

13. 12 − 16 −4   **14.** 0 − 18 −18   **15.** −15 − 8 −23   **16.** −15 − 15 −30

17. −4⅙ − 1⅙ −5⅓   **18.** −7¼ − 2¼ −9½   **19.** −7.3 − 2.7 −10   **20.** −2.7 − 7.3 −10

(Example 2)

21. 6 − (−10) 16   **22.** 18 − (−3) 21   **23.** −24 − (−10) −14   **24.** 0 − (−32) 32

Additional Answers
Classroom Exercises
4. 18 + 5
8. −13 + 19
12. −2.6 + (−3.4)

WRITTEN EXERCISES

Goal: To subtract positive and negative numbers

Sample Problems: a. −11 − (−11)   **b.** −11 − 11

Answers: a. 0   **b.** −22

Subtract. (Example 1)

1. 9 − 5 4   **2.** 18 − 6 12   **3.** 24 − 18 6   **4.** 16 − 9 7

5. 0 − 5 −5   **6.** 6 − 13 −7   **7.** 14 − 23 −9   **8.** 28 − 35 −7

9. 0 − 22 −22   **10.** 0 − 36 −36   **11.** −9.3 − 1.5 −10.8   **12.** −7.6 − 18.1

13. −18.9 − 5.8 −24.7 **14.** −0.14 − 0.58 −0.72 **15.** ⅛ − ⅞ −¾   **16.** −⅚ − ⅙ −1

Assignment Guide
Basic
Day 1 pp. 9–10: 1–43 odd
Day 2 pp. 9–10: 2–44 even
Average pp. 9–10: 1–27, 41, 43; p. 7: 1–27 odd, 31
Above average p. 10: 29–40, 41, 43; p. 7: 19–32

Additional Answers
Written Exercises
12. −25.7

(Example 2)

17. 25 − (−6) 31

18. 19 − (−7) 26

19. 13 − (−19) 32

20. 26 − (−11) 37

21. −16 − (−18) 2

22. −14 − (−20) 6

23. 0 − (−42) 42

24. 0 − (−96) 96

25. 2.7 − (−5.6) 8.3

26. 4.3 − (−8.9) 13.2

27. $-8\frac{5}{6} - (-4\frac{5}{6})$ −4

28. $-10\frac{3}{8} - (-4\frac{3}{8})$ −6

MIXED PRACTICE

29. 11 − 24 −13

30. 11 − (−24) 35

31. 0 − 18 −18

32. 0 − (−18) 18

33. −4 − (−12) 8

34. −4 − 12 −16

35. 7.8 − 3.4 4.4

36. 3.4 − 7.8 −4.4

37. −3.2 − (−4.5) 1.3

38. −0.71 − 0.82 −1.53

39. $\frac{5}{7} - \frac{6}{7}$ $-\frac{1}{7}$

40. $-\frac{1}{3} - (-\frac{4}{3})$ 1

APPLICATIONS

(Example 3)

Problem-Solving Skills
Choosing the operation
(Ex. 41–44)

41. When mixed equally with water in an automobile's radiator, antifreeze has a freezing point of −17°F. By itself, antifreeze has a freezing point of −37°F. How much lower is the freezing point for the pure antifreeze than for the mixture? 20°F

42. The record low temperature for Houston, Texas, is 5°F. The record low temperature for Buffalo, New York, is −21°F. How much lower is the record low temperature for Buffalo than for Houston? 26°F

43. A new business had a net income of −$582 in August. Its net income for September was −$620. How much lower was the business' September net income than its August net income? $38

44. On May 31, Jill's checking account balance was $265. On June 31, her checking account balance was −$14. How much lower was Jill's account balance on June 31 than on May 31? $279

REVIEW CAPSULE FOR SECTION 1.4

Write a fraction for each of the following.

1. $1\frac{1}{4}$ $\frac{5}{4}$

2. 9 $\frac{9}{1}$

3. $1\frac{5}{6}$ $\frac{11}{6}$

4. 13 $\frac{13}{1}$

5. $5\frac{2}{5}$ $\frac{27}{5}$

Multiply.

6. (10)(3) 30

7. (4)(1.3) 5.2

8. $(\frac{1}{2})$(18) 9

9. $(\frac{1}{2})(\frac{2}{3})$ $\frac{1}{3}$

10. (1.5)(1.5) 2.25

11. (0.8)(0.7) 0.56

12. (0.9)(1.6) 1.44

13. (5)(0) 0

14. (0)(28) 0

15. $1\frac{1}{2} \times \frac{2}{3}$ 1

16. $\frac{3}{4} \times \frac{8}{9}$ $\frac{2}{3}$

17. $2\frac{1}{4} \times 3\frac{1}{5}$ $7\frac{1}{5}$

10 / *Chapter 1*

10

1.4 Multiplication

OBJECTIVES: To multiply 0 and any number
To multiply a positive number and a negative number
To multiply two negative numbers

In the following multiplication problems, one factor, 8, stays the same. The other factor decreases by 1.

$$8(\ 3\ ) = \ 24$$
$$8(\ 2\ ) = \ 16$$
$$8(\ 1\ ) = \ \ 8$$
$$8(\ 0\ ) = \ \underline{\ ?\ }$$
$$8(-1) = \ \underline{\ ?\ }$$
$$8(-2) = \ \underline{\ ?\ }$$
$$8(-3) = \ \underline{\ ?\ }$$

◄ **Each product is eight less than the preceding one.**

Think of the positions of 24, 16, and 8 on the number line. Start at 24 and move to the left in steps of 8 units.

◄ **The product of a positive number and a negative number is a negative number.**

P–1 **If the pattern continues, what is the answer to 8(0)?** 0

Multiplication Property of Zero	$3 \cdot 0 = 0$
For any number a, $a \cdot 0$ or $0 \cdot a = 0$.	$0 \cdot \frac{1}{2} = 0$

P–2 **If the pattern above continues, what are the other missing numbers?**

$-8, -16, -24$

The following rule tells how to find the product of a positive and a negative number by using absolute value.

Rule for Multiplying a Positive Number and a Negative Number	
1. Multiply the absolute values.	$(3)(-7) = -21$
2. Write the opposite of the result.	$(-8)(4) = -32$

EXAMPLE 1   Multiply: $(9)(-8)$

Solution: ☐1 Multiply the absolute values. ⟶ $|9| \cdot |-8| = 9 \cdot 8$
$= 72$

☐2 Write the opposite of the result. ⟶ $(9)(-8) = -72$

P–3 **Multiply.**

a. $(-6)(7)$ -42 b. $(4)(-12)$ -48 c. $(-\frac{1}{2})(0)$ 0 d. $(-\frac{3}{4})(\frac{4}{3})$ -1

Operations on Numbers / 11

Lesson Resources
Warm–Up: p. M–19
Maintenance: See below.
Practice Worksheet 1.4
Exploration 1

Maintenance
1. Multiply: $\frac{3}{5} \cdot \frac{10}{9}$
   Ans: $\frac{2}{3}$
2. Add: $0.03 + 0.21 + 0.004$
   Ans: 0.244
3. Locate 2 on a number line.
   Ans:

4. Change $3\frac{7}{20}$ to a decimal.
   Ans: 3.35
5. Jack has $3 more than Joni. Jack has $12. How much money does Joni have?
   Ans: $9

Additional Examples
Example 1
Multiply.
1. $(-15)(6)$
   Ans: -90
2. $(-6.3)(100)$
   Ans: -630
3. $(6)(-\frac{1}{3})$
   Ans: -2

In the following multiplication problems, one factor, −6, stays the same. The other factor decreases by 1.

Think of the positions of −18, −12, −6, and 0 on the number line. Start at −18 and move to the right in steps of 6 units.

$$-6(\ 3\ ) = -18$$
$$-6(\ 2\ ) = -12$$
$$-6(\ 1\ ) = -6$$
$$-6(\ 0\ ) = \ \ \ 0$$
$$-6(-1) = \underline{\ \ ?\ \ }$$
$$-6(-2) = \underline{\ \ ?\ \ }$$
$$-6(-3) = \underline{\ \ ?\ \ }$$

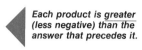 Each product is greater (less negative) than the answer that precedes it.

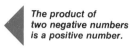 The product of two negative numbers is a positive number.

P-4 **If the pattern continues, what are the missing numbers?** 6, 12, 18

The following rule tells you how to find the product of two negative numbers by using absolute value.

> **Rule for Multiplying Two Negative Numbers**
> 1. Multiply the absolute values.
> 2. Write the result. $(-7)(-6) = 42$

EXAMPLE 2   Multiply: $(-8)(-5)$

Solution: ☐1 Multiply the absolute values. ───────→ $|-8| \cdot |-5| = 8 \cdot 5$
                                                                    $= 40$

          ☐2 Write the result. ───────→ $(-8)(-5) = 40$

P-5 **Multiply.**
  a. $(-5)(-9)$ 45 **b.** $(-7)(-10)$ 70 **c.** $(-2)(-0.13)$ 0.26 **d.** $(-6)(-0.5)$
                                                                                 3

EXAMPLE 3   At 7:00 P.M. on a winter day, the wind chill temperature was six times the thermometer reading of −5°F. What was the wind chill temperature?

Solution: $(6)(-5) = \underline{\ \ ?\ \ }$

          ☐1 Multiply the absolute values. ───────→ $|6| \cdot |-5| = 6 \cdot 5$
                                                                    $= 30$

          ☐2 Write the opposite of the result. ───────→ $(6)(-5) = -30$

          The wind chill temperature at 7:00 P.M. was −30°F.

Additional Examples
Example 2
Multiply.
1. $(-30)(-4)$
   Ans: 120
2. $(-8.2)(-11)$
   Ans: 90.2
3. $\left(-\frac{1}{2}\right)(-40)$
   Ans: 20

Example 3
1. Rudolph's net income in October was $\frac{1}{5}$ of his net income of −$300 in September. Find his net income in October.
   Ans. −$60
2. On a winter day, the wind chill temperature was three times the thermometer reading of −3°C. Find the wind chill temperature.
   Ans: −9°C

CLASSROOM EXERCISES

Multiply. (Example 1)

1. $(-4)(9)$ -36
2. $(10)(-6)$ -60
3. $(-12)(8)$ -96
4. $(-4)(12)$ -48

5. $(-\frac{1}{3})(9)$ -3
6. $(-15)(\frac{1}{5})$ -3
7. $(-37)(0)$ 0
8. $(0)(-75)$ 0

9. $(4)(-2.4)$ -9.6
10. $(-2)(1.8)$ -3.6
11. $(-3)(0.7)$ -2.1
12. $(-10)(2.6)$
-26

Multiply. (Example 2)

13. $(-11)(-7)$ 77
14. $(-5)(-9)$ 45
15. $(-10)(-15)$ 150
16. $(-14)(-6)$ 84

17. $(-7)(-21)$ 147
18. $(-43)(-3)$ 129
19. $(-5)(-1.2)$ 6
20. $(-0.7)(-30)$

21. $(-\frac{1}{4})(-4)$ 1
22. $(-\frac{1}{8})(-8)$ 1
23. $(-\frac{3}{8})(-\frac{8}{3})$ 1
24. $(-\frac{2}{5})(-\frac{3}{5})$ 21
$\frac{6}{25}$

WRITTEN EXERCISES

Goal: To multiply positive and negative numbers

Sample Problems: a. $(-7)(-8)$ **b.** $(-9)(12)$

Answers: a. 56   **b.** -108

Multiply. (Example 1)

1. $(-14)(5)$ -70
2. $(-16)(6)$ -96
3. $(0)(-63)$ 0
4. $(-30)(0)$ 0

5. $(22)(-3)$ -66
6. $(18)(-7)$ -126
7. $(-2)(24)$ -48
8. $(-4)(32)$

9. $(15)(-\frac{2}{3})$ -10
10. $(20)(-\frac{3}{4})$ -15
11. $(\frac{3}{5})(-\frac{4}{9})$ $-\frac{4}{15}$
12. $(-\frac{2}{9})(\frac{3}{4})$ $-\frac{1}{6}$

13. $(-12)(1.7)$ -20.4
14. $(-9)(3.4)$ -30.6
15. $(-1)(8.36)$ -8.36
16. $(9.01)(-1)$
-9.01

(Example 2)

17. $(-24)(-3)$ 72
18. $(-16)(-8)$ 128
19. $(-\frac{4}{5})(-\frac{5}{4})$ 1
20. $(-\frac{7}{4})(-\frac{4}{7})$ 1

21. $(-15)(-5.8)$ 87
22. $(-24)(-2.7)$ 64.8
23. $(-160)(-20)$ 3200
24. $(-180)(-99)$

25. $(-1.65)(-0.2)$ 0.33
26. $(-1.25)(-0.8)$ 1
27. $(-1)(-99)$ 99
28. $(-86)(-1)$
86

MIXED PRACTICE

29. $(36)(-12)$ -432
30. $(-42)(8)$ -336
31. $(-18)(-11)$ 198
32. $(-23)(-5)$
115

33. $(-240)(15)$ -3600
34. $(-325)(7)$ -2275
35. $(817)(0)$ 0
36. $(0)(-341)$ 0

37. $(-1.2)(-1.3)$ 1.56
38. $(-2.4)(-5.1)$ 12.24
39. $(-1\frac{2}{3})(2\frac{4}{5})$ $-4\frac{2}{3}$
40. $(-1\frac{3}{3})(-3\frac{1}{3})$
$5\frac{1}{3}$

Operations on Numbers / **13**

Common Error
When multiplying a positive number and a negative number, students use the sign of the number having the greater absolute value for the answer.

Example
$7 (-3) = 21$

Prescription
Have students determine the sign of the answer before performing the multiplication. Remind students that the product of a positive number and a negative number is always negative.

Assignment Guide
Basic pp. 13–14: 1–41 odd
Average pp. 13–14: 1–45 odd
Above Average: pp. 13–14:
1–41 odd, 43–47

Additional Answers
Written exercises
8. -128
24. $17,820$

Problem Solving Skills
Choosing the operation
(Ex. 41–42)
Solving a multi–step problem
(Ex. 43–47)
Using logical reasoning (Ex. 47)

APPLICATIONS   (Example 3)

41. The net change in the price of a stock for five successive days was $-\frac{3}{4}$ of a point for each day. Find the net change for the five days. $-3\frac{3}{4}$

42. On a winter day, the wind chill temperature was eight times the thermometer reading of $-4°C$. Find the wind chill temperature. $-32°C$

MORE CHALLENGING EXERCISES

Let $*$ mean: "Double the number with the larger absolute value. Then multiply by the remaining number." Compute each value.

EXAMPLE $3 * (-5)$ **SOLUTION** $3 * (-5) = 3 \cdot 2(-5)$
$= 3 \cdot (-10)$
$= -30$

Critical Thinking
See Exercises 43–47.

43. $-6 * 2$ -24 **44.** $-4 * (-8)$ 64 **45.** $1.2 * 3.6$ 8.64 **46.** $\frac{1}{2} * \left(-\frac{3}{8}\right)$
$-\frac{3}{8}$

47. Compute the value for $2 * (-4)$ and for $-4 * 2$. Is $*$ a commutative operation? Explain. Yes, because $2 * (-4) = -16$ and $-4 * 2 = -16$.

Quiz Sections 1.1–1.4
After completing this Mid-Chapter Review, you may want to administer a quiz covering the same sections. See the quiz provided on page 33 in the *Teacher's ResourceBank*.

███ **MID-CHAPTER REVIEW** ███████████

Write the opposite of each number in simplest form. (Section 1.1)

1. 17 -17 **2.** 32 -32 **3.** $-\frac{1}{2}$ $\frac{1}{2}$ **4.** $-1\frac{1}{2}$ $1\frac{1}{2}$ **5.** -0.06 0.06

Simplify. (Section 1.1)

6. $|20|$ 20 **7.** $|0|$ 0 **8.** $|-3.7|$ 3.7 **9.** $|2.5|$ 2.5 **10.** $\left|-1\frac{3}{5}\right|$ $1\frac{3}{5}$

Which number has the greater absolute value? (Section 1.1)

11. -17 or 12 -17 **12.** -8 or -12 -12 **13.** 10 or -5 10 **14.** 9 or -38
-38

Add. (Section 1.2)

15. $-20 + (-13)$ -33 **16.** $-15 + 35$ 20 **17.** $-18.1 + 19.1$ 1

18. $-6.8 + 6.8$ 0 **19.** $-12\frac{1}{2} + (-13\frac{1}{2})$ -26 **20.** $8\frac{1}{4} + (-12\frac{3}{4})$
$-4\frac{1}{2}$

Solve. (Section 1.2)

21. The temperature of a solution was $-5°C$. It was heated until its temperature rose $11°C$. What was its temperature after being heated? $6°C$

22. The thermometer reading at 4:00 A.M. was $-8°F$. By 1:00 P.M., it had risen $17°F$. Find the thermometer reading at 1:00 P.M. $9°F$

14 / *Chapter 1*

Write a sum for each difference. (Section 1.3)

23. $11 - 7$ $11 + (-7)$

24. $-8.7 - 9.6$ $-8.7 + (-9.6)$

25. $12.1 - (-7.9)$ $12.1 + 7.9$

26. $3.6 - 5.4$ $3.6 + (-5.4)$

27. $-7 - 4$ $-7 + (-4)$

28. $-8 - (-10)$ $-8 + 10$

Subtract. (Section 1.3)

29. $13 - 28$ -15

30. $-4 - (-12)$ 8

31. $0 - (-9.25)$ 9.25

32. $2.6 - (-3.3)$ 5.9

33. $\frac{1}{2} - 7$ $-6\frac{1}{2}$

34. $-6\frac{1}{2} - 6\frac{1}{2}$ -13

Solve. (Section 1.3)

35. The high temperature on a winter day was 14°F. The low temperature was −11°F. How much higher was the high temperature than the low temperature?   25°F

36. The wind chill temperature on a winter day was −17°F. The thermometer reading was 6°F. How much lower was the wind chill temperature than the thermometer reading?   23°F

Multiply. (Section 1.4)

37. $(-7)(8)$ -56

38. $(-4)(13)$ -52

39. $(9)(-\frac{1}{3})$ -3

40. $(12)(-\frac{1}{4})$ -3

41. $(-1)(-1)$ 1

42. $(-4)(-5)$ 20

43. $(-4)(-8.4)$ 33.6

44. $(-10)(-4.5)$ 45

Solve. (Section 1.4)

45. On a winter day, the low temperature was −2°F. The high temperature was −11 times the low temperature. Find the day's high temperature.   22°F

46. The net change in the price of a stock for four successive days was $-1\frac{1}{2}$ points each day. Find the total change in the price of the stock for the four days.   −6 points

REVIEW CAPSULE FOR SECTION 1.5

Write a fraction for each of the following.

1. $1\frac{1}{2}$ $\frac{3}{2}$

2. 12 $\frac{12}{1}$

3. $1\frac{3}{5}$ $\frac{8}{5}$

4. $2\frac{1}{3}$ $\frac{7}{3}$

5. $3\frac{1}{7}$ $\frac{22}{7}$

6. $4\frac{2}{5}$ $\frac{22}{5}$

7. $5\frac{2}{3}$ $\frac{17}{3}$

8. $2\frac{1}{6}$ $\frac{13}{6}$

9. 7 $\frac{7}{1}$

10. $3\frac{3}{4}$ $\frac{15}{4}$

Multiply or divide as indicated.

11. $\frac{7}{2} \times \frac{2}{7}$ 1

12. $-\frac{3}{5} \times (-\frac{5}{3})$ 1

13. $1\frac{2}{3} \times \frac{3}{5}$ 1

14. $(-6) \times (-\frac{1}{6})$ 1

15. $24 \div 8$ 3

16. $\frac{24}{8}$ 3

17. $24 \times \frac{1}{8}$ 3

18. $4\frac{1}{2} \div 6$ $\frac{3}{4}$

Maintenance

1. Add: $3\frac{1}{3} + \frac{1}{2}$

Ans: $3\frac{5}{6}$

2. Divide: $(3.6) \div (1.2)$

Ans: 3

3. Locate $-4\frac{1}{2}$ on a number line.

Ans:

4. Change $\frac{1}{5}$ to a decimal

Ans: 0.2

5. John earns $6.20 an hour. He works 40 hours. How much money has he earned?

Ans: $248.00

Additional Examples
Example 1
Divide.

1. $(-176) \div (-11)$

Ans: 16

2. $(-24) \div (-\frac{1}{4})$

Ans: 96

3. $\frac{-39}{-3}$

Ans: 13

OBJECTIVES: To divide two negative numbers

1.5 Division

To divide a positive number and a negative number
To divide 0 by any number

Each of the following products equals 1.

$$\frac{7}{2} \times \frac{2}{7} = 1 \qquad -\frac{3}{5} \times \left(-\frac{5}{3}\right) = 1 \qquad \frac{5}{3} \times \frac{3}{5} = 1$$

Definition

> Two numbers are *reciprocals* of each other if their product is 1.
>
> $a \cdot \frac{1}{a} = 1$ *(a is not zero.)*
>
> $\frac{7}{2}$ and $\frac{2}{7}$
>
> $-\frac{5}{3}$ and $-\frac{3}{5}$

P–1 **What is the reciprocal of each number below?**

The reciprocal of any number n, except 0, can be expressed as $\frac{1}{n}$.
Thus, the reciprocal of $\frac{2}{3}$ is $\frac{1}{\frac{2}{3}}$, or $\frac{3}{2}$.

a. $\frac{3}{10}$ $\frac{10}{3}$ **b.** -3 $-\frac{1}{3}$ **c.** $1\frac{1}{3}$ $\frac{3}{4}$ **d.** 1 1 **e.** -1 -1

The following illustrates that

dividing by a number is the same as multiplying by its reciprocal.

$$36 \div 9 = 4 \qquad 36 \times \frac{1}{9} = 4$$

9 and $\frac{1}{9}$ are reciprocals.

> To divide by a number, multiply by its reciprocal.
>
> $a \div b = a \cdot \frac{1}{b}$, $b \neq 0$ *(b is not zero.)*
>
>

Any number <u>except zero</u> can be a divisor. Thus, "$\frac{6}{2}$" is permitted because "$6 = 3 \cdot 2$" is true. However, "$\frac{6}{0}$" is not permitted because "$6 = a \cdot 0$" is false no matter what number a represents. $a \cdot 0$ always equals 0.

EXAMPLE 1   Divide: $(-12) \div \left(-\frac{1}{3}\right)$

Solution: The reciprocal of $-\frac{1}{3}$ is $-\frac{3}{1}$, or -3.

1 Write a product for the quotient. ⟶ $(-12) \div \left(-\frac{1}{3}\right) = (-12) \times (-3)$

2 Multiply. ⟶ $= 36$

P–2 **Divide.**

a. $(-16) \div \left(-\frac{1}{4}\right)$ 64 **b.** $\left(-\frac{5}{9}\right) \div (-5)$ $\frac{1}{9}$ **c.** $0 \div \left(-\frac{1}{8}\right)$ 0

0 · a = 0

16 / *Chapter 1*

EXAMPLE 2   Divide: $4\frac{1}{2} \div (-6)$

Solution: The reciprocal of -6 is $-\frac{1}{6}$.

1 Write a product for the quotient. ⟶ $4\frac{1}{2} \div (-6) = \frac{9}{2} \times (-\frac{1}{6})$

2 Multiply. ⟶ $= -\frac{9}{12}$

3 Write the answer in lowest terms. ⟶ $= -\frac{3}{4}$

P–3 **Divide.**

a. $5\frac{1}{6} \div (-3)$ $-1\frac{13}{18}$ b. $(-18) \div 1\frac{1}{6}$ $-15\frac{3}{7}$ c. $\dfrac{-2.4}{8}$

-0.3

◄ *This represents the quotient of -2.4 and 8.*

Rules for Division

1. If a positive number is divided by a positive number, the quotient is a positive number.

   $30 \div 3 = 10$

2. If a negative number is divided by a negative number, the quotient is a positive number.

   $-4\frac{1}{2} \div -2\frac{1}{4} = 2$

3. If a negative number is divided by a positive number, the quotient is a negative number.

   $-25 \div 5 = -5$

4. If a positive number is divided by a negative number, the quotient is a negative number.

   $35 \div -7 = -5$

5. If zero is divided by a positive or a negative number, the quotient is zero.

   $0 \div -6 = 0$

6. Zero can never be a divisor.

   $-5 \div 0$ is undefined.

EXAMPLE 3   A city's deficiency in rainfall for one year (12 months) was -36 centimeters. Find the average deficiency per month.

Solution:

1 Write the division expression. ⟶ $-36 \div 12 = \underline{\ ?\ }$

2 Divide. ⟶ $-36 \div 12 = -3$

The average deficiency in rainfall per month was -3 centimeters.

Operations on Numbers / 17

Additional Examples
Example 2
Divide.
1. $(-42) \div 3$
   Ans: -14
2. $(17.5) \div (-5)$
   Ans: -3.5
3. $(-6\frac{4}{5}) \div 2$
   Ans: $-3\frac{2}{5}$

Example 3
A company had a net loss for the year of $-\$15,408$. Find the average net loss per month.
Ans: $-\$1284$

Write a product for each quotient. (Step 1 in Examples 1 and 2)

1. $-8 \div \frac{1}{4}$ -8×4 **2.** $-12 \div 3$ $-12 \times \frac{1}{3}$ **3.** $28 \div (-7)$ $28 \times (-\frac{1}{7})$ **4.** $\frac{-3.2}{4}$ $-3.2 \times \frac{1}{4}$

Divide. (Example 1)

5. $-35 \div (-\frac{1}{5})$ 175 **6.** $-32 \div (-\frac{1}{2})$ 64 **7.** $-30 \div (-10)$ 3 **8.** $\frac{-44}{-11}$ 4

(Example 2)

9. $4\frac{1}{2} \div (-9)$ $-\frac{1}{2}$ **10.** $3\frac{1}{5} \div (-8)$ $-\frac{2}{5}$ **11.** $0 \div (-10\frac{3}{4})$ 0 **12.** $0 \div (-\frac{1}{5})$ 0

Goal: To divide positive and negative numbers
Sample Problems: a. $125 \div (-5)$ **b.** $(-1\frac{1}{2}) \div (-6\frac{1}{4})$
Answers: a. -25 **b.** $\frac{6}{25}$

Divide. (Example 1)

1. $-480 \div (-20)$ 24 **2.** $(-325) \div (-25)$ 13 **3.** $\frac{-27}{-9}$ 3 **4.** $\frac{-132}{-12}$ 11
5. $-16 \div (-1\frac{1}{2})$ $10\frac{2}{3}$ **6.** $-25 \div (-1\frac{1}{4})$ 20 **7.** $-1\frac{1}{4} \div (-7\frac{1}{2})$ $\frac{1}{6}$ **8.** $-4\frac{1}{2} \div (-\frac{3}{4})$ 6

(Example 2)

9. $16 \div (-1\frac{1}{2})$ $-10\frac{2}{3}$ **10.** $10 \div (-\frac{1}{3})$ -30 **11.** $3\frac{3}{4} \div (-\frac{3}{4})$ -5 **12.** $2\frac{2}{5} \div (-\frac{2}{5})$ -6
13. $\frac{24}{-8}$ -3 **14.** $\frac{18}{-9}$ -2 **15.** $108 \div (-12)$ -9 **16.** $176 \div (-16)$
17. $0 \div 17$ 0 **18.** $0 \div (-81)$ 0 **19.** $-121 \div 11$ -11 **20.** $-216 \div 12$
21. $-6.3 \div 7$ -0.9 **22.** $-1.8 \div 0.3$ -6 **23.** $-2\frac{1}{2} \div \frac{1}{4}$ -10 **24.** $-3\frac{1}{6} \div \frac{1}{3}$ $-9\frac{1}{2}$

MIXED PRACTICE

25. $\frac{36}{-9}$ -4 **26.** $\frac{-54}{-6}$ 9 **27.** $-25 \div 5$ -5 **28.** $72 \div (-12)$ -6
29. $-13 \div (-\frac{1}{4})$ 52 **30.** $-18 \div \frac{2}{3}$ -27 **31.** $\frac{0}{-3}$ 0 **32.** $\frac{0}{-21.2}$ 0
33. $\frac{-39}{13}$ -3 **34.** $\frac{17}{-2}$ $-8\frac{1}{2}$ **35.** $-\frac{7}{9} \div 2\frac{1}{3}$ $-\frac{1}{3}$ **36.** $-\frac{1}{2} \div 1\frac{1}{8}$ $-\frac{4}{9}$
37. $\frac{110}{-11}$ -10 **38.** $\frac{-23}{10}$ -2.3 **39.** $20.5 \div (-0.5)$ -41 **40.** $-3.6 \div (-1.2)$ 3

APPLICATIONS   (Example 3)

41. A company had a net loss for the year of $-\$28,200$. Find the average net loss per month. $-\$2350$

42. Over a period of ten days, a stock showed a net change in price of $-3\frac{3}{4}$ points. Find the average change per day. $-\frac{3}{8}$ point per day

Assignment Guide
Basic p. 18: 1–41 odd
Average p. 18: 1–41 odd
Above Average p. 18: 1–41 odd

Additional Answers
Written Exercises
16. -11
20. -18

Problem-Solving Skills
Choosing the operation
(Ex. 41–42)

1.6 Order of Operations

To evaluate an expression involving addition and multiplication, addition and division, subtraction and multiplication, or subtraction and division

An **expression,** such as $12 - 18$ or $2 \cdot 5 + 7$, includes at least one of the operations of addition, subtraction, multiplication, or division.

P–1 **Which of the following answers are correct?**

                                                           Correct

a. $2 \cdot 5 + 7 = 10 + 7$ **Ans. 17** Correct   **c.** $10 \div 5 - 3 = 2 - 3$ **Ans. −1**

b. $2 \cdot 5 + 7 = 2 \cdot 12$ **Ans. 24** Incorrect   **d.** $10 \div 5 - 3 = 10 \div 2$ **Ans. 5**

                                                              Incorrect

Rules are needed so that a given expression has only one value.

> If only addition and subtraction are involved, $8 - 2 - 1 = 5$
> the operations are performed from left to right. $1 - 2 + 8 = 7$

EXAMPLE 1   Evaluate: **a.** $12 + 5 - 4$ **b.** $4 - 9 + 3$ ◀ *Evaluate means "find the value of."*

Solutions: **a.** $12 + 5 - 4 = 17 - 4 = 13$ **b.** $4 - 9 + 3 = -5 + 3 = -2$

> If only multiplication and division are involved, $8 \times 2 \div 4 = 4$
> the operations are performed from left to right. $27 \div 9 \div 3 = 1$

EXAMPLE 2   Evaluate: **a.** $28 \div 7(-9)$ **b.** $(-2)(-13) \div (-4)$

Solutions: **a.** $28 \div 7(-9) = 4(-9) = -36$ **b.** $(-2)(-13) \div (-4) = 26 \div (-4) = -6\frac{1}{2}$

A common error is to perform the operation first that appears first in the expression.

> Multiplication or division is performed $2 \cdot 5 + 7 = 17$
> before addition or subtraction. $10 \div 5 - 3 = -1$

EXAMPLE 3   Evaluate.

    a. $2 \cdot 5 + 7$ **b.** $4 + 3 \cdot 6$ **c.** $10 - 6 \cdot 2$ **d.** $10 \div 5 - 3$

Solutions: **a.** $2 \cdot 5 + 7 = 10 + 7 = 17$ **b.** $4 + 3 \cdot 6 = 4 + 18 = 22$

              c. $10 - 6 \cdot 2 = 10 - 12 = -2$ **d.** $10 \div 5 - 3 = 2 - 3 = -1$

Maintenance
1. Divide: $\frac{1}{2} \div \frac{1}{4}$
  Ans: 2
2. Multiply: $(0.1)(0.1)$
  Ans: 0.01
3. Locate -1.25 on a number line.
  Ans:

4. Change $\frac{1}{4}$ to a decimal.
  Ans: 0.25
Paul earns $8.30 an hour. He works 30 hours. How much money has he earned?
Ans: $249.00

Additional Examples
Example 1
Evaluate.
1. $17 - 10 + 5$
  Ans: 12
2. $-11 + 1 - 9$
  Ans: −19
3. $-7.6 - 9 + 1.4$
  Ans: −15.2

Example 2
Evaluate.
1. $-16 \div (8)(2)$
  Ans: −4
2. $-11 (-8) \div 2$
  Ans: 44
3. $32 \div (-2) \div (-2)$
  Ans: 8

Example 3
Evaluate.
1. $-4 + (8)(-2)$
  Ans: −20
2. $16 \div 8 - (-2)$
  Ans: 4
3. $5(-6) - 18 \div (-3)$
  Ans: −24

CLASSROOM EXERCISES

Evaluate. (Example 1)

1. $12 - 3 + 7$ 16

2. $14 + 1 - 9$ 6

3. $7 - 10 + 3$ 0

4. $5 - 8 - 11$ −14

(Example 2)

5. $2 \cdot 15 \div 6$ 5

6. $42 \div 7 \cdot 3$ 18

7. $-8 \div 2(-3)$ 12

8. $-10 \div (-2) \div (\overline{}5)$ $\frac{-1}{5}$

(Example 3)

9. $3 + 10 \cdot 4$ 43

10. $20 - 4 \cdot 2$ 12

11. $21 - 27 \div 3$ 12

12. $3 + 5(-7)$ −32

13. $(-3)\,(5) - (-2)\,(2)$ −11

14. $-9 \div 3 - 7$ −10

15. $11 \div (-\frac{1}{2}) + 6$ −16

16. $-8 + (-4) \div (-\frac{1}{4})$ 8

WRITTEN EXERCISES

Goal: To evaluate an expression that involves order of operations
Sample Problems: a. Evaluate: $5 + 8 \cdot 3$ **b.** Evaluate: $-2(-3) + 3 \div (-1)$
Answers: a. 29 **b.** 3

Evaluate. (Example 1)

1. $5 - 8 + 12$ 9

2. $10 - 14 + 7$ 3

3. $5 - 10 - 7$ −12

4. $-6 + 7 + 8$ 9

5. $-0.4 + 0.7 - 0.8$ −0.5

6. $1.2 - 2.3 + 3.6$ 2.5

7. $8 + 9 - 10 + 6$ 13

8. $17 - 13 - 2 + 10$ 12

(Example 2)

9. $8 \cdot 2 \div 4$ 4

10. $20 \div 5 \cdot 4$ 16

11. $-240 \div 8 \cdot 2$ −60

12. $9(-8) \div (-12)$ 6

13. $-3 \cdot 12 \div (-4)$ 9

14. $-6(-8) \div 3(4)$ 64

15. $15 \div \frac{1}{2}(-3)$ −90

16. $(-27) \div \frac{1}{3}(-5)$ 405

(Example 3)

17. $6 - 5 \cdot 4$ −14

18. $12 - 3 \cdot 9$ −15

19. $-17 + (-9)2$ −35

20. $-21 + 5(-3)$ −36

21. $12 - 32 \div (-4)$ 20

22. $15 - 39 \div (-3)$ 28

23. $6.3 + 4.8 \div (-8)$ 5.7

24. $1\frac{1}{8} + 3\frac{3}{8} \div (-18)$ $\frac{15}{16}$

MIXED PRACTICE

25. $8 - 5 + 9\,(-2)$ −15

26. $9 + 2 - 15$ −4

27. $20 - 6 \cdot 7$ −22

28. $-3 \cdot 5 - 21$ −36

29. $16 + 14 \div (-2)$ 9

30. $21 \div 7 + 4$ 7

31. $-1(8) - 2(-3)$ −2

32. $4(3) + 9(8)$ 84

33. $-3 + 18 \div (-9)$ −5

34. $-5 - 12 \div 6$ −7

35. $-66 \div (-2)(-4)$ −132

36. $-200 \div (-2)\,(-3)$ −300

NON-ROUTINE PROBLEMS

37. Suppose a person was born on the 25th day of the year 35 B.C. and died on the 25th day of the year 15 A.D. How many years did the person live?

38. Use four 4's and one or more of the operation symbols, $+$, $-$, $\div$, or $\times$, to express the whole numbers from 1 to 4.
Example: $(4 \div 4) + 4 - 4 = 1$

1.7 Powers

An expression such as $(-2)^3$ is called a **power.** The 3 is the **exponent** and the -2 is the **base.** The exponent tells how many times to use the base as a factor.

Teaching Suggestions p. M–18

EXAMPLE 1   Evaluate: $(-2)^3$

Solution: Use -2 as a factor 3 times ———▸ $(-2)^3 = (-2)(-2)(-2) = -8$

Powers are evaluated before multiplication and division are performed <u>and</u> before addition and subtraction are performed.

EXAMPLE 2   Evaluate.

a. $2 \cdot 3^2$ **b.** $-8 \div 4^3$ **c.** $-6 + 2^3$ **d.** $(-5)^2 - 6(-2)$

Solutions: **a.** $2 \cdot 3^2 = 2 \cdot 9 = 18$ **b.** $-8 \div 4^3 = (-8) \div 64 = \frac{-8}{64} = -\frac{1}{8}$

c. $-6 + 2^3 = -6 + 8$ **d.** $(-5)^2 - 6(-2) = 25 - (-12)$
$= 2$ $= 25 + 12 = 37$

P–1 **Evaluate:** **a.** $3 \cdot 2^2$ 12   **b.** $(-4)^2 \div 2^3$ 2   **c.** $-10 - 5(-\frac{1}{2})$ $-7\frac{1}{2}$

A **variable** is a letter such as x that represents one or more numbers. An **algebraic expression** is an expression with one or more variables, such as the following.

A variable can represent any number from its replacement set. $x + 2$ $5n$ $p^2 - 7y$ z^3

To evaluate an algebraic expression, replace each variable with its value. Then perform the operations.

EXAMPLE 3   Evaluate $3n^2 - 4$ when $n = 5$.

Solution: 1 Replace n with 5. ———▸ $3n^2 - 4 = 3(5)^2 - 4$
2 Evaluate the power. Then multiply. ———▸ $= 3(25) - 4$
3 Subtract. ———▸ $= 75 - 4 = 71$

Lesson Resources
Warm–Up: p. M–19
Maintenance: See below.
Practice Worksheet 1.7
Transparencies 1, 2, 3

Maintenance
1. Divide: $\frac{1}{8} \div \frac{3}{4}$
   Ans: $\frac{1}{6}$
2. Multiply: $(2.8)(3.4)$
   Ans: 9.52
3. Locate -1 on a number line.
   Ans:

   <table><tr><td>-3</td><td>-1</td><td>0</td><td></td><td>3</td></tr></table>

4. Change $1\frac{3}{8}$ to a decimal.
   Ans: 1.375
5. Della sold 30 hotdogs and 75 hamburgers. The hotdogs sold for $0.90 each, and the hamburgers sold for $1.15 each. How much money did Della collect?
   Ans: $113.25

Additional Examples
Example 1
Evaluate.
1. $(-3)^2$
   Ans: 9
2. $(-1)^7$
   Ans: −1
3. $(-2)^4$
   Ans: 16

Example 2
Evaluate.
1. $10 + (-2)^3$
   Ans: 2
2. $-3 \cdot 5^2$
   Ans: −75
3. $(-4)^3 \div (-16)$
   Ans: 4

Example 3
Evaluate.
1. $7x^2 - 19$ when $x = 2$
   Ans: 9
2. $5 - 3y^3$ when $y = -2$
   Ans: 29

Additional Examples
Example 4
Evaluate.
1. $n^2 + 8n$ when $n = -5$
   Ans: -15
2. $3a^2 - 7a$ when $a = 2$
   Ans: -2
3. $-x^2 + 6x - 11$ when
   $x = -3$
   Ans: -38

Common Error
Students incorrectly evaluate powers by multiplying the base times the exponent.

Example
$2^3 = 2 \cdot 3$
$ = 6$

Prescription
In Classroom Exercises 1–6, have students name the base and tell how many times the base will be used as a factor. Then have students write each power as a product and evaluate.

Assignment Guide
Basic
Day 1 pp. 22–23: 1–29 odd
Day 2 p. 23: 31–67 odd
Average pp. 22–23: 1–67 odd
Above Average pp. 22–23: 1–67 odd

P–2 Evaluate each expression when $p = 3$.

a. $p^2 - 5$ 4 **b.** $5 - p^2$ −4 **c.** $4p^2 - p$ 33

EXAMPLE 4   Evaluate $x^2 - 3x$ when $x = -2$.

Solution: ☐1 Replace x with -2. ⟶ $x^2 - 3x = (-2)^2 - 3(-2)$
          ☐2 Evaluate the power. Then multiply. ⟶ $= 4 - 3(-2)$
          ☐3 Subtract. ⟶ $= 4 - (-6) = 10$

CLASSROOM EXERCISES

Evaluate each expression in Exercises 1–14. (Example 1)

1. 6^2 36   **2.** 2^3 8   **3.** 5^3 125   **4.** $(-3)^2$ 9   **5.** $(-4)^3$ −64   **6.** $(-1)^4$ 1

(Example 2)

7. $12 - 4^2$ −4   **8.** $3 \cdot 7^2$ 147   **9.** $2 \cdot 3^2 - 10$ 8   **10.** $4 - 3 \cdot 2^2$ −8
11. $8^2 \div (-4)$ −16   **12.** $-9 \div 3^3$ $-\frac{1}{3}$ **13.** $-10 - 10^2$ −110   **14.** $(-4)^2 - 3(-4)$ 28

Evaluate each expression when $t = 4$ (Example 3)

15. $t^2 - 5$ 11   **16.** $2 - t^2$ −14   **17.** $-4t - 7$ −23   **18.** $-13 + 7t$ 15

Evaluate each expression when $y = -3$. (Example 3)

19. $2y^2$ 18   **20.** $y^3 + 6$ −21   **21.** $-8 - y^2$ −17   **22.** $-2y + 13$ 19

Evaluate each expression when $n = 5$. (Example 4)

23. $n^2 + n$ 30   **24.** $n^3 - 7n$ 90   **25.** $2n^2 - 5n$ 25   **26.** $-2n^2 - 4n + 5$ −65

Evaluate each expression when $x = -2$. (Example 4)

27. $3x - x^2$ −10   **28.** $x^3 - 7x$ 6   **29.** $3x^2 + 6x$ 0   **30.** $5x^3 - x^2 - x$ −42

WRITTEN EXERCISES

Goal: To evaluate an expression
Sample Problems: Evaluate: **a.** $5 + 9 \cdot 3^2$ **b.** $n^2 - 6$ when $n = 3$
Answers: **a.** 86   **b.** 3

Evaluate each expression in Exercises 1–14. (Example 1)

1. 11^2 121   **2.** 6^3 216   **3.** 10^3 1000   **4.** $(-5)^2$ 25   **5.** $(-7)^3$ −343   **6.** $(-3)^3$ −27

(Example 2)

7. $6(-3)^2$ 54 **8.** $-4 \cdot (-1)^3$ 4 **9.** $5 + 2 \cdot 4^2$ 37 **10.** $9 + 3 \cdot 5^2$ 84

11. $4 \cdot 2^3 - 50$ –18 **12.** $(-5)3^2 + 37$ –8 **13.** $5^2 - 4^2 + 3^2$ 18 **14.** $5^2 + 2^2 - 3^2$
20

Evaluate each expression when n = 3. (Example 3)

15. $n^2 + 4$ 13 **16.** $n^2 - 7$ 2 **17.** $8n^2 - 12$ 60 **18.** $-4n^2 + 9$ –27

Evaluate each expression when r = –1. (Example 3)

19. $7 + r^2$ 8 **20.** $13 - r^3$ 14 **21.** $4r^2 - 6$ –2 **22.** $-5r^2 + 10$ 5

Evaluate each expression when s = 4 or when w = –10. (Example 4)

23. $s^2 - 4s$ 0 **24.** $10s^2 - s$ 156 **25.** $-5s^2 - s$ –84 **26.** $s^2 + s + 1$ 21

27. $3s^2 - 5s + 2$ 30 **28.** $6s^2 + 3s - 5$ 103 **29.** $s^3 + 2s^2 - 3s$ –183 **30.** $3s^3 - 4s^2 + 2s$ 131 – 5

31. $3w^2$ 300 **32.** $-4w^2$ –400 **33.** $2w^2 + w$ 190 **34.** $5w - w^2$ –150

35. $w^2 + w - 30$ 60 **36.** $2w^2 - 4w + 20$ 260 **37.** $w^3 + 2w^2 - w$ –790 **38.** $3w^3 - 4w^2 - w + 60$ –3330

MIXED PRACTICE *Evaluate each expression.*

39. 4^4 256 **40.** $3 \cdot 2^3$ 24 **41.** $(-6)^3$ –216 **42.** $-2(-4)^2$ –32

43. $2^2 + 3^2$ 13 **44.** $4 \cdot 7 - 5^2$ 3 **45.** $-6 \div 4^2$ $-\frac{3}{8}$ **46.** $(-1)^5 - 4(-3)$ 11

Evaluate each expression when m = –5 or when t = 6.

47. $1 - 5m$ 26 **48.** $12m \div (-8)$ $7\frac{1}{2}$ **49.** $t^3 - 2t^2 \div (-4)$ 234 **50.** $-8t \div 4(-7)$ 84

51. $-3t^2 - 4t$ –132 **52.** $-4m - 6$ 14 **53.** $2m - m^2$ –35 **54.** $-t^2 - 5t$ –66

55. $2t^2$ 72 **56.** $-3t^3$ –648 **57.** $2m^2 \div (-4)$ $-12\frac{1}{2}$ **58.** $-t^3 - 10t$ –276

59. $m^2 + 3m - m$ 15 **60.** $-5 + m^2$ 20 **61.** $8(-4t^2)$ –1152 **62.** $m^2 - m + (-6)$ 24

APPLICATIONS

The Example below shows how to use a calculator to evaluate a power of a number. Note that you press "*" once for the second power, twice for the third power, three times for the fourth power, and so on.*

EXAMPLE Evaluate 16^3.

SOLUTION

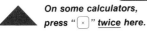
On some calculators, press "×*" twice here.*

Since the power is three, press "=*" twice.*

Use a calculator to do the following exercises.

63. 7^2 49 **64.** 9^3 729 **65.** 5^4 625 **66.** 21^4 194,481 **67.** 7^6 117,649 **68.** $(-4)^3$ –64 (HINT: First press "4 +/–.")

Quiz Sections 1.5–1.7
After completing Sections 1.5–1.7, you may want to administer a quiz covering the same sections. See the quiz provided on page 34 in the *Teacher's ResourceBank*.

Methods of Computation

Making a Choice

In everyday situations, you often have to consider whether you need an **exact answer** to a problem or whether an **estimate** will give an answer that is close enough. Then you have to decide whether to use **mental computation,** a **calculator, paper and pencil,** or some combination of these to solve the problem efficiently.

EXERCISES

1. Robert went to the hardware store to buy several items for a project. He was on his way to the check-out counter when he discovered that he had exactly $16.55 in his wallet. Robert thought: "I'll have to find the total cost to be sure I have enough money to pay for everything."

   a. Does Robert need to find the exact total or will an estimate be close enough? Explain.

   b. Which would be the more efficient way to estimate the sum, using a calculator or using mental computation? Explain. Answers will vary.

Paint (1 quart)	$4.95
Bolts	$1.90
Nails	$1.05
Drill bit	$2.19
Extension cord	$4.49

Robert used **front-end estimation** in this way.

$\boxed{1}$ Add the dollars first. $4 + $1 + $1 + $2 + $4 = $12

$\boxed{2}$ Estimate the cents.   0.95 ◄——— *About $1*
0.90 ◄——— *About $1*
0.05 ⎫
0.19 ⎬ ◄——— *About 75¢*
0.49 ⎭

Estimate: $12 + $2 + 75¢ = **$14.75**

   c. How far was Robert's estimate from the actual cost? Estimate was 17¢ higher.

   d. Did Robert have enough money to purchase all the items? Yes

2. The manager of Bargain Books listed one week's sales of the store's top selling books. The books are to be ranked in order, starting with the greatest number of sales.

Book Title	Sales	Rank
Yesterday	59	?
Disbelief	43	?
Of the Heart	60	?
My Life	48	?
Remember the Day	51	?
Last Boat Home	52	?
Faithful	49	?
Springtime	44	?

a. Is an exact sales listing or an approximate sales listing appropriate? Why? exact; Reasons will vary.

b. Do you think the manager will use paper and pencil or mental computation to do the rankings? Explain your answer.
Answers will vary.

3. The table at the right gives the scores awarded by five judges for the final dives of the top scorers at Uptown High School's swimming and diving meet. The judges use this formula to award points.

Diver	Degree of Difficulty	Judges' Scores	Points
Jones	2.3	7 7 8 8 9	?
Lang	2.1	7 7 8 8 8	?
Grimm	2.4	6 6 7 7 8	?
Mills	2.6	7 8 8 8 9	?
Curry	2.5	8 8 8 9 9	?

$$\frac{\text{Number of}}{\text{Points}} = \frac{\text{Sum of Middle}}{\text{Three Scores}} \times \frac{\text{Degree of}}{\text{Difficulty}}$$

a. Should an exact answer be computed or is an estimate sufficient? Why?

b. Would it be more efficient to use paper and pencil or a calculator to award points for each dive? Why? Answers will vary.

c. Rank the divers according to points, from highest to lowest.

d. What computation method did you use to arrange the rankings? Why?
Answers will vary.

4. Maria Stern must compute one week's earnings for five of her employees. This table shows the number of hours worked and the hourly pay rate for each employee.

Employee	Hours	Rate	Earnings
Wayne	20.5	$4.30	?
Eva	24.75	$5.20	?
Gloria	17.25	$4.20	?
Bill	18.5	$4.10	?
Carl	16.75	$3.80	?

a. Should Maria estimate each employee's earnings or is an exact answer needed? Why?

b. What method or methods of computation would you use to compute the earnings? Give reasons for your answer. Answers will vary.

c. Find the earnings for each employee.

CHAPTER SUMMARY

IMPORTANT TERMS	Opposite *(p. 2)*	Exponent *(p. 21)*
	Absolute value *(p. 2)*	Base *(p. 21)*
	Reciprocal *(p. 16)*	Variable *(p. 21)*
	Power *(p. 21)*	Algebraic expression *(p. 21)*

IMPORTANT IDEAS

1. The absolute value of a positive number equals that number. The absolute value of a negative number equals the opposite of that number. The absolute value of 0 equals 0.

2. Rules for Adding Positive and Negative Numbers: See pages 4 and 5.

3. The sum of opposites is 0.

4. To subtract a number, add its opposite.

5. *Multiplication Property of Zero:* $a \cdot 0 = 0 \cdot a = 0$.

6. Rules for Multiplying Positive and Negative Numbers: See pages 11–12.

7. To divide by a number, multiply by its reciprocal.

8. If a is any number, $\frac{a}{0}$ does not represent a number.

9. Rules for Dividing Positive and Negative Numbers: See page 17.

10. Rules for Order of Operations
    a. Multiplication or division is performed before addition or subtraction.
    b. If only addition and subtraction are involved, the operations are performed from left to right.
    c. If only multiplication and division are involved, the operations are performed from left to right.
    d. Powers are evaluated before multiplication and division are performed *and* before addition and subtraction are performed.

Chapter Test

Two Chapter Tests (Form A and Form B) are provided on pages 35–38 in the *Teacher's ResourceBank*.

CHAPTER REVIEW

SECTION 1.1

Write the opposite of each number in simplest form.

1. −1.5 1.5   **2.** 27 −27   **3.** $-\frac{3}{8}$ $\frac{3}{8}$ **4.** $-2\frac{1}{2}$ $2\frac{1}{2}$ **5.** 0 0

Simplify.

6. |24| 24   **7.** |−2.6| 2.6   **8.** $-|-\frac{3}{8}|$ $-\frac{3}{8}$ **9.** |−35| 35   **10.** −|−9| −9

Which number has the greater absolute value?

11. 0 or −3 −3   **12.** 0 or 1 1   **13.** −2 or 4 4   **14.** −5 or 3 −5

Add.

15. $-13 + 24$ 11

16. $53 + (-29)$ 24

17. $28 + (-42)$ -14

18. $-61 + 39$ -22

19. $-5.8 + (-7.7)$ -13.5

20. $-0.29 + (-0.58)$
-0.87

21. On Thursday, the change in a stock's price was $-1\frac{7}{8}$ points. On Friday, the change was $-1\frac{3}{8}$ points. Find the total change in price for the two days.
$-3\frac{1}{4}$ points

22. In February, two new record low temperatures were set in a city. The new records were $-12°F$ and $-19°F$. What is the sum of these two new record low temperatures?
$-31°F$

SECTION 1.3

Write each difference as a sum. Then compute.

23. $17 - 29$ $17 + (-29)$; -12

24. $42 - 68$ $42 + (-68)$; -26

25. $-27 - (-18)$ $-27 + 18$; -9

26. $-53 - (-26)$ $-53 + 26$; -27

27. $19.2 - (-4.7)$ $19.2 + 4.7$; 23.9

28. $42.9 - (-18.5)$
$42.9 + 18.5$; 61.4

29. The freezing point of the coolant in an automobile's radiator was $-8°F$. More antifreeze was added which made the freezing point $-27°F$. How much was the freezing point lowered? $19°F$

30. The high temperature on a December day was $4°C$. The low temperature was $-12°C$. How much higher was the high temperature than the low temperature? $16°C$

SECTION 1.4

Multiply.

31. $(12)(-17)$ -204

32. $(-15)(21)$ -315

33. $(-18)(-25)$ 450

34. $(-52)(-14)$ 728

35. $(-\frac{1}{5})(45)$ -9

36. $(\frac{1}{4})(-28)$ -7

37. On a winter day, the low temperature for Helena, Montana, was -4 times the low temperature of $3°F$ for Boise, Idaho. What was the low temperature for Helena?
$-12°F$

38. At 11:00 P.M., the thermometer reading was $-2°F$. The wind chill temperature was $9\frac{1}{2}$ times the thermometer reading. What was the wind chill temperature? $-19°F$

SECTION 1.5

Divide.

39. $\frac{-63}{9}$ -7

40. $\frac{84}{-7}$ -12

41. $\frac{-144}{-12}$ 12

42. $\frac{-128}{-8}$ 16

43. $13 \div (-\frac{1}{4})$ -52

44. $\frac{2}{3} \div (-12)$ $-\frac{1}{18}$

45. $(-1\frac{1}{4}) \div (-1\frac{7}{8})$ $\frac{2}{3}$

46. $(-3\frac{1}{2}) \div (-2\frac{5}{8})$
$1\frac{1}{3}$

Operations on Numbers / 27

Solve.

47. The deficiency in rainfall in a farming region for January through June was −8.4 inches. Find the average deficiency per month for that period. −1.4 inches per month

48. Over a period of five days, the net change in the price of a company's stock was $-3\frac{1}{8}$ points. Find the average change in the stock's price per day. $-\frac{5}{8}$ point per day

SECTION 1.6

Evaluate.

49. $12 - 27 + 8$ −7

50. $23 - 38 + 6$ −9

51. $16 + 4(-8)$ −16

52. $(-13) + 12 \cdot 3$ 23

53. $15 - 42 \div (-6)$ 22

54. $24 - (-16) \div 4$ 28

SECTION 1.7

Evaluate.

55. 4^3 64

56. $(-2)^3$ −8

57. $(6)^2 - 3(-5)$ 51

58. $(-12)(8) \div 24 + 3^2$ 5

59. $4(-3)^2 \div (-12) - 25$ −28

60. $-4(3)^2 \div 12 - (-15)$ 12

Evaluate each expression when n = 3.

61. $n^2 - 13$ −4

62. $4n^2 - 45$ −9

63. $-2n^2 - n$ −21

64. $-n - 5n^2$ −48

65. $n^3 - n^2 + n$ 21

66. $n^3 - 3n^2 + 1$ 1

67. $-n^3 + 3n^2 - 1$ −1

68. $-n^3 + n^2 - n$ −21

69. $-3n^3 + 2n^2 - 2n$ −69

Real Numbers

In this chapter you will study properties of numbers. Scientists study objects to determine if they exhibit mathematical properties. Crystals, for example, exhibit the mathematical property called *symmetry*. Similarities between crystals are revealed by their symmetry.

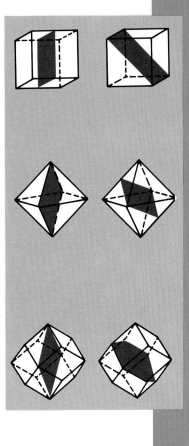

Background

The photo above shows the following crystals: top left—quartz; top right—galena; bottom left—pyrite; bottom right—sphalerite.

The diagram at the right shows planes of symmetry in a cube (6 faces), an octahedron (8 faces), and a dodecahedron (12 faces).

Lesson Resources
Warm–Up: p. M–20
Maintenance: See below.
Practice Worksheet 2.1
Exploration 2
Transparency 4

Maintenance
1. Simplify: $|-10|$
   Ans: 10 (Section 1.1)
2. Subtract: $24 - (-12)$
   Ans: 36 (Section 1.3)
3. Divide: $-9.6 \div (-0.3)$
   Ans: 32 (Section 1.5)
4. Evaluate: $3 \cdot 2^2 - 4$
   Ans: 8 (Section 1.7)
5. The thermometer reading at
   5:30 A.M. was $-5°C$. By
   11:00 A.M., it had risen $12°C$.
   Find the thermometer reading
   at 11:00 A.M.
   Ans: 7°C (Section 1.2)

Additional Examples
Example 1
Write as a quotient of two integers in two ways. Sample answers are given. Other answers
are possible.
1. 5
   Ans: $\frac{5}{1}$; $\frac{10}{2}$
2. -0.8
   Ans: $\frac{-8}{10}$; $\frac{4}{5}$
3. 1
   Ans: $\frac{7}{7}$; $\frac{-5}{-5}$

Example 2
Express each number as a repeating decimal.
1. $-\frac{2}{3}$
   Ans: $-0.\overline{6}$
2. $\frac{5}{9}$
   Ans: $0.\overline{5}$
3. $-5\frac{3}{4}$
   Ans: $-5.75000 \cdots$

OBJECTIVES: To write a rational number as a quotient of two integers
To express a fraction as a repeating decimal

2.1 Real Numbers
To identify a number as <u>Rational</u> or <u>Irrational</u>

The **set of whole numbers** consists of zero and the <u>counting numbers</u>.

Counting Numbers	**Whole Numbers**
$\{1, 2, 3, 4, 5, \cdots\}$	$\{0, 1, 2, 3, 4, 5, \cdots\}$

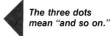
*The three dots
mean "and so on."*

The **set of integers** consists of the set of whole numbers and their
opposites. For example, 2 and -2 are opposites.

$$\{\cdots, -3, -2, -1, 0, 1, 2, 3, \cdots\}$$

P–1 **What is the opposite of each of the following?**
a. -2 2   **b.** 6 -6 **c.** 0 0   **d.** 213 -213

The number $-2\frac{1}{2}$ is a <u>rational number</u>. Here are other names for $-2\frac{1}{2}$.

$$-\frac{5}{2} \qquad \frac{-5}{2} \qquad \frac{5}{-2} \qquad \frac{-10}{4} \qquad \frac{20}{-8} \qquad -2.5 \qquad -2\frac{5}{10}$$

Definition

> A **rational number** is a number that can be
> expressed in the form $\frac{a}{b}$, where a and b
> are integers, but b is not 0.
>
> $-5 = -\frac{5}{1}$
> $3\frac{1}{3} = \frac{10}{3}$

EXAMPLE 1 Write each number as a quotient of two integers
in two ways.
  a. 4   **b.** 0   **c.** $3\frac{3}{4}$ **d.** $-\frac{5}{8}$ **e.** 0.3

Solutions: **a.** $\frac{4}{1}; \frac{8}{2}$ **b.** $\frac{0}{1}; \frac{0}{5}$ **c.** $\frac{15}{4}; \frac{45}{12}$ **d.** $\frac{-5}{8}$ or $\frac{5}{-8}; \frac{-10}{16}$ **e.** $\frac{3}{10}; \frac{30}{100}$

(Other solutions are possible.)

Rational numbers can also be expressed as <u>repeating decimals</u>.
Also, <u>all</u> repeating decimals are names for rational numbers.

EXAMPLE 2 Express $\frac{3}{11}$ as a repeating decimal.

Solution: $\frac{3}{11}$ means $3 \div 11$.

$$11\overline{)3.0000 \cdots} \quad 0.2727 \cdots$$

*A repeating decimal has
a digit or sequence
of digits that is
infinitely repeated.*

Express each rational number as a repeating decimal.

a. $\frac{2}{9}$ **b.** $\frac{4}{11}$ **c.** $\frac{1}{6}$ $0.1666\cdots$

0.222 $\cdots$ 0.3636 $\cdots$

The repeating decimal $0.2727\cdots$ can be abbreviated as $0.\overline{27}$. The bar indicates the digits that repeat.

The decimal form of many rational numbers is a <u>terminating decimal</u>. For example, $\frac{1}{4} = 0.25$, $-3\frac{1}{2} = -3.5$, and $1\frac{7}{8} = 1.875$. However, you can consider 0 as the repeating digit for these numbers.

$$\frac{1}{4} = 0.25000\cdots \qquad -3\frac{1}{2} = -3.5000\cdots \qquad 1\frac{7}{8} = 1.875000\cdots$$

EXAMPLE 3   Choose from the box the set of numbers that corresponds to each word description.

a. The attendance at a concert

b. The scale on a weather thermometer

c. The rise and fall of stock prices

> Rational numbers
> Counting numbers
> Integers

Solutions: **a.** Counting numbers   As their name implies, they are used for counting.

b. Integers   A weather thermometer's scale includes 0, negative numbers, and positive numbers.

c. Rational numbers   The changes in stock prices are expressed as positive and negative fractions.

Positive numbers such as 1, 4, 9, 16, 25, and 36 are **perfect squares,** because each is the square of a counting number.

$$1^2 = 1 \qquad 2^2 = 4 \qquad 3^2 = 9 \qquad 4^2 = 16 \qquad 5^2 = 25 \qquad 6^2 = 36$$

Thus, square roots of perfect squares are <u>rational</u>.

$$\sqrt{1} = 1 \qquad \sqrt{4} = 2 \qquad \sqrt{9} = 3 \qquad \sqrt{16} = 4 \qquad \sqrt{25} = 5 \quad\cdots$$

Square roots of counting numbers that are not perfect squares are <u>irrational</u>.

Not all irrational numbers are radicals. π is an irrational number.

An **irrational number** <u>cannot</u> be expressed as a quotient of integers. Thus, the decimal form of an irrational number is <u>nonrepeating</u>.

$$\sqrt{2} = 1.4142135\cdots \qquad \sqrt{3} = 1.7320508\cdots \qquad \pi = 3.1415926\cdots$$

Real Numbers / **31**

Additional Examples
Example 3
Choose from the following sets of numbers the set of numbers that corresponds to each word description: Integers, Rational numbers, Whole numbers.
1. Scores of a basketball game
    Ans: Whole numbers
2. Lap times of a race car
    Ans: Rational numbers
3. Distance in feet above or below sea level
    Ans: Integers

Additional Examples
Example 4
Write Rational or Irrational to de-scribe each number.
1. 3.876876 ···
   Ans: Rational
2. 3.878778777 ···
   Ans: Irrational
3. $-\sqrt{17}$
   Ans: Irrational
4. $\sqrt{121}$
   Ans: Rational

Example 5
Refer to the set below. Then list the numbers named.
$\{-8, -\sqrt{10}, -1.6, 0, \frac{10}{2}, \sqrt{49}, \pi\}$
1. Rational numbers
   Ans: $\{-8, -1.6, 0, \frac{10}{2}, \sqrt{49}\}$
2. Integers
   Ans: $\{-8, 0, \frac{10}{2}, \sqrt{49}\}$
3. Irrational numbers
   Ans: $\{-\sqrt{10}, \pi\}$

Common Error
When writing a fraction as a re-peating decimal, students rou-tinely divide the smaller number into the larger number.

Example
Write $\frac{3}{5}$ as a repeating decimal.

$$\begin{array}{r} 1.66 \\ 3\overline{)5.00} \end{array} \qquad \frac{3}{5} = 1.\overline{6}$$

Prescription
In Example 2, point out that $\frac{3}{11}$ means 3 divided by 11. Then give students several fraction ex-amples and have them write the fractions in other division for-mats. For example,
$\frac{2}{3}$ $2 \div 3$ $3\overline{)2}$

Additional Answers
Classroom Exercises
11. 0.571428571428 ···
13. −0.375000 ···
14. −1.4000 ···

32

EXAMPLE 4

Write Rational or Irrational to describe each number.

a. 12.326326 ···   **b.** 0.050050005 ···
c. $\sqrt{14}$ **d.** $-\sqrt{9}$

Ask students to explain how they know that b is nonrepeating.

Solutions: **a.** Rational, because 326 repeats.

b. Irrational, because the decimal is nonrepeating.

c. Irrational, because 14 is not a perfect square.

d. Rational, because 9 is a perfect square: $-\sqrt{9} = -3$.

P–3 **Describe each number as Rational or Irrational.**

a. 0.121314 ···   **b.** $\sqrt{24}$ **c.** 4.4444 ···
Irrational   Irrational   Rational

Definition | The **set of real numbers** contains all the rational numbers and all the irrational numbers.

EXAMPLE 5

Refer to the set below to list the numbers named.

$$\{-\tfrac{13}{4}, -\tfrac{12}{4}, -2.7, -\sqrt{2}, 0, \sqrt{3}, \pi, 4.9, \sqrt{49}\}$$

a. Integers   **b.** Rational numbers
c. Irrational numbers   **d.** Real numbers

Solutions: **a.** $-\tfrac{12}{4}, 0, \sqrt{49}$ **b.** $-\tfrac{13}{4}, -\tfrac{12}{4}, -2.7, 0, 4.9, \sqrt{49}$

c. $-\sqrt{2}, \sqrt{3}, \pi$ **d.** All are real numbers.

CLASSROOM EXERCISES

Write each number as a quotient of two integers in two ways. (Example 1)

1. $3\frac{1}{2}$ $\frac{7}{2}$; $\frac{14}{4}$ **2.** -15 **3.** 0 $\frac{0}{1}$; $\frac{0}{3}$ **4.** -0.9 **5.** $-\frac{1}{4}$ **6.** $\sqrt{9}$ $\frac{3}{1}$; $\frac{6}{2}$ **7.** $-\sqrt{25}$
        $\frac{-15}{1}$; $\frac{-30}{2}$ $\frac{-9}{10}$; $\frac{-18}{20}$ $\frac{-2}{8}$; $\frac{-4}{16}$ $\frac{-5}{1}$; $\frac{-10}{2}$

Write each rational number as a repeating decimal. (Example 2)

8. 5 **9.** $\frac{1}{2}$ **10.** $\frac{1}{3}$ **11.** $\frac{4}{7}$ **12.** $\frac{1}{9}$ **13.** $-\frac{3}{8}$ **14.** $-1\frac{2}{5}$
  5.000 ···   0.5000 ···   0.333 ···   0.111 ···

For Exercises 15–17, choose from the box the set of numbers that corresponds to each word description. (Example 3)

<table>
<tr><td></td><td>Positive rational
numbers</td></tr>
<tr><td></td><td>Counting numbers</td></tr>
<tr><td></td><td>Integers</td></tr>
</table>

15. The profit or loss in dollars of a business during one year Integers
16. The counter on a service station pump showing the number of gallons purchased Positive rational numbers
17. The page numbers in your textbook Counting numbers

Describe each number as Rational *or* Irrational. (Example 4)

18. $1.123123 \cdots$ Rational
19. $6.121231234 \cdots$ Irrational
20. 19   Rational
21. -16 Rational

Refer to the set of numbers below. Then list the numbers named in each exercise. (Example 5)

$$\{28, -1.8, 3\tfrac{1}{4}, 0.1010010001 \cdots, -124, 17, -\tfrac{22}{7}, 5.436436436 \cdots\}$$

22. Whole numbers 28, 17
23. Integers $28, -124, 17$
24. Negative numbers $-1.8, -124, -\tfrac{22}{7}$
25. Real numbers
26. Rational numbers
27. Irrational numbers

![bar] **WRITTEN EXERCISES** ![bar]

Goal: To identify rational and irrational numbers
Sample Problem: List the rational and irrational numbers in
$$\{\tfrac{4}{2}, \sqrt{16}, -3.5, -\sqrt{27}, 5.222 \cdots, \sqrt{33}\}.$$
Answer: Rationals: $\tfrac{4}{2}, \sqrt{16}, -3.5, 5.222 \cdots$; Irrationals: $-\sqrt{27}, \sqrt{33}$

Write each number as a quotient of two integers in two ways. (Example 1)

1. $-\tfrac{7}{8}$ $\tfrac{-14}{16}, \tfrac{-28}{32}$
2. $-\tfrac{13}{16}$ $\tfrac{-26}{32}, \tfrac{-39}{48}$
3. $4\tfrac{3}{8}$ $\tfrac{35}{8}, \tfrac{70}{16}$
4. $-5\tfrac{3}{4}$ $\tfrac{-23}{4}, \tfrac{-46}{8}$
5. -1.83

6. 3.47 $\tfrac{347}{100}, \tfrac{694}{200}$
7. 20 $\tfrac{20}{1}, \tfrac{40}{2}$
8. -46 $\tfrac{-46}{1}, \tfrac{-92}{2}$
9. $-\sqrt{9}$ $\tfrac{-3}{1}, \tfrac{-6}{2}$
10. $-\sqrt{64}$

Write each rational number as a repeating decimal. (Example 2)

11. -24 $-24.000 \cdots$
12. $\tfrac{5}{9}$ $0.555 \cdots$
13. $\tfrac{7}{11}$ $0.6363 \cdots$
14. $-3\tfrac{5}{8}$ $-3.625000 \cdots$
15. $\tfrac{3}{7}$
16. $\tfrac{11}{12}$

For Exercises 17–19, choose from the box the set of numbers that corresponds to each word description. (Example 3)

<table>
<tr><td></td><td>Positive rational
numbers</td></tr>
<tr><td></td><td>Whole numbers</td></tr>
<tr><td></td><td>Integers</td></tr>
</table>

17. A table of record high and low temperatures expressed to the nearest degree Integers
18. The odometer on a car Positive rational numbers
19. The attendance at a basketball game Whole numbers

Write Rational or Irrational to describe each number. (Example 4)

20. -3.7 Rational **21.** 7.9 Rational **22.** $\sqrt{29}$ Irrational **23.** $-\sqrt{13}$

24. $-1\frac{5}{8}$ Rational **25.** $3\frac{3}{4}$ Rational **26.** $\sqrt{49}$ Rational **27.** $-\sqrt{36}$

28. $-\pi$ **29.** $-7.1000\cdots$ **30.** $0.24681012\cdots$ **31.** $-0.3691215\cdots$
Irrational   Rational   Irrational

Refer to the set of numbers below. Then list the numbers named in Exercises 32–37. (Example 5)

$$\{\frac{10}{2}, -2, 4.7, 0, \pi, -1\frac{3}{4}, \frac{7}{8}, 19, -\frac{17}{4}, -6, -0.9\}$$

32. Rational numbers **33.** Irrational numbers π **34.** Negative numbers

35. Positive integers $\frac{10}{2}$, 19 **36.** Positive irrational numbers π **37.** Real numbers

Select one or more letters below to describe each of the following numbers. (Examples 4 and 5)

a. Whole numbers **b.** Integers **c.** Rational numbers
d. Irrational numbers **e.** Real numbers **f.** Counting numbers

38. 7 a, b, c, e, f **39.** -4 b, c, e **40.** $\frac{5}{8}$ c, e **41.** -20 b, c, e **42.** -1.7 c, e **43.** $7\frac{1}{5}$ c, e

44. $-6\frac{1}{9}$ c, e **45.** $\frac{22}{7}$ c, e **46.** $1.4238\cdots$ d, e **47.** $-\frac{28}{7}$ b, c, e **48.** 101
a, b, c, e, f **49.** -8.0
b, c, e

REVIEW CAPSULE FOR SECTION 2.2

Rewrite the following as sums. (Pages 8–10)

1. $6 - 4$ 6 + (−4) **2.** $7 - 6$ 7 + (−6) **3.** $15 - 8$ 15 + (−8) **4.** $4 - 9$
4 + (−9)

5. $5 - 12$ 5 + (−12) **6.** $23 - 19$ 23 + (−19) **7.** $14 - 18$ 14 + (−18) **8.** $2 - 25$
2 + (−25)

Add. (Pages 4–7)

9. $4 + 13$ 17 **10.** $8 + 9$ 17 **11.** $\frac{5}{16} + (-\frac{1}{4})$ $\frac{1}{16}$ **12.** $\frac{1}{12} + (-\frac{2}{5})$

13. $1.3 + (-0.2)$ 1.1 **14.** $2.3 + (-1.1)$ 1.2 **15.** $-4 + (-5)$ −9 **16.** $-8 + (-1)$

17. $-\frac{1}{3} + \frac{2}{9}$ $-\frac{1}{9}$ **18.** $\frac{1}{2} + 3\frac{2}{3}$ $4\frac{1}{6}$ **19.** $-1.5 + (-5.6)$ −7.1 **20.** $0.8 + (-5.9)$
−5.1

Subtract. (Pages 8–10)

21. $5 - 3$ 2 **22.** $12 - 9$ 3 **23.** $\frac{1}{3} - \frac{1}{4}$ $\frac{1}{12}$ **24.** $\frac{3}{4} - \frac{2}{15}$ $\frac{37}{60}$

25. $-0.6 - 1.2$ −1.8 **26.** $-0.2 - 0.7$ −0.9 **27.** $-4 - (-5)$ 1 **28.** $-6 - (-1)$

29. $1\frac{2}{3} - \frac{3}{4}$ $\frac{11}{12}$ **30.** $-\frac{5}{6} - (-\frac{11}{12})$ $\frac{1}{12}$ **31.** $0.5 - 3.1$ −2.6 **32.** $2.4 - (-8.3)$
10.7

OBJECTIVES: To find the value of the variable that will make a given equation
true based on one of the Addition Properties

2.2 Addition Properties

To simplify an arithmetic expression involving addition and subtraction
To simplify an algebraic expression involving addition and subtraction

The following properties apply to addition of real numbers.

Teaching Suggestions p. M–20

Lesson Resources
Warm–Up: p. M–21
Maintenance: See below.
Practice Worksheet 2.2

Table

Addition Property of Zero

For any real number a,

$$a + 0 = a \text{ and } 0 + a = a.$$

$-3.4 + 0 = -3.4$

$0 + 6\frac{1}{2} = 6\frac{1}{2}$

Addition Property of Opposites

For any real number a,

$$a + (-a) = 0.$$

$9.1 + (-9.1) = 0$

$8\frac{1}{3} + (-8\frac{1}{3}) = 0$

Commutative Property of Addition

Any two real numbers can be
added in either order.
For any real numbers a and b,

$$a + b = b + a.$$

$\left(-\frac{3}{4}\right) + 7 = 7 + \left(-\frac{3}{4}\right)$

$y + 5 = 5 + y$

Associative Property of Addition

The way real numbers are grouped
for addition does not affect
their sum.
For any real numbers a, b, and c,

$$(a + b) + c = a + (b + c).$$

$\left(9 + \frac{7}{10}\right) + \frac{3}{10} = 9 + \left(\frac{7}{10} + \frac{3}{10}\right)$

$(6 + x) + x = 6 + (x + x)$

Maintenance:
1. Simplify: $|0.003|$
   Ans: 0.003 (Section 1.1)
2. Subtract: $8 - 15$
   Ans: -7 (Section 1.3)
3. Divide: $-\frac{3}{4} \div \left(-\frac{2}{3}\right)$
   Ans: $1\frac{1}{8}$ (Section 1.5)
4. Evaluate $x^2 - 6$ when $x = 3$.
   Ans: 3 (Section 1.7)
5. On June 1, Jane's checking
   account balance was $345.
   On July 1, her balance was
   $-$8. How much lower was
   Jane's account balance on
   July 1 than on June 1?
   Ans: $353 (Section 1.3)

EXAMPLE 1

*An equation is a sentence
that contains the
equality symbol "=."*

Find a value of the variable that will make each
equation true. Name the property that gives the
reason for your choice.

a. $2.9 + x = 0$ **b.** $3 + n = 8\frac{1}{2} + 3$

c. $y + 0 = 4$ **d.** $(35 + 67) + 72 = 35 + (67 + a)$

Solutions:

	Equation	Value	Property
a.	$2.9 + x = 0$	$x = -2.9$	Addition Property of Opposites
b.	$3 + n = 8\frac{1}{2} + 3$	$n = 8\frac{1}{2}$	Commutative Property
c.	$y + 0 = 4$	$y = 4$	Addition Property of Zero
d.	$(35 + 67) + 72 = 35 + (67 + a)$	$a = 72$	Associative Property

Additional Examples
Example 1
Find the value of the variable
that will make each equation
true. Name the property that
gives the reason for your choice.
1. $0 + x = -5.7$
   **Ans: $x = -5.7$; Add.
   Property of Zero**
2. $11 + (-8) = -8 + y$
   **Ans: $y = 11$; Comm.
   Property of Addition**
3. $a = 17 + (-17)$
   **Ans: $a = 0$, Add.
   Property of Opposites**

Real Numbers / **35**

What value of *y* makes each equation true?   by which property?

a. $8.2 + 7.1 = 7.1 + y$ **b.** $(7 + 3\frac{1}{2}) + y = 7 + (3\frac{1}{2} + 6)$
8.2; Comm. Prop. of Add.   6; Assoc. Prop. of Add.

The addition properties can be used to simplify expressions. The first step is to write the expression as a sum.

EXAMPLE 2   Simplify: $-12 + 19 - 13 + 15$ Explain that the square brackets are just like parentheses.

Solution: Use the Commutative and Associative Properties.

① Write as a sum. ⟶ $-12 + 19 - 13 + 15 = -12 + 19 + (-13) + 15$
② Regroup. ⟶ $= [-12 + (-13)] + (19 + 15)$
③ Add. ⟶ $= \quad -25 \quad + \quad 34$
$= 9$

EXAMPLE 3   Simplify: $12 + t - 17$

Solution:

① Write as a sum. ⟶ $12 + t - 17 = 12 + t + (-17)$
② Regroup. ⟶ $= t + [12 + (-17)]$ *By the Commutative and Associative Properties*
③ Simplify. ⟶ $= t + (-5)$
$= t - 5$

EXAMPLE 4   Simplify: $5.2 + x - 8.3 + y$

Solution: $5.2 + x - 8.3 + y = [5.2 + (-8.3)] + x + y$
$= -3.1 + x + y$ The result can also be written as $x + y - 3.1$.

P–2   **Simplify each of the following.**

a. $3 + 4 - 7 + 99$ **b.** $9 - 2\frac{1}{2} - 7\frac{3}{8} - \frac{7}{8}$ **c.** $1\frac{1}{2} + y - \frac{2}{3}$ **d.** $3.2 + b - 4.0$
$\frac{5}{6} + y$ $b - 0.8$

CLASSROOM EXERCISES

Add. (Table)

1. $6 + (-6)$ 0   **2.** $-12 + 0$ −12   **3.** $4\frac{1}{3} + 0$ $4\frac{1}{3}$ **4.** $(-\frac{1}{3}) + \frac{1}{3}$ 0

5. $0 + (-0.7)$ −0.7   **6.** $(-0.4) + 0.4$ 0   **7.** $1\frac{1}{2} + (-1\frac{1}{2})$ 0   **8.** $0 + \frac{3}{8}$ $\frac{3}{8}$

Write the value of the variable that makes each equation true. (Example 1)

9. $k = 4 + (-4)$ 0

10. $3 + (4 + b) = (3 + 4) + 5$ 5

11. $3\frac{1}{2} + (4 + \frac{1}{3}) = (a + 4) + \frac{1}{3}$ $3\frac{1}{2}$

12. $6\frac{1}{2} + 0 = n$ $6\frac{1}{2}$

13. $1.4 + x = 0.6 + 1.4$ 0.6

14. $4.3 + d = 4.3$ 0

Simplify. (Example 2)

15. $4 - 5 - 2 + 6$ 3

16. $-11 + 2 - 7 + 3$ -13

17. $-3\frac{1}{2} + 1\frac{2}{3} - \frac{1}{6} + 2$ 0

18. $4\frac{1}{5} - 5\frac{8}{10} + 2\frac{3}{10} - \frac{1}{2}$ $\frac{1}{5}$

(Example 3)

19. $31 - a + 23$ $54 - a$

20. $-d + 21 - 32$ $-d - 11$

21. $2.6 + x - 0.5$ $2.1 + x$

22. $y - 3.2 + 1.7$ $y - 1.5$

(Example 4)

23. $x - 4 + y - 2$ $x + y - 6$

24. $-z + 6 - 9 - p$ $-z - p - 3$

25. $12.0 - a - b - 6.9$ $5.1 - a - b$

26. $-3.9 - 9.3 - m - k$ $-13.2 - m - k$

WRITTEN EXERCISES

Goal: To simplify expressions using the addition properties

Sample Problem: $-3\frac{1}{3} - a + \frac{2}{3} + c$ **Answer:** $-a + c - 2\frac{2}{3}$

Write the value of the variable that makes each equation true. Name the property that gives the reason for your choice. (Example 1)

1. $8 + r = 0$ -8; Add. Property of Opposites

2. $0 + 6 = t$ 6; Add. Property of Zero

3. $15 + 7 = x + 15$ 7; Comm. Property of Add.

4. $y + 0 = -12$ -12; Add. Prop. of Zero

5. $4\frac{1}{2} + (-4\frac{1}{2}) = b$ 0; Add. Property of Opposites

6. $(a + \frac{3}{8}) + 1\frac{1}{2} = \frac{2}{3} + (\frac{3}{8} + 1\frac{1}{2})$ $\frac{2}{3}$; Assoc. Prop. of Add.

Simplify. (Example 2)

7. $7 + 5 - 13 + 6$ 5

8. $12 - 10 + 23 - 15$ 10

9. $\frac{1}{2} - \frac{1}{4} + 2 - \frac{3}{4}$ $1\frac{1}{2}$

10. $-3\frac{5}{8} + \frac{1}{4} - \frac{1}{8} - \frac{3}{8}$ $-3\frac{7}{8}$

11. $2.5 - 1.2 - 0.8 + 2.0$ 2.5

12. $-0.3 - 5.9 + 2.8 + 12.9$ 9.5

13. $2.7 + 3.6 - 4.8 + 0.2$ 1.7

14. $-8.9 - 2.6 + 11.2 - 1.2$ -1.5

(Example 3)

15. $5 - 1 - 2 - u$ $2 - u$

16. $s - 16 - 0 + 9$ $s - 7$

17. $1\frac{7}{8} + p - \frac{1}{2} - \frac{3}{8}$ $p + 1$

18. $4\frac{1}{4} - y - 7\frac{1}{3} - 5\frac{3}{4}$ $-y - 8\frac{5}{6}$

Real Numbers / **37**

Assignment Guide
Basic pp. 37–38; 1–41 odd
Average pp. 37–38; 1–29 odd, 43; p. 41: 1–23 odd
Above Average p. 38: 31–44; p. 41: 7–23 odd, 25–28

19. $-a - 0.5 + 0.2$ $-a - 0.3$

20. $3.8 + b - 6.4$ $b - 2.6$

21. $-3.9 + f - 0.5$ $f - 4.4$

22. $-r + 6.8 - 8.1$ $-r - 1.3$

(Example 4)

23. $6 + k - 23 - b$ $-17 + k - b$

24. $-d + 56 - 39 + w$ $-d + w + 17$

25. $d - 2\frac{1}{2} + \frac{3}{8} - f$ $d - 2\frac{1}{8} - f$

26. $h - 0.7 + 2.4 - m$ $h + 1.7 - m$

27. $a - 5 + 14 - b$ $a + 9 - b$

28. $r - 14 - g - 16$ $r - g - 30$

29. $s + 5.9 - t - 5.9$ $s - t$

30. $-19.1 + k + 19.1 - x$ $k - x$

MIXED PRACTICE

31. $16 + 34 - 21 + 10$ 39

32. $16 - 35 + 17 + 35$ 33

33. $3\frac{1}{2} - 5\frac{3}{4} + 2\frac{1}{4} - 4\frac{1}{2}$ $-4\frac{1}{2}$

34. $4\frac{1}{4} - 6\frac{1}{2} - 2\frac{3}{4} + 1\frac{1}{4}$ $-3\frac{3}{4}$

35. $3 + 6.1 - 0.8 + 2.3$ 10.6

36. $0.9 - 1.3 - 4.8 + 3.7$ -1.5

37. $-18 - y + 18$ $-y$

38. $31 + x - 17$ $x + 14$

39. $-63 + 15 + q - 34$ $-82 + q$

40. $44 + j - 17 + w$ $27 + j + w$

41. $c - 7\frac{5}{6} - d - 3\frac{2}{3}$ $c - d - 11\frac{1}{2}$

42. $7\frac{5}{8} + a - b - 13\frac{1}{4}$ $a - b - 5\frac{5}{8}$

NON-ROUTINE PROBLEMS

43. Steve forgot his locker number at school. The school secretary reminds him that the digits in his locker number are all odd numbers and that, when added, they total 11. There are 120 lockers. What is Steve's locker number?

44. In the addition problem below, the letters P, K, and N represent three different digits. Identify the digits.

$$\begin{array}{r} P\,K \\ +\ \ K\,N \\ \hline K\,N\,K \end{array}$$

REVIEW CAPSULE FOR SECTION 2.3

Multiply. (Pages 11–14)

1. $(5)(-6)$ -30

2. $3 \cdot 15$ 45

3. $\left(\frac{1}{3}\right)\left(-\frac{3}{4}\right)$ $-\frac{1}{4}$

4. $\frac{1}{5} \cdot \frac{6}{5}$ $\frac{6}{25}$

5. $-0.4 \cdot 6.2$ -2.48

6. $(-1.8)(-3.9)$ 7.02

7. $(-14)(5)$ -70

8. $(-8)(-11)$ 88

9. $\left(-\frac{3}{8}\right)\left(-1\frac{2}{3}\right)$ $\frac{5}{8}$

10. $-1\frac{1}{2} \cdot 3\frac{1}{5}$ $-4\frac{4}{5}$

11. $1.9 \cdot 2.6$ 4.94

12. $6.8(-5.3)$ -36.04

Write the reciprocal of the following. (Pages 16–18)

13. 5 $\frac{1}{5}$

14. $\frac{2}{5}$ $\frac{5}{2}$

15. -4 $-\frac{1}{4}$

16. $\frac{1}{3}$ $\frac{3}{1}$, or 3

17. $-\frac{3}{8}$ $-\frac{8}{3}$

18. 10 $\frac{1}{10}$

19. 1 1

20. $-\frac{1}{12}$ $-\frac{12}{1}$, or -12

21. -13 $-\frac{1}{13}$

22. $11\frac{1}{11}$

23. -1 -1

24. $\frac{8}{5}$ $\frac{5}{8}$

38 / Chapter 2

Problem–Solving Skills
Using guess and check (Ex. 43–44)
Using logical reasoning (Ex. 43–44)
Working backwards (Ex. 44)

Non–Routine Problems
For a description of non-routine problems, see page M–13.

Additional Answers
Written Exercises
43. Steve's locker number must be less than 120. Since the sum of two odd numbers is an even number, his locker number must have 3 digits. The number 119 has all odd digits and the sum of the digits is 11. His locker number is 119.
44. Since the sum K + N = K, N represents zero. Since N represents 0, P + K must be 10. Thus, K represents 1 and P represents 9.

$$\begin{array}{r} 91 \\ +\ 10 \\ \hline 101 \end{array}$$

OBJECTIVES: To find the value of the variable that will make an equation true based
on one of the Multiplication Properties
To simplify an arithmetic expression involving multiplication

2.3 Multiplication Properties

To simplify an algebraic expression involving multiplication

The following properties apply to multiplication of real numbers.

Teaching Suggestions p. M–20

Lesson Resources
Warm–Up: p. M–21
Maintenance: See below.
Practice Worksheet 2.3

Table

Multiplication Property of One

For any real number a,

$a \cdot 1 = a$ and $1 \cdot a = a$.

$(-4)(1) = -4$

$1(0.3x) = 0.3x$

The Multiplication Property of Reciprocals corresponds to the Addition Property of Opposites except that 0 is excluded from the Multiplication Property.

Multiplication Property of Reciprocals

The product of any nonzero real number and its reciprocal is 1.

$a \cdot \frac{1}{a} = 1, a \neq 0$

$\frac{3}{5} \cdot \frac{5}{3} = 1$

$(-7)(-\frac{1}{7}) = 1$

Commutative Property of Multiplication

Any two real numbers can be multiplied in either order.
For any real numbers a and b,

$a \cdot b = b \cdot a$.

$(-3)(2.5) = (2.5)(-3)$

$(4\frac{1}{5})(-8) = (-8)(4\frac{1}{5})$

Associative Property of Multiplication

The way real numbers are grouped for multiplication does not affect their product.
For any real numbers a, b, and c,

$(ab)c = a(bc)$.

$(9 \cdot \frac{2}{3})5 = 9(\frac{2}{3} \cdot 5)$

$5.2(3x) = (5.2 \cdot 3)x$

Maintenance
1. Which number has the greater absolute value, -26 or 36?
   Ans: 36 (Section 1.1)
2. Add: $-5 + (-3)$
   Ans: -8 (Section 1.2)
3. Subtract: $6.5 - (-2.1)$
   Ans: 8.6 (Section 1.3)
4. Evaluate $3x^2 - 2x$ when $x = -1$.
   Ans: 5 (Section 1.7)
5. The net change in the price of a stock for each of five successive days was $-\frac{3}{8}$ of a point. Find the net change for the five days.
   Ans: $-1\frac{7}{8}$ point (Section 1.4)

EXAMPLE 1

Find the value of the variable that makes each equation true. Name the property that gives the reason for your choice.

a. $\frac{1}{3} \cdot y = 1$ **b.** $(4.1) \cdot m = (-3.5)(4.1)$

c. $n \cdot 1 = 6$ **d.** $(\frac{2}{3}d)4\frac{1}{2} = \frac{2}{3}(5 \cdot 4\frac{1}{2})$

Solutions:

	Equation	Value	Property
a.	$\frac{1}{3} \cdot y = 1$	$y = \frac{3}{1}$, or 3	Multiplication Property of Reciprocals
b.	$(4.1) \cdot m = (-3.5)(4.1)$	$m = -3.5$	Commutative Property
c.	$n \cdot 1 = 6$	$n = 6$	Multiplication Property of One
d.	$(\frac{2}{3}d)4\frac{1}{2} = \frac{2}{3}(5 \cdot 4\frac{1}{2})$	$d = 5$	Associative Property

Additional Examples
Example 1
Find the value of the variable that makes each equation true. Name the property that gives the reason for your choice.
1. $(\frac{3}{5}y)\,17 = 1 \cdot 17$
   Ans: $y = \frac{5}{3}$; Mult. Property of Reciprocals
2. $1 \cdot 9.2 = x$
   Ans: $x = 9.2$; Mult. Property of One
3. $17(8 \cdot 3) = (17 \cdot 8)a$
   Ans: $a = 3$; Assoc. Property of Multiplication

P–1 What value of x makes each equation true? by which property?

a. $(8.1)(6.3) = (6.3)x$ **b.** $(7 \cdot 9)\frac{1}{2} = x(9 \cdot \frac{1}{2})$ **c.** $(-9)x = 1$

The multiplication properties can be used to simplify expressions.

EXAMPLE 2 Multiply: $(-\frac{1}{2})(\frac{1}{4})(-4)(-12)$

Solution: Use the Associative and Commutative Properties.

☐1 Rewrite the expression ⟶ $(-\frac{1}{2})(\frac{1}{4})(-4)(-12) = (-\frac{1}{2})(-12)(\frac{1}{4})(-4)$

☐2 Multiply. ⟶ $(-\frac{1}{2})(-12)(\frac{1}{4})(-4) = (6)(-1)$

$= -6$

EXAMPLE 3 Multiply: $(9)(x)(-\frac{1}{3})(y)$.

Solution:

☐1 $(9)(x)(-\frac{1}{3})(y) = (9)(-\frac{1}{3})(x)(y)$ ◀ 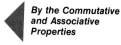 *By the Commutative and Associative Properties*

☐2 $(9)(-\frac{1}{3})(x)(y) = (-3)(x \cdot y)$

$= -3xy$ This is the simplest form of the product.

P–2 Multiply in each expression.

a. $(0.6)(3.2)(-5)$ –9.6 **b.** $(\frac{1}{8})(n)(24)(16)$ 48n **c.** $(-25)(p)(q)(-4)$100pq

---

████ **CLASSROOM EXERCISES** ████

Multiply. (Table)

1. $(\frac{1}{4})(4)$ 1
2. $-16(-\frac{1}{16})$ 1
3. $62 \cdot 1$ 62
4. $(1)(-23)$ -23
5. $(\frac{10}{9})(\frac{9}{10})$ 1
6. $(-\frac{2}{3})(-\frac{3}{2})$ 1
7. $1(17.8)$ 17.8
8. $(0.03)(1)$ 0.03

Write the value of the variable that makes each equation true. (Example 1)

9. $(12.8)(3.04) = n(12.8)$ 3.04
10. $(1)(-0.24) = d$ -0.24
11. $7(16) = y(7)$ 16
12. $(17 \cdot 3)b = 17(3 \cdot 5)$ 5
13. $54 \cdot 5 = 5q$ 54
14. $(4.1 \cdot 0.2)a = 4.1(0.2 \cdot 5)$ 5
15. $\frac{3}{8}(\frac{1}{3} \cdot k) = (\frac{3}{8} \cdot \frac{1}{3})\frac{4}{5}$ $\frac{4}{5}$
16. $1 = \frac{5}{12} \cdot z$ $\frac{12}{5}$
17. $\frac{2}{5}(\frac{5}{7} \cdot m) = (\frac{2}{5} \cdot \frac{5}{7})\frac{1}{2}$ $\frac{1}{2}$

Multiply. (Examples 2 and 3)

18. $(4)(17)(5)(-3)$ -1020

19. $(-3)(19)(5)(-2)$ 570

20. $(\frac{1}{3})(\frac{4}{5})(-6)(-\frac{5}{4})$ 2

21. $(-\frac{3}{2})(-\frac{1}{8})(48)(-\frac{4}{3})$ -12

22. $(14)(-4)(9)(-11)$ 5544

23. $(35)(\frac{1}{7})(10)(\frac{2}{5})$ 20

24. $(a)(0.25)(b)(0.4)$ $0.1ab$

25. $(1.6)(y)(-0.5)(z)$ $-0.8yz$

26. $(-j)(-3)(6)(k)$ $18jk$

27. $(-12)(m)(-n)(-10)$ $-120mn$

28. $(x)(-6.1)(1.3)(y)$ $-7.93xy$

29. $(-b)(1.8)(h)(0.6)$ $-1.08bh$

WRITTEN EXERCISES

Goal: To multiply in expressions using the multiplication properties

Sample Problem: $(-\frac{5}{8})(y)(\frac{8}{5})(-z)$ **Answer:** yz

Write the value of the variable that makes each equation true. Name the property that gives the reason for your choice. (Example 1)

1. $-\frac{5}{7} = f \cdot 1$

2. $\frac{6}{11} \cdot \frac{11}{6} = h$

3. $(-2.3 \cdot 6.4)8 = -2.3(6.4 \cdot s)$

4. $1 \cdot (-9) = x$

5. $a \cdot 3 = 3 \cdot (-\frac{1}{2})$

6. $(-6.8)(3.5) = m(-6.8)$

Multiply. (Examples 2 and 3)

7. $(-2)(-5)4$ 40

8. $(-1)(-8)(-3)$ -24

9. $(-\frac{1}{3})(12)(-9)$ 36

10. $(-4)(-\frac{3}{2})(-\frac{2}{3})(-\frac{1}{4})$ 1

11. $(0.6)(-p)(0.7)(-q)$ $0.42pq$

12. $(-3.5)(c)(-d)(-40)$ $-140cd$

13. $(-8)(18)(m)(-2)$ $288m$

14. $(d)(4)(39)(f)(2)$ $312df$

15. $(-3)(-3)(t)(-3)$ $-27t$

16. $(5)(p)(-6)(-2)$ $60p$

17. $(1.8)(m)(p)(-3)$ $-5.4mp$

18. $(-0.3)(a)(-0.7)(5)(b)$ $1.05ab$

19. $(-5)(x)(9)(y)(-5)$ $225xy$

20. $(b)(4)(-3)(c)(-5)$ $60bc$

21. $(1\frac{3}{10})(\frac{5}{6})(-r)(\frac{6}{5})(d)$ $-1\frac{3}{10}rd$

22. $(-\frac{1}{8})(-y)(\frac{3}{5})(-8)(m)$ $-\frac{3}{5}ym$

23. $(w)(16)(-5)(\frac{1}{8})(z)$ $-10wz$

24. $(-k)(14)(12)(-\frac{1}{7})(t)$ $24kt$

MORE CHALLENGING EXERCISES

For Exercises 25–28, a represents a positive integer and b represents a negative integer. Identify whether the value of each expression is positive, p, or negative, n.

25. $(-5)(b)(a)(a)$ p

26. $(-3)(a)(b)(-2)$ n

27. $(b)(-4)(-1)(b)(-3)$ n

28. $(2)(a)(b)(b)(b)$ n

Assignment Guide
Basic p. 41: 1–23 odd
Average p. 41: 1–23 odd;
p. 37: 1–29 odd, 43
Above Average p. 41:
7–23 odd, 25–28; p. 37: 31–44 all

Additional Answers
Written Exercises
1. $-\frac{5}{7}$; Multiplication Property of One
2. 1; Multiplication Property of Reciprocals
3. 8; Associative Property of Multiplication
4. -9; Multiplication Property of One
5. $-\frac{1}{2}$; Commutative Property of Multiplication
6. 3.5; Commutative Property of Multiplication

Critical Thinking
See Exercises 25–28

Problem–Solving Skills
Solving a multi–step problem
(Ex. 25–28)
Using logical reasoning
(Ex. 25–28)

Additional Answers
Mid–Chapter Review
12. $-6.142857142857\cdots$
23. $\frac{2}{3}, -\sqrt{5}, -1\frac{2}{7}, \sqrt{36}, -9.3$
28. $k - j - 5\frac{1}{4}$
29. $r - s - 4\frac{3}{8}$

MID-CHAPTER REVIEW

Write each number as a quotient of two integers in two ways. (Section 2.1)

1. $-\frac{5}{6}$ $\frac{-10}{12}; \frac{-15}{18}$ **2.** $-\frac{14}{17}$ $\frac{-28}{34}; \frac{-42}{51}$ **3.** $2\frac{4}{5}$ $\frac{14}{5}; \frac{28}{10}$ **4.** $-3\frac{1}{7}$ $\frac{-22}{7}; \frac{-44}{14}$ **5.** 3.5 $\frac{35}{10}; \frac{70}{20}$ **6.** -4.8 $\frac{-48}{10}; \frac{-96}{20}$

Write each rational number as a repeating decimal. (Section 2.1)

7. $\frac{4}{9}$ $0.444\cdots$ **8.** $\frac{2}{3}$ $0.666\cdots$ **9.** $\frac{8}{11}$ **10.** $-\frac{4}{7}$ **11.** $5\frac{1}{4}$ $5.2500\cdots$ **12.** $-6\frac{1}{7}$

$0.7272\cdots$ $0.571428571428\cdots$

Write <u>Rational</u> *or* <u>Irrational</u> *to describe each number.* (Section 2.1)

13. -2.8 Rational **14.** 8.1 Rational **15.** 31 Rational **16.** $\sqrt{81}$ Rational

17. $-4\frac{2}{3}$ Rational **18.** $\frac{\pi}{4}$ Irrational **19.** $6.13000\cdots$ Rational **20.** $2.323323332\cdots$
Irrational

Refer to the set of numbers below. Then list the numbers named in each exercise. (Section 2.1)

$$\{\tfrac{2}{3}, -\sqrt{5}, -1\tfrac{2}{7}, \sqrt{36}, -9.3\}$$

21. Integers $\sqrt{36}$ **22.** Irrational numbers $-\sqrt{5}$ **23.** Real numbers

Simplify. (Section 2.2)

24. $18 + x - 18$ x **25.** $-32 - y + 32$ $-y$ **26.** $-43 + w - 82 + 37$ $w - 88$

27. $y - 12 + 8 + z$ $y + z - 4$ **28.** $-12\frac{3}{4} + k + 7\frac{1}{2} - j$ **29.** $5\frac{3}{8} + r - 9\frac{3}{4} - s$

30. $t - 0.25 - w - 0.75$ $t - w - 1$ **31.** $-6.3 - a + b - 9.2$ $-15.5 - a + b$ **32.** $-8.4 - 2.5 - 0.6 - 1.3$ -12.8

Multiply. (Section 2.3)

33. $(6)(-2)(-\frac{1}{3})$ 4 **34.** $(-\frac{1}{4})(-24)(8)$ 48 **35.** $(-2)(-3)(h)(-\frac{1}{2})$ $-3h$

36. $(5)(-3)(x)(\frac{1}{5})$ $-3x$ **37.** $(-40)(-k)(-10)(4)$ $-1600k$ **38.** $(n)(-2)(-2)(-2)(-2)$ $16n$

39. $(6.3)(a)(-1.4)(y)$ $-8.82ay$ **40.** $(2.2)(-0.4)(d)(-s)$ $0.88ds$ **41.** $(p)(8)(-3)(f)$ $-24pf$

REVIEW CAPSULE FOR SECTION 2.4

Simplify.

1. $(-3 \cdot 4)(x \cdot y \cdot y)$ $-12xy^2$ **2.** $(5 \cdot 7)(a \cdot a \cdot b \cdot b)$ $35a^2b^2$ **3.** $(9 \cdot \frac{1}{3})(p^2 \cdot q \cdot r)$ $3p^2qr$

4. $(-\frac{1}{5} \cdot 10)(m \cdot n^2 \cdot s)$ $-2mn^2s$ **5.** $x + (-8)$ $x - 8$ **6.** $-t + (-1)$ $-t - 1$

7. $(1.3 \cdot 5)(f^2 \cdot g \cdot t^2)$ $6.5f^2gt^2$ **8.** $w + (-16)$ $w - 16$ **9.** $-22 + (-p)$ $-22 - p$

OBJECTIVES: To simplify expressions of the form $-(a + b)$
To simplify expressions of the form $-1(a)$ or $-1(a + b)$
To simplify expressions of the form $-a(b)$ or $(-a)(-b)$

2.4 Special Properties

To simplify expressions of the form $a - (b + c)$

The following special properties are helpful when simplifying an expression that contains parentheses.

Opposite of a Sum

$$-(a + b) = (-a) + (-b) \qquad -(4 + x) = (-4) + (-x)$$
$$= -a - b \qquad\qquad\qquad = -4 - x$$

Multiplication Property of −1

$$-1(a) = -a \qquad -1(7.2) = -7.2; \; -1(-y) = y$$
$$-1(a + b) = -a - b \qquad -1(4 + x) = -4 - x; \; -1(-t - s) = t + s$$

Opposites and Products

$$(-a)b = -ab \qquad (-6)(m) = -6m; \; (8)(-z) = -8z$$
$$(-a)(-b) = ab \qquad (-4)(-y) = 4y; \; -(5x)(-y) = 5xy$$

Subtraction of a Sum or Difference

$$a - (b + c) = a - b - c \qquad 5 - (m + q) = 5 - m - q$$
$$a - (b - c) = a - b + c \qquad 6 - (c - 2) = 6 - c + 2$$

Emphasize that *a*, *b*, and *c* can represent 0 or negative numbers as well as positive numbers.

Note that the first two special properties above are equivalent. That is,

$$-(a + b) = -1(a + b).$$

Finding the opposite of a sum or difference is the same as multiplying the sum or difference by -1.

P–1 **Simplify.**

a. $-1(x + 2)$ **b.** $-(4 - t)$ **c.** $-1(\sqrt{2})$ **d.** $(-m)(-k)$ *mk*
$\quad$ *−x − 2* $\quad$ *−4 + t* $\quad$ *−√2*

EXAMPLE 1   Simplify: $(-rs)(5rt)$

Solution:

1 Use the Multiplication Property of −1. $\longrightarrow (-rs)(5rt) = (-1rs)(5rt)$

2 Use the Commutative and Associative Properties. $\longrightarrow = (-1 \cdot 5)(r \cdot r \cdot s \cdot t)$

3 Multiply. $\longrightarrow = (-5)(r^2st)$
$\qquad\qquad\qquad\qquad\qquad\qquad\qquad\qquad = -5r^2st$

Real Numbers / 43

Teaching Suggestions p. M–20

Lesson Resources
Warm–Up: p. M–21
Maintenance: See below.
Practice Worksheet 2.4

Maintenance
1. Which number has the greater absolute value, -5 or 3?
   Ans: –5 (Section 1.1)
2. Multiply: $(-3)(6)$
   Ans: –18 (Section 1.4)
3. Evaluate: $5 - 3 + 2$
   Ans: 4 (Section 1.6)
4. Evaluate $5x^2 - x$ when $x = -2$.
   Ans: 22 (Section 1.7)
5. A company had a net loss for the year of –$14,808. Find the average net loss per month.
   Ans: –$1234 (Section 1.5)

Additional Examples
Example 1
Simplify.
1. $(8x)(-11xy)$
   Ans: $-88x^2y$
2. $(-2.1ab)(-3ab)$
   Ans: $6.3a^2b^2$
3. $(-25h)(-hk)$
   Ans: $25h^2k$

43

Simplify.
1. $(6a)(-2b)(\frac{1}{2}b)(5a)$
   Ans: $-30a^2b^2$
2. $(11x)(-\frac{1}{11}x)(5y)(-\frac{1}{5})$
   Ans: x^2y
3. $(-0.5cd)(5d)(-2cd)$
   Ans: $5c^2d^3$

EXAMPLE 2   Simplify: $(-2a)(-a)(-x)(-\frac{1}{2}x)$

Solution:

1 Multiply from the left to right. ⟶ $(-2a)(-a)(-x)(-\frac{1}{2})x = (2a^2)(\frac{1}{2}x^2)$

2 Use the Commutative and Associative Properties. ⟶ $= (2 \cdot \frac{1}{2})(a^2 \cdot x^2)$

3 Use the Multiplication Property of 1. ⟶ $= 1 \cdot a^2x^2$

$= a^2x^2$

In the following special properties, note that no divisor is 0.

Special Properties of Division

$$\frac{a}{1} = a \qquad\qquad \frac{7qr}{1} = 7qr$$

$$\frac{a}{a} = 1, a \neq 0 \qquad\qquad \frac{-3s + 4t}{-3s + 4t} = 1$$

Opposites and Quotients

$$\frac{-a}{b} = -\frac{a}{b}; \frac{a}{-b} = -\frac{a}{b}; b \neq 0 \qquad \frac{-2k}{m} = -\frac{2k}{m}; \frac{7y}{-9x} = -\frac{7y}{9x}$$

$$\frac{-a}{-b} = \frac{a}{b}, b \neq 0 \qquad\qquad \frac{-3c}{-5d^2} = \frac{3c}{5d^2}$$

P–2   **Simplify.**

a. $\frac{-5k}{7} - \frac{5k}{7}$ b. $\frac{-3cd}{-3cd}$ 1   c. $\frac{6t + 4m}{6t + 4m}$ 1   d. $\frac{2xy}{-2xy}$ -1

Example 3
Simplify.
1. $\frac{-(3x - 5)}{5 - 3x}, x \neq \frac{5}{3}$
   Ans: 1
2. $\frac{-(7 - y)}{7 - y}, y \neq 7$
   Ans: -1
3. $\frac{-4 - 9x}{-(4 + 9x)}, x \neq -\frac{4}{9}$
   Ans: 1

EXAMPLE 3   Simplify: $\frac{-(4 - t)}{t - 4}, t \neq 4$ ◀ **If t is not 4, then $t - 4$ is not zero.**

Solution:

1 Simplify the numerator. ⟶ $\frac{-(4 - t)}{t - 4} - \frac{-4 + t}{t - 4}$ Note that the numerator $-(4 - t)$ is changed to the equivalent expression

2 Use the Commutative Property. ⟶ $= \frac{t + (-4)}{t - 4}$ $t - 4$ in step 3. In general, $-(a - b) = b - a$.

3 Simplify. ⟶ $= \frac{t - 4}{t - 4} = 1$

Simplify. (Examples 1–3)

1. $3(-k)$ $-3k$

2. $-1(-x)$ x

3. $(-8)(-6y)$ $48y$

4. $(-5p)(-3)$ $15p$

5. $-(x + y)$ $-x - y$

6. $-(-x - y)$ $x + y$

7. $(x - 2) - (y - a)$ $x - 2 - y + a$

8. $a - (6 - b)$ $a - 6 + b$

9. $(-20h)(\frac{1}{5})$ $-4h$

10. $(-\frac{1}{2})(-a)$ $\frac{1}{2}a$

11. $(-x)(2y)$ $-2xy$

12. $(xy)(-7)$ $-7xy$

13. $(-xy)(3x)$ $-3x^2y$

14. $(-a)(-a)$ a^2

15. $(-y)(-y)(z)(-z)$ $-y^2z^2$

16. $(-a)(-b)(-c)$ $-abc$

17. $(-\frac{2}{3}c)(-3ac)(-a)$ $-2a^2c^2$

18. $(\frac{1}{5})(-\frac{7}{8}g)(10h)(-4g)$

19. $(-m)(-4n)(-4n)(m)$ $-16m^2n^2$

20. $(2r)(-3s)(-r)(-6)$ $-36r^2s$

21. $(-9) \div f,\ f \neq 0$ $-\frac{9}{f}$

22. $\frac{a}{-b}$ $-\frac{a}{b}$

23. $\frac{-cd}{-cd}$ 1

24. $\frac{9a}{-9a}$ -1

25. $\frac{x - 7}{x - 7},\ x \neq 7$ 1

26. $\frac{-(a - b)}{b - a},\ b \neq a$ 1

27. $\frac{16 - c}{-(c - 16)},\ c \neq 16$ 1

■■■ **WRITTEN EXERCISES** ■■■

Goal: To simplify expressions using special properties

Sample Problems. **a.** $-6r - (r + 3)$ **b.** $(-3.2)(-x)(y)$

c. $\frac{2a - 5}{-(5 - 2a)},\ a \neq 2\frac{1}{2}$

Answers: a. $-7r - 3$ **b.** $3.2xy$ **c.** 1

Simplify. (Examples 1–3)

1. $(-a)(-c)$ ac

2. $x - (y + 10)$ $x - y - 10$

3. $2t - (r - 1)$ $2t - r + 1$

4. $(-1)(f)$ $-f$

5. $(-d)(4)$ $-4d$

6. $-(z + 11)$ $-z - 11$

7. $(m + n)(-1)$ $-m - n$

8. $5d - (6c - 7)$ $5d - 6c + 7$

9. $2a - (11 + b)$

10. $(8a)(-ab)$ $-8a^2b$

11. $(-4c)(-4d)$ $16cd$

12. $(-\frac{1}{3}t)(-15)$ $5t$

13. $(-16)(\frac{1}{4}q)$ $-4q$

14. $(-2.3x)(-2y)$ $4.6xy$

15. $(-5a)(1.8a)$ $-9a^2$

16. $(-3s)(-t)(s)(2t)$ $6s^2t^2$

17. $(2x)(-y)(-x)(3y)$ $6x^2y^2$

18. $(-\frac{1}{2}xy)(-\frac{1}{3}xy)(3)$

19. $(4b)(-\frac{2}{3}ab)(-\frac{1}{2}a)$ $\frac{4}{3}a^2b^2$

20. $(-5ab)(1.8ac)(-\frac{1}{5}b)$

21. $(1.3r)(-0.6s)(2rs)$

22. $\frac{-7}{a}$ $-\frac{7}{a}$

23. $\frac{3}{-p}$ $-\frac{3}{p}$ $1.8a^2b^2c$

24. $\frac{-4f}{-4f}$ 1

25. $\frac{bc}{bc}$ 1

26. $\frac{-(h + 3)}{k}$ $\frac{-h - 3}{k}$

27. $\frac{-(2 + y)}{x}$ $\frac{-2 - y}{x}$

Real Numbers / **45**

Common Error
Students incorrectly simplify expressions involving the opposite of a sum or difference.

Examples
$-(4 - a) = -4 - a$
$-(b + 3) = -b + 3$

Prescription
Emphasize that finding the opposite of a sum or difference is the same as multiplying by -1. Have students show each term multiplied by -1. Then simplify. For example,
$-(4 - a) = (-4) - (-a) = -4 + a$.

Additional Answers
Classroom Exercises
18. $7g^2h$

Assignment Guide
Basic pp. 45–46: 1–45 odd
Average pp. 45–46: 1–47 odd
Above Average pp. 45–46: 1–45 odd, 46, 47

Additional Answers
Written Exercises
9. $2a - 11 - b$
18. $\frac{1}{2}x^2y^2$
21. $-1.56r^2s^2$

Simplify. (Examples 1–3)

28. $\dfrac{-(2-x)}{x-2}$, $x \neq 2$ 1

29. $\dfrac{f-g}{f-g}$, $f \neq g$ 1

30. $\dfrac{a-7}{-(a-7)}$, $a \neq 7$ −1

31. $\dfrac{-(k-g)}{-(k-g)}$, $k \neq g$ 1

32. $(\frac{1}{2}xy)(-4x)(-y)$ $2x^2y^2$

33. $(-7)(-fg)$ $7fg$

34. $\dfrac{abc}{abc}$ 1

35. $\dfrac{mp}{-(mp)}$ −1

36. $(-1)(0.3c+4d)$ $-0.3c-4d$

37. $(1.3r)(-0.6r)$ $-0.78r^2$

38. $(-3a)(5b)(\frac{1}{10}a)(-4b)$ $6a^2b^2$

39. $(\frac{1}{5}x)(\frac{2}{3}xy)$ $\frac{2}{15}x^2y$

40. $(p-q)-(r-s)$ $p-q-r+s$

41. $12-(r-3s+15)-3$ $-r+3s$

42. $-(18+t)$ $-18-t$

43. $-(-r+2)$ $r-2$

44. $\dfrac{-2b+3}{-(-3+2b)}$, $b \neq \frac{3}{2}$ 1

45. $\dfrac{-(y-2)}{y-2}$, $y \neq 2$ −1

NON-ROUTINE PROBLEMS

46. Sarah doubled the amount in her
checking account during the month
of January. On the last day of the
month, she withdrew $16. Each
month after that, she repeated this
procedure. However, after with-
drawing $16 on the 30th of April,
Sarah had no money left in her
account. How much did Sarah have
in her account at the beginning of
January?

47. How many triangles are shown in
the figure below?

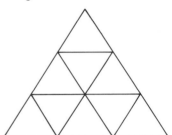

REVIEW CAPSULE FOR SECTION 2.5

Simplify.

1. $(2 \cdot 7)a+(2)(8)$ $14a+16$

2. $3 \cdot 7+(3 \cdot 8)x$ $21+24x$

3. $(-5 \cdot 2)+(-5 \cdot 6)b$ $-10-30b$

4. $(-8)y+(-8 \cdot 4)$ $-8y-32$

5. $(4p)(3q)+(4p)(-2)$ $12pq-8p$

6. $(\frac{1}{2}x)(-y)+(8x^2)(-y)$ $-\frac{1}{2}xy-8x^2y$

Write Yes or No to indicate whether 3xy is a factor of both given expressions.

7. $-6xy, 3x^2y$ Yes

8. $2xy, 6xy^2$ No

9. $-18x^3y, 3xy$ Yes

10. $4x^2y^2, 6xy$ No

11. $9x^2y^2, -13xy$ No

12. $21yx, -15x^2y^2$ Yes

OBJECTIVES: To apply the Distributive Property in writing an expression of the form $a(b + c)$ or $a(b - c)$ as a sum or difference in simplest form

2.5 Multiplying and Factoring

To apply the Distributive Property in factoring expressions of the form $ab + ac$ or $ab - ac$

The Distributive Property can be used to express a product as a sum.

Teaching Suggestions p. M–20

Lesson Resources
Warm–Up: p. M–21
Maintenance: See below.
Practice Worksheet 2.5
Transparency 5

These are sometimes called the "lefthand" and "righthand" distributive properties.

> ### Distributive Property
>
> For any real numbers a, b, and c:
>
> 1. **$a(b + c) = ab + ac$** $3(2c + 5b) = 3(2c) + 3(5b)$
> 2. **$(b + c)a = ba + ca$** $(-5a - 7b)2c = (-5a)2c + (-7b)2c$

Maintenance
1. Add: $3\frac{1}{2} + (-2\frac{1}{2})$
   Ans: 1   (Section 1.2)
2. Multiply: $(0)(-3.612)$
   Ans: 0   (Section 1.4)
3. Evaluate: $12 - 10 \div (-2)$
   Ans: 17   (Section 1.6)
4. Evaluate $2k^2 + 3k + 1$
   When $k = -4$.
   Ans: 21   (Section 1.7)
5. On Thursday, the change in the price of a stock was $-1\frac{1}{8}$ points. On Friday, the change was $-2\frac{5}{8}$ points. Find the total change in the price for the two days:
   Ans: $-3\frac{3}{4}$ points
   Section 1.2)

EXAMPLE 1   Multiply and simplify: $-3(4x + 7)$

Solution:

1. Use the Distributive Property. ⟶ $-3(4x + 7) = -3(4x) + (-3)(7)$

2. Use the Associative Property. ⟶ $= (-3 \cdot 4)x + (-3)(7)$

   $= -12x + (-21)$

3. Simplify. ⟶ $= -12x - 21$

Additional Examples
Example 1
Multiply and simplify.
1. $2(7a + 5)$
   Ans: $14a + 10$
2. $-6(6 - 5y)$
   Ans: $-36 + 30y$
3. $(10a + 2)(-\frac{1}{2})$
   Ans: $-5a - 1$

P–1 **Multiply and simplify.**

a. $5(-3a + 2)$ b. $(6m + 4)(-\frac{1}{2})$ c. $-3r(-2s + 5)$ $6rs - 15r$
   $-15a + 10$ $-3m - 2$

Note in Example 2 that $(3x - 2y)$ is written as the sum $[3x + (-2y)]$ before multiplying.

EXAMPLE 2   Multiply and simplify: $(3x - 2y)5x$

Solution: $(3x - 2y)5x = [3x + (-2y)]5x$

$= (3x)(5x) + (-2y)(5x)$

$= 15x^2 + (-10xy)$

$= 15x^2 - 10xy$

Example 2
Multiply and simplify.
1. $-4x(2x - 3)$
   Ans: $-8x^2 + 12x$
2. $(a - 13)(-11a)$
   Ans: $-11a^2 + 143a$
3. $2.1y(10y - 1)$
   Ans: $21y^2 - 2.1y$

The Distributive Property can be used to express a sum as a product.

$$7x + 14 = 7(x + 2)$$

This process is called **factoring**. The number 7 is the common factor of $7x$ and 14 because each is divisible by 7.

*Real Numbers   /   **47***

Find the common factors of each pair.

a. $3x^2$ and $3xy$ $3x$ **b.** $12ac$ and $4bc$ $4c$ **c.** $-21rt$ and $-7t$ $-7t$

Additional Examples
Example 3
Factor.
1. $12x^2 + 8x$
  Ans: $4x(3x + 2)$
2. $10ac + 30ad$
  Ans: $10a(c + 3d)$
3. $30h^2k + 24hk^2$
  Ans: $6hk(5h + 4k)$

EXAMPLE 3   Factor: $3x^2 + 3xy$

Solution: ☐1 Find the common factor. ⟶ $3x^2 + 3xy = (3x)\,(x) + (3x)\,(y)$

☐2 Rewrite using the common factor
as one factor of the product. ⟶ $= (3x)\,(x + y)$

Note in Example 4 that $12ac - 4bc$ and $-21rt - 7t$ are written as sums
<u>before</u> factoring.

Example 4
Factor.
1. $-15x^2 - 10xy$
  Ans: $-5x(3x + 2y)$
2. $8a^2 - 8a$
  Ans: $8a(a - 1)$
3. $-4h + 28h^2$
  Ans: $-4h(1 - 7h)$

EXAMPLE 4   Factor: **a.** $12ac - 4bc$ **b.** $-21rt - 7t$

Compare the factors of **a.** and **b.**
Solutions: Usually, factors with the least
number of opposite symbols are

a. $12ac - 4bc = 12ac + (-4bc)$ used. Otherwise, **b.** $-21rt - 7t = -21rt + (-7t)$
the factors of **a.** might be
$= 3a(4c) + (-b)\,(4c)$ $(-3a + b)(-4c)$ and of **b.** $= (-7t)\,(3r) + (-7t)\,(1)$
$7t(-3r - 1)$
$= [3a + (-b)]4c$ $= (-7t)\,(3r + 1)$
$= (3a - b)4c$ $= -7t(3r + 1)$

P-3   **What is the factored form of each of the following?**

a. $6y + 24$ **b.** $10mn - 4n$ **c.** $15x^2 - 3x$ **d.** $-9v - 3vw$
$6(y + 4)$ $2n(5m - 2)$ $3x(5x - 1)$ $-3v(3 + w)$

CLASSROOM EXERCISES

Common Error
Students forget to multiply by the
second term when using the Distributive Property.

Example
$-2(3x + 4) = -6x + 4$

Prescription
Have students look at examples
worked both ways as follows.
$-2(5 + 4) = -2(9) = -18$
$-2(5 + 4) = (-2)5 + (-2)4$
$= -18$
This will show that $-2(5) + 4$ is
wrong.

Multiply and simplify. (Examples 1 and 2)

1. $10(p + q)$ $10p + 10q$ **2.** $12(k - 1)$ $12k - 12$ **3.** $\frac{1}{2}(4k - 12)$ $2k - 6$ **4.** $(r + s)8$ $8r + 8s$

5. $-4x(-5x - 7)$ **6.** $(a - b)14a$ **7.** $(9m - 3)\,(-\frac{1}{3}m)$ **8.** $(3m - 5n)\,(-m)$
$20x^2 + 28x$ $14a^2 - 14ab$ $-3m^2 + m$ $-3m^2 + 5mn$

Factor. (Examples 3 and 4)
$a(3x + 5y)$ $2a(1 + 5a)$
9. $8m + 8n$ $8(m + n)$ **10.** $3ax + 5ay$ **11.** $6x + 10y$ $2(3x + 5y)$ **12.** $2a + 10a^2$

13. $5x - 5y$ $5(x - y)$ **14.** $4 - 6b$ $2(2 - 3b)$ **15.** $4mt - 9nt$ $t(4m - 9n)$ **16.** $x^2 - 3x$ $x(x - 3)$

17. $-3a - 3b$ $-3(a + b)$ **18.** $-5x - 10y$ **19.** $-6r^2 - r$ $-r(6r + 1)$ **20.** $-4st + 8s^2t$
                               $-5(x + 2y)$ $-4st(1 - 2s)$

48 / Chapter 2

WRITTEN EXERCISES

Goal: To use the Distributive Property to multiply and factor expressions

Sample Problem: a. $-5t(3r - 4s)$ **b.** $21mt - 9nt$

Answers: a. $-15tr + 20ts$ **b.** $3t(7m - 3n)$

Multiply and simplify. (Examples 1 and 2)

1. $10(a + b)$ $10a + 10b$ **2.** $5(x + y)$ $5x + 5y$ **3.** $-3(x + 2)$ $-3x - 6$ **4.** $-5(t + 4)$ $-5t - 20$

5. $7(r - 8)$ $7r - 56$ **6.** $12(s - 5)$ $12s - 60$ **7.** $8(3p + 2)$ $24p + 16$ **8.** $6(4r + 5)$ $24r + 30$

9. $(r + 3)4r$ $4r^2 + 12r$ **10.** $(m + 6)3m$ **11.** $(y - 6)(-2y)$ **12.** $(n - 10)(-4n)$

13. $(6t - 5u)4t$ **14.** $(8k - 7j)5k$ **15.** $(-8t + 12)(\frac{1}{4}t)$ **16.** $(2.8b - 1.4c)(-0.2ab)$

$-2t^2 + 3t$ $-0.56ab^2 + 0.28abc$

Factor. (Examples 3 and 4)

17. $5a + 5b$ $5(a + b)$ **18.** $14m + 7n$ **19.** $4r + 20s$ $4(r + 5s)$ **20.** $3ax + 6ay$ $3a(x + 2y)$

21. $2km + 4kn$ **22.** $y + 13y^2$ **23.** $p^2q + pq^2$ $pq(p + q)$ **24.** $6m + 18n$ $6(m + 3n)$

25. $3r - 3s$ $3(r - s)$ **26.** $4a - 4$ $4(a - 1)$ **27.** $8x - 8$ $8(x - 1)$ **28.** $12rk - 18sk$

29. $8kx^2 - 16xy^2$ **30.** $13x^2 - 26xy^2$ **31.** $\frac{3}{4}rx - \frac{3}{4}ry$ $\frac{3}{4}r(x - y)$ **32.** $\frac{1}{3}am - \frac{1}{3}an$ $\frac{1}{3}a(m - n)$

33. $-14ab - 7b$ **34.** $-24rs - 8s$ **35.** $-3x^2 + 6x$ $-3x(x - 2)$ **36.** $-5n^2 + 15n$

37. $-2xy + 5xy$ **38.** $-10a - 2b$ **39.** $-4x^2 - x$ $-x(4x + 1)$ **40.** $-11rs^2 + 33r^2s$

$-11rs(s - 3r)$

MORE CHALLENGING EXERCISES

Factor.

EXAMPLE: $6a^3 + 9a^2 - 3a$

SOLUTION: $6a^3 + 9a^2 - 3a = (3a)(2a^2) + (3a)(3a) + (3a)(-1)$

$= 3a(2a^2 + 3a - 1)$

$3x(2x^2 + x - 6)$

41. $5a - 15b + 10c$ $5(a - 3b + 2c)$ **42.** $4r + 8s - 24t$ $4(r + 2s - 6t)$ **43.** $6x^3 + 3x^2 - 18x$

44. $7y^3 - 14y^2 + 21y$ **45.** $30r^2 - 36x - 138x^2$ **46.** $-2x^2 - 4x^2y - 8x^2y^2$

In the expression $(ax + by)$, the value of a is less than the value of b.
The common factor of ax and by is an integer c.

47. Can $c = a$? Explain, using examples to illustrate your answer.

Yes. Let $a = 2$ and $b = 4$. $(2x + 4y) = 2(x + 2y)$. Thus, $c = 2$.

48. Can $c = b$? Explain, using examples to illustrate your answer.

No. Let $a = 3$ and $b = 6$. $(3x + 6y) = 3(x + 2y)$. Thus $c \neq 6$.

REVIEW CAPSULE FOR SECTION 2.6

Add or subtract, as indicated. (Pages 4–10)

1. $12 + (-5)$ 7 **2.** $-8 + (-2)$ -10 **3.** $7 + (-13)$ -6 **4.** $-4 + (-9)$ -13

5. $8 - 14$ -6 **6.** $3 - (-2)$ 5 **7.** $-5 - 15$ -20 **8.** $-1 - (-6)$ 5

Assignment Guide
Basic
Day 1 p. 49: 1–16
Day 2 p. 49: 17–39 odd
Average p. 49: 1–47 odd
Above Average p. 49:
1–39 odd, 41–48

Additional Answers
Written Exercises
10. $3m^2 + 18m$
11. $-2y^2 + 12y$
12. $-4n^2 + 40n$
13. $24t^2 - 20ut$
14. $40k^2 - 35jk$
18. $7(2m + n)$
21. $2k(m + 2n)$
22. $y(1 + 13y)$
28. $6k(2r - 3s)$
29. $8x(kx - 2y^2)$
30. $13x(x - 2y^2)$
33. $-7b(2a + 1)$
34. $-8s(3r + 1)$
36. $-5n(n - 3)$
37. $3xy$ or $xy(-2 + 5)$
38. $-2(5a + b)$
44. $7y(y^2 - 2y + 3)$
45. $6(5r^2 - 6x - 23x^2)$
46. $-2x^2(1 + 2y + 4y^2)$

Problem–Solving Skills
Using logical reasoning
(Ex. 47–48)

Critical Thinking
See Exercises 47 and 48.

Lesson Resources
Warm–Up: p. M–21
Maintenance: See below.
Practice Worksheet 2.6
Manipulative 3, p. 544
Teaching Aids 1, 3, 4
Transparencies 50, 51, 53, 54,
55, 56

Maintenance
1. Add: $-17 + (-12)$
   Ans: -29 (Section 1.2)
2. Multiply: $(-2)(-3)(-5)$
   Ans: -30 (Section 1.4)
3. Evaluate: $-3(-4) \div 12$
   Ans: 1 (Section 1.6)
4. Evaluate: $(-4)^2 - 4$
   Ans: 12 (Section 1.7)
5. The thermometer reading at noon on a winter day was $-10°C$. The wind chill temperature was $-22°C$. How much lower was the wind chill temperature than the thermometer reading?
   Ans: 12°C (Section 1.3)

Additional Examples
Example 1
Simplify.
1. $13a^2 - a^2$ **Ans: $12a^2$**
2. $7xy + (-5xy)$ **Ans: $2xy$**
3. $-3.9h - 5.1h$ **Ans: $-9h$**

Example 2
Simplify.
1. $-3 + 5y + y - 23$
   Ans: $6y - 26$
2. $5a - 8b + 6b - 19a$
   Ans: $-14a - 2b$
3. $9.2x + 17 - 7.8x - 17$
   Ans: $1.4x$

Example 3
Simplify.
1. $x^2 + 7x - 9 - 3x + 5x^2$
   Ans: $6x^2 + 4x - 9$
2. $5a - 7a^2 + 9a^2 - a - 6$
   Ans: $2a^2 + 4a - 6$
3. $8 + 7.1y^2 - 3.4 + 5y - 3.2y^2 - y$
   Ans: $3.9y^2 + 4y + 4.6$

OBJECTIVES: To combine like terms in an expression of two terms

2.6 Combining Like Terms

To combine like terms in an expression of three or more terms and then simplify

The first and third terms in $3x + 4x^2 - 2x$ are <u>like terms</u>.

Definition	*Like terms* have exactly the same variables and the same powers of these variables.

P–1 **Which of the following are like terms?**

It is important that students be able to identify like terms.

a. $3x$ and $-5x$ Like **b.** $5y$ and $2y^2$ Unlike **c.** a^2b and $-2ab^2$ Unlike

You can add and subtract only <u>like terms</u>. When you add or subtract like terms, you are *combining* like terms.

EXAMPLE 1 Simplify: $x^2 - 8x^2$

Solution: ① Factor. ⟶ $x^2 - 8x^2 = (1 - 8)x^2$
② Add. ⟶ $= (-7)x^2$ ◀ *$1 - 8 = 1 + (-8)$*
$= -7x^2$

It may be helpful to write the expression as a sum.

EXAMPLE 2 Simplify: $x - 5 - 11x + 7$

Solution:
① Write as a sum. ⟶ $x - 5 - 11x + 7 = 1x + (-5) + (-11x) + 7$
② Group like terms. ⟶ $= [1x + (-11x)] + (-5 + 7)$
③ Factor. ⟶ $= [1 + (-11)]x + (-5 + 7)$
$= -10x + 2$

It is not necessary to show detailed steps as in Examples 1 and 2.

The detailed steps are helpful when a student has trouble with the short-cut method shown in Example 3.

EXAMPLE 3 Simplify: $-2x^2 - 3x + 5 - 5x + 11x^2$

Solution: 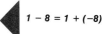 $-2x^2 - 3x + 5 - 5x + 11x^2 = 9x^2 - 8x + 5$ ◀ *$-2x^2 + 11x^2 = 9x^2$*
$-3x - 5x = -8x$

Simplify. (Examples 1–3)

1. $3y + 8y$ $11y$

2. $2ax + 7ax$ $9ax$

3. $7x - 5x$ $2x$

4. $2r - 2r$ 0

5. $ab + ab$ $2ab$

6. $2z + 10z$ $12z$

7. $6y^2 - 9y^2$ $-3y^2$

8. $-3b^2 + b^2$ $-2b^2$

9. $-t^2 - t^2$ $-2t^2$

10. $2a^2b + 3ba^2$ $5a^2b$

11. $3x - 1 + 5x$ $8x - 1$

12. $5a - a + 2$ $4a + 2$

13. $-7 + 2x - 2x$ -7

14. $-3t + t + 5$ $-2t + 5$

15. $-x^2 - 5x^2 - 3$ $-6x^2 - 3$

16. $2a^2 - 3a - 7a^2$ $-5a^2 - 3a$

17. $a^2 - 3a + 2a + 5$

18. $d^2 + 9 + d + 3d^2$

19. $x^2 + x + 1 - x + x^2$ $2x^2 + 1$

20. $5d - 16f + 23d - 41 - 12f$

21. $\frac{1}{9}a^2 + \frac{1}{3} + \frac{1}{2}a + \frac{1}{6} + 2a$ $\frac{1}{9}a^2 + 2\frac{1}{2}a + \frac{1}{2}$

22. $\frac{1}{4}x + 2 - 4 - \frac{1}{2}x + 2x^2 + 1$

23. $1.5m + 0.9n + 2.1m^2 + 4.2m + 0.1n$

24. $0.6y^2 - 0.5y - 0.3y + y^2 + 1$

25. $15p^2 - 31 - 81p^2 - 42 - p$ $-66p^2 - p - 73$

26. $14n + 5n^2 + 5n^2 + 16m^2 - 16n^2$

Goal: To combine like terms

Sample Problems: a. $-3y + 5y$ **b.** $-3x^2 + 6 - x^2 - 4$

Answers: a. $2y$ **b.** $2 - 4x^2$

Simplify. (Examples 1–3)

1. $13y + 7y$ $20y$

2. $2x - x - 6x$ $-5x$

3. $17t + 2t$ $19t$

4. $5a - 11a$ $-6a$

5. $-2x - 15x$ $-17x$

6. $-m + 5m$ $4m$

7. $3t^2 - 5t^2$ $-2t^2$

8. $2r^2 - 11r^2$ $-9r^2$

9. $-4z^2 + 11z^2$ $7z^2$

10. $-11d^2 + 20d^2$ $9d^2$

11. $-3a - 6 + 9a$ $6a - 6$

12. $9m + 2n - 3m - 5n$

13. $0.3t^2 + 2t - 0.5t^2$

14. $3c^2 - 3x - 3c + 5$

15. $-2.1g + 5.3 - 8.7 - 0.9g$

16. $5.4d - 0.6 - 8.9d - 8.5$

17. $\frac{1}{4}a - b + \frac{3}{5}a - \frac{1}{2}b$ $\frac{17}{20}a - 1\frac{1}{2}b$

18. $\frac{2}{3}r + \frac{5}{3}s - \frac{1}{3}s + \frac{2}{3}r$ $1\frac{1}{3}r + 1\frac{1}{3}s$

19. $50b + 200z + 5bz - 300bz - 112b$

20. $15m - 17 - 17m - 4n - 11n$

21. $0.9t + 1.6t^2 + 0.3 + 1.7t^2 + 1.9t$

22. $0.05 + 13n^2 + 3.15 - 2.1n^2 - 3.4n$

23. $\frac{1}{3}r^2 - \frac{1}{5}s^2 - \frac{1}{3}s^2 + \frac{1}{8}r^2 + \frac{1}{3}$ $\frac{11}{24}r^2 - \frac{8}{15}s^2 + \frac{1}{3}$

24. $\frac{2}{3} - \frac{4}{5}b - \frac{3}{4}b^2 - \frac{7}{8}b^2 - \frac{3}{10}$

25. $12f^2 - 15f - 62f^2 - 45 - 2f$ $-50f^2 - 17f - 45$

26. $a^2 - 6 - 3a^2 - 12 - 5a$

27. $-0.3t + 0.7t + 0.1t$ $0.5t$

28. $1.8ab - 0.7ab + 2.1ab$ $3.2ab$

29. $ab^2 - a^2b + ab - a^2b^2 - ba^2$

30. $-5y^2 - 3y + y^2 - y$ $-4y^2 - 4y$

31. $8v^2 + 8w^2 + 2w + 2w^2 - 6w$ $8v^2 + 10w^2 - 4w$

32. $12k^2 + 9k - 12k + 6k^2 + 6$

33. $77a^2 + 35 - 2a^2 - 10$ $75a^2 + 25$

34. $17a^2 + 19a^2$ $36a^2$

35. $\frac{1}{2}x - x^2 - \frac{3}{4}x - 3x^2 - \frac{1}{3}x^2 + \frac{1}{5}$ $-4\frac{1}{3}x^2 - \frac{1}{4}x + \frac{1}{5}$

36. $\frac{1}{3}d + \frac{1}{8}d^2 - \frac{1}{2}d$ $\frac{1}{8}d^2 - \frac{1}{6}d$

Real Numbers / **51**

Additional Answers
Classroom Exercises
17. $a^2 - a + 5$
18. $4d^2 + d + 9$
20. $28d - 28f - 41$
22. $2x^2 - \frac{1}{4}x - 1$
23. $2.1m^2 + 5.7m + n$
24. $1.6y^2 - 0.8y + 1$
26. $-6n^2 + 14n + 16m^2$

Assignment Guide
Basic p. 51: 1–35 odd
Average p. 51: 1–35 odd
Above Average p. 51:
1–35 odd

Additional Answers
Written Exercises
12. $6m - 3n$
13. $-0.2t^2 + 2t$
14. $3c^2 - 3x - 3c + 5$
15. $-3g - 3.4$
16. $-3.5d - 9.1$
19. $-62b - 295bz + 200z$
20. $-2m - 15n - 17$
21. $3.3t^2 + 2.8t + 0.3$
22. $10.9n^2 - 3.4n + 3.2$
24. $-1\frac{5}{8}b^2 - \frac{4}{5}b + \frac{11}{30}$
26. $-2a^2 - 5a - 18$
29. $ab^2 - 2a^2b + ab - a^2b^2$
32. $18k^2 - 3k + 6$

Quiz Sections 2.4–2.6
After completing Sections 2.4–2.6, you may want to administer a quiz covering the same sections. See the quiz provided on page 54 in the *Teacher's ResourceBank*.

Using Statistics

Comparing Averages

The field of statistics involves collecting, organizing, and interpreting data. **Quality control engineers** collect statistics on the quality of a product by devising and carrying out testing procedures.

Acme Quality Control, Inc. conducts gasoline mileage tests on automobiles. Each of three cars was tested five times. The data is shown in the table.

EXAMPLE: Find the mean, median, and mode of the gasoline mileages for Car A.

GASOLINE MILEAGE IN MILES PER GALLON (mpg)			
	Car A	Car B	Car C
Test 1	26.7	30.4	30.1
Test 2	28.4	27.5	28.2
Test 3	28.3	30.4	28.2
Test 4	28.6	27.9	28.1
Test 5	26.7	27.4	31.3

SOLUTION:

1. The **mean** is the sum of the mileages divided by the number of mileages.

$$\frac{26.7 + 28.4 + 28.3 + 28.6 + 26.7}{5} = \frac{138.7}{5} = 27.74, \text{ or } 27.7$$

 Rounded to the nearest tenth

2. Write the mileages in order from least to greatest. The **median** is the middle number.

    26.7   26.7   28.3   28.4   28.6
    _____2 numbers_____/   ↑median   _____2 numbers_____/

    If there are two middle numbers, the median is the mean of these two numbers.

3. The **mode** is the mileage which occurs most often. ⟶ 26.7

    There can be no mode, or more than one mode.

The mean is **27.7 mpg**; the median is **28.3 mpg**; and the mode is **26.7 mpg**.

EXERCISES

1. Which statistic gives the greatest average gasoline mileage for Car A, the mean, median, or mode? Median

2. If you were in charge of the advertising for Car A, would you give the mean, median, or mode as the *average* mileage? Explain. Median; It is greater than the mean and mode.

52 / *Chapter 2*

3. Find the mean, median, and mode for Car B.

4. Which average would you use to advertise Car B? Explain.

5. Find the mean, median, and mode for Car C.

6. Which average would you use to advertise Car C? Explain.

The manufacturer of each car would like to claim that the test results show that its car got the best average gasoline mileage.

7. Which manufacturer do you think would use the mean to support a claim that its car gets the best average mileage? Explain.

8. Do you think the manufacturer of Car B would use the median or the mode to support its claim for the best average mileage? Explain.

9. Which average do you think the manufacturer of Car A would use to claim the best average mileage? Explain.

*The **mode** is a useful average when analyzing data on clothing or package sizes. It indicates the most frequently used size or sizes.*
*The **median** is useful when discussing salaries or home prices. It is not affected by a few extremely large or small values in the group.*
*The **mean** is affected by extremely large or small values in a group. The mean is useful when discussing scores or sports performances, where all the values are equally important.*
For Exercises 10-14, reasons will vary.
Choose the average (mean, median, or mode) you feel would be the best to use in each case. Give a reason for your choice.

10. A dress manufacturer wants to know the "average" size that women will buy. Mode

11. A real estate developer wants to know the "average" family income in a community. Median

12. A teacher wants to know the "average" score on an exam. Mean

13. A homebuyer wants to know the "average" selling price for houses in a neighborhood. Median

14. A canning company wants to know the "average" can size that shoppers will buy. Mode

*Real Numbers / **53***

CHAPTER SUMMARY

IMPORTANT TERMS

Counting numbers *(p. 30)*
Whole numbers *(p. 30)*
Integers *(p. 30)*
Rational numbers *(p. 30)*
Repeating decimal *(p. 30)*
Terminating decimal *(p. 31)*
Perfect squares *(p. 31)*

Irrational numbers *(p. 31)*
Nonrepeating decimal *(p. 31)*
Real numbers *(p. 32)*
Equation *(p. 35)*
Factoring *(p. 47)*
Like terms *(p. 50)*
Combining like terms *(p. 50)*

IMPORTANT IDEAS

1. Rational numbers can be expressed as repeating decimals.

2. The decimal forms of irrational numbers are nonrepeating.

3. The following addition properties are true for any real numbers.
   a. Addition Property of Zero: $a + 0 = a$
   b. Addition Property of Opposites: $a + (-a) = 0$
   c. Commutative Property of Addition: $a + b = b + a$
   d. Associative Property of Addition: $(a + b) + c = a + (b + c)$

4. The following multiplication properties are true for any real numbers.
   a. Multiplication Property of One: $a \cdot 1 = a$
   b. Multiplication Property of Reciprocals: $a \cdot \dfrac{1}{a} = 1,\ a \neq 0$
   c. Commutative Property of Multiplication: $a \cdot b = b \cdot a$
   d. Associative Property of Multiplication: $(ab)c = a(bc)$

5. The following special properties are true for any real numbers.
   a. Property of the Opposite of a Sum: $-(a + b) = (-a) + (-b)$
   b. Multiplication Property of -1: $-1(a) = -a;\ -1(a + b) = -a - b$
   c. Opposites and Products: $(-a)(b) = -ab;\ (-a)(-b) = ab$
   d. Subtraction of a Sum or Difference:
      $a - (b + c) = a - b - c;\ a - (b - c) = a - b + c$
   e. Special Division Properties: $\dfrac{a}{1} = a;\ \dfrac{a}{a} = 1,\ a \neq 0$
   f. Opposites and Division: $\dfrac{-a}{b} = \dfrac{a}{-b};\ \dfrac{a}{-b} = -\dfrac{a}{b};\ \dfrac{-a}{-b} = \dfrac{a}{b};\ b \neq 0$

6. The opposite of a sum or difference is the same as multiplying the sum or difference by -1.

7. Distributive Property: For any real numbers a, b, and c,
   a. $a(b + c) = ab + ac$, and
   b. $(b + c)a = ba + ca$.

8. Steps for Combining Like Terms
   a. Add or subtract the numerical factors.
   b. Use the same variable(s) and exponent(s) to write one term.

54 / *Chapter 2*

CHAPTER REVIEW

SECTION 2.1

Write each number as a quotient of two integers in two ways.

1. $-\frac{13}{16}$ $\frac{-26}{32}$; $\frac{-39}{48}$ 2. $-\frac{4}{5}$ $\frac{-8}{10}$; $\frac{-12}{15}$ 3. $3\frac{7}{8}$ $\frac{31}{8}$; $\frac{62}{16}$ 4. $-2\frac{3}{4}$ $\frac{-11}{4}$; $\frac{-22}{8}$ 5. -37 $\frac{-37}{1}$; $\frac{-74}{2}$ 6. 16 $\frac{16}{1}$; $\frac{32}{2}$

Write each rational number as a repeating decimal.

7. -21 $-21.000\cdots$ 8. $\frac{4}{5}$ $0.8000\cdots$ 9. $-\frac{1}{3}$ $-0.333\cdots$ 10. $\frac{7}{12}$ $0.58333\cdots$ 11. $4\frac{1}{2}$ $4.5000\cdots$ 12. $8\frac{1}{7}$

For Exercises 13–15, choose from the box the set of numbers that corresponds to each word description.

13. The counter on a service station pump showing the cost of the gasoline dispensed Positive rational numbers

14. An electronic traffic counter Whole numbers

15. The numbers used by a TV weatherperson to describe the excess or deficiency of rainfall Rational numbers

> Rational numbers
> Whole numbers
> Positive rational numbers

Refer to the set of numbers below. Then list the numbers named in Exercises 16–21.

$$\{13\tfrac{1}{2},\ -5,\ 0,\ -\sqrt{5},\ 4.1,\ 13,\ -\sqrt{9},\ \tfrac{5}{8},\ \pi,\ -0.5\}$$ $13\frac{1}{2}$, 4.1, 13, $\frac{5}{8}$, π

16. Rational numbers 17. Irrational numbers 18. Positive real numbers

19. Negative real numbers $-5, -\sqrt{5}, -\sqrt{9}, -0.5$ 20. Integers $-5, 0, 13 -\sqrt{9}$ 21. Whole numbers 0, 13

SECTION 2.2

Write the value of the variable that makes each equation true. Name the property that gives the reason for your choice.

22. $44 + (63 + 2) = (44 + 63) + d$

23. $h + 0 = 5\frac{1}{2}$

24. $x + 5.7 = 5.7 + 3.9$

25. $14 + m = 0$

Simplify.

26. $23 - 15 - 18 + 5$ -5

27. $-37 + 14 - 9 + 21$ -11

28. $-12.9 + 8.6 + 13.7 - 18.3$ -8.9

29. $16.8 - 13.4 - 25.7 + 17.8$ -4.5

30. $f - 3\frac{1}{4} + 2\frac{1}{2} - d$ $f - d - \frac{3}{4}$

31. $9\frac{5}{8} - 6\frac{1}{2} - a + 2\frac{7}{8}$ $6 - a$

32. $56 + r - 92 + 26 - s - 18$ $-28 + r - s$

33. $-73 - r + 39 + w - 14 + 58$ $10 - r + w$

Real Numbers / **55**

Chapter Test
Two Chapter Tests (Form A and Form B) are provided on pages 55–58 in the *Teacher's ResourceBank*.

Additional Answers

12. $8.142857142857\cdots$

16. $13\frac{1}{2}$, -5, 0, 4.1, 13, $-\sqrt{9}$, $-\frac{5}{8}$, -0.5

17. $-\sqrt{5}$, π

22. 2; Associative Property of Addition

23. $5\frac{1}{2}$; Addition Property of Zero

24. 3.9; Commutative Property of Addition

25. -14; Addition Property of Opposites

SECTION 2.3

Multiply.

34. $(-4)(7)(-3)$ 84

35. $(-5)(20)(-4)$ 400

36. $(-30)(-2)(-4)$ −240

37. $(-10)(-40)(-5)$ −2000

38. $(12)(-3)(\frac{1}{4})(-\frac{1}{9})(-17)$ −17

39. $(\frac{1}{6})(-8)(3)(-\frac{1}{4})(-63)$ −63

40. $(-2)(-2)(-3)(-1)$ 12

41. $(-1)(-2)(-2)(-3)(-3)$ −36

42. $(-6)(12)(x)(-4)$ $288x$

43. $(\frac{1}{4})(y)(8)(-5)$ $-10y$

44. $(7)(-c)(-\frac{1}{3})(21)$ $49c$

45. $(-18)(5)(-d)(-\frac{1}{9})$ $-10d$

SECTION 2.4

Simplify.

Additional Answers
51. $r + s - p + q$
57. $72q^3r^2$
64. $-8b - 72$
65. $-7r - 70$
72. $10(t - w)$
83. $-1.4q - 12.1$

46. $-(s - 5)$ $-s + 5$

47. $-(12m - 2n - p)$ $-12m + 2n + p$

48. $-5a - (-3b + c)$ $-5a + 3b - c$

49. $w - (k + 12)$ $w - k - 12$

50. $(m - n) - (a + b)$ $m - n - a - b$

51. $(r + s) - (p - q)$

52. $(-4s)(-t)$ $4st$

53. $(-k)(-12n)$ $12kn$

54. $(\frac{1}{4}x)(-12x)$ $-3x^2$

55. $(-\frac{1}{3}t)(-21t)$ $7t^2$

56. $(-10ab)(5b)(-2ac)$ $100a^2b^2c$

57. $(-4qr)(6q)(-3qr)$

58. $\frac{-xy}{-xy}$ 1

59. $\frac{-8t}{8t}$ −1

60. $\frac{5 - s}{5 - s}$, $s \neq 5$ 1

61. $\frac{a - 4}{a - 4}$, $a \neq 4$ 1

62. $\frac{t - 1}{1 - t}$, $t \neq 1$ −1

63. $\frac{-(7 - x)}{x - 7}$, $x \neq 7$ 1

SECTION 2.5

Multiply and simplify.

64. $-8(b + 9)$

65. $(r + 10)(-7)$

66. $(-5 + 6t)(-3)$ $15 - 18t$

67. $-9(2a + 1)$ $-18a - 9$

68. $3r(-2r + 5)$ $-6r^2 + 15r$

69. $6m(-3m + 5)$ $-18m^2 + 30m$

70. $(-9t - 12)(\frac{1}{4}t)$ $-2\frac{1}{4}t^2 - 3t$

71. $(3.4a + 2.6)(-0.3b)$ $-1.02ab - 0.78b$

Factor.

72. $10t - 10w$

73. $\frac{1}{2}a - \frac{1}{2}b$ $\frac{1}{2}(a - b)$

74. $2x + 6y$ $2(x + 3y)$

75. $15r + 10s$ $5(3r + 2s)$

76. $6x^2 + 3x$ $3x(2x + 1)$

77. $-4y^2 + 2y$ $-2y(2y - 1)$

78. $-12x - 3y$ $-3(4x + y)$

79. $-22xy^2 + 44xy$ $-22xy(y - 2)$

SECTION 2.6

Simplify.

80. $5k - 8k$ $-3k$

81. $-12t^2 - 15t^2 - 5$ $-27t^2 - 5$

82. $6.3n - 4.7n - 3.2 - 3.9n$ $-2.3n - 3.2$

83. $7.2q - 5.9q - 12.1 - 2.7q$

84. $r^2s - 4s^2 + r^2s + s^2$ $2r^2s - 3s^2$

85. $-p^2q + 2pq^2 - qp^2 - pq^2$ $-2p^2q + pq^2$

56 / Chapter 2

Equations and Problem Solving

Many events in nature can be examined using mathematics. For example, mathematics can be used to determine the effects of a drop of liquid falling into a liquid, the speed of a flash of lightning, and the size and shape of the hexagonal-patterned snowflake.

Lesson Resources
Warm–Up: p. M–22
Maintenance: See below.
Practice Worksheet 3.1
Manipulative 4, p. 546
Teaching Aids 1, 3
Transparencies 50, 51, 53, 54

Maintenance
1. Simplify: $-|-36.2|$
   Ans: -36.2 (Section 1.1)
2. Evaluate: $(6)(-3) - 2$
   Ans: -20 (Section 1.6)
3. Simplify: $-10 + 12 - 13 + 8$
   Ans: -3 (Section 2.2)
4. Simplify: $(2x)(-3y)(-x)(-y)$
   Ans: $-6x^2y^2$ (Section 2.4)
5. Factor: $-7t - 14t^2$
   Ans: $-7t(1 + 2t)$
   (Section 2.5)

Additional Examples
Example 1
Solve.
1. $a = 29 + 29$
   Ans: $a = 58$
2. $-11 = h - 9$
   Ans: $h = -2$
3. $x - 5.7 = 18$
   Ans: $x = 23.7$

3.1 Addition and Subtraction

OBJECTIVE: To solve and check equations of the form $x + a = b$ or $x - a = b$ in which a and b are rational numbers

Recall from Section 2.2 that an **equation** is a sentence that contains the equality symbol "=." A **solution** or **root** of an equation is a number that makes the equation true.

P–1 **Which equations are true when x is replaced by 19?**

a. $x - 9 = 10$ **b.** $x - 9 + 9 = 10 + 9$ **c.** $x = 19$
    True   True   True

Equations **a, b,** and **c** are <u>equivalent</u>. **Equivalent equations** have the same solution. Note that 9 was added to each side of Equation **a** to get Equation **b** and Equation **c**.

"$x - 6 + 6 = 10 + 6$" and "$x = 16$" are both equivalent to "$x - 6 = 10$."

> **Addition Property for Equations**
>
> Adding the same real number to each side of an equation forms an equivalent equation.
>
> $x - 6 = 10$
> $x - 6 + 6 = 10 + 6$
> $x = 16$

The goal in finding the solution of an equation is to get the variable alone on one side of the equation. To **solve** an equation means to find its solution.

EXAMPLE 1   Solve: **a.** $n - 8 = -15$ **b.** $73 = t - 9$

Solution: a.
$\boxed{1}$ Add 8 to each side. $\longrightarrow$

$$n - 8 = -15$$
$$n - 8 + 8 = -15 + 8$$
$$n = -7$$

Check: $n - 8 = -15$
$-7 - 8$
-15

Solution: b.
$\boxed{1}$ Add 9 to each side. $\longrightarrow$

$$73 = t - 9$$
$$73 + 9 = t - 9 + 9$$
$$82 = t$$

Check: $73 = t - 9$
$82 - 9$
73

As Example 1 shows, you check the answer by replacing the variable with your answer to see if it is the solution.

P–2 **Solve.**

a. $y - 12 = -13$ **b.** $14 = x - 15$ **c.** $t - 18 = -18$ **d.** $11 = k - 11$
    $y = -1$ $x = 29$ $t = 0$ $k = 22$

As you would expect, there is a subtraction property for equations.

"$x + 7 - 7 = -2 - 7$"
and "$x = -9$" are
both equivalent to
"$x + 7 = -2$."

> **Subtraction Property for Equations**
>
> Subtracting the same real number from each side of an equation forms an equivalent equation.

$$x + 7 = -2$$
$$x + 7 - 7 = -2 - 7$$
$$x = -9$$

The checks are left for you to do in Example 2.

EXAMPLE 2   Solve: **a.** $3.4 + x = -7.9$.   **b.** $\frac{2}{5} = y + \frac{3}{5}$

Solution:
$$3.4 + x = -7.9$$
1 Subtract 3.4 from each side. ⟶ $3.4 + x - 3.4 = -7.9 - 3.4$ ◀ $\begin{array}{l} -7.9 - 3.4 = \\ -7.9 + (-3.4) \end{array}$
$$x = -11.3$$

Solution:
$$\frac{2}{5} = y + \frac{3}{5}$$
1 Subtract $\frac{3}{5}$ from each side. ⟶ $\frac{2}{5} - \frac{3}{5} = y + \frac{3}{5} - \frac{3}{5}$ ◀ $\begin{array}{l} \frac{2}{5} - \frac{3}{5} = \\ \frac{2}{5} + (-\frac{3}{5}) \end{array}$
$$-\frac{1}{5} = y$$

P–3 **Solve.**

a. $n + 13 = 12$
$n = -1$

b. $17 + k = 18$
$k = 1$

c. $24 = x + 24$
$x = 0$

CLASSROOM EXERCISES

What number must be added to or subtracted from each side of each equation? (Step 1, Examples 1 and 2)

1. $k + 12 = -9$ Subtract 12.
2. $x - 8 = 14$ Add 8.
3. $8 + t = 29$

4. $m - 12.9 = -4.7$ Add 12.9.
5. $9.2 = w - 13.7$ Add 13.7.
6. $8\frac{3}{4} + r = 1\frac{1}{8}$

7. $\frac{13}{8} = p - \frac{5}{2}$ Add $\frac{5}{2}$.
8. $8 + m = 3$ Subtract 8.
9. $14 = 23 + s$

Solve. Check each answer. (Examples 1 and 2)

10. $x - 7 = 14$ $x = 21$
11. $y - 12 = -7$ $y = 5$
12. $p - 23 = -18$

13. $17 = q - 9$ $q = 26$
14. $-0.8 = r - 0.8$ $r = 0$
15. $2.5 = b - 1.5$

16. $f + \frac{3}{8} = 2\frac{1}{4}$ $f = 1\frac{7}{8}$
17. $h + 6\frac{1}{2} = 3\frac{2}{3}$ $h = -2\frac{5}{6}$
18. $12 + v = -9$

19. $32 = y + 15$ $y = 17$
20. $-12 = 20 + x$ $x = -32$
21. $-6 = 11 + c$

Equations and Problem Solving / **59**

Assignment Guide
Basic p. 60: 1–41 odd
Average p. 60: 1–41 odd
Above Average p. 60: 1–41 odd

Additional Answers
Written Exercises
  6. $c = 46.4$
  9. $f = 24$
 12. $h = -2\frac{1}{3}$
 18. $t = 0.5$
 21. $k = 2$
 24. $z = 7\frac{1}{4}$
 30. $t = 1$
 33. $x = -3\frac{1}{2}$
 36. $m = -11.4$
 39. $z = -62$
 42. $k = 29\frac{7}{10}$

Review Capsule
16. 117
20. -102.5
24. -52
28. $-\frac{1}{20}$

■■■■ **WRITTEN EXERCISES** ■■■■

Goal: To solve equations by using the Addition and Subtraction Properties
Sample Problems: a. $x - 19 = 12$ **b.** $-30 = 27 + n$
Answers: a. $x = 31$ **b.** $-57 = n$

Solve. Check each answer. (Example 1)

1. $x - 9 = 24$ $x = 33$
2. $y - 15 = 27$ $y = 42$
3. $z - 21 = -9$ $z = 12$
4. $a - 32 = -15$ $a = 17$
5. $b - 1.3 = -5.5$ $b = -4.2$
6. $c - 27.1 = 19.3$
7. $23 = d - 15$ $d = 38$
8. $-46 = r - 31$ $r = -15$
9. $23 = f - 1$
10. $-236 = p - 407$ $p = 171$
11. $\frac{1}{2} = g - \frac{1}{2}$ $g = 1$
12. $-5\frac{1}{6} = h - 2\frac{5}{6}$

(Example 2)

13. $j + 11 = 28$ $j = 17$
14. $14 + k = 33$ $k = 19$
15. $n + 15 = -9$ $n = -24$
16. $38 + q = 26$ $q = -12$
17. $0.6 + x = -0.4$ $x = -1$
18. $t + 0.25 = 0.75$
19. $10.0 = b + 5.2$ $b = 4.8$
20. $30.7 = x + 27.6$ $x = 3.1$
21. $9 = 7 + k$
22. $413 = w + 198$ $w = 215$
23. $-\frac{4}{5} = \frac{2}{5} + y$ $y = -1\frac{1}{5}$
24. $9\frac{3}{4} = z + 2\frac{1}{2}$

MIXED PRACTICE

25. $x + 20 = -13$ $x = -33$
26. $a + 2.75 = -7.48$ $a = -10.23$
27. $7.5 = b - 2.6$ $b = 10.1$
28. $c - 59.2 = -34.5$ $c = 24.7$
29. $3.4 = d + 6.3$ $d = -2.9$
30. $\frac{5}{6} = t - \frac{1}{6}$
31. $10.5 + f = 19.3$ $f = 8.8$
32. $-59.6 = 80.7 + g$ $g = -140.3$
33. $1\frac{3}{4} + x = -\frac{7}{4}$
34. $h + 26 = 405$ $h = 379$
35. $k - 60.2 = -47.4$ $k = 12.8$
36. $-7.3 = 4.1 + m$
37. $16.2 = p + 14.9$ $p = 1.3$
38. $s - 59 = -102$ $s = -43$
39. $74 + z = 12$
40. $3\frac{3}{4} = v - \frac{7}{8}$ $v = 4\frac{5}{8}$
41. $9.3 = -24.1 + b$ $b = 33.4$
42. $k - 12\frac{1}{5} = 17\frac{1}{2}$

REVIEW CAPSULE FOR SECTION 3.2

Perform the indicated operations. (Pages 11–14 and 16–18)

1. $3\left(\frac{1}{3}\right)$ 1
2. $\frac{1}{8} \cdot 8$ 1
3. $\left(-\frac{1}{7}\right)(-7)$ 1
4. $(-16)\left(-\frac{1}{16}\right)$ 1
5. $\left(\frac{3}{4}\right)\left(\frac{4}{3}\right)$ 1
6. $\left(-\frac{7}{2}\right)\left(-\frac{2}{7}\right)$ 1
7. $\left(-\frac{5}{16}\right)\left(-\frac{16}{5}\right)$ 1
8. $\left(\frac{8}{3}\right)\left(\frac{3}{8}\right)$ 1
9. $33 \div 33$ 1
10. $(-4.5) \div (-4.5)$ 1
11. $\frac{5}{8} \div \frac{5}{8}$ 1
12. $312 \div 312$ 1
13. $15 \cdot \frac{3}{5}$ 9
14. $(-1.8)(3.9)$ -7.02
15. $\left(\frac{5}{8}\right)(-12)$ $-7\frac{1}{2}$
16. $(-9)(-13)$
17. $\left(-\frac{3}{4}\right)\left(-\frac{7}{2}\right)$ $2\frac{5}{8}$
18. $(13)(-76)$ -988
19. $(58)(8)$ 464
20. $-12.5 \cdot 8.2$
21. $-7.2 \div -0.9$ 8
22. $\frac{1}{6} \div \frac{2}{3}$ $\frac{1}{4}$
23. $135 \div 9$ 15
24. $-13 \div \frac{1}{4}$
25. $-15.5 \div 0.6$ $-25\frac{5}{6}$
26. $0.1 \div 0.25$ 0.4
27. $\left(-\frac{2}{3}\right) \div \left(\frac{4}{15}\right)$ $-2\frac{1}{2}$
28. $-\frac{4}{5} \div 16$

3.2 Multiplication and Division

To solve and check an equation of the form $ax = b$, showing all steps

Equations involving multiplication or division are solved somewhat like equations that involve addition or subtraction.

Teaching Suggestions p. M–22

Lesson Resources
Warm–Up: p. M–23
Maintenance: See below.
Practice Worksheet 3.2
Manipulative 5, p. 548
Teaching Aids 1, 3
Transparencies 50, 51, 53, 54

P–1 **Which equations are true when n is replaced by 27?**

a. $\frac{n}{9} = 3$ **b.** $\frac{n}{9} \cdot 9 = 3 \cdot 9$ **c.** $n = 27$
   True   True   True

Equations **a**, **b**, and **c** are equivalent. Note that each side of Equation **a** was multiplied by 9 to get Equation **b** and Equation **c**.

All three of these equations are equivalent.

> **Multiplication Property for Equations**
>
> Multiplying each side of an equation by the same nonzero real number forms an equivalent equation.

$\frac{x}{5} = 1.6$

$\frac{x}{5}(5) = 1.6(5)$

$x = 8$

Maintenance
1. Add: $(0.34) + (-0.26)$
   Ans: 0.08 (Section 1.2)
2. Multiply: $(-13)(-5)$
   Ans: 65 (Section 1.4)
3. Multiply: $(\frac{5}{12})(\frac{12}{5})(-2)(-\frac{1}{2})$
   Ans: 1 (Section 2.3)
4. Simplify: $\frac{-6t}{-6t}$
   Ans: 1 (Section 2.4)
5. The thermometer reading at sunrise was $-8°C$. By noon, it had risen 10°C. Find the thermometer reading at noon.
   Ans: 2°C (Section 1.2)

Recall that $\frac{x}{4}$ means $\frac{1}{4}x$. Thus, in the example below,

$\frac{x}{4}(4)$ means $(\frac{1}{4}x)(4)$ or, $1 \cdot x$.

EXAMPLE 1   Solve: **a.** $\frac{x}{4} = -12$ **b.** $40 = \frac{y}{-8}$

Solution: **a.** $\frac{x}{4} = -12$ **Check:** $\frac{x}{4} = -12$

☐1 Multiply each side by 4. ⟶ $\frac{x}{4}(4) = -12(4)$

$x = -48$ $\frac{-48}{4}$
                                      -12

Solution: **b.** $40 = \frac{y}{-8}$ **Check:** $40 = \frac{y}{-8}$

☐1 Multiply each side by -8. ⟶ $40(-8) = \frac{y}{-8}(-8)$ $\frac{-320}{-8}$

$-320 = y$ 40

Additional Examples
Example 1
Solve.

1. $\frac{a}{7} = -11$
   Ans: $a = -77$

2. $5 = \frac{x}{-3}$
   Ans: $x = -15$

3. $\frac{t}{-1.6} = -10$
   Ans: $t = 16$

P–2 **Solve.** $b = 256$

a. $\frac{b}{6} = -6$ $b = -36$ **b.** $20 = \frac{x}{-5}$ $x = -100$ **c.** $\frac{n}{3} = \frac{1}{3}$ $n = 1$ **d.** $-32 = \frac{b}{-8}$

As you would expect, there is a division property for equations.

Equations and Problem Solving / **61**

Division Property for Equations

Dividing each side of an equation by the same nonzero real number forms an equivalent equation.

$$7x = -3$$
$$\frac{7x}{7} = \frac{-3}{7}$$
$$x = -\frac{3}{7}$$

EXAMPLE 2

Solve: $-6x = -8.4$

Solution:

$$-6x = -8.4$$

1️⃣ Divide each side by -6. ⟶ $\dfrac{-6x}{-6} = \dfrac{-8.4}{-6}$

$$x = 1.4$$

Check: $-6x = -8.4$

$$-6(1.4)$$

$$-8.4$$

P–3 **Solve.**

a. $3y = 0.3$ $\quad y = 0.1$

b. $21 = -7x$ $\quad x = -3$

c. $-5n = -5$ $\quad n = 1$

d. $-12 = 4t$ $\quad t = -3$

Recall that <u>dividing by a number is the same as multiplying by its reciprocal</u>. This idea is used in the next example.

EXAMPLE 3

Solve: $12 = -\frac{3}{5}y$

Solution:

$$12 = -\frac{3}{5}y$$

1️⃣ Multiply each side by $-\frac{5}{3}$. ⟶ $12\left(-\frac{5}{3}\right) = \left(-\frac{3}{5}y\right)\left(-\frac{5}{3}\right)$

 Since $-\frac{5}{3} \cdot \left(-\frac{3}{5}\right) = 1$, they are reciprocals.

2️⃣ Rewrite the equation. ⟶ $12\left(-\frac{5}{3}\right) = \left(-\frac{3}{5} \cdot -\frac{5}{3}\right)y$

3️⃣ Multiply. ⟶ $\left(\frac{\overset{4}{12}}{1}\right)\left(-\frac{5}{\underset{1}{3}}\right) = 1 \cdot y$

$$-20 = y$$

The check is left for you.

P–4 **What number should you multiply each side of the equation by to get the variable alone?**

a. $\frac{1}{2}a = 19$ 2

b. $12 = -\frac{3}{4}m$ $-\frac{4}{3}$

c. $-1\frac{2}{3}x = 16$ $-\frac{3}{5}$

d. $-9 = \frac{4}{5}y$ $\frac{5}{4}$

62 / *Chapter 3*

See the Answers to Selected Exercises for answers to Ex. 1-7 odd.

CLASSROOM EXERCISES

What number should you multiply by or divide by to get the variable alone?
(Step 1, Examples 1, 2, and 3)

1. $\frac{n}{12} = -89$ **2.** $\frac{r}{-1.9} = 3.4$ **3.** $14a = -72$ **4.** $-23m = -59$

5. $5.6 = -8.9k$ **6.** $-8.65 = 0.29q$ **7.** $127 = -\frac{2}{3}w$ **8.** $-28 = \frac{15}{4}x$

Solve. Check each answer (Examples 1, 2, and 3) $a = -60$

9. $\frac{j}{7} = 4$ $j = 28$ **10.** $\frac{y}{-6} = -11$ $y = 66$ **11.** $\frac{z}{7} = 20$ $z = 140$ **12** $\frac{a}{12} = -5$

13. $-4 = \frac{r}{1.2}$ $r = -4.8$ **14.** $8 = \frac{s}{-7}$ $s = -56$ **15.** $-9 = \frac{t}{-9}$ $t = 81$ **16.** $0 = \frac{w}{32}$ $w = 0$

17. $5t = -30$ $t = -6$ **18.** $-8y = 56$ $y = -7$ **19.** $8a = -2.4$ $a = -0.3$ **20.** $-0.6x = -5.4$

21. $-24 = \frac{3}{2}k$ $k = -16$ **22.** $-15 = -\frac{1}{2}x$ $x = 30$ **23.** $-24 = -\frac{2}{3}k$ $k = 36$ **24.** $-27 = \frac{3}{4}p$
   $p = -36$

See the Answers to Selected Exercises for answers to Ex. 11, 15, 21, 23, and 25.

WRITTEN EXERCISES

Goal: To solve equations using the Multiplication and Division Properties

Sample Problems: a. $\frac{w}{5} = -7.2$ **b.** $4\frac{2}{3} = -3x$

Answers: a. $w = -36$ **b.** $-\frac{14}{9} = x$

Solve. Check each answer. (Examples 1, 2, and 3) $k = -108$

1. $\frac{x}{-5} = 13$ $x = -65$ **2.** $\frac{g}{-3} = 30$ $g = -90$ **3.** $\frac{y}{8} = -12$ $y = -96$ **4.** $\frac{k}{12} = -9$

5. $\frac{x}{8.3} = -14$ $x = -116.2$ **6.** $\frac{s}{0.2} = 1000$ $s = 200$ **7.** $\frac{t}{-4} = -15$ $t = 60$ **8.** $\frac{m}{-5} = 75$

9. $23 = \frac{v}{8}$ $v = 184$ **10.** $-14 = \frac{w}{-7}$ $w = 98$ **11.** $-16 = \frac{x}{5.2}$ **12.** $-2.08 = \frac{m}{-12}$

13. $8\frac{1}{4} = \frac{s}{-4}$ $s = -33$ **14.** $5\frac{1}{8} = \frac{z}{16}$ $z = 82$ **15.** $0.34 = \frac{r}{-1.6}$ **16.** $-70 = \frac{b}{0.5}$

17. $4a = -28$ $a = -7$ **18.** $6c = -54$ $c = -9$ **19.** $-9d = 72$ $d = -8$ **20.** $-11e = 77$

21. $-1.8x = 18.36$ **22.** $-5b = 3.5$ $b = -0.7$ **23.** $-108 = 12x$ **24.** $-72 = -0.9y$

25. $-140 = -16x$ **26.** $-y = 99$ $y = -99$ **27.** $-62r = 0$ $r = 0$ **28.** $-90 = -6y$

29. $\frac{1}{2}z = -44$ $z = -88$ **30.** $\frac{2}{3}r = -54$ $r = -81$ **31.** $30 = \frac{2}{5}k$ $k = 75$ **32.** $15 = \frac{3}{4}c$

33. $-\frac{2}{3}w = 20$ $w = -30$ **34.** $-\frac{1}{4}a = 8$ $a = -32$ **35.** $36 = \frac{3}{4}m$ $m = 48$ **36.** $24 = \frac{5}{6}e$

37. $\frac{2}{5}x = -14$ $x = -35$ **38.** $-100 = \frac{1}{5}y$ $y = -500$ **39.** $-\frac{7}{8}p = -49$ **40.** $56 = -\frac{1}{4}s$
   $p = 56$ $s = -224$

MORE CHALLENGING EXERCISES

41. $6|x| = 1.2$ **42.** $-\frac{1}{3}|p| = -5$ **43.** $\frac{|-n|}{-7} = -3$ **44.** $\frac{|-t|}{5} = 3.2$ $t = 16,$
   $x = 0.2$, or $x = -0.2$ $p = 15$, or $p = -15$ $n = 21$, or $n = -21$ or $t = -16$

Equations and Problem Solving / 63

Common Error
When solving equations, students use the operation indicated in the equation to get the variable alone.

Example
$-4n = 8$
$n = 8(-4)$
$n = -32$

Prescription
Insist that students check each solution by substituting in the original equation. Keep emphasizing that the same operation must be performed on both sides.

Additional Answers
Classroom Exercises
 2. Multiply by -1.9.
 4. Divide by -23.
 6. Divide by 0.29.
 8. Multiply by $\frac{4}{15}$.
 20. $x = 9$

Assignment Guide
Basic p. 63: 1–39 odd
Average p. 63: 1–43 odd
Above Average p. 63:
1–39 odd, 41–44

Additional Answers
Written Exercises
 8. $m = -375$
 12. $m = 24.96$
 16. $b = -35$
 20. $e = -7$
 24. $y = 80$
 28. $y = 15$
 32. $c = 20$
 36. $e = 28\frac{4}{5}$

Problem–Solving Skills
Solving a multi–step problem
(Ex. 41–44)
Using logical reasoning
(Ex. 41–44)

Critical Thinking
See Exercises 41–44.

REVIEW CAPSULE FOR SECTION 3.3

Simplify. (Pages 58–60)

1. $5n + 10 - 10$ $5n$

2. $3r - 7 + 7$ $3r$

3. $6b \div 6$ b

4. $(12t + 1.8 - 1.8) \div 12$ t

5. $(18x - 3\frac{1}{3} + 3\frac{1}{3}) \div 18$ x

6. $(\frac{y}{4} + 4 - 4)4$ y

Teaching Suggestions p. M–22

Lesson Resources
Warm–Up: p. M–23
Maintenance: See below.
Practice Worksheet 3.3
Manipulative 6, p. 550
Teaching Aids 1, 3
Transparencies 6, 7, 50, 51, 53, 54
Exploration 3

Maintenance
1. Subtract: $(-21) - (-13)$
   Ans: -8   (Section 1.3)
2. Evaluate: $(-27) \div (3)(-3)$
   Ans: 27   (Section 1.6)
3. Express $\frac{5}{11}$ as a repeating
   decimal.
   Ans: $0.\overline{45}$ (Section 2.1)
4. Simplify: $-(-3x - 5)$
   Ans: $3x + 5$ (Section 2.4)
5. Factor: $-6p^2r^2 - 3pr^3$
   Ans: $-3pr^2(2p + r)$
   (Section 2.5)

Additional Examples
Example 1
Solve.
1. $3t + 6 = 15$
   Ans: $t = 3$
2. $7x - 9 = -37$
   Ans: $x = -4$
3. $-11.3 + 2y = 15.1$
   Ans: $y = 13.2$

Example 2
Solve.
1. $-11 = -6t + 19$
   Ans: $t = 5$
2. $11 = 1 - 5y$
   Ans: $y = -2$
3. $24 = -3x - 9$
   Ans: $x = -11$

OBJECTIVES: To solve and check an equation with the variable on one side and involving multiplication and either addition or subtraction, showing all steps

3.3   Two or More Operations

To solve and check an equation with the variable on one side and involving division and either addition or subtraction, showing all steps

To solve an equation that involves more than one operation, use the Addition or Subtraction Property for Equations before using the Multiplication or Division Property.

Think of the Properties of Equations as "undoing" the operations in the equation.

EXAMPLE 1   Solve: $5n - 7 = 14$

Solution: $5n - 7 = 14$ **Check:** $5n - 7 = 14$

⬜1 Add 7 to each side. ⟶ $5n - 7 + 7 = 14 + 7$ $5(4.2) - 7$

                           $5n = 21$ $21 - 7$

⬜2 Divide each side by 5. ⟶ $\dfrac{5n}{5} = \dfrac{21}{5}$ 14

                            $n = 4.2$

Sometimes the variable is on the right side of an equation. The steps for solving are still the same.

EXAMPLE 2   Solve: $-40 = 8 - 4y$

Solution: $-40 = 8 - 4y$ **Check:** $-40 = 8 - 4y$

⬜1 Subtract 8 from each side. ⟶ $-40 - 8 = 8 - 4y - 8$ $8 - 4(12)$

                             $-48 = -4y$ $8 - 48$

⬜2 Divide each side by -4. ⟶ $\dfrac{-48}{-4} = \dfrac{-4y}{-4}$ -40

                         $12 = y$

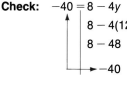

P–1   **Solve.**

a. $3t - 7 = 14$ $t = 7$

b. $-k + 16 = 8$ $k = 8$

c. $29 = -3 - 8r$ $r = -4$

Recall that $\frac{x}{12}$ means $\frac{1}{12}x$. Thus, in the example below,

$$\frac{x}{12}(12) \text{ means } (\tfrac{1}{12}x)\,(12) \text{ or, } 1 \cdot x.$$

EXAMPLE 3   Solve: $\frac{x}{12} - 14 = -8$

Solution: $\frac{x}{12} - 14 = -8$

☐1 Add 14 to each side. ⟶ $\frac{x}{12} - 14 + \mathbf{14} = -8 + \mathbf{14}$

$$\frac{x}{12} = 6$$

☐2 Multiply each side by 12. ⟶ $\frac{x}{12}(\mathbf{12}) = 6(\mathbf{12})$ ◀ $\frac{x}{12}(\mathbf{12}) = (\tfrac{1}{12})\,(\tfrac{12}{1})x$

$$x = 72$$

The check is left for you.

P–2 **Solve.**

a. $\frac{b}{7} + 6 = 5$ **b.** $11 + \frac{s}{5} = 16$ **c.** $-33 = -3 - \frac{d}{2}$ $d = 60$
$\quad b = -7$ $\quad s = 25$

CLASSROOM EXERCISES

What number must be added to or subtracted from each side of each equation? (Step 1, Examples 1–3)

1. $2n + 5 = 8$ **2.** $7w - 12 = -3$ **3.** $6p - 5.6 = 7.3$ **4.** $1.4 + 0.9k = 0.5$

5. $3 - 2t = -9$ **6.** $12 = 8w + 7$ **7.** $-25 = 5 + 7q$ **8.** $16 = 2 - 12x$

9. $\frac{a}{4} - 6 = 23$ **10.** $\frac{m}{-5} + 3 = -7$ **11.** $6.5 - 1.3r = -3.7$ **12.** $0.5 + 0.8w = -2.7$

Write <u>Yes</u> *or* <u>No</u> *after checking whether the solution of each equation below is correct.* (Checks in Examples 1–3)

13. $4x - 9 = -21$ Yes   **14.** $-3y + 7 = -6$ No   **15.** $12 - 3c = -15$
$\quad x = -3$ $\quad y = 4$ $\quad c = 1$ No

16. $16 = -8 + 3f$ Yes   **17.** $25 - b = 17$ No   **18.** $12 + \frac{1}{2}n = 18$
$\quad f = 8$ $\quad b = -8$ $\quad n = 3$ No

19. $-\frac{1}{3}r + 5 = -2$ Yes   **20.** $-19 + \frac{a}{5} = -23$ Yes   **21.** $-5.9 = 1.3p - 2$
$\quad r = 21$ $\quad a = -20$ $\quad p = 3$ No

Equations and Problem Solving / 65

Assignment Guide
Basic
Day 1 p. 66: 1–21 odd
Day 2 p. 66: 23–41 odd
Average p. 66: 1–43 odd
Above Average p. 66:
1–43 odd, 42

Problem–Solving Skills
Using logical reasoning
(Ex. 43–44)
Solving a multi–step problem
(Ex. 43)
Using guess and check
(Ex. 44)

Non–Routine Problems
For a description of non–routine
problems, see page M–13.

Additional Answers
10. $x = -3\frac{2}{3}$
16. $k = -37.8$
19. $x = -5.5$
22. $k = 6.5$
34. $p = -22.5$
43. Since $25 \div 3 = 8$ R1, the
clerk measured 8 times using the short yardstick. The
wire was missing 8×2, or
16 inches from the 25–foot
length the customer wanted.
44. Use the clues with guess
and check to find the correct
number. The hundreds digit
must be a 3 or a 4. List the
19 3–digit numbers begin-
ning with a 3 or a 4 in which
the sum of the digits is 12.
Circle the ones in which the
tens digit is the sum of the
other two digits. The correct
numbers are 363 and 462.

■■■■■ **WRITTEN EXERCISES** ■■■■■

Goal: To solve equations using the Addition, Subtraction, Multiplication,
and Division Properties for Equations
Sample Problems: a. $-8 + 4r = -20$ **b.** $8.5 = 5x + 6$
Answers: a. $r = -3$ **b.** $0.5 = x$

Solve. Show all steps. Check each answer. (Example 1)

1. $3x - 8 = 13$ $x = 7$ **2.** $4a - 12 = 16$ $a = 7$ **3.** $2b + 9 = -17$ $b = -13$

4. $6x + 5 = 25$ $x = 3\frac{1}{3}$ **5.** $16 + 3c = -50$ $c = -22$ **6.** $-56 + 24d = -254$
$d = -8\frac{1}{4}$

(Example 2)

7. $-31 = 2e + 7$ $e = -19$ **8.** $17 = -4x + 3$ $x = -3\frac{1}{2}$ **9.** $-52 = 12 + 8y$ $y = -8$

10. $-16\frac{1}{2} = -5\frac{1}{2} + 3x$ **11.** $-17\frac{1}{4} = 4y - 2\frac{1}{4}$ $y = -3\frac{3}{4}$ **12.** $250 = 5f - 100$ $f = 70$

(Example 3)

13. $\frac{x}{7} - 4 = -12$ $x = -56$ **14.** $\frac{h}{10} - 6 = -15$ $h = -90$ **15.** $-26 + \frac{x}{1.2} = -34$ $x = -9.6$

16. $-19 + \frac{k}{2.7} = -33$ **17.** $20 = \frac{a}{5} - 10$ $a = 150$ **18.** $70 = 8 + \frac{b}{2}$ $b = 124$

MIXED PRACTICE

19. $3x - 4.8 = -21.3$ **20.** $-32 + 30t = -287$ $t = -8\frac{1}{2}$ **21.** $\frac{h}{4} + 15 = -60$ $h = -300$

22. $-2.4 = 10.6 - 2k$ **23.** $\frac{2}{3}g - \frac{1}{3} = \frac{2}{3}$ $g = 1\frac{1}{2}$ **24.** $50 = \frac{1}{2}d + 30$ $d = 40$

25. $12 - 5z = 48$ $z = -7\frac{1}{5}$ **26.** $1.5f + 8.5 = 11.5$ $f = 2$ **27.** $125 = 10c - 25$ $c = 15$

28. $\frac{n}{6} - 13 = 20$ $n = 198$ **29.** $4x - 12.3 = -18.5$ $x = -1.55$ **30.** $12 - \frac{3}{4}m = 15$ $m = -4$

31. $\frac{j}{3} - 4 = 7$ $j = 33$ **32.** $2a - 9.3 = -6.7$ $a = 1.3$ **33.** $\frac{1}{5}b + 15 = 25$ $b = 50$

34. $-10.4 - p = 12.1$ **35.** $\frac{r}{8} + 15 = 12$ $r = -24$ **36.** $\frac{2}{7}w + \frac{1}{7} = -\frac{5}{7}$ $w = -3$

37. $86 = 9 + 3x$ $x = 25\frac{2}{3}$ **38.** $12 - \frac{t}{5} = 7$ $t = 25$ **39.** $-36 = -4x + 6$ $x = 10\frac{1}{2}$

40. $-\frac{p}{6} + 7 = -2.5$ $p = 57$ **41.** $5\frac{1}{3} = 12 - 8k$ $k = \frac{5}{6}$ **42.** $6.2 = -2x - 5.4$ $x = -5.8$

NON-ROUTINE PROBLEMS

43. A customer wanted a 25-foot piece
of wire. The clerk incorrectly
measured the wire with a yardstick
that was 2 inches too short. How
many inches were missing from the
customer's length of wire?

44. A three-digit number is between
300 and 500. The sum of the digits
is 12. The sum of the hundred's
digit and the one's digit equals the
ten's digit. Find the number.

OBJECTIVES: Given a word problem describing values of two of the variables n the formula
$d = rt$, to solve for the third variable

■■■■■ **PROBLEM SOLVING AND APPLICATIONS** ■■■■■

3.4 **Using Formulas** Given a word problem describing values of p and
either ℓ or w in the formula $p = z(\ell + w)$, to solve for the third variable

The following word rule and <u>formula</u> relate distance, rate, and time.

Word Rule: <u>Distance</u> equals <u>rate</u> multiplied by <u>time</u>.

Formula: $d = rt$

In a formula, variables and symbols are used to represent words. If
you know the values of two of the variables in $d = rt$, you use the
techniques for solving equations to find the third value.

EXAMPLE 1 In 1903, Orville Wright flew his airplane a distance of
37 meters at an <u>average speed</u> (rate) of 3.08 meters
per second. To the nearest tenth, how long did the
flight last?

Solution: ☐1 Write the formula. ──────▶ $d = rt$ $d = distance$
$r = rate$
☐2 Identify the known values. ──────▶ $d = 37$ meters; $t = time$

$r = 3.08$ meters per second

☐3 Replace the variables. ──────▶ $37 = 3.08t$

☐4 Solve the equation. ──────▶ $\dfrac{37}{3.08} = t$ or, $12.01 = t$

To the nearest tenth, the flight lasted 12.0 seconds.

Word Rule: The <u>perimeter</u> of a rectangle equals the
sum of the <u>lengths of the sides</u>.

Formula: $p = \ell + \ell + w + w$ 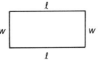 **Combine like terms.
Then factor.**

$p = 2\ell + 2w$ or, $p = 2(\ell + w)$

EXAMPLE 2 The perimeter of this postage stamp
is 9.4 centimeters (cm). The width is
2.2 centimeters. Find the length.

Solution: ☐1 Write the formula. ──────▶ $p = 2(\ell + w)$
☐2 Identify the known values. ──────▶ $p = 9.4$ cm; $w = 2.2$ cm
☐3 Replace the variables. ──────▶ $9.4 = 2(\ell + 2.2)$ Remind students
☐4 Solve the equation. ──────▶ $9.4 = 2\ell + 4.4$ that *both* ℓ and
2.2 must be
$5.0 = 2\ell$ multiplied by 2.
$2.5 = \ell$ The length is 2.5 cm.

Equations and Problem Solving / **67**

Teaching Suggestions p. M–22

Lesson Resources
Warm–Up: p. M–23
Maintenance: See below.
Practice Worksheet 3.4

Maintenance
1. Subtract: $38 - 46$
   Ans: −8 (Section 1.3)
2. Evaluate: $3^3 - 2 \cdot 2^3$
   Ans: 11 (Section 1.7)
3. Write True or False.
   22.3 is a whole number.
   Ans: False (Section 2.1)
4. Simplify: $9 + x - 15$
   Ans: x − 6 (Section 2.2)
5. At midnight, the wind chill
   temperature was four times
   the thermometer reading of
   −3°F. What was the wind chill
   temperature?
   Ans: −12°F (Section 1.4)

Additional Examples
Example 1
For each problem, find the rate,
r, or the time, t. Round your answer to the nearest tenth.
1. Dwight drove his car a distance of 528 miles in 11
   hours. Find his average
   speed.
   Ans: 48 miles per hour
2. During the first 20,000 feet after liftoff, a rocket's average
   speed is 1200 feet per second. How long does it take a
   rocket to fly that distance?
   Ans: 16.7 seconds

Example 2
The perimeter of a rectangular
rug is 50 feet. The length of the
rug is 17 feet. Find the width.
Ans: 8 feet

These exercises provide practice on solving for each of the variables in the formulas.

CLASSROOM EXERCISES

Find the value of the unknown variable. (Example 1)

1. $d = 1200$ kilometers; $r = 400$ kilometers per hour; $t = \underline{}$ 3 hours

2. $d = \underline{}$; $r = 25$ meters per minute; $t = 30$ minutes   750 meters

3. $d = 280$ feet; $r = \underline{}$; $t = 4$ seconds   70 feet per second

(Example 2)

4. $p = \underline{}$; $\ell = 18$ meters; $w = 12$ meters   60 meters

5. $p = 150$ kilometers; $\ell = \underline{}$; $w = 30$ kilometers   45 kilometers

6. $p = 29.2$ millimeters; $\ell = 6.4$ millimeters; $w = \underline{}$ 8.2 millimeters

Assignment Guide
Basic p. 68: 1–6
Average pp. 68–69: 1–7, 9
Above Average pp. 68–69: 1–9

WRITTEN EXERCISES

Sample Problems: a. Given: $d = 60.2$ m and $r = 7$ meters per second. Find t.
b. Given: $p = 37.0$ mm and $\ell = 12.7$ mm. Find w.
Answers: a. 8.6 seconds   **b.** $w = 5.8$ mm

Problem–Solving Skills
Using a formula (Ex. 1–6)
Drawing a diagram (Ex. 5–6)

For each problem, find the rate, r, or the time, t. Round your answer to the nearest tenth. (Example 1)

1. On a certain day, it took 1.2 seconds for light to travel from the moon to the earth 360,000 kilometers away. Find the speed of light in kilometers per second.   300,000 kilometers per second

2. During a thunderstorm, lightning was observed 5100 feet away. The thunderclap followed 4.5 seconds later. Find the speed of sound in feet per second.   1133.3 ft per second

3. A cheetah ran a distance of 80 yards traveling at an average speed of 35 yards per second. How long did it take the cheetah to run that distance?   2.3 seconds

4. In 1932, Amelia Earhart flew a plane from Newfoundland to Ireland, a distance of 3260 kilometers, at an average speed of 217 kilometers per hour. How long did her trip take?   15.0 hours

For each problem, find the width, w, or length, ℓ. Round your answer to the nearest tenth. (Example 2)

5. The perimeter of a picture frame is 3.8 meters. The length of the frame is 1.2 meters. Find the width.   0.7 meters

6. The perimeter of a hand calculator is 12.6 inches. The width is 2.5 inches. Find the length.   3.8 inches

APPLICATIONS

You can use a calculator to check an equation. However, some calculators do not automatically follow the rules for the order of operations. You may have to rewrite the equation for the calculator.

EXAMPLE   Check the answer to Sample Problem **a** on page 66.

SOLUTION   Equation: $-8 + 4r = -20$ Answer to be checked: -3
First rewrite the left side for the calculator: $4r - 8$

Multiplication is performed before addition.

Note: For "-3" press "③ ⟨⁺/₋⟩."

$$-20.$$

Since both sides have the same value, -20, the answer, -3, is correct.

Check the given "answer" to each equation below.

7. $5n - 7 = 14$; 4.2
Correct

8. $24 = 6 + 2x$; 3
Incorrect

9. $4 + 3x = -11$; -5
Correct

▰▰▰ MID-CHAPTER REVIEW ▰▰▰

Solve. Check each answer. (Section 3.1)

1. $b + 9 = -3$ $b = -12$

2. $10 + w = 40$ $w = 30$

3. $y - 1.5 = 3$ $y = 4.5$

4. $-3\frac{1}{8} = k - 2\frac{1}{4}$ $k = -\frac{7}{8}$

5. $t - 67 = 83$ $t = 150$

6. $0.75 = a + 0.25$ $a = 0.5$

Solve. Check each answer. (Section 3.2)

7. $-39 = 13a$ $a = -3$

8. $\frac{m}{-2} = -3$ $m = 6$

9. $-9z = 30$ $z = -3\frac{1}{3}$

10. $\frac{r}{0.5} = 16.5$ $r = 8.25$

11. $6\frac{1}{2} = \frac{x}{2}$ $x = 13$

12. $-42 = -13h$ $h = 3\frac{3}{13}$

Solve. Check each answer. (Section 3.3)

13. $3y + 7 = 28$ $y = 7$

14. $40 = \frac{h}{3} - 10$ $h = 150$

15. $15 = 7\frac{1}{2} - \frac{1}{2}t$ $t = -15$

16. $-20 = 15 + 15n$ $n = -2\frac{1}{3}$

17. $1.7 + \frac{j}{6} = 3.7$ $j = 12$

18. $8v - 17 = -33$ $v = -2$

Solve. Round your answer to the nearest tenth. (Section 3.4)

19. A coyote ran a distance of 50 yards traveling at an average speed of 20 yards per second. How long did it take the coyote to run that distance? 2.5 seconds

20. The perimeter of a park is 196 yards. The width is 46 yards. Find the length. 52 yards

Equations and Problem Solving / 69

Quiz Sections 3.1–3.4
After completing this Mid-Chapter Review, you may want to administer a quiz covering the same sections. See the quiz provided on page 71 in the *Teacher's ResourceBank.*

Using a Formula

Engineering

It is important for an **engineer** to know the greatest safe load that a beam can bear. The following formula applies to a steel beam with a rectangular cross section. Its load is in pounds, distributed along the beam.

Greatest Safe Load

$$S = \frac{1780Ad}{l}$$

◀ *A = cross-sectional area (in²)*
d = depth of beam (in)
l = distance between supports (ft)

EXAMPLE: Compute the greatest safe load in pounds for a solid steel beam 20 feet long with a cross section 8 inches wide and 6 inches deep.

SOLUTION: $S = \dfrac{1780Ad}{l}$

$A = 48$ in² $(6'' \times 8'')$, $l = 20$ ft, $d = 6$ in

$S = \dfrac{(1780)(48)(6)}{20}$

1 7 8 0 ⊠ 4 8 ⊠ 6 ⊡ 2 0 ⊟

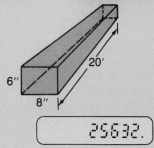

The greatest safe load is **25,632 pounds**.

EXERCISES

1. The distance between supports of a solid rectangular steel beam is 16 feet. The width of the beam is 2 inches and the depth is 4 inches. Find the greatest safe load. 3560 pounds

2. The width of a solid rectangular steel beam is 6 inches and the depth is 6 inches. The distance between supports is 18 feet. Find the greatest safe load. 21,360 pounds

3. If the distance between the supports of a solid steel rectangular beam is increased, does the greatest safe load increase or decrease? decreases

4. Two solid rectangular steel beams have the same depth and distance between supports. The width of one beam is twice that of the other. Compare the greatest safe loads.

5. The formula $S = \dfrac{1795Ad}{l}$ is used to find the greatest safe load of a steel I-beam. A steel I-beam has a length of 24 feet, a depth of 4 inches, and a cross-sectional area of 16 square inches. Find the greatest safe load.
4786.67 pounds

70 / Chapter 3

Simplify. (Pages 47–51)

1. $5n + 3 - 2n$ $3n + 3$

2. $2.4t - t + 1.3t$ $2.7t$

3. $k + 1\frac{1}{4} - 3\frac{1}{2}k - 1\frac{3}{4}$ $-2\frac{1}{2}k - \frac{1}{2}$

4. $16m - 14 - 6 - 3m + m$ $14m - 20$

5. $5(6 - x)$ $30 - 5x$

6. $15(3 + 2n)$ $45 + 30n$

7. $-8(k - 6)$ $-8k + 48$

8. $-12(-5 - 3x)$ $60 + 36x$

9. $\frac{1}{4}(2y + 3)$ $\frac{1}{2}y + \frac{3}{4}$

OBJECTIVES: To solve and check an equation with the variable in two or more
like terms on the same side

3.5   Like Terms in Equations

To solve and check an equation with the variable on one side that
requires use of the Distributive Property and combining like terms

In solving some equations, you should combine like terms first.

EXAMPLE 1   Solve and check: $5x - 7 - 3x = -19$

Solution:

$5x - 7 - 3x = -19$

Check: $5x - 7 - 3x = -19$

1 Combine like terms. ⟶ $5x - 7 - 3x = -19$

$5(-6) - 7 - 3(-6)$

$2x - 7 = -19$

$-30 - 7 + 18$

2 Add 7 to each side. ⟶ $2x - 7 + 7 = -19 + 7$

$-37 + 18$

$2x = -12$

-19

3 Divide each side by 2. ⟶ $\frac{2x}{2} = \frac{-12}{2}$

$x = -6$

P–1 **Solve.**

a. $7m - 3 + 4m = 30$ $m = 3$

b. $9k + 6 - 3k = -36$ $k = -7$

In solving some equations, you have to use the Distributive Property
first to remove parentheses. The check in the following examples is
left for you to do.

EXAMPLE 2   Solve and check: $26 = 5 - 3(2t + 3)$

Solution: 1 Use the Distributive Property. ⟶ $26 = 5 - 3(2t + 3)$

2 Simplify. ⟶ $26 = 5 - 6t - 9$

$26 = -4 - 6t$

Equations and Problem Solving / **71**

Lesson Resources
Warm–Up: p. M–23
Maintenance: See below.
Practice Worksheet 3.5
Manipulative 7, p. 552
Teaching Aids 1, 3
Transparencies 5, 6, 7, 50, 51,
  53, 54

Maintenance
1. Multiply: $(-12)(11)$
   Ans: −132 (Section 1.4)
2. Evaluate $-2n^2 + n - 1$
   when $n = 3$.
   Ans: −16 (Section 1.7)
3. Multiply: $(-4)(3)(-2)(n)$
   Ans: 24n (Section 2.3)
4. Simplify: $\frac{-(2 - n)}{n - 2}$
   Ans: 1 (Section 2.4)
5. Multiply and simplify:
   $2x(3x + 1)$
   Ans: $6x^2 + 2x$
   (Section 2.5)

Additional Examples
Example 1
Solve and check.
1. $-6t + 3 + 2t = 7$
   Ans: t = −1
2. $-23 = 12 - 8x + x$
   Ans: x = 5
3. $3a - a - 1.6 = -9.8$
   Ans: a = −4.1

Example 2
Solve and check.
1. $3(h + 1) - 2 = -5$
   Ans: h = −2
2. $y - 2(1 - y) = 34$
   Ans: y = 12
3. $8 = -6(2x + 1) + 4x - 2$
   Ans: x = −2

③ Add 4 to each side.	$\longrightarrow$	$26 + \mathbf{4} = -4 - 6t + \mathbf{4}$

$$30 = -6t$$

④ Divide each side by -6.	$\longrightarrow$	$\dfrac{30}{\mathbf{-6}} = \dfrac{-6t}{\mathbf{-6}}$

$$-5 = t$$

P–2 **Solve.**

a. $24 = 8 - 4(3y - 1)$ **b.** $17 = 12 + 5(r - 4)$
   $y = -1$ $r = 5$

EXAMPLE 3   Solve: $2(3x - 4) - 4(x + 1) = -15$

Solution:

① Use the Distributive Property twice.	$\longrightarrow$	$2(3x - 4) - 4(x + 1) = -15$

Step 2 may require special attention.

② Combine like terms.	$\longrightarrow$	$6x - 8 - 4x - 4 = -15$

Show that $-4(x + 1)$ equals $-4x - 4$.

$$2x - 12 = -15$$

③ Add 12 to each side.	$\longrightarrow$	$2x - 12 + \mathbf{12} = -15 + \mathbf{12}$

$$2x = -3$$

④ Divide each side by 2.	$\longrightarrow$	$\dfrac{2x}{2} = \dfrac{-3}{2}$

$$x = -1\tfrac{1}{2}$$

CLASSROOM EXERCISES

Write the equation that results after combining like terms. (Step 1, Example 1)

1. $5y + 6 - y = 13$ $4y + 6 = 13$

2. $-10 = 8t - 3t - 7$ $-10 = 5t - 7$

3. $-4r + 8 - r = 17$ $-5r + 8 = 17$

4. $10q - 6 - 2q - 3q = 7$

5. $-9.7 = 5.8m - 2.7m - 4.8$ $-9.7 = 3.1m - 4.8$

6. $-3.2a - 1.6 + 4.1a = 9.3$

a. *Write the equation that results after removing parentheses.*
b. *Then write the equation that results after combining like terms.*
(Steps, 1 and 2 in Examples 2 and 3)

7. $3(4x - 1) + 6 = 10$

8. $-5(2x + 3) - 7 = 12$

9. $\frac{1}{3} - 2(q - 5) + \frac{1}{4}q = 6$

10. $-7 = -\frac{1}{2}(4r - 6) + 13$

11. $\frac{3}{4}(x - 2) + 2(3x + \frac{1}{4}) = -11$

12. $5(k - \frac{1}{3}) - 3(2k + \frac{1}{3}) = 19$

13. $-6 = 4(3 - 2t) + 5(t - 4)$

14. $-6 + 5w - 3(4 - w) = 25$

72 / *Chapter 3*

Additional Examples
Example 3
Solve.
1. $3(v + 2) + 2(v - 3) = -5$
   Ans: $v = -1$
2. $5(2x - 7) - 3(x + 8) = -31$
   Ans: $x = 4$
3. $-43 = -2(5 - 8y) + \frac{1}{2}(6y + 10)$
   Ans: $y = -2$

Additional Answers
Classroom Exercises
 4. $5q - 6 = 7$
 6. $0.9a - 1.6 = 9.3$
 7. a. $12x - 3 + 6 = 10$
    b. $12x + 3 = 10$
 8. a. $-10x - 15 - 7 = 12$
    b. $-10x - 22 = 12$
 9. a. $\frac{1}{3} - 2q + 10 + \frac{1}{4}q = 6$
    b. $-1\frac{3}{4}q + 10\frac{1}{3} = 6$
10. a. $-7 = -2r + 3 + 13$
    b. $-7 = -2r + 16$
11. a. $\frac{3}{4}x - \frac{3}{2} + 6x + \frac{1}{2} = -11$
    b. $6\frac{3}{4}x - 1 = -11$
12. a. $5k - \frac{5}{3} - 6k - 1 = 19$
    b. $-k - 2\frac{2}{3} = 19$
13. a. $-6 = 12 - 8t + 5t - 20$
    b. $-6 = -8 - 3t$
14. a. $-6 + 5w - 12 + 3w = 25$
    b. $-18 + 8w = 25$

Evaluate each expression for the given value of the variable.
(Check, Example 1)

15. $3y - 8$ when $y = 5$ 7

16. $12 - 2p$ when $p = -3$ 18

17. $z - 5z - 3$ when $z = 2$ -11

18. $3r - 5 - 5r$ when $r = -1$ -3

19. $2(d - 3) - 5$ when $d = 4$ -3

20. $-3(c - 2) - 4$ when $c = -2$ 8

████████ **WRITTEN EXERCISES** ████████

Goal: To solve equations having like terms
Sample Problem: $-5(2x + 1) - 7x = 29$
Answer: $x = -2$

Solve. Check each answer. (Examples 1, 2, and 3)

1. $3x + 5 + x = 33$ $x = 7$

2. $2y - 8 - 5y = -47$ $y = 13$

3. $-z + 1.2 - 3z = 2.8$ $z = -0.4$

4. $4.6a - 5.8a - 30 = -12$

5. $37 = 8x - 19 - 15x$ $x = -8$

6. $-26 = 22 - c - 5c$ $c = 8$

7. $\frac{1}{4}(b + 3) = 2\frac{1}{2}$ $b = 7$

8. $15 = 3(\frac{2}{3}d - \frac{1}{3})$ $d = 8$

9. $-2(x + 1) - 7 = 15$ $x = -12$

10. $-3(4n + 1) + 8n = -35$

11. $-32 = -5(f - 2) + 2f$ $f = 14$

12. $15 - 4(2g - 3) + 4g = 39$

13. $(4k - 3) - (7k + 8) = 6\frac{1}{2}$ $k = -5\frac{5}{6}$

14. $17\frac{1}{2} = \frac{1}{3}(y + 1) - (2y - \frac{1}{2})$

15. $92 = (m + 4) - (3m - 12)$ $m = -38$

16. $2(3 - 2x) - 4(x + 4) = -118$

17. $-0.5(3n + 3) - 2(n + 0.6) = 0.8$ $n = -1$

18. $-7 = 6(4 + 0.3) + 0.5(2r - 5)$

MIXED PRACTICE

19. $\frac{1}{2}x + 6 + 4x = 5\frac{1}{2}$ $x = -\frac{1}{9}$

20. $7(n + 1) = 21$ $n = 2$

21. $-12 + 2(5 - h) - 3h = -57$ $h = 11$

22. $-3.6b + 26 - 0.8b = -84$

23. $19.2 - 23.8 = 5d - 14.8 - 7d$ $d = -5.1$

24. $-2(5 - x) - 6(x + 3) = -102$

25. $0.9 - 5.7 = 10.4 - e - 3e$ $e = 3.8$

26. $16 = 2r - 4(r - 6) + 6r$

27. $-85 = 5t - 9t - (14 - 2t)$ $t = 35\frac{1}{2}$

28. $4z + 6z - (20 - 8z) = 25$

29. $2\frac{1}{2}y - 3(y + \frac{1}{4}) + 6y = 13$ $y = 2\frac{1}{2}$

30. $70 = 6(2t - 1) - 3(2t + 1)$

MORE CHALLENGING EXERCISES

Find each answer without solving the equation.

31. If $2x + 3x = 15$, what is the value of $10x$?   30

32. If $3(x - 4) - (x - 4) = -3$, what is the value of $6(x - 4)$? -9

Equations and Problem Solving / **73**

Assignment Guide
Basic p. 73: 1–29 odd
Average p. 73: 1–31 odd
Above Average p. 73:
1–31 odd, 32

Additional Answers
Written Exercises
 4. $a = -15$
10. $n = 8$
12. $g = -3$
14. $y = -10$
16. $x = 13\frac{1}{2}$
18. $r = -30.3$
22. $b = 25$
24. $x = 18.5$
26. $r = -2$
28. $z = 2\frac{1}{2}$
30. $t = 13\frac{1}{6}$

Problem–Solving Skills
Making a comparison
(Ex. 31–32)
Using logical reasoning
(Ex. 31–32)

Critical Thinking
See Exercises 31 and 32.

Maintenance
1. Divide: $(-16) \div \left(\frac{2}{5}\right)$
   Ans: -40   (Section 1.5)
2. Simplify: $17 - y + 12$
   Ans: $29 - y$
   (Section 2.2)
3. Simplify: $(-3p)(-2m)(k)$
   Ans: $6pmk$ (Section 2.4)
4. Factor: $5rt - rk$
   Ans: $r(5t - k)$
   (Section 2.5)
5. Simplify: $5n - 2m + 3n - 7m$
   Ans: $8n - 9m$
   (Section 2.6)

Additional Examples
Example 1
Write an algebraic expression for each word expression. Use the variable n.
1. The sum of the height of the tower and 35 feet.
   Ans: $n + 35$
2. $12 less than the list price of the cabinet
   Ans: $n - 12$
3. The product of the width and 17.2
   Ans: $17.2n$

OBJECTIVES: To write an algebraic expression for a word expression involving only one operation

PROBLEM SOLVING AND APPLICATIONS

3.6   From Words to Symbols

To write and solve an equation for a word sentence involving only one operation

In order to solve word problems by algebra it is necessary to translate key words of the problem to algebraic symbols. Table 1 shows some word expressions and the corresponding algebraic expressions.

Table 1

Operation	Word Expression	Algebraic Expression
Addition	The <u>sum</u> of the number of years, y, and 12	$y + 12$
	Thirteen <u>plus</u> the number of centimeters, c	$13 + c$
	$6 <u>more than</u> the cost of a bus ticket, b	$b + 6$
	Some number, n, <u>increased by</u> $3\frac{2}{3}$	$n + 3\frac{2}{3}$
See NOTE below. Subtraction	The number of hours, h, <u>decreased by</u> $5\frac{1}{2}$	$h - 5\frac{1}{2}$
	The <u>difference</u> between some number, x, and 9.4	$x - 9.4$
	$54 <u>minus</u> the amount earned, a	$54 - a$
	The number of students, s, <u>less</u> 5	$s - 5$
	Four <u>less than</u> the number of lemons, ℓ	$\ell - 4$
Multiplication	The <u>product</u> of the number of feet, f, and 4.9	$4.9f$
	Some number, q, <u>multiplied by</u> $\frac{2}{3}$	$\frac{2}{3}q$
	Twenty-four <u>times</u> the number of meters, m	$24m$
	The number of people, p, <u>doubled</u>	$2p$
Division	The <u>quotient</u> of some number, k, and 15	$\frac{k}{15}$
	The number of miles, m, <u>divided by</u> 2.9	$\frac{m}{2.9}$

"4 less the number of lemons" is $4 - \ell$.

NOTE: The usual agreement is to write "difference between" in the order in which the numbers occur.

EXAMPLE 1   Write an algebraic expression for this word expression.

The record number scored decreased by 14 points

Solution: [1] Choose a variable. ⟶ Let $n =$ the record number scored

[2] Identify the operation. ⟶ subtraction ("decreased by")

[3] Write an algebraic expression.

Think: **The record number scored decreased by 14 points**

Translate: n $-$ 14

Table 2 shows some sentences and their corresponding equations.

Table 2	Word Sentence	Equation
	19.8 **is** the <u>sum</u> of some number and 4.3.	$19.8 = t + 4.3$
	An unknown number of cars <u>decreased by</u> 37 **equals** 196.	$n - 37 = 196$
	The <u>product</u> of the number of hours worked and the rate per hour, $5.25, **is** $183.75	$5.25w = 183.75$
	The <u>quotient</u> of the total cost and the number of payments, 36, **equals** $173.50.	$\frac{c}{36} = 173.50$

EXAMPLE 2   Write an equation for this sentence. Then solve the equation.
The list price of a lamp less a discount of $3.00 equals the sale price of $26.95.

Solution: 1. Choose a variable. ————————————→ Let p = the list price.
2. Identify the operation. ————————————→ subtraction ("less")
3. Write an equation.   Note how the word sentence directly translates to an equation.

Think: **The list price less the discount equals the sale price.**

Translate: p $-$ 3.00 $=$ 26.95

4. Solve the equation. ————————————→ $p - 3.00 + 3.00 = 26.95 + 3.00$
$p = 29.95$

The price of the lamp is $29.95.

Note carefully how these two subtraction expressions differ.

<u>y less 5</u>   **is not the same as**   <u>y less than 5</u>
$y - 5$ **is not the same as** $5 - y$.

CLASSROOM EXERCISES

For Exercises 1–10, write an algebraic expression for each word expression.
Use the variable n. (Table 1 and Example 1)

1. Twelve <u>more than</u> a number of stamps $n + 12$

2. Four dollars <u>less than</u> the cost of two tickets $n - 4$

3. The total number of points <u>divided by</u> 8 $\frac{n}{8}$

4. A savings account balance <u>increased by</u> $1200 $n + 1200$

Equations and Problem Solving / 75

5. Eighteen hundred entries in the race <u>times</u> the entry fee $1800n$

6. The number of swimmers <u>decreased by</u> eighteen $n - 18$

7. The <u>quotient</u> of 1200 miles and the number of gallons of fuel $\frac{1200}{n}$

8. The number of shoppers over the last year <u>tripled</u> $3n$

9. The <u>difference</u> between twenty-six and the number of fish caught $26 - n$

10. Six thousand <u>plus</u> the number of records sold this year $6000 + n$

Choose a variable and write an equation for each word sentence. Then solve the equation. (Table 2 and Example 2) For Ex. 11-18, the choice of variable may vary.

11. The low temperature for the month increased by 27° is 50°.

12. The high temperature for the day decreased by 18° is 78°.

13. The total number of points divided by 20 games equals 18 points per game.

14. The cost of an unknown number of tickets multiplied by 4 dollars equals 940 dollars.

15. The total number of persons on the flight, 104, is the number of passengers plus a flight crew of eleven.

16. The number of kilometers driven divided by 75 liters equals 10.2 kilometers per liter.

17. The average number of centimeters of rainfall less 9.4 centimeters is 69.3 centimeters, the amount of rainfall for the past year.

18. The total amount of this year's sales tripled would represent $372,000 in sales next year.

WRITTEN EXERCISES

Goal: To write and solve an algebraic equation for a word sentence involving one operation

Sample Problem: The low temperature for one day of −35° equals the high temperature decreased by 12°.

Answer: $-35 = h - 12$; $h = -23°$

Choose a variable and write an expression for each word expression.
(Example 1) For Ex. 1-20, the choice of variable may vary.

1. Twice the number of taxicabs $2t$

2. Three times the number of musicians $3m$

3. The number of TV viewers increased by 25,000 $v + 25,000$

4. Twenty-nine less than the number of pro golfers $g - 29$

5. The number of meters decreased by 15.8 $m - 15.8$

6. Five hundred sixty dollars more than the amount of a loan $a + 560$

7. A bowler's total score divided by 28 games $\frac{b}{28}$

8. The quotient of the number of kilometers sailed and 3.2 hours $\frac{k}{3.2}$

9. The difference between the number of accidents reported and 512 $a - 512$

10. The product of the number of cartons shipped and 14.3 kilograms $14.3c$

Choose a variable and write an equation for each word sentence. Then solve the equation. (Example 2)

11. The quotient of the total cost and 24 payments is 86 dollars.

12. The product of the number of tickets sold and $6.50 equals $6916.

13. The width of a rectangle, 14.2 meters, is 5.8 meters less than the length.

14. The quotient of the total cost and 288 cans of tennis balls equals the cost per can of $1.75.

15. The low temperature of −5.3°C is 7.9 degrees less than the high temperature for the day.

16. The high temperature of −12.6°F is 8.7 degrees more than the low temperature for the day.

17. The amount of the sale multiplied by 0.06 equals the sales tax of $33.60.

18. Fifty-four dollars is $7.75 more than the cost of the items one year ago.

19. The sum of Tim's weekly salary and his commission of $283 is $561.

20. The difference between the number of balloons purchased and the 283 balloons sold is 104.

MIXED PRACTICE

Match each word expression or word sentence in Exercises 21–28 with the corresponding expression or equation from **a–t.**

21. Thirty-five more than the number of horses   l

22. The number of shoppers increased by 4500 equals 12,700.   p

23. The number of books in a collection less 485 books is 714.   n

24. The paper carrier's daily supply of papers less 17, the number not sold   r

25. Three times the number of state senators is 72.   o

26. The number of players on the squad who played plus 24, the number who did not play   i

27. Twice the number of planes landing each day   k

28. A golfer's score of 86 is 3 strokes less than his handicap score.   m

a. $17 - w$

b. $2p = 58$

c. $28.6 + 2.3 = m$

d. $4500n = 12,700$

e. $n + 24 = 56$

f. $86 - 3$

g. $485 - r = 714$

h. $3 + n = 72$

i. $r + 2.3 = 28.6$

j. $24 + q$

k. $2(y)$

l. $t + 35$

m. $s - 3 = 86$

n. $714 = k - 485$

o. $72 = 3s$

p. $12,700 = 4500 + x$

q. $r + 2\frac{2}{3} = 84\frac{1}{8}$

r. $p - 17$

s. $2\frac{3}{4} + b$

t. $76\frac{5}{8} + 2\frac{3}{4}$

Equations and Problem Solving / **77**

Lesson Resources
Warm–Up: p. M–23
Maintenance: See below.
Practice Worksheet 3.7
Transparencies 6, 7, 9

Maintenance
1. Subtract: $7 - 9$
   Ans: -2 (Section 1.3)
2. Evaluate: $12 + 8 \div 2^2$
   Ans: 14 (Section 1.7)
3. Write True or False.
   6.8 is a rational number.
   Ans: True (Section 2.1)
4. Factor: $16ac - 4a$
   Ans: $4a (4c - 1)$
   (Section 2.5)
5. On a winter day, the low temperature was $-6°F$. The high temperature was -8 times the low temperature. Find the day's high temperature.
   Ans: 48°F (Section 1.4)

Additional Examples
Example 1
1. The Tigers' score of 12 was 9 less than the Mustangs' score. Find the Mustangs' score.
   Ans: 21
2. The sum of Tracie's first jump, 8.7 feet, and her second jump was 17.5 feet. Find the length of her second jump.
   Ans: 8.8 feet

OBJECTIVES: To write an algebraic expression for a word expression involving more than one operation

PROBLEM SOLVING AND APPLICATIONS

3.7 Using Real Numbers

To write and solve an equation for a word sentence involving more than one operation

Thus far in the text, you have solved word problems involving positive and negative numbers and by replacing variables in formulas. Many word problems in algebra can be solved by writing an equation. The following steps for doing this are illustrated in the examples.

1. **Understand** the problem: Read the problem carefully to determine what information is given and what you are asked to find. Use this information to choose a variable to represent the unknown(s).

2. Make a **plan:** In algebra, the plan usually involves writing an equation.

3. **Solve** the problem.

4. **Check** your results with the statements in the problem. Write the answer in a complete sentence.

EXAMPLE 1   A meteorologist reports that the high temperature for yesterday increased by 6.2 degrees equals the high temperature for today. The high temperature for today is $-3°C$. Find the high temperature for yesterday.

Solution:

1. Choose a variable to represent the unknown. ——————→ Let $n =$ yesterday's high temperature.

2. Write an equation.

   Think: <u>Yesterday's high</u> <u>increased by</u> <u>6.2°</u> <u>equals</u> <u>today's high</u>.

   Translate: n $+$ 6.2 $=$ -3

3. Solve the equation. ——→ $n + 6.2 - 6.2 = -3 - 6.2$
                                        $n = -9.2$

4. **Check:** When -9.2 is increased by 6.2, is the sum -3?
        Does $-9.2 + 6.2 = -3$?   Yes ✔

   The high temperature for yesterday was $-9.2°C$.

78 / *Chapter 3*

Sometimes you use more than one operation to solve a word problem.

EXAMPLE 2   Twice the cost of a pair of skis decreased by $40 equals $246. Find the cost of the skis.

Solution:

1. Choose a variable to represent the unknown. ⟶ Let s = the cost of the skis.

2. Write an equation.

   Think: Twice the cost of the skis   decreased by $40   equals $246.

   Translate: $2s$ —   40   =   246

3. Solve the equation. ⟶ $2s - 40 + 40 = 246 + 40$

   $2s = 286$

   $\dfrac{2s}{2} = \dfrac{286}{2}$

   $s = 143$

   Note how the word sentence from the problem can be directly translated to an algebraic equation.

4. The check is left for you.

   The cost of the skis is $143.

Using Algebra to Solve Word Problems

1. Choose a variable. Use the variable to represent the unknowns.
2. Write an equation for the problem.
3. Solve the equation.
4. Check your answer with the statements in the original problem. Answer the question.

CLASSROOM EXERCISES

For Exercises 1–6:   For Ex. 1-6, the choice of variable may vary.

a. *Choose a variable to represent the unknowns.*
b. *Write an equation for each problem.* (Example 1)

1. On a certain weekend, the Waldron family drove 12 miles less on Sunday than on Saturday. They drove 75 miles on Sunday. How many miles did they drive on Saturday?

2. Apex stock's net change on Tuesday was $\frac{3}{4}$ of a point more than its net change on Monday. Its net change on Tuesday was $-\frac{1}{8}$ of a point. Find the net change on Monday.

Equations and Problem Solving / **79**

Additional Examples
Example 2
1. Three times the cost of a small pizza less $10 equals $2. Find the cost of the small pizza.
   Ans: $4
2. The quotient of the hourly wage and 5 increased by $16 equals $18.40. Find the hourly wage.
   Ans: $12

Common Error
Students have difficulty getting started with the solutions to the word problems.

Example
Written Exercises 1–11

Prescription
Remind students to read the problem carefully and decide what the unknown is and what has to be done to find it before they try to write an equation.

Additional Answers
1. a. Let m = number of miles driven on Saturday.
   b. $75 = m - 12$
2. a. Let n = net change on Monday.
   b. $-\frac{1}{8} = \frac{3}{4} + n$

3. A company's loss (−) for August divided by −1.2 equals the company's profit of $5000 for September. Find the loss for August.

4. The record high temperature for Moline, Illinois, multiplied by −0.125 equals the record low temperature of −12°F. Find the record high temperature.

(Example 2)

5. The number of letters mailed on Tuesday was 16 less than twice the number of letters mailed on Monday. There were 132 letters mailed on Tuesday. How many letters were mailed on Monday?

6. The number of tourists visiting Cosmic Caves this year is 5 more than one-half the number of tourists that visited last year. The number of tourists this year is 2410. How many tourists visited last year?

WRITTEN EXERCISES

Goal: To use equations to solve word problems

Sample Problem: A new car dealer sold 8 more cars in April than in March. The dealer sold 34 cars in April. Find the number of cars sold in March.

Answer: $n + 8 = 34$; $n = 26$ cars

For Exercises 1–11:   For Ex. 1-11, the choice of variable may vary.
a. *Choose a variable to represent the unknowns.*
b. *Write an equation for each problem.*
c. *Solve the problem.* (Example 1)

1. On Saturday, the members of the Math Club washed 7 more cars in the afternoon than in the morning. They washed 18 cars in the afternoon. Find the number of cars they washed in the morning.

2. Mt. McKinley in Alaska has an elevation of 20,320 feet above sea level. The difference in elevation between Mt. McKinley and Death Valley in California is 20,602 feet. Find the elevation of Death Valley.

80 / *Chapter 3*

80

3. The record low temperature for Denver, Colorado, multiplied by −3.5 equals the record high temperature of 105°F. Find the record low temperature.

(Example 2)

5. On a winter day, four times the reading on a thermometer increased by 5° equals the wind chill temperature. The wind chill temperature is −23°F. What is the reading on the thermometer?

MIXED PRACTICE

7. The number of people attending graduation this year is 12 less than twice the number that attended last year. This year, there are 326 people. Find how many people attended last year.

9. The number of hours needed to complete a job divided by 6 employees equals an average workday of 7.5 hours for each employee. Find the number of hours needed to complete the job.

10. One half the cost of a new car decreased by $1027 equals the cost of the same car after three years. The car will cost $3134 after three years. What is the cost of the new car?

11. A company's loss (−) for 1985 multiplied by −2.5 equals the company's profit of $30,000 for 1986. Find the loss for 1985.

4. The quotient of the number of tourists and 9 tour buses equals the number of tourists on each bus. There are 35 tourists on each bus. How many tourists are there in all?

6. The number of books in the main library is 15 less than 3 times the number of books in the branch library. The main library has 12,000 books. Find the number of books in the branch library.

8. On Tuesday, a nursery received a shipment of 42 new trees. The number of trees the nursery had increased by the number of new trees equals a total of 74 trees. How many trees did the nursery have before the shipment?

Quiz Sections 3.5–3.7
After completing Sections 3.5–3.7, you may want to administer a quiz covering the same sections. See the quiz provided on page 72 in the *Teacher's ResourceBank*.

Focus on Reasoning: Logic Tables

Logical reasoning is used to solve many types of problems.

EXAMPLE   Jeff, Cathy, and Amy work on the school newspaper. One is the editor, one is a reporter, and one is a word processor. Use the clues below to find each person's job.

CLUE A Amy's only exercise is jogging.
CLUE B Jeff and the editor play tennis together.
CLUE C Amy and the reporter are cousins.

Solution: Make a table to show all the possibilities. Use an X to show that a possibility cannot be true. Use a ✔ when you are certain that a possibility is true.

1 Read Clue A. Can you reach any conclusion about Amy's job?

2 Read Clue B. Is Jeff the editor?
Place an X next to Jeff's name in the editor column.
Read Clue A and Clue B. Is Amy the editor?
Place an X next to Amy's name in the editor column.
Who must be the editor?
Place a ✔ next to Cathy's name in the editor column. Place X's next to Cathy's name in the reporter and word processor columns.

	Editor	Reporter	Word Processor
Jeff	X		
Cathy	✔	X	X
Amy	X		

3 Read Clue C. Is Amy the reporter?
Place an X in the reporter column and a ✔ in the word processor column next to Amy's name.
Who must be the reporter?

**Jeff: reporter   Cathy: editor
Amy: word processor**

	Editor	Reporter	Word Processor
Jeff	X	✔	X
Cathy	✔	X	X
Amy	X	X	✔

Check the answer with the clues.

EXERCISES

Copy the table. Use the clues to complete the table to solve the problem.

1. Alice, Nathan, and Marie play in the school band. One plays the drum, one plays the saxophone, and one plays the flute.

CLUE A Alice is a senior.
CLUE B Alice and the saxophone player practice together after school.
CLUE C Nathan and the flute player are sophomores.

	Drum	Saxophone	Flute
Alice	✓	X	X
Nathan	X	✓	X
Marie	X	X	✓

Who plays each instrument? Alice: drum; Nathan: saxophone; Marie: flute

2. Brenda has three cats named Tiki, Moby, and Copper. One is a Persian, one is a Siamese, and one is a Himalayan.

CLUE A Copper's favorite food is Fish Treats.
CLUE B The Siamese will only eat Liver Bites.
CLUE C Tiki will not eat with the Himalayan.
CLUE D The Himalayan does not have a favorite food.

	Persian	Siamese	Himalayan
Tiki	X	✓	X
Moby	X	X	✓
Copper	✓	X	X

What is the name of Brenda's Siamese cat? Tiki

Solve. Make a table to show all the possibilities.

3. Alex, Dee, Rod, and Sue are athletes. One is on the baseball team, one is on the soccer team, one is on the track team, and one is on the golf team.

CLUE A Alex is taller than the soccer player.
CLUE B Dee and Rod do not know how to play golf.
CLUE C Neither Alex nor Dee has time to play baseball.
CLUE D Sue's sport does not use a ball.

Who plays on the soccer team?
Dee

4. Chuck, Lillian, Denise, and Brad each have a different hobby. The hobbies are stamp collecting, oil painting, kite flying, and photography.

CLUE A Chuck and Denise are older than the stamp collector.
CLUE B Denise and Brad do not paint or take photographs.
CLUE C Chuck and the photographer are cousins.

What is Lillian's hobby? Photography

Additional Answers

3.

	Baseball	Soccer	Track	Golf
Alex	X	X	X	✓
Dee	X	✓	X	X
Rod	✓	X	X	X
Sue	X	X	✓	X

4.

	Stamps	Painting	Kites	Photography
Chuck	X	✓	X	X
Lillian	X	X	X	✓
Denise	X	X	✓	X
Brad	✓	X	X	X

IMPORTANT TERMS	Equation (p. 58)	Root (p. 58)
	Solution (p. 58)	Equivalent equations (p. 58)

IMPORTANT IDEAS

1. *Addition Property for Equations:* Adding the same real number to each side of an equation forms an equivalent equation.

2. *Subtraction Property for Equations:* Subtracting the same real number from each side of an equation forms an equivalent equation.

3. *Multiplication Property for Equations:* Multiplying each side of an equation by the same nonzero real number forms an equivalent equation.

4. *Division Property for Equations:* Dividing each side of an equation by the same nonzero real number forms an equivalent equation.

5. To solve an equation that involves more than one operation, the Addition or Subtraction Property for Equations is generally used before the Multiplication or Division Property.

6. In equations that contain parentheses, use the Distributive Property first.

7. The words <u>sum</u>, <u>plus</u>, <u>more than</u>, and <u>increased by</u> suggest the operation of addition.

8. The words <u>decreased by</u>, <u>difference</u>, <u>minus</u>, <u>less</u>, and <u>less than</u> suggest the operation of subtraction.

9. The words <u>product</u>, <u>multiplied by</u>, <u>times</u>, and <u>doubled</u> suggest the operation of multiplication.

10. The words <u>quotient</u> and <u>divided by</u> suggest the operation of division.

11. Using Algebra to Solve Word Problems
   a. Choose a variable. Use the variable to represent the unknowns.
   b. Write an equation for the problem.
   c. Solve the equation.
   d. Check your answer with the statements in the original problem. Answer the question.

Chapter Test

Two Chapter Tests (Form A and Form B) are provided on pages 73–76 in the *Teacher's ResourceBank*.

CHAPTER REVIEW

For Exercises 1–24, solve each equation. Check each answer.

SECTION 3.1

1. $x + 23 = 16$ $x = -7$

2. $y + 29 = -18$ $y = -47$

3. $b - 47 = -28$ $b = 19$

4. $d - 37 = 49$ $d = 86$

5. $-28 = 14 + f$ $f = -42$

6. $76 = -32 + g$ $g = 108$

7. $5.6 = t - 24.7$ $t = 30.3$

8. $-12.3 = x + 19.8$ $x = -32.1$

84 / Chapter 3

SECTION 3.2

9. $-7k = 63$ $k = -9$ **10.** $-96 = 12n$ $n = -8$ **11.** $-8 = \frac{m}{14}$ $m = -112$ **12.** $\frac{x}{-8} = 19$ $x = -152$

13. $-\frac{3}{4}t = -27$ $t = 36$ **14.** $60 = -\frac{12}{4}z$ $z = -20$ **15.** $12y = -99.6$ $y = -8.3$ **16.** $131.4 = -9r$ $r = -14.6$

SECTION 3.3

17. $4z - 3 = 42$ $z = 11\frac{1}{4}$ **18.** $6t - 14 = 38$ $t = 8\frac{2}{3}$ **19.** $-68 = 8m + 12$ $m = -10$ **20.** $-54 = 18 + 12p$ $p = -6$

21. $78 - 6x = 34$ $x = 7\frac{1}{3}$ **22.** $69 = 102 - 8n$ $n = 4\frac{1}{8}$ **23.** $-22 + \frac{v}{2.4} = -46$ $v = -57.6$ **24.** $-9 = \frac{w}{1.9} + 48$ $w = -108.3$

SECTION 3.4

For each problem, find the rate, r, or the time, t. Round your answer to the nearest tenth.

25. A wildebeest was observed racing a distance of 86.4 meters in 4.5 seconds. Find the speed of the wildebeest in meters per second. 19.2 meters per second

26. The speed of a radio signal is 186,000 miles per second. How long does it take for a radio signal to travel from the earth to Mars, which is about 35,000,000 miles away? 188.2 seconds or about 3.1 minutes

For each problem, find the width, w, or length, ℓ. Round your answer to the nearest tenth.

27. The perimeter of a wall mirror is 204 inches. The width is 18.1 inches. Find the length of the mirror. 83.9 inches

28. A steel ribbon 3.8 meters long is exactly long enough to wrap around a packing carton once. The base of the carton is 1.2 meters long. How wide is the base of the carton? 0.7 meters

SECTION 3.5

Solve each equation. Check each answer.

29. $8x + 6 - 5x = 39$ $x = 11$

30. $56 = 12k - 8k$ $k = 14$

31. $-19.3y - 16 + 5.9y = -83$ $y = 5$

32. $-55 = 2.4n + 74 - 15.3n$ $n = 10$

33. $35 = 4(t - 2) - 6t$ $t = -21.5$

34. $7z - 3(z + 7) = -67$ $z = -11.5$

35. $-28 = (3p - 2) - 2(5p + 12)$ $p = \frac{2}{7}$

36. $(18 - x) - (3x + 4) = -48$ $x = 15\frac{1}{2}$

Equations and Problem Solving / **85**

SECTION 3.6

For Exercises 37–40, choose a variable and write an equation for the word sentence. Then solve the equation. Variables may vary.

37. The total distance traveled less 25.6 kilometers equals 278.3 kilometers.

38. Twenty-four times the cost of each box equals $30.72.

39. A bowler's score divided by 5 games equals her average game score of 168.

40. The low temperature for one day increased by 24° equals the high temperature of 30.5°.

SECTION 3.7

For Exercises 41–44:
a. *Choose a variable to represent the unknowns.*
b. *Write an equation for each problem.*
c. *Solve.* Variables may vary.

41. The number of rainy days last year increased by 6 equals the previous recorded high of 82 rainy days in a year. Find the number of rainy days last year.

42. The record low temperature for Raleigh, North Carolina, divided by 0.04 equals the record high temperature of 100°F. Find the record low temperature.

43. Three times the reading on a thermometer decreased by 7 degrees equals the wind chill temperature of −31°F. Find the thermometer reading.

44. Three less than two times the number of pages in the first edition of a book equals the number of pages in the second edition of the book. The second edition contains 210 pages. Find the number of pages in the first edition.

CUMULATIVE REVIEW: CHAPTERS 1–3

Simplify. (Section 1.1)

1. $|-0.07|$ 0.07   **2.** $-|2\frac{1}{5}|$ $-2\frac{1}{5}$ **3.** $-|-12|$ -12 **4.** $|-(-134)|$ 134

Add. (Section 1.2)

5. $-8 + (-14)$ -22 **6.** $-17 + (-35)$ -52 **7.** $-139 + 39$ -100 **8.** $3.7 + (-1.6)$ 2.1

Subtract. (Section 1.3)

9. $16 - 28$ -12 **10.** $-10 - 10$ -20 **11.** $5\frac{2}{3} - (-1\frac{1}{3})$ 7   **12.** $-19 - (-21)$ 2

Multiply. (Section 1.4)

13. $(-4)(7)$ -28 **14.** $(2)(-26)$ -52 **15.** $(-1.7)(-100)$ 170   **16.** $(-5)(-16)$ 80

Divide. (Section 1.5)

17. $\frac{-20}{-5}$ 4   **18.** $\frac{-72}{9}$ -8 **19.** $-330 \div 22$ -15 **20.** $-18 \div (-1\frac{1}{2})$ 12

Evaluate. (Section 1.6)

21. $4 - 7 + 5$ 2   **22.** $-32 \div 2(-4)$ 64   **23.** $-11 + (-5)(-2)$ -1 **24.** $12 \div (-3) - 18$ -22

Evaluate each expression. (Section 1.7)

25. $2n^2 - 3$ when $n = 4$ 29   **26.** $y^2 + 7y$ when $y = -3$ -12 **27.** $-2x + 3x^2$ when $x = -5$ 85

Refer to the set of numbers below. Then list the numbers named in Exercises 28–30. (Section 2.1)

$$\{-5, -\sqrt{3}, -1\frac{2}{3}, 0, 4, \pi, \frac{18}{3}\}$$

28. Rational numbers $-5, -1\frac{2}{3}, 0, 4, \frac{18}{3}$ **29.** Integers $-5, 0, 4, \frac{18}{3}$ **30.** Real numbers $-5, -\sqrt{3}, -1\frac{2}{3}, 0, 4, \pi, \frac{18}{3}$

Simplify. (Section 2.2)

31. $-11 + 18 - 19 + 12$ 0   **32.** $6.4 - 8.2 + p - 0.4$ $p - 2.2$ **33.** $-44 + k - w + 62$ $k - w + 18$

Multiply. (Section 2.3)

34. $(-3)(6)(-2)$ 36   **35.** $(-2)(-2)(x)(-2)$ $-8x$ **36.** $(-\frac{2}{3})(t)(18)(p)$ $-12tp$

Simplify. (Section 2.4)

37. $(8xy)(-4x)$ $-32x^2y$ **38.** $4x - (y - 3)$ $4x - y + 3$ **39.** $(2a)(-3ab)(-5b)$ $30a^2b^2$

Cumulative Review / **87**

Cumulative Test
A cumulative test covering Chapters 1–3 is provided on pages 77–80 in the *Teacher's ResourceBank*.

Multiply and simplify. (Section 2.5)

40. $3(x + y)$ $3x + 3y$ **41.** $-6(8a - 1)$ $-48a + 6$ **42.** $(-2x + 7)5x$ $-10x^2 + 35x$ **43.** $(-16x + 8y)(-\frac{1}{2})$ $8x - 4y$

Factor. (Section 2.5)

44. $5x - 5y$ $5(x - y)$ **45.** $12r - 9$ $3(4r - 3)$ **46.** $21ab + 7a$ $7a(3b + 1)$ **47.** $-6x^2 - 18x$ $-6x(x + 3)$

Simplify. (Section 2.6)

48. $23x^2 - x^2$ $22x^2$ **49.** $9a - 12 + 7a + 4$ $16a - 8$ **50.** $3y^2 - 7y + 2 + y^2 - y$ $4y^2 - 8y + 2$

Solve. Check each answer. (Sections 3.1 through 3.3)

51. $x - 3 = -7$ $x = -4$ **52.** $0.2 = k + 0.9$ $k = -0.7$ **53.** $-2 = \dfrac{h}{-2}$ $h = 4$ **54.** $-8x = 32$ $x = -4$

55. $-36 = \frac{2}{3}y$ $y = -54$ **56.** $5a - 3 = -7$ $a = -\frac{4}{5}$ **57.** $-4 - 3x = 13$ $x = -\frac{17}{3}$ **58.** $\dfrac{m}{6} + 19 = -2$ $m = -126$

Solve each problem. (Section 3.4)

59. The perimeter of a picture frame is 37 inches. The length of the frame is 11 inches. Find the width. $7\frac{1}{2}$ inches

60. A certain car traveled a distance of 1120 miles at an average speed of 40 miles per hour. How long did it take the car to travel that distance? 28 hours

Solve. Check each answer. (Section 3.5)

61. $-n + 8 + 4n = 10$ $n = \frac{2}{3}$

62. $18 = 5x - (2x + 6)$ $x = 8$

63. $10 = 4 - 2(3 - 2x)$ $x = 3$

64. $3(x - 5) + 2(3 - 2x) = 18$ $x = -27$

Choose a variable and write an equation for each word sentence. Then solve the equation. (Section 3.6) Variables may vary.

65. The quotient of the number of apples and 17 is 5.

66. The product of 9 and the width of a rectangle is equal to 23.4 units.

Choose a variable to represent the unknowns. Write an equation for each problem. Solve. (Section 3.7) Variables may vary.

67. Lincoln High School Cheerleaders sold 54 fewer football programs this week than they sold last week. They sold 192 football programs this week. How many programs did they sell last week?

68. The number of trees planted by C. M. Grow's Nursery in April was 3 more than twice the number of trees planted by C. M. Grow's Nursery in March. Seventy-one trees were planted in April. Find the number of trees planted in March.

Additional Answers

65. Let n = the number of apples; $\frac{n}{17} = 5$; $n = 85$ apples

66. Let w = width of rectangle; $9w = 23.4$; $w = 2.6$ units

67. Let p = programs sold last week; $192 = p - 54$; $p = 246$ programs

68. Let n = number of trees planted in March; $71 = 2n + 3$; $n = 34$ trees

Problem Solving: One Variable

In nature, patterns can be found in the waves made by drops of rain in a still pool, in the veins of a leaf, and in the rings of a tree trunk. In mathematics, finding patterns is one method used in problem solving.

Maintenance
1. Add: $81.3 + (-26.1)$
   Ans: 55.2   (Section 1.2)
2. Simplify: $\frac{1}{3} + j - \frac{2}{3} + 3j$
   Ans: $4j - \frac{1}{3}$
   (Section 2.2)
3. Multiply: $(c)(-4)(2)(a)$
   Ans: $-8ac$ (Section 2.3)
4. Solve and check:
   $n - 8 = 26$
   Ans: $n = 34$ (Section 3.1)
5. A city's deficiency in rainfall
   for one year was -24 centimeters. Find the average deficiency per month.
   Ans: -2 centimeters
   (Section 1.5)

Additional Examples
Example 1
1. The cost of a bologna sandwich is four times the cost of
   a glass of milk (Condition 1).
   The total cost of the sandwich
   and glass of milk is $3.25
   (Condition 2). Find the cost of
   each.
   Ans: sandwich: $2.60
   milk: $0.65
2. The gravel in a truck weighs
   25,000 pounds more than the
   weight of the truck alone
   (Condition 1). The total weight
   of the truck and the gravel is
   61,000 pounds (Condition 2).
   Find the weight of each.
   Ans: truck: 18,000 pounds
   gravel: 43,000 pounds

OBJECTIVES: To solve and check word problems with two conditions and involving
two or more operations

PROBLEM SOLVING AND APPLICATIONS

4.1   Using Conditions

To help students
achieve success,
Condition 1 appears
before Condition 2
in each stated
problem of this
lesson.

To solve word problems, it is helpful to identify the conditions in the
problem. One condition is used to represent the unknowns of the
problem. Call this Condition 1. Another condition is used to write the
equation. Call this Condition 2. Then solve the equation.

EXAMPLE 1

One model of subcompact car has 71 cubic feet more
passenger space than luggage space (Condition 1).
The total amount of passenger and luggage space is
87 cubic feet (Condition 2). Find the amount of both
types of space.

Solution:

1. Use Condition 1 to represent
   the unknowns. $\longrightarrow$ Let $s =$ the luggage space.
                                          Then $s + 71 =$ the passenger space.

2. Use Condition 2 to write an equation.
   Think: Luggage space + Passenger space = Total space
   Translate: s $+$ $s + 71$ $=$ 87

3. Solve the equation. $\longrightarrow$ $2s + 71 = 87$
                                            $2s = 16$
                                            $s = 8$ *Don't forget to
                                                            find $s + 71$.*
                                            $s + 71 = 79$

4. **Check:** Condition 1 Are there 71 more cubic feet of passenger space than
                           luggage space?   Does $79 = 8 + 71$?   Yes ✔
             Condition 2 Does the total amount of space equal 87 cubic feet?
                          Does $8 + 79 = 87$?   Yes ✔

The passenger space is 79 cubic feet, and the luggage space is 8 cubic feet.

Steps for Solving Word Problems

1. Identify Condition 1. Use Condition 1 to represent the
   unknowns.
2. Identify Condition 2. Use Condition 2 to write an equation.
3. Solve the equation.
4. Check the results with the conditions. Answer the question.

90 / Chapter 4

EXAMPLE 2

The number of eleventh grade members of the Ski Club this year is 5 fewer than twice the number of twelfth grade members (Condition 1). The total membership is 82 (Condition 2). How many eleventh grade members are there?

Solution:

1. Use Condition 1 to represent the unknowns. ⟶ Let n = the number of twelfth graders. Then $2n - 5$ = the number of eleventh graders.

2. Use Condition 2 to write an equation.

Think: $\dfrac{\text{Number of}}{\text{twelfth graders}}$ + $\dfrac{\text{Number of}}{\text{eleventh graders}}$ = $\dfrac{\text{Total}}{\text{membership}}$

Translate: n + $2n - 5$ = 82

3. Solve the equation. ⟶ $3n - 5 = 82$

$$3n = 87$$
$$n = 29$$
$$2n - 5 = 53$$

4. **Check:** Condition 1 Is the number of eleventh grade members 5 less than twice the number of twelfth grade members?

Does $53 = 2(29) - 5$? Yes ✔

Condition 2 Is the total membership 82?

Does $29 + 53 = 82$? Yes ✔

There are 29 twelfth grade members and 53 eleventh grade members.

See the Answers to Selected Exercises for answers to Ex. 1 and 3.

CLASSROOM EXERCISES

For Exercises 1–10, Condition 1 is underscored once. Condition 2 is underscored twice. (Example 1)
a. Use Condition 1 to represent the unknowns. Use p as the variable.
b. Use Condition 2 to write an equation for the problem.

1. There were 186 more seats for tourist class than for first class on one flight. The total number of seats was 234.

2. The number of parking spaces for small cars in a lot is 75 less than the number for standard cars. The total number of spaces is 425.

3. The parts for a repair job on a car cost $3.58 more than the labor. The total bill was $75.62.

4. One model of compact car has 69 cubic feet more passenger space than cargo space. The total amount of space is 101 cubic feet.

Additional Examples
Example 2
1. The base price of a car is $1100 more than three times the cost of the options (Condition 1). The total cost of the car is $12,300 (Condition 2). Find the base price and the cost of the options.
   **Ans: base price: $9500
   options: $2800**

Common Error
Students often pick the numbers out of the problem and set up the equation randomly.

Example
Written Exercise 1
$x + 448 = 3640$

Prescription
Require students to represent the unknowns as shown in Step 1 of Examples 1 and 2, and to write the equation in words and in symbols as shown in Step 2. Then students should check their solution with the conditions.

Additional Answers
2. **a.** Let p = number of parking spaces for standard cars. Then $p - 75$ = number of parking spaces for small cars.
   b. $p + p - 75 = 425$
4. **a.** Let p = cubic feet of cargo space. Then $p + 69$ = cubic feet of passenger space.
   b. $p + p + 69 = 101$

See the Answers to Selected Exercises for answers to Ex. 5-9 odd.

6. a. Let p = amount of gas bill.
Then $2p - 0.35$ = amount of electric bill.
b. $p + 2p - 0.35 = 76.30$

8. a. Let p = number of teachers.
Then $10p + 153$ = number of students.
b. $p + 10p + 153 = 3244$

10. a. Let p = mass of shipping crate.
Then $20p - 7.6$ = mass of machine.
b. $p + 20p - 7.6 = 239.4$

Assignment Guide
Basic
Day 1 pp. 92–94: 1–15 odd
Day 2 pp. 92–94: 2–16 even
Average pp. 92–94: 1–19 odd
Above Average pp. 92–94:
1–15 odd, 17–20; p. 98: 9–16

Problem-Solving Skills
Writing and solving an equation
(Ex. 1–16)

Additional Answers
Written Exercises

1. a. Let x = number of visiting runners.
Then $x + 448$ = number of local runners.
b. $x + x + 448 = 3640$
c. visiting runners: 1596; local runners: 2044

2. a. Let x = number of games lost.
Then $x + 121$ = number of games won.
b. $x + x + 121 = 531$
c. games lost: 205; games won: 326

(Example 2)

5. A city's rainfall in June was $4\frac{3}{5}$ inches less than twice the amount in May. The total rainfall for the two months was $14\frac{9}{10}$ inches.

6. A family's electric bill one month was $0.35 less than twice the gas bill. The total amount of the two bills was $76.30.

7. A professional golfer earned $5,000 less than twice the amount earned the previous year. The total amount earned in two years was $93,000.

8. The number of students in school is 153 more than ten times the number of teachers. The total number of students and teachers is 3244.

9. The number of multiple–choice questions on a test was seven less than three times the number of true–false questions. The total number of questions was 105.

10. The mass of a machine is 7.6 kilograms less than 20 times the mass of its shipping crate. The total mass of the shipment is 239.4 kilograms.

WRITTEN EXERCISES

Goal: To represent conditions of a word problem and to solve the problem

Sample Problem: The attendance at the first performance of an orchestra was 560 more than at the second performance (Condition 1). The total attendance for the two performances was 5320 (Condition 2). Find the number attending each performance.

Answers: 1st performance: 2940; 2nd performance: 2380

For some problems, Condition 1 is underscored once. Condition 2 is underscored twice.
a. Use Condition 1 to represent the unknowns.
b. Use Condition 2 to write an equation for the problem.
c. Solve. (Example 1)

1. A marathon race had 448 more local participants than visiting runners. The total number in the race was 3640. Find the number of local runners and the number of visiting runners.

2. A coach's teams during his career won 121 more games than they lost. The total number of games was 531. Find the number of wins and the number of losses.

3. A tennis match ending in two sets had 4 more games in the second set than in the first. The total number of games was 20. Find the number played in each set.

(Example 2)

5. In one year, a retired couple spent $150 less on housing and transportation than twice the amount spent on food. The total amount spent on housing, transportation, and food was $8500. Find the amount spent for food.

7. A family's large car holds 2.3 gallons less fuel than twice the capacity of its compact car. The total fuel capacity of the two cars is 35.2 gallons. Find the fuel capacity of each car.

8. An art book contains 16 fewer pages of illustrations than three times the number of printed pages. The total number of pages is 512. Find the number of pages of illustrations.

MIXED PRACTICE

9. The passenger space in one car contains 5 cubic feet more than five times the cargo space. The total passenger and cargo space is 131 cubic feet. Find the size of each space.

11. Simon has a part-time job. In one two-week period he worked $2\frac{3}{4}$ hours more the second week than the first. His total time for the two weeks was $45\frac{3}{4}$ hours. How many hours did he work each week?

4. A collector has paintings and sculptures. The paintings are valued at $350,000 more than the sculptures. The total collection is valued at $1,260,000. Find the value of the paintings and the value of the sculptures.

6. At one point of a season the leading scorer in pro basketball had scored four fewer points on field goals than four times the number of points on free throws. Her total number of points was 1776. Find the number of points scored on field goals and the number scored on free throws.

10. A bedroom requires 120 square meters more carpeting than a living room. The total amount of carpeting needed is 968 square meters. Find the amount needed for the bedroom and the living room.

12. A trip from a family's home to the coast required 4.5 gallons more fuel returning than going. The total amount needed for the round-trip was 180.3 gallons. How much fuel was needed for each leg of the trip?

Problem Solving: One Variable / **93**

Additional Answers

4. a. Let x = value of sculptures.
Then $x + 350,000$ = value of paintings.
b. $x + x + 350,000 = 1,260,000$
c. sculptures: $455,000
paintings: $805,000

6. a. Let x = number of points scored on free throws. Then $4x - 4$ = number of points scored on field goals.
b. $x + 4x - 4 = 1776$
c. free throws: 356 points
field goals: 1420 points

8. a. Let x = number of printed pages. Then $3x - 16$ = number of illustrated pages.
b. $x + 3x - 16 = 512$
c. illustrated pages: 380

10. a. Let x = amount of carpeting for living room. Then $x + 120$ = amount of carpeting for bedroom.
b. $x + x + 120 = 968$
c. living room: 424 m^2
bedroom: 544 m^2

12. a. Let x = fuel required going. Then $x + 4.5$ = fuel required returning.
b. $x + x + 4.5 = 180.3$
c. going: 87.9 gallons
returning: 92.4 gallons

See the Answers to Selected Exercises for answers to Ex. 13 and 15.

Problem-Solving Skills
Using logical reasoning
(Ex. 17–20)
Making a list (Ex. 17–18)
Using guess and check (Ex. 19)
Drawing a diagram (Ex. 20)

Non-Routine Problems
For a description of non-routine
problems, see page M-13.

Additional Answers
Written Exercises
14. a. Let x = amount of rainfall
   Then $2x - 5.4 =$
   amount of snowfall.
   b. $x + 2x - 5.4 = 100.9$
   c. snowfall: 65.47 inches
16. a. Let x = first year sales.
   Then $2x - 50,000 =$
   second year sales.
   b. $x + 2x - 50,000 =$
   980,000
   c. second year: $636,666.67
17. In order to make change for
   any purchase from 1¢ to
   99¢, the cashier will need 9
   coins: 4 pennies, 1 nickel, 2
   dimes, 1 quarter, and 1 half–
   dollar.
18. List the possible ways 4
   people can be arranged
   when A is always at the left.
   ABCD   ACBD   ADBC
   ABDC   ACDB   ADCB
   The 4 people can be arranged 6 ways for each person at the left. The total is 6
   × 4, or 24 ways.
19. The smallest number divisible by 2, 3, 4, 5, or 6 is 60.
   Thus, when 61 is divided by
   2, 3, 4, 5, or 6, the remainder is 1. The store owner
   has 61 fish.
20. The front cover of Math I is
   against the back cover of
   Math II. Thus, the distance
   separating the first page of
   Math I and the last page of
   Math II is $2 \times \frac{1}{8}$, or $\frac{1}{4}$ inch.

13. Alfredo has a coin collection. He has 26 fewer U.S. coins than three times the number of foreign coins. The collection consists of 998 coins. Find the number of each type of coin.

15. Mary Ann enters customer records on a computer. In a two-day period she entered 12 more records the second day than the first day. In the two days she entered 256 records. Find the number of records entered each day.

NON-ROUTINE PROBLEMS

17. What is the smallest number of coins that a cashier needs in order to make change for a purchase of less than one dollar?

19. A pet store owner wanted to separate his fish so that the same number of fish were in each tank. When he separated the fish by twos, by threes, by fours, by fives, or by sixes, there was always one fish left over. What is the least number of fish the pet owner could have had?

14. The average snowfall for a city is 5.4 inches less than twice the amount of rainfall. The total amount of precipitation is 100.9 inches. Find the average amount of snowfall.

16. An insurance analyst set a goal of doubling her sales from the previous year. Her sales actually were $50,000 less than twice the previous year's sales. The total for the two years was $980,000. Find her sales for the second year.

18. At a banquet, four people are to be seated on one side of the head table. In how many different ways can these four people be seated?

20. Two books, Math I and Math II, are placed side by side on a bookshelf so that Math I is to the left of Math II. Each book cover is $\frac{1}{8}$ inch thick, and the pages, not counting the covers, are 2 inches thick. What is the distance from the first page of Math I to the last page of Math II?

REVIEW CAPSULE FOR SECTION 4.2

Solve and check each equation. (Pages 71–73)

1. $2t + 2(t - 10) = 84$ $t = 26$ **2.** $256 = 2w + 2(w + 5)$ $w = 61.5$ **3.** $2(5n) + 2(13n) = 151.2$ $n = 4.2$
4. $143 = 2(7p) + 2(15p)$ $p = 3.25$ **5.** $3x + 8x + 13x = 163.2$ $x = 6.8$ **6.** $185.4 = 4r + 5r + 9r$ $r = 10.3$

Represent the unknowns in each problem. Use the variable d.

EXAMPLE: The ratio of rainy days to sunny days is 1:6.   **ANSWER:** 1d; 6d

7. The ratio of tin to copper in an alloy is 1:4. d; 4d
8. The ratio of gaining stocks to losing stocks is 7:2. 7d; 2d
9. The ratio of orange, grapefruit, and pineapple sections in a fruit cup is 5:4:2. 5d; 4d; 2d
10. Stainless steel contains steel, chromium, and nickel in the ratio 37:9:4. 37d; 9d; 4d

OBJECTIVES: To solve word problems involving perimeters of triangles and rectangles
To solve word problems involving perimeters of triangles and rectangles
with the lengths of sides described by ratios

▮▮▮▮▮ PROBLEM SOLVING AND APPLICATIONS ▮▮▮▮▮

4.2   Using Geometry Formulas: Perimeter

The **perimeter** of a geometric figure such as a triangle, square, or rectangle is the sum of the lengths of its sides.

P–1   **What is the perimeter of each figure below?**

a.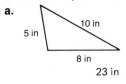

b.

c.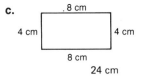

23 in   24 m   24 cm

EXAMPLE 1   The length of a drawing board is 5 inches more than the width (Condition 1). The perimeter is 94 inches (Condition 2). Find the length and width.

Solution:

$\boxed{1}$ Use Condition 1 to represent the unknowns. ——▶ Let z = the width.
Then $z + 5$ = the length.

$\boxed{2}$ Use Condition 2 to write an equation.

$$\underset{\downarrow}{p} = \underset{\downarrow}{2\ell} + \underset{\downarrow}{2w}$$

◀ **Formula for the perimeter of a rectangle**

$94 = 2(z + 5) + 2z$

$\boxed{3}$ Solve the equation. ——▶ $94 = 2z + 10 + 2z$

Note that $(z + 5)$ has been substituted for ℓ and z has been substituted for w.

$94 = 4z + 10$

$84 = 4z$

$21 = z$ ◀ **Don't forget to find $z + 5$.**

$z + 5 = 26$

$\boxed{4}$ **Check:**   Condition 1   Is the length 5 inches more than the width?
Does $26 = 21 + 5$?   Yes ✔

Condition 2   Is the perimeter 94 inches?
Does $2(26) + 2(21) = 94$?   Yes ✔

The length is 26 inches and the width is 21 inches.

Problem Solving: One Variable   /   95

Teaching Suggestions p. M–24

Lesson Resources
Warm–Up: p. M–25
Maintenance: See below.
Practice Worksheet 4.2
Transparencies 10, 11

Maintenance
1. Subtract: $6\frac{1}{2} - (-2)$
   Ans: $8\frac{1}{2}$ (Section 1.3)
2. Simplify: $2 - (x + y)$
   Ans: $2 - x - y$
   (Section 2.4)
3. Multiply and simplify:
   $4n(-2n + 3)$
   Ans: $-8n^2 + 12n$
   (Section 2.5)
4. Solve and check:
   $-4 = \frac{1}{5}y - 3$
   Ans: $y = -5$
   (Section 3.3)
5. A train traveled a distance of 300 miles at an average speed of 50 miles per hour. How long did the trip last?
   Ans: 6 hours   (Section 3.4)

Additional Examples
Example 1
1. A yard is in the shape of a rectangle. The length of the yard is 50 feet more than the width (Condition 1). The perimeter is 380 feet (Condition 2). Find the length and the width.
   **Ans: length: 120 feet
   width: 70 feet**
2. An advertising sign is triangular in shape. One side of the sign is 2.5 meters long. The other two sides are of equal length (Condition 1). The perimeter of the sign is 5.5 meters (Condition 2). Find the two unknown lengths.
   Ans: 1.5 meters; 1.5 meters

You may want to explain to students that the sum of the lengths of any two sides of a triangle is greater than the length of the third side.

Instead of the lengths of the sides of a figure, sometimes the ratio of the lengths of the sides is given. For example, if the lengths of a particular triangle are in the ratio 5:8:11, the lengths can be represented by 5x, 8x, and 11x.

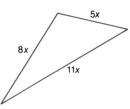

Additional Examples
Example 2
1. The ratio of the length to the width of a flag is 9:5 (Condition 1). The perimeter is 336 inches (Condition 2). Find the length and the width.
   Ans: length: 108 inches
   width: 60 inches
2. The three sides of a triangle are in the ratio of 4 : 5 : 7 (Condition 1). The perimeter of the triangle is 38.4 centimeters (Condition 2). Find the lengths of the sides.
   Ans: 9.6 centimeters; 12 centimeters; 16.8 centimeters

EXAMPLE 2

A cross country ski course is along a triangular route. The three legs of the course in order from the starting point have lengths in the ratio 2:3:4 (Condition 1). The length of the entire course is 45 kilometers (Condition 2). Find the length of each leg.

Solution:

1. Use Condition 1 to represent the unknowns. ⟶ Let $2x$ = the length of the first leg.
   Then $3x$ = the length of the second leg.
   And $4x$ = the length of the third leg.

2. Use Condition 2 to write an equation.

   Think: $\dfrac{\text{Length of}}{\text{first leg}} + \dfrac{\text{Length of}}{\text{second leg}} + \dfrac{\text{Length of}}{\text{third leg}} = \dfrac{\text{Total}}{\text{length}}$

   Translate: $2x + 3x + 4x = 45$

3. Solve the equation. ⟶ $9x = 45$
   $$x = 5$$
   $$2x = 10$$
   $$3x = 15$$
   $$4x = 20$$

4. **Check:** Condition 1 Are lengths of the legs in the ratio 2:3:4?

   Does $\dfrac{10}{15} = \dfrac{2}{3}$? (Does $10 \cdot 3 = 15 \cdot 2$?)   Yes ✓

   Does $\dfrac{15}{20} = \dfrac{3}{4}$?   Yes ✓

   Does $\dfrac{10}{20} = \dfrac{2}{4}$?   Yes ✓

   Condition 2 Does the sum of the lengths of the legs equal 45 km?
   Does $10 + 15 + 20 = 45$?   Yes ✓

The lengths of the legs of the course are 10 kilometers, 15 kilometers, and 20 kilometers.

▓▓▓▓ CLASSROOM EXERCISES ▓▓▓▓

Find the perimeter of each figure below. (P-1)

1.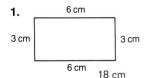

2.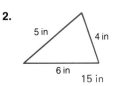

3.

For Exercises 4–15, Condition 1 is <u>underscored once</u>. Condition 2 is <u>underscored twice</u>.

a. *Use Condition 1 to represent the unknowns. Use the variable m.*

b. *Use Condition 2 to write an equation for the problem.* (Example 1)

4. The width of a screen for viewing slides is 10 inches less than the length. The perimeter is 140 inches.

5. The length of a flag is 27 inches more than its width. The perimeter is 174 inches.

6. The top of a stereo cabinet is a rectangle with its length 16 inches more than its width. The perimeter is 102 inches.

7. The length of one size of photo slide is 1.1 centimeters more than the width. The perimeter is 11.4 centimeters.

8. A scout troop hikes along a triangular route. The first leg is twice as long as the second leg. The third leg is 17 miles long. The entire hike is for 50 miles.

9. A sailboat race is along a triangular route. The first leg is 25.3 kilometers. The third leg is three times as long as the second leg. The race has a length of 67.3 kilometers.

(Example 2)

10. One special triangle has the lengths of its sides in the ratio 3:4:5. The perimeter is 38.4 centimeters.

11. Three cities form a triangle and the distances between the cities are in the ratio 4:5:6. The perimeter of the triangle is 184.5 kilometers.

12. The ratio of the width to the length of a tennis court is 6:13. The perimeter is 228 feet.

13. The ratio of the length to the width of a door is 20:9. The perimeter is 232 inches.

14. A surveyor working on a highway locates three stakes to form a triangle. One side of the triangle measures 34.3 meters. The ratio of the lengths of the other sides is 5:9. The perimeter is 70.7 meters.

15. A guy wire supports a pole and is anchored 9.8 meters from the base of the pole. The ratio of the length of the guy wire to the height of the pole is 7:5. The perimeter of the triangle formed is 33.8 meters.

Problem Solving: One Variable / **97**

97

Assignment Guide
Basic
Day 1 pp. 98–99: 1–13 odd
Day 2 pp. 98–99: 2–14 even
Average pp. 98–99: 1–15 odd
Above Average p. 99:
9–16 odd; p. 92: 1–15 odd, 17–20

Problem–Solving Skills
Drawing a diagram (Ex. 1–16)
Writing and solving an equation
(Ex. 1–16)
Solving a multi-step problem
(Ex. 15–16)

Additional Answers
1. **a.** Let x = height.
   Then $x + 6$ = length.
   b. $2x + 2(x + 6) = 84$
   c. height: 18 feet;
   length: 24 feet
2. **a.** Let x = width.
   Then $x + 3$ = length.
   b. $2x + 2(x + 3) = 36$
   c. width: 7.5 inches;
   length: 10.5 inches
3. **a.** Let x = length of shorter
   side.
   Then $x + 1.7$ = length of
   longer side.
   b. $x + x + 1.7 + 8.5 = 51.2$
   c. shorter side: 20.5 m
   longer side: 22.2 m
4. **a.** Let x = length of longer
   side.
   Then $x - 6.6$ = length of
   shorter side.
   b. $12.5 + x + x - 6.6 = 36.1$
   c. longer side: 15.1 cm;
   shorter side: 8.5 cm
6. **a.** Let $15x$ = width and
   $22x$ = length.
   b. $2(15x) + 2(22x) = 370$
   c. width: 75 m; length:
   110 m
8. **a.** Let $50x$ = length and
   $31x$ = width.
   b. $2(50x) + 2(32x) = 243$
   c. length: 75 cm; width:
   46.5 cm

WRITTEN EXERCISES

Goal: To solve word problems involving perimeters of triangles and rectangles.

Sample Problem: The ratio of the length to the width of a flag is 19:10 (Condition 1). Find the length and width of a flag having a perimeter of 174 feet (Condition 2).

Answer: 57 feet by 30 feet

For some problems, Condition 1 is <u>underscored once</u>. Condition 2 is <u>underscored twice</u>.
a. Use Condition 1 to represent the unknowns.
b. Use Condition 2 to write an equation for the problem.
c. Solve. (Example 1)

1. A sign is rectangular in shape. Its length is 6 feet greater than its height. Its perimeter is 84 feet. Find its length and height.

2. A mailing carton has a rectangular base. The length is 3 inches greater than the width. The perimeter is 36 inches. Find the length and width.

3. A sculptor designs a monument that includes a triangle with a base of 8.5 meters. The longer of the other two sides has a length 1.7 meters greater than the other. The perimeter of the triangle is 51.2 meters. Find the two unknown lengths.

4. A company's trademark is triangular. One side of the original design is 12.5 centimeters long. The shorter of the other two sides has a length 6.6 centimeters less than the length of the other. The perimeter of the triangle is 36.1 centimeters. Find the two unknown lengths.

See the Answers to Selected Exercises for answers to Ex. 5 and 7.
(Example 2)

5. One special triangle has sides in the ratio 5:12:13. Find the lengths of the sides if the perimeter is 204.0 meters.

6. The ratio of the width to the length of a soccer field is 15:22. The perimeter is 370 meters. Find the length and width.

7. A pilot flies a plane on a triangular route. The ratio of the lengths of the three legs of the flight is 4:7:9. The entire route is 684 kilometers long. Find the lengths of the legs of the route.

8. A picture is to be framed with its length and width in the ratio 50:31. The length of the frame needed (perimeter) is 243 centimeters. Find the length and width of the picture.

MIXED PRACTICE

9. A house has a rectangular floor plan. The length is 4 feet less than twice the width. The perimeter is 148 feet. Find the length and width.

11. The width of a computer terminal screen is $2\frac{1}{2}$ inches more than the height. The perimeter of the screen is 44 inches. Find the height and width.

12. A computer printout sheet is 9.9 centimeters longer than it is wide. The perimeter is 131.4 centimeters. Find the length and width.

13. A cable attached to the top of a tower is anchored 35.6 meters from the base of the tower. The perimeter of the triangle formed is 108.1 meters. The ratio of the cable's length to the tower's height is 18:11. Find the tower's height.

MORE CHALLENGING EXERCISES

15. A triangle has sides in the ratio 3:4:5. A square has sides equal in length to the smallest side of the triangle. The sum of the perimeters of the two figures is 96 centimeters. Find the lengths of the sides in each figure.

10. An official U.S. flag must have a ratio of length to width of 19:10. Find the length and width of a flag with a perimeter of 87 inches.

14. A punch press stamps a rectangular opening in a metal part. The length of the opening is 4.5 centimeters more than the width. The perimeter is 37.8 centimeters. Find the length and width of the opening.

16. Lot A is rectangular in shape. It is 8 meters longer than it is wide. Lot B is in the shape of a square. Each side of Lot B has the same measure as the width of Lot A. The sum of the perimeters of the two lots is 296 meters. Find the length and width of each lot.

REVIEW CAPSULE FOR SECTION 4.3

For Exercises 1–6, evaluate both expressions for the given value of the variable. (Pages 21–23)

1. $\left.\begin{array}{l}5x+3 \\ 4x-9\end{array}\right\}$ $x=6$ 33, 15

2. $\left.\begin{array}{l}20-2y \\ 56-5y\end{array}\right\}$ $y=12$ $-4, -4$

3. $\left.\begin{array}{l}4-3z \\ z+22\end{array}\right\}$ $z=-4\frac{1}{2}$ $17\frac{1}{2}, 17\frac{1}{2}$

4. $\left.\begin{array}{l}21+m \\ -19-7m\end{array}\right\}$ $m=-5$ 16, 16

5. $\left.\begin{array}{l}5r+2(r-1) \\ 3(r+1)\end{array}\right\}$ $r=1\frac{1}{4}$ $6\frac{3}{4}, 6\frac{3}{4}$

6. $\left.\begin{array}{l}8p-2(2p-1) \\ 6(p+2)\end{array}\right\}$ $p=-5$ $-18, -18$

Problem Solving: One Variable / **99**

**Additional Answers
Written Exercises**

9. a. Let x = width.
Then $2x - 4$ = length.
b. $2x + 2(2x - 4) = 148$
c. width: 26 feet;
length: 48 feet

10. a. Let $19x$ = length and $10x$ = width.
b. $2(19x) + 2(10x) = 87$
c. length: 28.5 inches;
width: 15 inches

11. a. Let x = height.
Then $x + 2\frac{1}{2}$ = width.
b. $2x + 2(x + 2\frac{1}{2}) = 44$
c. height: $9\frac{3}{4}$ inches;
width: $12\frac{1}{4}$ inches

12. a. Let x = width.
Then $x + 9.9$ = length.
b. $2x - 2(x + 9.9) = 131.4$
c. width: 27.9 cm;
length: 37.8 cm

13. a. Let $18x$ = cable length and $11x$ = tower height.
b. $35.6 + 18x + 11x = 108.1$
c. height: 27.5 m

14. a. Let x = width.
Then $x + 4.5$ = length.
b. $2x + 2(x + 4.5) = 37.8$
c. width: 7.2 cm;
length: 11.7 cm

15. a. Let $3x$, $4x$, and $5x$ be the lengths of the sides of the triangle. Let $3x$ = length of a side of the square.
b. $3x + 4x + 5x + 4(3x) = 96$
c. triangle: $3x = 12$ cm; $4x = 16$ cm; $5x = 20$ cm; square: 12 cm

16. a. Let x = width of Lot A. Then $x + 8$ = length of Lot A and x = length of a side of Lot B.
b. $2x + 2(x + 8) + 4(x) = 296$
c. width, Lot A: 35 m; length, Lot A: 43 m; side of Lot B: 35 m

99

4.3 Variable on Both Sides

When the variable appears on both sides of an equation, the first step is to eliminate the variable from one side.

EXAMPLE 1   Solve: $5x + 12 = 3x - 6$

Solution: $5x + 12 = 3x - 6$

1 Subtract 3x from each side. ⟶ $5x + 12 - \mathbf{3x} = 3x - 6 - \mathbf{3x}$

$2x + 12 = -6$

2 Subtract 12 from each side. ⟶ $2x + 12 - \mathbf{12} = -6 - \mathbf{12}$ ◀ $-6 - 12 = -6 + (-12)$

$2x = -18$

3 Divide each side by 2. ⟶ $\dfrac{2x}{2} = \dfrac{-18}{2}$ You will want to demonstrate optional ways to solve the equations of the examples.

$x = -9$ For example, you could subtract 5x from each side as the first step.

Check: $5x + 12 = 3x - 6$

$5(-9) + 12 \mid 3(-9) - 6$

$-45 + 12 \mid -27 - 6$ ◀ $-27 - 6 = -27 + (-6)$

$-33 \mid -33$

P–1   **Solve.**

a. $11t + 16 = 4t + 2$ **b.** $9j + 6 = 4j - 14$ $_{j\,=\,-4}$
   $t = -2$

EXAMPLE 2   Solve: $12 - 7r = 42 - 3r$

Solution: $12 - 7r = 42 - 3r$

1 Add 7r to each side. ⟶ $12 - 7r + \mathbf{7r} = 42 - 3r + \mathbf{7r}$

$12 = 42 + 4r$

2 Subtract 42 from each side. ⟶ $12 - \mathbf{42} = 42 + 4r - \mathbf{42}$

$-30 = 4r$

3 Divide each side by 4. ⟶ $\dfrac{-30}{4} = \dfrac{4r}{4}$

$-7\frac{1}{2} = r$ ◀ *The check is left for you.*

P–2   **Solve.**
                          $p = -14$
a. $3 - 2p = 17 - p$ **b.** $15 - 3f = 3 + 6f$ $_{f\,=\,1\frac{1}{3}}$

In equations that contain parentheses, use the Distributive Property first. Then eliminate the variable from one side of the equation.

EXAMPLE 3   Solve: $5 + 2x = 2(x - 3) + x$

Solution:

$\boxed{1}$ Use the Distributive Property. $\longrightarrow$ $5 + 2x = 2(x - 3) + x$

$\boxed{2}$ Combine like terms. $\longrightarrow$ $5 + 2x = 2x - 6 + x$

$$5 + 2x = 3x - 6$$

$\boxed{3}$ Subtract $2x$ from each side. $\longrightarrow$ $5 + 2x - \mathbf{2x} = 3x - 6 - \mathbf{2x}$

$$5 = x - 6$$

$\boxed{4}$ Add 6 to each side. $\longrightarrow$ $5 + \mathbf{6} = x - 6 + \mathbf{6}$

$$11 = x$$

Check:
$$5 + 2x = 2(x - 3) + x$$

$5 + 2(11)$	$2(11 - 3) + 11$
$5 + 22$	$2(8) + 11$
27	27

P–3 Solve.

a. $9 + 4y = 3(y + 4) - y$
  $y = 1\frac{1}{2}$

b. $2 + 10g = 6(2g + 3) + 2g$
  $g = -4$

CLASSROOM EXERCISES

Write the two equations that are formed in eliminating the variable from each side of the following equations. (Step 1, Examples 1–2)

1. $3x - 2 = 2x + 5$ 2. $3 - 4y = 7 - 5y$ 3. $8 + a = -3a + 6$

4. $10 + 5g = -8 + 3g$ 5. $-4d - 6 = 12 - d$ 6. $10b - 7 = 9 - 4b$

For each equation, write an equivalent equation that has the variable on one side. (Steps 1 and 2, Examples 1–2)

EXAMPLE: $3x - 7 = 2x + 8$ ANSWER: $x - 7 = 8$ or $-7 = -x + 8$

7. $4p + 2 = 6 - p$ 8. $5 - 2r = 3r - 2$ 9. $-3z - 8 = z + 5$

10. $3m + 2 = -2m - 8$ 11. $10.8 - 3.4j = 9.2 + 7.1j$ 12. $8.2 - h = 3h - 3.4$

Problem Solving: One Variable / **101**

Additional Examples
Example 3
Solve.
1. $3(2 - y) + 5y = 3y + 8$
   Ans: $y = -2$
2. $-4(3t - 20) = 70 - 10t$
   Ans: $t = 5$
3. $6y - 17 = \frac{1}{2}(8y + 2) - y$
   Ans: $y = 6$

Common Error
When solving equations with the variable on both sides, students incorrectly combine like terms.

Example
$$9x + 4 = 3x + 20$$
$$12x + 4 = 20$$
$$12x = 24$$
$$x = 2$$

Prescription
Remind students that terms can be added only if they are on the same side of the equation. Insist that students check the solution in the original equation.

Additional Answers
Classroom Exercises
1. $x - 2 = 5$; $-2 = -x + 5$
2. $3 = 7 - y$; $3 + y = 7$
3. $8 = -4a + 6$; $8 + 4a = 6$
4. $10 + 2g = -8$;
   $10 = -8 - 2g$
5. $-3d - 6 = 12$;
   $-6 = 12 + 3d$
6. $14b - 7 = 9$;
   $-7 = 9 - 14b$
7. $5p + 2 = 6$ or $2 = 6 - 5p$
8. $5 - 5r = -2$ or
   $5 = 5r - 2$
9. $-4z - 8 = 5$ or
   $-8 = 4z + 5$
10. $5m + 2 = -8$ or
    $2 = -5m - 8$
11. $10.8 - 10.5j = 9.2$ or
    $10.8 = 9.2 + 10.5j$
12. $8.2 - 4h = -3.4$ or
    $8.2 = 4h - 3.4$

101

Assignment Guide
Basic
Day 1 p. 102: 1–15 odd
Day 2 p. 102: 17–35 odd
Average
Day 1 p. 102: p. 102: 1–23 odd
Day 2 p. 102: 25–37 odd
Above Average p. 102:
1–37 odd, 38

Additional Answers
6. $d = -0.8$
14. $x = -72$
18. $d = -7$
20. $g = -2\frac{2}{3}$
34. $x = -16$

WRITTEN EXERCISES

Goal: To solve equations that have a variable on both sides
Sample Problem: Solve: $5 + 4x = 3x - 10 - x$
Answer: $x = -7\frac{1}{2}$

Solve. Check each answer. (Example 1)

1. $3x - 5 = 2x + 12$ $x = 17$
2. $5x - 3 = 4x + 9$ $x = 12$
3. $12a - 9 = 10a + 17$ $a = 13$
4. $8b - 14 = 6b + 28$ $b = 21$
5. $1.6c - 7.5 = 14.1 + 3.4c$ $c = -12$
6. $0.5d - 3 - d = -1.8 + d$
7. $3\frac{1}{2}k + 17 = 2\frac{3}{4}k + 12$ $k = -6\frac{2}{3}$
8. $\frac{1}{20} + \frac{1}{5}s - 11 = \frac{3}{5}s - 11$ $s = \frac{1}{8}$

(Example 2)

9. $12 - x = 26 - 3x$ $x = 7$
10. $20 - 2y = 56 - 5y$ $y = 12$
11. $2 - 3z = z + 22$ $z = -5$
12. $5 - 2m = -22 - 4m$ $m = -13\frac{1}{2}$
13. $2\frac{1}{4} - 1\frac{4}{5}p - 19 = \frac{2}{3}p + 38 - 2\frac{1}{3}$ $p = -21\frac{1}{4}$
14. $9 - 5\frac{1}{4}x = 27 - 2\frac{1}{4}x - 2\frac{3}{4}x$
15. $2t + 1.2 - 5t = 2.8 - t$ $t = -0.8$
16. $3.2 + z = -6.7 - 1.2z$ $z = -4.5$

(Example 3)

17. $3(4 - x) - 22 = x + 46$ $x = -14$
18. $-4(d - 2) - 14 = 36 + 2d$
19. $14 - 3y = -3(4 + 2y) - 7$ $y = -11$
20. $-6g - 8 = 12 - 4(3g + 9)$
21. $0.2(n + 1) = 0.6(n - 1) - 0.7$ $n = 3.75$
22. $3(y + 0.5) = 0.5y + 1.5$ $y = 0$
23. $\frac{4}{5}y - \frac{1}{2}(y - 1) = \frac{3}{4}(y + 1)$ $y = -\frac{5}{9}$
24. $\frac{1}{2}(z + 1) - 2z = 6z + \frac{1}{2}$ $z = 0$

MIXED PRACTICE

25. $10 - 5y = y + 61$ $y = -8\frac{1}{2}$
26. $14 - 3b = -16 - 7b$ $b = -7\frac{1}{2}$
27. $6 - 15q = 11q - 46$ $q = 2$
28. $21 + m = -19 - 7m$ $m = -5$
29. $3(2 - k) - 7 = 4(2k + 8)$ $k = -3$
30. $5z + 2(z - 1) = 3(z + 1)$ $z = 1\frac{1}{4}$
31. $-10 - 18d = 26 - 12d$ $d = -6$
32. $-4f - 19 = 17 + f$ $f = -7\frac{1}{5}$
33. $4.4 + 2.1d = 5.7d - 17.2$ $d = 6$
34. $2.3x - 31.5 = 14.9 + 5.2x$
35. $8p - 2(2p - 1) = 6(p + 2)$ $p = -5$
36. $5(2r + 3) - 4r = 20r + 1$ $r = 1$

MORE CHALLENGING EXERCISES

For Exercises 37–38, a represents an even integer and b represents an odd integer. Solve for x. Identify x as an even integer or an odd integer.

37. $2a + 2x = x + b$ $x = b - 2a$; x is odd
38. $4x - a = 3\frac{1}{2}x - 3b$ $x = 2(-3b + a)$; x is even

Problem–Solving Skills
Solving a multi–step problem
(Ex. 37–38)
Using logical reasoning
(Ex. 37–38)

Critical Thinking
See Exercises 37 and 38.

MID-CHAPTER REVIEW

For Exercises 1–8, Condition 1 is <u>underscored once</u>. Condition 2 is <u><u>underscored twice</u></u>. Use Condition 1 to represent unknowns. Use Condition 2 to write an equation for the problem. Solve. (Section 4.1)

1. A tumbling class for 5-6 year olds at the Y had seven more girls than boys. The total number enrolled for the class was 45. Find the number of girls and the number of boys.

2. Joel bought a sports coat for $15 less than five times the cost of a pair of slacks. The total cost of the coat and slacks was $273. Find the cost of each item.

3. A traffic count between 6:00 a.m. and 8:00 a.m. on a highway showed 2900 fewer cars the first hour than the second hour. The total number of cars for the two-hour period was 24,500. Find the number for each hour.

4. A school club's receipts from special projects was $5.00 more than three times the receipts from dues. The amount of total receipts was $485. Find the amounts received from dues and from special projects.

(Section 4.2)

5. A rectangular lot has a length that is 6.2 meters greater than twice its width. The perimeter is 106.6 meters. Find the length and width.

6. An Olympic-size swimming pool has a length and width in the ratio 25:12. The pool's perimeter is 148 meters. Find the length and width.

7. A stone monument has a triangular base. The lengths of the sides of the base are in the ratio 5:4:2. The perimeter is 264 inches. Find the length of each side.

8. A rectangular desk top has a length that is 2 centimeters less than twice its width. The perimeter is 460 centimeters. Find the length and width of the desk top.

For Exercises 9–14, solve each equation. Check each answer. (Section 4.3)

9. $6d - 5 = 9d + 37$ $d = -14$

10. $4r + 11 = 8r + 43$ $r = -8$

11. $16 - 3p = 32 - 7p$ $p = 4$

12. $23 - 5y = 89 - 11y$ $y = 11$

13. $-3(x - 2) + x = 10 - 6x$ $x = 1$

14. $14 + 5c = 2c - 7(c + 8)$ $c = -7$

REVIEW CAPSULE FOR SECTION 4.4

Compute.

EXAMPLE: 8.5% of 350   **ANSWER:** $8.5\% = 0.085; 0.085 \cdot 350 = 29.75$

1. 12% of $960 $115.20
2. 8% of $87.50 $7
3. 5% of 2780 139
4. 40% of $29.95 $11.98
5. 75% of 56 42
6. 17% of 250 42.5
7. 4.5% of $750 $33.75
8. $33\frac{1}{3}\%$ of 915 305

Problem Solving: One Variable / 103

Quiz Sections 4.1–4.3
After completing this Mid-Chapter Review, you may want to administer a quiz covering the same sections. See the quiz provided on page 91 in the *Teacher's ResourceBank.*

Additional Answers
Mid–Chapter Review

1. **a.** Let x = number of boys. Then $x + 7$ = number of girls.
   b. $x + x + 7 = 45$
   c. boys: 19; girls: 26
2. **a.** Let x = cost of slacks. Then $5x - 15$ = cost of coat.
   b. $x + 5x - 15 = 273$
   c. slacks: $48; coat: $225
3. **a.** Let x = count second hour. Then $x - 2900$ = count first hour.
   b. $x + x - 2900 = 24,500$
   c. second hour: 13,700 cars; first hour: 10,800 cars
4. **a.** Let x = receipts from dues. Then $3x + 5$ = receipts from special projects.
   b. $x + 3x + 5 = 485$
   c. dues: $120; special projects: $365
5. **a.** Let x = width. Then $2x + 6.2$ = length.
   b. $2x + 2(2x + 6.2) = 106.6$
   c. width: 15.7 m; length: 37.6 m
6. **a.** Let $25x$ = length. Then $12x$ = width.
   b. $2(25x) + 2(12x) = 148$
   c. length: 50 m; width: 24 m
7. **a.** Let $5x$, $4x$, and $2x$ be the lengths of the sides.
   b. $5x + 4x + 2x = 264$
   c. $5x = 120$ inches; $4x = 96$ inches; $2x = 48$ inches
8. **a.** Let x = width. Then $2x - 2$ = length.
   b. $2x + 2(2x - 2) = 460$
   c. width: $77\frac{1}{3}$ cm; length: $152\frac{2}{3}$ cm

103

Maintenance
1. Divide: $-4.34 \div (-0.7)$
   Ans: 6.2   (Section 1.5)
2. Simplify: $5 + n - 12 - m$
   Ans: $-7 + n - m$
   (Section 2.2)
3. Simplify:
   $-4x^2 + 6x + 12x^2 - 2x$
   Ans: $8x^2 + 4x$
   (Section 2.6)
4. Solve and check:
   $5x + 2x - 17 = 53$
   Ans: $x = 10$ (Section 3.5)
5. Write an algebraic expression
   for this word expression. Use
   the variable n.
   The number of swim suits decreased by 25
   Ans: $n - 25$ (Section 3.6)

Additional Examples
Example 1
1. A new car was on sale at a
   rate of discount of 5%. The
   sale price was $8075. Find
   the list price and the discount.
   Ans: list price: $8500
   discount: $425
2. A dress was on sale at a rate
   of discount of 20%. The sale
   price was $30.40. Find the list
   price and the discount.
   Ans: list price: $38.00
   discount: $7.60

OBJECTIVES: To solve word problems involving prices, discounts, and rates
of discount expressed as per cents

PROBLEM SOLVING AND APPLICATIONS

4.4 Using Per Cent

To solve sports problems involving averages expressed as per cents
To solve miscellaneous word problems involving per cents

Merchandise is often offered on sale to consumers at a reduction in
price. The price before the reduction is called the **list price** or **regular
price.** The amount that the item is reduced is called the **discount.** The
reduced price is called the **sale price.**

The following word rule and formula relate sale price, list price, and
discount.

Word Rule: The <u>sale price</u> equals the <u>list price</u> less the <u>discount</u>.

Formula: $s = p - d$ s = sale price
p = list price
d = discount

You should emphasize the difference between
discount and *rate of discount*. Discount is an
amount in dollars. Rate of discount is expressed
as a per cent.

Discount expressed as a per cent of the list price is called the **rate of
discount.**

The following word rule and formula relate discount, rate of discount,
and list price.

Word Rule: <u>Discount</u> equals the <u>rate of discount</u> times the <u>list price</u>.

Formula: $d = r \cdot p$ ◀ r = rate of discount

List Price	Rate of Discount	Discount
$60	20%	$(0.20)\,(60) = \$12.00$
$150	10%	$(0.10)\,(150) = 15.00$
$225	8%	$(0.08)\,(225) = 18.00$

EXAMPLE 1   A lamp was on sale at a rate of discount of 20%. The
sale price was $25.00. Find the list price of the lamp
and the discount.

Solution:

☐1 Represent the list price ——→ Let p = the list price.
and the discount.   Then $0.20p$ = the discount.

☐2 Use the formula to $s = p - d$
write an equation. ——————→ $25.00 = p - 0.20p$

104 / *Chapter 4*

3 Solve the equation. ⟶ $25.00 = 0.80p$

$31.25 = p$

$0.20p = 6.25$

> **Don't forget to find 0.20p.**

4 **Check:** Was the discount 20% of the list price?
Does $6.25 \div 31.25 = 0.2$ or 20%?   Yes ✓

Did the list price less the discount equal the sale price?
Does $31.25 - 6.25 = 25.00$?   Yes ✓

The list price was $31.25 and the discount was $6.25.

In Example 2, you are asked to organize the information given so that you can answer the question that is asked. Making a table will help you to do this and to write an equation for the problem.

EXAMPLE 2

You might ask, for discussion purposes, what per cent each team lost of the games played.

During a softball tournament, Team B played 2 more games than Team A (Condition 1). Team A won 60% of the games it played, and Team B won 50% of the games it played. The two teams won the same number of games (Condition 2). Find the number of games played by each team.

Solution:

	Games Played (Condition 1)	Per Cent Won	Games Won (Condition 2)
1 Team A	x	60%	$0.60x$
Team B	$x + 2$	50%	$0.50(x + 2)$

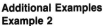

2 Use Condition 2 to write an equation.

Think: Games won by Team A = Games won by Team B

Translate: $0.60x$ = $0.50(x + 2)$

3 Solve the equation. ⟶ $0.60x = 0.50x + 1$

$0.10x = 1$

$x = 10$

$x + 2 = 12$

4 **Check:** Condition 1   Did Team B play 2 more games than Team A?
Does $12 = 10 + 2$?   Yes ✓

Condition 2   Did Team A and Team B win the same number of games?
Does $0.60(10) = 0.50(12)$?   Yes ✓

Team A played 10 games and Team B played 12 games.

Problem Solving: One Variable / **105**

Find the discount on each item. (Table)

1. A dining table listed at $640 with a discount rate of 20% $128

2. A shirt listed at $36 with a discount rate of 25% $9

3. A record album listed at $9.50 with a discount rate of 40% $3.80

4. A pick-up truck listed at $8500 with a discount rate of 6% $510

Solve. Refer to the formulas on page 104. (Step 2, Example 1)

5. The rate of discount of a radio on sale was 35%. The discount was $31.50. Find the list price. $90

6. The rate of discount of a pair of skis on sale was 20%. The discount was $75. Find the list price. $375

(Example 1)

7. A typewriter was on sale with a discount rate of 12%. The sale price was $396. Find the list price and the discount. $450; $54

8. A sweater was on sale with a discount rate of 35%. The sale price was $39. Find the list price and the discount. $60; $21

For Exercises 9–10, Condition 1 is underscored once. Condition 2 is underscored twice.

a. *Use the given information (Condition 1 and Condition 2) to complete the table.*

b. *Use Condition 2 to write an equation for the problem.* (Example 2)

9. Julia's monthly income is $240 greater than Emily's monthly income. Emily spends 30% of her monthly income on rent, and Julia spends 25% of her monthly income on rent. Julia and Emily spend the same amount each month on rent. Find their monthly incomes.

$0.30 x = 0.25(x + 240)$

	Monthly Income	Per Cent Spent	Amount Spent
Emily	x	30%	? $0.30x$
Julia	? $x + 240$	25%	? $0.25(x + 240)$

10. Before a special sale, the Sports Shop had 20 more baseball gloves in stock than baseball bats. During the sale, the shop sold 70% of its stock of bats and 50% of its stock of gloves. The same number of bats and gloves were sold. Find the number of bats and gloves in stock before the sale.

$0.70 n = 0.50(n + 20)$

Item	Number in Stock	Per Cent Sold	Number Sold
Bats	n	70%	? $0.70n$
Gloves	? $n + 20$	50%	? $0.50(n + 20)$

WRITTEN EXERCISES

Goal: To solve problems involving per cent

Sample Problem: A stereo was on sale for $328 at a discount of 18%. Find the list price and the discount.

Answers: List Price: $400; Discount: $72

Solve. Refer to the formulas on page 104. (Example 1)

1. The rate of discount of a refrigerator on sale was 20%. The discount was $115. Find the list price. $575

2. A color TV was on sale for $331.50. The discount rate was 15%. Find the list price and the discount.
   $p = $390; d = 58.50

3. A woman's warmup suit was on sale at a rate of discount of 20%. The sale price was $48. Find the list price and the discount. $p = $60; d = 12

4. A store offers customers a 2% discount if bills are paid within 15 days. This discount would save one customer $5.12. Find the original amount of the customer's bill. $256

For Exercises 5–7, Condition 1 is <u>underscored once</u>. Condition 2 is <u>underscored twice</u>.
a. *Use Condition 1 to represent the unknowns.*
b. *Use Condition 2 to write an equation for the problem.*
c. *Solve.*
(Example 2)

5. A basketball team played 4 more games this season than last season. Last season, the team won 55% of its games and this season it won 50% of its games. The team won the same number of games this season as last season. Find the number of games played each season.

6. In August, U-Fix Auto Parts sold 40% of its stock of batteries and 30% of its stock of oil filters. At the beginning of August, there were 40 more oil filters in stock than batteries. The same number of batteries and oil filters were sold. How many of each item were in stock at the beginning of August?

7. The number of players selected for the girls' tennis team was 75% of the number who tried out. There were 4 more who tried out for the boys' team than for the girls' team, but only 60% of the boys were selected. The two teams had the same number of members. Find how many tried out for each team.

*Problem Solving: One Variable / **107***

Assignment Guide
Basic p. 107: 1–7
Average p. 107: 1–7
Above Average p. 107: 1–7

Problem–Solving Skills
Using a formula (Ex. 1–4)
Making a table (Ex. 5–7)
Writing and solving an equation (Ex. 5–7)

Additional Answers
5. a. Let x = number of games last season.
   Then $x + 4$ = number of games this season.
   b. $0.55x = 0.50(x + 4)$
   c. last season: 40 games; this season: 44 games
6. a. Let x = batteries in stock.
   Then $x + 40$ = oil filters in stock.
   b. $0.40x = 0.30(x + 40)$
   c. batteries: 120; oil filters: 160
7. a. Let x = number trying out for girl's team.
   Then $x + 4$ = number trying out for boy's team.
   b. $0.75x = 0.60 (x + 4)$
   c. girl's team: 16; boy's team: 20

Teaching Suggestions p. M–24

Lesson Resources
Warm–Up: p. M–25
Maintenance: See below.
Practice Worksheet 4.5
Transparencies 10, 12

Maintenance
1. Evaluate: $-26 + 3(-4)$
   Ans: -38 (Section 1.6)
2. Simplify: $\dfrac{-3x + 7}{-(-7 + 3x)}$
   Ans: 1 (Section 2.4)
3. Solve and check:
   $-64 = k - 13$
   Ans: $k = -51$
   (Section 3.1)
4. Solve and check:
   $8w - 21 - 5w = -15$
   Ans: $w = 2$ (Section 3.5)
5. The high temperature on a winter day was 5°C. The low temperature was -11°C. How much higher was the high temperature than the low temperature?
   Ans: 16°C (Section 1.3)

Additional Examples
Example 1
1. A camper walked to town and then jogged back. The first half of the trip took 1.5 hours and the return trip took 0.9 hours (Condition 2). The average rate returning was 2 miles per hour more than the average rate going (Condition 1). Find the average speed returning.
   Ans: 5 miles per hour
2. A shuttle bus makes a round-trip run. The first half of the trip takes 20 minutes and the return trip takes 25 minutes (Condition 2). The average rate going is 5 miles per hour more than the average rate returning (Condition 1). Find the average speed on the first half of the trip.
   Ans: 25 miles per hour

■■■■■ **PROBLEM SOLVING AND APPLICATIONS** ■■■■■

4.5 Using Distance Formulas

To solve word problems involving cruising range, fuel mileage, and fuel tank capacity

The following word rule and formula are used to solve time/rate/distance problems.

Word Rule: <u>Distance</u> equals <u>rate</u> multiplied by <u>time</u>.

Formula: $\boldsymbol{d = r \cdot t}$ or $\boldsymbol{d = rt}$ ◀
$d = distance$
$r = rate$
$t = time$

In solving these problems, make a table showing the rate, time and distance for each trip. For <u>round trips</u>,

distance going equals distance returning.

This is Condition 2 in the problems for this section. Note that Condition 2 is sometimes stated before Condition 1.

Remind students that Condition 2 has been used to write an equation.
Condition 1 has been used to represent the unknowns.

EXAMPLE 1

A salesperson took a round-trip flight to another city. The first half of the trip took 5 hours and the return trip took 4.5 hours (Condition 2). The average rate returning was 54 miles per hour (abbreviated mi/hr) more than the average rate going (Condition 1). Find the average speed returning.

1 **Solution:** Make a table.

	Rate (Condition 1)	Time	Distance (Condition 2) $d = rt$
Going	x mi/hr	5 hr	$x(5)$, or $5x$ mi
Returning	$(x + 54)$ mi/hr	4.5 hr	$(x + 54)4.5$, or $4.5(x + 54)$ mi

2 Use Condition 2 to write an equation.

Think: <u>Distance going</u> = <u>Distance returning</u>

Translate: $5x$ = $4.5(x + 54)$

3 Solve the equation. ⟶ $5x = 4.5x + 243$

$0.5x = 243$

$\dfrac{0.5x}{0.5} = \dfrac{243}{0.5}$

$x = 486$ ◀ ***Don't forget to find x + 54.***

$x + 54 = 540$

108 / *Chapter 4*

$\boxed{4}$ **Check:** Condition 1   Was the average rate returning 54 miles per hour more than the average rate going?

Does $540 = 486 + 54$?   Yes ✔

Condition 2   Does the distance going equal the distance returning?

Does $486(5) = 540(4.5)$?   Yes ✔

The average rate returning was 540 miles per hour.

The following word rule and formula are similar to the distance formula. The **cruising range** of a car is the distance it can travel on a full tank of fuel. **Mileage** means miles per gallon (mi/gal) or kilometers per liter (km/L).

Word Rule: The <u>cruising range</u> (distance) of a car equals the <u>fuel tank capacity</u> multiplied by the <u>mileage</u>.

Formula: $r = m \cdot c$ or $r = mc$ ◀ r = cruising range
                                        m = mileage
                                        c = fuel tank capacity

EXAMPLE 2

The rate of use of fuel is sometimes referred to as "fuel economy."

The fuel tank capacity of Car B is 36 liters less than that of Car A (Condition 1). Car A averages 7 kilometers per liter of fuel. Car B averages 9 kilometers per liter. The range of Car B is 132 kilometers less than that of Car A (Condition 2). Find the fuel tank capacity of each car.

Solution:

	Mileage	Fuel Tank Capacity (Condition 1)	Range (Condition 2) $r = mc$
$\boxed{1}$ Car A	7 km/L	x L	$7x$ km
Car B	9 km/L	$(x - 36)$ L	$9(x - 36)$ km

$\boxed{2}$ **Think:**   <u>Range of Car B</u> = <u>Range of Car A</u> − 132   ◀ *Condition 2*

Translate: $9(x-36)$ = $7x$ − 132

$\boxed{3}$
$$9x - 324 = 7x - 132$$
$$9x - 7x - 324 = 7x - 132 - 7x$$
$$2x - 324 + 324 = -132 + 324$$
$$2x = 192$$
$$x = 96$$
$$x - 36 = 60$$

◀ *Tank capacity of Car A: 96 L*
Tank capacity of Car B: 60 L

Problem Solving: One Variable   /   **109**

4 Check: Condition 1 Is the fuel tank capacity of Car B 36 liters less than that of Car A? Does $60 = 96 - 36$? Yes ✔

  Condition 2 Is the range of Car B 132 kilometers less than that of Car A?

  Does $9(60) = 7(96) - 132$ Yes ✔

The fuel tank capacity of Car A is 96 liters.
The fuel tank capacity of Car B is 60 liters.

CLASSROOM EXERCISES

For Exercises 1–3, Condition 1 is underscored once. Condition 2 is underscored twice. (Examples 1 and 2)
a. *Use the given information to complete the table.*
b. *Use Condition 2 to write an equation for the problem.*

1. Sarah traveled to New Orleans and back by train. The trip to New Orleans took 6 hours and the trip returning took 6.4 hours. The average rate returning was 6 kilometers per hour less than the average rate going.

$6x = 6.4(x - 6)$

	Rate (Condition 1)	Time	Distance (Condition 2) $d = rt$
Going	x km/hr	6 hr	? $6x$
Returning	? $x - 6$	6.4 hr	? $6.4(x - 6)$

2. On a round trip to another city, José spent 3 hours going and $2\frac{3}{4}$ hours returning. The average rate returning was 4 miles per hour faster than the average rate going.

$3r = \frac{11}{4}(r + 4)$

	Rate (Condition 1)	Time	Distance (Condition 2) $d = rt$
Going	r mi/hr	?3	? $3r$
Returning	? $r + 4$	?$\frac{11}{4}$	? $\frac{11}{4}(r + 4)$

3. The fuel tank capacity of Car B is 2 liters less than that of Car A. Car A averages 10 km/L of fuel and Car B averages 12 km/L. The range of Car B is 112 kilometers more than that of Car A.

$10x + 112 = 12(x - 2)$

	Mileage	Fuel Tank Capacity (Condition 1)	Range (Condition 2) $r = mc$
Car A	10 km/L	x L	? $10x$
Car B	12 km/L	? $x - 2$	? $12(x - 2)$

WRITTEN EXERCISES

Goal: To solve distance/rate/time problems

Sample Problem: A round-trip jet flight takes 2.25 hours to reach Miami and 2 hours to return (Condition 2). The rate returning is 80 kilometers per hour greater than the rate going (Condition 1). Find the average rate going and the average rate returning.

Answers: Average rate going: 640 km/hr; Average rate returning: 720 km/hr

For Exercises 1–7, Condition 1 is <u>underscored once</u>. *Condition 2 is* <u>underscored twice</u>.
a. Use Condition 1 and Condition 2 to make a table.
b. Use Condition 2 to write an equation for the problem.
c. Solve.
(Examples 1 and 2)

See the Answers to Selected Exercises for answers to Exercises 1, 3, and 5.

1. A bus makes a round-trip run. <u>The average rate going is 15 miles per hour more than the average rate returning.</u> <u><u>The first half of the trip takes $2\frac{4}{5}$ hours and the return trip takes 4 hours.</u></u> Find the average speed on the first half of the trip.

2. An excursion boat makes a round-trip to a park. <u>The return trip is 6 kilometers per hour slower than the rate going to the park.</u> <u><u>The trip to the park takes 4 hours and the return trip takes 5 hours.</u></u> Find the average speed returning.

3. In going back and forth to work every day, Marge finds that the <u>average rate returning is 7 miles per hour less than the rate going.</u> <u><u>It takes Marge 1 hour to go to work and $1\frac{1}{4}$ hours to return home.</u></u> Find the average rate going and the average rate returning.

4. A jet makes a round-trip to San Francisco. <u><u>The first half of the trip takes 5 hours and the return trip takes 4 hours.</u></u> <u>The average rate returning is 80 kilometers per hour more than the average rate going.</u> Find the average rate going and returning.

5. <u><u>Two cars have the same cruising range.</u></u> <u>Car B's fuel tank holds 35 fewer liters than Car A.</u> <u>Car B averages 8.5 kilometers per liter of fuel. Car A averages 5 kilometers per liter.</u> Find the fuel tank capacity of each car.

6. The Lee family decided to drive to the beach and back. <u><u>The trip going took 4 hours. The trip returning took 5 hours.</u></u> <u>The average rate returning was 10 miles per hour slower than the rate going.</u> Find the average rate returning.

*Problem Solving: One Variable / **111***

Assignment Guide
Basic p. 111: 1–6
Average p. 111: 1–6
Above Average p. 111: 1–6

Problem–Solving Skills
Making a table (Ex. 1–6)
Using a formula (Ex. 1–6)
Writing and solving an equation (Ex. 1–6)

Additional Answers

2. a.

	Going	Returning
r(km/hr)	x	$x - 6$
t(hrs)	4	5
d(km)	$4x$	$5(x - 6)$

b. $4x = 5(x - 6)$
c. Returning: 24 km/hr

4. a.

	Going	Returning
r(km/hr)	x	$x + 80$
t(hrs)	5	4
d(km)	$5x$	$4(x + 80)$

b. $5x = 4(x + 80)$
c. Going: 320 km/hr
   Returning: 400 km/hr

6. a.

	Going	Returning
r(mi/hr)	x	$x - 10$
t(hrs)	4	5
d(mi)	$4x$	$5(x - 10)$

b. $4x = 5(x - 10)$
c. Returning: 40 mi/hr

Quiz Sections 4.4–4.5
After completing Sections 4.4–4.5, you may want to administer a quiz covering the same sections. See the quiz provided on page 92 in the *Teacher's ResourceBank.*

111

Making Tables   Surface Area/Volume

An important goal in industrial production is to keep the amount of raw materials to a minimum. For example, a food processor wants containers that provide the desired **capacity** but with dimensions that require the least amount of material **(surface area)**. The formula below gives S, the surface area of a cylindrical container in square centimeters.

$$S = \frac{(3.14r^3 + c)2}{r}$$ $c = $ capacity in milliliters
$r = $ radius in centimeters

EXAMPLE: Find, to the nearest whole number, the value of r that will require a minimum amount of material for a cylindrical can with a capacity of 354 milliliters.

SOLUTION: $S = \dfrac{(3.14r^3 + 354)2}{r}$

Use a calculator to complete the table shown at the right.

r	3.0	3.5	4.0	4.5
S	?	?	?	?

By calculator: When $r = 3.0$, $S = (3^3 \times 3.14 + 354) \times 2 \div 3$.

Solve also for $r = 3.5$, 4.0, and 4.5. ▶

In the table, the smallest value for S is 277.5. The corresponding value of r is about **4 centimeters.**

r	3.0	3.5	4.0	4.5
S	292.5	279.2	277.5	284.5

EXERCISES

1. In the Example, show that 3.8 is the value of r to the nearest *tenth*.
   (HINT: Find the values of S for $r = 3.7$, 3.8, and 3.9)

2. In the Example, the formula for the height of the can is $h = \dfrac{c}{3.14r^2}$. Find, using $r = 3.8$, the value of h that requires a minimum amount of material.

CHAPTER SUMMARY

IMPORTANT TERMS	Perimeter *(p. 95)* Rate of discount *(p. 104)* List price *(p. 104)* Mileage *(p. 109)* Discount *(p. 104)* Cruising range *(p. 109)* Sale price *(p. 104)*

IMPORTANT IDEAS	*1.* Steps for Solving Word Problems
	     a. Identify Condition 1. Use Condition 1 to represent the unknowns.
	     b. Identify Condition 2. Use Condition 2 to write an equation.
	     c. Solve the equation.
	     d. Check the results with the two conditions. Answer the question.
	 2. In equations that contain parentheses, use the Distributive Property first.

CHAPTER REVIEW

For Exercises 1–8, Condition 1 is <u>underscored once</u>. Condition 2 is <u>underscored twice</u>. Use Condition 1 to represent the unknowns. Use Condition 2 to write an equation for the problem. Solve.

SECTION 4.1

1. A librarian issued 82 more fiction books in one day than nonfiction books. The total number of books issued was 372. How many fiction books and how many non-fiction books were issued?

2. In a 2-day period the amount of rainfall the first day was 3.7 centimeters less than the amount the second day. In two days it rained 15.3 centimeters. Find the amount it rained each day.

3. On a family trip Alice drove 100 miles more than twice the number of miles her father drove. The total number of miles was 1150. Find how many miles Alice and her father each drove.

4. The labor cost of producing a machine part is $0.05 less than twice the cost of materials. The total unit production cost of the part is $1.72. Find the cost of materials and the cost of labor.

SECTION 4.2

5. A helicopter pad on top of a building is rectangular in shape. The length is 11.8 meters more than the width. The perimeter is 80.0 meters. Find the length and width.

6. A ceiling of a room is covered with rectangular panels. The length of each panel is twice the width. The perimeter is 365.76 centimeters. Find the length and width.

Problem Solving: One Variable / **113**

Chapter Test
Two Chapter Tests (Form A and Form B) are provided on pages 93–96 in the *Teacher's ResourceBank*.

Additional Answers
1. **a.** Let x = number of non-fiction books.
   Then $x + 82$ = number of fiction books.
   b. $x + x + 82 = 372$
   c. nonfiction: 145 books; fiction: 227 books
2. **a.** Let x = amount second day.
   Then $x - 3.7$ = amount first day.
   b. $x + x - 3.7 = 15.3$
   c. second day: 9.5 cm; first day: 5.8 cm
3. **a.** Let x = number of miles father drove.
   Then $2x + 100$ = number of miles Alice drove.
   b. $x + 2x + 100 = 1150$
   c. father: 350 miles; Alice: 800 miles
4. **a.** Let x = cost of materials.
   Then $2x - 0.05$ = cost of labor.
   b. $x + 2x - 0.05 = 1.72$
   c. materials: $0.59; labor: $1.13
5. **a.** Let x = width.
   Then $x + 11.8$ = length.
   b. $2x + 2(x + 11.8) = 80.0$
   c. width: 14.1 miles; length: 25.9 miles
6. **a.** Let x = width.
   Then $2x$ = length.
   b. $2x + 2(2x) = 365.76$
   c. width: 60.96 cm; length: 121.92 cm

7. **a.** Let $7x$ = length and $3x$ = width.
   b. $2(7x) + 2(3x) = 166.0$
   c. length: 58.1 mm; width: 24.9 mm

8. **a.** Let $5x$, $9x$, and $13x$ = lengths of legs of the route.
   b. $5x + 9x + 13x = 3681$
   c. $5x = 681\frac{2}{3}$ km; $9x = 1227$ km; $13x = 1772\frac{1}{3}$ km

19. **a.** Let x = Russell's monthly income. Then $x + 200$ = Brad's monthly income.
    b. $0.32x = 0.28(x + 200)$
    c. Russell's income: $1400 Brad's income: $1600

20. **a.** Let x = number of games played.
    b. $0.90x = 0.85x + 1$
    c. Each team played 20 games.

21. **a.** Let x = rate returning. Then $x - 5$ = rate going.
    b. $4.25(x - 5) = 4x$
    c. returning: 85 km/hr; going: 80 km/hr

22. **a.** Let x = Car A fuel tank capacity. Then $x - 5$ = Car B fuel tank capacity.
    b. $11.9x = 9.8(x - 5) + 154$
    c. Car A: 50 L; Car B: 45 L

---

7. The ratio of the length to the width of a rectangle is 7:3. The perimeter is 166.0 millimeters. Find the length and width.

8. A triangular plane route has legs in the ratio 5:9:13. The total route has a length of 3681 kilometers. Find the length of each leg.

SECTION 4.3

For Exercises 9–16, solve each equation. Check each answer.

9. $8h + 3 = 5h - 18$ $h = -7$

10. $2x + 5 = 6x + 29$ $x = -6$

11. $28 - 7y = y - 48$ $y = 9\frac{1}{2}$

12. $44 - a = -18 + 2a$ $a = 20\frac{2}{3}$

13. $\frac{1}{2}(4t - 6) - 3t = 2t - 7$ $t = 1\frac{1}{3}$

14. $-2v + 6 = 8 - \frac{1}{3}(9 - 3v)$ $v = \frac{1}{3}$

15. $4.3g - 28 = -1.9g + 65$ $g = 15$

16. $62 - 3.7x = 1.9x + 34$ $x = 5$

SECTION 4.4

Solve. Use the formulas $d = r \cdot p$ and $s = p - d$.

17. A tape deck is on sale at a discount rate of 40%. The sale price is $274.50. Find the list price. $457.50

18. An airline offered a special fare at a discount of 18%. The special fare was $479.70. Find the regular fare for this flight. $585

For Exercises 19–22, Condition 1 is underscored once. Condition 2 is underscored twice. Use Condition 1 to represent the unknowns. Use Condition 2 to write an equation for the problem. Solve.

19. Brad and Russell spend the same amount each month for rent. Brad's monthly income is $200 greater than Russell's monthly income. Brad spends 28% of his monthly income on rent and Russell spends 32% of his monthly income on rent. Find their monthly incomes.

20. The first and second place teams in a division won 90% and 85% of their games respectively. The first place team won one more game than the second place team. They played the same number of games. Find the number of games played by each team.

SECTION 4.5

21. Juan and Maria computed the average speed of the family car on a vacation trip. The trip to their vacation spot took 4.25 hours and the return trip took 4 hours. The average speed going was 5 kilometers per hour less than the speed returning. Find the average speed each direction.

22. The fuel tank capacity of Car B is 5 liters less than that of Car A. Car A averages 11.9 kilometers per liter and Car B averages 9.8 kilometers per liter. The range of Car A is 154 kilometers more than that of Car B. Find the fuel tank capacity of each car.

CHAPTER 5

Inequalities

Water will freeze when the temperature is less than or equal to 32°F, or 0°C. To prevent the water in a car radiator from freezing, antifreeze is added to the water because it lowers the freezing point.

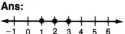
5.1   Graphs of Inequalities

OBJECTIVES: To write the solution set of an inequality from its graph
To draw a graph of an inequality based on its given or implied domain

The sentence "$x = 2$" means

"x is equal to two".

The sentence "$x < 2$" means "x is less than two". The solutions of these sentences are illustrated in the <u>graphs</u> below. The set of replacements for x is {integers}.

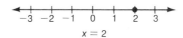   

P–1   **Why is "etc." necessary in the second graph?** To show that all integers less than 2 are solutions

The set of replacements for a variable is the **domain.** In the second graph above the domain of x is {integers}. If the domain is {real numbers}, then the graph of $x < 2$ is as shown below.

   *The point corresponding to 2 is circled in order to show that 2 is not a solution.*

Sentences such as $x < 2$ are called **inequalities.** Other kinds of inequalities are shown below.

Inequality	Meaning
$x > 2$	"x is greater than two."
$x \neq 2$	"x is not equal to two."
$x \leq 2$	"x is less than or equal to two."

P–2   **Why is 2 a solution of $x \leq 2$?** 2 is equal to 2.

P–3   **Why is 3 a solution of $x \neq 2$?** 3 is not equal to 2.

EXAMPLE 1   Draw a graph of the solutions of $x < 3$ on a number line. Domain = {counting numbers}. Examples 1 and 2 show the importance of defining the domain of the variable.

Solution: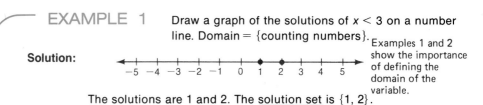

The solutions are 1 and 2. The <u>solution set</u> is {1, 2}.

EXAMPLE 2   Draw a graph of the solution set of $x < 3$.
             Domain = {integers}.

Solution:

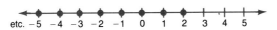

etc. -5 -4 -3 -2 -1 0 1 2 3 4 5

Solution set: {all integers less than 3}

If no domain is stated, it is always assumed in this text to be {real numbers}. *Underscore the importance of this statement.*

EXAMPLE 3   Draw a graph of $x \leq 3$.

"Graph of $x \leq 3$"
means "graph of
the solution set
of $x \leq 3$."

Solution:

-5 -4 -3 -2 -1 0 1 2 3 4 5

Solution set: {all real numbers less than or equal to 3}

P–4 **Why is the point corresponding to 3 darkened?** To show that 3 is a solution of $x \leq 3$

EXAMPLE 4   Graph $x > -\frac{1}{2}$.

"Graph" means
"draw a graph."

Solution:

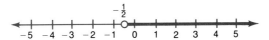

$-\frac{1}{2}$

-5 -4 -3 -2 -1 0 1 2 3 4 5

Solution set: {all real numbers greater than $-\frac{1}{2}$}

P–5 **What solution sets are graphed below?**
What domain is assumed? {all real numbers greater than or equal to $3\frac{1}{2}$}

a.

$3\frac{1}{2}$

-5 -4 -3 -2 -1 0 1 2 3 4 5

domain: {real numbers}

b.

$-2\frac{1}{2}$

-5 -4 -3 -2 -1 0 1 2 3 4 5

{all real numbers less than $-2\frac{1}{2}$}

domain: {real numbers}

Inequalities / **117**

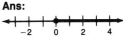
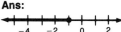
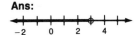

117

CLASSROOM EXERCISES

Find the solution set that is graphed. If necessary, name the domain that you are assuming. (P-5)

1.

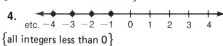

$-4\ -3\ -2\ -1\ \ 0\ \ 1\ \ 2\ \ 3\ \ 4$ etc.

$\{$all integers greater than $-2\}$

2.

$-4\ -3\ -2\ -1\ \ 0\ \ 1\ \ 2\ \ 3\ \ 4$

$\{$all real numbers less than $1\}$

3.

$2\frac{1}{2}$

$-4\ -3\ -2\ -1\ \ 0\ \ 1\ \ 2\ \ 3\ \ 4$

$\{$all real numbers less than $2\frac{1}{2}\}$

4.

etc. $-4\ -3\ -2\ -1\ \ 0\ \ 1\ \ 2\ \ 3\ \ 4$

$\{$all integers less than $0\}$

Name three solutions of each inequality. (Examples 1–4) Answers will vary.

5. $x \le 10$; Domain = $\{$whole numbers$\}$ 0; 4; 9

6. $x < -9$; Domain = $\{$integers$\}$ $-15, -13, -10$

7. $x \le 3.9$ $-1.2; 0; 2.1$

8. $x > -\pi$ $-3; \sqrt{2}; \pi$

9. $x \ge -5$; Domain = $\{$rational numbers$\}$ $-5; 3\frac{1}{4}; \frac{7}{8}$

10. $x \ge -2$; Domain = $\{$integers$\}$ $-2; 0; 16$

WRITTEN EXERCISES

Goals: To write the solution set of an inequality from a graph and to graph an inequality

Sample Problem: Graph $x < 6$. Domain = $\{$integers$\}$.

Answer:

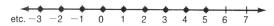

etc. $-3\ -2\ -1\ \ 0\ \ 1\ \ 2\ \ 3\ \ 4\ \ 5\ \ 6\ \ 7$

Write the solution set of each graph. If necessary, write the domain you are assuming. (P-5)

1. $\{-2, -1, 0, 1\}$

$-4\ -3\ -2\ -1\ \ 0\ \ 1\ \ 2\ \ 3\ \ 4$

$\{$all integers less than $1\}$

2. $\{-1, 0, 1, 2, 3\}$

$-4\ -3\ -2\ -1\ \ 0\ \ 1\ \ 2\ \ 3\ \ 4$

$\{$all integers greater than $-2\}$

3.

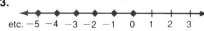

etc. $-5\ -4\ -3\ -2\ -1\ \ 0\ \ 1\ \ 2\ \ 3$

$\{$all real numbers greater than $-2\}$

4.

$-3\ -2\ -1\ \ 0\ \ 1\ \ 2\ \ 3\ \ 4\ \ 5$ etc.

$\{$all real numbers less than $-1\}$

5.

$-4\ -3\ -2\ -1\ \ 0\ \ 1\ \ 2\ \ 3\ \ 4$

6.

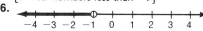

$-4\ -3\ -2\ -1\ \ 0\ \ 1\ \ 2\ \ 3\ \ 4$

118 / *Chapter 5*

Assignment Guide
Basic pp. 118–119: 1–18
Average pp. 118–119: 1–18, 19–25 odd
Above Average p. 118–119: 1–17 odd, 19–26; p. 122: 9–31 odd

7. {all real numbers less than or equal to -1}

-4 -3 -2 -1 0 1 2 3 4

8. {all real numbers greater than or equal to -2}

-4 -3 -2 -1 0 1 2 3 4

9.

$-1\frac{1}{2}$

-4 -3 -2 -1 0 1 2 3 4

{all real numbers greater than or equal to $-1\frac{1}{2}$}

10.

π

-4 -3 -2 -1 0 1 2 3 4

{all real numbers less than π}

MIXED PRACTICE

Draw a graph of each inequality.

11. $x \geq 2$; Domain = {whole numbers}   **12.** $x < 6$; Domain = {whole numbers}

13. $x < -2$; Domain = {integers}   **14.** $x \geq 5$; Domain = {integers}

15. $x < -\frac{3}{4}$ **16.** $x \geq -3\frac{1}{2}$ **17.** $x \geq 2.7$ **18.** $x < -1.9$

MORE CHALLENGING EXERCISES

Write Yes or No to show whether each graph below describes the solution set of the sentence. Domain = {real numbers}.

19. $-x = 2$

-2 -1 0 1 2 3

No

20. $-x = -3$

-4 -3 -2 -1 0 1

No

21. $-x > -3$

-1 0 1 2 3 4

Yes

22. $-x \geq -2\frac{1}{2}$

$-2\frac{1}{2}$

-4 -3 -2 -1 0 1

No

Does the graph of $x < a$ contain the point 0? Domain = {real numbers}.

23. The variable a represents a positive integer.   Yes

24. The variable a represents a negative integer.   No

25. The variable a represents a real number greater than 100.   Yes

26. The variable a represents zero.   No

REVIEW CAPSULE FOR SECTION 5.2

Change each fraction to decimal form. Then identify the greater number of each pair.

EXAMPLES: a. $\frac{2}{5}$ and $\frac{3}{5}$ **b.** $-\frac{7}{10}$ and $-\frac{8}{10}$

SOLUTIONS: a. $\frac{2}{5} = 0.4$; $\frac{3}{5} = 0.6$; $0.6 > 0.4$ Thus, $\frac{3}{5} > \frac{2}{5}$.

b. $-\frac{7}{10} = -0.7$; $-\frac{8}{10} = -0.8$; $-0.7 > -0.8$ Thus, $-\frac{7}{10} > -\frac{8}{10}$.

1. $\frac{3}{10}$ and $\frac{5}{10}$ **2.** $\frac{3}{4}$ and $\frac{1}{4}$ **3.** $\frac{2}{5}$ and $-\frac{3}{5}$ **4.** $\frac{6}{8}$ and $\frac{7}{8}$

5. $-\frac{4}{10}$ and $-\frac{5}{10}$ **6.** $-\frac{3}{8}$ and $-\frac{5}{8}$ **7.** $\frac{4}{6}$ and $\frac{5}{6}$ **8.** $-\frac{1}{3}$ and $-\frac{2}{3}$

*Inequalities / **119***

Additional Answers
Written Exercises

11. On a number line, the points 2, 3, 4, 5, etc.

12. On a number line, the points 5, 4, 3, 2, 1, 0

13. On a number line, the points -3, -4, -5, etc.

14. On a number line, the points 5, 6, 7, 8, etc.

15. On a number line, all points to the left of and not including $-\frac{3}{4}$

16. On a number line, all points to the right of and including $-3\frac{1}{2}$

17. On a number line, all points to the right of and including 2.7

18. On a number line, all points to the left of and not including -1.9

Problem–Solving Skills
Using logical reasoning
(Ex. 23–26)

Critical Thinking
See Exercises 23–26.

Additional Answers
Review Capsule

1. 0.3, 0.5; $\frac{5}{10} > \frac{3}{10}$

2. 0.75, 0.25, $\frac{3}{4} > \frac{1}{4}$

3. 0.4, -0.6; $\frac{2}{5} > -\frac{3}{5}$

4. 0.75, 0.875; $\frac{7}{8} > \frac{6}{8}$

5. -0.4, -0.5, $-\frac{4}{10} > -\frac{5}{10}$

6. -0.375, -0.625, $-\frac{3}{8} > -\frac{5}{8}$

7. 0.666 ···, 0.833 ···; $\frac{5}{6} > \frac{4}{6}$

8. -0.333 ···, -0.666 ···; $-\frac{1}{3} > -\frac{2}{3}$

119

5.2 Properties of Order

To write an equivalent inequality of the opposites of two real numbers from a given inequality
To write an inequality for x and y from two inequalities such as $x < a$ and $a < y$

Teaching Suggestions p. M–26

Lesson Resources
Warm–Up: p. M–27
Maintenance: See below.
Practice Worksheet 5.2

Maintenance

1. Evaluate: $6 - 4 + 3$
   Ans: 5 (Section 1.6)
2. Simplify: $\dfrac{-(5 - p)}{p - 5}$
   Ans: 1 (Section 2.4)
3. Multiply and simplify:
   $5(4y + 10)$
   Ans: 20y + 50
   (Section 2.5)
4. Solve and check:
   $x - 15.3 = -20.1$
   Ans: x = −4.8
   (Section 3.1)
5. A high school football team has 13 more senioir players than junior players (Condition 1). The total number of junior and senior players on the team is 57 (Condition 2). Find the number of juniors and the number of seniors on the team.
   Ans: juniors: 22
   seniors: 35
   (Section 4.1)

Additional Examples
Example 1
Write a new inequality by taking the opposite of each side.
1. $6 > -2$
   Ans: −6 < 2
2. $-4 < -3$
   Ans: 4 > 3
3. $-5.6 < 3.9$
   Ans: 5.6 > −3.9

Comparison Property

If a and b are real numbers, then exactly one of the following is true.

1. $a > b$ 2. $a = b$ 3. $a < b$

P–1 If $a = -1$ and $b = -2$, what inequality can you form by use of the Comparison Property? $a > b$

P–2 If $a = -1.3$ and $b = \dfrac{-13}{-10}$, what inequality can you form by use of the Comparison Property? $a < b$

EXAMPLE 1 Write an inequality by taking the opposite of each side of $-3 < 5$.

Solution:
$$-3 < 5$$
$$-(-3) \ \underline{?} \ -(5)$$
$$3 > -5$$

Taking the opposite of each side reverses the order.

You will need to emphasize again and again that taking the opposite of both sides reverses the inequality.

Order Property of Opposites

If a and b are real numbers and $a < b$, then $-a > -b$.

$-5.2 < -4.7$
$5.2 > 4.7$

You can write $-3 < 5$ as $5 > -3$ and $3 > -5$ as $-5 < 3$. Similarly, you can write $a < b$ as $b > a$ and $-a > -b$ as $-b < -a$.

P–3 What two inequalities can you form by applying the Order Property of Opposites to each of the following inequalities?

a. $y > -6$ b. $-13 < t$ c. $-r < s$ $r > -s; -s < r$
$-y < 6; 6 > -y$ $13 > -t; -t < 13$

P–4 What two inequalities relate r and 2 in the graph below?

$r < 2$
$2 > r$

120 / Chapter 5

P-5 What two inequalities relate 2 and *t* in the graph below?

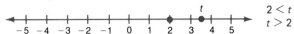

$2 < t$
$t > 2$

P-6 What two inequalities relate *r* and *t* in P-4 and P-5?

$r < t; t > r$

Note, if $a < b$ and $c < b$, you cannot determine the order of *a* and *c*.

> **Transitive Property of Order**
>
> If $a < b$ and $b < c$, then $a < c$.
>
> $-3.4 < 0$ and $0 < 1.9$.
> Then $-3.4 < 1.9$.

P-7 If $x < -2$ and $-2 < y$, what two inequalities relate *x* and *y*?

$x < y; y > x$

EXAMPLE 2   Suppose that "$-3 > q$" and "$-3 < p$" are both true. Write an inequality for *p* and *q*. Use "$<$."

Solution: $q < -3$ and $-3 < p$ ◀ **Transitive Property of Order**

$q < p$

You know that you can write $a < b$ as $b > a$, $b < c$ as $c > b$, and $a < c$ as $c > a$.

P-8 How would you state a Transitive Property of Order using "$>$"?

If $a > b$ and $b > c$, then $a > c$.

CLASSROOM EXERCISES

Write an inequality for each pair of numbers. (P-1, P-2)

1. $-5; 2$ $\quad -5 < 2$ **2.** $15; -13$ $\quad 15 > -13$ **3.** $-12; -9$ $\quad -12 < -9$ **4.** $14; -14$ $\quad 14 > -14$

5. $-16; 6$ $\quad -16 < 6$ **6.** $-21; -20$ $\quad -21 < -20$ **7.** $1.3; 1.03$ $\quad 1.3 > 1.03$ **8.** $-1.6; -1.7$ $\quad -1.6 > -1.7$

Complete each statement. Use $<$, $>$, or $=$. a, b, and c are real numbers.
(Example 1, P-7)

9. If $-c > -1$, then $c \underline{\ ? \ } 1$. $<$

10. If $-a < 4$, then $a \underline{\ ? \ } -4$. $>$

11. If $a > b$ and $b > 5$, then $a \underline{\ ? \ } 5$. $>$

12. If $c < -5$ and $a > -5$, then $c \underline{\ ? \ } a$. $<$

*Inequalities / **121***

121

Assignment Guide
Basic p. 122: 1–31 odd
Average p. 122: 1–31 odd
Above Average p. 122:
9–31 odd; p. 118: 1–17 odd, 19–26

WRITTEN EXERCISES

Goal: To write inequalities using the Comparison Property, the Order Property of Opposites, and the Transitive Property of Order

Sample Problem: Suppose that "$x < -31$" and "$y > -31$" are both true. Write an inequality for x and y. Use "$<$".

Answer: $x < y$

Write an inequality for the two numbers. Use "$<$". (P-1, P-2)

1. -5; 4 $\quad -5 < 4$
2. 17; -19 $\quad -19 < 17$
3. -12; -11 $\quad -12 < -11$
4. -5; -15 $\quad -15 < -5$

5. 1.7; 1.6 $\quad 1.6 < 1.7$
6. 0.7; -0.9 $\quad -0.9 < 0.7$
7. $-\frac{3}{4}$; $-\frac{7}{8}$ $\quad -\frac{7}{8} < -\frac{3}{4}$
8. $-\frac{1}{2}$; $-\frac{1}{3}$ $\quad -\frac{1}{2} < -\frac{1}{3}$

Write a new inequality by applying the Order Property of Opposites. (Example 1)

9. $3 < -b$ $\quad -3 > b$
10. $-a < 5$ $\quad a > -5$
11. $-a < -2$ $\quad a > 2$
12. $-7 > -b$ $\quad 7 < b$

13. $-(a+b) < 2$ $\quad a + b > -2$
14. $-10 < -(b+a)$ $\quad 10 > b + a$
15. $-b > a$ $\quad b < -a$
16. $-a < b$ $\quad a > -b$

Write True or False for each sentence. a, b, and c are real numbers. (Example 2)

17. If $c < -2$ and $-2 < a$, then $c < a$. True
18. If $b > 5$ and $5 > c$, then $b > c$. True
19. If $a < -2$, then $a < -5$ is true. False
20. If $c > -3$, then $-1 < c$ is true. False
21. If $a < 4$ and $b < 4$, then $a < b$. False
22. If $a > 5$ and $b > 5$, then $b > a$. False
23. If $b < c$ and $a < b$, then $a < c$. True
24. If $7 > b$ and $a > 7$, then $a > b$. True

Write an inequality for x and y. Use "$<$". (Example 2)

25. $y < -2$ and $-2 < x$ $\quad y < x$
26. $y < 5$ and $5 < x$ $\quad y < x$
27. $x < -3.8$ and $y > -3.8$ $\quad x < y$
28. $y < 0.9$ and $x > 0.9$ $\quad y < x$
29. $213 > -x$ and $y < -213$ $\quad y < x$
30. $-56 > y$ and $-x < 56$ $\quad y < x$
31. $-x > 19.3$ and $19.3 > -y$ $\quad x < y$
32. $-y < -10.3$ and $-x > -10.3$ $\quad x < y$

REVIEW CAPSULE FOR SECTION 5.3

Solve each equation. (Pages 58–60)

1. $64 = t - 71$ $\quad t = 135$
2. $x + 7 = -8$ $\quad x = -15$
3. $y - 5 = 8$ $\quad y = 13$
4. $3.7 = w - 4.2$ $\quad w = 7.9$
5. $t + 0.9 = -5.2$ $\quad t = -6.1$
6. $-\frac{3}{4} = \frac{7}{8} - r$
7. $-d - 1\frac{2}{3} = -\frac{5}{6}$ $\quad d = -\frac{5}{6}$
8. $-2.07 = p - 5.93$ $\quad p = 3.86$
9. $s + 0.87 = 0.34$ $\quad s = -0.53$

5.3 Addition and Subtraction

OBJECTIVES: To solve inequalities of the forms $x - a > b$ or $x - a < b$
To solve inequalities of the forms $x + a > b$ or $x + a < b$

Compare the solution sets and graphs for each pair of inequalities in the table below. The domain is $\{-1, 0, 1, 2, 3, 4, 5, 6, 7, 8\}$.

Inequalities	Solution Sets	Graphs
$x < 5$ $x + 3 < 5 + 3$ or $x + 3 < 8$	$\{-1, 0, 1, 2, 3, 4\}$	
$x > 5$ $x + 3 > 5 + 3$ or $x + 3 > 8$	$\{6, 7, 8\}$	

In the table above,

$$x < 5 \text{ is } \textit{equivalent} \text{ to } x + 3 < 8$$
and
$$x > 5 \text{ is } \textit{equivalent} \text{ to } x + 3 > 8,$$

because each pair has the **same solution set.** This suggests that adding the same number to each side of an inequality results in a new inequality that is equivalent to the first.

EXAMPLE 1   Solve: $x - 2 > -1$ ◀ **Solve means to find the solution set.**

Solution: $x - 2 > -1$

☐1 Add 2 to each side. ────────▶ $x - 2 + 2 > -1 + 2$

☐2 Simplify. ──────────────▶ $x > 1$

Solution set: {all real numbers greater than 1}

Point out the analogy between this property and the Addition Property for Equations.

> **Addition Property for Inequalities**
>
> Adding the same real number to each side of an inequality does not change the order of the inequality. The new inequality is equivalent to the first.
>
> $x - 12 > 3$
> $x - 12 + 12 > 3 + 12$
> $x > 15$

Inequalities / **123**

Teaching Suggestions p. M–26

Lesson Resources
Warm–Up: p. M–27
Maintenance: See below.
Practice Worksheet 5.3
Exploration 5

Maintenance
1. Evaluate: $(3)(4) + 6$
   Ans: 18 (Section 1.6)
2. Factor: $-8x^2 - 24x$
   Ans: $-8x(x + 3)$
   (Section 2.5)
3. Solve and check: $-6y = 57$
   Ans: $y = -9.5$
   (Section 3.2)
4. Solve and check:
   $4(2x - 5) = -2(x + 3)$
   Ans: $x = \frac{7}{5}$
   (Section 4.3)
5. A limousine makes a round-trip run (Condition 2). The average rate going is 6 miles per hour less than the average rate returning (Condition 1). The first half of the trip takes 5 hours and the return takes 4.5 hours (Condition 2). Find the average rate on the first half of the trip.
   Ans: 54 mi/hr (Section 4.5)

Additional Examples
Example 1
Solve.
1. $x - 4 > 2$
   Ans: {all real numbers greater than 6}
2. $x - 8 > -3$
   Ans: {all real numbers greater than 5}
3. $x - 11.3 > -5.7$
   Ans: {all real numbers greater than 5.6}

Additional Examples
Example 2
Solve.
1. $x - 1.3 < 1.3$
   Ans: {all real numbers less than 2.6}
2. $15 > x - 11$
   Ans: {all real numbers less than 26}
3. $-8 > x - 5.2$
   Ans: {all real numbers less than −2.8}

Example 3
Solve.
1. $x + 4 < 7.6$
   Ans: {all real numbers less than 3.6}
2. $x + 2\frac{1}{2} > -6\frac{1}{2}$
   Ans: {all real numbers greater than −9}
3. $-0.9 < x + 7.2$
   Ans: {all real numbers greater than −8.1}

Common Error
Students incorrectly solve inequalities involving addition or subtraction.

Example
$$x - 5 > 7 \qquad x + 3 < 12$$
$$x > 2 \qquad\quad x < 15$$

Prescription
Insist that students show all steps in the solution, especially the step showing addition of a number to each side or subtraction of a number from each side.

Additional Answers
7. {all real numbers less than 11}
8. {all real numbers greater than 6}
9. {all real numbers less than 8}

EXAMPLE 2 — Solve: $x - 5 < -1$

Solution:

$$x - 5 < -1$$

1. Add 5 to each side. → $x - 5 + \mathbf{5} < -1 + \mathbf{5}$
2. Simplify. → $x < 4$

Solution set: {all real numbers less than 4}

The inequality $x + 7 + (\mathbf{-7}) < 12 + (\mathbf{-7})$ is equivalent to $x + 7 - \mathbf{7} < 12 - \mathbf{7}$. This suggests a subtraction property.

> **Subtraction Property for Inequalities**
>
> Subtracting the same real number from each side of an inequality does not change the order of the inequality. The new inequality is equivalent to the first.
>
> $$x + \tfrac{1}{2} < \tfrac{3}{4}$$
> $$x + \tfrac{1}{2} - \tfrac{1}{2} < \tfrac{3}{4} - \tfrac{1}{2}$$
> $$x < \tfrac{1}{4}$$

EXAMPLE 3 — Solve: $x + 5.3 < 2.9$

Example 3 may also be solved by adding −5.3 to each side.

Solution:

$$x + 5.3 < 2.9$$

1. Subtract 5.3 from each side. → $x + 5.3 - \mathbf{5.3} < 2.9 - \mathbf{5.3}$
2. Simplify. → $x < -2.4$

Solution set: {all real numbers less than −2.4}

CLASSROOM EXERCISES

Write what number to add to, or subtract from, each side of the inequality to get the variable by itself. (Step 1 in Examples 1–3)

1. $r + 5 < 3$ Subtract 5.
2. $w - \frac{1}{2} > 2$ Add $\frac{1}{2}$.
3. $-13 > a + 7$ Subtract 7.
4. $-5.6 < 1.8 + c$ Subtract 1.8.
5. $-0.8 + x < 5.9$ Add 0.8.
6. $-3\frac{1}{4} > t - 5\frac{1}{8}$ Add $5\frac{1}{8}$.

Solve. (Examples 1–3)

7. $x - 1 < 10$
8. $y + 2 > 8$
9. $a - 3 < 5$

See the Answers to Selected Exercises for answers to Ex. 13-17 odd.

10. $t + 7 > -2$ **11.** $w - 5 < -1$ **12.** $10 > -3 + s$

13. $-5 + r < -1$ **14.** $p - 2 > -3$ **15.** $-0.5 + x > -8.5$

16. $3.6 < -1.4 + y$ **17.** $2x - 3 - x < 5$ **18.** $-7 > 3a + 5 - 2a$

▓▓▓▓ WRITTEN EXERCISES ▓▓▓▓

See the Answers to Selected Exercises for answers to Ex. 11-27 odd.

Goal: To solve an inequality using the Addition or Subtraction Property for Inequalities

Sample Problems: Solve: **a.** $x - 5.3 < 2.9$ **b.** $x + 5.3 < 2.9$

Answers: a. {all real numbers less than 8.2}
             b. {all real numbers less than -2.4}

Write Yes or No to show whether the two inequalities are equivalent.
(Examples 1–3)

1. $x + 2 < 3; x < 5$ No

2. $-5 < y - 2; -7 < y$ No

3. $2a - 7 < -2; 2a < 5$ Yes

4. $5r + 6 > -1; 5r > -7$ Yes

5. $x + 2 < 5; x + 1 < 6$ No

6. $y - 3 < 4; y - 2 < 5$ Yes

7. $2x + 2 < 5x; 2x < 5x - 2$ Yes

8. $3t < 4t - 1; 3t - 1 < 4t$ No

9. $3w - 1 > w + 7; 3w - 2 > w + 8$ No

10. $9g - 5 > g + 3; 9g - 3 > g + 5$ Yes

MIXED PRACTICE

Solve.

11. $y + 2 > -3$ **12.** $x + 5 < 8$ **13.** $-3 + t < -10$

14. $-7 + r > -13$ **15.** $x - 1 < -2$ **16.** $p - 3 > -6$

17. $-3.4 > -0.3 + q$ **18.** $-4.5 < b - 1.2$ **19.** $-2 + w > 4$

20. $-5 + y > -12$ **21.** $4.2 < y - 3.7$ **22.** $3.9 < y + 1.6$

MORE CHALLENGING EXERCISES

Solve. Domain = {integers}.

23. $2 - x < 3$ **24.** $-y - 7 > -2$ **25.** $-12 > -5 - a$

26. $15 < -b + 12$ **27.** $-p + \frac{1}{2} \leq 5\frac{1}{2}$ **28.** $\frac{2}{3} - q \geq -4\frac{1}{3}$

For Exercises 29–30, use the given inequalities to write an inequality for a and b. Use "$<$."

29. $a - y > b$ and $y > 0$
         $b < a$

30. $y - a > c$ and $y - b < c$
               $a < b$

NON-ROUTINE PROBLEMS

31. Dan has only a 9-quart pail, a 3-quart pail, and a 2-quart pail to carry water. How can Dan measure exactly eight quarts of water if he completely fills each of the pails just once?

32. Two shirts and one jacket cost the same amount as two sweaters. One sweater and two shirts cost the same amount as one jacket. Which costs the least—a shirt, a jacket, or a sweater?

33. Susan dug up 100 flower bulbs and separated them into five paper bags. The number of bulbs put in the first and second bags totaled 48. In the second and third bags, the bulbs totaled 32; in the third and fourth bags, the bulbs totaled 27; and in the fourth and fifth bags, the bulbs totaled 41. How many bulbs were put in each bag?

34. Trace the figure below. Connect all the points in the figure by drawing exactly four line segments without lifting your pencil off the paper. Do not draw through any one point more than once.

MID-CHAPTER REVIEW

Draw a graph of each inequality. (Section 5.1)

1. $x > -4$; Domain = {real numbers}

2. $x \leq 3$; Domain = {real numbers}

3. $x < 5$; Domain = {integers}

4. $x \geq -2$; Domain = {integers}

5. $x \leq 6$; Domain = {whole numbers}

6. $x > 4$; Domain = {real numbers}

Write an inequality for a and b. Use "<." (Section 5.2)

7. $a < 3$ and $b > 5$ $a < b$

8. $a > 2$ and $b < -3$ $b < a$

9. $-3.8 < a$ and $-4.3 > b$ $b < a$

10. $4 < -a$ and $b > -4$ $a < b$

11. $a > -3\frac{1}{2}$ and $b < -4$ $b < a$

12. $-a < 0$ and $-5 > b$ $b < a$

Solve. (Section 5.3)

13. $t - 8 < -3$

14. $-2 + w > 5$

15. $8.2 < x + 9.4$

16. $-4.7 > -2.3 + q$

17. $r + 1.5 < 1.2$

18. $-1.7 < y - 8.3$

REVIEW CAPSULE FOR SECTION 5.4

Solve each equation. (Pages 71–73 and 100–102)

1. $6n + 5 - n = 14$ $n = 1\frac{4}{5}$

2. $-12t + 16 = -8 - 3t$ $t = 2\frac{2}{3}$

3. $4(w - 8) = 9w + 3$ $w = -7$

4. $1 - 5y = -3(y - 7)$ $y = -10$

5. $-6(2 - p) = 2(-3p + 4)$ $p = 1\frac{2}{3}$

6. $-\frac{1}{4}(8m - 4) = -\frac{1}{3}(12 + 3m)$ $m = 5$

OBJECTIVES: To solve an inequality in which the variable appears more than once but only on one side

5.4 Solving Inequalities

To solve an inequality in which the variable appears on both sides
To solve an inequality in which the variable appears on both sides and requires use of the Distributive Property

Teaching Suggestions p. M–26

Lesson Resources
Warm–Up: p. M–27
Maintenance: See below.
Practice Worksheet 5.4

EXAMPLE 1   Solve: $4x + 2 - 5x < -12$

Solution:
$$4x + 2 - 5x < -12$$

1	Combine like terms.	$\longrightarrow$	$-x + 2 < -12$
2	Subtract 2 from each side.	$\longrightarrow$	$-x + 2 - 2 < -12 - 2$
3	Simplify.	$\longrightarrow$	$-x < -14$
4	Order Property of Opposites	$\longrightarrow$	$x > 14$

Solution set: {all real numbers greater than 14}

The variable often appears on both sides of an inequality.

EXAMPLE 2   Solve: $2x - 5 > 3x + 2$

Solution:

1	Subtract 2x from each side.	$\longrightarrow$	$2x - 5 - \mathbf{2x} > 3x + 2 - \mathbf{2x}$
2	Simplify.	$\longrightarrow$	$-5 > x + 2$
3	Subtract 2 from each side.	$\longrightarrow$	$-5 - 2 > x + 2 - 2$
4	Simplify.	$\longrightarrow$	$-7 > x$ or $x < -7$

Solution set: {all real numbers less than −7}

EXAMPLE 3   Solve: $3(x - 3) < 2x + 1$

Solution:

1	Distributive Property	$\longrightarrow$	$3x - 9 < 2x + 1$
2	Subtract 2x from each side.	$\longrightarrow$	$3x - 9 - \mathbf{2x} < 2x + 1 - \mathbf{2x}$
3	Simplify.	$\longrightarrow$	$x - 9 < 1$
4	Add 9 to each side.	$\longrightarrow$	$x - 9 + 9 < 1 + 9$
5	Simplify.	$\longrightarrow$	$x < 10$

Solution set: {all real numbers less than 10}

EXAMPLE 4   Solve: $2x - 5 > 2(x + 3)$

Solution:

1	Distributive Property	$\longrightarrow$	$2x - 5 > 2x + 6$
2	Subtract 2x from each side.	$\longrightarrow$	$2x - 5 - \mathbf{2x} > 2x + 6 - \mathbf{2x}$
3	Simplify.	$\longrightarrow$	$-5 > 6$

"−5 > 6" is _false_. The solution set is _empty_.

Inequalities / 127

Maintenance
1. Evaluate: $2 - 4 \cdot 3^2$
   Ans: −34 (Section 1.7)
2. Factor: $2a^2 + 6a$
   Ans: 2a(a + 3)
   (Section 2.5)
3. Solve and check:
   $18 = x + 20$
   Ans: x = −2 (Section 3.1)
4. Solve and check:
   $2.6x - 4 = 1.8x + 12$
   Ans: x = 20 (Section 4.3)
5. The rate of discount of a vase on sale was 15%. The discount was $4.50. Find the list price.
   Ans: $30.00 (Section 4.4)

Additional Examples
Example 1
Solve:
$7x + 8 - 8x > 20$
Ans: {all real numbers less than −12}

Example 2
Solve:
$4x < 11x - 35$
Ans: {all real numbers greater than 5}

Example 3
Solve:
$7(x - 3) < 6x + 5$
Ans: {all real numbers less than 26}

Example 4
Solve:
$3x - 7 > 3(x + 4)$
Ans: ϕ

Since the inequality $-5 > 6$ is false, there is no replacement for x that will make $2x - 5 > 2(x + 3)$ true. The solution set is empty. The symbol for the empty set is "ϕ". *Have students substitute various values for x in the original equation to show that it will always be false.*

EXAMPLE 5   Solve: $3x + 7 - 2x > x - 3$

Solution:

1. Combine like terms. ⟶ $x + 7 > x - 3$
2. Subtract x from each side. ⟶ $x + 7 - x > x - 3 - x$
3. Simplify. ⟶ $7 > -3$

Have students substitute various values for x in the original equation to show that it will always be true.

The inequality $7 > -3$ is true. Thus, $3x + 7 - 2x > x - 3$ is true for any number that replaces x. The solution set is {real numbers}.

CLASSROOM EXERCISES

See the Answers to Selected Exercises for answers to Ex. 1-21 odd.

Write the inequality you get as the first step in solving each inequality in the Written Exercises. (Step 1 of Examples 1–5)

WRITTEN EXERCISES

See the Answers to Selected Exercises for answers to Ex. 1-21 odd.

Goal: To solve an inequality in which the variable appears more than once
Sample Problem: Solve: $4x - 5 < 3(x + 2)$
Answer: {all real numbers less than 11}

Solve each inequality. (Examples 1–5)

1. $-3 < -7x + 2 + 8x$
2. $8 > 4x - 3 - 3x$
3. $3\frac{1}{2}a - 2 - 2\frac{1}{2}a < -7$
4. $-5\frac{1}{4}t + 7 + 6\frac{1}{4}t < -9$
5. $12y + \frac{1}{2} - 13y > -12\frac{1}{4}$
6. $3\frac{1}{8} > -7r - \frac{1}{4} + 6r$
7. $3w - 5 < 2w + 12$
8. $s - 2 < 2s - 8$
9. $3\frac{1}{2}x - 2 > 2\frac{1}{2}x + 10$
10. $\frac{2}{3}x - 5 > -\frac{1}{3}x - 2$
11. $0.6m - 0.7 < -0.4m + 0.2$
12. $1.2n - 3.6 < 0.2n + 5.1$
13. $3(-x + 2) < -5 - 2x$
14. $x - 6 > 2(x - 5)$
15. $5 - x < -2(x - 3) - 2$
16. $\frac{1}{2}(4x - 3) + \frac{1}{2} < x - 6$
17. $5y - 2 - 3y < \frac{1}{2}(6y - 8)$
18. $5(2 - a) > -3a - 7 - a$
19. $3a - 5 - 2a < a + 2$
20. $3(n - 4) > n - 2 + 2n$
21. $5(x - 1) < 2x - 6 + 3x$

REVIEW CAPSULE FOR SECTION 5.5

Solve each equation. (Pages 61–63)

1. $\frac{3}{4}x = 27$ $x = 36$
2. $\frac{8}{7}n = -96$ $n = -84$
3. $-12 = -\frac{1}{5}t$ $t = 60$
4. $\frac{r}{-6} = 2\frac{2}{3}$ $r = -16$
5. $-0.8 = \frac{w}{7}$ $w = -5.6$
6. $4p = -22$ $p = -5\frac{1}{2}$
7. $-3q = -12\frac{3}{5}$ $q = 4\frac{1}{5}$
8. $0.3s = -7.8$ $s = -26$

OBJECTIVES: To solve inequalities of the form $\frac{a}{b} \cdot x > c$ or $\frac{a}{b} \cdot x < c$ by the
Multiplication Property for Inequalities

5.5 **Multiplication and Division**

To solve inequalities of the form $ax > b$ or $ax < b$ by the
Division Property of Inequalities

Teaching Suggestions p. M–26

Lesson Resources
Warm–Up: p. M–27
Maintenance: See below.
Practice Worksheet 5.5

The table below shows that multiplying each side of an inequality by the same positive number does not change the direction of the inequality.

Multiplying each side by a negative number, however, reverses the direction of the inequality.

Table	Inequality	Multiply by 1.	Multiply by 2.	Multiply by −1.	Multiply by −2.
	$4 < 9$	$4 < 9$	$8 < 18$	$-4 > -9$	$-8 > -18$
	$-3 < 6$	$-3 < 6$	$-6 < 12$	$3 > -6$	$6 > -12$
	$5 > -7$	$5 > -7$	$10 > -14$	$-5 < 7$	$-10 < 14$

This table suggests the following property.

Maintenance
1. Multiply: $(2x - 5y)7x$
   Ans: $14x^2 - 35xy$
   (Section 2.5)
2. Solve and check:
   $\frac{1}{3}n + 6 = 12$
   Ans: $n = 18$ (Section 3.3)
3. Solve and check:
   $4 - 3x + 2 - 7x = -24$
   Ans: $x = 3$ (Section 3.5)
4. Solve and check:
   $10 - 6x = 25 - 3x$
   Ans: $x = -5$ (Section 4.3)
5. The temperature at midday was $-1°C$. By midafternoon it had risen $6°C$. Find the temperature at midafternoon.
   Ans: $5°C$ (Section 1.2)

> **Multiplication Property for Inequalities**
>
> If each side of an inequality is multiplied by the same nonzero real number, an equivalent inequality is obtained as follows.
> 1. The direction of the inequality is unchanged when each side is multiplied by the same positive number.
> 2. The direction of the inequality is reversed when each side is multiplied by the same negative number.
>
> $\frac{1}{2}x > 12$
> $2(\frac{1}{2}x) > 2(12)$
> $x > 24$
>
> $-\frac{2}{3}x < 6$
> $-\frac{3}{2}(-\frac{2}{3}x) > -\frac{3}{2}(6)$
> $x > -9$

Students tend to forget to reverse the inequality when multiplying each side by a negative number.

P–1 **What do you multiply each side of these inequalities by to get the variable alone?**

a. $\frac{1}{2}x > 3$ 2   **b.** $-\frac{1}{3}x < 8$ −3   **c.** $\frac{4}{5}x > 9$ $\frac{5}{4}$ **d.** $-\frac{2}{3}x > 10$ $-\frac{3}{2}$

EXAMPLE 1   Solve: $-\frac{2}{3}x > 10$

Solution:
1. Multiply each side by $-\frac{3}{2}$. ⟶ $-\frac{3}{2}(-\frac{2}{3}x) < -\frac{3}{2}(10)$ ◀ **Direction of the inequality is reversed.**
2. Simplify. ⟶ $x < -15$

Solution set: {all real numbers less than -15}

Additional Examples
Example 1
Solve.
1. $\frac{3}{5}x < -15$
   Ans: {all real numbers less than -25}
2. $-\frac{1}{2}x > 10$
   Ans: {all real numbers less than -20}
3. $-24 < -\frac{2}{3}x$
   Ans: {all real numbers less than 36}

For each inequality below, what inequality do you get if each side is divided by 3? by −3?

$$4x < -1; -4x > 1$$

a. $3x < 9$ **b.** $-6x > 6$ **c.** $12x < -3$ **d.** $-15x < 1$

$x < 3; -x > -3$ $-2x > 2; 2x < -2$ $-5x < \frac{1}{3}; 5x > -\frac{1}{3}$

Division Property for Inequalities

If each side of an inequality is divided by the same nonzero real number, an equivalent inequality is obtained as follows.

1. The direction of the inequality is unchanged when each side is divided by the same positive number.
2. The direction of the inequality is reversed when each side is divided by the same negative number.

$$5x < 15$$
$$\frac{5x}{5} < \frac{15}{5}$$
$$x < 3$$

$$-3x > 9$$
$$\frac{-3x}{-3} < \frac{9}{-3}$$
$$x < -3$$

Additional Examples

Example 2
Solve.
1. $5x > -30$
   Ans: {all real numbers greater than −6}
2. $4x < 18$
   Ans: {all real numbers less than 4.5}
3. $-20 < 2x$
   Ans: {all real numbers greater than −10}

Example 3
Solve.
1. $-8x > 40$
   Ans: {all real numbers less than −5}
2. $-0.7x > -21$
   Ans: {all real numbers less than 30}
3. $55 > -5x$
   Ans: {all real numbers greater than −11}

Common Error
Students do not change the direction of an inequality when multiplying or dividing each side by a negative number.

Example
$-2x > 12$
$x > -6$

Prescription
Review the Multiplication Property for Inequalities on page 129. Have the students choose one number from their solution set and check that number in the original inequality.

EXAMPLE 2   Solve: $3x < 15$ This can also be solved by *multiplying* each side by $\frac{1}{3}$.

Solution:

1. Divide each side by 3.   ⟶ $\dfrac{3x}{3} < \dfrac{15}{3}$

2. Simplify.   ⟶ $x < 5$

Solution set: {all real numbers less than 5}

This can also be solved by *multiplying* each side by $-\frac{1}{5}$.

EXAMPLE 3   Solve: $-5x < 20$

Solution:

1. Divide each side by −5.   ⟶ $\dfrac{-5x}{-5} > \dfrac{20}{-5}$

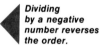
Dividing by a negative number reverses the order.

2. Simplify.   ⟶ $x > -4$

Solution set: {all real numbers greater than −4}

CLASSROOM EXERCISES

Multiply each side by 5. Then write an equivalent inequality.
(Table, Multiplication Property for Inequalities)

1. $x > 2$ $5x > 10$ **2.** $y < -3$ $5y > -15$ **3.** $2a > 10$ $10a > 50$ **4.** $\dfrac{t < 3}{\frac{1}{5}t < \frac{3}{5}}$

Multiply each side by −3. Then write an equivalent inequallty.
(Step 1 of Example 1, Multiplication Property for Inequalities)

5. $y > 2$ $-3y < -6$ **6.** $s < -5$ $-3s > 15$ **7.** $-\frac{1}{3}r < 2$ $r > -6$ **8.** $-\frac{1}{3}y > -10$ $y < 30$

Write the number you would multiply or divide each side by to get the variable by itself. (Step 1 of Examples 1–3)

9. $2x < 1$ Divide by 2. **10.** $-\frac{1}{2}x > \frac{2}{3}$ Multiply by −2. **11.** $-\frac{1}{2}x < -3$ Multiply by −2. **12.** $-\frac{1}{3}y > -2$ Multiply by −3.

To form pairs of equivalent inequalities, indicate if $<$ or $>$ should be used as the missing symbol. (Table, P-2, Multiplication or Division Property for Inequalities)

13. $5y > 35$; y __?__ 7 $>$ **14.** $\frac{1}{2}t < -1$; -2 __?__ t $>$
15. $-3 > -\frac{1}{3}x$; 9 __?__ x $<$ **16.** $2 < -0.1y$; y __?__ -20 $<$

WRITTEN EXERCISES

Goal: To solve an inequality using the Multiplication or Division Property for Inequalities
Sample Problems: a. $-\frac{3}{4}x > 12$ **b.** $4x < -8$
Answers: a. {all real numbers less than −16}
b. {all real numbers less than −2}

Solve. (Examples 1–3)   See the Answers to Selected Exercises for answers to Ex. 1-19 odd.

1. $3y < 15$ **2.** $5y > 30$ **3.** $\frac{1}{4}t > -5$ **4.** $-10t > -20$
5. $-7r < 21$ **6.** $\frac{1}{3} < -\frac{1}{6}y$ **7.** $\frac{1}{2} > -\frac{1}{4}x$ **8.** $\frac{1}{2}x < -8$
9. $\frac{2}{3}w < -4$ **10.** $\frac{3}{5}p > -9$ **11.** $|-\frac{1}{2}|x < |-3|$ **12.** $|-\frac{1}{3}|y > |-2|$
13. $-\frac{4}{3}q > 12$ **14.** $-\frac{8}{7}b < 16$ **15.** $-3x > 11$ **16.** $-5t < -12$
17. $0.6n < -6$ **18.** $-0.9m > 18$ **19.** $-4.5 < 1.5x$ **20.** $3.4 > -17y$

NON-ROUTINE PROBLEM

21. Each of the three boxes shown in the figure at the right contains socks. The socks are either gray, G, or black, B. One box contains two gray socks; one box contains two black socks; and one box contains one gray sock and one black sock. None of the boxes is labeled correctly. From which box can you draw exactly one sock to determine how all three boxes should be labeled?

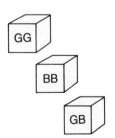

Assignment Guide
Basic p. 131: 1–19 odd
Average p. 131: 1–21 odd
Above Average p. 131:
1–21 odd

Problem–Solving Skills
Using a diagram (Ex. 21)
Using guess and check (Ex. 21)
Using logical reasoning (Ex. 21)

Non–Routine Problems
For a description of non–routine problems, see page M–13.

Additional Answers
Written Exercises
2. {all real numbers greater than 6}
4. {all real numbers less than 2}
6. {all real numbers less than −2}
8. {all real numbers less than −16}
10. {all real numbers greater than −15}
12. {all real numbers greater than 6}
14. {all real numbers greater than −14}
16. {all real numbers greater than $\frac{12}{5}$}
18. {all real numbers less than −20}
20. {all real numbers greater than −0.2}
21. Since none of the boxes is labeled correctly, draw a sock from the box labeled GB. If you draw a black sock, the box should be labeled BB, the box labeled BB should be labeled GG, and the box labeled GG can only be correctly labeled GB.

Teaching Suggestions p. M–26

Lesson Resources
Warm–Up: p. M–27
Maintenance: See below.
Practice Worksheet 5.6
Transparency 14

Maintenance
1. Simplify: $-(-x + 3y)$
   Ans: $x - 3y$ (Section 2.4)
2. Solve and check:
   $4(2m - 3) = 44$
   Ans: $m = 7$ (Section 3.5)
3. Solve and check:
   $3x - 3 = 2x + 5$
   Ans: $x = 8$ (Section 4.3)
4. Solve and check
   $4(n + 3) - 7 = 2(n + 6) + 2$
   Ans: $n = 4.5$ (Section 4.3)
5. A poster is rectangular in shape. Its length is 12 inches greater than its height (Condition 1). Its perimeter is 76 inches (Condition 2). Find its length and height.
   Ans: length: 25 inches; height: 13 inches
   (Section 4.2)

Additional Examples
Example 1
Solve.
1. $5x - 4 > 21$
   Ans: {all real numbers greater than 5}
2. $-14 > -6 + 8x$
   Ans: {all real numbers less than -1}

Example 2
Solve.
1. $-12x > -5x - 14$
   Ans: {all real numbers less than 2}
2. $7 - 5x < 4x + 34$
   Ans: {all real numbers greater than -3}

OBJECTIVES: To solve inequalities of the forms $ax + b < c$, $ax + b > c$, $ax - b < c$, and $ax - b > c$

5.6 Inequalities: More Than One Step

To solve inequalities in which more than one step is required by using the Addition, Subtraction, Multiplication, and Division Properties of Inequalities and the Distributive Property

When you multiply each side of an inequality by a positive **number to** get the variable alone, you are using the same steps as you **used to** solve an equation.

The methods used in solving inequalities are similar to those used in solving equations

EXAMPLE 1   Solve: $3x - 2 < 10$

Solution:

1. Add 2 to each side. $\longrightarrow$ $3x - 2 + 2 < 10 + 2$
2. Simplify. $\longrightarrow$ $3x < 12$
3. Multiply by the reciprocal of 3. $\longrightarrow$ $\frac{1}{3}(3x) < \frac{1}{3}(12)$
4. Simplify. $\longrightarrow$ $x < 4$

Solution set: {all real numbers less than 4}

P–1 **Why is the order symbol "<" not reversed in any of the steps in Example 1?** *The order of an inequality is unchanged if each side is multiplied by a positive number.*

In solving inequalities, use the Addition and Subtraction Properties for Inequalities before the Multiplication and Division Properties.

EXAMPLE 2   Solve: $-4x - 11 > 5 - 2x$

Solution:

1. Add $2x$ to each side. $\longrightarrow$ $-4x - 11 + 2x > 5 - 2x + 2x$
2. Simplify. $\longrightarrow$ $-2x - 11 > 5$
3. Add 11 to each side. $\longrightarrow$ $-2x - 11 + 11 > 5 + 11$
4. Simplify. $\longrightarrow$ $-2x > 16$
5. Multiply each side by $-\frac{1}{2}$. $\longrightarrow$ $-\frac{1}{2}(-2x) < -\frac{1}{2}(16)$
6. Simplify. $\longrightarrow$ $x < -8$

The order of the inequality is reversed.

Solution set: {all real numbers less than -8}

P–2 **Why must the order be reversed in Step 5 ?** *Each side is multiplied by a negative number.*

In inequalities that contain parentheses, use the Distributive Property first. Then eliminate the variable from one side of the inequality.

EXAMPLE 3   Solve: $-2x + 3(x + 3) > -4x - 6$

Solution:

This type of inequality is a good test of students' skills, because so many steps are involved.

☐1 Use the Distributive Property. ⟶ $-2x + 3x + 9 > -4x - 6$

☐2 Combine like terms. ⟶ $x + 9 > -4x - 6$

☐3 Add $4x$ to each side. ⟶ $x + 9 + \mathbf{4x} > -4x - 6 + \mathbf{4x}$

☐4 Simplify. ⟶ $5x + 9 > -6$

☐5 Subtract 9 from each side. ⟶ $5x + 9 - \mathbf{9} > -6 - \mathbf{9}$

☐6 Simplify. ⟶ $5x > -15$

☐7 Multiply each side by $\frac{1}{5}$. ⟶ $\frac{1}{5}(5x) > \frac{1}{5}(-15)$

☐8 Simplify. ⟶ $x > -3$

◀ Solution set: {**all real numbers greater than −3**}

P-3   **Solve:** $3y - 4 < 2(-2y + 4) + 5y$ {all real numbers less than 6}

P-4   **What is the solution set of the inequality in a below? in b?**

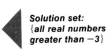

a	b
$2(3 - 2x) > x - 1 - 5x$	$-\frac{1}{2}x - \frac{1}{3} + \frac{3}{4}x > \frac{3}{8}x + 1 - \frac{1}{8}x$
$6 - 4x > -1 - 4x$	$-\frac{1}{3} + \frac{1}{4}x > \frac{2}{8}x + 1$
$6 - 4x + 4x > -1 - 4x + 4x$	$-\frac{1}{3} + \frac{1}{4}x - \frac{1}{4}x > \frac{1}{4}x + 1 - \frac{1}{4}x$
$6 > -1$ {real numbers}	$-\frac{1}{3} > 1$ ϕ

The steps in **a** lead to a true sentence. The solution set is {real numbers}. The steps in **b** lead to a false sentence. Its solution set is the empty set, ϕ.

▨ **CLASSROOM EXERCISES** ▨

Write the equivalent sentence you get as the first step in solving each inequality. (Step 1 of Examples 1, 2, and 3)

1. $2x + 1 > 3$ **2.** $-3x + 4 < -2$ **3.** $\frac{1}{2}x - 3 < -7$

4. $3x + 2 < x - 3$ **5.** $5 - 2x > 4x - 3$ **6.** $\frac{1}{2}x - 6 > \frac{3}{2}x + 4$

7. $5(2x - 6) < -10$ **8.** $3(x - 1) - 5 < x + 4$ **9.** $-5x + 3 > -2(x + 3) - 6$

*Inequalities / **133***

Additional Examples
Example 3
Solve.
1. $3x - 5 < 6(x - 1) + 10$
   Ans: {all real numbers greater than −3}
2. $-4(a - 5) > 7(a - 2) - 32$
   Ans: {all real numbers less than 6}

Additional Answers
Classroom Exercises
1. $2x + 1 - 1 > 3 - 1$
2. $-3x + 4 - 4 < -2 - 4$
3. $\frac{1}{2}x - 3 + 3 < -7 + 3$
4. $3x + 2 - x < x - 3 - x$
5. $5 - 2x - 4x > 4x - 3 - 4x$
6. $\frac{1}{2}x - 6 - \frac{3}{2}x > \frac{3}{2}x + 4 - \frac{3}{2}x$
7. $10x - 30 < -10$
8. $3x - 3 - 5 < x + 4$
9. $-5x + 3 > -2x - 6 - 6$

Additional Answers
Classroom Exercises

10. {all real numbers less than −1}
12. {all real numbers greater than 2}
14. {all real numbers greater than −6}
16. {all real numbers greater than −2}
18. {all real numbers greater than −5.6}
20. φ
22. {all real numbers greater than −5}
24. {all real numbers less than 2}

Solve. (Example 1)

10. $2x - 2 < -4$
11. $3x - 6 > -12$
12. $3x - 1 > 5$
13. $2x + 3 < 5$
14. $-\frac{1}{2}x + 2 < 5$
15. $\frac{1}{3}x + 1 > -2$

(Example 2)

16. $1 - 2x < -x + 3$
17. $6x - 1 > 4x - 3$
18. $2.2 + 1.4x > 0.4x - 3.4$
19. $-6x + 2 > 14 - 2x$
20. $3x + 4 < 3x - 2$
21. $2.3x - 3 < 5 + 2.3x$

(Example 3)

22. $1 + 2(x - 1) > -11$
23. $9 > -7(x + 2) - 5$
24. $-1 + n < 3(1 - n) + 2n$
25. $5(x + 3) - 3x > -3 + 5x$

Assignment Guide
Basic
Day 1 p. 134: 1–11
Day 2 p. 134: 12–22
Average p. 134: 1–21 odd
Above Average p. 134: 1–21 odd

Additional Answers
Written Exercises

2. {all real numbers less than 5}
4. {all real numbers less than −$\frac{7}{2}$}
6. {all real numbers less than 33}
8. {all real numbers greater than 2}
10. {all real numbers less than −$\frac{7}{2}$}
12. {all real numbers less than −$\frac{1}{4}$}
14. {all real numbers less than −5}
16. {all real numbers greater than 0}
18. {all real numbers less than 3}
20. {real numbers}
22. φ

Goal: To solve an inequality using at least two of the Properties of Inequalities
Sample Problem: $2x - 1 < 5$
Answer: {all real numbers less than 3}

Solve. Show the steps. (Examples 1, 2, and 3)

1. $3x - 1 < 7$ {all real numbers less than $\frac{8}{3}$}
2. $5x + 3 < 28$
3. $-10 > 2x + 3$ {all real numbers less than $-\frac{13}{2}$}
4. $-12 > 2x - 5$
5. $-\frac{1}{2}y - 3 < -7$ {all real numbers greater than 8}
6. $-\frac{1}{3}a + 1 > -10$
7. $3b + 7 < 5b$ {all real numbers greater than $\frac{7}{2}$}
8. $2 - 2n < -n$
9. $2m - 5 > 5m + 1$
10. $5t + 3 < 3t - 4$
11. $3x + 2 - 8x < 3$
12. $-p - 2 - 3p > -1$
13. $\frac{4}{3}q - \frac{1}{2} > \frac{7}{2} + \frac{2}{3}q$
14. $\frac{3}{5}r + \frac{5}{3} < -\frac{1}{3} + \frac{1}{5}r$
15. $-2(3x - 3) < 6$
16. $-3 > \frac{1}{2}(-4x - 6)$
17. $-4(x - 1) + 2x > 3x - 6$
18. $x - 6 < 3(4 - x) - 2x$
19. $-3x + \frac{1}{2} + x < -\frac{1}{2}(4x + 1)$
20. $-5(2x - 3) > 3 - 9x - x$
21. $-2(\frac{1}{2}x - \frac{1}{4}) < -\frac{1}{4}x + 1 - \frac{3}{4}x$
22. $6(3 - \frac{1}{3}x) < -\frac{1}{4}(8x + 1)$

OBJECTIVES: To write an inequality given a word sentence

PROBLEM SOLVING AND APPLICATIONS

5.7 Using Inequalities

To solve a word problem by writing and solving a related inequality

The steps for solving word problems involving inequalities are similar to those for solving word problems involving equations.

EXAMPLE 1   Rosa plans to spend less than $50 for a pair of shoes and a sweater (Condition 2). The cost of the pair of shoes she wants to buy is $5 more than twice the cost of a sweater (Condition 1). What is the greatest amount she can spend for the sweater?

Solution:

1. Let c = the cost of the sweater.
   Then $2c + 5$ = the cost of the shoes. ◀ **Condition 1**

2. Write an inequality for the problem.

   Think: | Cost of the sweater | + | Cost of the shoes | is less than | $50. |

   Translate: c + $2c + 5$ < 50   ◀ **Condition 2**

3. Solve the inequality. ───▶
   $$3c + 5 < 50$$
   $$3c + 5 - 5 < 50 - 5$$
   $$3c < 45$$
   $$\frac{3c}{3} < \frac{45}{3}$$
   $$c < 15$$

   Since $c < 15$, the greatest amount Rosa can spend for a sweater is $14.99.

4. **Check:** Is $14.99 + 2($14.99) + $5 < 50?
   Is $49.97 < $50?   Yes ✓

Recall that the *mean* of two or more scores is the sum of the scores divided by the number of scores.

In Example 2, Condition 1 is underscored once. Condition 2 is underscored twice.

Inequalities / 135

Lesson Resources
Warm–Up: p. M–27
Maintenance: See below.
Practice Worksheet 5.7

Maintenance
1. Simplify: $-7x - 13x$
   Ans: $-20x$ (Section 2.6)
2. Solve and check: $\frac{1}{2}c = 20$
   Ans: $c = 40$ (Section 3.2)
3. Evaluate $8x + 6 - 4x = 34$
   Ans: $x = 7$ (Section 3.5)
4. Solve and check:
   $-12.4x + 18.3 = -6.3x$
   Ans: $x = 3$ (Section 4.3)
5. The number of fiction books in a library is 10 less than 3 times the number of science books. The library contains 890 fiction books. Find the number of science books.
   Ans: 300 (Section 3.7)

Additional Examples
Example 1
1. Bertha estimates that the headphones she wants to buy cost $38 less than a certain cassette player (Condition 1). She plans to spend less than $58 for both items (Condition 2). Find the greatest amount she can spend for the cassette players.
   Ans: $47.99
2. John and Wayne own shares of the same stock. John owns 4 less than twice the number of shares that Wayne owns (Condition 1). They own more than 152 shares in all (Condition 2). Is it possible that John owns only 99 shares?
   Ans: No

1. Fay's mean bowling score for five games is greater than 123. Her scores for 4 games of bowling were 142, 106, 128, and 111 (Condition 2). Is it possible that her score on the fifth game was 130 (Condition 1)?
   Ans: Yes
2. Ray's scores on three quizzes are 58, 63, and 81. He wants his mean score on four quizzes to be greater than 75 (Condition 2). Find the lowest score he can get on the fourth quiz (Condition 1).
   Ans: 99

Common Error
Students incorrectly give the solution to the inequality as the answer for a word problem.

Example
Written Exercise 1
Answer: $x < 85$

Prescription
Remind students that the final answer must answer the question asked in the problem. If $x < 85$, then the greatest amount he can spend on the racket is $84.99.

Additional Answers
3. $2d - 3 > 20$
4. $3m + 8 < 40$
5. $\dfrac{32 + 45 + x}{3} < 42$
6. $\dfrac{82 + 87 + 91 + y}{4} > 86$
7. $\dfrac{65 + 70 + 75 + x}{4} > 72$
8. $\dfrac{92 + 94 + 98 + 100 + y}{5} < 95$

EXAMPLE 2

Note that Condition 2 appears first in the problem.

Felipe's scores on three tests were 86, 92, and 72. He wants his mean score on four tests to be greater than 85. Find the lowest score Felipe can get on the fourth test.

Solution:

1. Let x = Felipe's score on the fourth test.
2. Write an inequality for the problem.

   Think: The mean of the four tests must be greater than 85.

   Translate: $\dfrac{86 + 92 + 72 + x}{4} \qquad > \qquad 85$

3. Solve the inequality. $\longrightarrow \quad 4\left(\dfrac{86 + 92 + 72 + x}{4}\right) > 4(85)$

   $$86 + 92 + 72 + x > 340$$
   $$250 + x > 340$$
   $$250 + x - 250 > 340 - 250$$
   $$x > 90$$

   Since $x > 90$, Felipe's score must be at least 91.

4. **Check:** Is $\dfrac{86 + 92 + 72 + 91}{4} > 85$? Is $\dfrac{341}{4} > 85$?   Yes ✔

Steps for Solving Word Problems Involving Inequalities

1. Identify Condition 1. Use Condition 1 to represent the unknowns.
2. Identify Condition 2. Use Condition 2 to write an inequality.
3. Solve the inequality. Answer the question.

CLASSROOM EXERCISES

Write an inequality for each sentence. (Step 2, Examples 1 and 2)

1. The cost of a radio, c, is less than $60. $c < 60$

2. The number of tickets, t, increased by 12 is greater than 70. $t + 12 > 70$

3. Twice the number of days, d, decreased by 3 is greater than 20.

4. Eight plus 3 times the number of meters, m, is less than 40.

5. The mean of 32, 45, and x is less than 42.

6. The mean of 82, 87, 91, and y is greater than 86.

7. The mean of 65, 70, 75, and x is greater than 72.

8. The mean of 92, 94, 98, 100, and y is less than 95.

WRITTEN EXERCISES

Goal: To solve word problems involving inequalities
Sample Problem: The Stinson family plans to spend less than $1000 for a new sofa and chair (Condition 2). They estimate that the sofa will cost $300 more than the chair (Condition 1). What is the greatest amount they can spend for the chair?

Answer: $349.99

For Exercises 1–8, Condition 1 is <u>underscored once</u>. Condition 2 is <u>underscored twice</u>.
a. Use Condition 1 to represent the unknowns.
b. Use Condition 2 to write an inequality for the problem.
c. Solve. (Example 1)

1. Brian estimates that a pair of tennis shoes he wants to buy costs $50 less than a certain tennis racket. <u><u>He plans to spend less than $120 for both items.</u></u> Find the greatest amount he can spend for the racket.

2. <u><u>The Williams family plans to spend less than $800 each month for rent and food.</u></u> The amount budgeted for rent is $65 more than twice the amount budgeted for food. What is the greatest amount that can be budgeted for food?

3. <u><u>Andrew Klein spent more than $525 for a television and a radio.</u></u> The cost of a television was $30 more than four times the cost of the radio. Is it possible that the radio cost $100?

4. At a certain concert, the number of orchestra seat tickets sold was 20 less than three times the number of balcony tickets sold. <u><u>More than 400 tickets were sold in all.</u></u> Is it possible that only 90 balcony tickets were sold?

See the Answers to Selected Exercises for answers to Ex. 5 and 7.
    (Example 2)

5. Flora's scores on three math quizzes were 85, 72, and 81. <u><u>She wants her mean score on four quizzes to be greater than 80.</u></u> Find the lowest score she can get on the fourth quiz.

6. <u><u>Cara wants her mean score on five tests to be greater than 85.</u></u> Her scores on four tests are 85, 92, 70, and 86. What is the lowest score she can get on the fifth test?

7. Ryan's scores for three games of bowling were 95, 130, and 125. <u><u>In order to have a mean score greater than 110,</u></u> what is the lowest score he can bowl on the fourth game?

8. <u><u>Tyrone's mean bowling score for five games is greater than 120.</u></u> His scores for 4 games of bowling were 125, 100, 115, and 130. Is it possible that his score on the fifth game was 132?

Inequalities / 137

Assignment Guide
Basic p. 137: 1–7 odd
Average p. 137: 1–8 all
Above Average p. 137: 1–8

Problem–Solving Skills
Writing and solving an equation (Ex. 1–8)

Additional Answers
1. a. Let x = cost of tennis racket. Then $x - 50$ = cost of tennis shoes.
   b. $x + x - 50 < 120$
   c. $x < 85$; $84.99
2. a. Let x = amount budgeted for food. Then $2x + 65$ = amount budgeted for rent.
   b. $x + 2x + 65 < 800$
   c. $x < 245$; $244.99
3. a. Let x = cost of radio. Then $4x + 30$ = cost of television.
   b. $x + 4x + 30 > 525$
   c. $x > 99$; Yes
4. a. Let x = number of balcony tickets. Then $3x - 20$ = number of orchestra seat tickets.
   b. $x + 3x - 20 > 400$
   c. $x > 105$; No
6. a. Let x = score on fifth test.
   b. $\frac{85 + 92 + 70 + 86 + x}{5} > 85$
   c. $x > 92$; 93
8. a. Let x = score in fifth game.
   b. $\frac{125 + 100 + 115 + 130 + x}{5} > 120$
   c. $x > 130$; Yes

Quiz Sections 5.4–5.7
After completing Sections 5.4–5.7, you may want to administer a quiz covering the same sections. See the quiz provided on page 110 in the *Teacher's ResourceBank*.

137

Using Statistics

Making Diagrams

One important task of **statisticians** is to display data so that it can be interpreted more easily. For example, the median and the mode can often be read directly from a **stem–leaf plot.**

The Bright Light Bulb Company conducts tests on the light bulbs it manufactures. Thirty-five sample light bulbs are tested to find the number of hours each one lasts before burning out. The results of one test are shown at the right.

Lifetime of Thirty-six Light Bulbs (to the nearest hour)					
52	81	50	62	71	70
59	53	65	75	76	92
93	77	95	68	83	73
53	88	80	74	57	87
68	73	61	71	66	89
71	79	66	65	88	85

EXAMPLE: Construct a stem–leaf plot for the data in the table.

SOLUTION:

1. Since the smallest entry is 50 and the largest is 95, the stems can be the ten's digits, 5 through 9.

   Write the stems vertically. Draw a vertical line to the right.

   The **leaves** are the one's digits of the 36 entries. Write each one's digit to the right of its ten's digit stem.

```
5 | 2, 9, 3, 3, 0, 7
6 | 8, 5, 1, 6, 2, 8, 5, 6
7 | 1, 7, 3, 9, 5, 4, 1, 1, 6, 0, 3
8 | 1, 8, 0, 3, 8, 7, 9, 5
9 | 3, 5, 2
```

5 | 2 represents the entry 52.

2. Rewrite the leaves for each stem in increasing order from left to right. This is an **ordered stem-leaf plot.**

```
5 | 0, 2, 3, 3, 7, 9
6 | 1, 2, 5, 5, 6, 6, 8, 8
7 | 0, 1, 1, 1, 3, 3, 4, 5, 6, 7, 9
8 | 0, 1, 3, 5, 7, 8, 8, 9
9 | 2, 3, 5
```

EXERCISES

For Exercises 1–2, draw a stem–leaf plot for the given data.

1. **Number of Students on Fifteen School Buses**

40	36	35	28	40
32	35	32	27	25
27	31	40	31	33

2. **Number of Points Scored in Eighteen Basketball Games**

70	57	49	80	74	77
66	52	84	63	76	83
72	64	74	76	66	59

For Exercises 3–6, use the ordered stem-leaf plot in the Example.

3. Find the median of the data.
   (HINT: The mean of the 18th and
   19th leaves is $\frac{1+3}{2}$, or 2.) 72

4. Find the mode of the data. 71

5. The mean of the data is 72.7 hours.

   a. Which statement would you use
      if you were in charge of the
      company's advertising?
      Lasts Up to 95 Hours, or
      Lasts an Average of 70 Hours

   b. Give a reason for your choice.

6. You purchase a Bright Light bulb.

   a. Is it more likely to last 75
      hours or 90 hours? 75 hours

   b. Could it last 105 hours?
      45 hours? Explain. Yes; Yes;
      The lifetime of a single bulb could vary
      considerably from the averages.

BRIGHT LIGHT
Bulbs

Last Up to 95 Hours

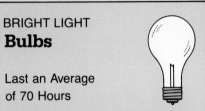

BRIGHT LIGHT
Bulbs

Last an Average
of 70 Hours

*A two-sided stem-leaf plot can be used to compare two sets of data.
The leaves of the second set of data are written to the left of the stems,
and they are read from right to left.*

*Two local movie theaters have both
shown the same movie for the past 15
days. The diagram at the right shows
each theater's daily attendance for that
movie. The **stems** are the hundred's and
ten's digits. The **leaves** are the one's
digits.*

Theater A		Theater B
4, 3, 0	15	
4, 2	16	9
5	17	2, 3, 3, 5, 6, 7
8, 1	18	1, 2, 5, 5, 5, 9
7, 6, 6, 0	19	6, 7
5, 4, 0	20	

0 | 20 **represents
the number 200.**

7. Find the median and mode for the
   attendance at Theater A.

8. Find the median and mode for the
   attendance at Theater B.

9. Which theater had the greater
   median? the greater mode?

10. Which theater do you think had the
    greater mean attendance? Find the
    mean attendance for each theater.
    Was your guess correct?

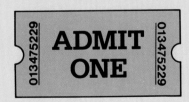

Inequalities / **139**

CHAPTER SUMMARY

IMPORTANT TERMS		
	Graph *(p. 116)*	Inequality *(p. 116)*
	Solution set *(p. 116)*	Empty set *(p. 128)*
	Domain of a variable *(p. 116)*	

IMPORTANT IDEAS

1. *Comparison Property:* If a and b are real numbers, then exactly one of the following is true: $a > b$, $a = b$, or $a < b$.

2. *Order Property of Opposites:* If a and b are real numbers and $a < b$, then $-a > -b$.

3. *Transitive Property of Order:* If $a < b$ and $b < c$, then $a < c$.

4. *Addition Property for Inequalities:* Adding the same real number to each side of an inequality does not change the direction of the inequality. The new inequality is equivalent to the first.

5. *Subtraction Property for Inequalities:* Subtracting the same real number from each side of an inequality does not change the direction of the inequality. The new inequality is equivalent to the first.

6. *Multiplication Property for Inequalities:* If each side of an inequality is multiplied by the same nonzero real number, an equivalent inequality is obtained as follows.
   a. The direction of the inequality is unchanged if each side is multiplied by the same positive number.
   b. The direction of the inequality is reversed if each side is multiplied by the same negative number.

7. *Division Property for Inequalities:* If each side of an inequality is divided by the same nonzero real number, an equivalent inequality is obtained as follows.
   a. The direction of the inequality is unchanged if each side is divided by the same positive number.
   b. The direction of the inequality is reversed if each side is divided by the same negative number.

8. Steps for Solving Word Problems Involving Inequalities
   See page 136.

CHAPTER REVIEW

SECTION 5.1

Write the solution set of each graph. If necessary, write the domain you are assuming.

1. 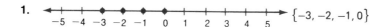 $\{-3, -2, -1, 0\}$

140 / *Chapter 5*

2. 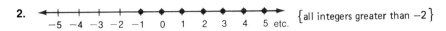 {all integers greater than -2}

3. {all real numbers less than $1\frac{1}{2}$}

4. {all real numbers greater than or equal to $-\frac{1}{2}$}

Draw a graph of each inequality.

5. $x < 5$; Domain $= $ {whole numbers}

6. $x < -2$; Domain $=$ {integers}

7. $x > -3$ **8.** $x > 1$ **9.** $x < 4$ **10.** $x < -1$

SECTION 5.2

Write a new inequality by applying the Order Property of Opposites.

11. $r < -12$ $-r > 12$ **12.** $x > -8$ $-x < 8$ **13.** $a > -3$ $-a < 3$ **14.** $-4.5 < t$

15. $-x > -4.2$ $x < 4.2$ **16.** $3\frac{1}{4} > -t$ $-3\frac{1}{4} < t$ **17.** $-1 > -x$ $1 < x$ **18.** $k < -n$

19. If $a > -5$ and $-5 > c$, what two inequalities relate a and c? $c < a;\, a > c$

20. If $-b < c$ and $-c > a$, what two inequalities relate a and b? $a < b;\, b > a$

Write True or False for each sentence.

21. If $a < -5$ and $-5 < b$, then $a < b$. True

22. If $a > 6$ and $6 > b$, then $a < b$.   False

SECTION 5.3

Write Yes or No to show whether the two inequalities are equivalent.

23. $x + 3 < 5$; $x + 1 < 7$ No

24. $x - 4 < 10$; $x - 5 < 9$ Yes

25. $-5 > 2x - 2$; $-7 > 2x$ No

26. $5x - 8 > 2$; $5x > 10$ Yes

27. $2a - 7 > -2$; $2a > 5$ Yes

28. $3b - 1 < b + 7$; $3b - 2 < b + 8$ No

Solve.

29. $x - 3 < -2$ **30.** $x + 9 > -3$ **31.** $-4 + x < -6$

32. $-8 + x > -10$ **33.** $-8.3 > 1.4 + x$ **34.** $5.6 < x - 3.9$

Inequalities / 141

Solve.

35. $4x + 3 - 3x < -5$ $\{$all real numbers less than $-8\}$

36. $3 > 5x + 7 - 4x$ $\{$all real numbers less than $-4\}$

37. $3x - 5 > 2(x + 3)$
$\{$all real numbers greater than $11\}$

38. $3(2x - 1) < x - 2 + 4x$
$\{$all real numbers less than $1\}$

SECTION 5.5

Solve.

39. $4x < 30$

40. $8x > -20$

41. $-9x > 45$

42. $-5x < -60$

43. $-6 < -\frac{3}{2}x$

44. $\frac{1}{2} > -\frac{1}{4}x$

SECTION 5.6

Solve.

45. $0.3 - 0.6x < 1.4x - 2.9$

46. $3.2x - 1.1 > 0.2x + 1.3$

47. $x - 19 > 12 - 2x$

48. $2x - 5 + 3x < -x + 7$

49. $-2 + 3x > 3(1 + x)$ ϕ

50. $2x - 4 + x < -1 + 3x$
$\{$real numbers$\}$

SECTION 5.7

For Exercises 51–54, Condition 1 is <u>underscored once</u>. Condition 2 is <u>underscored twice</u>.

Use Condition 1 to represent the unknowns. Use Condition 2 to write an inequality for the problem. Solve.

51. <u>The cost of a coat Emilio would like to buy is $20 more than three times the cost of a pair of pants.</u> <u><u>He plans to spend less than $105 for both items.</u></u> What is the greatest amount he can spend for the pants?

52. <u><u>Julia spent more than $25 for a book and a compact disc.</u></u> <u>The cost of the book was $7 less than the cost of the compact disc.</u> Is it possible that she spent only $15 for the compact disc?

53. <u>Suzanne's scores on three tests were 85, 93, and 78.</u> <u><u>She wants her mean score on four tests to be greater than 85.</u></u> Find the lowest score she can get on the fourth test.

54. <u><u>Alan's mean bowling score for five games is greater than 110.</u></u> <u>His scores for 4 games of bowling were 105, 120, 88, and 108.</u> Is it possible that his score on the fifth game was 135?

Additional Answers

39. $\{$all real numbers less than $7\frac{1}{2}\}$

40. $\{$all real numbers greater than $-2\frac{1}{2}\}$

41. $\{$all real numbers less than $-5\}$

42. $\{$all real numbers greater than $12\}$

43. $\{$all real numbers less than $4\}$

44. $\{$all real numbers greater than $-2\}$

45. $\{$all real numbers greater than $1.6\}$

46. $\{$all real numbers greater than $0.8\}$

47. $\{$all real numbers greater than $10\frac{1}{3}\}$

48. $\{$all real numbers less than $2\}$

51. a. Let x = cost of pants. Then $3x + 20$ = cost of coat.
b. $x + 3x + 20 < 105$
c. $x < 21.25$; $21.24

52. a. Let x = cost of compact disc. Then $x - 7$ = cost of book.
b. $x + x - 7 > 25$
c. $x > 16$; No

53. a. Let x = score on fourth test.
b. $\frac{85 + 93 + 78 + x}{4} > 85$
c. $x > 84$; 85

54. a. Let x = score in fifth game.
b. $\frac{105 + 120 + 88 + 108 + x}{5} > 110$
c. $x > 129$; Yes

CHAPTER 6

Powers

The *spiral galaxy* shown below is about 10^7 light-years away from Earth. In scientific notation, this is about 5.86×10^9 miles. The microscopic *diatom,* a one-celled plant, ranges in size from 5×10^{-4} inch to 8×10^{-2} inch.

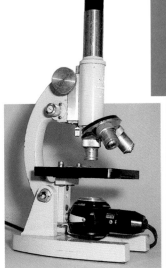

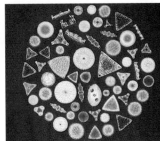

Lesson Resources
Warm–Up: p. M–28
Maintenance: See below.
Practice Worksheet 6.1
Exploration 6
Transparency 2

Maintenance
1. Divide: $\frac{3}{4} \div \left(-\frac{1}{6}\right)$
   Ans: $-4\frac{1}{2}$ (Section 1.5)
2. Solve and check:
   $16 + x = -12$
   Ans: $x = -28$
   (Section 3.1)
3. Solve and check:
   $3x + 8 = 2x - x$
   Ans: $x = -4$ (Section 4.3)
4. Solve: $6 < 4 + y$
   **Ans: {all real numbers
   greater than 2}**
   (Section 5.3)
5. The perimeter of a desk top is
   14 feet. The width is 4 feet.
   Find the length of the desk
   top.
   Ans: 3 feet (Section 3.4)

Additional Examples
Example 1
Write as one power.
1. $3^4 \cdot 3^9$
   Ans: 3^{13}
2. $x^8 \cdot x$
   Ans: x^9
3. $5^3 \cdot 5^6$
   Ans: 5^9

Example 2
Write as one power.
1. $(-6)^2(-6)^5$
   Ans: $(-6)^7$
2. $(-a)^5(-a)^6$
   Ans: $(-a)^{11}$
3. $(-3)^2(-3)^5$
   Ans: $(-3)^7$

144

OBJECTIVES: To write a product of powers of the same base as one power

6.1 Products of Powers

To write a product of one power and the sum or difference of two powers
all with the same base as a sum or difference using the Distributive Property
and the Product Property of Powers
To solve a special equation by applying the Product Property of Powers

P–1 **Write each power below as a product.**

a. 2^4 **b.** x^3 **c.** y^5 **d.** 4^3 **e.** z^2 $z \cdot z$
$2 \cdot 2 \cdot 2 \cdot 2$ $x \cdot x \cdot x$ $y \cdot y \cdot y \cdot y \cdot y$ $4 \cdot 4 \cdot 4$

2 is a factor of 2^4 four times. x is a factor of x^3 three times.

P–2 **How many times is y a factor of y^5?** 5

If a is a positive integer, then
x^a means that x is a factor a times.

$$x^a = \underbrace{x \cdot x \cdot x \cdots x}_{a \text{ factors}}$$

EXAMPLE 1   Write $2^3 \cdot 2^5$ as one power.

Solution: $2^3 \cdot 2^5 = (2 \cdot 2 \cdot 2)(2 \cdot 2 \cdot 2 \cdot 2 \cdot 2)$
$= (2 \cdot 2 \cdot 2 \cdot 2 \cdot 2 \cdot 2 \cdot 2 \cdot 2)$
$= 2^8$

P–3 **How can you express $(-3)^2$ as a product? $(-3)^4$ as a product?**
$(-3)(-3)$ $(-3)(-3)(-3)(-3)$

EXAMPLE 2   Write $(-3)^2(-3)^4$ as one power.

Solution: $(-3)^2(-3)^4 = (-3 \cdot -3)(-3 \cdot -3 \cdot -3 \cdot -3)$
$= (-3 \cdot -3 \cdot -3 \cdot -3 \cdot -3 \cdot -3)$
$= (-3)^6$

P–4 **How many times is x used as a factor in $x^a \cdot x^b$ if a and b are
positive integers?** $a + b$ times

$$x^a \cdot x^b = \underbrace{(x \cdot x \cdot x \cdots x)}_{a \text{ factors}}\underbrace{(x \cdot x \cdot x \cdots x)}_{b \text{ factors}}$$

144 / Chapter 6

Two common pitfalls are: 1) to multiply exponents and 2) to apply this property to powers with unlike bases.

Product Property of Powers

If a and b are integers and x is any nonzero real number, then $x^a \cdot x^b = x^{a+b}$.

$13^4 \cdot 13^5 = 13^9$

$(-5.6)^2(-5.6) = (-5.6)^3$

P–5 Write each product below as one power.

a. $(-1.9)^2(-1.9)^3$
$(-1.9)^5$

b. $(\frac{1}{2})^4(\frac{1}{2})^3$
$(\frac{1}{2})^7$

c. $t^2 \cdot t^7$
t^9

◀ **Keep the same base. Add the exponents.**

EXAMPLE 3   Write the product as a sum: $r^3(r^2 + r)$

Solution:

1 Distributive Property ⟶ $r^3(r^2 + r) = r^3 \cdot r^2 + r^3 \cdot r$

2 Product Property of Powers ⟶ $= r^5 + r^4$

P–6 Write each product below expressed as a sum or difference.

a. $t^2(t^4 + t^3)$
$t^6 + t^5$

b. $k^3(3k^2 - 2k + 3)$ $3k^5 - 2k^4 + 3k^3$

EXAMPLE 4   Solve: $12^4 \cdot 12^x = 12^9$

Solution: $12^4 \cdot 12^x = 12^9$

$12^{4+x} = 12^9$

$4 + x = 9$

$x = 5$

◀ **Since the bases are equal, the exponents are also equal.**

P–7   **Solve: $(-4.8)^3 \cdot (-4.8)^x = (-4.8)^7$** $x = 4$

CLASSROOM EXERCISES

Write each product as one power. (Examples 1–2)

1. $2^3 \cdot 2^2$ 2^5
2. $a^5 \cdot a^2$ a^7
3. $(-5)^2(-5)^4$ $(-5)^6$
4. $(-1.8)^5(-1.8)^2$ $(-1.8)^7$
5. $(\frac{1}{2})^2(\frac{1}{2})^3$ $(\frac{1}{2})^5$
6. $n \cdot n^4$ n^5
7. $k^{10} \cdot k^{12}$ k^{22}
8. $a^3 \cdot a \cdot a^2$ a^6
9. $7 \cdot 7^2 \cdot 7^3$ 7^6
10. $(-1)^3(-1)^5$ $(-1)^8$
11. $2^4 \cdot 2^3$ 2^7
12. $m^5 \cdot m^4$ m^9

Powers / **145**

Additional Examples
Example 3
Write the product as a sum or difference.
1. $y(y^3 + y^4)$
   Ans: $y^4 + y^5$
2. $x^2(3x^3 - x)$
   Ans: $3x^5 - x^3$
3. $h^5(h^7 - h^6 + h^2)$
   Ans: $h^{12} - h^{11} + h^7$

Example 4
Solve.
1. $11^x \cdot 11^7 = 11^{10}$
   Ans: $x = 3$
2. $(\frac{1}{3})^9 \cdot (\frac{1}{3})^x = (\frac{1}{3})^{15}$
   Ans: $x = 6$
3. $(-2)^x \cdot (-2) = (-2)^5$
   Ans: $x = 4$

Common Error
When multiplying powers of the same base, students multiply exponents.

Example
$3^2 \cdot 3^4 = 3^8$

Prescription
If students are unsure of the reason for the rule, have them work several examples by listing the factors and then counting them.
$3^2 \cdot 3^4 = (3 \cdot 3) \cdot (3 \cdot 3 \cdot 3 \cdot 3)$
     $= 3^6$

145

Solve each equation. (Example 4)

13. $10^x = 10^5$ $x = 5$ **14.** $10^x \cdot 10^2 = 10^9$ $x = 7$ **15.** $3^x \cdot 3^5 = 3^{11}$ $x = 6$

16. $(\frac{1}{2})^3(\frac{1}{2})^5 = (\frac{1}{2})^x$ $x = 8$ **17.** $3^x \cdot 3^2 = 3^6$ $x = 4$ **18.** $5^3 \cdot 5^x = 5^4$ $x = 1$

▓▓▓▓ **WRITTEN EXERCISES** ▓▓▓▓

Goal: To write a product of powers as one power
Sample Problem: $r^4 \cdot r \cdot r^3$ **Answer:** r^8

Write each product as one power. (Examples 1–2)

1. $12^3 \cdot 12^7$ 12^{10} **2.** $10^5 \cdot 10^4$ 10^9 **3.** $5 \cdot 5^3$ 5^4 **4.** $6 \cdot 6^4$ 6^5

5. $(3.2)^5(3.2)^7$ $(3.2)^{12}$ **6.** $(3.14)^3(3.14)^2$ $(3.14)^5$ **7.** $(-15)^4(-15)^5$ $(-15)^9$ **8.** $(-10)^4(-10)^2$

9. $(-6)^2(-6)^9$ $(-6)^{11}$ **10.** $(-4)^3(-4)^5$ $(-4)^8$ **11.** $(-\frac{2}{3})^2(-\frac{2}{3})^5$ $(-\frac{2}{3})^7$ **12.** $(-\frac{1}{4})^3(-\frac{1}{4})^2$

13. $a^4 \cdot a^3$ a^7 **14.** $x^2 \cdot x^5$ x^7 **15.** $r^8 \cdot r$ r^9 **16.** $y^6 \cdot y^3$ y^9

17. $a^3 \cdot a^2 \cdot a^6$ a^{11} **18.** $b^4 \cdot b \cdot b^3$ b^8 **19.** $5^2 \cdot 5^7$ 5^9 **20.** $12^3 \cdot 12^3$ 12^6

Write each product as a sum or difference. (Example 3)

21. $y^2(y + y^2)$ $y^3 + y^4$ **22.** $a^3(a^2 + a)$ $a^5 + a^4$ **23.** $x^4(x^5 + x^3)$ $x^9 + x^7$ **24.** $m^5(m^2 + m^3)$ $m^7 + m^8$

25. $(2^3 - 2^5)2^2$ $2^5 - 2^7$ **26.** $(10^4 - 10^2)10^3$ $10^7 - 10^5$ **27.** $n(n^3 - n^2 + n)$ $n^4 - n^3 + n^2$ **28.** $s^3(s^5 - s^3 - s)$ $s^8 - s^6 - s^4$

Solve each equation. (Example 4)

29. $10^5 \cdot 10^6 = 10^x$ $x = 11$ **30.** $8^3 \cdot 8^9 = 8^x$ $x = 12$

31. $(0.5)^x(0.5)^2 = (0.5)^7$ $x = 5$ **32.** $(1.2)^5(1.2)^x = (1.2)^9$ $x = 4$

33. $3^x \cdot 3^5 = 3^{10}$ $x = 5$ **34.** $5^5 \cdot 5^x = 5^{12}$ $x = 7$

35. $(\pi)^x(\pi)^3 = (\pi)^9$ $x = 6$ **36.** $(-\frac{2}{3})^2(-\frac{2}{3})^x = (-\frac{2}{3})^{11}$ $x = 9$

REVIEW CAPSULE FOR SECTION 6.2

Evaluate both sides of each equation. Then write True or False for the equation. (Pages 21–23)

1. $\frac{2^5}{2^2} = 2^3$ True   **2.** $\frac{3^5}{3^3} = 3^2$ True   **3.** $\frac{2^6}{2^2} = 2^3$ False   **4.** $\frac{x^5}{y^3} = \left(\frac{x}{y}\right)^2$ when $x = -2$, $y = 1$ False

Assignment Guide
Basic p. 146: 1–35 odd
Average p. 146: 3, 6, 9, ⋯, 36;
p. 149: 3, 6, 9, ⋯, 45, 47
Above Average p. 146:
3, 6, 9, ⋯, 36; p. 149: 3, 6,
9, ⋯, 45, 47

Additional Answers
Written Exercises
8. $(-10)^6$
12. $(-\frac{1}{4})^5$

OBJECTIVES: To write a quotient of two powers of the same base as one power

6.2 Quotients of Powers and Zero Exponents

To identify whether or not an equation involving a quotient of powers is true
To solve a special equation by applying the Quotient Property of Powers
To simplify an expression involving one or more powers with 0 as an exponent

P–1 How can you express x^5 as a product? x^3 as a product?

$x \cdot x \cdot x \cdot x \cdot x$ $\qquad$ $x \cdot x \cdot x$

EXAMPLE 1 Write $\dfrac{x^5}{x^3}$ as one power.

Solution:

1. Meaning of a power $\longrightarrow$ $\dfrac{x^5}{x^3} = \dfrac{x \cdot x \cdot x \cdot x \cdot x}{x \cdot x \cdot x}$

2. Product Rule for Fractions $\longrightarrow$ $= \dfrac{x \cdot x \cdot x}{x \cdot x \cdot x} \cdot \dfrac{x \cdot x}{1}$

   Remember that $\dfrac{x \cdot x \cdot x}{x \cdot x \cdot x} = 1$.

3. Multiplication Property of 1 $\longrightarrow$ $= \dfrac{x \cdot x}{1}$

   Try to get students to suggest the relation between this result and the original quotient of the two powers.

4. Simplify. $\longrightarrow$ $= x^2$

Quotient Property of Powers

If a and b are integers and x is a nonzero real number, then $\dfrac{x^a}{x^b} = x^{a-b}$.

$\dfrac{5^{12}}{5^3} = 5^{12-3}$

$= 5^9$

P–2 Write each quotient below as one power.

a. $\dfrac{17^7}{17^4}$ $\qquad$ **b.** $(1.8)^8 \div (1.8)^3$ $\qquad$ **c.** $\dfrac{m^{12}}{m^5}$

17^3 $\qquad\qquad$ $(1.8)^5$ $\qquad\qquad$ m^7

Keep the same base. Subtract the exponents.

EXAMPLE 2 Solve: $\dfrac{12^t}{12^3} = 12^4$

Solution: $\dfrac{12^t}{12^3} = 12^4$

$12^{t-3} = 12^4$

$t - 3 = 4$

Since the bases are equal, the exponents are also equal.

$t = 7$

Teaching Suggestions p. M–28

Lesson Resources
Warm–Up: p. M–29
Maintenance: See below.
Practice Worksheet 6.2

Maintenance
1. Evaluate: $16 \div 4 + 2 \cdot 2$
   Ans: 8 (Section 1.6)
2. Solve and check: $-\frac{2}{3}x = 18$
   Ans: $x = -27$
   (Section 3.2)
3. Solve and check:
   $27 - 6y = 4y - 7y$
   Ans: $y = 9$ (Section 4.3)
4. Solve: $x - 1.6 < 2.1$
   Ans: {all real numbers less than 3.7}
   (Section 5.3)
5. Write an equation for this sentence. Then solve. The length multiplied by 6 equals 180.
   Ans: $6\ell = 180$; $\ell = 30$
   (Section 3.6)

Additional Examples
Example 1
Write as one power.
1. $\dfrac{a^7}{a^2}$
   Ans: a^5
2. $\dfrac{y^4}{y}$
   Ans: y^3
3. $\dfrac{5^{11}}{5^3}$
   Ans: 5^8

Example 2
Solve.
1. $\dfrac{8^x}{8^5} = 8^8$
   Ans: $x = 13$
2. $\dfrac{10^x}{10} = 10^{10}$
   Ans: $x = 11$
3. $\dfrac{6^{4x}}{6^{2x}} = 6^{10}$
   Ans: $x = 5$

P–3 How can you express $\dfrac{113^5}{113^5}$ as one power by using the Quotient Property of Powers? $\quad 113^0$

P–4 Simplify $\dfrac{a}{a}$ if a is any nonzero real number. $\quad 1$

$$\frac{113^5}{113^5} = 113^{5-5}$$ ◀ **By the Quotient Property of Powers**

$$= 113^0$$

$$\frac{113^5}{113^5} = 1$$ ◀ **By the special property:** $\dfrac{a}{a} = 1$

This discussion inductively establishes the Property of a Zero Exponent stated below.

Thus, it is clear that 113^0 equals 1.

P–5 What is the value of any nonzero real number raised to the 0 power? $\quad 1$

Consider $\dfrac{x^m}{x^m}$, in which m is any integer and x is any nonzero real number.

$$\frac{x^m}{x^m} = x^{m-m}$$ ◀ **By the Quotient Property of Powers**

$$= x^0$$

$$\frac{x^m}{x^m} = 1$$ ◀ **By the special property:** $\dfrac{a}{a} = 1$

This shows that x^0 equals 1. Thus, the following property has been proved.

> **Property of a Zero Exponent**
>
> If x is any nonzero real number, then $x^0 = 1$.
>
> $(-27)^0 = 1$
>
> $(\sqrt{5})^0 = 1$

P–6 Simplify each expression below.

a. $\left(\dfrac{3}{4}\right)^0 + 12^0 \quad 2$ **b.** $(-5.2)^0 - (\sqrt{3})^0 \quad 0$ **c.** $\left(\dfrac{t}{25}\right)^0 \quad 1$

Write each quotient as one power. No divisor equals 0. (Example 1)

1. $\dfrac{3^5}{3^2}$ 3^3 **2.** $\dfrac{10^8}{10^3}$ 10^5 **3.** $\dfrac{(-3)^6}{(-3)^2}$ $(-3)^4$ **4.** $\dfrac{x^{10}}{x^7}$ x^3

5. $\dfrac{a^7}{a^5}$ a^2 **6.** $\dfrac{b^4}{b}$ b^3 **7.** $2^7 \div 2^4$ 2^3 **8.** $y^{12} \div y^7$ y^5

9. $\left(\dfrac{1}{2}\right)^{10} \div \left(\dfrac{1}{2}\right)^3$ $\left(\dfrac{1}{2}\right)^7$ **10.** $12^3 \div 12^k$ 12^{3-k} **11.** $\dfrac{(-5)^t}{(-5)^4}$ $(-5)^{t-4}$ **12.** $\dfrac{(0.5)^p}{(0.5)^q}$

                                                                                          $(0.5)^{p-q}$

Write True or False for each sentence. (Example 1)

13. $\dfrac{6^4}{3^2} = 2^2$ False   **14.** $\dfrac{10^{10}}{10^5} = 1^5$ False   **15.** $\dfrac{2^7}{3^2} = \left(\dfrac{2}{3}\right)^5$ False

16. $\dfrac{5^4}{5^3} = \dfrac{1}{5}$ False   **17.** $\dfrac{2^4}{2^2} = 2^2$ True   **18.** $\dfrac{13}{7^2} = (13 - 7)^3$

                                                                                                            False

Simplify. (P-5, P-6)

19. 2^0 1   **20.** 100^0 1   **21.** $(-5)^0$ 1   **22.** $\left(-\dfrac{2}{3}\right)^0$ 1

23. $(0.5)^0$ 1   **24.** $(-\pi)^0$ 1   **25.** $w^0; w \neq 0$ 1   **26.** $\left(\dfrac{1}{t}\right)^0; t \neq 0$ 1

Goal: To write a quotient of powers as one power
Sample Problem: $y^{12} \div y^4$
Answer: y^8

Write each quotient as one power. (Example 1)

1. $\dfrac{10^5}{10^3}$ 10^2 **2.** $\dfrac{8^6}{8^3}$ 8^3 **3.** $\left(\dfrac{1}{2}\right)^3 \div \left(\dfrac{1}{2}\right)^2$ $\left(\dfrac{1}{2}\right)^1$, or $\left(\dfrac{1}{2}\right)$ **4.** $\left(\dfrac{1}{4}\right)^5 \div \left(\dfrac{1}{4}\right)^3$ $\left(\dfrac{1}{4}\right)^2$

5. $\dfrac{x^{12}}{x^8}$ x^4 **6.** $\dfrac{a^{15}}{a^{12}}$ a^3 **7.** $\dfrac{2^{12}}{2^9}$ 2^3 **8.** $\dfrac{10^7}{10^3}$ 10^4

9. $\dfrac{(0.6)^3}{(0.6)^2}$ $(0.6)^1$, or 0.6   **10.** $\dfrac{(1.3)^6}{(1.3)^5}$ $(1.3)^1$, or 1.3   **11.** $\dfrac{\pi^5}{\pi^2}$ π^3 **12.** $\dfrac{(-\pi)^5}{(-\pi)^3}$ $(-\pi)^2$

Assignment Guide
Basic pp. 149–150: 1–45 odd
Average pp. 149–150: 3, 6, 9, ⋯, 45, 47; p. 146: 3, 6, 9, ⋯, 36
Above Average pp. 149–150: 3, 6, 9, ⋯, 36, 45, 47; p. 146: 3, 6, 9, ⋯, 36

Write <u>True</u> *or* <u>False</u> *for each equation.* (Example 1)

13. $\dfrac{10^{11}}{10^{10}} = 10$ True

14. $\dfrac{100^{98}}{100^{96}} = 10{,}000$ True

15. $\dfrac{10^5}{5^2} = 2^3$ False

16. $\dfrac{12^7}{3^4} = 4^3$ False

17. $\dfrac{3^8}{3^4} = 3^2$ False

18. $\dfrac{5^6}{5^2} = 5^3$ False

19. $\dfrac{(\sqrt{2})^7}{(\sqrt{2})^2} = (\sqrt{2})^5$ True

20. $\dfrac{(-\sqrt{3})^{10}}{(-\sqrt{3})^6} = (-\sqrt{3})^4$ True

21. $\dfrac{10^3 \cdot 10^5}{10^4} = 10^4$ True

22. $\dfrac{10^9}{10^2 \cdot 10^3} = 10{,}000$ True

Solve each equation. (Example 2)

23. $\dfrac{2^x}{2^2} = 2^3$ $x = 5$

24. $\dfrac{3^y}{3^5} = 3^2$ $y = 7$

25. $\dfrac{x^5}{10^3} = 10^2$ $x = 10$

26. $\dfrac{5^9}{a^3} = 5^6$ $a = 5$

27. $\dfrac{10^7}{10^2} = 10^x$ $x = 5$

28. $\dfrac{10^{10}}{10^3} = 10^y$

29. $\dfrac{11^{2x}}{11^x} = 11^4$ $x = 4$

30. $\dfrac{5^{3x}}{5^x} = 5^{12}$ $x = 6$

31. $10^0 = x$

32. $7^0 = n$ $n = 1$

33. $\left(\dfrac{1}{2}\right)^0 = y$ $y = 1$

34. $\left(-\dfrac{1}{4}\right)^0 = a$

35. $(0.7)^a = 1$ $a = 0$

36. $(-1.3)^x = 1$ $x = 0$

37. $7^x - 5^x = 0$

38. $\left(\dfrac{\sqrt{2}}{\sqrt{7}}\right)^x = 1$ $x = 0$

39. $1 = p^0$ p = any non-zero real number

40. $y^0 = 1$

Simplify. (P-6)

41. $93^0 - 2$ -1

42. $(-114)^0 + 5$ 6

43. $3x^0 + 2y^0$ 5

44. $5m^0 + 3n^0$ 8

45. $-2r^0 + t^0$ -1

46. $p^0 - 4q^0$ -3

NON-ROUTINE PROBLEM

47. Copy the addition problem at the right and find the sum. Then replace six of the digits in the problem with a zero so that the sum is 1,111.

$$\begin{array}{r} 1\ 1\ 1 \\ 3\ 3\ 3 \\ 5\ 5\ 5 \\ 7\ 7\ 7 \\ +\ 9\ 9\ 9 \\ \hline \end{array}$$

6.3 Negative Exponents

To simplify products or quotients of powers involving negative exponents

The Product and Quotient Properties of Powers apply to negative exponents as well as to positive and zero exponents.

EXAMPLE 1 Simplify $\dfrac{x^3}{x^6}$.

Solution:

1. Meaning of a power $\longrightarrow$ $\dfrac{x^3}{x^6} = \dfrac{x \cdot x \cdot x}{x \cdot x \cdot x \cdot x \cdot x \cdot x}$

2. Product Rule for Fractions $\longrightarrow$ $= \dfrac{x \cdot x \cdot x}{x \cdot x \cdot x} \cdot \dfrac{1}{x \cdot x \cdot x}$

3. Multiplication Property of 1 $\longrightarrow$ $= \dfrac{1}{x \cdot x \cdot x}$

4. Simplify. $\longrightarrow$ $= \dfrac{1}{x^3}$

The Quotient Property of Powers is used for the same expression.

$$\frac{x^3}{x^6} = x^{3-6}$$

◀ *Exponent of numerator less the exponent of denominator*

$$= x^{-3}$$

Thus, the following sentence is true for all nonzero real numbers.

$$\frac{1}{x^3} = x^{-3}$$

◀ $\frac{1}{x^3}$ *is considered to be the simplest form.*

Since x cannot equal 0 in $\frac{1}{x^3}$, it cannot equal 0 in x^{-3}.

You may want to discuss why $x^{-a} = \frac{1}{x^a}$ is also true when a is *any* integer.

> **Property of Negative Exponents**
>
> If $-a$ represents a negative integer and x is a nonzero real number, then $x^{-a} = \dfrac{1}{x^a}$.
>
> $3^{-2} = \dfrac{1}{3^2}$
>
> $(-1.3)^{-5} = \dfrac{1}{(-1.3)^5}$

P–1 Simplify.

a. r^{-3} $\dfrac{1}{r^3}$

b. m^{-6} $\dfrac{1}{m^6}$

c. 3^{-2} $\dfrac{1}{3^2}$ or $\dfrac{1}{9}$

d. 2^{-4} $\dfrac{1}{2^4}$ or $\dfrac{1}{16}$

e. 10^{-3} $\dfrac{1}{10^3}$ or $\dfrac{1}{1000}$

Powers / 151

Teaching Suggestions p. M–28

Lesson Resources
Warm–Up: p. M–29
Maintenance: See below.
Practice Worksheet 6.3
Transparencies 2, 15

Maintenance
1. Evaluate: $-12 \div 2^2$
   Ans: −3 (Section 1.7)
2. Solve and check:
   $4y + 2 = -34$
   Ans: y = −9
   (Section 3.3)
3. Solve and check:
   $4z - 5z = -28 + 3z$
   Ans: z = 7 (Section 4.3)
4. Solve: $3 - 6x > 8 - 7x$
   Ans: {all real numbers greater than 5}
   (Section 5.4)
5. On Friday, the number of customers of a barber shop was equal to 10 more than twice the number of customers on Saturday. There were 36 customers on Friday. Find the number of customers on Saturday.
   Ans: 13 (Section 3.7)

Additional Examples
Example 1
Simplify.
1. $\dfrac{y^8}{y^{10}}$

   Ans: $\dfrac{1}{y^2}$

2. $\dfrac{x}{x^5}$

   Ans: $\dfrac{1}{x^4}$

3. $\dfrac{4^3}{4^7}$

   Ans: $\dfrac{1}{4^4}$

151

What is the rule for multiplying powers of the same base?

Add the exponents. Keep the same base.

Additional Examples
Example 2
Write as one power.
1. $a^5 \cdot a^{-2}$

   Ans: a^3
2. $x^{-7} \cdot x^3$

   Ans: $\dfrac{1}{x^4}$
3. $3^{-2} \cdot 3^{-3}$

   Ans: $\dfrac{1}{3^5}$

EXAMPLE 2   Write $y^{-5} \cdot y^3$ as one power. Then simplify.

Solution:

1. Product Property of Powers ⟶ $y^{-5} \cdot y^3 = y^{-5+3}$

2. Add the exponents. ⟶ $= y^{-2}$

3. Simplify. ⟶ $= \dfrac{1}{y^2}$

P-3 **Write each product below as one power.**

a. $17^{-2} \cdot 17 \, 17^{-1}$ b. $(1.2)^{-2}(1.2)^6 (1.2)^4$ c. $r^{-1} \cdot r^3 \cdot r^{-5} \, r^{-3}$

P-4 **What is the rule for dividing powers of the same base?**

Subtract the exponents. Keep the same base.

Example 3
Write as one power.
1. $\dfrac{y^{-4}}{y^{-2}}$

   Ans: $\dfrac{1}{y^2}$
2. $\dfrac{x^5}{x^{-3}}$

   Ans: x^8
3. $\dfrac{h^{-6}}{h^3}$

   Ans: $\dfrac{1}{h^9}$

EXAMPLE 3   Write $\dfrac{x^{-5}}{x^{-2}}$ as one power.

Solution:

1. Quotient Property of Powers ⟶ $\dfrac{x^{-5}}{x^{-2}} = x^{-5-(-2)}$

2. Subtract the exponents. ⟶ $= x^{-3}$

 The simplest form is $\dfrac{1}{x^3}$.

P-5 **Write each quotient below as one power.**

a. $\dfrac{5^{-1}}{5^{-3}} \, 5^2$ b. $\dfrac{(3.7)^2}{(3.7)^5}$ c. $\dfrac{y^0}{y^{-2}} \, y^2$ d. $\dfrac{x^0}{x^{-5}} x^5$ e. $\dfrac{r^{-4}}{r^0} \, r^{-4}$

   $(3.7)^{-3}$

Example 4
Write as one power and simplify.
1. $\dfrac{x \cdot x^{-4}}{x^{-2}}$

   Ans: $\dfrac{1}{x}$
2. $\dfrac{y^{-3} \cdot y^{-5}}{y^{-8}}$

   Ans: 1
3. $\dfrac{a^{11} \cdot a^{-21}}{a^{-5}}$

   Ans: $\dfrac{1}{a^5}$

EXAMPLE 4   Write as one power and simplify: $\dfrac{m^6 \cdot m^{-3}}{m^4}$

Solution:

1. Product Property of Powers ⟶ $\dfrac{m^6 \cdot m^{-3}}{m^4} = \dfrac{m^3}{m^4}$

2. Quotient Property of Powers ⟶ $= m^{-1}$

3. Simplify. ⟶ $= \dfrac{1}{m}$

Write each expression as a power with a positive exponent. (Property of Negative Exponents)

1. 5^{-10} $\quad \frac{1}{5^{10}}$ **2.** $(-3)^{-4}$ $\quad \frac{1}{(-3)^4}$ **3.** $(5.2)^{-2}$ $\quad \frac{1}{(5.2)^2}$ **4.** $(325)^{-5}$ $\quad \frac{1}{(325)^5}$ **5.** r^{-8} $\quad \frac{1}{r^8}$ **6.** t^{-1} $\quad \frac{1}{t^1}$

Write each product or quotient as one power. (Examples 2–3)

7. $x^2 \cdot x^5$ $\quad x^7$ **8.** $(2^{-3})(2^5)$ $\quad 2^2$ **9.** $\dfrac{2^5}{2^3}$ $\quad 2^2$ **10.** $\dfrac{10^{-2}}{10^{-1}}$ $\quad 10^{-1}$

WRITTEN EXERCISES

Goal: To simplify powers and products and quotients of powers involving positive and negative exponents

Sample Problem: $\dfrac{a^2 \cdot a^{-7}}{a^{-10}}$ **Answer:** a^5

Simplify each expression. (Examples 2–3)

1. r^{-4} $\quad \frac{1}{r^4}$ **2.** t^{-8} $\quad \frac{1}{t^8}$ **3.** $2y^{-5}$ $\quad \frac{2}{y^5}$ **4.** $6m^{-2}$ $\quad \frac{6}{m^2}$

5. 4^{-2} $\quad \frac{1}{4^2}$, or $\frac{1}{16}$ **6.** 3^{-3} $\quad \frac{1}{3^3}$, or $\frac{1}{27}$ **7.** $(-3)^{-3}$ $\quad \frac{1}{(-3)^3}$, or $-\frac{1}{27}$ **8.** $(-2)^{-3}$

9. ab^{-3} $\quad \frac{a}{b^3}$ **10.** rs^{-2} $\quad \frac{r}{s^2}$ **11.** $\dfrac{2^3}{2^{-2}}$ $\quad 2^5$, or 32 **12.** $\dfrac{10^{-1}}{10^{-3}}$

13. $\dfrac{a^2}{a^5}$ $\quad \frac{1}{a^3}$ **14.** $\dfrac{y}{y^3}$ $\quad \frac{1}{y^2}$ **15.** $(2^{50})(2^{-49})$ $\quad 2$ **16.** $(10^{100})(10^{-99})$ $\quad 10$

Write each product or quotient as one power. Then simplify where necessary. (Examples 2–4)

17. $a^5 \cdot a^{-3}$ $\quad a^2$ **18.** $b^{-2} \cdot b^7$ $\quad b^5$ **19.** $x^{-3} \cdot x^{-2}$ $\quad x^{-5}; \frac{1}{x^5}$ **20.** $y^{-1} \cdot y^{-2}$

21. $\dfrac{n^2}{n^3}$ $\quad n^{-1}; \frac{1}{n}$ **22.** $\dfrac{m^5}{m^7}$ $\quad m^{-2}; \frac{1}{m^2}$ **23.** $\dfrac{r^{-2}}{r^{-5}}$ $\quad r^3$ **24.** $\dfrac{s^{-6}}{s^{-3}}$

25. $\dfrac{t^0}{t^{-3}}$ $\quad t^3$ **26.** $\dfrac{w^0}{w^{-5}}$ $\quad w^5$ **27.** $\dfrac{p^{-2}}{p^0}$ $\quad p^{-2}; \frac{1}{p^2}$ **28.** $\dfrac{g^5}{g^0}$ $\quad g^5$

29. $(10^2)(10^7)(10^{-5})$ $\quad 10^4$ **30.** $(5^{-2})(5^4)(5^{-3})$ $\quad 5^{-1}; \frac{1}{5}$ **31.** $(y^{-5})(y^6)(y^{-3})$

32. $(b^7)(b^{-3})(b^{-2})$ $\quad b^2$ **33.** $\dfrac{r^3 \cdot r^{-5}}{r^{-2}}$ $\quad r^0; 1$ **34.** $\dfrac{s^{-2} \cdot s^{-3}}{s^{-7}}$ $\quad s^2$

Powers / **153**

Assignment Guide
Basic pp. 153–154: 1–41 odd
Average pp. 153–154: 1–45 odd
Above Average pp. 153–154: 1–41 odd, 43–46

Additional Answers
Written Exercises
8. $\dfrac{1}{(-2)^3}$, or $-\dfrac{1}{8}$
12. 10^2, or 100
20. $y^{-3}; \dfrac{1}{y^3}$
24. $s^{-3}; \dfrac{1}{s^3}$
31. $y^{-2}; \dfrac{1}{y^2}$

153

APPLICATIONS

You can use a calculator to evaluate a fraction with a power in the denominator.

EXAMPLE   Evaluate $\dfrac{1}{(2.9)^3}$

SOLUTION

On some calculators press "=" twice here.

Evaluate.

35. $\dfrac{1}{2^{10}}$ **36.** $\dfrac{1}{3^{12}}$ **37.** $\dfrac{1}{(4.8)^5}$ **38.** $\dfrac{1}{(0.9)^4}$

39. $\dfrac{1}{(0.86)^{12}}$ **40.** $\dfrac{1}{(0.92)^{15}}$ **41.** $\dfrac{1}{(3^8)(3.14)}$ **42.** $\dfrac{1}{(0.6)^7(3140)}$

MORE CHALLENGING EXERCISES

Let a be a positive integer and let b be a negative integer. Determine whether each expression is positive or negative.

43. ab^{-2} positive   **44.** $a^{-2} \cdot b$ negative   **45.** $\dfrac{a}{a^2}$ positive   **46.** $\dfrac{b^{-3}}{b^{-1}}$

positive

MID-CHAPTER REVIEW

Write each product as one power. (Section 6.1)

$(-3)^8$

1. $(12)^5(12)^8$ 12^{13} **2.** $(4)^3(4)^7$ 4^{10} **3.** $(-6)^4(-6)^{10}$ $(-6)^{14}$ **4.** $(-3)^2(-3)^6$

5. $(\frac{3}{4})^8 \cdot (\frac{3}{4})^5$ $(\frac{3}{4})^{13}$ **6.** $(-\frac{5}{4})^4 \cdot (-\frac{5}{4})^7$ $(-\frac{5}{4})^{11}$ **7.** $(1.4)^2(1.4)^5(1.4)^7$ **8.** $(3.7)^4(3.7)^{24}$

$(3.7)^{28}$

Write each quotient as one power (Section 6.2)

9. $\dfrac{(21)^5}{(21)^3}$ 21^2 **10.** $\dfrac{(-7)^8}{(-7)^3}$ $(-7)^5$ **11.** $(\frac{3}{4})^6 \div (\frac{3}{4})^3$ $(\frac{3}{4})^3$ **12.** $\dfrac{(5.3)^{10}}{(5.3)^5}$

Simplify.

-7

13. $(0.9)^0$ 1 **14.** $(-\frac{5}{8})^0$ 1 **15.** $5^0 - 144^0$ 0 **16.** $-7t^0$, $t \neq 0$

Write each product or quotient as one power. Then simplify. (Section 6.3)

17. $(0.6)^{-7}(0.6)^4$ **18.** $(8.1)^{-2}(8.1)^{-3}$ **19.** $w^{-1} \cdot w^{-5}$ **20.** $p^8 \cdot p^{-3}$

21. $(12)^3 \div (12)^5$ **22.** $(59)^{-2} \div (59)^4$ **23.** $\dfrac{m^2}{m^5}$, $m \neq 0$ **24.** $\dfrac{t^{-1}}{t^4}$, $t \neq 0$

Using Formulas   **Distance/Rate/Time**

Speeds and distances in space are so great that **scientific notation** is useful in representing them.

Minimum Distance from Earth to:
Mars: 56,000,000 (5.6×10^7) kilometers
Sun: 147,000,000 (1.47×10^8) kilometers
Saturn: 1,200,000,000 (1.2×10^9) kilometers
Pluto: 4,270,000,000 (4.27×10^9) kilometers
Proxima Centauri (nearest star):
   40,000,000,000,000 (4.0×10^{13}) kilometers

Speed of Light or of a Radio Signal:
300,000 (3.0×10^5) kilometers per second

Voyager 2 took this photograph of Saturn on July 21, 1981, at a distance of 33.9 million kilometers from the planet.

This special topic covers applications of scientific notation in the field of astronomy. Students are challenged to compare some measurements in various metric units.

EXAMPLE: Find the approximate time in seconds needed for a radio signal to reach the earth from Saturn.

SOLUTION: Use $d = rt$.

$$d \;=\; r \;\times t$$
$$1.2 \times 10^9 = (3.0 \times 10^5) \times t \text{ or,}$$
$$(3.0 \times 10^5)t = 1.2 \times 10^9$$

$d = 1.2 \times 10^9$ kilometers
$r = 3.0 \times 10^5$ km per second

$$t = \frac{1.2 \times 10^9}{3.0 \times 10^5}$$

Use mental math.
$1.2 \div 0.3 = 0.4$
$10^9 \div 10^5 = 10^4$

$$= 0.4 \times 10^4, \text{ or } \textbf{4000 seconds}$$

EXERCISES

For Exercises 1–4, refer to the speed and distances given above to find each of the following. Round your answer to the nearest whole number.

1. The approximate time in seconds for light from the sun to reach the earth
   490 seconds

2. The approximate time in minutes (60 seconds) for a radio signal from Mars to reach the earth   3 minutes

3. The approximate time in hours for a radio signal from Pluto to reach the earth   4 hours

4. The approximate time in years (365 days) for light to reach the earth from the nearest star   4 years

5. Which is the greatest distance? c
   a. 1.48×10^6 meters
   b. 14.8×10^3 kilometers
   c. 1.48×10^5 kilometers
   d. 1480×10^2 meters

6. Which is the fastest speed? a
   a. 2.74×10^5 kilometers per hour
   b. 2.74×10^6 meters per hour
   c. 27.4×10^3 kilometers per hour
   d. 27.4×10^4 meters per hour

Maintenance
1. Simplify: $k - 19 + 0 + 19$
   Ans: k (Section 2.2)
2. Solve and check:
   $5t - 9t + 5 = -11$
   Ans: $t = 4$ (Section 3.5)
3. Solve and check:
   $2y + 5y + 3 = 10y$
   Ans: $y = 1$ (Section 4.3)
4. Solve: $-22 + 10x > -14 + 8x$
   Ans: {all real numbers greater than 4}
   (Section 5.6)
5. The number of multiple choice questions on a test is 26 more than three times the number of essay questions (Condition 1). The total number of questions on the test is 38 (Condition 2). Find the number of each type of question.
   Ans: multiple choice: 35; essay: 3 (Section 4.1)

Additional Examples
Example 1
Simplify.
1. $(5y)^3$
   Ans: $125y^3$
2. $(-2a)^4$
   Ans: $16a^4$

Example 2
Simplify.
1. $(2x)^{-3}$
   Ans: $\dfrac{1}{8x^3}$
2. $(ab)^{-4}$
   Ans: $\dfrac{1}{a^4b^4}$

6.4 Powers of Products and Quotients

OBJECTIVES: To simplify a power of a product by use of the Property of the Power of a Product
To simplify a power of a quotient by use of the Property of the Power of a Quotient

P–1 What is the value of $2x^3$ if x equals 1? ◀ *Raise 1 to the third power. Then multiply by 2.* 2

P–2 What is the value of $(2x)^3$ if x equals 1? ◀ *Multiply $2 \cdot 1$. Then raise to the third power.* 8

You can see that $2x^3$ does not equal $(2x)^3$.

EXAMPLE 1 Simplify $(2x)^3$.

Solution:
1. Meaning of a power ⟶ $(2x)^3 = (2x)(2x)(2x)$
2. Commutative and Associative Properties of Multiplication ⟶ $= (2 \cdot 2 \cdot 2)(x \cdot x \cdot x)$ ◀ *Thus, $(2x)^3 = 2^3x^3$.*
3. Simplify. ⟶ $= 8x^3$

Explain to students that each factor is raised to the power.

Property of the Power of a Product	$(-3t)^3 = (-3)^3t^3$
If x and y are nonzero real numbers and a is any integer, then $(xy)^a = x^ay^a$.	$= -27t^3$ $(pq)^{-2} = p^{-2}q^{-2}$

P–3 Simplify.

a. $(xy)^2$ b. $(3a)^2$ c. $(2r)^3$ d. $(-3r)^2$
   x^2y^2 $9a^2$ $8r^3$ $9r^2$

EXAMPLE 2 Simplify $(rs)^{-3}$.

Show that it can also be done by these steps:

Solution:
1. Property of the Power of a Product ⟶ $(rs)^{-3} = r^{-3}s^{-3}$ $(rs)^{-3} = \dfrac{1}{(rs)^3}$
2. Property of Negative Exponents ⟶ $= \dfrac{1}{r^3} \cdot \dfrac{1}{s^3}$ $= \dfrac{1}{r^3s^3}$
3. Product Rule for Fractions ⟶ $= \dfrac{1}{r^3s^3}$

EXAMPLE 3   Simplify $\left(\dfrac{x}{3}\right)^3$.

Solution:

1 Meaning of a power ──────────▶ $\left(\dfrac{x}{3}\right)^3 = \left(\dfrac{x}{3}\right)\left(\dfrac{x}{3}\right)\left(\dfrac{x}{3}\right)$

2 Product Rule of Fractions ──────────▶ $= \dfrac{x \cdot x \cdot x}{3 \cdot 3 \cdot 3}$

3 Meaning of a power ──────────▶ $= \dfrac{x^3}{3^3},$ or $\dfrac{x^3}{27}$ ◀ *Simplest form*

Explain that both dividend and divisor are raised to the power.

> **Property of the Power of a Quotient**
>
> If x and y are nonzero real numbers and
> a is any integer, then $\left(\dfrac{x}{y}\right)^a = \dfrac{x^a}{y^a}$
>
> $\left(\dfrac{2}{3}\right)^2 = \dfrac{2^2}{3^2}$
>
> $\left(\dfrac{r}{s}\right)^{-3} = \dfrac{r^{-3}}{s^{-3}}$

P-4   **Simplify.**

a. $\left(\dfrac{3}{t}\right)^2$ $\dfrac{9}{t^2}$ b. $\left(\dfrac{p}{q}\right)^4$ $\dfrac{p^4}{q^4}$ c. $\left(\dfrac{2r}{s}\right)^3$ $\dfrac{8r^3}{s^3}$ d. $\left(\dfrac{1}{r}\right)^4$ $\dfrac{1}{r^4}$

Expression **c** above involves both properties.

$$\left(\dfrac{2r}{s}\right)^3 = \dfrac{(2r)^3}{s^3} = \dfrac{2^3 r^3}{s^3} = \dfrac{8r^3}{s^3}$$

EXAMPLE 4   Simplify $\left(\dfrac{2}{x}\right)^{-3}$.

Solution:

1 Property of the Power of a Quotient ──────────▶ $\left(\dfrac{2}{x}\right)^{-3} = \dfrac{2^{-3}}{x^{-3}}$

2 Property of a Negative Exponent ──────────▶ $= \dfrac{1}{2^3} \div \dfrac{1}{x^3}$

3 Meaning of division ──────────▶ $= \dfrac{1}{2^3} \cdot \dfrac{x^3}{1}$

4 Simplify. ──────────▶ $= \dfrac{x^3}{8}$

CLASSROOM EXERCISES

Simplify. (Examples 1–3)

1. $(2a)^2$ $4a^2$
2. $(3x)^2$ $9x^2$
3. $(\frac{1}{2}y)^2$ $\frac{1}{4}y^2$
4. $(2y)^3$

5. $(5n)^2$ $25n^2$
6. $(4t)^2$ $16t^2$
7. $(-2b)^2$ $4b^2$
8. $(-x)^2$

9. $(0.5a)^2$ $0.25a^2$
10. $(-3m)^2$ $9m^2$
11. $(-2r)^3$ $-8r^3$
12. $(-2a)^4$

13. $(xy)^2$ x^2y^2
14. $(mn)^3$ m^3n^3
15. $(ab)^4$ a^4b^4
16. $(st)^5$

17. $\left(\dfrac{3}{x}\right)^2$ $\dfrac{9}{x^2}$
18. $\left(\dfrac{y}{2}\right)^3$ $\dfrac{y^3}{8}$
19. $\left(\dfrac{m}{n}\right)^5$ $\dfrac{m^5}{n^5}$
20. $\left(\dfrac{2a}{b}\right)^2$

21. $\left(\dfrac{-x}{y}\right)^2$ $\dfrac{x^2}{y^2}$
22. $\left(\dfrac{r}{-3s}\right)^2$ $\dfrac{r^2}{9s^2}$
23. $\left(\dfrac{-2}{x}\right)^3$ $\dfrac{-8}{x^3}$
24. $\left(\dfrac{-1}{y}\right)^4$

WRITTEN EXERCISES

Goal: To simplify a power of a product or a power of a quotient

Sample Problems: a. $(4xy)^3$ **b.** $\left(\dfrac{4x}{y}\right)^3$ **Answers: a.** $64x^3y^3$ **b.** $\dfrac{64x^3}{y^3}$

Simplify. (Examples 1–4)

1. $(3r)^3$ $27r^3$
2. $(4t)^3$ $64t^3$
3. $(-2a)^2$ $4a^2$
4. $(-3y)^2$ $9y^2$
5. $(-3p)^3$

6. $(-2q)^3$ $-8q^3$
7. $(-ab)^2$ a^2b^2
8. $(-xy)^2$ x^2y^2
9. $(-rs)^3$ $-r^3s^3$
10. $(-mn)^3$

11. $(2xy)^2$ $4x^2y^2$
12. $(3ab)^2$ $9a^2b^2$
13. $(-3pq)^2$ $9p^2q^2$
14. $(-2mn)^2$ $4m^2n^2$
15. $(-2ab)^3$

16. $(-4cd)^3$ $-64c^3d^3$
17. $\left(\dfrac{2}{x}\right)^3$ $\dfrac{8}{x^3}$
18. $\left(\dfrac{3}{y}\right)^2$ $\dfrac{9}{y^2}$
19. $\left(\dfrac{a}{-2}\right)^3$ $-\dfrac{a^3}{8}$
20. $\left(\dfrac{b}{-3}\right)^2$

21. $\left(\dfrac{-x}{-y}\right)^2$ $\dfrac{x^2}{y^2}$
22. $\left(\dfrac{-2a}{-2b}\right)^3$ $\dfrac{a^3}{b^3}$
23. $\left(\dfrac{2}{y}\right)^{-1}$ $\dfrac{y}{2}$
24. $\left(\dfrac{3}{r}\right)^{-2}$ $\dfrac{r^2}{9}$
25. $\left(\dfrac{pq}{4}\right)^{-1}$

26. $\left(\dfrac{cd}{3}\right)^{-3}$ $\dfrac{27}{c^3d^3}$
27. $(-3t)^{-1}$ $-\dfrac{1}{3t}$
28. $(-4k)^{-2}$ $\dfrac{1}{16k^2}$
29. $\left(\dfrac{3a}{b}\right)^3$ $\dfrac{27a^3}{b^3}$
30. $\left(\dfrac{2r}{3s}\right)^3$

REVIEW CAPSULE FOR SECTION 6.5

Write each product as one power. (Pages 144–146)

1. $m^4 \cdot m^4 \cdot m^4$ m^{12}
2. $(-3)^2 \cdot (-3)^2 \cdot (-3)^2$ $(-3)^6$
3. $t^{-3} \cdot t^{-3} \cdot t^{-3}$ t^{-9}

4. $(2a)^{-2} \cdot (2a)^{-2} \cdot (2a)^{-2}$ $(2a)^{-6}$
5. $g^{-5} \cdot g^{-5} \cdot g^{-5} \cdot g^{-5}$ g^{-20}
6. $(-0.5)^3 \cdot (-0.5)^3$ $(-0.5)^6$

OBJECTIVES: To simplify a power of a power
To simplify a power of an expression involving a product having one or more powers

6.5   Powers of Powers

To simplify a power of an expression involving a quotient where each term has one or more powers

P–1   Write $(x^2)^3$ as a product of three factors.

EXAMPLE 1   Simplify $(x^2)^3$. $(x^2)(x^2)(x^2)$

Solution:

① Meaning of a positive integral power ⟶ $(x^2)^3 = x^2 \cdot x^2 \cdot x^2$

② Product Property of Powers ⟶ $= x^6$

P–2   How is the exponent 6 related to the exponents 2 and 3? $2 \cdot 3 = 6$

EXAMPLE 2   Simplify $(y^{-4})^{-3}$.

Solution:

① Property of a Negative Exponent ⟶ $(y^{-4})^{-3} = \dfrac{1}{(y^{-4})^3}$

Students may want to write $(y^{-4})^{-3}$ as $\dfrac{1}{(y^4)^3}$.

② Meaning of a positive integral power ⟶ $= \dfrac{1}{(y^{-4})(y^{-4})(y^{-4})}$

③ Product Property of Powers ⟶ $= \dfrac{1}{y^{-12}}$

④ Property of a Negative Exponent ⟶ $= y^{12}$

$(-4)(-3) = 12$

P–3   How is the exponent 12 related to the exponents −4 and −3?

To raise a power to a power, multiply exponents.

> **Property of the Power of a Power**
>
> If x is a nonzero real number and a and b are integers, then $(x^a)^b = x^{ab}$
>
> $(r^{-2})^4 = r^{(-2)(4)}$
> $= r^{-8}$

P–4   Simplify.

a. $(m^3)^5$ b. $(k^{-4})^{-2}$ c. $(t^4)^{-1}$ ◀ **Keep the same base. Multiply the exponents.**
m^{15} k^8 t^{-4}

Powers / **159**

Teaching Suggestions p. M–28

Lesson Resources
Warm–Up: p. M–29
Maintenance: See below.
Practice Worksheet 6.5
Transparencies 2, 15

Maintenance
1. Simplify: $-(x - 3)$
   Ans: $-x + 3$ (Section 2.4)
2. Solve and check:
   $7p - (6p - 5) = -4$
   Ans: $p = -9$ (Section 3.5)
3. Write an inequality for 10 and −9. Use "<".
   Ans: $-9 < 10$
   (Section 5.2)
4. Solve: $2(3 + 2x) < 3x - 8$
   Ans: {all real numbers less than −14}
   (Section 5.4)
5. The length and the width of a rectangle are in the ratio 3:2 (Condition 1). The perimeter is 70 meters (Condition 2). Find the length and the width of the rectangle.
   Ans: length: 21 meters; width: 14 meters
   (Section 4.2)

Additional Examples
Example 1
Simplify.
1. $(y^4)^5$
   Ans: y^{20}
2. $(x^3)^3$
   Ans: x^9
3. $(10^5)^2$
   Ans: 10^{10}

Example 2
Simplify.
1. $(x^{-3})^4$
   Ans: $\dfrac{1}{x^{12}}$
2. $(y^{-1})^{-5}$
   Ans: y^5
3. $(3^4)^{-3}$
   Ans: $\dfrac{1}{3^{12}}$

159

Additional Examples
Example 3
Simplify.

1. $(3x^4y^{-2})^2$

Ans: $\dfrac{9x^8}{y^4}$

2. $(-2a^{-2}b^{-3})^{-2}$

Ans: $\dfrac{a^4b^6}{4}$

3. $(4h^{-1}k^4)^{-3}$

Ans: $\dfrac{h^3}{64k^{12}}$

EXAMPLE 3   Simplify $(2x^2y^3)^{-2}$

Solution:

1	Property of the Power of a Product	⟶	$(2x^2y^3)^{-2} = 2^{-2}(x^2)^{-2}(y^3)^{-2}$
2	Property of the Power of a Power	⟶	$= 2^{-2}x^{-4}y^{-6}$
3	Property of a Negative Exponent	⟶	$= \dfrac{1}{2^2} \cdot \dfrac{1}{x^4} \cdot \dfrac{1}{y^6}$
4	Product Rule for Fractions	⟶	$= \dfrac{1}{4x^4y^6}$

The table below shows powers of 2.

Base	Exponent	Power
2	12	4096
2	11	2048
2	10	1024
2	9	512
2	8	256
2	7	128
2	6	64
2	5	32
2	4	16
2	3	8
2	2	4
2	1	2
2	0	1
2	-1	$\frac{1}{2}$
2	-2	$\frac{1}{4}$
2	-3	$\frac{1}{8}$
2	-4	$\frac{1}{16}$
2	-5	$\frac{1}{32}$
2	-6	$\frac{1}{64}$
2	-7	$\frac{1}{128}$
2	-8	$\frac{1}{256}$
2	-9	$\frac{1}{512}$

P-5 **Simplify.**

a. 2^9 **b.** 2^{-11} $\frac{1}{2048}$ **c.** $(2^{-5})^2$ $\frac{1}{1024}$ **d.** $(m^{-1})^{-5}$ m^5
   512

EXAMPLE 4   Simplify $\left(\dfrac{2^4 s^3 t^2}{r^4}\right)^3$

Solution:

1 Property of the Power of a Quotient $\longrightarrow$ $\left(\dfrac{2^4 s^3 t^2}{r^4}\right)^3 = \dfrac{(2^4 s^3 t^2)^3}{(r^4)^3}$

2 Property of the Power of a Product $\longrightarrow$ $= \dfrac{(2^4)^3 (s^3)^3 (t^2)^3}{(r^4)^3}$

3 Property of the Power of a Power $\longrightarrow$ $= \dfrac{2^{12} s^9 t^6}{r^{12}}$

4 Simplify. $\longrightarrow$ $= \dfrac{4096 s^9 t^6}{r^{12}}$ ◀ **Use the table of powers of 2.**

Additional Examples
Example 4
Simplify.

1. $\left(\dfrac{2^3 a b^4}{c^2}\right)^4$

   Ans: $\dfrac{4096 a^4 b^{16}}{c^8}$

2. $\left(\dfrac{2^2 x^4 y^{-2}}{z^3}\right)^5$

   Ans: $\dfrac{1024 x^{20}}{y^{10} z^{15}}$

3. $\left(\dfrac{a^{-3} b^4}{2^4 c^{-2}}\right)^2$

   Ans: $\dfrac{b^8 c^4}{256 a^6}$

CLASSROOM EXERCISES

Simplify. (Examples 1-2)

1. $(3^2)^3$ 3^6
2. $(10^4)^3$ 10^{12}
3. $(a^{-1})^3$ $\frac{1}{a^3}$

4. $(b^2)^{-3}$ $\frac{1}{b^6}$
5. $(y^{-2})^{-3}$ y^6
6. $(t^{-5})^0$ 1

7. $(0.5^2)^4$ $(0.5)^8$
8. $(100^{-2})^{10}$ $\frac{1}{100^{20}}$
9. $(-3^4)^6$ $(-3)^{24}$

Simplify. (Examples 2-4)

10. $(r^2)^{-1}$ $\frac{1}{r^2}$
11. $(t^{-3})^2$ $\frac{1}{t^6}$
12. $(m^2 n^{-1})^3$

13. $(3q^{-2})^{-2}$ $\frac{q^4}{9}$
14. $(r^{-10})^{-2}$ r^{20}
15. $\left(\dfrac{x^2}{x^3}\right)^2$

16. $\left(\dfrac{a^3}{b^4}\right)^3$ $\dfrac{a^9}{b^{12}}$
17. $\left(\dfrac{r^{-2}}{s^3}\right)^3$ $\dfrac{1}{r^6 s^9}$
18. $\left(\dfrac{m^{-1}}{m^{-2}}\right)^{-5}$

19. $\left(\dfrac{x^2}{y^{-3}}\right)^{-3}$ $\dfrac{1}{x^6 y^9}$
20. $\left(\dfrac{10^{-2}}{y^{-3}}\right)^{-1}$ $\dfrac{10^2}{y^3}$
21. $\left(\dfrac{-2^3}{-5^2}\right)^0$

22. $\left(\dfrac{6^{-2}}{8^{-3}}\right)^0$ 1
23. $\left(\dfrac{s^{-4}}{m^{-2}}\right)^{-2}$ $\dfrac{s^8}{m^4}$
24. $\left(\dfrac{-x^2}{y^{-2}}\right)^{-3}$

Common Error
Students add the exponents when applying the Property of the Power of a Power.

Example
$(x^2)^3 = x^{2+3} = x^5$

Prescription
Have students write the factors indicated by the outside exponent.
$(x^2)^3 = x^2 \cdot x^2 \cdot x^2 = x^6$

Additional Answers
Classroom Exercises

12. $\dfrac{m^6}{n^3}$

15. $\dfrac{1}{x^2}$

18. $\dfrac{1}{m^5}$

21. 1

24. $\dfrac{1}{(-x)^6 y^6}$

See the Answers to Selected Exercises for answers to Ex. 3, 9, 15, 21 and 27.

Goal: To simplify powers of powers
Sample Problem: $(t^{-2})^{-5}$ **Answer:** t^{10}

Simplify. (Examples 1–4)

1. $(a^2)^4$ a^8
2. $(p^3)^2$ p^6
3. $(y^{-2})^5$

4. $(r^2)^{-4}$ $\dfrac{1}{r^8}$
5. $(m^{-1})^{-5}$ m^5
6. $(n^{-2})^{-5}$

7. $(t^2)^{-2}$ $\dfrac{1}{t^4}$
8. $(w^{-3})^4$ $\dfrac{1}{w^{12}}$
9. $(3r^3)^2$

10. $(4s^5)^2$ $16s^{10}$
11. $(2a^2b)^3$ $8a^6b^3$
12. $(2ab^3)^2$

13. $(2xy^2)^5$ $32x^5y^{10}$
14. $(2x^3y^2)^4$ $16x^{12}y^8$
15. $\left(\dfrac{x^3}{2}\right)^5$

16. $\left(\dfrac{y^2}{2}\right)^6$ $\dfrac{y^{12}}{64}$
17. $\left(\dfrac{2a^2b^3}{c^{21}}\right)^5$ $\dfrac{32a^{10}b^{15}}{c^{105}}$
18. $\left(\dfrac{2xy^3}{r^4}\right)^6$

19. $(-2t^3)^3$ $-8t^9$
20. $(-2m^{-2}n^3)^{-1}$ $-\dfrac{m^2}{2n^3}$
21. $(2xy^2)^{-3}$

22. $(3r^2s^3)^{-2}$ $\dfrac{1}{9r^4s^6}$
23. $(4m^{-2}n^3)^{-1}$ $\dfrac{m^2}{4n^3}$
24. $(10a^{-4}b^2)^{-1}$

25. $(2^{-3}r^{-1}s^3)^{-2}$ $\dfrac{64r^2}{s^6}$
26. $(2^{-4}p^4q^{-2})^{-3}$ $\dfrac{4096q^6}{p^{12}}$
27. $(-2x^2y^{-1}z^3)^{-5}$

28. $(-2r^{-2}s^4t^3)^{-6}$ $\dfrac{r^{12}}{64s^{24}t^{18}}$
29. $(0.2m^4n^5)^7$ $0.0000128m^{28}n^{35}$
30. $(-0.2x^5y^2)^5$

NON-ROUTINE PROBLEMS

31. Train A leaves Pittsburgh for Boston
traveling 70 miles per hour. At the
same time, Train B leaves Boston
for Pittsburgh traveling 50 miles per
hour. How far apart are the trains 1
hour before they pass each other?

32. Mr. Thomas plans to drive his car
45,000 miles. Each of the tires he
buys will last a maximum of 30,000
miles. Find the fewest number of
tires he will have to buy to travel
45,000 miles.

REVIEW CAPSULE FOR SECTION 6.6

*Multiply each number (**a**) by 1000 and (**b**) by 100,000.*

1. 0.0015678
2. 0.826605027
3. 48.2783
4. 5280.

*Divide each number (**a**) by 10,000 and (**b**) by 1,000,000.*

5. 675,928
6. 5,026,500
7. 0.18
8. 160,000,000

6.6 Scientific Notation

Teaching Suggestions p. M–28

Lesson Resources
Warm–Up: p. M–29
Maintenance: See below.
Practice Worksheet 6.6
Transparency 16

Scientists use very large and very small numbers. Such numbers are often expressed in <u>scientific notation</u>. Here are some examples.

a. 3×10^5 **b.** 1.6×10^2 **c.** 2.3×10^{-7}

Definition

This means that in expressing N, the decimal point is placed between the first two digits.

> **Scientific notation** is a numeral of the form $N \times 10^a$, in which a is an integer and N is a rational number such that $1 \leq N < 10$.
>
> $53,000. = 5.3 \times 10^4$
> $0.0048 = 4.8 \times 10^{-3}$

The sentence $1 \leq N < 10$ means that "$N \geq 1$ <u>and</u> $N < 10$."

P–1 **What is the value of N in each number below? the value of a?**

a. 5.8×10^3 **b.** 3.76×10^{-1} **c.** 1.0059×10^{-6}
$N = 5.8; a = 3$ $N = 3.76; a = -1$ $N = 1.0059; a = -6$

The table below shows some powers of 10.

Powers of Ten
$10^4 = 10,000$
$10^3 = 1,000$
$10^2 = 100$
$10^1 = 10$
$10^0 = 1$
$10^{-1} = 0.1$ or $\frac{1}{10}$
$10^{-2} = 0.01$ or $\frac{1}{100}$
$10^{-3} = 0.001$ or $\frac{1}{1000}$
$10^{-4} = 0.0001$ or $\frac{1}{10,000}$

Maintenance
1. Factor: $10 - 20c$
   Ans: $10(1 - 2c)$
   (Section 2.5)
2. Solve and check:
   $5n - 4 = 3n + 18$
   Ans: $n = 11$ (Section 4.3)
3. Solve: $y - 2 > 7$
   Ans: {all real numbers greater than 9}
   (Section 5.3)
4. Solve: $-3x < 27$
   Ans: {all real numbers greater than -9}
   (Section 5.5)
5. A radio was on sale with a discount rate of 40%. The sale price was $24. Find the list price and the discount.
   Ans: list price: $40; discount: $16
   (Section 4.4)

Additional Examples
Example 1
Write in decimal form.
1. 5.2×10^3
   Ans: 5200
2. 9.7×10^8
   Ans: 970,000,000
3. 4.3×10^{11}
   Ans: 430,000,000,000

EXAMPLE 1   Write 1.2×10^4 in decimal form.

Solution:
1. Find the value of 10^4 from the table. ——→ $10^4 = 10,000$
2. Substitute. ——→ $1.2 \times 10^4 = (1.2)(10,000)$
3. Multiply. ——→ $= 12,000.$

The result is the same as moving the decimal point four places to the right in 1.2.

Powers / 163

EXAMPLE 2   Write 5.3×10^{-3} in decimal form.

Solution:
1. Find the value of
   10^{-3} from the table. ⟶ $10^{-3} = 0.001$
2. Substitute. ⟶ $5.3 \times 10^{-3} = (5.3)(0.001)$
3. Multiply. ⟶ $= 0.0053$

The result is the same as moving the decimal point three places to the left in 5.3.

P–2 **How is the exponent of 10 in scientific notation related to the location of the decimal point in decimal form?** The exponent shows how many places to "move" the decimal point.

> To change scientific notation $N \times 10^a$ to decimal form:
> 1. move the decimal point of N to the right a places if a is positive, or
> 2. move the decimal point of N to the left a places if a is negative.

EXAMPLE 3   Write 186,000 in scientific notation.

Solution:
1. Place a caret to locate the decimal point in N. ⟶ 186000.
2. Count the number of places from the caret to
   the decimal point. ⟶ 186000.
   5 places
3. Write in scientific notation. ⟶ 1.86×10^5

Remind students to check by converting the result to ordinary decimal form. The result should equal the original number.

EXAMPLE 4   Write 0.0000053 in scientific notation.

Solution:
1. Place a caret to locate the decimal point in N. ⟶ 0.0000053
2. Count the number of places from the caret to
   the decimal point. ⟶ 0.0000053
   6 places
3. Write in scientific notation. ⟶ 5.3×10^{-6}

164 / *Chapter 6*

Write the decimal form for each number. (Examples 1–2)

1. 3×10^2 300   **2.** 2.0×10^{-2} 0.02   **3.** 7.03×10^3 7030   0.0039 **4.** 3.9×10^{-3}

5. 4.62×10^1 46.2   **6.** 8.1627×10^{-1} **7.** 5.0001×10^4 50,001   **8.** 9.99×10^{-4} 0.000999

0.81627

Write the scientific notation for each number. (Examples 3–4)

2.738 × 10⁻¹ written as 2.738×10^{-1}

9. 530 5.3×10^2 **10.** 0.0926 9.26×10^{-2} **11.** 48.6 4.86×10^1 **12.** 0.2738

13. 76,000. 7.6×10^4 **14.** 6.318 6.318×10^0 **15.** 0.000483 4.83×10^{-4} **16.** 0.008092 8.092×10^{-3}

WRITTEN EXERCISES

Goal: To write numbers in scientific notation
Sample Problems: a. 418,200   **b.** 0.00531
Answers: a. 4.182×10^5 **b.** 5.31×10^{-3}

For Exercises 1–16, write each number in decimal form. (Examples 1–2)

0.0091
1. 4.2×10^3 4200   **2.** 3.7×10^2 370   **3.** 2.9×10^{-4} 0.00029   **4.** 9.1×10^{-3}

5. 8.62×10^4 86,200   **6.** 6.01×10^5 601,000   **7.** 1.00726×10^{-1} **8.** 2.053×10^{-2}

9. 7.280×10^6 **10.** 3.1709×10^7 **11.** 5.607×10^{-5} **12.** 9.090×10^{-6}

13. The diameter of the earth is about 1.28×10^7 meters.   12,800,000

14. The diameter of the sun is about 1.39×10^9 meters.   1,390,000,000

15. A helium atom has a diameter of 2.2×10^{-8} centimeters.   0.000000022

16. If one foot of copper wire is heated 1° Celsius, it expands 1.6×10^{-5} feet. 0.000016

For Exercises 17–32, write each number in scientific notation. (Examples 3–4)

17. 0.0092 9.2×10^{-3} **18.** 0.000184 1.84×10^{-4} **19.** 2705.

20. 937. 9.37×10^2 **21.** 0.01734 1.734×10^{-2} **22.** 0.80672

23. 27,000,000. 2.7×10^7 **24.** 186,000. 1.86×10^5 **25.** 0.0000275

26. 0.00000031801 3.1801×10^{-7} **27.** 2 million 2×10^6 **28.** 3.5 billion

29. The speed of light is about 297,600 kilometers per second.

30. Our galaxy has about 100,000,000,000 stars. 1.0×10^{11}

31. The sun is about 150,000,000 kilometers from the earth. 1.5×10^8

32. Special balances can weigh something as small as 0.00000001 gram. 1.0×10^{-8}

Assignment Guide
Basic p. 165: 1–31 odd
Average p. 165: 1–31 odd
Above Average p. 165: 1–31 odd

Additional Answers
Written Exercises
7. 0.100726
8. 0.02053
9. 7,280,000
10. 31,709,000
11. 0.00005607
12. 0.000009090
19. 2.705×10^3
22. 8.0672×10^{-1}
25. 2.75×10^{-5}
28. 3.5×10^9
29. 2.976×10^5

Quiz Sections 6.4–6.6
After completing Sections 6.4–6.6, you may want to administer a quiz covering the same sections. See the quiz provided on page 128 in the *Teacher's ResourceBank.*

Focus on Reasoning: Number Patterns

A **number sequence** is a succession of numbers that follow a fixed pattern. Each number in the sequence is related to the preceding number according to a definite plan. **Identifying the plan** is the key to writing other numbers in each sequence.

EXAMPLE Write the next two numbers in each sequence.

a. $1, \frac{2}{3}, \frac{1}{3}, 0, -\frac{1}{3}, -\frac{2}{3}, \cdots$ **b.** $1, -2, -4, 8, 16, -32, -64, \cdots$

Solution: **a. Think:** $1, \quad \frac{2}{3}, \quad \frac{1}{3}, \quad 0, \quad -\frac{1}{3}, \quad -\frac{2}{3}$

$-\frac{1}{3} \quad -\frac{1}{3} \quad -\frac{1}{3} \quad -\frac{1}{3} \quad -\frac{1}{3}$

Rule: Subtract $\frac{1}{3}$ from the preceding number.

Next two numbers: $\mathbf{-1, -1\frac{1}{3}}$

b. Think: $1, \quad -2, \quad -4, \quad 8, \quad 16, \quad -32, \quad -64$

$\times(-2) \quad \times 2 \quad \times(-2) \quad \times 2 \quad \times(-2) \quad \times 2$

Rule: Alternate multiplying by -2 and 2.

Next two numbers: **128, 256**

EXERCISES

Write the next four numbers in each sequence.

1. $-11, -7, -3, 1, 5, 9, \cdots$ 13, 17, 21, 25

2. $-7, -4, -1, 2, 5, 8, \cdots$ 11, 14, 17, 20

3. $-\frac{2}{9}, -\frac{2}{3}, 2, 6, -18, -54, \cdots$

4. $-\frac{5}{4}, -\frac{5}{2}, 5, 10, -20, -40, \cdots$

5. $5.5, 1, -3.5, -8, -12.5, -17, \cdots$

6. $6.3, 3.2, 0.1, -3, -6.1, -9.2, \cdots$

7. $1, \frac{1}{3}, \frac{1}{5}, \frac{1}{7}, \frac{1}{9}, \frac{1}{11}, \cdots \frac{1}{13}, \frac{1}{15}, \frac{1}{17}, \frac{1}{19}$

8. $\frac{1}{2}, \frac{1}{4}, \frac{1}{6}, \frac{1}{8}, \frac{1}{10}, \frac{1}{12}, \cdots \frac{1}{14}, \frac{1}{16}, \frac{1}{18}, \frac{1}{20}$

9. $1, \frac{1}{4}, \frac{1}{9}, \frac{1}{16}, \frac{1}{25}, \frac{1}{36}, \cdots \frac{1}{49}, \frac{1}{64}, \frac{1}{81}, \frac{1}{100}$

10. $\frac{1}{2}, \frac{3}{4}, \frac{5}{6}, \frac{7}{8}, \frac{9}{10}, \frac{11}{12}, \cdots \frac{13}{14}, \frac{15}{16}, \frac{17}{18}, \frac{19}{20}$

11. $1, 9, 25, 49, 81, 121, \cdots$

12. $4, 16, 36, 64, 100, 144, \cdots$

13. $12, 10, 6, 0, -8, -18, \cdots$

14. $21, 18, 13, 6, -3, -14, \cdots$

15. $1, 4, 10, 22, 46, 94, \cdots$

16. $1, 3, 7, 15, 31, 63, \cdots$

Focus on Reasoning: Geometric Patterns

Sometimes a sequence is a succession of geometric figures. Look for what changes and what remains the same. This will help you to identify the pattern.

EXAMPLE   What is the next figure?

Think: The figures are in pairs.
The pairs have the same shape.
The second figure of each
pair is upside down.

Next figure: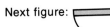

EXERCISES

Identify the pattern. Then draw the next figure in each sequence.

1. ⬭ ⚬⚬⚬ ☐ 🔲 ⬡
2. ⬡ 🔵🔵 △ ▲▲▲ ◇
3. ⬭ ◝ ⬡ ◞ ☐
4. ◔ ◔ ◰ ⊟ ◰
5. ◰ ◰ ◰ ◔ ◔
6. ◕ ◕ ◕ ▱ ▱
7. ◔ ◔ ◔ ◰ ◰
8. ◠ ◠ ◠ ◡ ▱
9. △ △ △ △ △
10. ⬡ ⬡ ⬡ ⬡ ⬡
11. ⊞ ⊞ ⊞ ⊞ ⊞
12. ◓ ◓ ◓ ◓ ◓

Additional Answers

1.
2.
3.
4.
5.
6.
7.
8.
9.
10.
11.
12.

Focus on Reasoning / **167**

167

CHAPTER SUMMARY

IMPORTANT TERMS	Power of a power *(p. 159)* Scientific notation *(p. 163)*

IMPORTANT IDEAS	*1.* If *a* is a positive integer, then x^a means that *x* is a factor *a* times.

2. *Product Property of Powers:* If *a* and *b* are integers and *x* is any nonzero real number, then $x^a \cdot x^b = x^{a+b}$.

3. *Quotient Property of Powers:* If *a* and *b* are integers and *x* is a nonzero real number, then $\frac{x^a}{x^b} = x^{a-b}$.

4. *Property of a Zero Exponent:* If *x* is any nonzero real number, then $x^0 = 1$.

5. *Property of Negative Exponents:* If $-a$ represents a negative integer and *x* is a nonzero real number, then $x^{-a} = \frac{1}{x^a}$.

6. *Property of the Power of a Product:* If *x* and *y* are nonzero real numbers and *a* is any integer, then $(xy)^a = x^a y^a$.

7. *Property of the Power of a Quotient:* If *x* and *y* are nonzero real numbers and *a* is any integer, then $\left(\frac{x}{y}\right)^a = \frac{x^a}{y^a}$.

8. *Property of the Power of a Power:* If *x* is a nonzero real number and *a* and *b* are integers, then $(x^a)^b = x^{ab}$.

9. To change scientific notation $N \times 10^a$ to decimal form:
   a. move the decimal point of *N* to the right *a* places if *a* is positive, or
   b. move the decimal point of *N* to the left *a* places if *a* is negative.

Chapter Test
Two Chapter Tests (Form A and Form B) are provided on pages 129–132 in the *Teacher's ResourceBank.*

CHAPTER REVIEW

SECTION 6.1

Write each product as one power.

1. $(27)^6(27)^3$ 27^9

2. $(-8.5)^4(-8.5)^4$ $(-8.5)^8$

3. $r^2 \cdot r^5$ r^7

4. $m^3 \cdot m^{12}$ m^{15}

5. $t^3 \cdot t \cdot t^5$ t^9

6. $p^4 \cdot p^5 \cdot p$ p^{10}

SECTION 6.2

Write each quotient as one power.

7. $\frac{19^8}{19^3}$ 19^5

8. $\frac{24^{12}}{24^7}$ 24^5

9. $\frac{(2.7)^{15}}{(2.7)^{13}}$ $(2.7)^2$

10. $\dfrac{(-3.9)^6}{(-3.9)^3}$ $(-3.9)^3$

11. $\dfrac{(\sqrt{7})^8}{(\sqrt{7})^5}$ $(\sqrt{7})^3$

12. $\dfrac{\pi^{19}}{\pi^{10}}$ π^9

Simplify.

13. $(35)^0$ $\quad 1$

14. $(242)^0$ $\quad 1$

15. $(-23.9)^0$ $\quad 1$

16. $\left(\dfrac{5}{4}\right)^0$ $\quad 1$

17. $(13.4)^0 - 2x^0$ $\quad -1$

18. $5n^0 + (-95)^0$
$\qquad\qquad\qquad 6$

SECTION 6.3

Simplify.

19. $(t)^{-10}$ $\dfrac{1}{t^{10}}$

20. $(2)^{-6}$ $\dfrac{1}{64}$

21. $5x^{-2}$ $\dfrac{5}{x^2}$

22. $100y^{-1}$ $\dfrac{100}{y}$

23. $p^{-2}q$ $\dfrac{q}{p^2}$

24. $m^{-3}n$ $\dfrac{n}{m^3}$

Write each product or quotient as one power. Then simplify where necessary.

25. $k^{-4} \cdot k^2$ $\;k^{-2};\,\dfrac{1}{k^2}$

26. $r^{-5} \cdot r^{-3}$ $\;r^{-8};\,\dfrac{1}{r^8}$

27. $\dfrac{m^{-2}\,m^{-5}}{m^3}$ $\;m^{-5};\,\dfrac{1}{m^5}$

28. $\dfrac{t^4}{t^{-3}}$ $\;t^7$

29. $\dfrac{p^2 \cdot p^{-3}}{p^4}$ $\;p^{-5};\,\dfrac{1}{p^5}$

30. $\dfrac{x^{-1} \cdot x^{-3}}{x^4}$
$\qquad\qquad x^{-8};\,\dfrac{1}{x^8}$

SECTION 6.4

Simplify.

31. $(4t)^2$ $\;16t^2$

32. $(2q)^4$ $\;16q^4$

33. $(-2r)^5$ $\;-32r^5$

34. $(-3y)^3$ $\;-27y^3$

35. $\left(\dfrac{5}{p}\right)^2$ $\dfrac{25}{p^2}$

36. $\left(\dfrac{-3}{s}\right)^3$ $-\dfrac{27}{s^3}$

37. $(-4m)^{-2}$ $\dfrac{1}{16m^2}$

38. $(-3n)^{-3}$ $-\dfrac{1}{27n^3}$

39. $\left(\dfrac{-1}{-y}\right)^3$ $\dfrac{1}{y^3}$

SECTION 6.5

Simplify.

40. $(d^3)^5$ $\;d^{15}$

41. $(g^5)^{-2}$ $\dfrac{1}{g^{10}}$

42. $\left(\dfrac{y^4}{2}\right)^5$ $\dfrac{y^{20}}{32}$

43. $\left(\dfrac{3m^2n^3}{p}\right)^3$ $\dfrac{27m^6n^9}{p^3}$

44. $(-2p^{-2}q^3r^{-4})^2$ $\dfrac{4q^6}{p^4r^8}$

45. $(2^{-2}x^2y^{-1}z^{-4})^3$
$\qquad\qquad \dfrac{x^6}{64y^3z^{12}}$

Powers / **169**

In Exercises 46–53, write each number in decimal form.

46. 1.83×10^{-2} 0.0183

47. 3.187×10^4 31,870

48. 2.1692×10^5 216,920

49. 5.279×10^{-3} 0.005279

50. 7.10036×10^{-4} 0.000710036

51. 8.00002×10^3 8000.02

52. The diameter of the planet Mars is about 4.214×10^3 miles.   4214

53. The orbit of the Space Shuttle is about 2.224×10^4 miles above the equator.   22,240

In Exercises 54–60, write each number in scientific notation.

54. 47,280. 4.728×10^4

55. 0.005219 5.219×10^{-3}

56. 0.00002009 2.009×10^{-5}

57. 512,000,000. 5.12×10^8

58. The speed of light is approximately 300,000 kilometers per second. 3.0×10^5 Write 300,000 in scientific notation.

59. There are about 31,536,000 seconds in one year. Write 31,536,000 in scientific notation. 3.1536×10^7

60. The average concentration of radium in seawater is about 0.00000000006 milligrams per liter. Write 0.00000000006 in scientific notation. 6.0×10^{-11}

CUMULATIVE REVIEW: CHAPTERS 1–6

Perform the indicated operations. (Sections 1.2 through 1.5)

1. $27 + (-19)$ 8
2. $-6.3 - 2.7$ -9
3. $\frac{2}{3}(-27)$ -18
4. $-65 \div (-5)$ 13

Evaluate each expression. (Section 1.7)

5. $3x^2 - 8$ when $x = 4$ 40
6. $m^2 - 3m$ when $m = -2$ 10
7. $5x^2 - 6x$ when $x = -3$ 63

Multiply and simplify. (Sections 2.3 through 2.5)

8. $(-7)(2)(x)(-5)$ $70x$
9. $(3x)(12xy)(-y)$ $-36x^2y^2$
10. $-3(2x - 5)$ $-6x + 15$
11. $(2y + 11)5y$ $10y^2 + 55y$

Factor. (Section 2.5)

12. $7a + 7q$ $7(a + q)$
13. $8c - 12$ $4(2c - 3)$
14. $9x - 3xy$ $3x(3 - y)$
15. $-4a^2 - 4a$ $-4a(a + 1)$

Simplify. (Section 2.6)

16. $-7w + w$ $-6w$
17. $4m - 2 - 8m - 1$ $-4m - 3$
18. $2a^2 - 7a - 3a + 8a^2$ $10a^2 - 10a$

Solve. Check each answer. (Sections 3.1 through 3.3)

19. $y + 1 = -8$ $y = -9$
20. $-\frac{3}{4}j = -30$ $j = 40$
21. $-22 = 3k + 5$ $k = -9$
22. $-3 = \frac{t}{4} - 7$ $t = 16$
23. $7h - 2 + h = 14$ $h = 2$
24. $11 = 2(3x - 2) - 3$ $x = 3$

Use Condition 1 to represent the unknowns. Use Condition 2 to write an equation for the problem. Solve. (Sections 4.1 and 4.2)

25. Joyce scored 15 fewer points in her first game of bowling than in her second game. Her total number of points for the two games was 309. Find the number of points Joyce scored in the first game.

26. One special triangle has sides in the ratio 3:5:7. The perimeter of the triangle is 165 centimeters. Find the lengths of the sides.

Solve. Check each answer. (Section 4.3)

27. $7x - 2 = 2x + 13$ $x = 3$
28. $2(2 + r) - 3r = r + 8$ $r = -2$

Solve. (Section 4.4)

29. O'Leary's Clothing Store is having a sale on winter coats. The rate of discount of a particular coat is 35%. The discount is $39.90. Find the list price of the coat. $114

30. All of the gas grills at Big John's Hardware Store were on sale at a 30% rate of discount. Sally paid the sale price of $126 for a grill. Find the list price of the grill and the discount. $p = \$180;\ d = \$54$

Cumulative Test
A cumulative test covering Chapters 4–6 is provided on pages 133–136 in the *Teacher's ResourceBank.*

Additional Answers
25. a. Let x = points scored in second game.
Then $x - 15$ = points scored in first game.
 b. $x + x - 15 = 309$
 c. first game: 147 points
26. a. Let $3x$, $5x$, and $7x$ be the lengths of the sides of the triangle.
 b. $3x + 5x + 7x = 165$
 c. $3x = 33$ cm; $5x = 55$ cm; $7x = 77$ cm

See the Answers to Selected Exercises for answer to Ex. 31.

Additional Answers

32. a.

	To work	Return
r(mi/hr)	$x + 15$	x
t(hrs)	0.6	0.9
d(mi)	$0.6(x + 15)$	$0.9x$

33. On a number line, the points to the left of, and not including, 4

34. On a number line, the points to the right of, and not including, -2

35. On a number line, the points to the right of, and including, -3.7

36. On a number line, the points to the left of, and including, $1\frac{1}{2}$

37. {all real numbers less than -6}

38. {all real numbers greater than 2.2}

39. {all real numbers greater than -4}

40. {all real numbers greater than -37}

41. {all real numbers less than -25}

42. {all real numbers less than $5\frac{1}{2}$}

43. {all real numbers greater than 4}

44. {all real numbers less than -3}

45. {all real numbers greater than 5}

46. a. Let x = cost of V.C.R. Then $2x - 100$ = cost of television.
b. $x + 2x - 100 < 950$
c. $599.99

47. a. Let x = score on fifth test.
b. $\dfrac{73 + 87 + 85 + 82 + x}{5} > 84$
c. 94

Use Conditions 1 and 2 to make a table. Use Condition 2 to write an equation for the problem. Solve. (Sections 4.4 and 4.5)

31. A football player attempted 6 more field goals this season than last season. This season he made 60% of his attempts. Last season he made 75% of his attempts. He made the same number of field goals this season as last season. How many did he attempt this season?

32. Anthony's average rate driving to work is 15 miles per hour faster than his average rate returning home. It takes him 0.6 hours to drive to work and 0.9 hours to drive home. Find Anthony's average rate driving to work and his average rate returning home.

Draw the graph of each inequality. (Section 5.1)

33. $x < 4$ **34.** $x > -2$ **35.** $x \geq -3.7$ **36.** $x \leq 1\frac{1}{2}$

Solve each inequality. (Sections 5.3 through 5.6)

37. $x + 9 < 3$ **38.** $a - 3.9 > -1.7$ **39.** $11 > 3y + 7 - 4y$

40. $3(x - 7) < 4x + 16$ **41.** $\frac{3}{5}a < -15$ **42.** $-11 < -2b$

43. $2g + 5 > 13$ **44.** $2 - k > 6k + 23$ **45.** $2(10 + t) - 5 < 5t$

Use Condition 1 to represent the unknowns. Use Condition 2 to write an equation for the problem. Solve. (Section 5.7)

46. Ryan estimates that a television costs $100 less than twice the cost of a V.C.R. He plans to spend less than $950 for both items. Find the greatest amount he can spend for the television.

47. Meghan's scores on four math tests were 73, 87, 85, and 82. She wants her mean score on five tests to be greater than 84. What is the lowest score she can get on the fifth test?

Write each product or quotient as one power. Then simplify where necessary. (Sections 6.1 through 6.3)

48. $x \cdot x^9$ x^{10} **49.** $b^2 \cdot b^7 \cdot b^7$ b^{16} **50.** $\dfrac{2^{19}}{2}$ 2^{18} **51.** $\dfrac{x^9}{x^2}$ x^7

52. $\dfrac{c^4}{c^7}$ $c^{-3}; \frac{1}{c^3}$ **53.** $\dfrac{r^0}{r^{-3}}$ r^3 **54.** $x^{11} \cdot x^{-17}$ $x^{-6}; \frac{1}{x^6}$ **55.** $\dfrac{m^2 \cdot m^{-4}}{m^{-8}}$ m^6

Simplify. (Sections 6.4 and 6.5)

56. $(2a)^3$ $8a^3$ **57.** $\left(\dfrac{4}{x}\right)^2$ $\dfrac{16}{x^2}$ **58.** $(4a^{-7})^{-2}$ $\dfrac{a^{14}}{16}$ **59.** $\left(\dfrac{3x^4w}{y^5}\right)^3$ $\dfrac{27x^{12}w^3}{y^{15}}$

Write each number in decimal form. (Section 6.6)

60. 9.2×10^3 9200   **61.** 4.12×10^{-4} .000412   **62.** 6.72×10^{-2} 0.0672   **63.** 7.3×10^5 730,000

Roots

On December 3, 1980, *Janice Brown* made the first solar-powered flight in the *Solar Challenger.* The formula $D = 3.56\sqrt{x}$ gives the approximate distance to the horizon in kilometers from an aircraft that is *x* meters above the ground (altitude). How far could Janice see to the horizon at an altitude of 30 meters?

Background

The Solar Challenger was designed by aerodynamicist Paul Beattie MacCready. The plane was powered by 16,128 solar cells connected to two electric motors. It weighed 210 pounds and had a wingspan of 47 feet.

At an altitude of 30 meters, Janice Brown could see approximately 20 kilometers to the horizon.

Teaching Suggestions p. M–30

Lesson Resources
Warm–Up: p. M–30
Practice Worksheet 7.1
Exploration 7

Maintenance

1. Simplify: $7m + 3t - 4m - 7t$
   Ans: $3m - 4t$
   (Section 2.6)
2. Solve and check:
   $2(y - 3) + 5(2y + 1) = 11$
   Ans: $y = 1$ (Section 3.5)
3. Solve: $6x - 18 < 5x - 6$
   Ans: {all real numbers less than 12} (Section 5.4)
4. Write $\dfrac{m^{-5}}{m^{-4}}$ as one power.
   Ans: $\dfrac{1}{m}$ (Section 6.3)
5. The net change in the price of a stock for each of ten successive business days was $-\frac{7}{8}$ of a point. Find the net change for the ten days.
   Ans: $-8\frac{3}{4}$ points
   (Section 1.4)

7.1 Numbers in Radical Form

To write the positive square root of a perfect square counting number by using prime factorization

P–1 **What is the simplest form of $\sqrt{36}$?** 6
of $-\sqrt{36}$? -6

◀ $\sqrt{}$ *is a radical symbol.*

Emphasize that $\sqrt{}$ is used to represent a nonnegative square root.
$-\sqrt{}$ is used for a negative square root.

A **square root** is one of two equal factors of a number. Both 6 and -6 are square roots of 36. Also, 36 is the **square** of both 6 and -6.

$$(6)(6) = 36 \qquad (-6)(-6) = 36$$

P–2 **What is the value of $\sqrt{0}$?** 0

P–3 **Why doesn't $\sqrt{-16}$ have a real-number value?**
There are no two equal factors of -16.

The square of a positive number is positive.
The square of a negative number is positive.
The square of 0 is 0.

◀ *A negative number cannot have a real number as a square root.*

P–4 **What is the square of each number below?**

a. -7 49 **b.** $\dfrac{1}{2}$ $\dfrac{1}{4}$ **c.** -0.3 0.09 **d.** 20 400 **e.** $-\dfrac{3}{4}$ $\dfrac{9}{16}$

P–5 **Simplify each expression below.**

a. $-\sqrt{49}$ -7 **b.** $\sqrt{\dfrac{1}{4}}$ $\dfrac{1}{2}$ **c.** $-\sqrt{0.09}$ -0.3 **d.** $\sqrt{10,000}$ 100

This is an appropriate subset of the prime numbers for students to memorize. Seldom will a prime factor greater than 23 be needed.

It may be helpful to review prime factorization at this time. The set of prime numbers less than 25 is shown below.

$$\{2, 3, 5, 7, 11, 13, 17, 19, 23\}$$

Definition

> A **prime number** is a counting number greater than 1 that has exactly two counting-number factors, 1 and the number itself.

P–6 **Why is 15 not a prime number?**
It has more than two counting number factors.

The set of counting-number factors of 15 is $\{1, 3, 5, 15\}$.

◀ *The prime factorization of 15 is 3 · 5.*

You will recall that *prime factorization* is writing a number as a product of its prime-number factors. The table below shows the prime factorization of some numbers. The numbers 4, 9, 25, and 36 in the table are *perfect squares*. Note that each prime factor of a perfect square occurs an even number of times.

Number	Prime Factorization	Number	Prime Factorization
4	2 · 2	20	2 · 2 · 5
6	2 · 3	25	5 · 5
9	3 · 3	27	3 · 3 · 3
12	2 · 2 · 3	36	2 · 2 · 3 · 3

EXAMPLE 1   Write the prime factorization of 150.

Solution:

1 Divide by 2 (the smallest prime number).
Write the quotient 75.

$$\begin{array}{r|r} 2 & 150 \\ \hline & 75 \end{array}$$

2 Check to see whether 2 is a factor of 75. It is not.

3 Select the next greater prime number 3. It is a factor of 75. Divide and write the quotient 25.

$$\begin{array}{r|r} 2 & 150 \\ \hline 3 & 75 \\ \hline & 25 \end{array}$$

4 Check again to see whether 3 is a factor of 25. It is not.

5 Select the next greater prime number 5. Divide by 5 and write the quotient 5. The quotient is prime.

$$\begin{array}{r|r} 2 & 150 \\ \hline 3 & 75 \\ \hline 5 & 25 \\ \hline & \mathbf{5} \end{array}$$

6 Write the prime factorization of 150.

$$150 = 2 \cdot 3 \cdot 5 \cdot 5$$

EXAMPLE 2   Simplify $\sqrt{441}$.

◀ *Prime factorization can help in simplifying radicals.*

Solution:

1 Write the prime factorization of 441. ⟶ $\sqrt{441} = \sqrt{3 \cdot 3 \cdot 7 \cdot 7}$

2 Write two equal factors of 441. ⟶ $= \sqrt{(3 \cdot 7)(3 \cdot 7)}$

$= \sqrt{21 \cdot 21}$

3 Simplify. ⟶ $= 21$

$$\begin{array}{r|r} 3 & 441 \\ \hline 3 & 147 \\ \hline 7 & 49 \\ \hline & 7 \end{array}$$

Roots / 175

Write the square of each number. (P-4)

1. -10 100 **2.** 11 121 **3.** -12 144 **4.** $\frac{1}{2}$ $\frac{1}{4}$ **5.** $-\frac{1}{3}$ $\frac{1}{9}$

6. -0.2 0.04 **7.** 0.4 0.16 **8.** -13 169 **9.** $\frac{5}{2}$ $\frac{25}{4}$ **10.** -1.2 1.44

Simplify. (P-1, P-5)

11. $\sqrt{9}$ 3 **12.** $-\sqrt{1}$ -1 **13.** $\sqrt{0}$ 0 **14.** $-\sqrt{36}$ -6 **15.** $\sqrt{100}$ 10

16. $\sqrt{\frac{9}{4}}$ $\frac{3}{2}$ **17.** $-\sqrt{\frac{1}{16}}$ $-\frac{1}{4}$ **18.** $\sqrt{\frac{9}{25}}$ $\frac{3}{5}$ **19.** $-\sqrt{0.01}$ -0.1 **20.** $\sqrt{0.0004}$ 0.02

Write the prime factorization. (Example 1)

21. 21 $3 \cdot 7$ **22.** 12 $2 \cdot 2 \cdot 3$ **23.** 8 $2 \cdot 2 \cdot 2$ **24.** 25

25. 26 $2 \cdot 13$ **26.** 35 $5 \cdot 7$ **27.** 49 $7 \cdot 7$ **28.** 77

29. 39 $3 \cdot 13$ **30.** 18 $2 \cdot 3 \cdot 3$ **31.** 45 $3 \cdot 3 \cdot 5$ **32.** 81

WRITTEN EXERCISES

Goal: To write square roots of numbers
Sample Problems: a. $\sqrt{225}$ **b.** $-\sqrt{2^2 \cdot 3^4}$
Answers: a. 15 **b.** -18

Simplify. (Steps 2 and 3 of Example 2)

1. $\sqrt{121}$ 11 **2.** $\sqrt{196}$ 14 **3.** $-\sqrt{100}$ -10 **4.** $-\sqrt{81}$ -9 **5.** $\sqrt{\frac{4}{25}}$ $\frac{2}{5}$

6. $\sqrt{\frac{9}{49}}$ $\frac{3}{7}$ **7.** $-\sqrt{\frac{1}{9}}$ $-\frac{1}{3}$ **8.** $-\sqrt{\frac{25}{36}}$ $-\frac{5}{6}$ **9.** $\sqrt{0.81}$ 0.9 **10.** $\sqrt{0.25}$

11. $\sqrt{(2 \cdot 5)(2 \cdot 5)}$ 10 **12.** $\sqrt{(5 \cdot 11)(5 \cdot 11)}$ 55 **13.** $\sqrt{2 \cdot 3 \cdot 3 \cdot 2 \cdot 2 \cdot 2}$

14. $\sqrt{3 \cdot 2 \cdot 5 \cdot 3 \cdot 2 \cdot 5}$ 30 **15.** $\sqrt{2^2 \cdot 5^4}$ 50 **16.** $\sqrt{3^4 \cdot 5^2 \cdot 11^2}$ 495

Write the prime factorization. (Example 1)

17. 65 $5 \cdot 13$ **18.** 99 $3 \cdot 3 \cdot 11$ **19.** 18 $2 \cdot 3 \cdot 3$ **20.** 400 **21.** 54 $2 \cdot 3 \cdot 3 \cdot 3$

22. 60 $2 \cdot 2 \cdot 3 \cdot 5$ **23.** 120 **24.** 108 **25.** 198 **26.** 264

Simplify. (Example 2)

27. $\sqrt{484}$ 22 **28.** $\sqrt{676}$ 26 **29.** $\sqrt{784}$ 28 **30.** $\sqrt{1225}$ 35 **31.** $\sqrt{1936}$ 44

32. $\sqrt{4096}$ 64 **33.** $\sqrt{3136}$ 56 **34.** $\sqrt{4624}$ 68 **35.** $\sqrt{5776}$ 76 **36.** $\sqrt{6561}$

37. $\sqrt{5184}$ 72 **38.** $\sqrt{7225}$ 85 **39.** $\sqrt{9025}$ 95 **40.** $\sqrt{4356}$ 66 **41.** $\sqrt{3844}$ 62

7.2 Products

OBJECTIVE: To simplify the product of two or more square root radicals

Radicals such as $\sqrt{9}$, $-\sqrt{49}$, and $\sqrt{64}$ represent rational numbers. **A square root of a perfect square is rational.**

Radicals such as $\sqrt{7}$, $\sqrt{15}$, $-\sqrt{34}$, and $\sqrt{50}$ represent irrational numbers.

> Any whole number that is not a perfect square has an irrational square root.

P–1 **Which numbers below are rational? Which are irrational?**

a. $\sqrt{19}$ **b.** $-\sqrt{144}$ **c.** $-\sqrt{80}$ **d.** $\sqrt{0}$ Rational
Irrational   Rational   Irrational

EXAMPLE 1   Multiply and simplify: $\sqrt{13} \cdot \sqrt{13}$ A square root of 13 is one of the two equal factors of 13 designated as $\sqrt{13}$. So, $\sqrt{13} \cdot \sqrt{13}$ must equal 13.

Solution: $\sqrt{13} \cdot \sqrt{13} = (\sqrt{13})^2$ **A square root is one of two equal factors of a number.**

$= 13$

> If a is any nonnegative real number, then $\sqrt{a} \cdot \sqrt{a} = a$.
>
> $\sqrt{20} \cdot \sqrt{20} = 20$
> $\sqrt{3.2} \cdot \sqrt{3.2} = 3.2$

P–2 **Multiply and simplify.**

a. $\sqrt{29} \cdot \sqrt{29}$ 29 **b.** $\sqrt{\frac{1}{4}} \cdot \sqrt{\frac{1}{4}}$ $\frac{1}{4}$ **c.** $\sqrt{1029} \cdot \sqrt{1029}$ 1029

EXAMPLE 2   Multiply and simplify: $(\sqrt{3} \cdot \sqrt{10})^2$

Solution:

[1] Property of the Power of a Product ⟶ $(\sqrt{3} \cdot \sqrt{10})^2 = (\sqrt{3})^2(\sqrt{10})^2$

[2] Simplify. ⟶ $= (3)(10)$ or 30

Teaching Suggestions p. M–30

Lesson Resources
Warm–Up: p. M–31
Maintenance: See below.
Practice Worksheet 7.2

Maintenance
1. Factor: $\frac{1}{2}a - \frac{1}{2}c$
   Ans: $\frac{1}{2}(a - c)$ (Section 2.5)
2. Solve and check:
   $-2(2 - d) - d = 3 + 4d$
   Ans: $d = -\frac{7}{3}$
   (Section 4.3)
3. Solve: $2(x + 5) > x + 15$
   Ans: {**all real numbers greater than 5**}
   (Section 5.4)
4. Write $\frac{t^3 \cdot t^4}{t^{-7}}$ as one power.
   Ans: t^{14} (Section 6.3)
5. Gloria rode her bicycle a distance of 22.5 miles at an average speed of 18 miles per hours. How long did she ride her bicycle?
   Ans: 1.25 hours
   (Section 3.4)

Additional Examples
Example 1
Multiply and simplify.
1. $\sqrt{28} \cdot \sqrt{28}$
   Ans: 28
2. $\sqrt{0.91} \cdot \sqrt{0.91}$
   Ans: 0.91

3. $\sqrt{\frac{3}{5}} \cdot \sqrt{\frac{3}{5}}$
   Ans: $\frac{3}{5}$

Example 2
Multiply and simplify.
1. $(\sqrt{7} \cdot \sqrt{11})^2$
   Ans: 77
2. $(\sqrt{0.6} \cdot \sqrt{60})^2$
   Ans: 36
3. $(\sqrt{12} \cdot \frac{1}{2})^2$
   Ans: 6

177

The product $\sqrt{3} \cdot \sqrt{10}$ is positive. Example 2 shows that $\sqrt{3} \cdot \sqrt{10}$ represents the positive square root of 30.

$$\sqrt{3} \cdot \sqrt{10} = \sqrt{30}$$

If a, for instance, were negative, then $\sqrt{a}$ would not be a real number. Such numbers are beyond the scope of this course.

Product Property of Radicals

If a and b are nonnegative real numbers, then $\sqrt{a} \cdot \sqrt{b} = \sqrt{ab}$.

$\sqrt{5} \cdot \sqrt{11} = \sqrt{55}$

$\sqrt{2} \cdot \sqrt{8} = \sqrt{16}$
$\qquad = 4$

P–3 **Multiply and simplify.**

a. $\sqrt{7} \cdot \sqrt{5}$ $\sqrt{35}$ b. $\sqrt{6} \cdot \sqrt{11}$ $\sqrt{66}$ c. $\sqrt{3} \cdot \sqrt{12}$ 6

EXAMPLE 3 Multiply and simplify: $-\sqrt{10} \cdot \sqrt{13}$

Solution:

1 Product of a positive and negative number ⟶ $-\sqrt{10} \cdot \sqrt{13} = -(\sqrt{10} \cdot \sqrt{13})$

2 Product Property of Radicals ⟶ $= -(\sqrt{130})$

3 Simplify. ⟶ $= -\sqrt{130}$

EXAMPLE 4 Multiply and simplify: $(-\sqrt{8})(-\sqrt{18})$

Solution: $(-\sqrt{8})(-\sqrt{18}) = (\sqrt{8})(\sqrt{18})$
$\qquad = \sqrt{8 \cdot 18}$
$\qquad = \sqrt{144}$
$\qquad = 12$

EXAMPLE 5 Multiply and simplify: $\sqrt{3} \cdot \sqrt{10} \cdot \sqrt{7}$

Solution: $\sqrt{3} \cdot \sqrt{10} \cdot \sqrt{7} = (\sqrt{3} \cdot \sqrt{10}) \cdot \sqrt{7}$

 You may group the factors as you please.

$\qquad = (\sqrt{30})(\sqrt{7})$
$\qquad = \sqrt{210}$

178 / *Chapter 7*

CLASSROOM EXERCISES

Multiply and simplify. (Examples 1–4)

1. $\sqrt{2} \cdot \sqrt{5}$ $\sqrt{10}$ **2.** $\sqrt{3} \cdot \sqrt{7}$ $\sqrt{21}$ **3.** $\sqrt{8} \cdot \sqrt{8}$ 8   **4.** $\sqrt{8} \cdot \sqrt{2}$ 4

5. $\sqrt{2} \cdot \sqrt{50}$ 10   **6.** $\sqrt{4} \cdot \sqrt{25}$ 10   **7.** $\sqrt{\frac{1}{5}} \cdot \sqrt{10}$ $\sqrt{2}$ **8.** $\sqrt{\frac{1}{2}} \cdot \sqrt{10}$

9. $\sqrt{\frac{1}{3}} \cdot \sqrt{\frac{1}{3}}$ $\frac{1}{3}$ **10.** $\sqrt{0.5} \cdot \sqrt{0.5}$ 0.5   **11.** $\sqrt{0} \cdot \sqrt{5}$ 0   **12.** $(-\sqrt{4})(\sqrt{9})$

13. $(-\sqrt{16})(-\sqrt{25})$ 20   **14.** $(-\sqrt{2})(-\sqrt{3})$ $\sqrt{6}$ **15.** $(-\sqrt{2})(\sqrt{2})$ -2 **16.** $(-\sqrt{3})^2$ 3

WRITTEN EXERCISES

Goal: To simplify the products of radicals
Sample Problem: $(\sqrt{2})(\sqrt{5})(-\sqrt{7})$ **Answer:** $-\sqrt{70}$

Multiply and simplify. (Examples 1–5)

1. $\sqrt{13} \cdot \sqrt{13}$ 13   **2.** $\sqrt{17} \cdot \sqrt{17}$ 17   **3.** $\sqrt{12} \cdot \sqrt{12}$ 12   **4.** $\sqrt{19} \cdot \sqrt{19}$ 19

5. $\sqrt{81} \cdot \sqrt{36}$ 54   **6.** $\sqrt{144} \cdot \sqrt{16}$ 48   **7.** $\sqrt{2} \cdot \sqrt{18}$ 6   **8.** $\sqrt{3} \cdot \sqrt{27}$ 9

9. $\sqrt{3} \cdot \sqrt{17}$ $\sqrt{51}$ **10.** $\sqrt{2} \cdot \sqrt{19}$ $\sqrt{38}$ **11.** $(-\sqrt{5})(\sqrt{2})$ $-\sqrt{10}$ **12.** $(\sqrt{7})(-\sqrt{2})$

13. $(-\sqrt{3})(-\sqrt{10})$ $\sqrt{30}$ **14.** $(-\sqrt{5})(-\sqrt{6})$ $\sqrt{30}$ **15.** $\sqrt{\frac{1}{3}} \cdot \sqrt{15}$ $\sqrt{5}$ **16.** $\sqrt{\frac{1}{2}} \cdot \sqrt{34}$

17. $\sqrt{\frac{1}{2}} \cdot \sqrt{\frac{1}{2}}$ $\frac{1}{2}$ **18.** $\sqrt{\frac{2}{3}} \cdot \sqrt{\frac{2}{3}}$ $\frac{2}{3}$ **19.** $\sqrt{\frac{1}{3}} \cdot \sqrt{48}$ 4   **20.** $\sqrt{\frac{1}{5}} \cdot \sqrt{20}$ 2

21. $(\sqrt{2} \cdot \sqrt{5})^2$ 10   **22.** $(\sqrt{3} \cdot \sqrt{14})^2$ 42   **23.** $(\sqrt{\frac{1}{2}} \cdot \sqrt{\frac{1}{3}})^2$ $\frac{1}{6}$ **24.** $(\sqrt{\frac{1}{4}} \cdot \sqrt{8})^2$ 2

25. $\sqrt{2} \cdot \sqrt{3} \cdot \sqrt{6}$ 6   **26.** $\sqrt{3} \cdot \sqrt{5} \cdot \sqrt{15}$ 15

27. $\sqrt{2} \cdot \sqrt{3} \cdot \sqrt{5}$ $\sqrt{30}$ **28.** $\sqrt{3} \cdot \sqrt{5} \cdot \sqrt{7}$ $\sqrt{105}$

29. $\sqrt{2} \cdot \sqrt{5} \cdot \sqrt{10}$ 10   **30.** $\sqrt{3} \cdot \sqrt{7} \cdot \sqrt{21}$ 21

MORE CHALLENGING EXERCISES

Solve each equation. $x = \sqrt{33}$

31. $\frac{1}{\sqrt{3}}x = \sqrt{12}$ $x = 6$ **32.** $\frac{1}{\sqrt{5}}x = \sqrt{20}$ $x = 10$ **33.** $\frac{1}{\sqrt{7}}x = \sqrt{2}$ $x = \sqrt{14}$ **34.** $\frac{1}{\sqrt{11}}x = \sqrt{3}$

$x = 6$ $x = 10$

35. $x^2 = 36$ or $x = -6$ **36.** $x^2 = 100$ or $x = -10$ **37.** $x^2 = 14$ **38.** $x^2 = 33$

Give an example to show that each of the following is a false statement.

39. If $\sqrt{a} \cdot \sqrt{b} = \sqrt{c}$, then $c > a$ and $c > b$. $a = b = c = 1$

40. If $\sqrt{a} \cdot \sqrt{\frac{1}{b}} = 1$, then $a > b$. $a = b = 1$

Roots / **179**

7.3 Simplifying Radicals

To simplify a cube root or a fourth root radical in which the radicand has a perfect cube factor or perfect fourth power factor, respectively

P–1 **Simplify each radical below.**

▶ $\sqrt{49}$ is a <u>radical</u> and 49 is the <u>radicand</u>.

a. $\sqrt{49}$ 7   **b.** $-\sqrt{81}$ –9   **c.** $\sqrt{5^2}$ 5   **d.** $-\sqrt{13^2}$ –13

EXAMPLE 1   Simplify $\sqrt{12}$.

Solution:

1 Write the prime factorization of 12. ⟶ $\sqrt{12} = \sqrt{2 \cdot 2 \cdot 3}$

2 Product Property of Radicals ⟶ $= \sqrt{2 \cdot 2} \cdot \sqrt{3}$

3 Meaning of square root ⟶ $= 2\sqrt{3}$

▶ $\sqrt{12}$ is irrational because 12 is not a perfect square.

The simplest form of $\sqrt{12}$ is $2\sqrt{3}$.

A square root radical is in <u>simplest</u> <u>form</u> if its radicand has no perfect square factor.

EXAMPLE 2   Simplify $\sqrt{200}$.

Solution:

1 Write the prime factorization. ⟶ $\sqrt{200} = \sqrt{2^3 \cdot 5^2}$

2 Product Property of Radicals ⟶ $= \sqrt{2^2 \cdot 5^2} \cdot \sqrt{2}$

3 Meaning of square root ⟶ $= 2 \cdot 5\sqrt{2}$

4 Simplify. ⟶ $= 10\sqrt{2}$

▶ *Even powers of factors are grouped under one radical symbol.*

The simplest form of $\sqrt{200}$ is $10\sqrt{2}$.

Other roots besides square roots are possible. A **cube root** of a number is one of its three equal factors. A **fourth root** of a number is one of its four equal factors.

$2 \cdot 2 \cdot 2 = 8$ ◀ **2 is a cube root of 8.**

$3 \cdot 3 \cdot 3 \cdot 3 = 81$ ◀ **3 is a fourth root of 81.**

The Product Property of Radicals can be extended to other roots.

$$\sqrt[3]{x} \cdot \sqrt[3]{y} = \sqrt[3]{xy}$$

$$\sqrt[4]{x} \cdot \sqrt[4]{y} = \sqrt[4]{xy}$$

 True for all nonnegative values of x and y

Here, for convenience, negative numbers are excluded as values of the radicand. You may want to explain that $\sqrt[3]{-8}$, for example, is a real number. That is, $\sqrt[3]{-8} = -2$ because $(-2)(-2)(-2) = -8$

EXAMPLE 3   Simplify $\sqrt[3]{54}$.

Solution:

1. Write the prime factorization. ⟶ $\sqrt[3]{54} = \sqrt[3]{2 \cdot 3 \cdot 3 \cdot 3}$

2. Product Property of Radicals ⟶ $= \sqrt[3]{3 \cdot 3 \cdot 3} \cdot \sqrt[3]{2}$

3. Simplify. ⟶ $= 3\sqrt[3]{2}$

The simplest form of $\sqrt[3]{54}$ is $3\sqrt[3]{2}$. **Read "3 times the cube root of 2."**

The table below shows the third and fourth powers of some counting numbers.

n	n^3	n^4	n	n^3	n^4
1	1	1	6	216	1296
2	8	16	7	343	2401
3	27	81	8	512	4096
4	64	256	9	729	6561
5	125	625	10	1000	10,000

P-2   **Simplify each radical below.**

a. $\sqrt[4]{2^4}$ 2   b. $\sqrt[3]{2^3 \cdot 3^3}$ 6   c. $\sqrt[3]{125}$ 5   d. $\sqrt[4]{256}$ 4

CLASSROOM EXERCISES

Simplify. (Examples 1–2)

1. $\sqrt{2^2}$ 2   2. $\sqrt{3^2}$ 3   3. $-\sqrt{5^2}$ −5   4. $\sqrt{7^2}$ 7   5. $\sqrt{2^4}$ 4

6. $\sqrt{3^4}$ 9   7. $\sqrt{5^4}$ 25   8. $\sqrt{7^4}$ 49   9. $\sqrt{11^2}$ 11   10. $\sqrt{3^6}$ 27

11. $\sqrt{6^4}$ 36   12. $-\sqrt{10^2}$ −10   13. $\sqrt{2^2 \cdot 5^2}$ 10   14. $\sqrt{3^2 \cdot 5^2}$ 15   15. $\sqrt{2^2 \cdot 7^2}$ 14

16. $\sqrt{2^4 \cdot 5^2}$ 20   17. $\sqrt{2^2 \cdot 3^2 \cdot 5^2}$ 30   18. $\sqrt{2 \cdot 5 \cdot 2}$ $2\sqrt{5}$ 19. $\sqrt{3 \cdot 3 \cdot 5}$ $3\sqrt{5}$ 20. $\sqrt{2 \cdot 3 \cdot 2}$ $2\sqrt{3}$

Roots / **181**

Additional Examples
Example 3
Simplify.
1. $\sqrt[3]{40}$
    Ans: $2\sqrt[3]{5}$
2. $\sqrt[3]{135}$
    Ans: $3\sqrt[3]{5}$
3. $\sqrt[3]{128}$
    Ans: $4\sqrt[3]{2}$

Common Error
Students do not simplify radicals completely.

Example
$\sqrt{72} = \sqrt{9 \cdot 8}$
$= \sqrt{9} \cdot \sqrt{8}$
$= 3\sqrt{8}$

Prescription
Instill in students the habit of checking the radicand to see if it contains any perfect square factors.

Simplify. (Example 3)

21. $\sqrt[3]{3^3}$ 3   22. $\sqrt[3]{8}$ 2   23. $\sqrt[4]{1296}$ 6

24. $\sqrt[3]{729}$ 9   25. $\sqrt[4]{2401}$ 7   26. $\sqrt[3]{2^3 \cdot 7}$ $2\sqrt[3]{7}$

27. $\sqrt[3]{3 \cdot 5 \cdot 3 \cdot 5 \cdot 5}$ $5\sqrt[3]{9}$ 28. $\sqrt[4]{5^4 \cdot 11^4}$ 55   29. $\sqrt[4]{2 \cdot 3 \cdot 2 \cdot 3 \cdot 5 \cdot 3 \cdot 3}$

                                                      $3\sqrt[4]{20}$

WRITTEN EXERCISES

Goal: To simplify radicals

Sample Problem: $\sqrt{288}$ **Answer:** $12\sqrt{2}$

Simplify. (Examples 1–2)

1. $\sqrt{6^2}$ 6   2. $\sqrt{4^2}$ 4   3. $\sqrt{38^2}$ 38   4. $\sqrt{19^2}$ 19

5. $\sqrt{2^2 \cdot 3^2} \cdot \sqrt{3}$ $6\sqrt{3}$ 6. $\sqrt{3^2 \cdot 5^2} \cdot \sqrt{5}$ $15\sqrt{5}$ 7. $\sqrt{5^2 \cdot 7^2} \cdot \sqrt{5 \cdot 7}$

8. $\sqrt{2^2 \cdot 11^2} \cdot \sqrt{2 \cdot 11}$ $22\sqrt{22}$ 9. $\sqrt{2^4} \cdot \sqrt{2 \cdot 3}$ $4\sqrt{6}$ 10. $\sqrt{3^4} \cdot \sqrt{3 \cdot 5}$

11. $\sqrt{2^4 \cdot 3^2} \cdot \sqrt{5 \cdot 2}$ $12\sqrt{10}$ 12. $\sqrt{3^4 \cdot 5^2} \cdot \sqrt{2 \cdot 11}$ $45\sqrt{22}$ 13. $\sqrt{2 \cdot 3 \cdot 3 \cdot 5 \cdot 2}$

14. $\sqrt{3 \cdot 5 \cdot 2 \cdot 3 \cdot 5}$ $15\sqrt{2}$ 15. $\sqrt{5 \cdot 7 \cdot 5 \cdot 3 \cdot 2 \cdot 2}$ $10\sqrt{21}$ 16. $\sqrt{3 \cdot 11 \cdot 7 \cdot 3 \cdot 7 \cdot 3}$

17. $\sqrt{8}$ $2\sqrt{2}$ 18. $\sqrt{18}$ $3\sqrt{2}$ 19. $\sqrt{27}$ $3\sqrt{3}$ 20. $\sqrt{20}$ $2\sqrt{5}$ 21. $\sqrt{24}$

22. $\sqrt{28}$ $2\sqrt{7}$ 23. $\sqrt{40}$ $2\sqrt{10}$ 24. $\sqrt{44}$ $2\sqrt{11}$ 25. $-\sqrt{48}$ $-4\sqrt{3}$ 26. $-\sqrt{60}$

27. $\sqrt{120}$ $2\sqrt{30}$ 28. $\sqrt{108}$ $6\sqrt{3}$ 29. $\sqrt{96}$ $4\sqrt{6}$ 30. $\sqrt{180}$ $6\sqrt{5}$ 31. $\sqrt{270}$

(Example 3)

32. $\sqrt[3]{64}$ 4   33. $\sqrt[4]{16}$ 2   34. $\sqrt[4]{625}$ 5   35. $\sqrt[3]{81}$ $3\sqrt[3]{3}$ 36. $\sqrt[3]{72}$

37. $\sqrt[4]{144}$ $2\sqrt[4]{9}$ 38. $\sqrt[3]{243}$ $3\sqrt[3]{3}$ 39. $\sqrt[3]{1250}$ $5\sqrt[3]{10}$ 40. $\sqrt[3]{2160}$ $6\sqrt[3]{10}$ 41. $\sqrt[4]{1250}$

MORE CHALLENGING EXERCISES

                                                            $-5\sqrt[3]{3}$

42. $\sqrt[3]{-512}$ -8 43. $\sqrt[3]{-1000}$ -10 44. $\sqrt[3]{-24}$ $-2\sqrt[3]{3}$ 45. $\sqrt[3]{-56}$ $-2\sqrt[3]{7}$ 46. $\sqrt[3]{-375}$

NON-ROUTINE PROBLEMS

47. Meg spent $\frac{1}{4}$ of the quarters she had saved. She gave her brother $\frac{3}{4}$ of the rest. Then she had 3 quarters left. How many quarters had she saved?

48. Two positive integers are each less than 10. The sum of their squares, added to their product, equals a perfect square. Find the integers.

182 / *Chapter 7*

7.4 Simplifying Radical Expressions

OBJECTIVE: To simplify a square root or a cube root radical in which the radicand is a product involving powers of one or more variables

P-1 **What is $\sqrt{x}$ for each value of x below?**

a. $x = 100$ $\;10$ **b.** $x = \frac{1}{4}$ $\;\frac{1}{2}$ **c.** $x = 17$ $\;\sqrt{17}$ **d.** $x = \frac{4}{9}$ $\;\frac{2}{3}$

P-2 **Why cannot x have a negative value in $\sqrt{x}$?** A negative number cannot have a real number as a square root.

The domain of each variable will be {nonnegative real numbers} unless stated otherwise.

EXAMPLE 1 Simplify $\sqrt{x^2y^4}$.

Solution:

$\boxed{1}$ Product Property of Radicals $\longrightarrow$ $\sqrt{x^2y^4} = \sqrt{x^2} \cdot \sqrt{y^4}$

$\boxed{2}$ Meaning of square root $\longrightarrow$ $= xy^2$

An even power of a variable is called a perfect square because each of its square roots can be expressed without the radical symbol.

P-3 **Simplify each radical.**

a. $\sqrt{a^6}$ $\;a^3$ **b.** $\sqrt{x^8}$ $\;x^4$ **c.** $\sqrt{b^{12}}$ $\;b^6$ **d.** $\sqrt{y^{28}}$ $\;y^{14}$ Warn students not to "take the square root of" the exponent.

$\sqrt{x^9} \ne x^3$

EXAMPLE 2 Simplify $\sqrt{x^3y^5}$.

Solution:

$\boxed{1}$ Product Property of Powers $\longrightarrow$ $\sqrt{x^3y^5} = \sqrt{(x^2 \cdot x)(y^4 \cdot y)}$ ◀ x^2 and y^4 are perfect squares.

$\boxed{2}$ Commutative and Associative Properties of Multiplication $\longrightarrow$ $= \sqrt{(x^2y^4)(xy)}$ ◀ Group the perfect square factors.

$\boxed{3}$ Product Property of Radicals $\longrightarrow$ $= \sqrt{x^2y^4} \cdot \sqrt{xy}$

$\boxed{4}$ Meaning of square root $\longrightarrow$ $= xy^2\sqrt{xy}$

The simplest form is $xy^2\sqrt{xy}$.

Roots / **183**

Teaching Suggestions p. M–30

Lesson Resources
Warm–Up: p. M–31
Maintenance: See below.
Practice Worksheet 7.4
Transparency 17

Maintenance
1. Solve and check:
   $-t - 13 = 29$
   Ans: $t = -42$ (Section 3.3)
2. Solve and check:
   $7c + 4 - c = 2c - 3$
   Ans: $c = -\frac{7}{4}$ (Section 4.3)
3. Solve: $-34 > 2p$
   Ans: {all real numbers less than -17}
   (Section 5.5)
4. Simplify: $(3xy)^3$
   Ans: $27x^3y^3$ (Section 6.4)
5. The low temperature for yesterday was 12°F lower than the low temperature for today. The low temperature for yesterday was 8°F. Find the low temperature for today.
   Ans: 20°F (Section 3.7)

Additional Examples
Example 1
Simplify.
1. $\sqrt{a^4b^{16}}$
   Ans: a^2b^8
2. $\sqrt{x^6y^{10}}$
   Ans: x^3y^5
3. $\sqrt{c^2d^{100}}$
   Ans: cd^{50}

Example 2
Simplify.
1. $\sqrt{h^6k^5}$
   Ans: $h^3k^4\sqrt{k}$
2. $\sqrt{a^7b^{11}}$
   Ans: $a^3b^5\sqrt{ab}$
3. $\sqrt{x^2y^3z^{15}}$
   Ans: $xyz^7\sqrt{yz}$

183

Additional Examples
Example 3
Simplify.

1. $\sqrt{20a^3b^4}$
   Ans: $2ab^2\sqrt{5a}$
2. $\sqrt{40x^5y^9}$
   Ans: $2x^2y^4\sqrt{10xy}$
3. $\sqrt{27a^4b^5c^6}$
   Ans: $3a^2b^2c^3\sqrt{3b}$

Example 4
Simplify.

1. $\sqrt[3]{x^5y^7}$
   Ans: $xy^2\ \sqrt[3]{x^2y}$
2. $\sqrt[3]{8a^9b^{11}}$
   Ans: $2a^3b^3\ \sqrt[3]{b^2}$
3. $\sqrt[3]{54x^4y^{13}}$
   Ans: $3xy^4\ \sqrt[3]{2xy}$

Common Error
When simplifying cube roots, students incorrectly treat them as square roots.

Example
$\sqrt[3]{x^6} = \sqrt{x^3 \cdot x^3}$
$\qquad = x^3$

Prescription
Remind students that when they are finding a cube root, the number or expression must have 3 equal factors.

EXAMPLE 3   Simplify $\sqrt{12c^3d^7}$.

Solution:
$$\sqrt{12c^3d^7} = \sqrt{(2 \cdot 2 \cdot 3)(c \cdot c^2)(d \cdot d^6)}$$
$$= \sqrt{(2^2 \cdot c^2 \cdot d^6)(3cd)}$$
$$= \sqrt{2^2 \cdot c^2 \cdot d^6} \cdot \sqrt{3cd}$$
$$= 2cd^3\sqrt{3cd}$$

EXAMPLE 4   Simplify $\sqrt[3]{r^4s^5}$.

Solution:

1. Find the greatest third power factors of r^4 and s^5. $\longrightarrow$ $r^4 = r^3 \cdot r;\ s^5 = s^3 \cdot s^2$

2. Product Property of Powers $\longrightarrow$ $\sqrt[3]{r^4s^5} = \sqrt[3]{(r^3 \cdot r)(s^3 \cdot s^2)}$

3. Associative and Commutative Properties of Multiplication $\longrightarrow$ $= \sqrt[3]{(r^3s^3)(rs^2)}$

   ◄ **Group the perfect cube factors.**

4. Product Property of Radicals $\longrightarrow$ $= \sqrt[3]{r^3s^3} \cdot \sqrt[3]{rs^2}$

5. Meaning of cube root $\longrightarrow$ $= rs\sqrt[3]{rs^2}$

P–4   What is $\sqrt[3]{a^3}$? $\sqrt[3]{x^6}$? $\sqrt[3]{y^9}$?
    a x^2

You know that a square root radical is in simplest form if its radicand has no perfect square factor. A cube root radical is in simplest form if its radicand has no perfect cube factor.

The following radicals are in simplest form.

   a. $\sqrt{x}$ **b.** $\sqrt{10}$ **c.** $\sqrt[3]{p^2q}$ **d.** $\sqrt{2t}$

P–5   **Which of the following radicals are in simplest form?**

   a. $\sqrt{28x^3}$ No   **b.** $\sqrt{21st}$ Yes   **c.** $\sqrt[3]{9m^2n^2}$ Yes   **d.** $\sqrt[3]{24c^5d^7}$ No

---

CLASSROOM EXERCISES

Simplify. (Examples 1–4)

1. $\sqrt{a^2}$ a **2.** $\sqrt{x^4}$ x^2 **3.** $\sqrt{a^2b^2}$ ab **4.** $\sqrt{x^6}$ x^3 **5.** $\sqrt{x^2y^4}$
                                                                       xy^2

6. $\sqrt{4x^2}$ $2x$ 7. $\sqrt{9y^2}$ $3y$ 8. $\sqrt{16a^2}$ $4a$ 9. $\sqrt{25n^2}$ $5n$ 10. $\sqrt{4x^4}$ $2x^2$

11. $\sqrt{9x^6}$ $3x^3$ 12. $\sqrt{16x^2y^4}$ $4xy^2$ 13. $\sqrt{x^4}\cdot\sqrt{x}$ $x^2\sqrt{x}$ 14. $\sqrt{a^6}\cdot\sqrt{a}$

15. $\sqrt{2^2\cdot a^2}\cdot\sqrt{2a}$ 16. $\sqrt{3^2\cdot x^4}\cdot\sqrt{2x}$ 17. $\sqrt{x^2y^2}\cdot\sqrt{2x}$ 18. $\sqrt{3^4\cdot x^4}\cdot\sqrt{2x}$

19. $\sqrt{6x^2}$ $x\sqrt{6}$ 20. $\sqrt{10y^4}$ $y^2\sqrt{10}$ 21. $\sqrt[3]{r^3t^3}$ rt 22. $\sqrt[3]{m^6n^3}$ m^2n

23. $\sqrt[3]{x^{12}}$ x^4 24. $\sqrt[3]{p^3q^9}$ pq^3 25. $\sqrt[3]{8k^6}$ $2k^2$ 26. $\sqrt[3]{27r^3}$ $3r$

14. $a^3\sqrt{a}$
15. $2a\sqrt{2a}$
16. $3x^2\sqrt{2x}$
17. $xy\sqrt{2x}$
18. $9x^2\sqrt{2x}$

▨▨▨ WRITTEN EXERCISES ▨▨▨

Goal: To simplify radicals containing variables
Sample Problem: $\sqrt[3]{16a^4b^6}$
Answer: $2ab^2\sqrt[3]{2a}$

Simplify. The domain is {nonnegative real numbers}. (Examples 1–4)

1. $\sqrt{36y^2}$ $6y$ 2. $\sqrt{9a^4}$ $3a^2$ 3. $\sqrt{y^2z^6}$ yz^3 4. $\sqrt{a^6b^8}$

5. $\sqrt{x^5}$ $x^2\sqrt{x}$ 6. $\sqrt{a^3}$ $a\sqrt{a}$ 7. $\sqrt{4x}$ $2\sqrt{x}$ 8. $\sqrt{9y}$

9. $\sqrt{16y^3}$ $4y\sqrt{y}$ 10. $\sqrt{25y^5}$ $5y^2\sqrt{y}$ 11. $\sqrt{8x^2}$ $2x\sqrt{2}$ 12. $\sqrt{8y^4}$

13. $\sqrt{8a^5}$ $2a^2\sqrt{2a}$ 14. $\sqrt{8x^3}$ $2x\sqrt{2x}$ 15. $\sqrt{12ab}$ $2\sqrt{3ab}$ 16. $\sqrt{18xy}$

17. $\sqrt{49a^3b^3}$ $7ab\sqrt{ab}$ 18. $\sqrt{36x^3y^3}$ $6xy\sqrt{xy}$ 19. $\sqrt{24a^2b^3}$ 20. $\sqrt{40x^3y^2}$

21. $\sqrt{75r^5s^6}$ $5r^2s^3\sqrt{3r}$ 22. $\sqrt{45c^4d^7}$ 23. $\sqrt{20r^2s^3t^5}$ 24. $\sqrt{27a^5b^4c^3}$

25. $\sqrt[3]{x^{12}}$ x^4 26. $\sqrt[3]{r^6}$ r^2 27. $\sqrt[3]{x^{13}}$ $x^4\sqrt[3]{x}$ 28. $\sqrt[3]{8y^7}$

29. $\sqrt[3]{27x^3y^6}$ $3xy^2$ 30. $\sqrt[3]{64r^6s^9}$ $4r^2s^3$ 31. $\sqrt[3]{16a^2b^3}$ 32. $\sqrt[3]{24m^4n^6}$

33. $\sqrt[3]{54m^5n^8}$ 34. $\sqrt[3]{32a^{11}b^5}$ 35. $\sqrt[3]{32r^7s^{10}}$ $2r^2s^3\sqrt[3]{4rs}$ 36. $\sqrt[3]{128a^4b^9}$ $4ab^3\sqrt[3]{2a}$

MORE CHALLENGING EXERCISES

Write each product as one radical. Then simplify.

EXAMPLE: $\sqrt{2x}\cdot\sqrt{2x^3}$ **SOLUTION:** $\sqrt{2x}\cdot\sqrt{2x^3}=\sqrt{4x^4}$
$$=2x^2$$

37. $\sqrt{x}\cdot\sqrt{x^3}$ x^2 38. $\sqrt{x}\cdot\sqrt{x^5}$ x^3 39. $\sqrt{2}\cdot\sqrt{2x}$ $2\sqrt{x}$

40. $\sqrt{3}\cdot\sqrt{3y}$ $3\sqrt{y}$ 41. $\sqrt{2x}\cdot\sqrt{2x^3}$ $2x^2$ 42. $\sqrt{3y^3}\cdot\sqrt{3y}$

43. $\sqrt[3]{4x}\cdot\sqrt[3]{3x^2}$ $x\sqrt[3]{12}$ 44. $\sqrt[3]{2r^4}\cdot\sqrt[3]{4r^2}$ $2r^2$ 45. $\sqrt[3]{9a}\cdot\sqrt[3]{3a^5}$

46. $\sqrt[3]{16r^2s^2}\cdot\sqrt[3]{4rs^4}$ $4rs^2$ 47. $\sqrt[3]{8b^2}\cdot\sqrt[3]{8b^7}$ $4b^3$ 48. $\sqrt[3]{27rs^{11}}\cdot\sqrt[3]{r^8s}$ $3r^3s^4$

Day 1 p. 185: 1–19 odd
Day 2 p. 185: 21–35 odd
Average p. 185: 1–47 odd
Above Average p. 185:
1–35 odd, 37–48

4. a^3b^4
8. $3\sqrt{y}$
12. $2y^2\sqrt{2}$
16. $3\sqrt{2xy}$
19. $2ab\sqrt{6b}$
20. $2xy\sqrt{10x}$
22. $3c^2d^3\sqrt{5d}$
23. $2rst^2\sqrt{5st}$
24. $3a^2b^2c\sqrt{3ac}$
28. $2y^2\sqrt[3]{y}$
31. $2b\sqrt[3]{2a^2}$
32. $2mn^2\sqrt[3]{3m}$
33. $3mn^2\sqrt[3]{2m^2n^2}$
34. $2a^3b\sqrt[3]{4a^2b^2}$
42. $3y^2$
45. $3a^2$

APPLICATIONS

You can use a calculator with a square root key, $\boxed{\sqrt{}}$ *, to find roots other than square roots. For example, you can think of a fourth root as a "square root of a square root."*

EXAMPLE 1 $\sqrt{39}$ SOLUTION $\boxed{3}\;\boxed{9}\;\boxed{\sqrt{}}$ $\fbox{6.2449979}$

EXAMPLE 2 $\sqrt[4]{36}$ SOLUTION Think of $\sqrt[4]{36}$ as $\sqrt{\sqrt{36}}$.

$\boxed{3}\;\boxed{6}\;\boxed{\sqrt{}}\;\boxed{\sqrt{}}$ $\fbox{2.4494897}$

Evaluate each of the following.

49. $\sqrt{64}$ 8   **50.** $\sqrt{121}$ 11   **51.** $\sqrt{43}$ 6.5574385   **52.** $\sqrt[4]{16}$ 2   **53.** $\sqrt[4]{1296}$ 6   **54.** $\sqrt[8]{6581}$ 3.0011416

Quiz Sections 7.1–7.4
After completing this Mid-Chapter Review, you may want to administer a quiz covering the same sections. See the quiz provided on page 151 in the *Teacher's ResourceBank*.

▰▰▰ MID-CHAPTER REVIEW ▰▰▰

Simplify. (Section 7.1)

1. $\sqrt{(13 \cdot 17)(13 \cdot 17)}$ 221   **2.** $\sqrt{2 \cdot 5 \cdot 7 \cdot 2 \cdot 7 \cdot 5}$ 70   **3.** $\sqrt{5^4 \cdot 7^2 \cdot 11^2}$ 1925   **4.** $\sqrt{3^2 \cdot 5^4 \cdot 13^2}$ 975

5. $\sqrt{729}$ 27   **6.** $\sqrt{1024}$ 32   **7.** $\sqrt{1296}$ 36   **8.** $\sqrt{2025}$ 45   **9.** $\sqrt{2304}$ 48

Multiply and simplify. (Section 7.2)

10. $\sqrt{23} \cdot \sqrt{23}$ 23   **11.** $\sqrt{9} \cdot \sqrt{49}$ 21   **12.** $\sqrt{12} \cdot \sqrt{3}$ 6   **13.** $(-\sqrt{7})(\sqrt{2})$ $-\sqrt{14}$

14. $\left(-\sqrt{\frac{1}{2}}\right)(-\sqrt{162})$ 9   **15.** $\sqrt{\frac{1}{5}} \cdot \sqrt{125}$ 5   **16.** $\sqrt{\frac{5}{6}} \cdot \sqrt{\frac{5}{6}}$ $\frac{5}{6}$ **17.** $\sqrt{5} \cdot \sqrt{7} \cdot \sqrt{6}$ $\sqrt{210}$

Simplify. (Section 7.3)

18. $\sqrt{360}$ $6\sqrt{10}$ **19.** $\sqrt{490}$ $7\sqrt{10}$ **20.** $\sqrt{320}$ $8\sqrt{5}$ **21.** $\sqrt{700}$ $10\sqrt{7}$ **22.** $\sqrt{98}$ $7\sqrt{2}$

23. $\sqrt{405}$ $9\sqrt{5}$ **24.** $\sqrt[3]{40}$ $2\sqrt[3]{5}$ **25.** $\sqrt[3]{108}$ $3\sqrt[3]{4}$ **26.** $\sqrt[4]{48}$ $2\sqrt[4]{3}$ **27.** $\sqrt[4]{162}$ $3\sqrt[4]{2}$

Simplify. The domain is {nonnegative real numbers}. (Section 7.4)

28. $\sqrt{81a^4b^2}$ $9a^2b$ **29.** $\sqrt{49m^6n^2}$ $7m^3n$ **30.** $\sqrt{24p^3}$ $2p\sqrt{6p}$ **31.** $\sqrt{27r^4s^5}$

32. $\sqrt{75p^2q^3r^6}$ $5pqr^3\sqrt{3q}$ **33.** $\sqrt{108a^3b^5c^7}$ $6ab^2c^3\sqrt{3abc}$ **34.** $\sqrt[3]{216m^3n^4}$ $6mn\sqrt[3]{n}$ **35.** $\sqrt[3]{54p^7q^6}$ $3p^2q^2\sqrt[3]{2p}$

Additional Answers
Mid–Chapter Review
31. $3r^2s^2\sqrt{3s}$

REVIEW CAPSULE FOR SECTION 7.5

Combine like terms. (Pages 50–51)

1. $9m + 8m$ 17m   **2.** $5.6t - 2.9t$ 2.7t   **3.** $76k + k$ 77k   **4.** $4.8w - 5.9w + 0.8w$ $-0.3w$

7.5 Sums and Differences

OBJECTIVES: To add or subtract radicals having equal radicands
To simplify radicals in a sum or difference and then write the sum or difference in simplest form by adding or subtracting like radicals

P–1 **How is each sum or difference below expressed as a product by the Distributive Property?**

> **Find a common factor.**

a. $7x + ax$ **b.** $ty - ry$ **c.** $11m + 3m$

$(7 + a)x$ $(t - r)y$ $(11 + 3)m$

The Distributive Property can also be applied to sums or differences of radicals.

EXAMPLE 1 Add: $3\sqrt{5} + 4\sqrt{5}$

Solution:

1. Distributive Property ⟶ $3\sqrt{5} + 4\sqrt{5} = (3 + 4)\sqrt{5}$

> **Adding $3\sqrt{5}$ and $4\sqrt{5}$ is much like adding $3x$ and $4x$.**

2. Simplify. ⟶ $= 7\sqrt{5}$

P–2 **Add: $6\sqrt{2} + 5\sqrt{2}$** $11\sqrt{2}$

Radicals such as $6\sqrt{2}$ and $5\sqrt{2}$ are called **like radicals.** They have equal radicands.

The Distributive Property is used in these Examples to show how like radicals are combined. Students will soon learn to combine like radicals just as they combine like terms.

P–3 **Add or subtract as indicated.**

a. $4\sqrt{5} + \sqrt{5}$ $5\sqrt{5}$ **b.** $9\sqrt{3} - 2\sqrt{3}$ $7\sqrt{3}$ **c.** $\sqrt{10} - 4\sqrt{10}$ $-3\sqrt{10}$

P–4 **What are the prime factors of 12?** 2, 2, and 3

EXAMPLE 2 Add: $\sqrt{12} + 7\sqrt{3}$

Solution:

1. Write the prime factorization of 12. ⟶ $\sqrt{12} + 7\sqrt{3} = \sqrt{2 \cdot 2 \cdot 3} + 7\sqrt{3}$

2. Product Property of Radicals ⟶ $= \sqrt{2^2} \cdot \sqrt{3} + 7\sqrt{3}$

3. Meaning of square root ⟶ $= 2\sqrt{3} + 7\sqrt{3}$

4. Distributive Property ⟶ $= (2 + 7)\sqrt{3}$

5. Simplify. ⟶ $= 9\sqrt{3}$

> **Simplest form**

Teaching Suggestions p. M–30

Lesson Resources
Warm–Up: p. M–31
Maintenance: See below.
Practice Worksheet 7.5
Transparency 17

Maintenance
1. Divide: $6\frac{1}{2} \div (-2)$
   Ans: $-3\frac{1}{4}$ (Section 1.5)
2. Solve and check:
   $5x + 9 - 3x = -9$
   Ans: $x = -9$ (Section 3.5)
3. Solve: $2(3 + 2x) < -3x - 8$
   Ans: {all real numbers less than -2} (Section 5.6)
4. Simplify: $(ab)^{-4}$
   Ans: $\frac{1}{a^4b^4}$ (Section 6.4)
5. Mark's monthly car payment is $20 less than half his monthly rent payment (Condition 1). The total amount of the two payments is $715 (Condition 2). Find the amount of each monthly payment.
   Ans: car: $225; rent: $490 (Section 4.1)

Additional Examples
Example 1
Add or subtract.
1. $8\sqrt{7} + 7\sqrt{7}$
   Ans: $15\sqrt{7}$
2. $9\sqrt{2} + 9\sqrt{2}$
   Ans: $18\sqrt{2}$
3. $4\sqrt{10} - 7\sqrt{10}$
   Ans: $-3\sqrt{10}$

Example 2
Add or subtract.
1. $9\sqrt{2} + \sqrt{8}$
   Ans: $11\sqrt{2}$
2. $\sqrt{50} - 8\sqrt{2}$
   Ans: $-3\sqrt{2}$
3. $-\sqrt{7} + \sqrt{63}$
   Ans: $2\sqrt{7}$

Simplify: a. $\sqrt{3}$ $\sqrt{3}$ b. $\sqrt{15}$ $\sqrt{15}$

EXAMPLE 3   Add: $\sqrt{3} + \sqrt{15}$

Solution: Each radical is in simplest form. Therefore, $\sqrt{3} + \sqrt{15}$ is in simplest form.

P–6 **What is the prime factorization of 8?   of 50?** $2 \cdot 5 \cdot 5$
$2 \cdot 2 \cdot 2$

EXAMPLE 4   Add: $\sqrt{8} + 3\sqrt{2} + \sqrt{50}$

Solution:
$$\sqrt{8} + 3\sqrt{2} + \sqrt{50} = \sqrt{2 \cdot 2 \cdot 2} + 3\sqrt{2} + \sqrt{2 \cdot 5 \cdot 5}$$
$$= \sqrt{2^2} \cdot \sqrt{2} + 3\sqrt{2} + \sqrt{5^2} \cdot \sqrt{2}$$
$$= 2\sqrt{2} + 3\sqrt{2} + 5\sqrt{2}$$
$$= (2 + 3 + 5)\sqrt{2}$$
$$= 10\sqrt{2} \qquad \text{◀ \textit{Simplest form}}$$

EXAMPLE 5   Subtract: $\sqrt{12n} - \sqrt{27n}$

Solution:

1. Write the prime factorization. ⟶ $\sqrt{12n} - \sqrt{27n} = \sqrt{2 \cdot 2 \cdot 3 \cdot n} - \sqrt{3 \cdot 3 \cdot 3 \cdot n}$

2. Product Property of Radicals ⟶ $= \sqrt{2^2} \cdot \sqrt{3n} - \sqrt{3^2} \cdot \sqrt{3n}$

3. Meaning of square root ⟶ $= 2\sqrt{3n} - 3\sqrt{3n}$

4. Distributive Property ⟶ $= (2 - 3)\sqrt{3n}$

5. Simplify. ⟶ $= -1\sqrt{3n}$ or $-\sqrt{3n}$

CLASSROOM EXERCISES

Add or subtract as indicated. (Example 1)

1. $2\sqrt{3} + 5\sqrt{3}$ $7\sqrt{3}$

2. $3\sqrt{5} + 2\sqrt{5}$ $5\sqrt{5}$

3. $3\sqrt{2} + \sqrt{2}$ $4\sqrt{2}$

4. $\sqrt{7} + \sqrt{7}$ $2\sqrt{7}$

5. $7\sqrt{10} + 5\sqrt{10}$ $12\sqrt{10}$

6. $8\sqrt{3} - 5\sqrt{3}$ $3\sqrt{3}$

Additional Examples
Example 3
Add or subtract.
1. $\sqrt{6} + \sqrt{7}$
   Ans: $\sqrt{6} + \sqrt{7}$
2. $\sqrt{15} - \sqrt{10}$
   Ans: $\sqrt{15} - \sqrt{10}$

Example 4
Add or subtract.
1. $4\sqrt{3} + \sqrt{27} + \sqrt{3}$
   Ans: $8\sqrt{3}$
2. $\sqrt{20} - \sqrt{45} + 7\sqrt{5}$
   Ans: $6\sqrt{5}$
3. $9\sqrt{7} - \sqrt{63} - \sqrt{28}$
   Ans: $4\sqrt{7}$

Example 5
Add or subtract.
1. $\sqrt{18a} - \sqrt{8a}$
   Ans: $\sqrt{2a}$
2. $5\sqrt{6x} + \sqrt{54x}$
   Ans: $8\sqrt{6x}$
3. $\sqrt{25x} - \sqrt{9x}$
   Ans: $2\sqrt{x}$

Common Error
Students incorrectly add or subtract radicals.

Examples
$2\sqrt{10} + 3\sqrt{5} = 5\sqrt{15}$
$6\sqrt{40} - 3\sqrt{10} = 3\sqrt{30}$

Prescription
Emphasize that only like radicals can be added or subtracted.

Add or subtract as indicated. Simplify where necessary. (Examples 1 and 3)

7. $4\sqrt{2} - \sqrt{2}$ $\quad 3\sqrt{2}$ **8.** $5\sqrt{5} - 4\sqrt{5}$ $\quad \sqrt{5}$ **9.** $3\sqrt{11} - 5\sqrt{11}$ **10.** $3\sqrt{13} - 4\sqrt{13}$

11. $2\sqrt{x} + 3\sqrt{x}$ $\quad 5\sqrt{x}$ **12.** $\frac{1}{2}\sqrt{y} + 2\sqrt{y}$ $\quad \frac{5}{2}\sqrt{y}$ **13.** $\frac{3}{4}\sqrt{a} - \frac{1}{4}\sqrt{a}$ $\quad \frac{1}{2}\sqrt{a}$ **14.** $\sqrt{n} + \sqrt{n}$ $\quad 2\sqrt{n}$

15. $9\sqrt{m} + \sqrt{m}$ **16.** $\sqrt{16} - \sqrt{1}$ $\quad 3$ **17.** $\sqrt{4} + \sqrt{9}$ $\quad 5$ **18.** $\sqrt{2} + \sqrt{5}$
$$\sqrt{2} + \sqrt{5}$$

Goal: To add and subtract radicals
Sample Problem: $3\sqrt{5} - 7\sqrt{5} + \sqrt{5}$ **Answer:** $-3\sqrt{5}$

Add or subtract as indicated. Simplify where necessary. (Examples 1–3, 5)

1. $10\sqrt{3} + 5\sqrt{3}$ $\quad 15\sqrt{3}$ **2.** $8\sqrt{5} + 3\sqrt{5}$ $\quad 11\sqrt{5}$ **3.** $7\sqrt{2} - 3\sqrt{2}$

4. $8\sqrt{7} - 3\sqrt{7}$ $\quad 5\sqrt{7}$ **5.** $2\sqrt{3} - 5\sqrt{3}$ $\quad -3\sqrt{3}$ **6.** $3\sqrt{14} - 7\sqrt{14}$

7. $5\sqrt{2} - \sqrt{2}$ $\quad 4\sqrt{2}$ **8.** $6\sqrt{6} - \sqrt{6}$ $\quad 5\sqrt{6}$ **9.** $\sqrt{11} + \sqrt{11}$

10. $\sqrt{15} + \sqrt{15}$ $\quad 2\sqrt{15}$ **11.** $\sqrt{5} + \sqrt{15}$ $\quad \sqrt{5} + \sqrt{15}$ **12.** $2\sqrt{3} + \sqrt{7}$

13. $\sqrt{19} - \sqrt{3}$ $\quad \sqrt{19} - \sqrt{3}$ **14.** $3\sqrt{5} - 2\sqrt{3}$ $\quad 3\sqrt{5} - 2\sqrt{3}$ **15.** $3\sqrt{x} + 9\sqrt{x}$

16. $5\sqrt{a} + 10\sqrt{a}$ $\quad 15\sqrt{a}$ **17.** $-2\sqrt{t} - 3\sqrt{t}$ $\quad -5\sqrt{t}$ **18.** $-\sqrt{r} - 5\sqrt{r}$

19. $\sqrt{s} + \sqrt{s}$ $\quad 2\sqrt{s}$ **20.** $\sqrt{12} + 3\sqrt{3}$ $\quad 5\sqrt{3}$ **21.** $3\sqrt{8} + 5\sqrt{2}$

22. $3\sqrt{12} + 5\sqrt{3}$ $\quad 11\sqrt{3}$ **23.** $\sqrt{20} - 2\sqrt{5}$ $\quad 0$ **24.** $\sqrt{18} - 6\sqrt{2}$

25. $\sqrt{27} + \sqrt{12}$ $\quad 5\sqrt{3}$ **26.** $\sqrt{32} + \sqrt{8}$ $\quad 6\sqrt{2}$ **27.** $2\sqrt{45} + \sqrt{5}$

28. $\sqrt{24} + \sqrt{96}$ $\quad 6\sqrt{6}$ **29.** $2\sqrt{72} - \sqrt{2}$ $\quad 11\sqrt{2}$ **30.** $4\sqrt{27} - \sqrt{12}$

31. $\sqrt{28a} + 3\sqrt{7a}$ $\quad 5\sqrt{7a}$ **32.** $\sqrt{44n} + 5\sqrt{11n}$ $\quad 7\sqrt{11n}$ **33.** $\sqrt{2r} + \sqrt{18r}$

34. $\sqrt{16x} - \sqrt{4x}$ $\quad 2\sqrt{x}$ **35.** $\sqrt{25y} + \sqrt{49y}$ $\quad 12\sqrt{y}$ **36.** $\sqrt{2x} - 3\sqrt{32x}$

(Example 4)

37. $2\sqrt{3} + \sqrt{3} + 4\sqrt{3}$ $\quad 7\sqrt{3}$ **38.** $7\sqrt{2} + \sqrt{2} + 2\sqrt{2}$ $\quad 10\sqrt{2}$

39. $3\sqrt{7} - 4\sqrt{7} + 2\sqrt{7}$ $\quad \sqrt{7}$ **40.** $4\sqrt{5} - 6\sqrt{5} + \sqrt{5}$ $\quad -\sqrt{5}$

41. $7\sqrt{y} - 2\sqrt{y} - 5\sqrt{y}$ $\quad 0$ **42.** $-3\sqrt{b} + 10\sqrt{b} - 7\sqrt{b}$ $\quad 0$

REVIEW CAPSULE FOR SECTION 7.6

Multiply and simplify. (Pages 177–179) $\frac{1}{3}x^2$

1. $\sqrt{3} \cdot \sqrt{12}$ $\quad 6$ **2.** $\sqrt{5x^3} \cdot \sqrt{5x}$ $\quad 5x^2$ **3.** $\sqrt{13x} \cdot \sqrt{13x}$ $\quad 13x$ **4.** $\sqrt{\frac{1}{9}x} \cdot \sqrt{x^3}$

Roots / **189**

Additional Answers
Classroom Exercises
 9. $-2\sqrt{11}$
10. $-\sqrt{13}$
15. $10\sqrt{m}$

Assignment Guide
Basic
Day 1 p. 189: 1–21 odd
Day 2 p. 189: 23–41 odd
Average
Day 1 p. 189: 1–21 odd
Day 2 p. 189: 23–41 odd
Above Average p. 189:
1–41 odd

Additional Answers
Written Exercises
 3. $4\sqrt{2}$
 6. $-4\sqrt{14}$
 9. $2\sqrt{11}$
12. $2\sqrt{3} + \sqrt{7}$
15. $12\sqrt{x}$
18. $-6\sqrt{r}$
21. $11\sqrt{2}$
24. $-3\sqrt{2}$
27. $7\sqrt{5}$
30. $10\sqrt{3}$
33. $4\sqrt{2r}$
36. $-11\sqrt{2x}$

Teaching Suggestions p. M–30

Lesson Resources
Warm–Up: p. M–31
Maintenance: See below.
Practice Worksheet 7.6
Transparency 17

Maintenance
1. Evaluate $m^3 - m$ when
   $m = -4$.
   Ans: −60 (Section 1.7)
2. Solve and check:
   $2(2s + 3) = -18$
   Ans: s = −6 (Section 4.3)
3. Solve:
   $5a + 3(a - 7) > a + 7$
   Ans: {all real numbers greater than 4}
   (Section 5.6)
4. Solve: $\dfrac{8^x}{8^3} = 8^5$

   Ans: x = 8 (Section 6.2)
5. The length of a rectangular shaped swimming pool is 12 feet more than the width (Condition 1). The perimeter is 60 feet (Condition 2). Find the length and the width of the pool.
   Ans: length: 21 feet; width: 9 feet (Section 4.2)

Additional Examples
Example 1
Simplify.

1. $\sqrt{\dfrac{x^3}{9}}$ **Ans:** $\dfrac{x\sqrt{x}}{3}$

2. $\sqrt{\dfrac{20}{a^2}}$ **Ans:** $\dfrac{2\sqrt{5}}{a}$

Example 2
Simplify.

1. $\dfrac{\sqrt{32}}{\sqrt{2}}$
   Ans: 4

2. $\dfrac{\sqrt{500y^3}}{\sqrt{5y}}$
   Ans: 10y

7.6 Quotients

To rationalize the denominator in the quotient of two square root radicals and then simplify

P–1 **Simplify:** **a.** $\sqrt{\dfrac{4}{9}}$ $\tfrac{2}{3}$ **b.** $\dfrac{\sqrt{4}}{\sqrt{9}}$ $\tfrac{2}{3}$

Explain that the numerator can equal 0 but the denominator cannot equal 0

> **Quotient Property of Radicals**
>
> If a is any nonnegative real number and b is any positive real number,
> then $\sqrt{\dfrac{a}{b}} = \dfrac{\sqrt{a}}{\sqrt{b}}$.
>
> $\sqrt{\dfrac{4}{9}} = \dfrac{\sqrt{4}}{\sqrt{9}}$
>
> $\sqrt{\dfrac{3}{7}} = \dfrac{\sqrt{3}}{\sqrt{7}}$

EXAMPLE 1   Simplify $\sqrt{\dfrac{8}{r^2}}$.

Solution:

1 Quotient Property of Radicals ⟶ $\sqrt{\dfrac{8}{r^2}} = \dfrac{\sqrt{8}}{\sqrt{r^2}}$

2 Product Property of Radicals ⟶ $= \dfrac{\sqrt{4} \cdot \sqrt{2}}{\sqrt{r^2}}$

3 Meaning of square root ⟶ $= \dfrac{2\sqrt{2}}{r}$ **Simplest form**

P–2 **Simplify each radical.**

a. $\sqrt{\dfrac{n^2}{9}}$ $\tfrac{n}{3}$ **b.** $\sqrt{\dfrac{2}{x^2}}$ $\dfrac{\sqrt{2}}{x}$ **c.** $\sqrt{\dfrac{r^3}{16}}$ $\dfrac{r\sqrt{r}}{4}$ **d.** $\sqrt{\dfrac{64}{a^2}}$ $\dfrac{8}{a}$

EXAMPLE 2   Simplify $\dfrac{\sqrt{18x^3}}{\sqrt{2x}}$.

Solution:

1 Quotient Property of Radicals ⟶ $\dfrac{\sqrt{18x^3}}{\sqrt{2x}} = \sqrt{\dfrac{18x^3}{2x}}$

2 Divide. ⟶ $= \sqrt{9x^2}$

3 Meaning of square root ⟶ $= 3x$ **Simplest form**

Sometimes it is inconvenient to have a radical in a denominator. The procedure used in Example 3 below is called **rationalizing the denominator.**

EXAMPLE 3   Rationalize the denominator of $\dfrac{\sqrt{3}}{\sqrt{2}}$.

Solution:

1. Multiplication Property of One ⟶ $\dfrac{\sqrt{3}}{\sqrt{2}} = \dfrac{\sqrt{3}}{\sqrt{2}} \cdot \dfrac{\sqrt{2}}{\sqrt{2}}$ 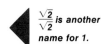 $\dfrac{\sqrt{2}}{\sqrt{2}}$ *is another name for 1.*

2. Product Property of Radicals ⟶ $= \dfrac{\sqrt{6}}{\sqrt{4}}$

3. Meaning of square root ⟶ $= \dfrac{\sqrt{6}}{2}$ $\dfrac{\sqrt{6}}{2}$ *has no radical in the denominator.*

P–3   **What name for 1 would be used to rationalize the denominator of each fraction below?**

a. $\dfrac{\sqrt{2}}{\sqrt{5}} \quad \dfrac{\sqrt{5}}{\sqrt{5}}$ b. $\dfrac{\sqrt{m}}{\sqrt{2n}} \quad \dfrac{\sqrt{2n}}{\sqrt{2n}}$ c. $\dfrac{\sqrt{5r}}{\sqrt{12s}}$ d. $\dfrac{\sqrt{3y}}{\sqrt{x^3}} \quad \dfrac{\sqrt{x}}{\sqrt{x}}$ ◀ *Remember that $\sqrt{x^3}\sqrt{x} = x^2$.*

$\dfrac{\sqrt{12s}}{\sqrt{12s}}$ or $\dfrac{\sqrt{3s}}{\sqrt{3s}}$

EXAMPLE 4   Rationalize the denominator of $\dfrac{\sqrt{5r}}{\sqrt{12s}}$.

Then simplify.

You may also want to show the method in this Example of multiplying by

Solution:

1. Multiplication Property of One ⟶ $\dfrac{\sqrt{5r}}{\sqrt{12s}} = \dfrac{\sqrt{5r}}{\sqrt{12s}} \cdot \dfrac{\sqrt{12s}}{\sqrt{12s}}$ $\dfrac{\sqrt{3s}}{\sqrt{3s}}$ *in Step* 1

2. Product Property of Radicals ⟶ $= \dfrac{\sqrt{60rs}}{12s}$

3. Product Property of Radicals ⟶ $= \dfrac{\sqrt{4} \cdot \sqrt{15rs}}{12s}$

4. Meaning of square root ⟶ $= \dfrac{2\sqrt{15rs}}{12s}$

5. Simplify. ⟶ $= \dfrac{\overset{1}{2}\sqrt{15rs}}{\underset{6}{12}s}$ or $\dfrac{\sqrt{15rs}}{6s}$

Simplify. (Example 1)

1. $\sqrt{\frac{1}{4}}$ $\frac{1}{2}$ 2. $\sqrt{\frac{1}{9}}$ $\frac{1}{3}$ 3. $\sqrt{\frac{9}{25}}$ $\frac{3}{5}$ 4. $\sqrt{\frac{36}{49}}$ $\frac{6}{7}$ 5. $\sqrt{\frac{64}{100}}$ $\frac{4}{5}$ 6. $\sqrt{\frac{100}{144}}$ $\frac{5}{6}$

Simplify. (Example 2)

7. $\frac{\sqrt{12}}{\sqrt{3}}$ 2 8. $\frac{\sqrt{8}}{\sqrt{2}}$ 2 9. $\frac{\sqrt{20}}{\sqrt{5}}$ 2 10. $\frac{\sqrt{24}}{\sqrt{6}}$ 2 11. $\frac{\sqrt{28}}{\sqrt{7}}$ 2

━━━━━ **WRITTEN EXERCISES** ━━━━━

Goal: To simplify quotients of radicals

Sample Problem: $\frac{\sqrt{132r^3}}{\sqrt{6r}}$ **Answer:** $r\sqrt{22}$

Simplify. (Example 1)

1. $\sqrt{\frac{x^2}{16}}$ $\frac{x}{4}$ 2. $\sqrt{\frac{y^2}{25}}$ $\frac{y}{5}$ 3. $\sqrt{\frac{49}{a^2}}$ $\frac{7}{a}$ 4. $\sqrt{\frac{64}{y^2}}$ $\frac{8}{y}$ 5. $\sqrt{\frac{y^4}{100}}$

6. $\sqrt{\frac{x^4}{121}}$ $\frac{x^2}{11}$ 7. $\sqrt{\frac{2}{x^2}}$ $\frac{\sqrt{2}}{x}$ 8. $\sqrt{\frac{3}{r^2}}$ $\frac{\sqrt{3}}{r}$ 9. $\sqrt{\frac{12}{a^2}}$ $\frac{2\sqrt{3}}{a}$ 10. $\sqrt{\frac{18}{t^2}}$

Simplify. (Example 2)

11. $\frac{\sqrt{50}}{\sqrt{2}}$ 5 12. $\frac{\sqrt{75}}{\sqrt{3}}$ 5 13. $\frac{\sqrt{108}}{\sqrt{3}}$ 6 14. $\frac{\sqrt{72}}{\sqrt{2}}$ 6 15. $\frac{\sqrt{98}}{\sqrt{2}}$

16. $\frac{\sqrt{180}}{\sqrt{5}}$ 6 17. $\frac{\sqrt{125}}{\sqrt{5}}$ 5 18. $\frac{\sqrt{147}}{\sqrt{3}}$ 7 19. $\frac{\sqrt{x^3}}{\sqrt{x}}$ x 20. $\frac{\sqrt{a}}{\sqrt{a^3}}$

21. $\frac{\sqrt{4y}}{\sqrt{y^3}}$ $\frac{2}{y}$ 22. $\frac{\sqrt{r^3}}{\sqrt{9r}}$ $\frac{r}{3}$ 23. $\frac{\sqrt{ab^2}}{\sqrt{a^3}}$ $\frac{b}{a}$ 24. $\frac{\sqrt{x^3y^3}}{\sqrt{xy}}$ xy 25. $\frac{\sqrt{10}}{\sqrt{2}}$

26. $\frac{\sqrt{14}}{\sqrt{7}}$ $\sqrt{2}$ 27. $\frac{\sqrt{18}}{\sqrt{3}}$ $\sqrt{6}$ 28. $\frac{\sqrt{22}}{\sqrt{11}}$ $\sqrt{2}$ 29. $\frac{\sqrt{33}}{\sqrt{3}}$ $\sqrt{11}$ 30. $\frac{\sqrt{38}}{\sqrt{2}}$

Rationalize each denominator. Then simplify. (Examples 3–4)

31. $\frac{\sqrt{3}}{\sqrt{5}}$ $\frac{\sqrt{15}}{5}$ 32. $\frac{\sqrt{2}}{\sqrt{7}}$ $\frac{\sqrt{14}}{7}$ 33. $\frac{\sqrt{a^2}}{\sqrt{b}}$ $\frac{a\sqrt{b}}{b}$ 34. $\frac{\sqrt{r^3}}{\sqrt{p}}$ $\frac{r\sqrt{rp}}{p}$ 35. $\frac{\sqrt{3s}}{\sqrt{5t}}$ $\frac{\sqrt{15st}}{5t}$ 36. $\frac{\sqrt{2x}}{\sqrt{3y}}$

REVIEW CAPSULE FOR SECTION 7.7

Write the prime factorization. (Pages 174–176)

1. 168   2. 180   3. 244 $2 \cdot 2 \cdot 61$ 4. 288   5. 207   6. 234
   $2 \cdot 2 \cdot 2 \cdot 3 \cdot 7$ $3 \cdot 3 \cdot 23$

Assignment Guide
Basic
Day 1 p. 192: 1–19 odd
Day 2 p. 192: 21–35 odd
Average p. 192: 1–35 odd
Above Average p. 192:
1–35 odd

Additional Answers
Written Exercises
5. $\frac{y^2}{10}$
10. $\frac{3\sqrt{2}}{t}$
15. 7
20. $\frac{1}{a}$
25. $\sqrt{5}$
30. $\sqrt{19}$
36. $\frac{\sqrt{6xy}}{3y}$

Review Capsule
2. $2 \cdot 2 \cdot 3 \cdot 3 \cdot 5$
4. $2 \cdot 2 \cdot 2 \cdot 2 \cdot 2 \cdot 3 \cdot 3$
6. $2 \cdot 3 \cdot 3 \cdot 13$

OBJECTIVES: To approximate positive square roots of certain counting numbers, not in a given Table of Squares and Square Roots, by forming a product involving one or more radicals that are within the range of the table

7.7 Approximations

To rationalize the denominator of a fraction and approximate the given radical by use of a Table

A Table of Squares and Square Roots is provided on page 194 for the positive integers from 1 to 150. Square roots in this table are approximated to three decimal places.

P–1 **What is each power below based on the table?**

a. 28^2 784 **b.** 76^2 5776 **c.** 26^2 676

Each number in the **Number** column is the positive square root of a number in the **Square** column.

P–2 **What is the value of each square root below?**

a. $\sqrt{784}$ 28 **b.** $-\sqrt{6084}$ −78 **c.** $\sqrt{729}$ 27

P–3 **What is the approximate value of each square root below?**

a. $\sqrt{26}$ **b.** $\sqrt{78}$ **c.** $-\sqrt{28}$ −5.292
   5.099   8.832

The process of simplifying radicals can sometimes be used to approximate the square root of numbers not in the table.

EXAMPLE 1   Approximate $\sqrt{160}$ to three decimal places.

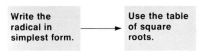

Write the radical in simplest form. → **Use the table of square roots.**

Solution:

1. Write the prime factorization. ——→ $\sqrt{160} = \sqrt{2 \cdot 2 \cdot 2 \cdot 2 \cdot 2 \cdot 5}$

2. Product Property of Radicals ——→ $= \sqrt{2^4} \cdot \sqrt{2 \cdot 5}$

3. Simplify. ——————————→ $= 4\sqrt{10}$

4. Use the table on page 194. ——→ $\approx 4(3.162)$

≈ 12.648

The approximate value of $\sqrt{160}$ is 12.648.

Teaching Suggestions p. M–30

Lesson Resources
Warm–Up: p. M–31
Maintenance: See below.
Practice Worksheet 7.7
Transparencies 17, 18

Maintenance
1. Multiply: $(-t)(6)(-r)(-2)$
   Ans: −12tr (Section 2.3)
2. Solve and check:
   $8 - 4(x - 1) = -36 + 2x$
   Ans: x = 8 (Section 4.3)
3. Write $6^3 \cdot 6^5$ as one power.
   Ans: 6^8 (Section 6.1)
4. Simplify: $(-2^3)^0$
   Ans: 1 (Section 6.5)
5. Plastic garbage cans are on sale with a discount rate of 25%. The sale price is $15. Find the list price and the discount.
   Ans: list price: $20; discount: $5 (Section 4.4)

Additional Examples
Example 1
Approximate to three decimal places.
1. $\sqrt{240}$
   Ans: 15.492
2. $\sqrt{990}$
   Ans: 31.464
3. $\sqrt{252}$
   Ans: 15.875

Table of Squares and Square Roots

No.	Square	Square Root	No.	Square	Square Root	No.	Square	Square Root
1	1	1.000	51	2601	7.141	101	10,201	10.050
2	4	1.414	52	2704	7.211	102	10,404	10.100
3	9	1.732	53	2809	7.280	103	10,609	10.149
4	16	2.000	54	2916	7.348	104	10,816	10.198
5	25	2.236	55	3025	7.416	105	11,025	10.247
6	36	2.449	56	3136	7.483	106	11,236	10.296
7	49	2.646	57	3249	7.550	107	11,449	10.344
8	64	2.828	58	3364	7.616	108	11,664	10.392
9	81	3.000	59	3481	7.681	109	11,881	10.440
10	100	3.162	60	3600	7.746	110	12,100	10.488
11	121	3.317	61	3721	7.810	111	12,321	10.536
12	144	3.464	62	3844	7.874	112	12,544	10.583
13	169	3.606	63	3969	7.937	113	12,769	10.630
14	196	3.742	64	4096	8.000	114	12,996	10.677
15	225	3.873	65	4225	8.062	115	13,225	10.724
16	256	4.000	66	4356	8.124	116	13,456	10.770
17	289	4.123	67	4489	8.185	117	13,689	10.817
18	324	4.243	68	4624	8.246	118	13,924	10.863
19	361	4.359	69	4761	8.307	119	14,161	10.909
20	400	4.472	70	4900	8.367	120	14,400	10.954
21	441	4.583	71	5041	8.426	121	14,641	11.000
22	484	4.690	72	5184	8.485	122	14,884	11.045
23	529	4.796	73	5329	8.544	123	15,129	11.091
24	576	4.899	74	5476	8.602	124	15,376	11.136
25	625	5.000	75	5625	8.660	125	15,625	11.180
26	676	5.099	76	5776	8.718	126	15,876	11.225
27	729	5.196	77	5929	8.775	127	16,129	11.269
28	784	5.292	78	6084	8.832	128	16,384	11.314
29	841	5.385	79	6241	8.888	129	16,641	11.358
30	900	5.477	80	6400	8.944	130	16,900	11.402
31	961	5.568	81	6561	9.000	131	17,161	11.446
32	1024	5.657	82	6724	9.055	132	17,424	11.489
33	1089	5.745	83	6889	9.110	133	17,689	11.533
34	1156	5.831	84	7056	9.165	134	17,956	11.576
35	1225	5.916	85	7225	9.220	135	18,225	11.619
36	1296	6.000	86	7396	9.274	136	18,496	11.662
37	1369	6.083	87	7569	9.327	137	18,769	11.705
38	1444	6.164	88	7744	9.381	138	19,044	11.747
39	1521	6.245	89	7921	9.434	139	19,321	11.790
40	1600	6.325	90	8100	9.487	140	19,600	11.832
41	1681	6.403	91	8281	9.539	141	19,881	11.874
42	1764	6.481	92	8464	9.592	142	20,164	11.916
43	1849	6.557	93	8649	9.644	143	20,449	11.958
44	1936	6.633	94	8836	9.695	144	20,736	12.000
45	2025	6.708	95	9025	9.747	145	21,025	12.042
46	2116	6.782	96	9216	9.798	146	21,316	12.083
47	2209	6.856	97	9409	9.849	147	21,609	12.124
48	2304	6.928	98	9604	9.899	148	21,904	12.166
49	2401	7.000	99	9801	9.950	149	22,201	12.207
50	2500	7.071	100	10,000	10.000	150	22,500	12.247

EXAMPLE 2   Approximate $\sqrt{\dfrac{3}{8}}$ to three decimal places.

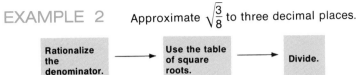

| Rationalize the denominator. | → | Use the table of square roots. | → | Divide. |

Solution:

1 Quotient Property of Radicals ⟶ $\sqrt{\dfrac{3}{8}} = \dfrac{\sqrt{3}}{\sqrt{8}}$

2 Multiplication Property of One ⟶ $= \dfrac{\sqrt{3}}{\sqrt{8}} \cdot \dfrac{\sqrt{8}}{\sqrt{8}}$

3 Product Property of Radicals ⟶ $= \dfrac{\sqrt{24}}{8}$

4 Simplify the numerator. ⟶ $= \dfrac{2\sqrt{6}}{8}$

5 Write in lowest terms. ⟶ $= \dfrac{\sqrt{6}}{4}$

6 Use the table on page 194. ⟶ $\approx \dfrac{2.449}{4}$

≈ 0.612 ◀ **Approximate value** of $\sqrt{\dfrac{3}{8}}$

Discuss the use of $\dfrac{\sqrt{2}}{\sqrt{2}}$ as a name for 1 in Step 2 . Also, note the importance of rationalizing the denominator. $\dfrac{\sqrt{6}}{4}$ is easier to approximate than $\dfrac{\sqrt{3}}{\sqrt{8}}$.

EXAMPLE 3   The area of a square plaque is 75 square inches. Find the length of each side of the plaque to the nearest tenth of an inch.

Solution:

1 Write the formula for the area of a square. ⟶ $A = s^2$

2 Replace A with 75. ⟶ $75 = s^2$

3 Solve for s. ⟶ $\sqrt{75} = s$

4 Simplify. ⟶ $\sqrt{25} \cdot \sqrt{3} = s$

$5\sqrt{3} = s$ ◀ **Use the table on page 194.**

5 Approximate the value of s. ⟶ $5(1.732) \approx s$

$8.660 \approx s$

The length of each side to the nearest tenth is 8.7 inches.

If you are sitting in a plane that is high in the sky, you can see the earth's horizon several miles away. The distance to the horizon depends on the plane's **altitude.** The altitude is the plane's distance above the ground.

The approximate distance to the horizon is given by the following formula, where D is the distance to the horizon in miles and A is the altitude of the plane in feet.

$$D = 1.22\sqrt{A}$$

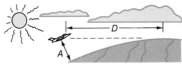

EXAMPLE 4   Find the approximate distance in miles to the horizon from a plane that is flying at an altitude of 25,000 feet. Round the distance to the nearest mile.

Solution:

1 Write the formula. ──────────▶ $D = 1.22\sqrt{A}$

2 Replace A with 25,000. ──────▶ $D = 1.22\sqrt{25,000}$

3 Simplify. ───────────────▶ $D = 1.22\sqrt{2500} \cdot \sqrt{10}$

◀ *Use the table on page 194.*

$D = (1.22)(50)\sqrt{10}$

$D \approx 61(3.162)$

$D \approx 192.882$

The distance to the horizon is approximately 193 miles.

CLASSROOM EXERCISES

Find each value by referring to the Table of Squares and Square Roots.
(P-1, P-2, P-3)

			13,689
1. 29^2 841	**2.** 56^2 3136	**3.** 93^2 8649	**4.** 117^2
5. $\sqrt{5184}$ 72	**6.** $\sqrt{961}$ 31	**7.** $-\sqrt{7744}$ −88	**8.** $\sqrt{13,225}$
9. $\sqrt{38}$ 6.164	**10.** $\sqrt{131}$ 11.446	**11.** $\sqrt{78}$ 8.832	**12.** $-\sqrt{43}$
13. $\sqrt{3481}$ 59	**14.** $\sqrt{13,456}$ 116	**15.** $-\sqrt{150}$ −12.247	**16.** $\sqrt{109}$
			10.440

WRITTEN EXERCISES

Goal: To approximate the values of square roots to two or three decimal places

Sample Problem: Approximate $\sqrt{304}$ to three decimal places.

Answer: 17.436

Approximate each square root to three decimal places. Use the table on page 194. (Examples 1–2)

1. $\sqrt{168}$ 12.962

2. $\sqrt{180}$ 13.416

3. $\sqrt{244}$ 15.620

4. $\sqrt{288}$

5. $\sqrt{207}$ 14.388

6. $\sqrt{234}$ 15.297

7. $\sqrt{275}$ 16.585

8. $\sqrt{425}$

9. $\sqrt{468}$ 21.636

10. $\sqrt{684}$ 26.154

11. $\sqrt{\frac{3}{4}}$ 0.866

12. $\sqrt{\frac{5}{16}}$

13. $\sqrt{\frac{2}{3}}$ 0.816

14. $\sqrt{\frac{5}{6}}$ 0.913

15. $\sqrt{\frac{7}{8}}$ 0.936

16. $\sqrt{\frac{5}{8}}$

17. $\sqrt{\frac{11}{12}}$ 0.958

18. $\sqrt{\frac{7}{12}}$ 0.764

19. $\sqrt{\frac{3}{32}}$ 0.306

20. $\sqrt{\frac{7}{18}}$

APPLICATIONS

Solve each problem. Use the table on page 194. Round the answers to the nearest tenth. (Example 3)

21. A homesite is in the shape of a square. It has an area of 4840 square yards. Find the length of each side of the homesite in yards.
  69.6 yards

22. The Andersens have a square-shaped backyard. It has an area of 396 square meters. Find the length of each side of the yard in meters.   19.9 meters

For Exercises 23–24, use the formula $D = 1.22\sqrt{A}$. (Example 4)

23. Find the approximate distance in miles to the horizon from a plane that is at an altitude of 5000 feet. Round the distance to the nearest mile. 86 miles

24. Find the approximate distance in miles to the horizon from a plane that is at an altitude of 15,000 feet. Round the distance to the nearest mile.   149 miles

For Exercises 25–26, use the formula $D = 3.56\sqrt{A}$ where A is the altitude in meters and D is the distance in kilometers.

25. Find the approximate distance in kilometers to the horizon from a plane that is at an altitude of 4500 meters. Round the distance to the nearest kilometer. 239 kilometers

26. Find the approximate distance in kilometers to the horizon from a plane that is at an altitude of 7500 meters. Round the distance to the nearest kilometer.   308 kilometers

Roots / **197**

Assignment Guide
Basic p. 197: 1–25 odd
Average p. 197: 1–25 odd
Above Average p. 197: 1–25 odd

Additional Answers
4. 16.968
8. 20.615
12. 0.559
16. 0.791
20. 0.624

Problem-Solving Skills
Reading a table (Ex. 21–26)
Using a formula (Ex. 21–26)
Drawing a diagram (Ex. 21–22)

198

Teaching Suggestions p. M–30

Lesson Resources
Warm–Up: p. M–31
Maintenance: See below.
Practice Worksheet 7.8
Transparencies 18, 19

Maintenance
1. Solve and check:
   $2(2 - 3x) = 8 - 2(4x + 5)$
   Ans: $x = -3$ (Section 4.3)
2. Solve: $2y - 6 > 3y + 1$
   Ans: {all real numbers less than -7} (Section 5.4)
3. Solve: $\frac{10^3}{10^t} = 10^2$
   Ans: $t = 1$ (Section 6.2)
4. Write 3,000,000 in scientific notation.
   Ans: 3.0×10^6
   (Section 6.6)
5. Margo's scores on three tests were 67, 91, and 84. She wants her mean score on four tests to be greater than 84 (Condition 2). Find the lowest score Margo can get on the fourth test (Condition 1).
   Ans: 95 (Section 5.7)

Additional Examples
Example 1
Use the Rule of Pythagoras to determine whether each triangle is a right triangle.
1.
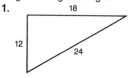

Ans: No

2.
3.5   9.1
8.4

Ans: Yes

PROBLEM SOLVING AND APPLICATIONS

7.8   The Rule of Pythagoras

To determine the length of one side of a right triangle given the lengths of the other two sides by using the Rule of Pythagoras

Example 1 and exercises based on it involve use of the converse and the contrapositive.
Converse: If $c^2 = a^2 + b^2$ is true, then $\triangle ABC$ is a right triangle.
Contrapositive: If $c^2 = a^2 + b^2$ is not true, then $\triangle ABC$ is not a right triangle.

A triangle with one angle of 90° is called a **right triangle.** The symbol ⌐ is used to indicate the 90° angle. The longest side of a right triangle is the **hypotenuse.** The two shorter sides are the **legs.** In right triangle ABC at the right, the hypotenuse is side AB and the legs are sides AC and BC.

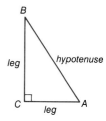

A rule about right triangles is named for Pythagoras, the Greek mathematician who developed it.

Rule of Pythagoras

In a right triangle, the square of the length of the hypotenuse, c, equals the sum of the squares of the lengths of the legs, a and b.

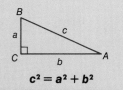

$$c^2 = a^2 + b^2$$

EXAMPLE 1   Use the Rule of Pythagoras to determine whether the triangle shown at the right is a right triangle.

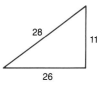

Solution: Let $a = 11$, $b = 26$, and $c = 28$.

1. Write the Rule of Pythagoras in symbols. ⟶ $c^2 = a^2 + b^2$
2. Replace the variables with their values. ⟶ $28^2 \overset{?}{=} 11^2 + 26^2$
3. Perform the operations. ⟶ $784 \overset{?}{=} 121 + 676$ **Not an equality**

   $784 \neq 797$

Since the result is not an equality, the triangle is not a right triangle.

P–1   **Which of the following are the lengths of the three sides of a right triangle?** a

   a. 3 meters, 4 meters, 5 meters   **b.** 8 feet, 12 feet, 16 feet

When the lengths of two sides of a right triangle are known, the Rule of Pythagoras can be used to find the length of the third side.

EXAMPLE 2

The legs of a right triangle have lengths of 9 centimeters and 12 centimeters. Find the length of the hypotenuse.

Solution: Let $a = 9$ and $b = 12$. Find c.

1 Write the Rule of Pythagoras in symbols. ⟶ $c^2 = a^2 + b^2$

2 Replace the known variables with their values. ⟶ $c^2 = 9^2 + 12^2$

3 Solve for c. ⟶ $c^2 = 81 + 144$

$c^2 = 225$

4 Find the square root. ⟶ $c = \sqrt{225}$ **Use the table on page 194.**

$c = 15$

The length of the hypotenuse is 15 centimeters.

Additional Examples
Example 2
Right triangle ABC has legs a and b and hypotenuse c. The lengths of two sides of the triangle are given. Find the length of the third side.
1. $a = 28$ feet; $b = 21$ feet
   Ans: $c = 35$ feet
2. $a = 4.2$ centimeters; $c = 15$ centimeters
   Ans: $b = 14.4$ centimeters

EXAMPLE 3

Find the approximate height of a 14-inch TV screen if the width is 11 inches. Round the height to the nearest tenth. (A 14-inch screen has a diagonal length of 14 inches.)

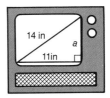

Solution: Let $b = 11$ and $c = 14$. Find a.

1 Write the Rule of Pythagoras in symbols. ⟶ $c^2 = a^2 + b^2$

2 Replace the known variables with their values. ⟶ $14^2 = a^2 + 11^2$

3 Solve for a. ⟶ $196 = a^2 + 121$

$196 - \mathbf{121} = a^2 + 121 - \mathbf{121}$

$75 = a^2$

4 Find the square root. ⟶ $\sqrt{75} = a$ **Use the table on page 194.**

$8.660 \approx a$

The height of the screen is 8.7 inches to the nearest tenth.

Example 3
A radio tower is 60 feet high and has a support wire which is 75 feet long. Find the distance from the base of the tower to the anchored wire.

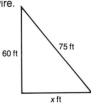

Ans: 45 feet

CLASSROOM EXERCISES

*Determine whether the given lengths are the lengths of the
sides of a right triangle.* (Example 1)

1. 2 feet, 3 feet, 4 feet   No

3. 3 inches, 5 inches, 6 inches No

2. 6 meters, 8 meters, 10 meters   Yes

4. 12 yards, 5 yards, 13 yards Yes

*Use the Rule of Pythagoras to write an equation for finding the unknown
length in each right triangle.* (Steps 1 and 2 in Examples 2 and 3)

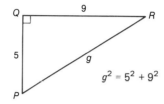

$$g^2 = 5^2 + 9^2$$

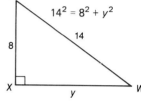

$$14^2 = 8^2 + y^2$$

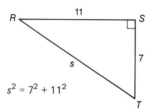

$$s^2 = 7^2 + 11^2$$

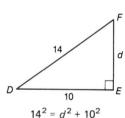

$$14^2 = d^2 + 10^2$$

WRITTEN EXERCISES

Goal: To solve for the length of one side of a right triangle when the
lengths of the other two sides are known

Sample Problem: Find the length of the hypotenuse of a right triangle when
the lengths of the legs are 10 inches and 24 inches.

Answer: 26 inches

*Determine whether the given lengths are the lengths of the sides of a
right triangle.* (Example 1)

1. 12 meters, 9 meters, 15 meters   Yes

3. 16 inches, 35 inches, 38 inches   No

2. 14 feet, 16 feet, 21 feet No

4. 15 feet, 36 feet, 39 feet Yes

200   /   *Chapter 7*

For Exercises 5–16, refer to triangle ABC at the right. Use the Rule of Pythagoras and the table on page 194 to find the missing length. Round to the nearest tenth where necessary.
(Examples 2 and 3)

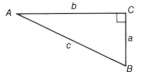

5. $a = 12$, $b = 16$ $c = 20$

6. $a = 10$, $b = 24$ $c = 26$

7. $a = 15$, $c = 25$

8. $a = 8$, $c = 10$ $b = 6$

9. $b = 7$, $c = 25$ $a = 24$

10. $b = 12$, $c = 13$

11. $c = 39$, $b = 36$ $a = 15$

12. $c = 50$, $b = 48$ $a = 14$

13. $c = 5$, $a = 2$

14. $c = 18$, $a = 12$ $b = 13.4$

15. $a = 8$, $b = 10$ $c = 12.8$

16. $a = 3$, $b = 5$

Additional Answers
 7. $b = 20$
10. $a = 5$
13. $b = 4.6$
16. $c = 5.8$

Solve each problem. Use the table on page 194. Round to the nearest tenth where necessary. (Example 3)

17. A TV with a 20-inch screen in on sale. The width of the screen is 16 inches. Find the height of the screen.   12 inches

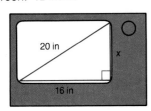

18. The distance between consecutive bases on a baseball diamond is 90 feet. Find the distance, d, from home plate to second base.

127.3 feet

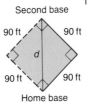

19. A utility pole has a support wire which is 36 feet long. The wire is anchored to the ground 20 feet from the base of the pole. Find the height of the pole.   29.9 feet

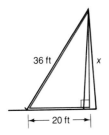

20. A boat sailed 20 kilometers due south from point A to point C. The boat then sailed 14 kilometers due west to point B. Find the direct sailing distance from point A to point B.   24.4 kilometers

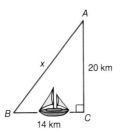

*Roots   /   **201***

Quiz Sections 7.5–7.8
After completing Sections 7.5–7.8, you may want to administer a quiz covering the same sections. See the quiz provided on page 152 in the *Teacher's ResourceBank.*

Two tables of interesting weather data are included in this special topic. One is for the temperature-humidity index and the other is for wind chill temperatures. Students are also shown a formula for computing wind chill temperature in Celsius degrees.

Using Tables

Meteorology

Meteorology is the study of the atmosphere. The best known activity of meteorologists is **weather forecasting.**

The table at the right shows the **Temperature–Humidity Index,** or **THI.** THI is a measure of how comfortable (or uncomfortable) people feel under given conditions of temperature and humidity during the warm season of the year.

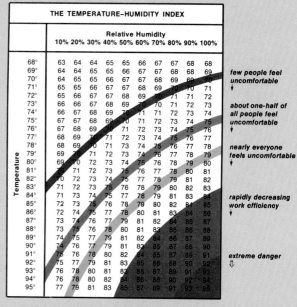

THE TEMPERATURE-HUMIDITY INDEX

Relative Humidity
10% 20% 30% 40% 50% 60% 70% 80% 90% 100%

few people feel uncomfortable

about one-half of all people feel uncomfortable

nearly everyone feels uncomfortable

rapidly decreasing work efficiency

extreme danger

EXAMPLE 1: Find the temperature–humidity index above which most people feel uncomfortable.

SOLUTION: Locate the curved band in the table for "nearly everyone feels uncomfortable." The band passes through the number 79 in each column.

Temperature–Humidity Index: **79**

In cold climates, the combined effect of high winds and cold temperatures can make a person feel colder than the actual thermometer reading. This is called **wind chill.** Weather forecasters obtain wind chill temperatures from tables such as the one in Example 2.

EXAMPLE 2: Find the wind chill temperature when the thermometer reading is −5°F and the wind speed is 10 miles per hour.

SOLUTION: Locate the wind speed, 10 miles per hour, along the left side of the table. Look horizontally to the right to the column headed "−5."

Wind Chill Temperature: **−27°F**

WIND CHILL TEMPERATURE TABLE									
Thermometer Reading (°F)									
−25	−20	−15	−10	−5	0	5	10	15	20
−31	−26	−21	−15	−10	−5	0	6	11	16
−52	−46	−40	−34	−27	−22	−15	−9	−3	3
−65	−58	−51	−45	−38	−31	−25	−18	−11	−5
−74	−67	−60	−53	−46	−39	−31	−24	−17	−10
−81	−74	−66	−59	−51	−44	−36	−29	−22	−15
−86	−79	−71	−64	−56	−49	−41	−33	−25	−18
−89	−82	−74	−67	−58	−52	−43	−35	−27	−20
−92	−84	−76	−69	−68	−53	−45	−37	−29	−21

Wind Speed (mph): 5, 10, 15, 20, 25, 30, 35, 40

The following formula can be used to approximate the wind chill temperature in degrees Celsius.

$$C = 33 - \frac{(10\sqrt{r} + 10.45 - r)(33 - t)}{22.1}$$

C: wind chill temperature in degrees Celsius
t: air temperature in degrees Celsius
r: wind speed in meters per second

EXAMPLE 3: On a fall day the air temperature is 5°C. The wind speed is 8 meters per second. Find the wind chill temperature.

SOLUTION: $C = 33 - \dfrac{(10\sqrt{r} + 10.45 - r)(33 - t)}{22.1}$

$C = 33 - \dfrac{(10\sqrt{8} + 10.45 - 8)(33 - 5)}{22.1}$ ◄ $r = 8$ $t = 5$

$= 33 - \dfrac{(10\sqrt{8} + 10.45 - 8)(28)}{22.1}$

8 [√] [×] 1 0 [+] 1 0 . 4 5 [−] 8 [×]

2 8 [÷] 2 2 . 1 [=] [+/−] [+] 3 3 [=]

$\boxed{-5.939347}$

The wind chill temperature is about −6°C.

EXERCISES

In Exercises 1–4, refer to the tables on page 202.

1. Find the temperature–humidity index above which about 50% of most people feel uncomfortable. 75

2. Find the temperature–humidity index when the temperature is 90°F and the relative humidity is 80%. 87

3. What wind speed produces a wind chill temperature of −53°F when the thermometer reading is −10°F?
   20 miles per hour

4. Does the wind chill temperature increase or decrease when the wind speed increases and the thermometer reading stays the same? Decreases

Use the formula above to find the wind chill temperature to the nearest degree Celsius (°C).

5. Air temperature: −3°C
   Wind speed: 7 meters per second
   −16°C

6. Air temperature: −12°C
   Wind speed: 6 meters per second
   −26°C

IMPORTANT TERMS		
Radical symbol *(p. 174)*	Fourth root *(p. 180)*	
Square root *(p. 174)*	Perfect cube *(p. 184)*	
Square *(p. 174)*	Like radicals *(p. 187)*	
Prime number *(p. 174)*	Rationalizing	
Prime-factorization *(p. 175)*	the denominator *(p. 191)*	
Perfect square *(p. 175)*	Right Triangle *(p. 198)*	
Radical *(p. 180)*	Hypotenuse *(p. 198)*	
Radicand *(p. 180)*	Legs *(p. 198)*	
Cube root *(p. 180)*		

IMPORTANT IDEAS

1. A negative number cannot have a real number as a square root.

2. Each prime factor of a perfect square occurs an even number of times.

3. A square root of a perfect square counting number is rational.

4. Any whole number that is not a perfect square has an irrational square root.

5. If a is any nonnegative real number, then $\sqrt{a} \cdot \sqrt{a} = a$.

6. *Product Property of Radicals:* If a and b are nonnegative real numbers, then $\sqrt{a} \cdot \sqrt{b} = \sqrt{ab}$.

7. A square root radical is in simplest form if its radicand has no perfect square factor.

8. A cube root radical is in simplest form if its radicand has no perfect cube factor.

9. *Quotient Property of Radicals:* If a is any nonnegative real number and b is any positive real number, then $\sqrt{\dfrac{a}{b}} = \dfrac{\sqrt{a}}{\sqrt{b}}$.

10. *Rule of Pythagoras:* In a right triangle, the square of the length of the hypotenuse equals the sum of the squares of the lengths of the legs.

Chapter Test
Two Chapter Tests (Form A and Form B) are provided on pages 153–156 in the *Teacher's ResourceBank.*

CHAPTER REVIEW

SECTION 7.1

Simplify.

1. $-\sqrt{49}$ -7

2. $\sqrt{\dfrac{1}{16}}$ $\dfrac{1}{4}$

3. $\sqrt{0.36}$ 0.6

4. $\sqrt{3^4 \cdot 7^2}$ 63

SECTION 7.2

Multiply and simplify.

5. $\sqrt{23} \cdot \sqrt{23}$ 23

6. $\sqrt{81} \cdot \sqrt{36}$ 54

7. $(\sqrt{3})(-\sqrt{11})$ $-\sqrt{33}$

8. $(\sqrt{2})(\sqrt{7})$ $\sqrt{14}$

9. $\left(\sqrt{\frac{1}{2}}\right)(\sqrt{72})$ 6

10. $\left(\sqrt{\frac{1}{3}}\right)(\sqrt{27})$ 3

SECTION 7.3

Simplify.

11. $\sqrt{2 \cdot 3 \cdot 2 \cdot 3 \cdot 3}$ $6\sqrt{3}$

12. $\sqrt{3 \cdot 5 \cdot 5 \cdot 2 \cdot 2 \cdot 2}$ $10\sqrt{6}$

13. $\sqrt{52}$ $2\sqrt{13}$ **14.** $\sqrt{60}$ $2\sqrt{15}$ **15.** $\sqrt{99}$ $3\sqrt{11}$ **16.** $\sqrt{63}$ $3\sqrt{7}$ **17.** $\sqrt[3]{24}$ $2\sqrt[3]{3}$ **18.** $\sqrt[3]{56}$ $2\sqrt[3]{7}$

SECTION 7.4

Simplify.

19. $\sqrt{r^6}$ r^3

20. $\sqrt{t^{10}}$ t^5

21. $\sqrt{9x^5}$ $3x^2\sqrt{x}$

22. $\sqrt{16y^3}$ $4y\sqrt{y}$

23. $\sqrt{28a^3b^5}$ $2ab^2\sqrt{7ab}$

24. $\sqrt{45r^4s^7}$ $3r^2s^3\sqrt{5s}$

25. $\sqrt[3]{16m^3n^4}$ $2mn\sqrt[3]{2n}$

26. $\sqrt[3]{27p^5q^6}$ $3pq^2\sqrt[3]{p^2}$

SECTION 7.5

Add or subtract as indicated. Simplify where necessary.

27. $8\sqrt{13} + 2\sqrt{13}$ $10\sqrt{13}$

28. $12\sqrt{3} - \sqrt{3}$ $11\sqrt{3}$

29. $2\sqrt{12} + 5\sqrt{3}$ $9\sqrt{3}$

30. $3\sqrt{8} + 5\sqrt{2}$ $11\sqrt{2}$

31. $\sqrt{24} - 9\sqrt{6}$ $-7\sqrt{6}$

32. $12\sqrt{5} - \sqrt{45}$ $9\sqrt{5}$

SECTION 7.6

Simplify.

33. $\sqrt{\frac{3}{24}}$ $\frac{\sqrt{2}}{4}$

34. $\sqrt{\frac{x}{81}}$ $\frac{\sqrt{x}}{9}$

35. $\sqrt{\frac{20}{t^2}}$ $\frac{2\sqrt{5}}{t}$

36. $\sqrt{\frac{54}{y^4}}$ $\frac{3\sqrt{6}}{y^2}$

Rationalize each denominator. Then simplify.

37. $\frac{\sqrt{7}}{\sqrt{10}}$ $\frac{\sqrt{70}}{10}$

38. $\frac{\sqrt{3}}{\sqrt{14}}$ $\frac{\sqrt{42}}{14}$

39. $\frac{\sqrt{3x^3}}{\sqrt{2y}}$ $\frac{x\sqrt{6xy}}{2y}$

40. $\frac{\sqrt{2a^5}}{\sqrt{5m}}$ $\frac{a^2\sqrt{10am}}{5m}$

SECTION 7.7

Approximate each square root to three decimal places. Use the table on page 194.

41. $\sqrt{153}$ 12.369

42. $\sqrt{200}$ 14.142

43. $\sqrt{\frac{11}{16}}$ 0.829

44. $\sqrt{\frac{7}{25}}$ 0.529

45. $\sqrt{\frac{17}{24}}$ 0.842

46. $\sqrt{\frac{11}{20}}$ 0.742

Solve each problem. Use the table on page 194. Round each answer to the nearest tenth.

47. A backyard patio is in the shape of a square. It has an area of 12 square meters. Find the length of each side of the patio in meters. 3.5 meters

48. The Mellott family has a square-shaped kitchen floor. The floor has an area of 108 square feet. Find the length of each side of the kitchen floor in feet. 10.4 feet

For Exercises 49–50, use the formula $D = 1.22\sqrt{A}$ where A is the altitude in feet and D is the distance in miles. Use the table on page 194. Round the answers to the nearest mile.

49. Find the approximate distance in miles to the horizon from a plane that is at an altitude of 4000 feet. 77 miles

50. Find the approximate distance in miles to the horizon from a plane that is at an altitude of 9000 feet. 116 miles

SECTION 7.8

Determine whether the given lengths are the lengths of the sides of a right triangle.

51. 6 meters, 8 meters, 10 meters Yes

52. 5 inches, 6 inches, 8 inches No

53. 8 feet, 9 feet, 12 feet No

54. 60 yards, 11 yards, 61 yards Yes

Solve each problem using the Rule of Pythagoras. Use the table on page 194. Round the length to the nearest tenth.

55. An appliance store is holding a sale on a TV with a 14-inch screen. The width of the screen is 10 inches. Find the height of the screen in inches. 9.8 inches

56. A construction worker is making a rectangular form for a patio. The length of the form is 20 feet and the width is 14 feet. He measures the diagonal to see if the form makes a right angle. What should be the length of the diagonal? 24.4 feet

CHAPTER 8

Polynomials

The photo below is the earth as seen from 400,000 kilometers in space. The formula $K = \frac{R^3}{T^2}$ says that for any planet in the solar system, the cube of its average distance from the sun divided by the square of its time for one revolution is a constant.

Background
The center photo of the earth was taken by the Apollo 17 spacecraft during the final lunar landing mission in NASA's Apollo program.
The top right photo shows the planet Saturn. NASA's Voyager 2 was 43 million kilometers (27 million miles) from Saturn when it took this photograph in 1981.
The photo of Mars at the bottom right was taken by the Viking aircraft.

Teaching Suggestions p. M–32

Lesson Resources
Warm–Up: p. M–32
Maintenance: See below.
Practice Worksheet 8.1
Transparencies 2, 3

Maintenance
1. Evaluate $m^2 - 2m + 3$ when $m = 2$.
   Ans: 3 (Section 1.7)
2. Solve and check:
   $4 + 7x = -(4 - 2x) - 3x$
   Ans: $x = -1$ (Section 4.3)
3. Write $\left(-\frac{1}{3}\right)^2 \cdot \left(-\frac{1}{3}\right)^3$ as one power.
   Ans: $\left(-\frac{1}{3}\right)^5$ (Section 6.1)
4. Simplify: $(x^{-3}y)^{-4}$
   Ans: $\dfrac{x^{12}}{y^4}$ (Section 6.5)

5. The perimeter of a rectangular shaped greeting card is 12.2 inches. The width is 2.6 inches. Find the length of the card.
   Ans: 3.5 inches
   (Section 3.4)

Additional Examples
Evaluate.
1. $7x^5$ if $x = -2$
   Ans: −224
2. $-5x^4$ if $x = 3$
   Ans: −405
3. $-2x^3$ if $x = -4$
   Ans: 128

OBJECTIVES: To determine whether a given numeral or expression is a monomial over {rational numbers}
To write the numerical coefficient of a given monomial

8.1 Monomials

To evaluate a monomial for a given value of its variable

The numerals and expressions below are monomials.

   a. $\frac{1}{3}x$ **b.** 5   **c.** $3x^2$ **d.** $-x$

Definition

Explain that the domains of a, x, and n are all different.

> A **monomial** of one variable over {rational numbers} is an expression of the form ax^n in which a is any rational number and n is any nonnegative integer.
>
> $-5x^3$
> $\frac{1}{4}x^5$
> $-3x^0$ or -3

P–1 **What are the values of a and n in each monomial below?**

   a. $3.7x^2$ **b.** $-x$ **c.** -15
   $a = 3.7; n = 2$ $a = -1; n = 1$ $a = -15;$ $n = 0$

Since $-15 = -15x^0$, -15 is a monomial.

If ax^n is a monomial, n cannot be negative. The following expressions are <u>not</u> monomials.

   a. $3x^{-2}$ **b.** $-\frac{1}{5}x^{-1}$ **c.** $0.7x^{-3}$

P–2 **Why is $\dfrac{3}{x^2}$ not a monomial?**

$\frac{3}{x^2}$ is equivalent to $3x^{-2}$.   *$n = -2$, which is a negative number.*

In a monomial of the form ax^n, the number represented by a is the **coefficient** or **numerical coefficient.**

P–3 **What is the coefficient of each monomial below?**

   a. $\frac{1}{3}x^2$ $\frac{1}{3}$ **b.** x^3 1   **c.** $-x$ -1 **d.** $-\dfrac{x}{5}$ $-\dfrac{1}{5}$

Monomials are classified by their numerical coefficients. The monomials below are monomials **over the set of integers.**

   a. $2x^3$ **b.** -10 **c.** $5x$ **d.** $-x^5$

EXAMPLE   Evaluate $-7x^3$ if x equals -2.

Solution: $-7x^3 = -7(-2)^3$ **Raise to a power first.**

                    $= -7(-2)(-2)(-2)$
                    $= (-7)(-8)$
                    $= 56$

CLASSROOM EXERCISES

Write Yes or No to tell whether each numeral or expression is a monomial over {rational numbers}. (P-2)

1. $5x^3$ Yes   2. $5x^{-1}$ No   3. $\dfrac{3}{x}$ No   4. $\dfrac{x}{3}$ Yes   5. $\sqrt{4} \cdot x^3$ Yes   6. $-7x^0$ Yes

Write the coefficient of each monomial. (P-3)

7. $2x^2$ 2   8. $-x^3$ -1 9. $-7\frac{1}{2}$ $-7\frac{1}{2}$ 10. $-2.8x^3$ -2.8 11. $3\frac{2}{3}x^0$ $3\frac{2}{3}$

WRITTEN EXERCISES

Goal: To evaluate monomials
Sample Problem: Evaluate $0.6x^3$ if $x = 20$.   **Answer:** 4800

Write Yes or No to show whether each numeral or expression is a monomial over {rational numbers}. (P-2)

1. $13x$ Yes   2. $27x^2$ Yes   3. $-5x^3$ Yes   4. $-3x^5$ Yes   5. $2x^{-2}$ No

6. $-3x^{-5}$ No   7. $\dfrac{x^2}{10}$ Yes   8. $\dfrac{-x^3}{7}$ Yes   9. $\dfrac{3}{x^2}$ No   10. $\dfrac{-2}{x^2}$ No

11. $0.6x^0$ Yes   12. $5.7x^0$ Yes   13. $\sqrt{7} \cdot x^2$ No   14. $-\sqrt{3} \cdot x^3$ No   15. $x + 3$ No

Write the coefficient of each monomial. (P-3)

16. $2x^2$ 2   17. $5x^4$ 5   18. $\frac{1}{3}x^2$ $\frac{1}{3}$ 19. $\frac{3}{5}x^3$ $\frac{3}{5}$ 20. $-0.3x$ -0.3 21. $1.8x^3$ 1.8

22. x 1   23. $-x^3$ -1 24. $\dfrac{x}{10}$ $\frac{1}{10}$ 25. $\frac{3}{5}$ $\frac{3}{5}$ 26. $2x^0$ 2   27. $\dfrac{3x}{5}$ $\frac{3}{5}$

Evaluate each monomial. (Example)

28. $-5x^2$ if $x = 3$ -45 29. $3x^3$ if $x = -1$ -3 30. $\frac{1}{4}x^4$ if $x = -2$ 4

31. $\frac{3}{4}x^5$ if $x = 2$ 24   32. $-0.5x^3$ if $x = 10$ -500 33. $-8.3x^0$ if $x = -19$ -8.3

REVIEW CAPSULE FOR SECTION 8.2

Write each product or quotient as one power. (Pages 144–150)

1. $m^2 \cdot m^4$ m^6 2. $t \cdot t^7$ t^8 3. $r^5 \cdot r^3 \cdot r^2$ r^{10} 4. $y^4 \cdot y^4 \cdot y^4$ y^{12}

5. $\dfrac{w^6}{w^2}$ w^4 6. $\dfrac{q^4}{q}$ q^3 7. $\dfrac{n^8}{n^4}$ n^4 8. $\dfrac{a^{10}}{a^5}$ a^5

*Polynomials / **209***

Review of Basic Algebra Skills
Before teaching Chapter 8, you may wish to use Survey Test 3 on pages 7–8 in the *Teacher's ResourceBank* to assess student competency with the basic algebra skills related to polynomials covered in *Holt Introductory Algebra 1.*

Additional practice with these skills can be found on pages 15–16 in the *Teacher's ResourceBank.*

Assignment Guide
Basic p. 209: 1–33 odd
Average p. 209: 1–33 odd
Above Average p. 209: 3, 6, 9, ⋯, 33;
p. 212: 3, 6, 9, ⋯, 27, 28

210

Lesson Resources
Warm–Up: p. M–33
Maintenance: See below.
Practice Worksheet 8.2

Maintenance
1. Simplify: $(-pq)(5pt)$
   Ans: $-5p^2qt$ (Section 2.4)
2. Solve and check:
   $3(c - 1) + 8 = -10 + 2c$
   Ans: $c = -15$
   (Section 4.3)
3. Solve: $2^3 \cdot 2^x = 2^9$
   Ans: $x = 6$ (Section 6.1)
4. Multiply and simplify:
   $\sqrt{2} \cdot \sqrt{5} \cdot \sqrt{10}$
   Ans: 10 (Section 7.2)
5. The French Club washed 16 more compact cars than full-sized cars. They washed 34 compact cars. Find the number of full-sized cars that were washed.
   Ans: 18 (Section 3.7)

Additional Examples
Example 1
Multiply.
1. $(-8y^4)(7y^2)$
   Ans: $-56y^6$
2. $(3x^2)(-9x)$
   Ans: $-27x^3$

Example 2
Multiply.
1. $(-4a^2bc^3)(-3abc^4)$
   Ans: $12a^3b^2c^7$
2. $(5xy^3)(-8x^2y^5)$
   Ans: $-40x^3y^8$

Example 3
Multiply.
1. $(-2xy^2)(-3x^2)(-5y^3)$
   Ans: $-30x^3y^5$
2. $(\frac{2}{3}a^3)(-12b^2)(3a^2b^2)$
   Ans: $-24a^5b^4$

8.2 Operations with Monomials

To divide two monomials

 Multiply.

Some students will forget and try to multiply exponents.

a. $x^3 \cdot x^4$ **b.** $y^2 \cdot y^9$ **c.** $r \cdot r^5$

$\qquad x^7 \qquad\qquad\quad y^{11} \qquad\qquad r^6$

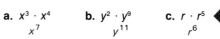

Remember to add exponents.

EXAMPLE 1   Multiply: $(3x^2)(-2x^3)$
Solution:
☐1 Commutative and Associative Properties
of Multiplication ⟶ $(3 \cdot -2)(x^2 \cdot x^3)$

☐2 Simplify. ⟶ $-6x^5$

Product Property of Powers

Monomials of more than one variable are shown below.

    a. $2x^3y$ **b.** $-3r^2st$ **c.** $\frac{1}{2}p^3q^2$

EXAMPLE 2   Multiply: $(8st^2)(-7r^2st)$
Solution:
☐1 Commutative and Associative Properties
of Multiplication ⟶ $(8 \cdot -7)(r^2)(s \cdot s)(t^2 \cdot t)$

☐2 Simplify. ⟶ $-56r^2s^2t^3$

P–2 **What is the simplest form of $4 \cdot -5 \cdot -\frac{1}{2}$?** 10

P–3 **What is the simplest form of $(rs)(s^2)(r)$?** r^2s^3

EXAMPLE 3   Multiply: $(4rs)(-5s^2)(-\frac{1}{2}r)$
Solution:
☐1 Commutative and Associative Properties
of Multiplication ⟶ $(4 \cdot -5 \cdot -\frac{1}{2})(r \cdot r)(s \cdot s^2)$

☐2 Simplify. ⟶ $10r^2s^3$

> **Steps for multiplying two or more monomials**
>
> 1. Multiply the numerical factors.
> 2. Add exponents of powers of the same base.

$$(6x^2y)(-7xy^3) = -42x^3y^4$$

NOTE: Remind students that monomials do not need to have common variables in order to multiply them.
e.g. $(-2x^2)(4y) = -8x^2y$

P-4   **Simplify each quotient.**

a. $\dfrac{x^5}{x^2}\ x^3$ **b.** $\dfrac{y^4}{y}\ y^3$ **c.** $\dfrac{r^6}{r^5}$
r

Remember to subtract exponents.

EXAMPLE 4   Divide: $\dfrac{-36n^8}{9n^5}$

Solution:

① Product Rule of Fractions   ⟶ $\dfrac{-36}{9} \cdot \dfrac{n^8}{n^5}$

② Simplify.   ⟶ $-4n^3$

Quotient Property of Powers

EXAMPLE 5   Divide: $\dfrac{-7r^3s^2}{-4rs}$

Solution:

① Product Rule of Fractions   ⟶ $\dfrac{-7}{-4} \cdot \dfrac{r^3}{r} \cdot \dfrac{s^2}{s}$

② Simplify.   ⟶ $\dfrac{7}{4}r^2s$ or $1\tfrac{3}{4}r^2s$

> **Steps for dividing monomials**
>
> 1. Divide the numerical factors.
> 2. Subtract exponents of powers of the same base.

$$\dfrac{56a^4bc^3}{-7a^2b} = -8a^2c^3$$

Additional Examples
Example 4
Divide.
1. $\dfrac{-24x^4}{3x}$
    Ans: $-8x^3$
2. $\dfrac{15a^6}{-3a^2}$
    Ans: $-5a^4$
3. $\dfrac{-3.6y^9}{1.2y^3}$
    Ans: $-3y^6$

Example 5
Divide.
1. $\dfrac{-15a^2b^3}{-5ab^3}$
    Ans: $3a$
2. $\dfrac{50x^4y^7}{30x^2y^2}$
    Ans: $\tfrac{5}{3}x^2y^5$ or $1\tfrac{2}{3}x^2y^5$

Common Error
Students multiply exponents when multiplying monomials.

Example
$(4st^2)(2s^2t^3) = 8s^2t^6$

Prescription
Review the Product Property of Powers on page 145 and the steps for multiplying two or more monomials on page 211. Emphasize that the exponents are to be added.

▨▨▨ CLASSROOM EXERCISES ▨▨▨

Multiply. (Examples 1–3)

1. $(4y^2)(5y)$ $20y^3$

2. $(-r^3)(2r^2)$ $-2r^5$

3. $(-6x^2)(7x^3)$ $-42x^5$

4. $(-8t^5)(-4t)$ $32t^6$

5. $(16xy)(-\tfrac{1}{2}x^2y)$ $-8x^3y^2$

6. $(2x)(3x)(5x)$ $30x^3$

7. $(-x)(2x^2)(3y)$ $-6x^3y$

8. $(r^2)(-4rs)(-s^3)$ $4r^3s^4$

9. $(2xy)(3x^2)(4xy^2)$ $24x^4y^3$

Polynomials / **211**

Divide. (Examples 4 and 5)

10. $\dfrac{12t^5}{-3t^2}$ $-4t^3$ 　　**11.** $\dfrac{-9x^4}{-x^2}$ $9x^2$ 　　**12.** $\dfrac{-15x^3}{5x}$ $-3x^2$ 　　**13.** $\dfrac{14b^6}{-7b^2}$ $-2b^4$

14. $\dfrac{-63r^5s^4}{-9rs^3}$ $7r^4s$ 　　**15.** $\dfrac{-54s^5t^3}{6st^2}$ $-9s^4t$ 　　**16.** $\dfrac{18r^2s^3t}{6rs}$ $3rs^2t$ 　　**17.** $\dfrac{-48p^3qr^5}{6pr^4}$ $-8p^2qr$

▨ WRITTEN EXERCISES ▨

Goals: To multiply and to divide monomials

Examples: a. $(-12r^3s^2)(5r^2s)$ 　　**b.** $\dfrac{-24x^3y}{-8xy}$ 　　**Answers: a.** $-60r^5s^3$ 　　**b.** $3x^2$

Multiply. (Examples 1–3)

1. $(8m^3)(9m^2)$ $72m^5$ 　　**2.** $(7a^4)(3a^3)$ $21a^7$ 　　**3.** $(-12r^3t^2)(8t^3)$ $-96r^3t^5$

4. $(15xy^4)(-4x^2)$ $-60x^3y^4$ 　　**5.** $(-\frac{1}{3}pq^3)(-15p^3q^2)$ $5p^4q^5$ 　　**6.** $(-20s^4t^2)(-\frac{1}{4}st^4)$ $5s^5t^6$

7. $(4t^2)(-3t)(2t)$ $-24t^4$ 　　**8.** $(5n)(3n^2)(-3n^3)$ $-45n^6$ 　　**9.** $(3mn)(-7n^2)(-m^2n)$

10. $(-4rt^2)(2rt)(-t^2)$ 　　**11.** $(-3ab)(-3ab^2)(-3a^2b^3)$ 　　**12.** $(-2q^2r)(-2qr^2)(-2qr^3)$

13. $(0.4rs^2t)(-3.8rst^3)$ 　　**14.** $(-2.8a^2bc^3)(7ab^2c)$ 　　**15.** $(-6xy^2z^3)(-1.4u^3x^2y)$

Divide. (Examples 4 and 5)

16. $\dfrac{-20y^5}{4y^2}$ $-5y^3$ 　　**17.** $\dfrac{12k^4}{-3k}$ $-4k^3$ 　　**18.** $\dfrac{-32r^2}{-8r}$ $4r$

19. $\dfrac{-36y^3}{-9y}$ $4y^2$ 　　**20.** $\dfrac{9.6p^6q^3}{-6p^2q}$ $-1.6p^4q^2$ 　　**21.** $\dfrac{-4.8m^8n^5}{8m^2n^3}$

22. $\dfrac{-13pq^3r}{-qr}$ $13pq^2$ 　　**23.** $\dfrac{-29a^4bc^3}{-a^3c}$ $29abc^2$ 　　**24.** $\dfrac{-17r^5s^6t^9}{-5r^2s^2t^3}$

25. $\dfrac{22x^{12}y^8z^6}{-7x^4y^2z^2}$ 　　**26.** $\dfrac{-5k^7p^{10}y^5}{12k^3py^4}$ $-\frac{5}{12}k^4p^9y$ 　　**27.** $\dfrac{-4m^{12}n^7q^3}{-15m^3n^5q}$

NON-ROUTINE PROBLEM

28. A five-digit number contains the digits shown at the right. Use the clues below to find the number.

Clue 1: The number is less than 30,000.
Clue 2: The number is even.
Clue 3: The 0 is next to the 1.
Clue 4: The 3 is not next to the 2 or 0.

1　　4
0
3　　2

Assignment Guide
Basic p. 212: 1–27 odd
Average p. 212: 1–27 odd, 28
Above Average p. 212:
3, 6, 9, ···, 27, 28;
p. 209: 3, 6, 9, ···, 33

Problem-Solving Skills
Using guess and check (Ex. 28)
Using logical reasoning (Ex. 28)

Non-Routine Problems
For a description of non-routine problems, see page M–13.

Additional Answers
Written Exercises
9. $21m^3n^4$
10. $8r^2t^5$
11. $-27a^4b^6$
12. $-8q^4r^6$
13. $-1.52r^2s^3t^4$
14. $-19.6a^3b^3c^4$
15. $8.4u^3x^3y^3z^3$
21. $-0.6m^6n^2$
24. $\frac{17}{5}r^3s^4t^6$
25. $-\frac{22}{7}x^8y^6z^4$
27. $\frac{4}{15}m^9n^2q^2$
28. Solutions may vary. Use the guess and check method to use the clues.
Clue 1: 21,340
Clue 2: 21,340
Clue 3: 23,410
Clue 4: 24,310

212

OBJECTIVES: To identify whether or not a given expression is a polynomial over {rational numbers}
To simplify a polynomial by combining like terms and arranging terms in descending order of powers

8.3 Polynomials

To evaluate a polynomial for values of its variable or variables

Lesson Resources
Warm–Up: p. M–33
Maintenance: See below.
Practice Worksheet 8.3
Manipulative 9, p. 556
Teaching Aids 1, 3, 4
Transparencies 20, 21, 46, 50, 51, 53, 54, 55, 56

P–1 **What monomials are in the following expression?**

$$2x^2, -9x, \frac{1}{4}$$

$2x^2 + (-9x) + \frac{1}{4}$ 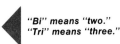 *A polynomial*

Definition

> A *polynomial* is a monomial or the sum of two or more monomials.
> $-1.8x^6$
> $-5x^4 + 2x^2 - 3x + 1$

The monomials that form a polynomial are called *terms*.

A *binomial* is a polynomial of two terms.

A *trinomial* is a polynomial of three terms.

 "Bi" means "two."
"Tri" means "three."

P–2 **Which polynomials below are binomials? trinomials?** b; c
a; d

a. $2x^2 + 6$ **b.** $x^2 - x + 1$

c. $\frac{1}{2} - \frac{1}{4}x^2 + \frac{3}{4}x$ **d.** $3.8x^2 - 1.3x$

The table below shows three polynomials expressed in simplest form.

Polynomials	Simplest Form
$3x^2 - 2x - x + 7$	$3x^2 - 3x + 7$
$5x - x^3 - 10 + 2x^2$	$-x^3 + 2x^2 + 5x - 10$
$1.3x^2 + 0.6 - x^4 + 5.6x^4 - x$	$4.6x^4 + 1.3x^2 - x + 0.6$

This order of terms is called "descending order of powers."

EXAMPLE 1   Simplify $2x^2 - 2x^3 + 3x - 5x - x^2 - 12 + x^4$.

Solution:
1. Combine like terms. $2x^2 - 2x^3 + 3x - 5x - x^2 - 12 + x^4 = x^2 - 2x^3 - 2x - 12 + x^4$
2. Arrange terms in order by exponents. ⟶ $= x^4 - 2x^3 + x^2 - 2x - 12$

Polynomials of more than one variable are shown below.

a. x^2y **b.** $3r^2st + 5rst^2$ **c.** $4x^3y^2 - 2x^2y + 5xy^3$

*Polynomials / **213***

Maintenance
1. Multiply and simplify:
   $-2(-3a + 2)$
   Ans: 6a − 4 (Section 2.5)
2. Solve: $-16 > -18 - t$
   Ans: {all real numbers greater than −2}
   (Section 5.2)
3. Solve: $12y - 8 > 5(11 + y)$
   Ans: {all real numbers greater than 9}
   (Section 5.6)
4. Solve: $\frac{7^5}{7^x} = 7^1$
   Ans: x = 4 (Section 6.2)
5. An estimate for building a sun deck showed that the cost of labor was 3 times the cost of the materials (Condition 1). The estimated total amount for labor and materials was $1,299.44 (Condition 2). Find the estimated cost for labor and for materials.
   Ans. materials: $324.86; labor: $974.58
   (Section 4.1)

Additional Examples
Example 1
Simplify.
1. $7x + x^2 - 9x + 6 + 5x^2$
   Ans: 6x² − 2x + 6
2. $4x^3 - 8x + 7x^3 - 11 - 8x + 5$
   Ans: 11x³ − 16x − 6
3. $2x^2 - x^5 + 7x^3 + 7x^2 + x^5 - x$
   Ans: 7x³ + 9x² − x

Additional Examples
Example 2
Simplify.
1. $2ab - 7a^2 + a - 6ab + 9a^2 - b$
   Ans: $2a^2 - 4ab + a - b$
2. $1.7x^2 - 4xy - 3.6y^2 + xy$
   Ans: $1.7x^2 - 3xy - 3.6y^2$

Example 3
Evaluate.
1. $x^2 - 6x - 9$ if $x = 3$
   Ans: -18
2. $5x^2 - x + 11$ if $x = -4$
   Ans: 95
3. $2x^3 + x^2 - 9x - 1$ if $x = -2$
   Ans: 5

Common Error
Students inadvertently omit one or more terms of a polynomial in the process of trying to simplify.

Example
$x^2 - 2x + 3x^2 - 5 + 2x^2 + 8 = 3x^2 - 2x + 3$

Prescription
Have students put a check mark over each item as it is used.

EXAMPLE 2   Simplify $1.3rs - 2.4r^2 + 3.8rs + 6.7s^2$.

Solution:

☐1 Combine like terms. ⟶ $1.3rs - 2.4r^2 + 3.8rs + 6.7s^2 = 5.1rs - 2.4r^2 + 6.7s^2$

☐2 Arrange in descending order of powers of r. ⟶ $= -2.4r^2 + 5.1rs + 6.7s^2$

OR

☐2 Arrange in descending order of powers of s. ⟶ $= 6.7s^2 + 5.1rs - 2.4r^2$

Since monomials represent real numbers in this course, polynomials also represent real numbers.

EXAMPLE 3   Evaluate $2x^2 - 5x + 7$ if $x = -2$.

Solution:
$$2x^2 - 5x + 7 = 2(-2)^2 - 5(-2) + 7$$
$$= 2(4) + 10 + 7$$
$$= 8 + 10 + 7$$
$$= 25$$

CLASSROOM EXERCISES

Write $\underline{Yes}$ or $\underline{No}$ to tell whether each numeral or expression is a polynomial over {rational numbers}. (Definition)

1. $3x + 5$ Yes
2. $-5x^2$ Yes
3. x^3 Yes
4. $\frac{1}{2}$ Yes
5. $-\frac{2}{3}x^3 + 3x^2$ Yes
6. $\frac{x}{2}$ Yes
7. $\frac{2}{x}$ No
8. $5x^0$ Yes
9. $3x^{-1}$ No
10. $100x^{100}$ Yes
11. $\sqrt{16} \cdot x - 1$ Yes
12. $0.3x^2$ Yes
13. $3x^3 + \frac{1}{2}x^2$ Yes
14. $\sqrt{2} \cdot x^2 + x$ No
15. 0 Yes

Simplify. (Examples 1 and 2)

16. $x^2 - 3x - 7x + 1$ $x^2 - 10x + 1$
17. $3 + 5x^2 - 2x$ $5x^2 - 2x + 3$
18. $x^2 - 3 + x - x^3 + 2x^4$ $2x^4 - x^3 + x^2 + x - 3$
19. $12 - 2x^3 + 5x^3$ $3x^3 + 12$

214 / *Chapter 8*

20. $8 - 3x + 7 - 2x^2$ $-2x^2 + 3x + 15$

21. $2x^2y + 3x^2y - 5$ $5x^2y - 5$

22. $4x^2y^2 - 1 - 3x^2y^2 - 8$ $x^2y^2 - 9$

23. $-3x^5 + 5x^8 - 3x^2 + x^{12}$

          $x^{12} + 5x^8 - 3x^5 - 3x^2$

Goals: To simplify and to evaluate polynomials

Sample Problems: a. $x^2 + y^2 + 5x^2$ **b.** Evaluate $-2x^3 + x$ if $x = -3$.

Answers: a. $6x^2 + y^2$ **b.** 51

Write Yes or No to show whether each numeral or expression is a polynomial over {rational numbers}. (Definition)

1. $2x - 5$ Yes

2. $4x + 3$ Yes

3. $x^2 + 3\pi$ No

4. $5x^2 - \sqrt{3}$ No

5. $\dfrac{x^2}{2} - \dfrac{3}{x} + 5$ No

6. $\dfrac{-x^2}{5} + \dfrac{x}{2} + 5$ Yes

7. $-3x^0$ Yes

8. $\frac{1}{3}x^5$ Yes

9. $1.3x^2 + 8.5x - 7.2$ Yes

10. $3x^{-2} + 4x^{-1} - 5^0$ No

Simplify. (Examples 1 and 2)

11. $3x^2 - x - 3x + 7$ $3x^2 - 4x + 7$

12. $-2x^3 + x^2 - 4x^2 + 5$

13. $5x - 3x^2 + x^5 - 1 + 4x^3$

14. $12x^2 - 5x + x^3 - 2x^4 + 3$

15. $7x^2 - 9x + 8 - 3x^2 - x^2$ $3x^2 - 9x + 8$

16. $15x^2 - x + 12 - 3x^2 - 6x$

17. $3x^3y - 5 + xy^3 + 5x^2y^2$

18. $6rs^3 - r^2s^2 + 3r^4s - 5s^4$

19. $m^2n + 5mn - 3m^2n + m^3 - 10$

20. $p^3q + 5q^3 + 3p^3q - pq^2$

21. $1.5x - 2.3x^4 + 0.9x - 1.9x^2 + 1.8x^4$

22. $0.3x^3 + 2.9x^5 - 5.2x + 6.3x - 1.8x^3$

Evaluate each polynomial. (Example 3)

23. $-x^2 + 3x - 5$ if $x = -3$ -23

24. $3x^2 - 4x + 5$ if $x = -4$ 69

25. $x^3 - x^2 + x - 1$ if $x = -2$ -15

26. $-x^3 + x^2 - x + 1$ if $x = -2$ 15

27. $2x^4 - x^3 + 3x + 2$ if $x = 2$ 32

28. $x^5 - 2x^4 - x^2 + 5$ if $x = 1$ 3

29. $x^2y^2 - 2xy + 1$ if $x = -2$ and $y = 3$ 49

30. $r^3s - r^2s^2 + 2$ if $r = -1$ and $s = 2$ -4

MORE CHALLENGING EXERCISES

Simplify.

31. $x^2 - (3x + 5)$ $x^2 - 3x - 5$

32. $2x^3 - (-x^2 + x)$ $2x^3 + x^2 - x$

33. $(x^2 + 4x) + (12 - 3x)$ $x^2 + x + 12$

34. $(x^3 - 2x^2) + (5x - 4x^2)$

          $x^3 - 6x^2 + 5x$

Polynomials   /   **215**

Assignment Guide
Basic p. 215: 1–29 odd
Average p. 215: 1–33 odd
Above Average p. 215:
1–29 odd, 31–34

Additional Answers
Written Exercises

12. $-2x^3 - 3x^2 + 5$

13. $x^5 + 4x^3 - 3x^2 + 5x - 1$

14. $-2x^4 + x^3 + 12x^2 - 5x + 3$

16. $12x^2 - 7x + 12$

17. $3x^3y + 5x^2y^2 + xy^3 - 5$

18. $3r^4s - r^2s^2 + 6rs^3 - 5s^4$

19. $m^3 - 2m^2n + 5mn - 10$

20. $4p^3q - pq^2 + 5q^3$

21. $-0.5x^4 - 1.9x^2 + 2.4x$

22. $2.9x^5 - 1.5x^3 + 1.1x$

APPLICATIONS

Polynomials can easily be evaluated with a calculator by grouping terms with parentheses in a special way.

$$5n^3 - 4n^2 + 12n - 17 = ((5n - 4)n + 12)n - 17$$

Have some students compare the efficiency of this method with evaluating the poly-nomial directly.

EXAMPLE Evaluate $5x^2 - 16x + 3$ if $x = 42$.

SOLUTION ☐1 Group terms in a special way. ⟶ $5x^2 - 16x + 3 = (5x - 16)x + 3$

☐2 Substitute and evaluate. ⟶ $= (5 \cdot 42 - 16)42 + 3$

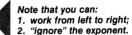

| 5 | × | 4 | 2 | − | 1 | 6 |

| × | 4 | 2 | + | 3 | = |

```
8151.
```

◄ Note that you can:
1. **work from left to right;**
2. **"ignore" the exponent.**

Evaluate each polynomial for the given value of x.

35. $120x^2 - 75x + 29$; $x = 24$ **36.** $76x^2 + 128x - 97$; $x = 29$ **37.** $x^3 - 2x^2 + 3x - 4$; $x = 12$
67, 349   67, 531   1472

■■■■ **MID-CHAPTER REVIEW** ■■■■

Evaluate each monomial. (Section 8.1)

1. $-4x^3$ if $x = -2$ **2.** $\frac{1}{2}x^2$ if $x = \frac{1}{2}$ $\frac{1}{8}$ **3.** $0.5x^4$ if $x = -2$ 8   **4.** $-(x^5)$ if $x = -1$ ^{1}
32

Multiply or divide. (Section 8.2)

5. $(-3rt)(2r^2)$ **6.** $(-10m^2n)(mn)$ **7.** $\dfrac{-30s^2t}{6st}$ $-5s$ **8.** $\dfrac{-48m^3n^2}{-8mn}$
$-6r^3t$ $-10m^3n^2$ $6m^2n$

Simplify. (Section 8.3)

9. $-x + 3 - 2x^2 + 7x - x^2 + 9$ **10.** $-5x^3 - 12 + 3x - x^3 + 2x^2 - 8x$

11. $-1.9x^2 + 0.2x^3 - 1.3x^2 + 2.8x - 1.4$ **12.** $0.9x - 1.8x^4 - 1.3x + 4.3x^3 - 2.7x^4$

Evaluate for the given value of the variable. (Section 8.3)

13. $r^2 - 2r + 5$; $r = -3$ **14.** $-5x^3 - x^2 + 2x - 3$; $x = 2$ **15.** $-p^2 + 7p - 10$; $p = -10$
20   −43   −180

REVIEW CAPSULE FOR SECTION 8.4

Arrange the terms of each polynomial in descending order of powers.
(Pages 213–215)

1. $2x - 3x^2$ **2.** $-4x - 3x^3$ **3.** $12 - 6x^2 + 4x$ **4.** $7 - 5x$ **5.** $3x - 5 + x^2$

216 / *Chapter 8*

Quiz Sections 8.1–8.3
After completing this Mid-Chapter Review, you may want to administer a quiz covering the same sections. See the quiz provided on page 171 in the *Teacher's ResourceBank.*

Additional Answers
Mid–Chapter Review
9. $-3x^2 + 6x + 12$
10. $-6x^3 + 2x^2 - 5x - 12$
11. $0.2x^3 - 3.2x^2 + 2.8x - 1.4$
12. $-4.5x^4 + 4.3x^3 - 0.4x$

Review Capsule
1. $-3x^2 + 2x$
2. $-3x^3 - 4x$
3. $-6x^2 + 4x + 12$
4. $-5x + 7$
5. $x^2 + 3x - 5$

OBJECTIVES: To add two or more polynomials arranged either vertically or horizontally

8.4 Addition and Subtraction

To subtract two polynomials arranged either vertically or horizontally

EXAMPLE 1   Add: $(x^2 - 3x + 4) + (3x^2 + 5x - 7)$

Solution:

① Write the sum
of all terms. ⟶ $(x^2 - 3x + 4) + (3x^2 + 5x - 7) = x^2 - 3x + 4 + 3x^2 + 5x - 7$

② Combine like terms. ⟶ $= 4x^2 + 2x - 3$ ◀ *Simplest form*

P-1 **Add: a. $(x^2 - 8x) + (7x + 3)$**   **b. $(3x^2 - 5) + (4x - x^2)$**
$x^2 - x + 3$ $2x^2 + 4x - 5$

Polynomials can be added by arranging them vertically in columns.

EXAMPLE 2   Add: $(3x^3 - x^2 + 4) + (3x^2 - 7x - 9) + (-x^3 + 3x - 6)$

Solution:

$$
\begin{array}{r}
3x^3 - x^2 \quad\ + 4 \\
3x^2 - 7x - 9 \\
(+) \underline{-x^3 \qquad + 3x - 6} \\
2x^3 + 2x^2 - 4x - 11
\end{array}
$$

◀ *Leave space for the missing term, and write in the + or − between terms.*

> The sum of any two polynomials is a polynomial.

EXAMPLE 3   Subtract: $(x^2 - 12x) - (3x^2 + 7x - 5)$

Solution:

① Meaning of
subtraction ⟶ $(x^2 - 12x) - (3x^2 + 7x - 5) = (x^2 - 12x) + -(3x^2 + 7x - 5)$

② Property of the Opposite of a Sum ⟶ $= x^2 - 12x - 3x^2 - 7x + 5$

③ Simplify. ⟶ $= -2x^2 - 19x + 5$

P-2 **Subtract: a. $(2x^2 + 6x) - (2x - 3)$**   **b. $(x^3 - 2x^2) - (-5x + 1)$**
$2x^2 + 4x + 3$ $x^3 - 2x^2 + 5x - 1$

P-3 **Subtract.**

a. $x^3 - (-3x^3)$ b. $0 - 2x^2$ c. $(-4x) - (-3x)$ d. $7 - (-1)$
$4x^3$ $-2x^2$ $-x$ 8

Remind students to subtract by adding the opposite of the subtrahend.

In adding or subtracting polynomials vertically, you arrange the terms in descending order of powers.

*Polynomials / **217***

Teaching Suggestions p. M–32

Lesson Resources
Warm–Up: p. M–33
Maintenance: See below.
Practice Worksheet 8.4
Manipulative 10, p. 558
Teaching Aids 1, 3, 4
Transparencies 20, 22, 47, 50, 51, 53, 54, 55, 56

Maintenance
1. Simplify:
   $-12y + 3x - 4y + 2x$
   Ans: 5x − 16y
   (Section 2.6)
2. Solve:
   $5x - 7 - 2x < 3x - 5$
   Ans: {real numbers}
   (Section 5.4)
3. Simplify: $10^0 - 5^0$
   Ans: 0   (Section 6.2)
4. Simplify: $-\sqrt{450}$
   Ans: −15 $\sqrt{2}$ (Section 7.3)
5. A jogging course is along a triangular route. The three legs of the course in order from the starting point have lengths in the ratio 2:4:5 (Condition 1). The total length of the course is 8.8 miles (Condition 2). Find the length of each leg.
   Ans: 1.6 miles; 3.2 miles; 4 miles   (Section 4.2)

Additional Examples
Example 1
Add:
$(7x^2 - 3x - 9) +$
$(x^2 + 6x - 1)$
Ans: 8x² + 3x − 10

Example 2
Add:
$(x^2 - 9) + (7x - 3) +$
$(4x^2 + 11x)$
Ans: 5x² + 18x − 12

Example 3
Subtract:
$(4x^2 - 8x) - (2x^2 - 3x + 7)$
Ans: 2x² − 5x − 7

217

Subtract.
1. $3x^2 + 9x + 2$
   $(-)-4x^2 - 3x - 3$
   Ans: $7x^2 + 12x + 5$
2. $7x^3\qquad + 6x - 3$
   $(-)\ x^3 + 2x^2 - 9x - 4$
   Ans: $6x^3 - 2x^2 + 15x + 1$

Example 5
Find the perimeter of each polygon.
1.

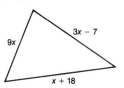

Ans: $13x + 11$
2.

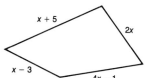

Ans: $8x + 1$

Common Error
When subtracting polynomials, students forget to distribute the subtraction to all of the terms in the subtrahend.

Example
$(2x^2 - 12x + 4) -$
$(x^2 + 3x + 5) =$
$x^2 - 12x + 4 - x^2 + 3x + 5$

Prescription
Have students practice writing the opposites of polynomials. For example,
$-(x^2 + 3x + 5) =$
$-x^2 - 3x - 5$

218

EXAMPLE 4   Subtract: $(7 + x^3 - 4x) - (2x^2 - 1 - 3x^3 - 3x)$

Solution:
Explain that
$(-)$ means to
subtract.

$$\begin{array}{r} x^3 \qquad - 4x + 7 \\ (-)\ \underline{-3x^3 + 2x^2 - 3x - 1} \\ 4x^3 - 2x^2 - \ x + 8 \end{array}$$

 Leave space for the x^2-term.

Compare the monomials in the answer to Example 4 with your answers to P-3.

> The difference of any two polynomials is a polynomial.

In Example 5, the lengths of the sides of a polygon are represented by polynomials.

EXAMPLE 5   Find the perimeter of the polygon at the left. Write the perimeter as a polynomial in simplest form.

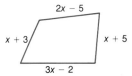

Solution: The perimeter of a polygon is the sum of the lengths of its sides.

[1] Write the sum of the polynomials. ⟶ $(x + 3) + (2x - 5) + (x + 5) + (3x - 2)$

[2] Combine like terms. ⟶ $(x + 2x + x + 3x) + (3 - 5 + 5 - 2)$

[3] Simplify. ⟶ $7x + 1$

The polynomial $7x + 1$ represents the perimeter.

▨▨ CLASSROOM EXERCISES ▨▨

Add. (Example 2)

1. $4x - 9$
   $\underline{3x + 2}$
   $7x - 7$

2. $-x + 15$
   $\underline{5x - \ 7}$
   $4x + 8$

3. $2x^2 - \ x + 3$
   $-x^2 \qquad - 4$
   $\underline{-3x^2 - 4x}$
   $-2x^2 - 5x - 1$

Subtract. (Example 4)

4. $6x + 12$
   $(-)\ \underline{5x + \ 3}$
   $x + 9$

5. $3x^2 + 6x + 4$
   $(-)\ \underline{\ x^2 + 2x + 3}$
   $2x^2 + 4x + 1$

6. $4x^2 \qquad - 1$
   $(-)\ \underline{6x^2 - x + 5}$
   $-2x^2 + x - 6$

Add or subtract as indicated. (Examples 1 and 3)

7. $(3x^2 - 2x) + (5x - 3)$
   $3x^2 + 3x - 3$

8. $(-7x^2 + 8x) - (x + 5)$
   $-7x^2 + 7x - 5$

9. $(2x^2 - 3x) - (3x^2 + x)$
   $-x^2 - 4x$

WRITTEN EXERCISES

Goals: To add and subtract polynomials

Sample Problem: $(-4x^2 + 7) + (9x - 12) - (x - x^2)$ **Answer:** $-3x^2 + 8x - 5$

Add. (Example 2)

1. $\begin{aligned}14x^2 - 9x + 4 \\ -3x^2 + 7x - 12\end{aligned}$ $11x^2 - 2x - 8$

2. $\begin{aligned}-10x^2 + 11x - 6 \\ 7x^2 - 9x + 13\end{aligned}$ $-3x^2 + 2x + 7$

3. $\begin{aligned}3.2x^2 + 0.5x - 1.8 \\ -1.7x^2 + 2.4x + 2.6\end{aligned}$ $1.5x^2 + 2.9x + 0.8$

4. $\begin{aligned}-4.1x^2 + 0.9x - 3.3 \\ 2.7x^2 - 2.4x - 1.6\end{aligned}$ $-1.4x^2 - 1.5x - 4.9$

5. $\begin{aligned}x^3 \qquad - 6x + 10 \\ -2x^3 + x^2 + 3x - 7 \\ -x^3 - 5x^2 \qquad + 6\end{aligned}$ $-2x^3 - 4x^2 - 3x + 9$

6. $\begin{aligned}-23x^3 + 14x^2 \qquad - 11 \\ x^3 - 9x^2 \qquad + 17 \\ -8x^3 \qquad + 15x - 2\end{aligned}$

$-30x^3 + 5x^2 + 15x + 4$

Subtract. (Example 4)

7. $\begin{aligned}x^2 + 5x - 10 \\ (-)\ 3x^2 - x + 8\end{aligned}$ $-2x^2 + 6x - 18$

8. $\begin{aligned}-4x^2 + x - 9 \\ (-)\ x^2 - 3x + 8\end{aligned}$ $-5x^2 + 4x - 17$

9. $\begin{aligned}2x^3 \qquad + 6x - 3 \\ (-)\ -x^3 + 5x^2 + 9x - 5\end{aligned}$ $3x^3 - 5x^2 - 3x + 2$

10. $\begin{aligned}3x^3 + 8x^2 \qquad - 5 \\ (-)\ 4x^3 \qquad + 12x - 13\end{aligned}$

$-x^3 + 8x^2 - 12x + 8$

MIXED PRACTICE   *Add or subtract as indicated.*

11. $(5x^2 - 3x + 12) + (3 - x^2 - 4x)$

12. $(8 - 6x - 4x^2) + (5x^2 - x - 13)$

13. $(x^3 - 2x + 5) - (4x^3 - 5x^2 + x - 2)$ $-3x^3 + 5x^2 - 3x + 7$

14. $(8x^3 - 2x^2 + 3x - 4) - (x^3 + 5x^2 - x + 7)$ $7x^3 - 7x^2 + 4x - 11$

15. $(4x^3 - 5x^2 - 6) - (2x^2 + 3x + 10) - (-3x^3 + x - 7)$ $7x^3 - 7x^2 - 4x - 9$

16. $(12x^2 - x - 10) - (x^3 + 4x^2 + 8) - (-3x^3 + 5x + 7)$ $2x^3 + 8x^2 - 6x - 25$

17. $(1.3x^2 - 2.7x - 0.5) + (2.5x^2 + 4.3x + 2.3) - (1.7x^2 - 2.8x - 4.9)$

18. $(0.9x^2 + 4.1x - 2.8) - (1.4x^2 - 2.7x + 5.1) + (-3.2x^2 + 4.5x - 1.8)$

APPLICATIONS   *Write a polynomial to represent each perimeter.* (Example 5)

19.
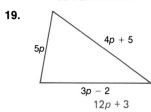
$12p + 3$

20.
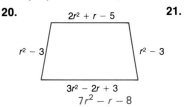
$7r^2 - r - 8$

21.
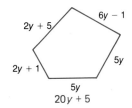
$20y + 5$

Polynomials / **219**

Assignment Guide
Basic p. 219: 1–21 odd
Average p. 219: 1–21 odd
Above Average p. 219: 1–21 odd

Additional Answers
11. $4x^2 - 7x + 15$
12. $x^2 - 7x - 5$
17. $2.1x^2 + 4.4x + 6.7$
18. $-3.7x^2 + 11.3x - 9.7$

Problem-Solving Skills
Using a diagram (Ex. 19–21)

220

Teaching Suggestions p. M–32

Lesson Resources
Warm–Up: p. M–33
Maintenance: See below.
Practice Worksheet 8.5
Manipulative 11, p. 560
Teaching Aids 1, 3, 4, 5
Transparencies 20, 23, 50, 51,
53, 54, 55, 57

Maintenance
1. Solve and check:
   $-7x - 3 + 4x = 12$
   Ans: $x = -5$ (Section 3.5)
2. Solve: $-18y < 27$
   **Ans: {all real numbers
   greater than $-\frac{3}{2}$}**
   (Section 5.5)
3. Write $\dfrac{x^3 \cdot x^{-4}}{x^6}$ as one power.
   Then simplify.
   Ans: $\dfrac{1}{x^7}$ (Section 6.3)
4. Simplify: $\sqrt{729}$
   Ans: 27 (Section 7.1)
5. Joel's scores on four tests
   were 75, 80, 85, and 90. He
   wants his mean score on five
   tests to be greater than 82
   (Condition 2). Find the lowest
   score Joel can get on the fifth
   test (Condition 1).
   Ans: 81 (Section 5.7)

Additional Examples
Example 1
Multiply.
1. $(x + 6)(x + 8)$
   Ans: $x^2 + 14x + 48$
2. $(2x - 1)(x + 7)$
   Ans: $2x^2 + 13x - 7$
3. $(4x - 3)(3x - 5)$
   Ans: $12x^2 - 29x + 15$

Example 2
Multiply.
1. $(x - 9)(x + 5)$
   Ans: $x^2 - 4x - 45$
2. $(4x + 3)(2x + 7)$
   Ans: $8x^2 + 34x + 21$
3. $(6x - 1)(2x - 3)$
   Ans: $12x^2 - 20x + 3$

8.5 Multiplication

OBJECTIVES: To multiply two binomials
To multiply a binomial and a quadratic trinomial
To multiply any two polynomials

P–1 Find each product.

a. $(-2x)(x)$
   $-2x^2$

b. $(3x)(-5)$
   $-15x$

c. $(-3)(-4)$
   12

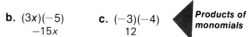 Products of monomials

EXAMPLE 1 Multiply: $(x + 3)(3x - 4)$  Two binomials

Solution:

1 Distributive Property $\longrightarrow$ $(x + 3)(3x - 4) = (x + 3)3x + (x + 3)(-4)$

2 Distributive Property $\longrightarrow$ $= x(3x) + 3(3x) + x(-4) + 3(-4)$

3 Simplest forms of products $\longrightarrow$ $= 3x^2 + 9x - 4x - 12$

4 Combine like terms. $\longrightarrow$ $= 3x^2 + 5x - 12$ ◀ Simplest form

In a polynomial, a term with no variables is called the **constant term.**

P–2 What is the constant term in each polynomial below?

a. $x^2 + 5$ 5   b. $3x^2 + 5x - 12$ -12 c. $x^3 - 7x + 9$ 9

Binomials can be multiplied by a short method.

The mnemonic device FOIL may help some students remember the method shown here.
F: First terms
O: Outside terms
I: Inside terms
L: Last terms

EXAMPLE 2 Multiply: $(5x - 2)(x - 3)$

| Multiply the first terms. | Multiply the outside terms. | Multiply the inside terms. | Multiply the last terms. |

Solution:

$(5x - 2)(x - 3)$ $(5x - 2)(x - 3)$ $(5x - 2)(x - 3)$ $(5x - 2)(x - 3)$

$5x^2$ $-15x$ $-2x$ 6

$(5x - 2)(x - 3) = 5x^2 - 15x - 2x + 6$
$= 5x^2 - 17x + 6$

P–3 **Find each product.**

a. $(2x + 1)(x + 4)$
$2x^2 + 9x + 4$

b. $(3x - 2)(2x + 3)$
$6x^2 + 5x - 6$

c. $(4x - 1)(3x - 5)$
$12x^2 - 23x + 5$

d. $(2x + 5)(x - 4)$
$2x^2 - 3x - 20$

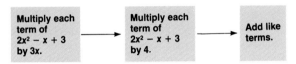
Try to combine like terms mentally.

Any two polynomials are multiplied by multiplying each term of one polynomial by each term of the other.

EXAMPLE 3   Multiply: $(3x + 4)(2x^2 - x + 3)$

| Multiply each term of $2x^2 - x + 3$ by $3x$. | → | Multiply each term of $2x^2 - x + 3$ by 4. | → | Add like terms. |

Solution: **Method 1**

$$(3x + 4)(2x^2 - x + 3) = 3x(2x^2) + 3x(-x) + 3x(3) + 4(2x^2) + 4(-x) + 4(3)$$
$$= 6x^3 - 3x^2 + 9x + 8x^2 - 4x + 12$$
$$= 6x^3 + 5x^2 + 5x + 12$$

Method 2

$$
\begin{array}{r}
2x^2 - x + 3 \\
3x + 4 \\
\hline
6x^3 - 3x^2 + 9x \\
8x^2 - 4x + 12 \\
\hline
6x^3 + 5x^2 + 5x + 12
\end{array}
$$

Multiply by 3x. *Multiply by 4.*

EXAMPLE 4   Multiply: $(x - 3)(5 + 3x^3 - x^2)$

Solution:
$$
\begin{array}{r}
3x^3 - x^2 + 5 \\
x - 3 \\
\hline
3x^4 - x^3 + 5x \\
- 9x^3 + 3x^2 - 15 \\
\hline
3x^4 - 10x^3 + 3x^2 + 5x - 15
\end{array}
$$

Arrange terms in order. Leave space for x term.

Columns of like terms

> The product of any two polynomials is a polynomial.

Polynomials / 221

EXAMPLE 5

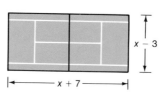

Write a polynomial to represent the area of the tennis court. Ask students what restrictions there would be on the domain of the variable in order for $x - 3$ and $x + 7$ to represent lengths.

Solution: The area of a rectangle equals the product of its length and width.

1. Represent the length and the width. ⟶ Length = $x + 7$; width = $x - 3$
2. Multiply. ⟶ $(x + 7)(x - 3) = x^2 - 3x + 7x - 21$
   $$= x^2 + 4x - 21 \quad \blacktriangleleft \; \textbf{Area}$$

CLASSROOM EXERCISES

Multiply. (Example 2)

1. $(x + 1)(x + 2)$ $x^2 + 3x + 2$
2. $(x + 3)(x + 4)$ $x^2 + 7x + 12$
3. $(x - 1)(x + 3)$ $x^2 + 2x - 3$
4. $(x + 2)(x - 3)$ $x^2 - x - 6$
5. $(x - 2)(x - 4)$ $x^2 - 6x + 8$
6. $(x - 3)(x - 7)$ $x^2 - 10x + 21$
7. $(2x + 1)(x + 2)$ $2x^2 + 5x + 2$
8. $(3x - 1)(x + 2)$ $3x^2 + 5x - 2$
9. $(4x + 1)(x - 3)$ $4x^2 - 11x - 3$
10. $(2x - 1)(2x - 1)$ $4x^2 - 4x + 1$

Write the missing row in each multiplication problem. (Examples 3–4)

11.
$$
\begin{array}{r}
2x + 5 \\
\underline{x - 6} \\
\boxed{} \quad 2x^2 + 5x \\
\underline{-12x - 30} \\
2x^2 - 7x - 30
\end{array}
$$

12.
$$
\begin{array}{r}
-3x - 4 \\
\underline{2x + 1} \\
-6x^2 - 8x \\
\underline{\boxed{}} \quad -3x - 4 \\
-6x^2 - 11x - 4
\end{array}
$$

13.
$$
\begin{array}{r}
x^2 - 2x + 3 \\
\underline{x - 2} \\
x^3 - 2x^2 + 3x \\
\underline{-2x^2 + 4x - 6} \\
\boxed{} \quad x^3 - 4x^2 + 7x - 6
\end{array}
$$

14.
$$
\begin{array}{r}
x^3 - 2x^2 + 5 \\
\underline{2x - 1} \\
2x^4 - 4x^3 + 10x \\
\underline{-x^3 + 2x^2 - 5} \\
2x^4 - 5x^3 + 2x^2 + 10x - 5
\end{array}
$$

WRITTEN EXERCISES

Goal: To multiply a polynomial by a polynomial

Sample Problem: $(5x^2 - 2x + 3)(x + 2)$ **Answer:** $5x^3 + 8x^2 - x + 6$

Multiply. (Examples 1 and 2)

1. $(x + 3)(x + 5)$
   $x^2 + 8x + 15$
2. $(x + 2)(x + 7)$
   $x^2 + 9x + 14$
3. $(x - 4)(x + 2)$
   $x^2 - 2x - 8$
4. $(x - 3)(x + 1)$
   $x^2 - 2x - 3$

222 / Chapter 8

5. $(x - 5)(x - 3)$ $x^2 - 8x + 15$

6. $(x - 2)(x - 4)$ $x^2 - 6x + 8$

7. $(x + 8)(x - 5)$ $x^2 + 3x - 40$

8. $(x + 10)(x - 6)$ $x^2 + 4x - 60$

9. $(2x + 3)(x + 5)$ $2x^2 + 13x + 15$

10. $(3x + 2)(x + 4)$ $3x^2 + 14x + 8$

11. $(3x - 1)(x + 4)$ $3x^2 + 11x - 4$

12. $(4x - 3)(x + 5)$ $4x^2 + 17x - 15$

13. $(5x - 2)(x - 3)$ $5x^2 - 17x + 6$

14. $(3x - 4)(x - 4)$ $3x^2 - 16x + 16$

15. $(4x + 5)(2x - 3)$ $8x^2 - 2x - 15$

16. $(2x + 5)(3x - 7)$ $6x^2 + x - 35$

Multiply. (Examples 3 and 4)

17. $x^2 - 2x + 3$
   $\underline{ x - 4}$

18. $x^2 + 3x - 2$
   $\underline{ x - 3}$

19. $2x^2 + 3x - 4$
   $\underline{ x + 5}$

20. $3x^2 - x + 4$
   $\underline{ x + 4}$

21. $x^2 - 5$
   $\underline{ x + 6}$

22. $x^2 - 4x$
   $\underline{ x - 5}$

23. $2x^2 - 3x + 7$
   $\underline{ 3x - 2}$

24. $3x^2 - 2x - 4$
   $\underline{ 2x + 3}$

25. $-5x^2 + 2x - 4$
   $\underline{ 4x + 5}$

26. $5x^3 + 4x^2 - 6$
   $\underline{ -2x + 5}$

27. $7x^3 + 6x^2 + 5$
   $\underline{ 6x + 3}$

28. $2x^3 -x + 2$
   $\underline{ 4x - 5}$

APPLICATIONS

Write a polynomial to represent each area.
(Example 5)

29. A rectangular parking lot has a length of $(x - 4)$ units and a width of $(x - 7)$ units. $x^2 - 11x + 28$

30. The length of a rectangular garden is $(2x + 3)$ units. The width of the garden is $(x + 4)$ units.

31. A rectangular swimming pool has a length of $(3x - 1)$ units and a width of $(x + 5)$ units. $3x^2 + 14x - 5$

32. Each side of a square piece of carpet has a length of $(x + 6)$ units.
$x^2 + 12x + 36$

REVIEW CAPSULE FOR SECTION 8.6

Subtract. (Pages 217–219)

1. $4x^2 - 3x + 2$
   $\underline{4x^2 + x }$
   $ -4x + 2$

2. $5x^2 - 4$
   $\underline{5x^2 + 2x }$
   $ -2x - 4$

3. $-2x^2 + 4x - 3$
   $\underline{-2x^2 - 4x }$
   $ 8x - 3$

4. $x^4 - 3x^2 + 6$
   $\underline{x^4 - 5x^2 - 2}$
   $ 2x^2 + 8$

Polynomials / **223**

Maintenance
1. Solve and check:
   $5x - 2x = 24 - 9x$
   Ans: $x = 2$ (Section 4.3)
2. Solve: $5x - 10 < 6x - 8$
   **Ans: {all real numbers
   greater than -2}**
   (Section 5.6)
3. Simplify: $\left(\dfrac{-a}{6}\right)^3$
   Ans: $-\dfrac{a^3}{6^3}$ (Section 6.4)
4. Add. $\sqrt{32} + 2\sqrt{18}$
   Ans: $10\sqrt{2}$ (Section 7.5)
5. A TV with a rectangular
   shaped 10–inch screen is on
   sale. The width of the screen
   is 8 inches. Find the height of
   the screen.
   Ans: 6 inches
   (Section 7.8)

Additional Examples
Example 1
Divide.
1. $(12x^3 - 3x^2 + 15x) \div 3x$
   Ans: $4x^2 - x + 5$
2. $(4x^3y + 12x^2y - 24xy) \div 4xy$
   Ans: $x^2 + 3x - 6$
3. $(14x^4 - 63x^3 - 42x^2) \div 7x^2$
   Ans: $2x^2 - 9x - 6$

Example 2
Divide.
1. $(3x^2 + 11x + 10) \div (x + 2)$
   Ans: $3x + 5$
2. $(2x^2 - 3x - 14) \div (x + 2)$
   Ans: $2x - 7$
3. $(6x^2 - 17x + 5) \div (2x - 5)$
   Ans: $3x - 1$

224

OBJECTIVES: To divide a polynomial of two or more terms by a monomial
and write the quotient and a zero remainder

8.6 Division

To divide a polynomial of two or more terms by a binomial and
write the quotient and a zero remainder

To divide a polynomial by a monomial, divide each term of the
polynomial by the monomial.

P–1 **Divide:** **a.** $\dfrac{12x^3}{2x}$ **b.** $\dfrac{8x^2}{2x}$ **c.** $\dfrac{4x}{2x}$ 2

$6x^2$ $4x$

EXAMPLE 1   Divide: $(12x^3 + 8x^2 - 4x) \div 2x$

*2x is the divisor.
$12x^3 + 8x^2 - 4x$ is
the dividend.*

Solution: $(12x^3 + 8x^2 - 4x) \div 2x = \dfrac{12x^3}{2x} + \dfrac{8x^2}{2x} - \dfrac{4x}{2x}$

$= 6x^2 + 4x - 2$

Examples 2 and 3 show division of a trinomial by a binomial.

EXAMPLE 2   Divide: $(2x^2 + x - 15) \div (x + 3)$

Explain that division
can be checked by
multiplying the
quotient and the
divisor.

Solution:

1. Divide $2x^2$, the first term of $2x^2 + x - 15$,
   by x, the first term of $x + 3$.

$$\begin{array}{r} 2x \\ x + 3 \overline{)2x^2 + x - 15} \end{array}$$

2. Multiply $(x + 3)$ by $2x$. Subtract
   the product from $2x^2 + x - 15$.

$$2x(x + 3) \longrightarrow \begin{array}{r} 2x \\ x + 3 \overline{)2x^2 + x - 15} \\ \underline{2x^2 + 6x} \\ -5x - 15 \end{array}$$

3. Divide $-5x$ by x. This is the
   second term of the quotient.

$$\begin{array}{r} 2x - 5 \\ x + 3 \overline{)2x^2 + x - 15} \\ \underline{2x^2 + 6x} \\ -5x - 15 \end{array}$$

4. Multiply $(x + 3)$ by -5. Subtract
   the product from $-5x - 15$.

$$-5(x + 3) \longrightarrow \begin{array}{r} 2x - 5 \\ x + 3 \overline{)2x^2 + x - 15} \\ \underline{2x^2 + 6x} \\ -5x - 15 \\ \underline{-5x - 15} \\ 0 \end{array}$$

The quotient is $2x - 5$.

P–2 **Divide:** a. $\dfrac{4x^3}{2x}$ b. $\dfrac{-2x^2}{2x}$ $-x$

$2x^2$

EXAMPLE 3   Divide: $(4x^3 - 3x - 1) \div (2x + 1)$

Solution:

1 Divide $4x^3$ by $2x$ to get the
first term of the quotient. ⟶

$$2x + 1 \overline{) \begin{array}{l} 2x^2 \\ 4x^3 - 3x - 1 \end{array}}$$

◀ Leave
space
for an
x^2 term.

2 Multiply $(2x + 1)$ by $2x^2$. Subtract
the product from $4x^3 - 3x - 1$. ⟶

$$2x + 1 \overline{) \begin{array}{l} 2x^2 \\ 4x^3 - 3x - 1 \end{array}}$$

$2x^2(2x + 1)$ ⟶ $\dfrac{4x^3 + 2x^2}{-2x^2 - 3x - 1}$

3 Divide $-2x^2$ by $2x$ for the second
term of the quotient. ⟶

$$2x + 1 \overline{) \begin{array}{l} 2x^2 - x \\ 4x^3 - 3x - 1 \end{array}}$$

$\dfrac{4x^3 + 2x^2}{-2x^2 - 3x - 1}$

4 Multiply $(2x + 1)$ by $-x$. Subtract
the product from $-2x^2 - 3x - 1$. ⟶

$$2x + 1 \overline{) \begin{array}{l} 2x^2 - x \\ 4x^3 - 3x - 1 \end{array}}$$

$\dfrac{4x^3 + 2x^2}{-2x^2 - 3x - 1}$

$-x(2x + 1)$ ⟶ $\dfrac{-2x^2 - x}{-2x - 1}$

5 Divide $-2x$ by $2x$ for the last term
of the quotient. Multiply
$(2x + 1)$ by -1 and subtract. ⟶

$$2x + 1 \overline{) \begin{array}{l} 2x^2 - x - 1 \\ 4x^3 - 3x - 1 \end{array}}$$

$\dfrac{4x^3 + 2x^2}{-2x^2 - 3x - 1}$

$\dfrac{-2x^2 - x}{-2x - 1}$

$-1(2x + 1)$ ⟶ $\dfrac{-2x - 1}{0}$

The quotient is $2x^2 - x - 1$.

Additional Examples
Example 3
Divide.
1. $(x^3 - 19x + 12) \div (x - 4)$
   Ans: $x^2 + 4x - 3$
2. $(4x^3 + x + 15) \div (2x + 3)$
   Ans: $2x^2 - 3x + 5$
3. $(8x^3 - 1) \div (2x - 1)$
   Ans: $4x^2 + 2x + 1$

Common Error
Students are confused because
division of polynominals appears
to involve only the first term of
the divisor and not the complete
divisor.

Example

$$x - 2 \overline{) \begin{array}{l} 2x \\ 2x^2 - 3x - 2 \end{array}}$$

Prescription
Show students this process is
similar to division of whole numbers. The procedure involves
discrete steps, and dividing by
the first term is similar to dividing
by the left-most digit with a
whole number divisor.

CLASSROOM EXERCISES

Divide. (Example 1)

1. $\dfrac{25x^4}{-5x^2}$ $-5x^2$ 2. $\dfrac{-20x^5}{4x}$ $-5x^4$ 3. $(15x^4 - 25x^2) \div 5x^2$ $3x^2 - 5$ 4. $(9x^3 + 6x^2) \div 3x^2$ $3x + 2$

Complete each division problem. (Examples 2–3)

5.
$$x - 2 \overline{)\, x^2 + 3x - 10}$$
$$\underline{x^2 - 2x}$$
quotient: $x + 5$

6.
$$x + 5 \overline{)\, x^2 - 7x - 60}$$
$$\underline{x^2 + 5x}$$
quotient: $x - 12$

7.
$$2x - 3 \overline{)\, 6x^2 - 5x - 6}$$
$$\underline{6x^2 - 9x}$$
quotient: $3x + 2$

8.
$$5x - 3 \overline{)\, 10x^2 - 31x + 15}$$
$$\underline{10x^2 - 6x}$$
quotient: $2x - 5$

Assignment Guide
Basic p. 226: 1–23 odd
Average p. 226: 1–25 odd
Above Average p. 226: 1–25 odd

Problem–Solving Skills
Using a diagram (Ex. 25)
Using guess and check (Ex. 25)
Using logical reasoning (Ex. 25)

Non–Routine Problems
For a description of non–routine problems, see page M–13.

Additional Answers
Written Exercises
12. $x^2 + 4x - 2$
14. $3x^2 + x + 1$
16. $2x - 5$
18. $x^2 + 2$
19. $2x^2 - 5x + 10$
20. $3x^2 + 4x - 7$
24. $x^3 + 2x^2 + 4x + 8$
25.

WRITTEN EXERCISES

Goal: To divide a polynomial by a binomial

Sample Problem: $(2x^3 + 3x^2 - 1) \div (2x - 1)$ **Answer:** $x^2 + 2x + 1$

Divide. (Example 1)

1. $(4x^2 + 6x) \div 2x$ $2x + 3$

2. $(9y^3 - 12y) \div 3y$ $3y^2 - 4$

3. $(18xy - 12x^2y^2) \div 6xy$ $3 - 2xy$

4. $(36x^3 + 12x) \div 12x$ $3x^2 + 1$

Divide. (Examples 2 and 3)

5. $x - 3 \overline{)\, 3x^2 - 5x - 12}$ $3x + 4$

6. $x + 2 \overline{)\, 4x^2 + 5x - 6}$ $4x - 3$

7. $2x - 1 \overline{)\, 10x^2 - 17x + 6}$ $5x - 6$

8. $3x - 1 \overline{)\, 12x^2 + 2x - 2}$ $4x + 2$

9. $4x + 1 \overline{)\, 12x^2 - 13x - 4}$ $3x - 4$

10. $2x + 3 \overline{)\, 8x^2 + 10x - 3}$ $4x - 1$

11. $x - 5 \overline{)\, x^3 - 8x^2 + 17x - 10}$ $x^2 - 3x + 2$

12. $x + 4 \overline{)\, x^3 + 8x^2 + 14x - 8}$

13. $2x + 3 \overline{)\, 4x^3 \qquad - 7x + 3}$ $2x^2 - 3x + 1$

14. $3x + 1 \overline{)\, 9x^3 + 6x^2 + 4x + 1}$

15. $x^2 - 2 \overline{)\, x^3 - 5x^2 - 2x + 10}$ $x - 5$

16. $x^2 + 3 \overline{)\, 2x^3 - 5x^2 + 6x - 15}$

17. $x^2 - 2x \overline{)\, x^4 - 2x^3 + 3x^2 - 6x}$ $x^2 + 3$

18. $x^2 - 5 \overline{)\, x^4 \qquad - 3x^2 \qquad - 10}$

19. $(10x^3 - 29x^2 + 60x - 20) \div (5x - 2)$

20. $(18x^3 + 9x^2 - 62x + 35) \div (6x - 5)$

21. $(x^3 - 8) \div (x - 2)$ $x^2 + 2x + 4$

22. $(x^3 + 27) \div (x + 3)$ $x^2 - 3x + 9$

23. $(x^4 - 81) \div (x - 3)$ $x^3 + 3x^2 + 9x + 27$

24. $(x^4 - 16) \div (x - 2)$

NON-ROUTINE PROBLEM

25. In a bowling alley, bowling pins are set up as shown in the figure at the right. Move exactly three of the pins to form a triangle exactly like the original triangle, but facing the opposite direction.

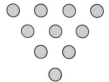

OBJECTIVE: To divide a polynomial of three or more terms by a binomial or a trinomial and write the quotient as a polynomial plus a fraction formed by the remainder and the divisor

8.7 Division with Remainders

Teaching Suggestions p. M–32

Lesson Resources
Warm–Up: p. 33
Maintenance: See below.
Practice Worksheet 8.7

| P-1 | **Divide:** $\dfrac{3x^2}{x}$. $3x$

EXAMPLE 1   Divide: $x - 4 \overline{\smash{)}3x^2 - 10x - 6}$

Solution:

To check this kind of division problem, multiply the quotient, $3x + 2$, and the divisor, $x - 4$, and add the remainder. The result should equal the dividend, $3x^2 - 10x - 6$.

1. Divide $3x^2$ by x. Multiply $(x - 4)$ by $3x$ and subtract.

$$\begin{array}{r} 3x \\ x - 4 \overline{\smash{)}3x^2 - 10x - 6} \\ 3x(x-4) \longrightarrow \underline{3x^2 - 12x} \\ 2x - 6 \end{array}$$

2. Divide $2x$ by x. Multiply $(x - 4)$ by 2 and subtract.

$$\begin{array}{r} 3x + 2 \\ x - 4 \overline{\smash{)}3x^2 - 10x - 6} \\ \underline{3x^2 - 12x} \\ 2x - 6 \\ 2(x-4) \longrightarrow \underline{2x - 8} \\ 2 \end{array}$$

The remainder is 2. The quotient can be written as $3x + 2 + \dfrac{2}{x - 4}$.

Since $\dfrac{2}{x - 4}$ is not a polynomial, the quotient in Example 1 is not a polynomial.

> The quotient of two polynomials is not necessarily a polynomial.

| P-2 | **Arrange $10x^2 - 12x - 8x^3 - 19$ in descending powers of x.**
$-8x^3 + 10x^2 - 12x - 19$

EXAMPLE 2   Divide: $(10x^2 - 12x - 8x^3 - 19) \div (2x^2 + 3)$

Solution:

1. Arrange terms in order. $\longrightarrow 2x^2 + 3 \overline{\smash{)}-8x^3 + 10x^2 - 12x - 19}$

2. Divide $-8x^3$ by $2x^2$. Multiply $(2x^2 + 3)$ by $-4x$ and subtract. $\longrightarrow$

$$\begin{array}{r} -4x \\ 2x^2 + 3 \overline{\smash{)}-8x^3 + 10x^2 - 12x - 19} \\ -4x(2x^2 + 3) \longrightarrow \underline{-8x^3 - 12x} \\ 10x^2 - 19 \end{array}$$

Polynomials / **227**

Maintenance
1. Solve and check:
   $x - 3 = 22 - 4x$
   Ans: $x = 5$ (Section 4.3)
2. Solve: $4(x + 2) > -9 + 5x$
   **Ans: {all real numbers
   less than 17}**
   (Section 5.6)
3. Simplify: $(y^{-3})^{-4}$
   Ans: y^{12} (Section 6.5)
4. Rationalize the
   denominator of $\dfrac{\sqrt{3}}{\sqrt{2}}$.
   Ans: $\dfrac{\sqrt{6}}{2}$ (Section 7.6)

5. A sailing course is in the shape of a right triangle. From the starting line the windward buoy is 48 miles due north. The reaching buoy is 14 miles due east of the windward buoy. Find the distance from the reaching buoy back to the starting line.
Ans: 50 miles
(Section 7.8)

Additional Examples
Example 1
Divide.
1. $x + 3 \overline{\smash{)}x^2 + 2x + 3}$
   Ans: $x - 1 + \dfrac{6}{x + 3}$
2. $2x + 1 \overline{\smash{)}10x^2 + x - 7}$
   Ans: $5x - 2 - \dfrac{5}{2x + 1}$

Example 2
Divide.
1. $x^2 + 4 \overline{\smash{)}3x^3 + 5x^2 + 12x + 21}$
Ans: $3x + 5 + \dfrac{1}{x^2 + 4}$
2. $2x^2 - 7 \overline{\smash{)}4x^3 - 6x^2 - 14x + 11}$
   Ans: $2x - 3 - \dfrac{10}{2x^2 - 7}$

3 Divide $10x^2$ by $2x^2$. Multiply $(2x^2 + 3)$ by 5 and subtract. $\longrightarrow$

$$2x^2 + 3 \overline{\smash{)}\, -8x^3 + 10x^2 - 12x - 19} \quad {}^{-4x+5}$$
$$\underline{-8x^3 \qquad\quad -12x}$$
$$10x^2 \qquad -19$$
$5(2x^2 + 3) \longrightarrow \quad \underline{10x^2 \qquad +15}$
$$-34$$

The quotient is written as $-4x + 5 - \dfrac{34}{2x^2 + 3}$. ◀ **Not a polynomial**

P–3 **Find the first term in the quotient.**

$$4x - 3y \overline{\smash{)}\, 8x^3 - 10x^2y - xy^2 + 2y^3} \quad 2x^2$$

Additional Examples
Example 3
Divide.
1. $3x^3 + 7x^2y - 2xy^2 - 3y^3 \div x + 2y$

   Ans: $3x^2 + xy - 4y^2 + \dfrac{5y^3}{x + 2y}$

2. $3x^3 - 17x^2y + 17xy^2 + 3y^3 \div 3x - 5y$

   Ans: $x^2 - 4xy - y^2 - \dfrac{2y^3}{3x - 5y}$

EXAMPLE 3   Divide: $4x - 3y \overline{\smash{)}\, 8x^3 - 10x^2y - xy^2 + 2y^3}$

Solution:

1 Divide $8x^3$ by $4x$. Multiply $(4x - 3y)$ by $2x^2$ and subtract. $\longrightarrow$

$$4x - 3y \overline{\smash{)}\, 8x^3 - 10x^2y - xy^2 + 2y^3} \quad {}^{2x^2}$$
$2x^2(4x - 3y) \longrightarrow \quad \underline{8x^3 - 6x^2y}$
$$-4x^2y - xy^2 + 2y^3$$

2 Divide $-4x^2y$ by $4x$. Multiply $(4x - 3y)$ by $-xy$ and subtract. $\longrightarrow$

$$4x - 3y \overline{\smash{)}\, 8x^3 - 10x^2y - xy^2 + 2y^3} \quad {}^{2x^2 - xy}$$
$$\underline{8x^3 - 6x^2y}$$
$$-4x^2y - xy^2 + 2y^3$$
$-xy(4x - 3y) \longrightarrow \quad \underline{-4x^2y + 3xy^2}$
$$-4xy^2 + 2y^3$$

3 Divide $-4xy^2$ by $4x$. Multiply $(4x - 3y)$ by $-y^2$ and subtract. $\longrightarrow$

$$4x - 3y \overline{\smash{)}\, 8x^3 - 10x^2y - xy^2 + 2y^3} \quad {}^{2x^2 - xy - y^2}$$
$$\underline{8x^3 - 6x^2y}$$
$$-4x^2y - xy^2 + 2y^3$$
$$\underline{-4x^2y + 3xy^2}$$
$$-4xy^2 + 2y^3$$
$-y^2(4x - 3y) \longrightarrow \quad \underline{-4xy^2 + 3y^3}$
$$-y^3$$

The quotient is $2x^2 - xy - y^2 - \dfrac{y^3}{4x - 3y}$.

228 / *Chapter 8*

CLASSROOM EXERCISES

Note: Q stands for quotient; R stands for remainder.

Write the quotient and remainder in each problem. (Examples 1 and 2)

1. $x + 2 \overline{)x^2 - 3x - 7}$ Q: $x - 5$

$\underline{x^2 + 2x}$ R: 3
$-5x - 7$
$\underline{-5x - 10}$

2. $x - 3 \overline{)x^2 + x + 5}$

$\underline{x^2 - 3x}$
$4x + 5$
$\underline{4x - 12}$

3. $3x - 1 \overline{)6x^2 + 4x - 3}$ Q: $2x + 2$

$\underline{6x^2 - 2x}$ R: -1
$6x - 3$
$6x - 2$

4. $2x + 1 \overline{)4x^2 - 6x + 1}$

$\underline{4x^2 + 2x}$
$-8x + 1$
$-8x - 4$

5. $x^2 - 1 \overline{)x^3 + x^2 - 3x + 1}$ Q: $x + 1$

$\underline{x^3 - x}$ R: $-2x + 2$
$x^2 - 2x + 1$
$x^2 - 1$

6. $x - 1 \overline{)x^3 - 6}$

$\underline{x^3 - x^2}$
$x^2 - 6$
$\underline{x^2 - x}$
$x - 6$
$\underline{x - 1}$

Write the polynomial missing from each row. (Example 2)

7. $2x - 3 \overline{)10x^2 - 9x - 11}$ quotient: $5x + 3$

$\boxed{}$ $10x^2 - 15x$
$6x - 11$
$\boxed{}$ $6x - 9$
-2

8. $4x + 1 \overline{)12x^2 - 17x - 1}$ quotient: $3x - 5$

$\boxed{}$
$-20x - 1$
$\boxed{}$
4

9. $x^2 - 3 \overline{)x^3 + 2x^2 - x + 5}$ quotient: $x + 2$

$\boxed{}$ $x^3 - 3x$
$2x^2 + 2x + 5$
$\boxed{}$ $2x^2 - 6$
$\boxed{}$ $2x + 11$

10. $x^2 - 2 \overline{)4x^3 - 3x^2 - 7x + 2}$ quotient: $4x - 3$

$\boxed{}$
$-3x^2 + x + 2$
$\boxed{}$

11. $x^2 + 2 \overline{)x^4 + x^3 - x^2 + 2x - 6}$ quotient: $x^2 + x - 3$

$\underline{x^4 + 2x^2}$
$\boxed{}$ $x^3 - 3x^2 + 2x - 6$
$\underline{x^3 + 2x}$
$-3x^2 - 6$
$\boxed{}$ $-3x^2 - 6$
0

12. $x^2 + 4 \overline{)3x^4 - x^3 + 10x^2 - 4x - 8}$ quotient: $3x^2 - x - 2$

$\underline{3x^4 + 12x^2}$
$\boxed{}$
$-x^3 - 4x$
$\boxed{}$
$-2x^2 - 8$
0

Additional Answers

2. Q: $x + 4$; R: 17
4. Q: $2x - 4$; R: 5
6. Q: $x^2 + x + 1$; R: -5
8. $12x^2 + 3x$;
$-20x - 5$
10. $4x^3 - 8x$;
$-3x^2 + 6$;
$x - 4$
12. $-x^3 - 2x^2 - 4x - 8$;
$-2x^2 - 8$

Polynomials / **229**

229

Assignment Guide
Basic
Day 1 p. 230: 1–17 odd
Day 2 p. 230: 2–18 even
Average p. 230: 1–27 odd
Above Average p. 230:
1–37 odd

Additional Answers

2. $4x + 1 + \dfrac{2}{x - 3}$

4. $4x - 3 - \dfrac{4}{3x + 2}$

6. $3x + 5 - \dfrac{6}{6x - 1}$

7. $5x - 6 - \dfrac{10}{2x + 5}$

8. $3x - 9 - \dfrac{6}{4x + 1}$

9. $6x - 5 + \dfrac{x}{2x^2 - 1}$

10. $-4x + 3 - \dfrac{x}{3x^2 + 5}$

11. $x^2 - 3x + 6 - \dfrac{8x - 8}{x^2 + x - 1}$

12. $x^2 - 2 - \dfrac{x - 1}{2x^2 - x + 1}$

20. $x^2 + 3x + 9 + \dfrac{54}{x - 3}$

21. $x^2 - 2xy + y^2 + \dfrac{y^3}{2x - 3y}$

22. $-3y^2 + 2xy - x^2 - \dfrac{2x^3}{3y - x}$

26. $x^2 + 2xy + y^2 + \dfrac{3y^3}{3x + 2y}$

27. $x^2 - 5x + 29 - \dfrac{148}{x + 5}$

28. $x^2 - 8x + 16 - \dfrac{31}{x + 2}$

30. $-6x + 33 - \dfrac{161}{x + 5}$

32. $a^2 - 5a + 6$

38. $x^2 - x + 1 + \dfrac{1}{x^2 + x + 1}$

Quiz Sections 8.4–8.7
After completing Sections 8.4–8.7, you may want to administer a quiz covering the same sections. See the quiz provided on page 172 in the *Teacher's ResourceBank*.

▮ WRITTEN EXERCISES ▮

Goal: To divide a polynomial by a polynomial

Sample Problem: $x + 2 \,)\overline{x^3 + 6x^2 + 11x + 10}$ **Answer:** $x^2 + 4x + 3 + \dfrac{4}{x + 2}$

Divide. (Examples 1–3)

1. $x + 4 \,)\overline{2x^2 + 3x - 17}$ $2x - 5 + \dfrac{3}{x + 4}$

2. $x - 3 \,)\overline{4x^2 - 11x - 1}$

3. $2x - 3 \,)\overline{4x^2 + 8x - 25}$ $2x + 7 - \dfrac{4}{2x - 3}$

4. $3x + 2 \,)\overline{12x^2 - x - 10}$

5. $5x - 6 \,)\overline{15x^2 + 2x - 29}$ $3x + 4 - \dfrac{5}{5x - 6}$

6. $6x - 1 \,)\overline{18x^2 + 27x - 11}$

7. $2x + 5 \,)\overline{10x^2 + 13x - 40}$

8. $4x + 1 \,)\overline{12x^2 - 33x - 15}$

9. $2x^2 - 1 \,)\overline{12x^3 - 10x^2 - 5x + 5}$

10. $3x^2 + 5 \,)\overline{-12x^3 + 9x^2 - 21x + 15}$

11. $x^2 + x - 1 \,)\overline{x^4 - 2x^3 + 2x^2 + x + 2}$

12. $2x^2 - x + 1 \,)\overline{2x^4 - x^3 - 3x^2 + x - 1}$

13. $x^2 + 8x + 2 \,)\overline{x^3 + 3x^2 - 38x - 16}$ $x - 5 - \dfrac{6}{x^2 + 8x + 2}$

14. $x^2 - x - 1 \,)\overline{2x^4 - 2x^3 - 5x^2 + 3x + 7}$ $2x^2 - 3 + \dfrac{4}{x^2 - x - 1}$

15. $2x^2 + 9x + 4 \,)\overline{2x^4 + 13x^3 + 16x^2 - 19x + 12}$ $x^2 + 2x - 3 + \dfrac{24}{2x^2 + 9x + 4}$

16. $a^2 - 4a + 4 \,)\overline{a^3 - 6a^2 + 13a - 9}$ $a - 2 + \dfrac{a - 1}{a^2 - 4a + 4}$

17. $x + y \,)\overline{x^3 + x^2y - xy^2 - 2y^3}$ $x^2 - y^2 - \dfrac{y^3}{x + y}$

18. $x - y \,)\overline{x^3 + x^2y - xy^2 - 2y^3}$ $x^2 + 2xy + y^2 - \dfrac{y^3}{x - y}$

MORE CHALLENGING EXERCISES

19. $x + 2 \,)\overline{x^3 - 8}$ $x^2 - 2x + 4 - \dfrac{16}{x + 2}$

20. $x - 3 \,)\overline{x^3 + 27}$

21. $2x - 3y \,)\overline{2x^3 - 7x^2y + 8xy^2 - 2y^3}$

22. $3y - x \,)\overline{-9y^3 + 9y^2x - 5yx^2 - x^3}$

23. $2x - 1 \,)\overline{8x^3 - 1}$ $4x^2 + 2x + 1$

24. $5y + 4 \,)\overline{125y^3 + 64}$ $25y^2 - 20y + 16$

25. $a^2 - 2 \,)\overline{a^3 + 6a^2 - 2a - 12}$ $a + 6$

26. $3x + 2y \,)\overline{3x^3 + 8x^2y + 7xy^2 + 5y^3}$

27. $x + 5 \,)\overline{x^3 + 4x - 3}$

28. $x + 2 \,)\overline{x^3 - 6x^2 + 1}$

29. $2a - 1 \,)\overline{5a + 6a^2 - 4}$ $3a + 4$

30. $x + 5 \,)\overline{4 - 6x^2 + 3x}$

31. $a - 3 \,)\overline{4a^3 - 24a - 3a^2 - 9}$ $4a^2 + 9a + 3$

32. $2a + 3 \,)\overline{2a^3 + 18 - 3a - 7a^2}$

33. $x + y \,)\overline{x^3 + y^3}$ $x^2 - xy + y^2$

34. $x - y \,)\overline{x^3 - y^3}$ $x^2 + xy + y^2$

35. $x^2 + xy + y^2 \,)\overline{x^3 - y^3}$ $x - y$

36. $a^2 - ab + b^2 \,)\overline{a^3 + b^3}$ $a + b$

37. $x - 1 \,)\overline{x^4 - 1}$ $x^3 + x^2 + x + 1$

38. $x^2 + x + 1 \,)\overline{x^4 + x^2 + 2}$

Using Formulas

Compound Interest

The focus of this special topic is the compound interest formula. Students will see the power of the calculator as it is applied to the computation needed in this formula.

Compound interest paid by a bank on deposits is computed on the total of the **principal** (amount deposited) <u>plus</u> the **interest** previously earned. The interest is usually paid more than once a year. This formula is used to compute compound interest.

$$A = p(1 + r)^n$$

A = amount of the new balance
r = rate of interest per period
p = principal
n = number of interest periods per year

EXAMPLE: The First City Bank pays 5% interest compounded quarterly (four times a year) on regular savings accounts. Anthony deposited $400 in a regular savings account. What will be the amount of his new balance after one year?

SOLUTION: $A = p(1 + r)^n$

$p = \$400$, $n = 4$, $r = 0.0125$ ◀ **r = 0.05 ÷ 4, or 0.0125**

$A = (400)(1 + 0.0125)^4$, or $400(1.0125)^4$

[1] [.] [0] [1] [2] [5] [×]

[=] [=] [=] [×] [4] [0] [0] [=]   ◀ **Round to the nearest cent.**

Anthony will have **$420.38** in his account after one year.

EXERCISES

For Exercises 1–3, find the new balance after one year.

1. $200; 6% interest
   Compounded
   quarterly $212.27

2. $500; 7% interest
   Compounded
   twice a year $535.61

3. $600; 6% interest
   Compounded
   monthly $637.01

4. Juan has $900 in a savings account. The bank pays 6% interest compounded twice a year. What will be the new balance after one year?
   $954.81

5. Cynthia has $300 in a savings account. The bank pays 5% interest compounded quarterly. What will be the new balance after one year?
   $315.28

Polynomials / **231**

CHAPTER SUMMARY

IMPORTANT TERMS	Monomial *(p. 208)* Coefficient *(p. 208)* Polynomial *(p. 213)* Term *(p. 213)* Binomial *(p. 213)*	Trinomial *(p. 213)* Constant term *(p. 220)* Divisor *(p. 224)* Dividend *(p. 224)*

IMPORTANT IDEAS

1. Monomials or polynomials are said to be over {integers} if all the coefficients are integers.

2. The sum of any two polynomials is a polynomial.

3. In adding or subtracting polynomials vertically, you arrange the terms in descending order of powers.

4. The difference of any two polynomials is a polynomial.

5. Any two polynomials are multiplied by multiplying each term of one polynomial by each term of the other.

6. The product of any two polynomials is a polynomial.

7. To divide a polynomial by a monomial, divide each term of the polynomial by the monomial.

8. The quotient of two polynomials is not necessarily a polynomial.

Chapter Test

Two Chapter Tests (Form A and Form B) are provided on pages 173–176 in the *Teacher's ResourceBank.*

CHAPTER REVIEW

SECTION 8.1

Write <u>Yes</u> *or* <u>No</u> *to show whether each numeral or expression is a monomial over* {*rational numbers*}.

1. $\frac{1}{3}x^2$ Yes

2. $\frac{3}{x^2}$ No

3. $-5x^{-3}$ No

4. $\sqrt{3} \cdot x$ No

Evaluate each monomial.

5. $4x^3$ if $x = -2$ −32

6. $-5x^2$ if $x = -3$ −45

7. $-0.8x^4$ if $x = 2$ −12.8

8. $1.2x^3$ if $x = -4$ −76.8

SECTION 8.2

Multiply.

9. $(3t^2)(7t^4)$ $21t^6$

10. $(-5xy^3)(13x^2y)$ $-65x^3y^4$

11. $(\frac{1}{3}ab)(-6ab^2)(-a^2b^2)$ $2a^4b^5$

12. $(1.4mn^2)(7m^2n)(-2n^2)$
 $-19.6m^3n^5$

232 / *Chapter 8*

Divide.

13. $\dfrac{-108t^5}{9t^3}$ $-12t^2$

14. $\dfrac{1.44r^3s^6t^2}{-12rs^5t}$ $-0.12r^2st$

15. $\dfrac{-3x^{12}y^7z^9}{-18x^3y^5z^7}$ $\frac{1}{6}x^9y^2z^2$

16. $\dfrac{23a^4bc^2}{a^3c}$ $23abc$

17. $\dfrac{-13x^4y^5}{-4x^2y}$ $\frac{13}{4}x^2y^4$

18. $\dfrac{16a^5c^6}{2a^4c^5}$ $8ac$

SECTION 8.3

Write Yes or No to show whether each numeral or expression is a polynomial over {rational numbers}.

19. $\dfrac{2}{x^2} - \dfrac{1}{x} + \dfrac{1}{2}$ No

20. $x^3 - 10$ Yes

21. $3\sqrt{x} - \sqrt{3}$ No

22. $x^{-3} + 2x^{-2} - 5x^{-1} - 7$ No

Simplify.

23. $7x^2 - 2x + 5x - 16$ $7x^2 + 3x - 16$

24. $4x^3 - 2x^2 - x^2 + 5x$ $4x^3 - 3x^2 + 5x$

25. $12x - x^2 + 5 - 3x^3$ $-3x^3 - x^2 + 12x + 5$

26. $3x^2 - 5 + 4x - 2x^3 + x^4$ $x^4 - 2x^3 + 3x^2 + 4x - 5$

SECTION 8.4

Add.

27. $\begin{aligned}-3x^2 + 2x - 5\\ \underline{7x^2 - 9x - 12}\end{aligned}$ $4x^2 - 7x - 17$

28. $\begin{aligned}4.3x^2 - 1.7x + 2.5\\ \underline{-3.9x^2 + 4.8x - 8.3}\end{aligned}$ $0.4x^2 + 3.1x - 5.8$

29. $(5x^3 - 3x + 7) + (2x^2 - 8x - 12)$ $5x^3 + 2x^2 - 11x - 5$

30. $(-2x^3 + x^2 - 5) + (5x^3 - x^2 + 7x - 12)$ $3x^3 + 7x - 17$

Subtract.

31. $\begin{aligned}x^2 - 3x + 4\\ \underline{-5x^2 + 4x + 9}\end{aligned}$ $6x^2 - 7x - 5$

32. $\begin{aligned}2x^3 \quad\;\; - 3x - 5\\ \underline{6x^3 - x^2 + 5x - 6}\end{aligned}$ $-4x^3 + x^2 - 8x + 1$

33. $(x^4 - 3x^2 + 7) - (5x^3 - x^2 + 2x - 5)$ $x^4 - 5x^3 - 2x^2 - 2x + 12$

34. $(3x^3 - x^2 - 4x) - (-x^3 + 4x^2 - x - 10)$ $4x^3 - 5x^2 - 3x + 10$

Write a polynomial to represent the perimeter of each polygon.

35.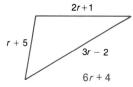
$2r + 1$
$r + 5$
$3r - 2$
$6r + 4$

36.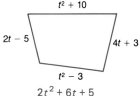
$t^2 + 10$
$2t - 5$
$4t + 3$
$t^2 - 3$
$2t^2 + 6t + 5$

37.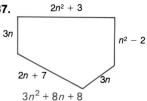
$2n^2 + 3$
$3n$
$n^2 - 2$
$2n + 7$
$3n$
$3n^2 + 8n + 8$

SECTION 8.5

Multiply.

38. $(2x + 7)(x - 3)$ $2x^2 + x - 21$

39. $(3x - 8)(x + 4)$ $3x^2 + 4x - 32$

40. $(5x - 2)(4x + 5)$ $20x^2 + 17x - 10$

41. $(6x - 5)(2x - 9)$ $12x^2 - 64x + 45$

42. $3x^3 - x^2 \qquad + 5$
$ 2x - 7$
$6x^4 - 23x^3 + 7x^2 + 10x - 35$

43. $-4x^3 \qquad -5x + 3$
$ 4x - 3$
$-16x^4 + 12x^3 - 20x^2 + 27x - 9$

Write a polynomial to represent each area.

44. The top of a rectangular table has a length of $(x + 3)$ units and a width of $(x - 5)$ units. $x^2 - 2x - 15$

45. The length of a soccer field is $(3x - 1)$ units and the width is $(2x + 8)$ units. $6x^2 + 22x - 8$

SECTION 8.6

Divide.

46. $(10x^3 + 6x^2 - 8x) \div 2x$ $5x^2 + 3x - 4$

47. $(24x^2y^2 - 16xy) \div 8xy$ $3xy - 2$

48. $2x - 5 \overline{)\, 2x^2 + 9x - 35}$ $x + 7$

49. $3x - 4 \overline{)\, 15x^2 + 4x - 32}$ $5x + 8$

50. $x^2 - 3 \overline{)\, x^4 - 3x^3 + x^2 + 9x - 12}$
$ x^2 - 3x + 4$

51. $x^2 + 2 \overline{)\, -x^4 + x^3 - 7x^2 + 2x - 10}$
$ -x^2 + x - 5$

SECTION 8.7

Divide.

52. $4x + 11 \overline{)\, -8x^2 - 2x + 60}$ $-2x + 5 + \dfrac{5}{4x + 11}$

53. $6x - 4 \overline{)\, -18x^2 + 36x - 35}$ $-3x + 4 - \dfrac{19}{6x - 4}$

54. $x^2 + 3 \overline{)\, x^4 - x^3 - 3x - 11}$ $x^2 - x - 3 - \dfrac{2}{x^2 + 3}$

55. $2x^2 - 5 \overline{)\, 2x^4 + 3x^2 - 26}$ $x^2 + 4 - \dfrac{6}{2x^2 - 5}$

Factoring Polynomials

This complicated formula is used to predict the weight of an average adult after a given number of days of dieting.

$$W = \frac{1}{a}\left[N + (aW_0 - N)\left(\frac{3500 - a}{3500}\right)^t\right]$$

Milk Group

Meat Group

Fruit and Vegetable Group

Bread and Cereal Group

Background

The variables in the formula above represent the following.

W: Weight in t days a: Required calories per pound of body weight

W_0: Initial weight N: Number of calories consumed daily

t: Time in days

9.1   Common Monomial Factors

OBJECTIVE: To factor a polynomial over $\{\text{integers}\}$ by common monomial factoring

P–1   **Express $4x^2 - 6x$ as a product.** $2x(2x - 3)$

Here are some ways to factor $4x^2 - 6x$.

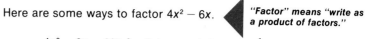 *"Factor" means "write as a product of factors."*

$$4x^2 - 6x = 2(2x^2 - 3x) \qquad 4x^2 - 6x = \tfrac{1}{2}x(8x - 12)$$
$$4x^2 - 6x = 4x(x - 1.5) \qquad 4x^2 - 6x = x^{-1}(4x^3 - 6x^2)$$
$$4x^2 - 6x = 2x(2x - 3)$$

The factors $2x$ and $(2x - 3)$ in the last factorization are special.

1. They are both polynomials.
2. Their coefficients are integers.
3. They are prime.

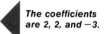 *The coefficients are 2, 2, and −3.*

There are only two ways to write $7x^2 - 2$ as a product of two or more polynomials that have integers as their coefficients.

$$7x^2 - 2 = 1(7x^2 - 2)$$
$$7x^2 - 2 = -1(-7x^2 + 2)$$

Either 1 times the polynomial or −1 times the opposite of the polynomial.

For this reason polynomials such as $7x^2 - 2$ are called **prime polynomials.**

EXAMPLE 1   Factor $6x^2 - 9x$.

 Always find the prime polynomial factors.

Solution:

[1] Find the greatest number that is a common factor of 6 and 9. ───────► 3 $6 = 3 \cdot 2$
$9 = 3 \cdot 3$

[2] Find the greatest power of x that is common to x^2 and x. ───────► x $x^2 = x \cdot x$
$x = x$

[3] Write the common monomial factor. ───────► $3x($ $)$

[4] Divide each term of $6x^2 - 9x$ by $3x$. ───────────────► $3x(2x - 3)$

Check your answer by multiplying the factors.

$$\begin{array}{r} 2x - 3 \\ 3x \\ \hline 6x^2 - 9x \end{array}$$

Explain that this is only a partial check. Refer to P-1 to show why.

Definition

> The **greatest common factor** of two or more counting numbers is the greatest number that is a factor of all the numbers.
>
> The greatest common factor of 6 and 9 is 3.

EXAMPLE 2   Factor $8x^3 + 12x^2$.

Solution:

1. Write the greatest common factor of 8 and 12. ──────➤ 4

2. Write the greatest power of x common to x^3 and x^2. ──────➤ x^2

3. Write the common monomial factor. ──────➤ $4x^2(\quad)$

4. Divide each term of $8x^3 + 12x^2$ by $4x^2$. ──────➤ $4x^2(2x + 3)$

P–2 **What is the common monomial factor of $8x^3 + 12x^2$?** $4x^2$

Any monomial factors do not need to be expressed as *prime* factors.

The prime factors of $8x^3 + 12x^2$ are $2 \cdot 2 \cdot x \cdot x(2x + 3)$. The simplest factored form is $4x^2(2x + 3)$. Expressing $8x^3 + 12x^2$ as $4x^2(2x + 3)$ is an example of **common monomial factoring.**

EXAMPLE 3   Factor $10x^4 - 15x^3 + 25x^2$.

Solution:

1. Find the greatest common factor of 10, 15, and 25. ──────➤ 5

2. Find the greatest power of x common to x^4, x^3, and x^2. ──────➤ x^2

3. Write the common monomial factor. ──────➤ $5x^2(\quad)$

4. Divide each term of $10x^4 - 15x^3 + 25x^2$ by $5x^2$. ──────➤ $5x^2(2x^2 - 3x + 5)$

Check:
$$\begin{array}{r} 2x^2 - 3x + 5 \\ 5x^2 \\ \hline 10x^4 - 15x^3 + 25x^2 \end{array}$$

◀ *Multiply the factors.*

In factoring, you should always check two conditions.

1. Any polynomial factors other than monomial factors should be prime.
2. The product of the factors must equal the given polynomial.

Factoring Polynomials / **237**

Additional Examples
Example 2
Factor.
1. $6x^3 + 10x^2$
   Ans: $2x^2(3x + 5)$
2. $18x^3 - 48x^2$
   Ans: $6x^2(3x - 8)$
3. $-20x^4 - 4x^3$
   Ans: $-4x^3(5x + 1)$

Example 3
Factor
1. $18x^3 + 6x^2 - 12x$
   Ans: $6x(3x^2 + x - 2)$
2. $5x^5 - 40x^3 - 15x^2$
   Ans: $5x^2(x^3 - 8x - 3)$
3. $16x^4 + 8x^3 + 28x^2$
   Ans: $4x^2(4x^2 + 2x + 7)$

Common Error
Students do not completely factor a polynomial.

Example
$8x^2 - 12x = 2x(4x - 6)$

Prescription
Have students identify the common monomial factors in Classroom Exercises 9–20.

Additional Answers
Classroom Exercises
12. $x(x - 3)$
16. $-x(x - 3)$
20. $7x^2(2x^3 - 3)$

Assignment Guide
Basic p. 238: 1–23 odd
Average p. 238: 1–23 odd
Above Average p. 238:
1–23, odd

Additional Answers
Written Exercises
4. $6(x - 3)$
8. $-4x(x - 5)$
9. $4x^2(2x - 3)$
10. $5x^2(2x + 5)$
11. $6x(2x^2 - 3)$
12. $9x(2x^2 - 3)$

Review Capsule
3. $x^2 + 3x + 7x + 3 \cdot 7$
6. $x^2 + 4x + 5x + 4 \cdot 5$

CLASSROOM EXERCISES

Write Yes or No to tell whether each polynomial is prime.

1. $x + 2$ Yes
2. $x - 5$ Yes
3. $2x + 4$ No
4. $5x + 10$ No

5. $x^2 + x$ No
6. $x^2 + 2$ Yes
7. $3x - 2$ Yes
8. $x^2 + 3x$ No

Factor. (Examples 1 and 2)

9. $4x + 2$ $2(2x + 1)$
10. $3x - 3$ $3(x - 1)$
11. $5x + 15$ $5(x + 3)$
12. $x^2 - 3x$

13. $2x^2 + 10$ $2(x^2 + 5)$
14. $3x^2 + 9x$ $3x(x + 3)$
15. $-3x + 6$ $-3(x - 2)$
16. $-x^2 + 3x$

17. $5x^3 - 10x$ $5x(x^2 - 2)$
18. $12x^2 + 18x$ $6x(2x + 3)$
19. $24 - 8x^2$ $8(3 - x^2)$
20. $14x^5 - 21x^2$

WRITTEN EXERCISES

Goal: To factor a polynomial by common monomial factoring
Sample Problem: Factor $9x^4 - 54x^3 + 18x^2$.   **Answer:** $9x^2(x^2 - 6x + 2)$

Factor. (Examples 1–3)

1. $4x - 4$ $4(x - 1)$
2. $3x + 3$ $3(x + 1)$
3. $5x + 10$ $5(x + 2)$
4. $6x - 18$

5. $2x^2 + 3x$ $x(2x + 3)$
6. $4x^2 - 5x$ $x(4x - 5)$
7. $-3x^2 + 9x$ $-3x(x - 3)$
8. $-4x^2 + 20x$

9. $8x^3 - 12x^2$
10. $10x^3 + 25x^2$
11. $12x^3 - 18x$
12. $18x^3 - 27x$

13. $x^3 - 2x^2 + 3x$ $x(x^2 - 2x + 3)$
14. $3x^3 + x^2 - 5x$ $x(3x^2 + x - 5)$

15. $2x^4 - 4x^2 - 10x$ $2x(x^3 - 2x - 5)$
16. $6x^4 + 8x^3 - 26x$ $2x(3x^3 + 4x^2 - 13)$

17. $-5x^2 - 10x$ $-5x(x + 2)$
18. $-4x^2 - 16x$ $-4x(x + 4)$

19. $8x^5 - 12x^3 + 20x^2$ $4x^2(2x^3 - 3x + 5)$
20. $18x^4 + 12x^3 - 30x^2$ $6x^2(3x^2 + 2x - 5)$

21. $5x^4 - 35x^2 - 10$ $5(x^4 - 7x^2 - 2)$
22. $3x^5 + 12x^3 - 9$ $3(x^5 + 4x^3 - 3)$

23. $21x^6 + 14x^5 - 10$ Prime polynomial
24. $18x^7 - 27x^4 + 9x^3$ $9x^3(2x^4 - 3x + 1)$

REVIEW CAPSULE FOR SECTION 9.2

Write each trinomial in the form $x^2 + rx + sx + rs$.

EXAMPLE: $x^2 + 7x + 10$ **SOLUTION:** $x^2 + 5x + 2x + 5 \cdot 2$

1. $x^2 + 4x + 4$ $x^2 + 2x + 2x + 2 \cdot 2$
2. $x^2 + 6x + 8$ $x^2 + 2x + 4x + 2 \cdot 4$
3. $x^2 + 10x + 21$

4. $x^2 + 5x + 6$ $x^2 + 2x + 3x + 2 \cdot 3$
5. $x^2 + 12x + 11$ $x^2 + 1x + 11x + 1 \cdot 11$
6. $x^2 + 9x + 20$

9.2 Factoring Trinomials

OBJECTIVES: To factor a polynomial of the form $x^2 + rx + sx + rs$ in which r and s are integers
To factor a quadratic trinomial of the form $x^2 + bx + c$ in which b and c are positive integers

A monomial such as $2x^5$ is said to be of the <u>fifth degree</u>.

> The **degree of a monomial** in one variable is determined by the exponent of the variable. $-8x^3$ is of the <u>third degree</u>.

P–1 **What is the degree of each of these monomials?**

a. $-3x^2$ b. $5x$ c. -16 d. $16x^8$
Second   First   Zero   Eighth

Definition

> The **degree of a polynomial** is the highest degree of the monomials of the polynomial. $5x^2 - 2x + 6$ is of <u>second degree</u>.

Second degree polynomials are called **quadratic polynomials.**

◄ $5x^2 - 2x + 6$ is a quadratic trinomial.

P–2 **What is the degree of each of these polynomials?**

a. $x - 5$ b. $x^3 - 2x^2 + 3x - 7$ c. $3x^2 - 2x + 2x^4 - 7$
First   Third   Fourth

EXAMPLE 1   Factor $x^2 + 2x + 3x + 6$.   ◄ *Remember always to write the <u>prime</u> factors.*

Solution:

1. Write as the sum of two binomials. ──► $x^2 + 2x + 3x + 6 = (x^2 + 2x) + (3x + 6)$

2. Factor each binomial. ────────────────────► $= x(x + 2) + 3(x + 2)$

3. Distributive Property ───────────────────► $= (x + 3)(x + 2)$

EXAMPLE 2   Factor $x^2 + rx + sx + rs$. (r and s represent integers.)

Solution:

1. Write as the sum of two binomials. ──► $x^2 + rx + sx + rs = (x^2 + rx) + (sx + rs)$

2. Factor each binomial. ────────────────────► $= x(x + r) + s(x + r)$

3. Distributive Property ───────────────────► $= (x + s)(x + r)$

*Factoring Polynomials / **239***

Teaching Suggestions p. M–34

Lesson Resources
Warm–Up: p. M–35
Maintenance: See below.
Practice Worksheet 9.2

Maintenance
1. Multiply:
   $(1.8)(-2.7)(1)(-1)$
   Ans: 4.86   (Section 2.3)
2. Solve: $3m - 4 < 2(m - 5)$
   Ans: {all real numbers less than −6}   (Section 5.4)
3. Write 0.00365 in scientific notation.
   Ans: 3.65×10^{-3}
   (Section 6.6)
4. Add: $3\sqrt{8} + 5\sqrt{2}$
   Ans: $11\sqrt{2}$ (Section 7.5)
5. A yearly subscription to a sports magazine cost $5.43 more than a subscription to a news magazine (Condition 1). The total cost of the two subscriptions is $63.33 (Condition 2). Find the cost of each subscription.
   Ans: news: $28.95; sports: $34.38
   (Section 4.1)

Additional Examples
Example 1
Factor.
1. $x^2 + 3x + 5x + 15$
   Ans: (x + 5)(x + 3)
2. $x^2 + 2x + 8x + 16$
   Ans: (x + 8)(x + 2)
3. $x^2 + x + 11x + 11$
   Ans: (x + 11)(x + 1)

Example 2
Factor.
1. $x^2 + ax + bx + ab$
   Ans: (x + b)(x + a)
2. $x^2 + hx + kx + hk$
   Ans: (x + k)(x + h)

| P-3 | **Factor each of the following.** |

a. $x(x + 3) + 7(x + 3)$ **b.** $x(x + 10) + 2(x + 10)$
$(x + 7)(x + 3)$ $(x + 2)(x + 10)$

| P-4 | **What number in $x^2 + 5x + 2x + 10$ corresponds to r in $x^2 + rx + sx + rs$? to s? to rs?** |

          5   2   10

EXAMPLE 3

Factor $x^2 + 5x + 2x + 10$. ◀ $x^2 + rx + sx + rs$

Solution: $x^2 + 5x + 2x + 10 = (x^2 + 5x) + (2x + 10)$

$$= x(x + 5) + 2(x + 5)$$

$$= (x + 2)(x + 5)$$

| P-5 | **What two integers have a sum of 13 and a product of 36?** 4 and 9 |

EXAMPLE 4

Write $x^2 + 13x + 36$ in the form $x^2 + rx + sx + rs$. Then factor.

Solution: $x^2 + 13x + 36 = x^2 + 4x + 9x + 36$ ◀ $r = 4$, $s = 9$

$$= (x^2 + 4x) + (9x + 36)$$

$$= x(x + 4) + 9(x + 4)$$

$$= (x + 9)(x + 4)$$

Values of r and s are to be found intuitively. An explicit rule will be discussed in Section 9.3.

CLASSROOM EXERCISES

Write the degree of each polynomial. (P-2)

1. $x^2 - 5$ Second
2. $3x^3 - 5x + 6$ Third
3. $x - 3x^0$ First
4. $5 + 2x - x^2$ Second
5. $x - x^5 + x^3 - x^2$ Fifth
6. $-3x + 7x^5 - 5x^{10}$ Tenth

Factor. (P-3)

7. $x(x + 3) + 4(x + 3)$ $(x + 4)(x + 3)$
8. $x(x + 2) + 7(x + 2)$ $(x + 7)(x + 2)$
9. $x(x + 5) + 3(x + 5)$ $(x + 3)(x + 5)$
10. $x(x + 2) + 8(x + 2)$ $(x + 8)(x + 2)$
11. $x(x + 4) + 5(x + 4)$ $(x + 5)(x + 4)$
12. $x(x + 7) + 3(x + 7)$ $(x + 3)(x + 7)$
13. $x(x + 8) + 4(x + 8)$ $(x + 4)(x + 8)$
14. $x(x + 9) + 5(x + 9)$ $(x + 5)(x + 9)$

Goal: To write two binomial factors of a quadratic trinomial
Sample Problem: Factor $x^2 + 16x + 63$.
Answer: $(x + 9)(x + 7)$

Factor. (Examples 1–3)

1. $x^2 + 3x + 4x + 12$ $(x + 3)(x + 4)$

2. $x^2 + 2x + 7x + 14$ $(x + 2)(x + 7)$

3. $x^2 + 5x + 3x + 15$ $(x + 5)(x + 3)$

4. $x^2 + 2x + 8x + 16$ $((x + 2)(x + 8)$

5. $x^2 + 4x + 5x + 20$ $(x + 4)(x + 5)$

6. $x^2 + 7x + 3x + 21$ $(x + 7)(x + 3)$

7. $x^2 + 8x + 4x + 32$ $(x + 8)(x + 4)$

8. $x^2 + 9x + 5x + 45$ $(x + 9)(x + 5)$

9. $x^2 + 3x + 11x + 33$ $(x + 3)(x + 11)$

10. $x^2 + 6x + 8x + 48$ $(x + 6)(x + 8)$

Write each trinomial in the form $x^2 + rx + sx + rs$. Then factor. (Example 4)

EXAMPLE: $x^2 + 18x + 77$

SOLUTION: $x^2 + 18x + 77 = x^2 + (7x + 11x) + 77$
$$= (x^2 + 7x) + (11x + 77)$$
$$= x(x + 7) + 11(x + 7)$$
$$= (x + 11)(x + 7)$$

11. $x^2 + 7x + 10$ $(x + 2)(x + 5)$

12. $x^2 + 9x + 14$ $(x + 2)(x + 7)$

13. $x^2 + 5x + 6$

14. $x^2 + 8x + 15$ $(x + 3)(x + 5)$

15. $x^2 + 10x + 21$ $(x + 3)(x + 7)$

16. $x^2 + 13x + 22$

17. $x^2 + 15x + 26$ $(x + 2)(x + 13)$

18. $x^2 + 14x + 33$ $(x + 3)(x + 11)$

19. $x^2 + 13x + 30$

20. $x^2 + 11x + 18$ $(x + 2)(x + 9)$

21. $x^2 + 11x + 28$ $(x + 4)(x + 7)$

22. $x^2 + 11x + 10$

MORE CHALLENGING EXERCISES

Find the value of k that will make the second polynomial a factor of the first.

23. $x^2 + 7x + k$; $x + 4$
$k = 12$

24. $x^2 + kx + 30$; $x + 5$
$k = 11$

25. $x^2 + 11x + k$; $x + 9$
$k = 18$

REVIEW CAPSULE FOR SECTION 9.3

Factor. (Pages 47–49)

1. $x(x + 5) - 2(x + 5)$

2. $x(x - 3) + 2(x - 3)$

3. $x(x - 6) - 4(x - 6)$

4. $x(x - 10) - 2(x - 10)$

5. $x(x - 5) - 5(x - 5)$

6. $x(x - 9) - 9(x - 9)$

Factoring Polynomials / **241**

Assignment Guide
Basic p. 241: 1–21 odd
Average p. 241: 1–25 odd
Above Average p. 241:
1–21 odd, 23–25

Additional Answers
Written Exercises
13. $(x + 2)(x + 3)$
16. $(x + 2)(x + 11)$
19. $(x + 3)(x + 10)$
22. $(x + 1)(x + 10)$

Critical Thinking
See Exercises 23–25.

Problem–Solving Skills
Using guess and check
(Ex. 23–25)
Using logical reasoning
(Ex. 23–25)
Working backwards (Ex. 23–25)

Review Capsule
1. $(x - 2)(x + 5)$
2. $(x + 2)(x - 3)$
3. $(x - 4)(x - 6)$
4. $(x - 2)(x - 10)$
5. $(x - 5)(x - 5)$
6. $(x - 9)(x - 9)$

Maintenance
1. Simplify: $(-12)(\frac{1}{4}p)$
   Ans: $-3p$ (Section 2.4)
2. Solve: $\frac{1}{6}t < -5$
   Ans: {all real numbers less than -30} (Section 5.5)
3. Write 2,003,000; in scientific notation.
   Ans: 2.003×10^6
   (Section 6.6)
4. Evaluate $\frac{2}{3}x^4$ if x equals -3.
   Ans: 54 (Section 8.1)
5. The length of a rectangular shaped window is 12 inches greater than its height (Condition 1). The perimeter of the window is 200 inches (Condition 2). Find the length and height.
   Ans: height: 44 inches; length: 56 inches
   (Section 4.2)

Additional Examples
Example 1
Factor.
1. $x^2 + 13x + 42$
   Ans: $(x + 7)(x + 6)$
2. $x^2 + 8x + 16$
   Ans: $(x + 4)(x + 4)$
3. $x^2 + 9x + 8$
   Ans: $(x + 1)(x + 8)$

OBJECTIVES: To write values of $r + s$ and $r \cdot s$ as a trinomial of the form $x^2 + bx + c$ as compared with $x^2 + (r + s)x + rs$

9.3 Factoring $x^2 + bx + c$

To write a quadratic trinomial in the form $x^2 + rx + sx + rs$
To write the binomial factors of a quadratic trinomial in the form $x^2 + bx + c$

The simplest polynomial form of $x^2 + rx + sx + rs$ is shown below.

$$x^2 + (r + s)x + rs$$

P–1 In the polynomial $x^2 + 10x + 16$, what integer corresponds to $r + s$? 10 What integer corresponds to rs? 16

P–2 What represents the value of b when $x^2 + (r + s)x + rs$ is compared with $x^2 + bx + c$? $r + s$

P–3 What represents the constant term c? rs

> To factor a trinomial of the form $x^2 + bx + c$, find two integers r and s such that
> $$r + s = b \quad \text{and} \quad r \cdot s = c.$$

Explain that if integers exist that make $r + s = b$ and $r \cdot s = c$ both true, there is just one such pair of integers.

EXAMPLE 1   Factor $x^2 + 10x + 16$.

Solution: $r + s = 10$ $rs = 16$ ◀ **Find two numbers whose sum is 10 and whose product is 16.**

1. Write all pairs of integers with product 16.

   (1)(16)   (−1)(−16)
   (2)(8)   (−2)(−8)
   (4)(4)   (−4)(−4)

2. Select the pair with a sum of 10. ⟶ $\begin{cases} r = 8 \\ s = 2 \end{cases}$

3. Write in the form $x^2 + rx + sx + rs$. ⟶ $x^2 + 10x + 16 = x^2 + 8x + 2x + 16$

4. Write as the sum of two binomials. ⟶ $= (x^2 + 8x) + (2x + 16)$

5. Factor each binomial. ⟶ $= x(x + 8) + 2(x + 8)$

6. Distributive Property ⟶ $= (x + 2)(x + 8)$

242 / Chapter 9

EXAMPLE 2   Factor $x^2 - 7x + 12$.

Solution:

1. Write all pairs of integers with product 12.

$$(1)(12) \qquad (-1)(-12)$$
$$(2)(6) \qquad (-2)(-6)$$
$$(3)(4) \qquad (-3)(-4)$$

2. Select the pair with a sum of -7. $\longrightarrow$ $\begin{cases} r = -3 \\ s = -4 \end{cases}$

3. Write in the form
$x^2 + rx + sx + rs$. $\longrightarrow$ $x^2 - 7x + 12 = x^2 - 3x - 4x + 12$

4. Write as a sum of two binomials. $\longrightarrow$ $= (x^2 - 3x) + (-4x + 12)$

5. Factor each binomial. $\longrightarrow$ $= x(x - 3) - 4(x - 3)$

6. Distributive Property $\longrightarrow$ $= (x - 4)(x - 3)$

Use $-4(x - 3)$, _not_ $4(-x + 3)$.

EXAMPLE 3   Factor $x^2 - x - 20$.

Solution:

1. Write all pairs of integers with product -20.

$$(1)(-20) \qquad (-1)(20)$$
$$(2)(-10) \qquad (-2)(10)$$
$$(4)(-5) \qquad (-4)(5)$$

2. Select the pair with a sum of -1. $\longrightarrow$ $\begin{cases} r = 4 \\ s = -5 \end{cases}$

3. $\qquad x^2 - x - 20 = x^2 + 4x - 5x - 20$

4. $\qquad\qquad\qquad = (x^2 + 4x) + (-5x - 20)$

5. $\qquad\qquad\qquad = x(x + 4) - 5(x + 4)$

6. $\qquad\qquad\qquad = (x - 5)(x + 4)$

Additional Examples
Example 2
Factor.
1. $x^2 - 10x + 24$
   Ans: $(x - 4)(x - 6)$
2. $x^2 - 6x + 9$
   Ans: $(x - 3)(x - 3)$
3. $x^2 - 12x + 11$
   Ans: $(x - 1)(x - 11)$

Example 3
Factor.
1. $x^2 - 2x - 24$
   Ans: $(x - 6)(x + 4)$
2. $x^2 + 3x - 10$
   Ans: $(x - 2)(x + 5)$
3. $x^2 + 7x - 8$
   Ans: $(x + 8)(x - 1)$

Additional Examples
Example 4
Factor.
1. $x^2 + 3x + 5$
    Ans: prime
2. $x^2 - 6x - 10$
    Ans: prime
3. $x^2 + 2x - 2$
    Ans: prime

EXAMPLE 4   Factor $x^2 + 5x + 10$.

Solution:

1. Write all pairs of integers with product 10.

         (1)(10)   (−1)(−10)   ◄ *Pairs of integers*
         (2)(5)   (−2)(−5)   *with product 10*

2. Select the pair with a sum of 5.   ◄ *None of the pairs*
                                           of integers meet
                                           this condition.

     Since no two integers meet both conditions, $x^2 + 5x + 10$ is a prime polynomial.

Additional Answers
2. 2, 6
4. 1, 2
6. None
8. 2, −4
10. −2, −4
12. None
15. $x^2 + 3x + 5x + 15$
18. $x^2 - 2x - 8x + 16$
21. $x^2 - x + 6x - 6$
24. $x^2 + x - 7x - 7$

CLASSROOM EXERCISES

Name two integers, if they exist, that have a product and a sum as given.
(Steps 1 and 2, Examples 1 and 2)

1. The product is 6 and the sum is 5.   2, 3   **2.** The product is 12 and the sum is 8.

3. The product is 18 and the sum is 9.   3, 6   **4.** The product is 2 and the sum is 3.

5. The product is 3 and the sum is 2.   None   **6.** The product is 8 and the sum is −2.

7. The product is −8 and the sum is 2.   −2, 4   **8.** The product is −8 and the sum is −2.

9. The product is 8 and the sum is 6.   2, 4   **10.** The product is 8 and the sum is −6.

11. The product is 10 and the sum is 3.   None   **12.** The product is 12 and the sum is 6.

Express each trinomial in the form $x^2 + rx + sx + rs$. (Examples 1–3)

13. $x^2 + 7x + 12$ $x^2 + 3x + 4x + 12$ **14.** $x^2 + 6x + 9$ $x^2 + 3x + 3x + 9$ **15.** $x^2 + 8x + 15$

16. $x^2 + 11x + 10$ $x^2 + x + 10x + 10$ **17.** $x^2 - 5x + 6$ $x^2 - 2x - 3x + 6$ **18.** $x^2 - 10x + 16$

19. $x^2 + 4x - 5$ $x^2 - x + 5x - 5$ **20.** $x^2 - 7x - 8$ $x^2 + x - 8x - 8$ **21.** $x^2 + 5x - 6$

22. $x^2 - 5x - 6$ $x^2 + x - 6x - 6$ **23.** $x^2 + 6x - 7$ $x^2 - x + 7x - 7$ **24.** $x^2 - 6x - 7$

WRITTEN EXERCISES

Goal: To factor quadratic trinomials of the form $x^2 + bx + c$
Sample Problems: a. $x^2 + 11x + 30$ **b.** $x^2 - 8x - 20$
Answers: a. $(x + 6)(x + 5)$ **b.** $(x - 10)(x + 2)$

For each trinomial, write the values of $r + s$ and $r \cdot s$. (P-1, P-4, P-5, P-6)

1. $x^2 + 7x + 12$ 7; 12
2. $x^2 + 14x + 15$ 14; 15
3. $x^2 + 14x + 33$

4. $x^2 + 9x + 20$ 9; 20
5. $x^2 + 4x - 32$ 4; −32
6. $x^2 + x - 30$

7. $x^2 - 8x - 48$ −8; −48
8. $x^2 - 10x - 64$ −10; −64
9. $x^2 - 17x + 72$

10. $x^2 + 5x - 36$ 5; −36
11. $x^2 - 3x + 2$ −3; 2
12. $x^2 + 7x - 30$

Write each trinomial in the form $x^2 + rx + sx + rs$. (Examples 1–3)

13. $x^2 + 13x + 12$ $x^2 + x + 12x + 12$
14. $x^2 + 16x + 15$ $x^2 + x + 15x + 15$
15. $x^2 + 11x + 30$

16. $x^2 + 13x + 42$ $x^2 + 6x + 7x + 42$
17. $x^2 - 8x + 15$ $x^2 - 3x - 5x + 15$
18. $x^2 - 9x + 14$

19. $x^2 - 11x + 24$ $x^2 - 3x - 8x + 24$
20. $x^2 - 14x + 45$ $x^2 - 5x - 9x + 45$
21. $x^2 - 3x - 40$

22. $x^2 - 2x - 35$ $x^2 + 5x - 7x - 35$
23. $x^2 + 4x - 21$ $x^2 - 3x + 7x - 21$
24. $x^2 + 4x - 32$

25. $x^2 - 9x - 36$ $x^2 + 3x - 12x - 36$
26. $x^2 - 13x + 36$ $x^2 - 4x - 9x + 36$
27. $x^2 - 7x + 6$

28. $x^2 - 7x - 8$ $x^2 + x - 8x - 8$
29. $x^2 - 11x - 42$ $x^2 + 3x - 14x - 42$
30. $x^2 - 3x + 2$

MIXED PRACTICE

Factor. When a polynomial cannot be factored, write <u>Prime</u>. (Examples 1–4)

31. $x^2 + 2x + 1$ $(x + 1)(x + 1)$
32. $x^2 + 4x + 4$ $(x + 2)(x + 2)$
33. $x^2 + 12x + 35$

34. $x^2 + 10x + 21$ $(x + 3)(x + 7)$
35. $x^2 - 7x + 12$ $(x - 3)(x - 4)$
36. $x^2 - 9x + 20$

37. $x^2 - 11x + 30$ $(x - 5)(x - 6)$
38. $x^2 - 10x + 24$ $(x - 4)(x - 6)$
39. $x^2 - x - 12$

40. $x^2 - x - 20$ $(x + 4)(x - 5)$
41. $x^2 - x - 30$ $(x + 5)(x - 6)$
42. $x^2 - 2x - 24$

43. $x^2 + x - 30$ $(x - 5)(x + 6)$
44. $x^2 + x - 24$ Prime
45. $x^2 - 9x + 14$

46. $x^2 - 11x + 24$ $(x - 3)(x - 8)$
47. $x^2 + 5x - 14$ $(x - 2)(x + 7)$
48. $x^2 + 5x - 24$

49. $x^2 - 3x + 36$ Prime
50. $x^2 - 20x + 64$ $(x - 4)(x - 16)$
51. $x^2 - 2x - 15$

52. $x^2 - 10x - 24$ $(x + 2)(x - 12)$
53. $x^2 + 19x + 34$ $(x + 2)(x + 17)$
54. $x^2 - 11x + 28$

55. $x^2 - 22x - 48$ $(x + 2)(x - 24)$
56. $x^2 - 26x + 48$ $(x - 2)(x - 24)$
57. $x^2 + 3x + 2$

58. $x^2 + 11x + 16$ Prime
59. $x^2 - x - 56$ $(x + 7)(x - 8)$
60. $x^2 - 10x - 56$

Factoring Polynomials / **245**

Assignment Guide
Basic
Day 1 p. 245: 1–29 odd
Day 2 p. 245: 31–59 odd
Average p. 245: 3, 6, 9, · · · 30, 31–59 odd
Above Average p. 245: 1–59 odd

Additional Answers
3. 14; 33
6. 1; −30
9. −17; 72
12. 7; −30
15. $x^2 + 5x + 6x + 30$
18. $x^2 - 2x - 7x + 14$
21. $x^2 + 5x - 8x - 40$
24. $x^2 - 4x + 8x - 32$
27. $x^2 - x - 6x + 6$
30. $x^2 - x - 2x + 2$
33. $(x + 5)(x + 7)$
36. $(x - 4)(x - 5)$
39. $(x + 3)(x - 4)$
42. $(x + 4)(x - 6)$
45. $(x - 2)(x - 7)$
48. $(x - 3)(x + 8)$
51. $(x + 3)(x - 5)$
54. $(x - 4)(x - 7)$
57. $(x + 1)(x + 2)$
60. $(x + 4)(x - 14)$

Lesson Resources
Warm–Up: p. M–35
Maintenance: See below.
Practice Worksheet 9.4
Transparencies 20, 23, 25

Maintenance
1. Factor: $-3y^2 - 9y$
   Ans: $-3y(y + 3)$
   (Section 2.5)
2. Solve: $-2a > \frac{4}{9}$
   Ans: {all real numbers less than $-\frac{2}{9}$}
   (Section 5.5)
3. Simplify: $\sqrt{256}$
   Ans: 16 (Section 7.1)
4. Divide: $\frac{-36a^4b^6}{9ab^3}$
   Ans: $-4a^3b^3$
   (Section 8.2)
5. There were three more who tried out for the women's volleyball team than for the men's volleyball team (Condition 1). Seventy-five per cent of those who tried out for the men's team were selected, but only 60 per cent of those who tried out for the women's team were selected. The two teams had the same number of members (Condition 2). Find how many tried out for each team.
   Ans: men's team: 12; women's team: 15
   (Section 4.4)

Additional Examples
Example 1
Multiply.
1. $(x + 5)(2x + 3)$
   Ans: $2x^2 + 13x + 15$
2. $(3x - 1)(x - 4)$
   Ans: $3x^2 - 13x + 4$
3. $(5x + 2)(x - 4)$
   Ans: $5x^2 - 18x - 8$

9.4 Factoring $ax^2 + bx + c$

OBJECTIVES: To multiply two binomials
To write the binomial factors of a quadratic trinomial of the form $ax^2 + bx + c$

In the last section you factored quadratic trinomials. Each of these trinomials had a leading coefficient of 1.

Definition

> The coefficient of the term of highest degree in a polynomial is called the **leading coefficient**.
>
> The leading coefficient of $x - 2x^2 - 5$ is -2.

P–1 **What is the leading coefficient in each polynomial below?**

a. $5x^3 - 2x^2 + 3x - 7$ 5   **b.** $x - 5$ 1   **c.** $2x^3 - x^4 + 3$ −1

Example 1 reviews multiplying two binomials.

EXAMPLE 1   Multiply: $(2x + 1)(x + 3)$

Solution:

1. Distributive Property ⟶ $(2x + 1)(x + 3) = (2x + 1)x + (2x + 1)3$
2. Distributive Property ⟶ $= 2x^2 + x + 6x + 3$
3. Combine like terms. ⟶ $= 2x^2 + 7x + 3$

Quadratic trinomials are factored by reversing steps 1, 2, and 3 above. The key step is in writing $2x^2 + 7x + 3$ as $2x^2 + x + 6x + 3$. You must find values for r and s in the polynomial form

$$2x^2 + rx + sx + 3$$
$$\text{or}$$
$$2x^2 + (r + s)x + 3.$$

◄ $2x^2 + x + 6x + 3$

Go over these two conditions carefully.

> To factor a quadratic trinomial of the form $ax^2 + bx + c$, you must find two integers, r and s, that meet these conditions.
> 1. The sum of r and s equals the coefficient of the first-degree term. ($r + s = b$)
> 2. The product of r and s equals the product of the leading coefficient and the constant term. ($r \cdot s = a \cdot c$)

246 / Chapter 9

EXAMPLE 2   Factor $6x^2 + 11x - 10$.

Solution: $r + s = 11$ $rs = -60$ **Find two numbers whose sum is 11 and whose product is −60.**

It may not be necessary for students to write all pairs. Have them check each pair of numbers as they write them.

[1] Write all pairs of integers with product −60.

$(-1)(60)$ $(1)(-60)$ $(-2)(30)$ $(2)(-30)$ $(-3)(20)$ $(3)(-20)$
$(-4)(15)$ $(4)(-15)$ $(-5)(12)$ $(5)(-12)$ $(-6)(10)$ $(6)(-10)$

[2] Select the pair with a sum of 11. ⟶ $r = -4$ and $s = 15$

[3] Write in the form $6x^2 + rx + sx - 10$. ⟶ $6x^2 + 11x - 10 = (6x^2 - 4x) + (15x - 10)$

[4] Factor each binomial. ⟶ $= 2x(3x - 2) + 5(3x - 2)$

[5] Distributive Property ⟶ $= (2x + 5)(3x - 2)$

EXAMPLE 3   Factor $2x^2 + 7x + 3$.

Solution: $r + s = 7$ $rs = 6$ **$2x^2 + 7x + 3 = 2x^2 + (r + s)x + 3$**

[1] Write all pairs of integers with product 6.

$(1)(6)$ $(2)(3)$ $(-1)(-6)$ $(-2)(-3)$

[2] Select the pair with a sum of 7. ⟶ $r = 1$ and $s = 6$

[3] $2x^2 + 7x + 3 = (2x^2 + x) + (6x + 3)$

[4] $= x(2x + 1) + 3(2x + 1)$

[5] $= (x + 3)(2x + 1)$

EXAMPLE 4   Factor $12x^2 - 11x + 2$.

Solution: $r + s = -11$ $rs = 24$

[1] $(1)(24)$ $(2)(12)$ $(3)(8)$ $(4)(6)$ **Pairs of integers with product 24**
$(-1)(-24)$ $(-2)(-12)$ $(-3)(-8)$ $(-4)(-6)$

[2] $r = -3$ and $s = -8$

[3] $12x^2 - 11x + 2 = (12x^2 - 3x) + (-8x + 2)$

[4] $= 3x(4x - 1) - 2(4x - 1)$

[5] $= (3x - 2)(4x - 1)$

Factoring Polynomials / **247**

CLASSROOM EXERCISES

Write the value of r + s. Then write the value of rs. (Examples 2–4)

1. $2x^2 + 3x + 1$ 3; 2
2. $2x^2 + 5x + 2$ 5; 4
3. $3x^2 + 10x + 3$
4. $3x^2 + 7x + 2$ 7; 6
5. $3x^2 + 11x + 6$ 11; 18
6. $3x^2 + 14x + 15$
7. $6x^2 + 11x + 3$ 11; 18
8. $6x^2 + 7x + 2$ 7; 12
9. $6x^2 + 17x + 5$
10. $6x^2 + 13x + 6$ 13; 36
11. $3x^2 - x - 2$ −1; −6
12. $3x^2 - 5x - 2$
13. $2x^2 - 5x - 25$ −5; −50
14. $2x^2 - 3x - 9$ −3; −18
15. $15x^2 - 7x - 2$
16. $15x^2 + 4x - 4$ 4; −60
17. $6x^2 - 17x + 12$ −17; 72
18. $10x^2 - 29x + 10$

Additional Answers
Classroom Exercises
3. 10; 9
6. 14; 45
9. 17; 30
12. −5; −6
15. −7; −30
18. −29; 100

Assignment Guide
Basic
Day 1 p. 248: 1–17 odd
Day 2 p. 248: 19–33 odd
Average
Day 1 p. 248: 1–17 odd
Day 2 pp. 248–249: 19–39 odd
Above Average pp. 248–249:
1–33 odd, 35–40

WRITTEN EXERCISES

Goal: To factor trinomials of the form $ax^2 + bx + c$
Sample Problem: $10x^2 - 11x - 6$
Answer: $(5x + 2)(2x - 3)$

Multiply. (Example 1)

1. $(2x - 1)(x - 1)$
   $2x^2 - 3x + 1$
2. $(2x - 1)(x - 2)$
   $2x^2 - 5x + 2$
3. $(3x - 2)(x + 1)$
   $3x^2 + x - 2$
4. $(3x - 1)(x + 2)$
   $3x^2 + 5x - 2$

Factor. (Examples 2–4)

5. $2x^2 + 3x + 1$ $(2x + 1)(x + 1)$
6. $2x^2 + 5x + 2$ $(2x + 1)(x + 2)$
7. $3x^2 + 10x + 3$
8. $3x^2 + 7x + 2$ $(3x + 1)(x + 2)$
9. $3x^2 + 11x + 6$ $(3x + 2)(x + 3)$
10. $3x^2 + 14x + 15$
11. $6x^2 + 11x + 3$ $(2x + 3)(3x + 1)$
12. $6x^2 + 7x + 2$ $(3x + 2)(2x + 1)$
13. $6x^2 + 17x + 5$
14. $6x^2 + 13x + 6$ $(3x + 2)(2x + 3)$
15. $3x^2 - x - 2$ $(3x + 2)(x - 1)$
16. $3x^2 - 5x - 2$
17. $2x^2 - 5x - 25$ $(2x + 5)(x - 5)$
18. $2x^2 - 3x - 9$ $(2x + 3)(x - 3)$
19. $15x^2 - 7x - 2$
20. $15x^2 + 4x - 4$ $(3x + 2)(5x - 2)$
21. $2x^2 - 7x + 6$ $(2x - 3)(x - 2)$
22. $3x^2 - 10x + 3$
23. $8x^2 - 14x + 3$ $(2x - 3)(4x - 1)$
24. $15x^2 - 13x + 2$ $(3x - 2)(5x - 1)$
25. $2x^2 + 5x - 3$
26. $3x^2 - 7x - 6$ $(3x + 2)(x - 3)$
27. $15x^2 + 14x - 8$ $(3x + 4)(5x - 2)$
28. $12x^2 - 7x - 5$
29. $18x^2 + 23x + 7$ $(2x + 1)(9x + 7)$
30. $15x^2 + 11x - 12$ $(3x + 4)(5x - 3)$
31. $12x^2 - 33x - 9$
32. $4x^2 - 29x + 7$ $(4x - 1)(x - 7)$
33. $8x^2 + 35x + 12$ $(8x + 3)(x + 4)$
34. $12x^2 + 25x + 12$

Additional Answers
Written Exercises
7. $(3x + 1)(x + 3)$
10. $(3x + 5)(x + 3)$
13. $(3x + 1)(2x + 5)$
16. $(3x + 1)(x - 2)$
19. $(3x - 2)(5x + 1)$
22. $(3x - 1)(x - 3)$
25. $(2x - 1)(x + 3)$
28. $(12x + 5)(x - 1)$
31. $(12x + 3)(x - 3)$ or
    $3(4x + 1)(x - 3)$
34. $(4x + 3)(3x + 4)$

NON-ROUTINE PROBLEMS

For Exercises 35–38, look for a pattern in the products in the box at the right. Then use the pattern to evaluate each expression without using pencil and paper.

$999 \times 2 = 1998$
$999 \times 3 = 2997$
$999 \times 4 = 3996$
$999 \times 5 = 4995$

35. 999×6 **36.** 999×8 **37.** $6993 \div 999$ **38.** $8991 \div 999$

39. In the figure below, a belt runs over six wheels in the direction of the arrows. How many wheels are turning clockwise?

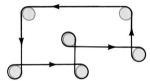

40. There are six water glasses in a row. Three of the glasses are empty and three are full. The arrangement of the glasses is empty, empty, empty, full, full, full. By moving one glass, change the arrangement to empty, full, empty, full, empty, full.

MID-CHAPTER REVIEW

Factor. (Section 9.1)

1. $12x^2 - 4x$
2. $-5x^3 - 10x^2$
3. $3x^3 + 9x^2 - 15x$
$3x(x^2 + 3x - 5)$
4. $4x^4 - 12x^2 - 2x$
$2x(2x^3 - 6x - 1)$

5. $2x^2y + 6xy$
6. $3xy^2 - 9x^2y$
7. $5x^3y - 10x^2y + 25x^2y^2$
8. $12rs^2t - 6r^2st^2$

Factor. (Section 9.2)

9. $x^2 + 12x + 35$
$(x + 5)(x + 7)$
10. $x^2 + 15x + 54$
$(x + 6)(x + 9)$
11. $x^2 + 15x + 50$
$(x + 5)(x + 10)$
12. $x^2 + 16x + 48$
$(x + 4)(x + 12)$

Factor. (Section 9.3)

13. $x^2 + 3x - 40$
14. $x^2 - 16x + 60$
15. $x^2 - 3x - 88$
$(x + 8)(x - 11)$
16. $x^2 - 2x - 63$
$(x + 7)(x - 9)$

17. $x^2 + 6x - 72$
$(x - 6)(x + 12)$
18. $x^2 + x - 72$
$(x - 8)(x + 9)$
19. $x^2 - 20x + 99$
$(x - 9)(x - 11)$
20. $x^2 - x - 99$
Prime

Factor. (Section 9.4)

21. $2x^2 + x - 6$
$(2x - 3)(x + 2)$
22. $8x^2 - 6x - 5$
$(2x + 1)(4x - 5)$
23. $20x^2 - 3x - 2$
$(4x + 1)(5x - 2)$
24. $6x^2 - 19x - 20$
$(6x + 5)(x - 4)$

REVIEW CAPSULE FOR SECTION 9.5

Factor. (Pages 239–241)

1. $(x^2 + 12x) - (12x + 144)$
$(x - 12)(x + 12)$
2. $(x^2 + 2xy) - (2xy + 4y^2)$
$(x - 2y)(x + 2y)$
3. $4x^2 - 6x + 6x - 9$
$(2x + 3)(2x - 3)$

*Factoring Polynomials / **249***

Using Probability

Making a Model

Life scientists are specialists in living things. Since biology is the science of life, many life scientists call themselves **biologists.** However, most are named by the type of living thing that they study or by the nature of the work that they perform. Some life scientist careers are agronomist, biochemist, botanist, horticulturist, microbiologist, nutritionist, and zoologist.

In the study of **genetics,** life scientists learn that the traits of parents are transmitted to their offspring by tiny particles called **genes.** Each living thing has two genes for each trait — one gene from each parent.

The genes for tall and short pea plants are represented below.

TT Pure tall
tt Pure short T = gene for tallness
Tt Hybrid tall t = gene for shortness

When a pure tall pea plant and a pure short pea plant are crossed, all offspring are hybrid tall.

Suppose that two hybrid tall pea plants are crossed. This problem is related to the algebra problem of squaring a binomial.

Let $(\frac{1}{2}T + \frac{1}{2}t)$ represent the two genes for tallness and shortness in each hybrid tall pea plant. Then

$$(\tfrac{1}{2}T + \tfrac{1}{2}t)(\tfrac{1}{2}T + \tfrac{1}{2}t) = \tfrac{1}{4}TT + \tfrac{1}{2}Tt + \tfrac{1}{4}tt.$$

This result indicates the following probabilities.

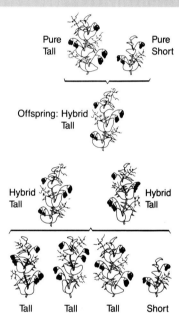

The probability (chances) that an offspring will be pure tall (TT) is 1 out of 4.

This means that, over a long enough period of time, you can expect about one fourth of all such offspring to be pure tall.

The probability (chances) that an offspring will be hybrid tall (Tt) is 1 out of 2.

That is, over a long period of time about one half of all such offspring can be expected to be hybrid tall.

The probability (chances) that an offspring will be pure short (tt) is 1 out of 4.

250 / *Chapter 9*

EXAMPLE 1: There are 24 offspring of two hybrid tall parent pea plants. Find the probable number of each type.

SOLUTION: Refer to the trinomial on page 250.

$$\frac{1}{4}TT + \frac{1}{2}Tt + \frac{1}{4}tt \quad \begin{cases} \frac{1}{4} \times 24 = 6 \text{ pure tall} \\ \frac{1}{2} \times 24 = 12 \text{ hybrid tall} \\ \frac{1}{4} \times 24 = 6 \text{ pure short} \end{cases}$$

Since this is merely the *probable* distribution, you would not always expect to obtain exactly these numbers of the various types.

In guinea pigs, a rough coat gene R is **dominant** to a smooth coat gene r. This means that guinea pigs will have coats as shown below.

RR: Pure rough coat Rr: Hybrid rough coat rr: Pure smooth coat

EXAMPLE 2: A guinea pig with a pure rough coat (RR) is mated with a guinea pig which has a hybrid rough coat (Rr). There are 36 offspring. Find the probable number of each type.

SOLUTION: $(\frac{1}{2}R + \frac{1}{2}R)(\frac{1}{2}R + \frac{1}{2}r) = \frac{1}{4}RR + \frac{1}{4}Rr + \frac{1}{4}RR + \frac{1}{4}Rr$
$$= \frac{1}{2}RR + \frac{1}{2}Rr$$

$\frac{1}{2} \times 36 = 18$ pure rough coat (RR) $\frac{1}{2} \times 36 = 18$ hybrid rough coat (Rr)

All of the offspring will have rough coats.

EXERCISES

There are 128 offspring of two pea plants. Find the probable number of pure tall (TT), pure short (tt), and hybrid tall (Tt) for each combination of parent plants.

1. Hybrid tall and pure short
2. Pure short and pure short
3. Pure tall and hybrid tall
4. Pure tall and pure short

There are 56 offspring of two guinea pigs. Find the probable number of offspring with pure rough coats (RR), hybrid rough coats (Rr), and pure smooth coats (rr) for each combination of parents.

5. Hybrid rough coat; hybrid rough coat
6. Pure rough coat; pure smooth coat
7. Pure rough coat; hybrid rough coat

Factoring Polynomials / 251

Additional Answers
1. 0 pure tall; 64 pure short; 64 hybrid tall
2. 0 pure tall; 128 pure short; 0 hybrid tall
3. 64 pure tall; 0 pure short; 64 hybrid tall
4. 0 pure tall; 0 pure short; 128 hybrid tall
5. 14 pure rough; 28 hybrid rough; 14 pure smooth
6. 0 pure rough; 56 hybrid rough; 0 pure smooth
7. 28 pure rough; 28 hybrid rough; 0 pure smooth

9.5 Difference of Two Squares

OBJECTIVES: To write the product of two binomials of the form $(a + b)$ and $(a - b)$
To factor a binomial that is the difference of two squares

A binomial such as $x^2 - 16$ is called a **difference of two squares.**

> **You can write** $x^2 - 16$ **as** $(x)^2 - (4)^2$.

The first term and the absolute value of the constant term are perfect squares.

P–1 **Which polynomials below represent the difference of two squares?**

a. $x^2 - 4$ **b.** $x^3 - 9$ **c.** $x^4 - 1$ **d.** $x^2 + 25$ *a and c*

The product of two special binomials suggests how to factor a difference of two squares.

EXAMPLE 1   Multiply: $(x + a)(x - a)$

Solution: $(x + a)(x - a) = x^2 + ax - ax - a^2$

$= x^2 - a^2$ ◀ **Difference of two squares**

P–2 **What are the binomial factors of $x^2 - a^2$?** $(x + a)(x - a)$

You may want to mention that the *sum* of two squares is *not* factorable.

> The prime factorization of a binomial such as $x^2 - a^2$ is $(x + a)(x - a)$. $x^2 - 49 = (x + 7)(x - 7)$

P–3 **What is the positive square root of 121?** 11

EXAMPLE 2   Factor $x^2 - 121$.

Solution: $x^2 - 121 = (x + 11)(x - 11)$ ◀ **(Sum of square roots) times (Difference of square roots)**

---

252 / *Chapter 9*

EXAMPLE 3   Factor $x^4 - 9$.

Solution: Both x^4 and 9 are perfect squares. You can write x^4 as $(x^2)^2$
and 9 as 3^2.

$$x^4 - 9 = (x^2 + 3)(x^2 - 3)$$

EXAMPLE 4   Factor $4x^2 - 9y^2$.

Solution: You can write $4x^2$ as $(2x)^2$ and $9y^2$ as $(3y)^2$.

$$4x^2 - 9y^2 = (2x + 3y)(2x - 3y)$$

P–4   Simplify.

a. $\sqrt{x^6}$ x^3 b. $\sqrt{r^8}$ r^4 c. $\sqrt{t^{12}}$
t^6

◄ Any power with an
even exponent is
a perfect square.

EXAMPLE 5   Factor $x^{10} - y^8$.

Write one factor
as a sum of square
roots.
→
Write one factor
as a difference
of square roots.

Solution: $x^{10} - y^8 = (x^5 + y^4)(x^5 - y^4)$

To factor the difference of squares, write one factor as the sum
of square roots and the other factor as the difference of
square roots.

■■■■■ **CLASSROOM EXERCISES** ■■■■■

Factor. (Examples 2–4)

1. $a^2 - 1$ 2. $y^2 - 4$

$(p + 3)(p - 3)$
3. $p^2 - 9$

$(r - 4)(r + 4)$
4. $r^2 - 16$

$(x + 5)(x - 5)$
5. $x^2 - 25$

6. $a^2 - 36$ 7. $t^2 - 49$ 8. $x^2 - 64$ 9. $b^2 - 81$ 10. $x^2 - 100$

11. $p^2 - q^2$ 12. $c^2 - d^2$ 13. $4x^2 - y^2$ 14. $x^2 - 9y^2$ 15. $x^4 - 25$
$(p + q)(p - q)$ $(c + d)(c - d)$

Factoring Polynomials / **253**

Factor.
1. $x^4 - 64$
   Ans: $(x^2 - 8)(x^2 + 8)$
2. $x^6 - 1$
   Ans: $(x^3 - 1)(x^3 + 1)$

Example 4
Factor.
1. $9x^2 - 16y^2$
   Ans: $(3x - 4y)(3x + 4y)$
2. $x^2 - 25y^2$
   Ans: $(x - 5y)(x + 5y)$

Example 5
Factor.
1. $x^8 - t^{10}$
   Ans: $(x^4 - t^5)(x^4 + t^5)$
2. $x^4 - h^2k^2$
   Ans: $(x^2 - hk)(x^2 + hk)$

Common Error
When factoring the difference of
two squares, students incorrectly
write the answer as the product
of the difference of two square
roots.

Example
$x^2 - 9 = (x - 3)(x - 3)$

Prescription
Have students find the product of
$(x - 3)(x - 3)$ and $(x - 3)$
$(x + 3)$. Point out that the first
product has three terms and is
not the difference of two squares.
Have students practice products
like Example 1 until they can see
the prime factorization.

Additional Answers
1. $(a + 1)(a - 1)$
2. $(y + 2)(y - 2)$
6. $(a + 6)(a - 6)$
7. $(t + 7)(t - 7)$
8. $(x + 8)(x - 8)$
9. $(b + 9)(b - 9)$
10. $(x + 10)(x - 10)$
13. $(2x + y)(2x - y)$
14. $(x + 3y)(x - 3y)$
15. $(x^2 + 5)(x^2 - 5)$

Assignment Guide
Basic p. 254: 1–35 odd
Average p. 254: 1–47 odd
Above Average p. 254:
1–47 odd

Additional Answers
Written Exercises
  6. $9a^2 - b^2$
  9. $25x^2 - 9$
16. $(3y + 1)(3y - 1)$
20. $(y^2 + 3)(y^2 - 3)$
21. $(2x^2 + 3)(2x^2 - 3)$
22. $(3a^2 + 1)(3a^2 - 1)$
24. $(4a + 3b)(4a - 3b)$
28. $(5r + 4)(5r - 4)$
31. $(4xy^2 + 3)(4xy^2 - 3)$
32. $(3ab^2 + 5)(3ab^2 - 5)$
44. 1584

▐ WRITTEN EXERCISES ▬▬▬▬▬▬

Goal: To factor the difference of two squares
Sample Problem: $25a^2 - 9b^2$
Answer: $(5a + 3b)(5a - 3b)$

Multiply. (Example 1)

1. $(x - 8)(x + 8)$ $x^2 - 64$ **2.** $(x - 9)(x + 9)$ $x^2 - 81$ $x^2 - 121$
3. $(x + 11)(x - 11)$

4. $(x + 12)(x - 12)$ $x^2 - 144$ **5.** $(2x - y)(2x + y)$ $4x^2 - y^2$ **6.** $(3a - b)(3a + b)$

7. $(xy - 3)(xy + 3)$ $x^2y^2 - 9$ **8.** $(rs - 5)(rs + 5)$ $r^2s^2 - 25$ **9.** $(5x + 3)(5x - 3)$

10. $(4x + 3)(4x - 3)$ $16x^2 - 9$ **11.** $(x^2 + 2)(x^2 - 2)$ $x^4 - 4$ **12.** $(y^2 + 3)(y^2 - 3)$
$y^4 - 9$

Factor. (Examples 2–5)

13. $a^2 - 25$ $(a+5)(a-5)$ **14.** $y^2 - 36$ $(y+6)(y-6)$ **15.** $4x^2 - 1$ $(2x+1)(2x-1)$ **16.** $9y^2 - 1$

17. $9a^2 - 16$ $(3a+4)(3a-4)$**18.** $4b^2 - 25$ $(2b+5)(2b-5)$**19.** $x^4 - 4$ $(x^2+2)(x^2-2)$ **20.** $y^4 - 9$

21. $4x^4 - 9$ **22.** $9a^4 - 1$ **23.** $4x^2 - y^2$ $(2x+y)(2x-y)$ **24.** $16a^2 - 9b^2$

25. $a^2b^2 - 1$ $(ab+1)(ab-1)$**26.** $x^2y^2 - 4$ $(xy+2)(xy-2)$ **27.** $16t^2 - 81$ $(4t+9)(4t-9)$ **28.** $25r^2 - 16$

29. $x^2y^2 - r^2$ $(xy+r)(xy-r)$**30.** $a^2b^2 - c^2$ $(ab+c)(ab-c)$**31.** $16x^2y^4 - 9$ **32.** $9a^2b^4 - 25$

33. $x^6 - 4$ $(x^3+2)(x^3-2)$ **34.** $a^6 - 9$ $(a^3+3)(a^3-3)$ **35.** $x^{10} - y^8$ $(x^5+y^4)(x^5-y^4)$**36.** $r^8 - s^{14}$
$(r^4+s^7)(r^4-s^7)$

MORE CHALLENGING EXERCISES

Multiply by the method shown in the example below.
Find the answer without using pencil and paper.

EXAMPLE: $(19)(21) = (20 - 1)(20 + 1)$
$= 400 - 1 = 399$

37. $(14)(16)$ 224   **38.** $(15)(17)$ 255   **39.** $(11)(15)$ 165   252
40. $(14)(18)$

41. $(22)(18)$ 396   **42.** $(23)(17)$ 391   **43.** $(26)(34)$ 884   **44.** $(36)(44)$

45. $(43)(37)$ 1591   **46.** $(54)(46)$ 2484   **47.** $(59)(61)$ 3599   **48.** $(89)(91)$
8099

REVIEW CAPSULE FOR SECTION 9.6

Multiply. (Pages 220–223)

1. $(x + 6)(x + 6)$ **2.** $(x - 3)(x - 3)$ **3.** $(2x - 5)(2x - 5)$ $4x^2 - 20x + 25$ **4.** $(5x + 1)(5x + 1)$ $25x^2 + 10x + 1$

5. $(2x + 7)(2x + 7)$ **6.** $(x - y)(x - y)$ **7.** $(3x + 2)(3x + 2)$ **8.** $(5x - 1)^2$
$9x^2 + 12x + 4$ $25x^2 - 10x + 1$

Review Capsule
1. $x^2 + 12x + 36$
2. $x^2 - 6x + 9$
5. $4x^2 + 28x + 49$
6. $x^2 - 2xy + y^2$

9.6 Trinomial Squares

OBJECTIVES: To identify whether or not a trinomial is a perfect square
To square a binomial
To factor a perfect square trinomial

A trinomial such as $x^2 + 6x + 9$ is called a **_trinomial square_**
because it has two equal binomial factors.

$$x^2 + 6x + 9 = (x + 3)(x + 3)$$

> A trinomial of the form $ax^2 + bx + c$ is a trinomial square if <u>both</u>
> of the following conditions are true.
>
> 1. ax^2 and c are perfect squares.
> 2. $(\frac{1}{2}b)^2 = ac$

In $x^2 + 6x + 9$, both x^2 and 9 are perfect squares.

P–1 What is the value of $\frac{1}{2}b$?^{3} of $(\frac{1}{2}b)^2$?^{9} of ac? ^{9}

You can see that $x^2 + 6x + 9$ meets both conditions.

EXAMPLE 1 Write <u>Yes</u> or <u>No</u> to show whether $4x^2 - 20x + 25$ is a
trinomial square.

Solution: ① $4x^2$ and 25 are both perfect squares.

$$4x^2 = (2x)(2x) \qquad 25 = (5)(5)$$

② "$(\frac{1}{2}b)^2 = ac$" is true.

$(\frac{1}{2}b)^2 = (\frac{1}{2} \cdot -20)^2 \qquad ac = 4 \cdot 25$ Explain that $4x^2 - 20x - 25$ is not
$\quad\quad\quad = (-10)^2 \qquad\qquad\quad = 100$ a trinomial square. Students often
$\quad\quad\quad = 100$ have trouble with this concept.

<u>Yes</u>, $4x^2 - 20x + 25$ is a trinomial square.

◀ A trinomial square is
also called a
"perfect square trinomial."

EXAMPLE 2 Factor $4x^2 - 20x + 25$. ◀ This is a trinomial square.

Solution:

① Write the first term of each binomial
as a square root of $4x^2$. ——————▶ 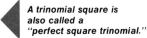 $4x^2 - 20x + 25 = (2x\quad)(2x\quad)$

② Write the second term as a square root of 25. ———▶ $= (2x\quad 5)(2x\quad 5)$

③ Write each binomial as a difference. ————————▶ $= (2x - 5)(2x - 5)$

Factoring Polynomials / **255**

Teaching Suggestions p. M–34

Lesson Resources
Warm–Up: p. M–35
Maintenance: See below.
Practice Worksheet 9.6
Manipulative 14, p. 566
Teaching Aids 1, 3, 4, 5
Transparencies 49, 50, 51, 53,
  54, 55, 57

Maintenance
1. Solve and check:
  $5d + 11 = 8d - 7$
  Ans: $d = 6$ (Section 4.3)
2. Simplify: $\left(\dfrac{10^{-3}}{m^{-4}}\right)^{-2}$
  Ans: $\dfrac{10^6}{m^8}$ (Section 6.5)
3. Multiply and simplify:
  $\sqrt{\dfrac{1}{2}} \cdot \sqrt{98}$
  Ans: 7 (Section 7.2)
4. Multiply: $(x - 4)(2x + 1)$
  Ans: $2x^2 -7x - 4$
  (Section 8.5)
5. A beach blanket has a length
  of $(x + 4)$ units and a width
  of $(x + 6)$ units. Find the area
  of the blanket.
  Ans: $x^2 + 10x + 24$
  square units
  (Section 8.5)

Additional Examples
Example 1
Write <u>Yes</u> or <u>No</u> to show whether
each trinomial is a trinomial
square.
1. $9x^2 + 12x + 4$
  Ans: Yes
2. $16x^2 - 8x + 1$
  Ans: Yes
3. $25x^2 + 30x - 9$
  Ans: No

Example 2
Factor.
1. $9x^2 + 6x + 1$
  Ans: $(3x + 1)(3x + 1)$
2. $x^2 - 10x + 25$
  Ans: $(x - 5)(x - 5)$
3. $4x^2 - 28x + 49$
  Ans: $(2x - 7)(2x - 7)$

255

P–2
Why is each factor of Example 2 a difference? The value of b is negative.

P–3 **Multiply.**

    a. $(2x + 5)(2x + 5)$ **b.** $(x - 7)(x - 7)$ **c.** $(3x - 1)(3x - 1)$
        $4x^2 + 20x + 25$ $x^2 - 14x + 49$ $9x^2 - 6x + 1$

Trinomials in more than one variable can be trinomial squares.

EXAMPLE 3   Factor $x^2y^2 + 8xy + 16$.

Solution: ☐1 x^2y^2 and 16 are perfect squares.

               ☐2   "$(\frac{1}{2} \cdot 8)^2 = 1 \cdot 16$" is true.

                       $x^2y^2 + 8xy + 16 = (xy + 4)(xy + 4)$

$x^2y^2 + 8xy + 16$ is a trinomial square.

To check, multiply the factors. The factors are sums because b is positive.

P–4 **Factor.**

    a. $x^2 + 2x + 1$ **b.** $x^2 - 12x + 36$ **c.** $9x^2 - 24x + 16$
        $(x + 1)(x + 1)$ $(x - 6)(x - 6)$ $(3x - 4)(3x - 4)$

CLASSROOM EXERCISES

Write Yes or No to tell whether each trinomial is a trinomial square.
(Example 1)

1. $16x^2 + 8x + 1$ Yes   **2.** $x^2 + 9x - 16$ No   **3.** $x^2 + 10x + 25$ Yes

4. $x^2 + 4x + 3$ No   **5.** $x^2 - 2x + 1$ Yes   **6.** $100 - 20x + x^2$ Yes

7. $x^2 - 49x + 14$ No   **8.** $x^2 + 20x + 100$ Yes   **9.** $2x^2 + 8x - 8$ No

10. $2x^2 - 8x + 8$ No   **11.** $81x^2 + 50x + 3$ No   **12.** $64x^2 + 48x + 9$ Yes

Factor each trinomial square. (Example 2)

13. $x^2 + 4x + 4$ $(x + 2)(x + 2)$ **14.** $x^2 - 2x + 1$ $(x - 1)(x - 1)$

15. $x^2 - 6x + 9$ $(x - 3)(x - 3)$ **16.** $x^2 - 10x + 25$ $(x - 5)(x - 5)$

17. $x^2 + 16x + 64$ $(x + 8)(x + 8)$ **18.** $x^2 + 18x + 81$ $(x + 9)(x + 9)$

19. $x^2 - 24x + 144$ $(x - 12)(x - 12)$ **20.** $x^2 - 14x + 49$ $(x - 7)(x - 7)$

21. $4x^2 + 4x + 1$ $(2x + 1)(2x + 1)$ **22.** $25x^2 - 20x + 4$ $(5x - 2)(5x - 2)$

Additional Examples
Example 3
Factor.
1. $x^2y^2 + 12xy + 36$
    Ans: (xy + 6)(xy + 6)
2. $x^4y^2 - 16x^2y + 64$
    Ans: (x²y − 8)(x²y − 8)
3. $9x^2y^2 - 30xy + 25$
    Ans: (3xy − 5)(3xy − 5)

Common Error
When factoring trinomial squares, students always write the factors as sums.

Example
$x^2 - 10x + 25 =$
$(x + 5)(x + 5)$

Prescription
Review Example 2 with the students. Explain that whenever the value of b is negative in a trinomial square, the factors are written as differences. Have students tell which of Classroom Exercises 13–22 would have factors written as differences.

WRITTEN EXERCISES

Goal: To factor trinomial squares
Sample Problem: $4x^2 - 20x + 25$ **Answer:** $(2x - 5)(2x - 5)$

Write Yes or No to show whether each trinomial is a trinomial square.
(Example 1)

1. $x^2 - 8x + 25$ No

2. $x^2 + 10x + 36$ No

3. $x^2 - 22x + 121$ Yes

4. $x^2 - 14x + 49$ Yes

5. $x^2 + 14x - 49$ No

6. $x^2 + 16x - 64$ No

7. $2x^2 - 12x + 25$ No

8. $3x^2 - 22x + 36$ No

9. $9x^2 + 18x + 1$ No

10. $16x^2 + 32x + 1$ No

11. $4x^2 - 4x + 1$ Yes

12. $5x^2 - 10x + 25$ No

Multiply. (P-3)

13. $(x - 5)(x - 5)$ $x^2 - 10x + 25$

14. $(x - 3)(x - 3)$ $x^2 - 6x + 9$

15. $(x + 7)(x + 7)$

16. $(x + 9)(x + 9)$ $x^2 + 18x + 81$

17. $(2x + 3)(2x + 3)$ $4x^2 + 12x + 9$

18. $(3x + 2)(3x + 2)$

19. $(5x - 1)(5x - 1)$ $25x^2 - 10x + 1$

20. $(4x - 3)(4x - 3)$ $16x^2 - 24x + 9$

21. $(4x + 5)^2$

22. $(6x + 5)^2$ $36x^2 + 60x + 25$

23. $(2x - 1)^2$ $4x^2 - 4x + 1$

24. $(3x - 7)^2$

Factor. (Examples 2–3)

25. $36x^2 - 12x + 1$ $(6x - 1)(6x - 1)$

26. $25x^2 - 10x + 1$ $(5x - 1)(5x - 1)$

27. $25x^2 + 20x + 4$

28. $16x^2 + 24x + 9$ $(4x + 3)(4x + 3)$

29. $9x^2 - 24x + 16$ $(3x - 4)(3x - 4)$

30. $4x^2 - 20x + 25$

31. $49x^2 + 140x + 100$

32. $36x^2 + 84x + 49$ $(6x + 7)(6x + 7)$

33. $49x^2 - 14x + 1$

34. $81x^2 + 18x + 1$ $(9x + 1)(9x + 1)$

35. $x^2 + 4xy + 4y^2$ $(x + 2y)(x + 2y)$

36. $x^2 + 6xy + 9y^2$

37. $4x^2 + 12xy + 9y^2$

38. $16x^2 + 40xy + 25y^2$

39. $x^2y^2 - 16xy + 64$

40. $4x^2y^2 - 20xy + 25$
$(2xy - 5)(2xy - 5)$

41. $9x^2y^2 + 42xy + 49$
$(3xy + 7)(3xy + 7)$

42. $x^2y^2 - 2xy + 1$
$(xy - 1)(xy - 1)$

REVIEW CAPSULE FOR SECTION 9.7

Match each polynomial with its prime factors. (Pages 236–254)

1. $4x^2 - 8x + 3$ B

2. $4x^2 - 10x - 6$ D

3. $4x^2 + 4x - 3$ A

4. $4x^2 - 9$ C

A. $(2x + 3)(2x - 1)$

B. $(2x - 3)(2x - 1)$

C. $(2x - 3)(2x + 3)$

D. $2(x - 3)(2x + 1)$

Factoring Polynomials / **257**

Assignment Guide
Basic p. 257: 1–41 odd
Average p. 257: 1–41 odd
Above Average p. 257:
1–41 odd

Additional Answers
Written Exercises
15. $x^2 + 14x + 49$
18. $9x^2 + 12x + 4$
21. $16x^2 + 40x + 25$
24. $9x^2 - 42x + 49$
27. $(5x + 2)(5x + 2)$
30. $(2x - 5)(2x - 5)$
31. $(7x + 10)(7x + 10)$
33. $(7x - 1)(7x - 1)$
36. $(x + 3y)(x + 3y)$
37. $(2x + 3y)(2x + 3y)$
38. $(4x + 5y)(4x + 5y)$
39. $(xy - 8)(xy - 8)$

Maintenance
1. Solve: $3y - 2(y + 7) > 5$
   Ans: {all real numbers greater than 19}
   (Section 5.4)
2. Simplify: $(trp)^{-3}$
   Ans: $\frac{1}{t^3 r^3 p^3}$
   (Section 6.5)
3. Simplify: $\sqrt[3]{1}$
   Ans: 1 (Section 7.3)
4. Divide:
   $(6x^2 - 7x + 2) \div (2x - 1)$
   Ans: $3x - 2$
   (Section 8.6)
5. A rectangular building has a length of $(x - 6)$ units and a width of $(x - 8)$ units. Find the area of the building.
   Ans: $x^2 - 14x + 48$ square units (Section 8.5)

Additional Examples
Example 1
Factor.
1. $5x^2 - 45$
   Ans: $5(x - 3)(x + 3)$
2. $12x^2 - 3$
   Ans: $3(2x + 1)(2x - 1)$
3. $2x^3 - 32x$
   Ans: $2x(x - 4)(x + 4)$

Example 2
Factor.
1. $5x^2 + 40x + 75$
   Ans: $5(x + 3)(x + 5)$
2. $x^3 - 12x^2 + 36x$
   Ans: $x(x - 6)(x - 6)$
3. $4x^3 - 32x^2 + 64x$
   Ans: $4x(x - 4)(x - 4)$

9.7 Combined Types of Factoring

OBJECTIVE: To write the prime factors of a polynomial by using one or more factoring methods

You have learned the following ways of factoring polynomials.

1. Common monomial factoring (Pages 236–238)
2. Factoring the difference of two squares (Pages 252–254)
3. Factoring quadratic trinomials (Pages 239–241)
4. Factoring trinomial squares (Pages 255–257)

When you factor a polynomial always write it as a product of prime polynomial factors.

In factoring a polynomial, always look for a common monomial factor as the first step.

P–1 **What is a common monomial factor of $2x^2 - 32$?** 2

EXAMPLE 1   Factor $2x^2 - 32$.

Solution:
1. Write the common monomial factor. ⟶ $2x^2 - 32 = 2(\quad)$
2. Divide each term of $2x^2 - 32$ by 2. ⟶ $= 2(x^2 - 16)$
3. Factor $x^2 - 16$ as a difference of two squares. ⟶ $= 2(x + 4)(x - 4)$

Remind students not to forget to write the factor 2 in the final form.

 *Prime polynomial factors*

When you write a polynomial as the product of prime polynomial factors, you are factoring completely. To "factor" always means to "factor completely."

P–2 **What is a common monomial factor of $3x^2 - 6x + 3$?** 3

EXAMPLE 2   Factor $3x^2 - 6x + 3$.

Solution:
1. Write the common factor. ⟶ $3x^2 - 6x + 3 = 3(\quad)$
2. Divide each term of $3x^2 - 6x + 3$ by 3. ⟶ $= 3(x^2 - 2x + 1)$
3. Factor $x^2 - 2x + 1$ as a trinomial square. ⟶ $= 3(x - 1)(x - 1)$

258 / *Chapter 9*

EXAMPLE 3   Factor $30x^2 - 5x - 10$.

Solution:

1. Use common monomial factoring. ⟶ $30x^2 - 5x - 10 = 5(6x^2 - x - 2)$

2. Determine whether $6x^2 - x - 2$ is a trinomial square. ⟶ *$6x^2 - x - 2$ is not a trinomial square.*

   Have students explain why it is not a trinomial square.

3. Write $-x$ as a sum. ⟶ $6x^2 - x - 2 = 6x^2 + rx + sx - 2$

4. Write all pairs of integers with product -12.

   $$(-1)(12) \qquad (-2)(6) \qquad (-3)(4)$$
   $$(1)(-12) \qquad (2)(-6) \qquad (3)(-4)$$

5. Select the pair with a sum of -1. ⟶ $\begin{cases} r = 3 \\ s = -4 \end{cases}$

6. $\qquad 6x^2 - x - 2 = 6x^2 + 3x - 4x - 2$
7. $\qquad\qquad\qquad = 3x(2x + 1) - 2(2x + 1)$ *Remember: 5 is also a factor of $30x^2 - 5x - 10$.*
8. $\qquad\qquad\qquad = (3x - 2)(2x + 1)$

Thus, $30x^2 - 5x - 10 = 5(6x^2 - x - 2)$
$$\qquad\qquad\qquad\qquad\quad = 5(3x - 2)(2x + 1)$$

EXAMPLE 4   Factor $4x^2 - 16y^2$.

Solution: You can see that $4x^2 - 16y^2$ is a difference of two squares.

$$4x^2 - 16y^2 = 4(x^2 - 4y^2)$$ *Common monomial factor*

$$\qquad\qquad = 4(x + 2y)(x - 2y)$$

CLASSROOM EXERCISES

Write the common factor. Then divide each term by the common factor.
(Steps 1 and 2 of Examples 1 and 2)

1. $4x^2 - 12$ $4; (x^2 - 3)$ **2.** $3x^2 + 6x + 6$ $3; (x^2 + 2x + 2)$

3. $4x^2 - 20x + 28$ $4; (x^2 - 5x + 7)$ **4.** $2x^2 - 8x + 12$ $2; (x^2 - 4x + 6)$

5. $15x^2 - 5x - 25$ $5; (3x^2 - x - 5)$ **6.** $16x^2 - 64$ $16; (x^2 - 4)$

Factoring Polynomials / **259**

Factor.
1. $4x^2 + 22x + 10$
   Ans: $2(2x + 1)(x + 5)$
2. $12x^2 + 69x - 18$
   Ans: $3(4x - 1)(x + 6)$
3. $7x^4 + 34x^3 - 5x^2$
   Ans: $x^2(x + 5)(7x - 1)$

Example 4
Factor.
1. $9x^2 - 36y^2$
   Ans: $9(x - 2y)(x + 2y)$
2. $16x^2 - 36y^2$
   Ans: $4(2x + 3y)(2x - 3y)$
3. $5x^2 - 45y^2$
   Ans: $5(x + 3y)(x - 3y)$

Common Error
Students forget to find a common monomial factor and incorrectly state that a polynomial cannot be factored.

Example
$3x^2 + 30x + 63$ cannot be factored.

Prescription
Before assigning the Written Exercises, have students look at Exercises 1–12 and name the common monomial factors. Remind students to look for the common monomial factor first.

Write the common factor. Then divide each term by the common factor.
(Steps 1 and 2 of Examples 1 and 2)

7. $x^2y^2 - 3xy + 5y$ $y; (x^2y - 3x + 5)$

8. $ax^2 - 3ax + 12a$ $a; (x^2 - 3x + 12)$

9. $-4x^2 - 20$ $-4; (x^2 + 5)$

10. $12ax^2 + 3a$ $3a; (4x^2 + 1)$

11. $3a^2b^2 - 9ab + 6b$ $3b; (a^2b - 3a + 2)$

12. $24 - 36x^2$ $12; (2 - 3x^2)$

13. $3x^2y - 30xy + 75y$ $3y; (x^2 - 10x + 25)$

14. $-12x^2y - 14xy + 6y$
$-2y; (6x^2 + 7x - 3)$

WRITTEN EXERCISES

Goal: To factor polynomials using more than one type of factoring
Sample Problem: $2a^2b^2 + 14a^2b + 24a^2$
Answer: $2a^2(b + 4)(b + 3)$

Factor. Remember factors should be prime polynomials. (Examples 1–4)

1. $2y^2 - 18$ $2(y + 3)(y - 3)$

2. $2a^2 - 50$ $2(a + 5)(a - 5)$

3. $2x^2 + 8x + 8$

4. $2x^2 + 4x + 2$ $2(x + 1)(x + 1)$

5. $3x^2 + 30x + 63$ $3(x + 7)(x + 3)$

6. $3x^2 + 36x + 105$

7. $4x^2 - 40x + 96$ $4(x - 4)(x - 6)$

8. $5x^2 - 55x + 150$ $5(x - 5)(x - 6)$

9. $5x^2 - 5x - 100$

10. $3x^2 - 3x - 36$ $3(x + 3)(x - 4)$

11. $4y^2 - 9$ $(2y + 3)(2y - 3)$

12. $x^2y^2 - 16$

13. $ax^2 + 4ax + 4a$ $a(x + 2)(x + 2)$

14. $x^2y + 6xy + 9y$ $y(x + 3)(x + 3)$

15. $6x^2 - 10x - 4$

16. $6x^2 - 2x - 4$ $2(3x + 2)(x - 1)$

17. $12x^2 - 40x + 12$ $4(3x-1)(x-3)$

18. $20x^2 - 70x + 60$

19. $-3x^2 - 12$ $-3(x^2 + 4)$

20. $-5a^2 - 125$ $-5(a^2 + 25)$

21. $4ax^2 - 9a$

22. $8ab^2 - 18a$ $2a(2b + 3)(2b - 3)$

23. $x^2y - 10xy + 25y$

24. $a^2b^2 - 6a^2b + 9a^2$

25. $3x^2 - 75$ $3(x + 5)(x - 5)$

26. $27y^2 - 3$ $3(3y + 1)(3y - 1)$

27. $32 - 16d + 2d^2$

28. $-36 + 15a^2 - 12a$

29. $3x^2 - 6x - 24$

30. $10y^2 - 26y + 12$

31. $2 - 128x^2$ $2(1 + 8x)(1 - 8x)$

32. $x^3 - x$ $x(x + 1)(x - 1)$

33. $ab^2 - ab - 20a$

34. $pq^2 + p^2q^2 - p$

35. $24a^2 - 30a + 9$
$3(2a - 1)(4a - 3)$

36. $30a^2 - 5a - 10$
$5(3a - 2)(2a + 1)$

MORE CHALLENGING EXERCISES

37. If $x - y = 6$ and $x + y = 1$,
find $x^2 - y^2$.   6

38. If $x \neq 0$ and $x^2 - x = x$, what is the
value of $x - 1$?   1

39. If $s^2 - t^2 = 6$ and $s - t = 3$,
find $s + t$.   2

40. If $m^2 - 4n^2 = 32$, and $m + 2n = 8$,
find $m - 2n$.   4

Additional Answers
Written Exercises
 3. $2(x + 2)(x + 2)$
 6. $3(x + 5)(x + 7)$
 9. $5(x + 4)(x - 5)$
12. $(xy + 4)(xy - 4)$
15. $2(3x + 1)(x - 2)$
18. $10(2x - 3)(x - 2)$
21. $a(2x + 3)(2x - 3)$
23. $y(x - 5)(x - 5)$
24. $a^2(b - 3)(b - 3)$
27. $2(4 - d)(4 - d)$
28. $3(5a + 6)(a - 2)$
29. $3(x + 2)(x - 4)$
30. $2(5y - 3)(y - 2)$
33. $a(b - 5)(b + 4)$
34. $p(q^2 + pq^2 - 1)$

Problem–Solving Skills
Using logical reasoning
(Ex. 37–40)

Quiz Sections 9.5–9.7
After completing Sections 9.5–
9.7, you may want to administer
a quiz covering the same sections. See the quiz provided on
page 190 in the *Teacher's
ResourceBank.*

Focus on Reasoning:
Mental Computation

Logical reasoning can help you to solve problems more efficiently. Sometimes it will enable you to solve problems using **mental computation** only. The ability to identify efficient strategies to solve problems is an important test-taking skill.

EXAMPLE 1   If $12 + y \cdot z = 9$, evaluate $(12 + y \cdot z)^2 - 20$.

   Solution: Since $12 + y \cdot z = 9$, $(12 + y \cdot z)^2 - 20 = (9)^2 - 20$
   $$= 81 - 20$$
   $$= \mathbf{61}$$

EXAMPLE 2   If $r + 7 - w = 36$, evaluate $15 + \sqrt{r + 7 - w}$.

   Solution: Since $r + 7 - w = 36$, $15 + \sqrt{r + 7 - w} = 15 + \sqrt{36}$
   $$= 15 + 6$$
   $$= \mathbf{21}$$

EXERCISES

1. If $3x - 2y = 4$, evaluate $(3x - 2y)^2$.   16

2. If $14p - 3k = -9$, evaluate $(14p - 3k)^2$.   81

3. If $9 + d \cdot f = -2$, evaluate $(9 + d \cdot f)^3 + 3$.   −5

4. If $11 - x \cdot y = -3$, evaluate $(11 - x \cdot y)^3 + 14$.   −13

5. If $r - w + 3 = 49$, evaluate $\sqrt{r - w + 3}$.   7

6. If $k + n + 2 = 64$, evaluate $\sqrt{k + n + 2}$.   8

7. If $g \cdot m - 6 = 81$, evaluate $\frac{1}{3}\sqrt{g \cdot m - 6}$.   3

8. If $q \cdot z - 7 = 144$, evaluate $\frac{1}{6}\sqrt{q \cdot z - 7}$.   2

9. If $6p + v = 27$, evaluate $\sqrt[3]{6p + v}$.   3

10. If $8f - u = 8$, evaluate $\sqrt[3]{8f - u}$.   2

11. If $a \cdot b - 5 = 7$, evaluate $(a \cdot b - 6)^2 - 10$.   26

12. If $t \cdot w + 4 = 9$, evaluate $(t \cdot w + 5)^2 - 18$.   82

13. If $d \div f + 11 = 35$, evaluate $2 + \sqrt{d \div f + 12}$.   8

14. If $h \div k - 9 = 17$, evaluate $5 + \sqrt{h \div k - 10}$.   9

This topic on mental computation is another good example of a special skill that students need to acquire in order to solve certain problems on college entrance exams. Getting some practice in solving problems using mental computation will increase their speed and confidence when they encounter problems of this type. This is a good example of a skill that, if not taught, students will probably not be able to apply even if they have an inherent ability to solve this type of problem.

Examples 3 and 4 apply logical reasoning to problems involving **factoring.** The key is in identifying a binomial product or the factors of a binomial product.

EXAMPLE 3   If $x + y = 6$ and $x - y = -3$, find the value of $x^2 - y^2$.

Think: $x^2 - y^2 = (x + y)(x - y)$

Solution: Since $x + y = 6$ and $x - y = -3$, $x^2 - y^2 = 6 \cdot (-3)$, or **−18.**

EXAMPLE 4   If $a^2 - b^2 = 10$ and $a + b = 5$, find the value of $a - b$.

Think: $a^2 - b^2 = (a + b)(a - b)$

Solution: Since $a^2 - b^2 = 10$ and $a + b = 5$, $5(a - b) = 10$.

So $$\dfrac{\overset{1}{\cancel{5}}(a - b)}{\underset{1}{\cancel{5}}} = \dfrac{\overset{2}{\cancel{10}}}{\underset{1}{\cancel{5}}}$$

$$a - b = \mathbf{2}$$

EXERCISES

15. If $x + y = -2$ and $x - y = 3$, find the value of $x^2 - y^2$.   −6

16. If $r + t = 4$ and $r - t = -3$, find the value of $r^2 - t^2$.   −12

17. If $c + 2d = -6$ and $c - 2d = -5$, find the value of $c^2 - 4d^2$.   30

18. If $n + 3p = -4$ and $n - 3p = -7$, find the value of $n^2 - 9p^2$.   28

19. If $3h - f = 2.4$ and $3h + f = 3.1$, find the value of $9h^2 - f^2$.   7.44

20. If $5k - m = 2.3$ and $5k + m = 1.8$, find the value of $25k^2 - m^2$.   4.14

21. If $2w - 5z = \dfrac{3}{5}$ and $2w + 5z = \dfrac{5}{12}$, find the value of $4w^2 - 25z^2$. $\dfrac{1}{4}$

22. If $3x - 4y = \dfrac{2}{3}$ and $3x + 4y = \dfrac{3}{4}$, find the value of $9x^2 - 16y^2$. $\dfrac{1}{2}$

23. If $x^2 - y^2 = 12$ and $x + y = 3$, find the value of $x - y$.   4

24. If $w^2 - z^2 = 24$ and $w + z = 8$, find the value of $w - z$.   3

25. If $4n^2 - p^2 = 1.8$ and $2n - p = 0.9$, find the value of $2n + p$.   2

26. If $9x^2 - y^2 = 3.6$ and $3x - y = 1.2$, find the value of $3x + y$.   3

27. If $3d + 5w = \dfrac{1}{6}$ and $9d^2 - 25w^2 = 4$, find the value of $3d - 5w$.   24

28. If $6r + 5t = \dfrac{1}{4}$ and $36r^2 - 25t^2 = 10$, find the value of $6r - 5t$.   40

29. If $c^2 - d^2 = 30$ and $c - d = -5$, find the value of $(c + d) - 11$.   −17

30. If $f^2 - h^2 = 40$ and $f - h = -8$, find the value of $(f + h) + 12$.   7

CHAPTER SUMMARY

IMPORTANT TERMS	Prime polynomial *(p. 236)* Quadratic polynomial *(p. 239)*
	Common monomial factor *(p. 236)* Leading coefficient *(p. 246)*
	Greatest common factor *(p. 237)* Difference of two squares *(p. 252)*
	Degree of a monomial *(p. 239)* Trinomial square *(p. 255)*
	Degree of a polynomial *(p. 239)*

IMPORTANT IDEAS

1. In a prime polynomial the only polynomial factors with coefficients that are integers are
  a. 1 and the polynomial, and
  b. -1 and the opposite of the polynomial.

2. To factor a quadratic trinomial you must find two integers r and s meeting the following conditions.
  a. The sum of r and s equals the coefficient of the first-degree term.
  b. The product of r and s equals the product of the leading coefficient and the constant term.

3. The prime factorization of $x^2 - a^2$ is $(x + a)(x - a)$.

4. For a quadratic trinomial such as $ax^2 + bx + c$ to be a trinomial square,
  a. ax^2 and c must be perfect squares, and
  b. "$(\frac{1}{2}b)^2 = ac$" must be true.

5. In factoring a polynomial, always look for a common monomial factor as the first step.

CHAPTER REVIEW

SECTION 9.1

Factor.

1. $4x^2 - 10x$ $2x(2x - 5)$

2. $3a^3 + 15a^2$ $3a^2(a + 5)$

3. $18a^2b + 30a^2b^2 - 42ab$ $6ab(3a + 5ab - 7)$

4. $24ax + 48ay - 96at$

5. $a^2bc^3 - ab^2c + a^2b^2c^2$ $abc(ac^2 - b + abc)$

6. $2x^5 - 4x^4 + 6x^3 - 8x^2 + 10x$
     $2x(x^4 - 2x^3 + 3x^2 - 4x + 5)$

SECTION 9.2

Factor.

7. $x^2 + 7x + 3x + 21$ $(x + 7)(x + 3)$

8. $x^2 + 8x + 5x + 40$ $(x + 8)(x + 5)$

9. $x^2 + 5x + 10x + 50$ $(x + 5)(x + 10)$

10. $x^2 + 9x + 6x + 54$ $(x + 9)(x + 6)$

11. $x^2 + 10x + 21$ $(x + 7)(x + 3)$

12. $x^2 + 13x + 22$ $(x + 2)(x + 11)$

Factoring Polynomials   /   **263**

Chapter Test
Two Chapter Tests (Form A and Form B) are provided on pages 191–194 in the *Teacher's ResourceBank*.

Additional Answer
4. $24a(x + 2y - 4t)$

SECTION 9.3

Factor.

13. $x^2 + 3x - 18$ $(x + 6)(x - 3)$ 14. $x^2 - x - 20$ $(x + 4)(x - 5)$ 15. $x^2 - 9x + 18$ $(x - 6)(x - 3)$

16. $x^2 - 14x + 33$ $(x - 3)(x - 11)$ 17. $x^2 - 5x - 81$ Prime 18. $x^2 - 5x - 36$ $(x + 4)(x - 9)$

SECTION 9.4

Factor.

19. $2x^2 - x - 3$ $(2x - 3)(x + 1)$ 20. $3x^2 - 5x - 2$ $(3x + 1)(x - 2)$ 21. $4x^2 + 7x - 2$ $(4x - 1)(x + 2)$

22. $5x^2 + 4x - 1$ $(5x - 1)(x + 1)$ 23. $3x^2 - 2x - 5$ $(3x - 5)(x + 1)$ 24. $4x^2 - 4x - 3$ $(2x - 3)(2x + 1)$

SECTION 9.5

Factor.

25. $9a^2 - 16b^2$ $(3a + 4b)(3a - 4b)$ 26. $4m^2 - 81n^2$ $(2m + 9n)(2m - 9n)$ 27. $r^2s^2 - x^2y^2$ $(rs + xy)(rs - xy)$

28. $4t^2 - 121x^2$ $(2t + 11x)(2t - 11x)$ 29. $x^4 - 169$ $(x^2 + 13)(x^2 - 13)$ 30. $25a^2b^4 - 144c^2$ $(5ab^2 + 12c)(5ab^2 - 12c)$

SECTION 9.6

Write Yes or No to show whether each trinomial is a trinomial square.

31. $x^2 - 12x + 36$ Yes   32. $x^2 + 3x + 9$ No

33. $4x^2 + 10x + 13$ No   34. $9x^2y^2 - 16xy + 25$ No

35. $8x^2 - 12x + 9$ No   36. $4x^2 - 20x + 25$ Yes

Factor.

37. $x^2 - 8x + 16$ $(x - 4)(x - 4)$ 38. $x^2 + 12x + 36$ $(x + 6)(x + 6)$ 39. $4x^2 - 12x + 9$ $(2x - 3)(2x - 3)$

40. $9x^2 - 6x + 1$ $(3x - 1)(3x - 1)$ 41. $x^2 - 4xy + 4y^2$ $(x - 2y)(x - 2y)$ 42. $16x^2 - 24x + 9$ $(4x - 3)(4x - 3)$

SECTION 9.7

Factor.

43. $5x^2 - 20$ $5(x + 2)(x - 2)$ 44. $28x^2 - 7$ $7(2x + 1)(2x - 1)$

45. $3x^2 - 12x + 12$ $3(x - 2)(x - 2)$ 46. $2x^2 + 4x - 70$ $2(x - 5)(x + 7)$

47. $3x^2 - 12x + 96$ $3(x^2 - 4x + 32)$ 48. $x^2y - 2xy - 24y$ $y(x + 4)(x - 6)$

CUMULATIVE REVIEW: CHAPTERS 1–9

Perform the indicated operations. (Sections 1.2 through 1.5)

1. $-31 + 18$ $\;_{-13}$ **2.** $-5.7 - (-8.3)$ $\;_{2.6}$ **3.** $(-4\frac{2}{3})(-6)$ $\;_{28}$ **4.** $64 \div (-4)$ $\;^{-16}$

Multiply and simplify. (Section 2.5)

5. $(3x - 7)8$ $\;_{24x - 56}$ **6.** $-4(6t - 2)$ $\;_{-24t + 8}$ **7.** $-3n(4n + 2)$ $\;^{-12n^2 - 6n}$ **8.** $(2k + 5)6k$ $\;_{12k^2 + 30k}$

Solve. Check your answer. (Section 3.5)

9. $x + 35 - 4x = -1$ $\;_{x = 12}$ **10.** $2a - 3(a + 1) = 5$ $\;_{a = -8}$

Choose a variable to represent the unknowns. Write an equation for each problem. Solve. (Section 3.7)

11. A particular house is worth $4000 more than the lot on which it is located. The house is worth $20,000. Find the worth of the lot.

12. The number of hours that Jeff worked was 2 less than the number of hours that Dan worked. Jeff worked 14 hours. How many hours did Dan work?

Use Condition 1 to represent the unknowns. Use Condition 2 to write an equation for the problem. Solve. (Section 4.1 and 4.2)

13. There are 80 more single-family homes in Fairfield Subdivision than multi-family homes. There are a total of 188 homes. Find the number of single-family homes.

14. The length of a transparency is 3 inches less than twice the width. The perimeter is 42 inches. Find the length and width.

Solve. Check your answer. (Section 4.3)

15. $5x + 3 = 3x - 5$ $\;_{x = -4}$ **16.** $3k - 2(5 - k) = 10k$ $\;_{k = -2}$

Draw the graph of each inequality. (Section 5.1)

17. $x \le 2$ **18.** $x \ge -3$ **19.** $x > -1\frac{1}{2}$ **20.** $x < 5.7$

Solve each inequality. (Sections 5.3 through 5.6)

21. $8x - 1 > 7x - 3$ **22.** $12 < -3x$ **23.** $2(2 + x) > 3x + 8 + x$

Simplify. (Sections 6.4 and 6.5)

24. $(-2hk)^3$ $\;_{-8h^3k^3}$ **25.** $(xy)^{-4}$ $\;_{\frac{1}{x^4y^4}}$ **26.** $(3a^2b^6)^4$ $\;_{81a^8b^{24}}$ **27.** $\left(\frac{5cd^2}{x^3}\right)^2$ $\;_{\frac{25c^2d^4}{x^6}}$

Cumulative Review / **265**

Cumulative Test
A cumulative test covering Chapters 7–9 is provided on pages 195–198 in the *Teacher's ResourceBank.*

Additional Answers
11. Let x = value of lot. Then $x + 4000$ = value of house. $x + 4000 = 20,000$ Lot: $16,000
12. Let x = number of hours Dan worked. Then $x - 2$ = number of hours Jeff worked. $x - 2 = 14$ Dan: 16 hours
13. Let x = number of multi-family homes. Then $x + 80$ = number of single-family homes. $x + x + 80 = 188$ Single-family homes: 134
14. Let x = width. Then $2x - 3$ = length. $2x + 2(2x - 3) = 42$ width: 8 inches; length: 13 inches
17. On a number line, all points to the left of, and including, 2.
18. On a number line, all points to the right of, and including, -3.
19. On a number line, all points to the right of, and not including, $-1\frac{1}{2}$.
20. On a number line, all points to the left of, and not including, 5.7.
21. {all real numbers greater than -2}
22. {all real numbers less than -4}
23. {all real numbers less than -2}

265

Simplify. (Sections 7.1 through 7.4)

28. $\sqrt{81}$ 9

29. $-\sqrt{\frac{9}{16}}$ $-\frac{3}{4}$

30. $\sqrt{2}\cdot\sqrt{7}\cdot\sqrt{14}$ 14

31. $\sqrt{12}\cdot\sqrt{27}$ 18

32. $\sqrt{500}$ $10\sqrt{5}$

33. $\sqrt[3]{24}$ $2\sqrt[3]{3}$

34. $\sqrt{a^4b^7}$ $a^2b^3\sqrt{b}$

35. $\sqrt{8x^7y^9}$

Add or subtract. Simplify where necessary. (Section 7.5)

36. $5\sqrt{2}+8\sqrt{2}$ $13\sqrt{2}$

37. $2\sqrt{3}-\sqrt{27}$ $-\sqrt{3}$

38. $\sqrt{5}+\sqrt{10}$

39. $\sqrt{50p}-\sqrt{18p}$ $2\sqrt{2p}$

Simplify. Rationalize the denominator where necessary. (Section 7.6)

40. $\sqrt{\frac{32}{w^2}}$ $\frac{4\sqrt{2}}{w}$

41. $\frac{\sqrt{a^5}}{\sqrt{16a^3}}$ $\frac{a}{4}$

42. $\frac{\sqrt{5}}{\sqrt{6}}$ $\frac{\sqrt{30}}{6}$

43. $\frac{\sqrt{2m}}{\sqrt{7p}}$ $\frac{\sqrt{14mp}}{7p}$

Approximate each square root to three decimal places. Use the table on page 194. (Section 7.7)

44. $\sqrt{80}$ 8.944

45. $\sqrt{150}$ 12.247

46. $\sqrt{\frac{3}{5}}$ 0.775

47. $\sqrt{\frac{2}{27}}$ 0.272

Solve each problem. (Section 7.8)

48. The legs of a right triangle have lengths of 15 inches and 36 inches. Find the length of the hypotenuse. 39 inches

49. Find the approximate width of a 20-inch TV screen if the height is 12 inches. 16 inches

Evaluate each expression. (Sections 8.1 and 8.3)

50. $-8x^3$ when $x=-2$ 64

51. y^2-3y when $y=2$ -2

52. $2a^2+5a-9$ when $a=-3$ -6

Multiply or divide as indicated. (Section 8.2)

53. $(12a^3)(-0.6a^2)$ $-7.2a^5$

54. $(5x^2y)(-9x^2y^4w)$ $-45x^4y^5w$

55. $\dfrac{-48c^7}{16c^2}$ $-3c^5$

56. $\dfrac{-24h^2k^5m}{-40hk^3m}$

Perform the indicated operation. (Sections 8.4 through 8.7)

57. $(a^2-3a+7)+(3a^2-a-19)$

58. $(y^2+y-11)-(7y^2+5)$

59. $(4w-1)(3w+5)$ $12w^2+17w-5$

60. $(g-3)(2g^2+3g-1)$

61. $3x+2\,)\,\overline{12x^2-x-6}$ $4x-3$

62. $n-4\,)\,\overline{n^3-19n+4}$

Factor. (Sections 9.1 through 9.7)

63. $10y^3-15y^2$ $5y^2(2y-3)$

64. $6x^4+21x^3-3x^2$

65. $x^2+7x+2x+14$

66. $h^2+19h+18$ $(h+1)(h+18)$

67. $g^2-10g-24$ $(g+2)(g-12)$

68. $3m^2-7m+2$

69. $6n^2+7n-3$ $(3n-1)(2n+3)$

70. c^2-9 $(c+3)(c-3)$

71. $4y^2-25d^2$

72. $9x^2-6x+1$ $(3x-1)(3x-1)$

73. $9x^2-36$ $(3x+6)(3x-6)$, or $9(x+2)(x-2)$

74. $24x^2+30x-9$

Products and Quotients of Rational Expressions

Glide ability is related to the ratio of the wing spread and the wing area. Electrical resistance in a wire is related to the ratio of its length and diameter. The velocity of sound waves is related to an increase or decrease in temperature.

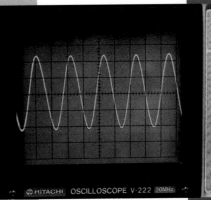

Lesson Resources
Warm–Up: p. M–36
Maintenance: See below.
Practice Worksheet 10.1

Maintenance
1. Subtract: $12 - (-20)$
   Ans: 32   (Section 1.3)
2. Solve: $\frac{2}{3}x < 15$
   Ans: {all real numbers less than $22\frac{1}{2}$}
   (Section 5.5)
3. Simplify: $\sqrt{3 \cdot 5 \cdot 7 \cdot 5 \cdot 3}$
   Ans: $15\sqrt{7}$ (Section 7.3)
4. Factor: $x^3 - 2x^2 + 5x$
   Ans: $x(x^2 - 2x + 5)$
   (Section 9.1)
5. The perimeter of a magazine cover is 38 inches. The length is 11 inches. Find the width.
   Ans: 8 inches
   (Section 3.4)

Additional Examples
Example 1
Evaluate each rational expression if x equals 2.
1. $\dfrac{x^2 + 6x + 5}{3x^2 + 9}$
   Ans: 1
2. $\dfrac{2x^2 - x + 3}{x^2 - 4x - 9}$
Ans: $-\dfrac{9}{13}$

10.1 Rational Expressions

OBJECTIVES: To identify whether or not an expression is a rational expression
To evaluate a rational expression for a given value of its variable
To write the number or numbers that cannot be in the domain of a rational expression

Definition

A *rational expression* is one that can be written as the quotient of two polynomials.

$$\frac{x^2 + x - 3}{x^2 - 3x + 5}; \frac{2}{x};$$

$$\frac{4}{3}; \frac{x^2 - 5}{1}$$

P–1 **Write $x^2 + x - 1$ as a quotient.** $\dfrac{x^2 + x - 1}{1}$

Any polynomial is also a rational expression since it can be expressed as a quotient with 1 as divisor.

P–2 **Which of the expressions below are rational expressions?**

a. $\dfrac{2x}{5}$ Yes **b.** $\dfrac{3x + 7}{x^2 - 3x + 10}$ Yes **c.** $\dfrac{|x^2 + 1|}{5}$ No

Remember that both dividend and divisor must be polynomials.

d. $\dfrac{-7}{10}$ Yes **e.** $\dfrac{\sqrt{x - 3}}{\sqrt{x + 3}}$ No **f.** 75 Yes

EXAMPLE 1

You may want to discuss how rational expressions can be evaluated by using a calculator. Refer to page 216.

Evaluate the following rational expression if x equals -2.

$$\frac{2x^2 - 3x + 1}{-x^2 + x - 3}$$

Solution:

① Replace x by -2. ⟶ $\dfrac{2x^2 - 3x + 1}{-x^2 + x - 3} = \dfrac{2(-2)^2 - 3(-2) + 1}{-(-2)^2 + (-2) - 3}$

② Simplify. ⟶ $= \dfrac{8 + 6 + 1}{-4 - 2 - 3}$

$= \dfrac{15}{-9}$ or $-\dfrac{15}{9}$

P–3 **Evaluate** $\dfrac{-x^2 + 5x - 1}{2x^2 - x + 4}$ **if x equals -3.** -1

268 / *Chapter 10*

> Any number that will result in a divisor of 0 must be restricted from the domain of the variable in a rational expression.

$$\frac{x+5}{x^2-1}$$

$$x \neq 1; \; x \neq -1$$

EXAMPLE 2   Write the number or numbers that cannot be in the domain of x in $\frac{x-2}{3x+4}$.

Solution: You can see that $3x + 4$ must not equal 0.

| Solve for the number that makes $3x + 4$ equal 0. | → | This number cannot be in the domain. |

$$3x + 4 = 0$$

$$3x + 4 - 4 = 0 - 4$$

$$3x = -4$$

$\frac{1}{3}(3x) = \frac{1}{3}(-4)$ Have students substitute $-\frac{4}{3}$ for x in $3x + 4$.

$x = -\frac{4}{3}$ The number $-\frac{4}{3}$ cannot be in the domain.

Additional Examples
Examples 2
Write the number or numbers that cannot be in the domain.

1. $\frac{2x^2 - 5x}{4x + 1}$

   Ans: $-\frac{1}{4}$

2. $\frac{x(x - 3)}{(x + 2)(x - 4)}$

   Ans: $-2; 4$

3. $\frac{3(x - 7)(x + 7)}{5x(x^2 - 9)}$

   Ans: $0; 3; -3$

■■■■ **CLASSROOM EXERCISES** ■■■■

Write Yes or No to tell whether each expression is a rational expression. (P-2)

1. $x^2 - 2x + 1$ Yes

2. $\frac{x^2 + 1}{5}$ Yes

3. $\sqrt{x + 1}$ No

4. 15 Yes

5. $\frac{2}{5}$ Yes

6. $\frac{2}{\sqrt{x}}$ No

7. $\frac{x - 3}{x + 1}$ Yes

8. $\frac{x(x + 3)}{x^2 + 2x + 1}$ Yes

Write the number or numbers that cannot be in the domain of each variable. (Example 2)

9. $\frac{5}{x}$ 0

10. $\frac{3}{x + 1}$ -1

11. $\frac{x + 3}{x - 3}$ 3

12. $\frac{x + 10}{-5x}$ 0

13. $\frac{x}{2x - 1}$ $\frac{1}{2}$

14. $\frac{3x + 1}{2x + 1}$ $-\frac{1}{2}$

15. $\frac{5}{x^2 - 1}$ 1, -1

16. $\frac{x}{(x - 3)(x + 2)}$ 3, -2

Products and Quotients of Rational Expressions / **269**

Assignment Guide
Basic p. 270: 1–31 odd
Average p. 270: 3, 6, 9, ⋯, 30,
32; p. 273: 3, 6, 9, ⋯, 30, 32
Above Average p. 270:
1–31 odd; p. 273: 3, 6, 9, ⋯, 30, 32

▓▓▓▓▓▓ **WRITTEN EXERCISES** ▓▓▓▓▓▓

Goal: To evaluate rational expressions

Sample Problem: Evaluate $\dfrac{x^2 - 4}{x + 3}$ if x equals 3.

Answer: $\dfrac{5}{6}$

Write <u>Yes</u> *or* <u>No</u> *to show whether each expression is a rational expression.*
(P-2)

1. $x^2 + 10$ Yes **2.** $x^2 - 3x + 5$ Yes **3.** $\dfrac{3}{4}$ Yes **4.** $-\dfrac{5}{6}$ Yes

5. $\dfrac{x - 2}{x + 2}$ Yes **6.** $\dfrac{x^2 - 5}{x^2 + 10}$ Yes **7.** $\dfrac{\sqrt{x} - 1}{\sqrt{x} + 1}$ No **8.** $\dfrac{x^2 + \sqrt{x}}{x^2 - \sqrt{x}}$ No

9. $\dfrac{|x|}{-x}$ No **10.** $\dfrac{x - 3}{|x + 2|}$ No **11.** $\dfrac{x^2 - 2x + 3}{3x^2 - 5x + 2}$ Yes **12.** $\dfrac{5 - x^2 + x}{x - 6 + 2x^2}$ Yes

Evaluate each rational expression if x equals 1. (Example 1)

13. $\dfrac{5}{-2x}$ $-\dfrac{5}{2}$ **14.** $\dfrac{-7}{5x}$ $-\dfrac{7}{5}$ **15.** $\dfrac{x + 3}{x + 2}$ $\dfrac{4}{3}$ **16.** $\dfrac{x - 2}{x + 2}$ $-\dfrac{1}{3}$

17. $\dfrac{x^2 - 2x + 5}{x^2 + 3x - 1}$ $\dfrac{4}{3}$ **18.** $\dfrac{x^2 - 1}{x^2 + 1}$ 0 **19.** $\dfrac{2x^2 - 3x + 1}{x^2 - 5x - 3}$ 0 **20.** $\dfrac{x^2 - x - 1}{x^2 - 7x + 3}$ $\dfrac{1}{3}$

Write the number or numbers that cannot be in the domain of each variable.
(Example 2)

21. $\dfrac{x^2 + 2}{x}$ 0 **22.** $\dfrac{x - 5}{-x}$ 0 **23.** $\dfrac{2x}{x - 10}$ 10 **24.** $\dfrac{3x}{x + 15}$ -15

25. $\dfrac{x^2 + 3}{5x}$ 0 **26.** $\dfrac{x^2 - 4}{-3x}$ 0 **27.** $\dfrac{2}{x^2 - 4}$ 2, -2 **28.** $\dfrac{5}{x^2 - 9}$ 3, -3

29. $\dfrac{2}{3x - 1}$ $\dfrac{1}{3}$ **30.** $\dfrac{5}{4x + 3}$ $-\dfrac{3}{4}$ **31.** $\dfrac{3}{(x - 1)(x + 7)}$ 1, -7 **32.** $\dfrac{7}{(x + 4)(x - 9)}$ $-4, 9$

REVIEW CAPSULE FOR SECTION 10.2

Write the prime factorization of each polynomial. (Pages 236–238)

1. $18x^3$ **2.** $12ab^2$ **3.** $x^2 - 25$ $(x + 5)(x - 5)$ **4.** $c^2 - 81$

5. $x^2 + 7x$ $x(x + 7)$ **6.** $a^2 - b^2$ $(a + b)(a - b)$ **7.** $x^2 - 6x + 9$ **8.** $3ab + 3b^2$

9. $3a^2 + 6ab$ $3a(a + 2b)$ **10.** $5a^2 - ab$ $a(5a - b)$ **11.** $x^2 + 4x + 4$ $(x + 2)(x + 2)$ **12.** $x^2 - y^2$ $(x + y)(x - y)$

10.2 Simplifying Rational Expressions

To simplify rational expressions involving monomials or trinomials

The Product Rule for Fractions can be applied to rational expressions.

Product Rule for Rational Expressions

If $\dfrac{a}{b}$ and $\dfrac{c}{d}$ are rational expressions,

then $\dfrac{a}{b} \cdot \dfrac{c}{d} = \dfrac{ac}{bd}$. (*b* and *d* are not zero.)

$$\frac{x}{x+1} \cdot \frac{x+3}{x-2} =$$

$$\frac{x^2 + 3x}{x^2 - x - 2}$$

You can simplify a rational expression by applying the Product Rule for Rational Expressions.

In working with rational expressions, it will be assumed that no divisor equals 0.

Examples 1, 2, and 3 demonstrate the detailed steps in simplifying rational expressions. Example 4 demonstrates a "shortcut" method which combines and disguises the explicit steps of the other examples.

EXAMPLE 1   Simplify $\dfrac{10x}{15x^2}$.

Solution:

1 Write the prime factorizations. $\longrightarrow \dfrac{10x}{15x^2} = \dfrac{2 \cdot 5 \cdot x}{3 \cdot 5 \cdot x \cdot x}$

2 Product Rule for Rational Expressions $\longrightarrow = \dfrac{5x}{5x} \cdot \dfrac{2}{3x}$ ◀ $\dfrac{5x}{5x}$ is a name for 1.

3 Multiplication Property of One $\longrightarrow = \dfrac{2}{3x}$ ◀ **Simplest form**

EXAMPLE 2   Simplify $\dfrac{x^2 + x}{x^2 + 2x + 1}$.

Solution:

1 $\dfrac{x^2 + x}{x^2 + 2x + 1} = \dfrac{x(x + 1)}{(x + 1)(x + 1)}$

2 $= \dfrac{x + 1}{x + 1} \cdot \dfrac{x}{x + 1}$

3 $= \dfrac{x}{x + 1}$ ◀ *x* is **not** a factor of *x* + 1.

*Products and Quotients of Rational Expressions / **271***

Teaching Suggestions p. M–36

Lesson Resources
Warm–Up: p. M–37
Maintenance: See below.
Practice Worksheet 10.2
Exploration 10
Transparencies 24, 26, 27, 28

Maintenance
1. Multiply:
   $(-\frac{1}{3})(-\frac{1}{4})(-2)(-6)$
   Ans: 1 (Section 2.3)
2. Solve: $-8t > 112$
   Ans: {all real numbers less than −14}
   (Section 5.5)
3. Simplify: $\sqrt[3]{512}$
   Ans: 8 (Section 7.3)
4. Factor: $a^2 - 7a - 18$
   Ans: (a + 2)(a − 9)
   (Section 9.3)
5. The length and width of a rectangle are in the ratio 4:3 (Condition 1). The perimeter is 42 meters (Condition 2). Find the length and the width of the rectangle.
   Ans: length: 12 meters; width: 9 meters
   (Section 4.2)

Additional Examples
Example 1
Simplify.
1. $\dfrac{21x^3}{14x}$
   Ans: $\dfrac{3x^2}{2}$
2. $\dfrac{18x^2}{20x^3}$
   Ans: $\dfrac{9}{10x}$

Example 2
Simplify.
1. $\dfrac{8x + 8}{x^2 - 1}$
   Ans: $\dfrac{8}{x - 1}$
2. $\dfrac{3x^2 - 15x}{x^2 - 4x - 5}$
   Ans: $\dfrac{3x}{x + 1}$

What are the prime factors of $x^2 - 4$? $(x + 3), (x - 2)$ **of $x^2 + x - 6$?** $(x + 2), (x - 2)$

$(x + 3), (x - 2)$

$(x + 2), (x - 2)$

Additional Examples
Example 3
Simplify.

1. $\dfrac{x^2 - 25}{x^2 - 6x + 5}$

   Ans: $\dfrac{x + 5}{x - 1}$

2. $\dfrac{x^2 - 2x - 8}{x^2 + 4x + 4}$

   Ans: $\dfrac{x - 4}{x + 2}$

3. $\dfrac{x^2 + 8x + 16}{x^2 - 16}$

   Ans: $\dfrac{x + 4}{x - 4}$

Example 4
Simplify.

1. $\dfrac{4x^2 + 4xy + y^2}{16x^2y + 8xy^2}$

   Ans: $\dfrac{2x + y}{8xy}$

2. $\dfrac{x^2 - 9y^2}{x^2 - 7xy + 12y^2}$

   Ans: $\dfrac{x + 3y}{x - 4y}$

3. $\dfrac{xy^2 + x}{xy + x}$

   Ans: $\dfrac{y^2 + 1}{y + 1}$

EXAMPLE 3 Simplify $\dfrac{x^2 - 4}{x^2 + x - 6}$.

Solution:

1 $\dfrac{x^2 - 4}{x^2 + x - 6} = \dfrac{(x + 2)(x - 2)}{(x - 2)(x + 3)}$

2 $= \dfrac{x - 2}{x - 2} \cdot \dfrac{x + 2}{x + 3}$

3 $= \dfrac{x + 2}{x + 3}$

EXAMPLE 4 Simplify $\dfrac{6x^2y - 3xy^2}{4x^2 - y^2}$.

Solution: The method shown below for identifying a name for 1 is often used to save work.

$$\dfrac{6x^2y - 3xy^2}{4x^2 - y^2} = \dfrac{3xy(2x - y)}{(2x + y)(2x - y)}$$ **Prime factorization**

This shortcut method sometimes encourages incorrect "cancellation," e.g. the two "x's" in $\dfrac{3xy}{2x + y}$.

$$= \dfrac{3xy\cancel{(2x - y)}^{1}}{(2x + y)\cancel{(2x - y)}_{1}}$$ **Note how a name for 1 is shown.**

$$= \dfrac{3xy}{2x + y}$$ **Simplest form**

Steps for simplifying rational expressions:

1. Write the prime factorizations of both polynomials.
2. Form a name for 1 from all factors common to the dividend and divisor.
3. Use this name for 1 and write an indicated product equivalent to the given rational expression.
4. Apply the Multiplication Property of One and write the result in the simplest form.

272 / *Chapter 10*

Simplify. No divisor equals 0. (Examples 1–3)

1. $\dfrac{2}{6}$ $\dfrac{1}{3}$

2. $\dfrac{2x}{3x}$ $\dfrac{2}{3}$

3. $\dfrac{x}{x^2}$ $\dfrac{1}{x}$

4. $\dfrac{3x}{3(x-2)}$ $\dfrac{x}{x-2}$

5. $\dfrac{3(x+1)}{x(x+1)}$ $\dfrac{\frac{3}{x}}{}$

6. $\dfrac{5(x^2+2)}{x^2+2}$ 5

7. $\dfrac{x+2}{x+3}$ $\dfrac{x+2}{x+3}$

8. $\dfrac{(x+1)(x-2)}{(x+1)}$ $x-2$

9. $\dfrac{x(x+2)}{x^2(x+2)}$ $\dfrac{1}{x}$

10. $\dfrac{x-1}{x^2-1}$ $\dfrac{1}{x+1}$

WRITTEN EXERCISES

Goal: To simplify rational expressions

Sample Problem: Simplify $\dfrac{9x^2+18x}{x^2+4x+4}$.   **Answer:** $\dfrac{9x}{x+2}$

Simplify. No divisor equals 0. (Examples 1–4)

1. $\dfrac{5x^2}{10x}$ $\dfrac{x}{2}$

2. $\dfrac{18x}{10x^2}$ $\dfrac{9}{5x}$

3. $\dfrac{12x^3}{28x^4}$ $\dfrac{3}{7x}$

4. $\dfrac{20x^3}{50x^2}$

5. $\dfrac{x(x-3)}{5(x-3)}$ $\dfrac{x}{5}$

6. $\dfrac{3(x+10)}{x(x+10)}$ $\dfrac{3}{x}$

7. $\dfrac{2x(x-5)}{3x(x-5)}$ $\dfrac{2}{3}$

8. $\dfrac{5x(x+7)}{4x(x+7)}$

9. $\dfrac{(x+2)(x-3)}{(x-3)(x+5)}$ $\dfrac{x+2}{x+5}$

10. $\dfrac{(x-4)(x+2)}{(x+2)(x-1)}$ $\dfrac{x-4}{x-1}$

11. $\dfrac{4x-4}{3x+9}$ $\dfrac{4x-4}{3x+9}$

12. $\dfrac{3x+3}{5x+5}$

13. $\dfrac{3x-6}{5x-10}$ $\dfrac{3}{5}$

14. $\dfrac{4x+12}{3x+9}$ $\dfrac{4}{3}$

15. $\dfrac{x^2+3x}{x^2+5x}$ $\dfrac{x+3}{x+5}$

16. $\dfrac{x^2-5x}{x^2+7x}$

17. $\dfrac{x+3}{x^2-9}$ $\dfrac{1}{x-3}$

18. $\dfrac{x-5}{x^2-25}$ $\dfrac{1}{x+5}$

19. $\dfrac{x^2-16}{x(x-4)}$ $\dfrac{x+4}{x}$

20. $\dfrac{x^2-36}{x(x+6)}$

21. $\dfrac{x^2+2x}{x^2-4}$ $\dfrac{x}{x-2}$

22. $\dfrac{x^2-49}{x^2+7x}$ $\dfrac{x-7}{x}$

23. $\dfrac{x(x-2)}{x^2-4x+4}$ $\dfrac{x}{x-2}$

24. $\dfrac{x^2+6x+9}{x(x+3)}$

25. $\dfrac{4x^2y}{10xy^2}$ $\dfrac{2x}{5y}$

26. $\dfrac{16a^2b}{12ab^2}$ $\dfrac{4a}{3b}$

27. $\dfrac{2a+2b}{2c}$ $\dfrac{a+b}{c}$

28. $\dfrac{3t}{3x-3y}$

29. $\dfrac{a^2-b^2}{3a+3b}$ $\dfrac{a-b}{3}$

30. $\dfrac{5x+5y}{x^2-y^2}$ $\dfrac{5}{x-y}$

31. $\dfrac{5a^2-ab}{3a^2+6ab}$ $\dfrac{5a-b}{3a+6b}$

32. $\dfrac{3a^2+3ab}{3ab+3b^2}$

REVIEW CAPSULE FOR SECTION 10.3

Write the prime factors of each expression. (Pages 236–238)

1. $12x^2y^3$ 2. $20r^3st^2$ 3. $18a^2b^4c^3$ 4. $36p^5q^2r^3$

Common Error
When simplifying rational expressions with binomial numerators and denominators, students try to simplify without factoring.

Example

$\dfrac{4x+12}{3x+9} = \dfrac{\overset{1}{\cancel{4x}}+\overset{4}{\cancel{12}}}{\underset{1}{\cancel{3x}}+\underset{3}{\cancel{9}}}$

$= \dfrac{4+4}{3+3}$

$= \dfrac{8}{6}$, or $\dfrac{4}{3}$

Prescription
Review Example 4. Show students that they must find the common factors of the numerator and denominator and not of separate terms.

Assignment Guide
Basic p. 273: 1–31 odd
Average p. 273: 3, 6, 9, ···, 30, 32; p. 270: 3, 6, 9, ···, 30, 32
Above Average p. 273: 3, 6, 9, ···, 30, 32; p. 270: 1–31

Additional Answers
Written Exercises
4. $\dfrac{2x}{5}$
8. $\dfrac{5}{4}$
12. $\dfrac{3}{5}$
16. $\dfrac{x-5}{x+7}$
20. $\dfrac{x-6}{x}$
24. $\dfrac{x+3}{x}$
28. $\dfrac{t}{x-y}$
32. $\dfrac{a}{b}$

Review Capsule
1. $2 \cdot 2 \cdot 3 \cdot x \cdot x \cdot y \cdot y \cdot y$
2. $2 \cdot 2 \cdot 5 \cdot r \cdot r \cdot r \cdot s \cdot t \cdot t$
3. $2 \cdot 3 \cdot 3 \cdot a \cdot a \cdot b \cdot b \cdot b \cdot b \cdot c \cdot c \cdot c$
4. $2 \cdot 2 \cdot 3 \cdot 3 \cdot p \cdot p \cdot p \cdot p \cdot p \cdot q \cdot q \cdot r \cdot r \cdot r$

273

Lesson Resources
Warm–Up: p. M–37
Maintenance: See below.
Practice Worksheet 10.3
Transparency 28

Maintenance
1. Solve and check:
   $3.5 = d + 2.3$
   Ans: $d = 1.2$
   (Section 3.1)

2. Solve: $\left(-\frac{2}{3}\right)^0 = a$
   Ans: $a = 1$ (Section 6.2)

3. Simplify: $\sqrt{25a^4 b^5}$
   Ans: $5a^2 b^2 \sqrt{b}$
   (Section 7.4)

4. Factor: $3x^2 + 10x + 7$
   Ans: $(3x + 7)(x + 1)$
   (Section 9.4)

5. A pair of tennis shoes was on sale with a discount rate of 20%. The sale price was $32. Find the list price and the discount.
   Ans: list price: $40;
   discount: $8 (Section 4.4)

Additional Examples
Example 1
Multiply and simplify.
1. $\dfrac{9}{28x} \cdot \dfrac{4x^2}{18x}$
   Ans: $\dfrac{1}{14}$
2. $\dfrac{30x^3}{21x} \cdot \dfrac{35x^2}{4x}$
   Ans: $\dfrac{25x^3}{2}$

Example 2
Multiply and simplify.
1. $\dfrac{9}{15x^3} \cdot \dfrac{10x^3}{3}$
   Ans: 2
2. $\dfrac{4x}{3x^2} \cdot \dfrac{6x^3}{8x^2}$
   Ans: 1

10.3 Multiplying Rational Expressions

OBJECTIVE: To multiply two or more rational expressions with monomial numerators and denominators, writing the product in simplest form

When multiplying fractions, you always simplify the product. This also applies to multiplying rational expressions.

P–1 Write each product in simplest form.

a. $\dfrac{10}{3} \cdot \dfrac{12}{5}$ 8
b. $\dfrac{2}{6} \cdot \dfrac{12}{5}$ $\dfrac{4}{5}$
c. $\dfrac{5}{6} \cdot \dfrac{21}{10}$ $1\dfrac{3}{4}$
d. $\dfrac{2}{1} \cdot \dfrac{3}{10} \cdot \dfrac{1}{12}$ $\dfrac{1}{20}$

Remind students that the Product Rule allows you to write a product as one rational expression or to write one rational expression as a product.

EXAMPLE 1 Multiply and simplify: $\dfrac{10x}{3} \cdot \dfrac{12}{5x^2}$

Solution:

1. Product Rule for Rational Expressions → $\dfrac{10x}{3} \cdot \dfrac{12}{5x^2} = \dfrac{(10x)(12)}{(3)(5x^2)}$

2. Factor. → $= \dfrac{(2 \cdot 5 \cdot x)(2 \cdot 2 \cdot 3)}{(3)(5 \cdot x \cdot x)}$

3. Product Rule for Rational Expressions → $= \dfrac{3 \cdot 5 \cdot x}{3 \cdot 5 \cdot x} \cdot \dfrac{2 \cdot 2 \cdot 2}{x}$

4. Multiplication Property of One → $= \dfrac{2 \cdot 2 \cdot 2}{x}$

5. Simplify. → $= \dfrac{8}{x}$

Steps can often be saved by identifying common factors first.

EXAMPLE 2 Multiply and simplify: $\dfrac{2x}{6x^2} \cdot \dfrac{12x}{5x^2}$

Solution:

$$\dfrac{2x}{6x^2} \cdot \dfrac{12x}{5x^2} = \dfrac{\overset{1}{2x} \cdot \overset{2 \cdot 1}{12x}}{\underset{1 \cdot x}{6x^2} \cdot \underset{x}{5x^2}}$$ x, 6, and x are common factors.

$$= \dfrac{4}{5x^2}$$ Simplest form

This demonstrates a shortcut method that combines many of the steps shown in Example 1.

274 / Chapter 10

EXAMPLE 3   Multiply and simplify: $\dfrac{5xy}{6a} \cdot \dfrac{21ab}{10y^2}$ 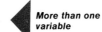 **More than one variable**

Solution:

$$\dfrac{5xy}{6a} \cdot \dfrac{21ab}{10y^2} = \dfrac{\overset{1 \cdot 1}{\cancel{5xy}}}{\underset{2 \cdot 1}{\cancel{6a}}} \cdot \dfrac{\overset{7 \cdot 1}{\cancel{21ab}}}{\underset{2 \cdot y}{\cancel{10y^2}}}$$

$$= \dfrac{1 \cdot x \cdot 1}{2 \cdot 1} \cdot \dfrac{7 \cdot 1 \cdot b}{2 \cdot y}$$

$$= \dfrac{x}{2} \cdot \dfrac{7b}{2y} = \dfrac{7bx}{4y}$$

EXAMPLE 4   Multiply and simplify: $\dfrac{2}{x} \cdot \dfrac{3x}{10} \cdot \dfrac{x^2}{12}$

Solution:

$$\dfrac{2}{x} \cdot \dfrac{3x}{10} \cdot \dfrac{x^2}{12} = \dfrac{\overset{1}{\cancel{2}}}{\underset{1}{\cancel{x}}} \cdot \dfrac{\overset{1}{\cancel{3x}}}{\underset{5}{\cancel{10}}} \cdot \dfrac{\overset{1}{\cancel{x^2}}}{\underset{4}{\cancel{12}}}$$

$$= \dfrac{x^2}{20}$$

P–2 **How do you know that** $\dfrac{x^2}{20}$ **is in simplest form?** x^2 and 20 have no common factors.

CLASSROOM EXERCISES

Multiply. Simplify where necessary. (Examples 1 and 2)

1. $\dfrac{2}{x} \cdot \dfrac{x}{5}$ $\dfrac{2}{5}$

2. $\dfrac{y}{3} \cdot \dfrac{2}{y}$ $\dfrac{2}{3}$

3. $\dfrac{x}{5} \cdot \dfrac{5}{x}$ 1

4. $\dfrac{3a}{5} \cdot \dfrac{7}{3a}$

5. $\dfrac{2}{3} \cdot \dfrac{x}{y}$ $\dfrac{2x}{3y}$

6. $\dfrac{7}{x} \cdot \dfrac{5}{x}$ $\dfrac{35}{x^2}$

7. $\dfrac{a}{3} \cdot \dfrac{5}{2a}$ $\dfrac{5}{6}$

8. $\dfrac{5x}{7} \cdot \dfrac{6}{x}$

9. $\dfrac{3}{a^2} \cdot \dfrac{a}{5}$ $\dfrac{3}{5a}$

10. $-\dfrac{x}{7} \cdot \dfrac{5}{x^2}$ $-\dfrac{5}{7x}$

11. $\dfrac{2}{x} \cdot \dfrac{x}{5}$ $\dfrac{2}{5}$

12. $\dfrac{x}{3} \cdot \dfrac{x^2}{5}$

Example 3
Multiply and simplify.

1. $\dfrac{14ax}{5b} \cdot \dfrac{3b^2y}{28xy}$
   Ans: $\dfrac{3ab}{10}$

2. $\dfrac{2x^3y}{3a^2} \cdot \dfrac{15ab}{4x^2}$
   Ans: $\dfrac{5bxy}{2a}$

Example 4
Multiply and simplify.

1. $\dfrac{3x}{8} \cdot \dfrac{4x}{6} \cdot \dfrac{16}{3x^3}$
   Ans: $\dfrac{4}{3x}$

2. $\dfrac{18x^2}{33a} \cdot \dfrac{20ab}{16x} \cdot \dfrac{11}{2b}$
   Ans: $\dfrac{15x}{4}$

3. $\dfrac{9}{28x} \cdot \dfrac{5x^3}{3x} \cdot \dfrac{4}{18x^2}$
   Ans: $\dfrac{5}{42x}$

Common Error
Students simplify incorrectly when multiplying rational expressions.

Example

$$\dfrac{3}{a^2} \cdot \dfrac{a}{5} = \dfrac{3}{\underset{a}{\cancel{a^2}}} \cdot \dfrac{\overset{1}{\cancel{a}}}{5}$$

$$= \dfrac{3}{10}$$

Prescription
Have students work Classroom Exercises 1-20 as shown in Example 1. Exercise 9, for example would be worked as follows.

$$\dfrac{3}{a^2} \cdot \dfrac{a}{5} = \dfrac{3 \cdot a}{a \cdot a \cdot 5}$$

$$= \dfrac{a}{a} \cdot \dfrac{3}{a \cdot 5} = \dfrac{3}{5a}$$

Additional Answers
Classroom Exercises

4. $\dfrac{7}{5}$

8. $\dfrac{30}{7}$

12. $\dfrac{x^3}{15}$

Multiply. Simplify where necessary. (Examples 1–4)

13. $\dfrac{2x}{3} \cdot \dfrac{1}{10x}$ $\dfrac{1}{15}$

14. $\dfrac{1}{5x} \cdot 15x$ 3

15. $\dfrac{3a}{b} \cdot 2b^2$ $6ab$

16. $\dfrac{3x^2}{2} \cdot \dfrac{2}{x}$ $3x$

17. $\left(-\dfrac{1}{3}\right)\left(-\dfrac{2}{y}\right)\left(-\dfrac{x}{2}\right)$ $-\dfrac{x}{3y}$

18. $\left(-\dfrac{1}{3}\right)\left(\dfrac{3}{x}\right)\left(-\dfrac{x}{5}\right)$ $\dfrac{1}{5}$

19. $\left(\dfrac{2a}{b}\right)\left(\dfrac{b}{3}\right)\left(\dfrac{3}{2a}\right)$ 1

20. $\dfrac{x}{y^2} \cdot \dfrac{y}{x^2}$ $\dfrac{1}{xy}$

WRITTEN EXERCISES

Goal: To multiply rational expressions

Sample Problem: $\dfrac{3x}{4a} \cdot \dfrac{5a^2}{9} \cdot \dfrac{6}{x^2}$ **Answer:** $\dfrac{5a}{2x}$

Multiply. Simplify where necessary. (Examples 1–4)

1. $\dfrac{3}{8} \cdot \dfrac{5}{9}$ $\dfrac{5}{24}$

2. $\dfrac{4}{5} \cdot \dfrac{3}{7}$ $\dfrac{12}{35}$

3. $\dfrac{5}{x} \cdot \dfrac{y}{3}$ $\dfrac{5y}{3x}$

4. $\dfrac{a}{7} \cdot \dfrac{4}{b}$ $\dfrac{4a}{7b}$

5. $\dfrac{2x}{5} \cdot \dfrac{3}{2x}$ $\dfrac{3}{5}$

6. $\dfrac{6}{7x} \cdot \dfrac{7x}{11}$ $\dfrac{6}{11}$

7. $\dfrac{4x}{3} \cdot \dfrac{7}{10x}$ $\dfrac{14}{15}$

8. $\dfrac{5}{12x} \cdot \dfrac{4x}{7}$ $\dfrac{5}{21}$

9. $\left(-\dfrac{x^2}{y}\right)\left(-\dfrac{2y}{x}\right)$

10. $\left(-\dfrac{a}{b^2}\right)\left(-\dfrac{2b}{a^2}\right)$ $\dfrac{2}{ab}$

11. $\dfrac{8xy}{3} \cdot \dfrac{15}{2x^2}$ $\dfrac{20y}{x}$

12. $\dfrac{3}{10ab} \cdot \dfrac{14b^2}{15}$

13. $-\dfrac{21b^2}{a^2} \cdot \dfrac{3ab}{15b}$ $-\dfrac{21b^2}{5a}$

14. $\left(\dfrac{5x^2y}{10x}\right)\left(-\dfrac{6y^2}{21xy}\right)$ $-\dfrac{y^2}{7}$

15. $\dfrac{5y}{12x}(15x^2)$

16. $(21a^2)\dfrac{5x}{6ab}$ $\dfrac{35ax}{2b}$

17. $\dfrac{36ab^2}{5c} \cdot \dfrac{20ac^2}{24b}$ $6a^2bc$

18. $\dfrac{4x}{18yz^2} \cdot \dfrac{12xyz}{30x^2}$

19. $\dfrac{a}{b} \cdot \dfrac{b}{c} \cdot \dfrac{c}{a}$ 1

20. $\dfrac{x}{y} \cdot \dfrac{z}{x} \cdot \dfrac{y}{z}$ 1

21. $\dfrac{2a^2}{5b} \cdot \dfrac{21b^2}{14c} \cdot \dfrac{11c^2}{10b}$

22. $\dfrac{4xy}{15z^2} \cdot \dfrac{3y}{10y^2} \cdot \dfrac{25yz}{7x^2}$ $\dfrac{2y}{7xz}$

23. $\dfrac{-5}{a} \cdot \dfrac{3a^2}{-b} \cdot \dfrac{-b^2}{10}$ $-\dfrac{3ab}{2}$

24. $\dfrac{-x}{-y^2} \cdot \dfrac{-3y}{4x} \cdot \dfrac{-8xy}{-z}$

NON-ROUTINE PROBLEMS

25. The time now is between 10:00 A.M. and 11:00 A.M. Fourteen minutes from now, it will be as many minutes before 12:00 noon as it is past 10:00 A.M. now. What time is it right now?

26. Sally's grandfather has lived a sixth of his life in Ohio, a fourth of his life in Iowa, a third of his life in Illinois, and the other 21 years of his life in Wisconsin. How old is Sally's grandfather?

276

10.4 Binomial Numerators and Denominators

OBJECTIVE: To multiply rational expressions that involve binomials, writing the product in simplest form

Teaching Suggestions p. M–36

Lesson Resources
Warm–Up: p. M–37
Maintenance: See below.
Practice Worksheet 10.4
Transparencies 23, 28

Binomials such as $x - 2$ and $x - 1$ are prime polynomials. The variable x is <u>not</u> a factor of $x - 2$ and $x - 1$. It is a term of each binomial.

EXAMPLE 1   Multiply and simplify: $\dfrac{x-2}{x+3} \cdot \dfrac{x+3}{x-1}$

Solution:

1. Product Property — $\dfrac{x-2}{x+3} \cdot \dfrac{x+3}{x-1} = \dfrac{(x-2)(x+3)}{(x+3)(x-1)}$

2. Commutative Property of Multiplication — $= \dfrac{(x+3)(x-2)}{(x+3)(x-1)}$

3. Product Property — $= \dfrac{x+3}{x+3} \cdot \dfrac{x-2}{x-1}$

4. Multiplication Property of 1 — $= \dfrac{x-2}{x-1}$ ◀ **Simplest form**

EXAMPLE 2   Multiply and simplify: $\dfrac{x(x+3)}{x-5} \cdot \dfrac{x-5}{2x}$

Solution:

$\dfrac{x(x+3)}{x-5} \cdot \dfrac{x-5}{2x} = \dfrac{\overset{1}{\cancel{x}}(x+3)(\overset{1}{\cancel{x-5}})}{(\underset{1}{\cancel{x-5}})2\underset{1}{\cancel{x}}}$ **A name for 1 is $\dfrac{x(x-5)}{(x-5)x}$.**

You may wish to show students the detailed steps first as in Example 1. Then show them this shortcut.

$= \dfrac{x+3}{2}$ ◀ **Simplest form**

P–1   **Write $-(a - b)$ without parentheses.** $-a + b$

You can use the Property of the Opposite of a Sum to write $-(a - b)$ without parentheses.

$-(a - b) = -(a + -b)$
$= (-a) + -(-b)$
$= (-a) + b$ Lead students through the steps of this "proof."
$= b + (-a)$
$= b - a$ **Thus, $a - b$ and $b - a$ are opposites.**

Products and Quotients of Rational Expressions / **277**

Maintenance
1. Solve and check:
   $-60 = 6 - 3y$
   Ans: $y = 22$ (Section 3.3)
2. Solve: $\dfrac{15^3}{x^2} = 15$
   Ans: $x = 15$ (Section 6.2)
3. Simplify:
   $3\sqrt{7} + 2\sqrt{28} - \sqrt{63}$
   Ans: $4\sqrt{7}$ (Section 7.5)
4. Factor: $5x^2 + 7x + 2$
   Ans: $(5x + 2)(x + 1)$
   (Section 9.4)
5. Fay's scores on four tests were 75, 72, 87, and 83. She wants her mean score on five tests to be greater than 79 (Condition 2). Find the lowest score she can get on the fifth test (Condition 1).
   Ans: 79   (Section 5.7)

Additional Examples
Example 1
Multiply and simplifly.
1. $\dfrac{x+7}{x-4} \cdot \dfrac{x+4}{x+7}$
   Ans: $\dfrac{x+4}{x-4}$
2. $\dfrac{x-9}{2x-3} \cdot \dfrac{2x-3}{x+6}$
   Ans: $\dfrac{x-9}{x+6}$

Example 2
Multiply and simplify.
1. $\dfrac{x+8}{x(x-8)} \cdot \dfrac{5x}{x+8}$
   Ans: $\dfrac{5}{x-8}$
2. $\dfrac{4x^2}{x(x+3)} \cdot \dfrac{(x+3)(x-4)}{6x(x-4)}$
   Ans: $\dfrac{2}{3}$

P–2 **Write the opposite of each expression.**

a. $x - 2$ **b.** $5 - t$ **c.** $r - m$

$2 - x$ $t - 5$ $m - r$

P–3 **What binomial is the opposite of $x - 2$?** $2 - x$

In Step ⬚1, point out that $(2 - x)$ has been replaced by
the equivalent $-(x - 2)$. This is a possible source of error.

Additional Examples
Example 3
Multiply and simplify.

1. $\dfrac{x + 6}{x - 1} \cdot \dfrac{1 - x}{x - 6}$

   Ans: $-\dfrac{x + 6}{x - 6}$

2. $\dfrac{x - 8}{x + 5} \cdot \dfrac{5 + x}{8 - x}$

   Ans: -1

EXAMPLE 3   Multiply and simplify: $\dfrac{x + 3}{x - 2} \cdot \dfrac{2 - x}{x - 5}$

Solution:

⬚1 Property of the Opposite of a Difference ⟶ $\dfrac{x + 3}{x - 2} \cdot \dfrac{2 - x}{x - 5} = \dfrac{(x + 3) \cdot -(x - 2)}{(x - 2)(x - 5)}$

⬚2 Multiplication Property of -1 ⟶ $= \dfrac{(x + 3) \cdot -1(x - 2)}{(x - 2)(x - 5)}$

⬚3 Product Rule for Rational Expressions ⟶ $= \dfrac{x - 2}{x - 2} \cdot \dfrac{-1(x + 3)}{x - 5}$

⬚4 Multiplication Property of 1 ⟶ $= \dfrac{-1(x + 3)}{x - 5}$

⬚5 Simplest form ⟶ $= -\dfrac{x + 3}{x - 5}$

Example 4
Multiply.

1. $\dfrac{x + 4}{x - 4} \cdot \dfrac{9}{x - 1}$

   Ans: $\dfrac{9x + 36}{x^2 - 5x + 4}$

2. $\dfrac{2x}{x + 10} \cdot \dfrac{x - 3}{x - 10}$

   Ans: $\dfrac{2x^2 - 6x}{x^2 - 100}$

EXAMPLE 4   Multiply: $\dfrac{x + 2}{x - 3} \cdot \dfrac{x + 1}{3}$

Solution: $\dfrac{x + 2}{x - 3} \cdot \dfrac{x + 1}{3} = \dfrac{(x + 2)(x + 1)}{(x - 3)3}$

$= \dfrac{x^2 + 3x + 2}{3x - 9}$

P–4 **What is the polynomial form of $(x + 2)(x + 1)$? of $(x - 3)3$?**

$x^2 + 3x + 2$ $3x - 9$

278 / *Chapter 10*

CLASSROOM EXERCISES

Multiply. Simplify where necessary. (Examples 1–4)

1. $\dfrac{2}{x-2} \cdot \dfrac{x-2}{3}$ $\dfrac{2}{3}$

2. $\dfrac{x+5}{5x} \cdot \dfrac{3x}{x+5}$ $\dfrac{3}{5}$

3. $\dfrac{x+1}{x+2} \cdot \dfrac{x+3}{x+4}$

4. $\dfrac{2(x-3)}{x+2} \cdot \dfrac{x+2}{6}$

5. $\dfrac{x+2}{x-1} \cdot \dfrac{x-1}{x+2}$ 1

6. $\dfrac{x+5}{x-2}(x-2)$ $x+5$

7. $\dfrac{1}{x-3} \cdot \dfrac{x-3}{x+2}$ $\dfrac{1}{x+2}$

8. $\dfrac{1}{x+2}(x+2)$ 1

9. $\dfrac{5}{x-5} \cdot \dfrac{5-x}{6}$ $-\dfrac{5}{6}$

10. $\dfrac{2}{-a-2} \cdot \dfrac{a+2}{5}$ $-\dfrac{2}{5}$

11. $\dfrac{x-y}{b-a} \cdot \dfrac{a-b}{y-x}$ 1

12. $\dfrac{2}{a-b} \cdot \dfrac{a+b}{3}$

13. $\dfrac{3(x+1)}{x-3} \cdot \dfrac{x-3}{3x}$ $\dfrac{x+1}{x}$

14. $\dfrac{y(x+y)}{x-y} \cdot \dfrac{2}{y(x+y)}$ $\dfrac{2}{x-y}$

15. $\dfrac{a-2}{3} \cdot \dfrac{2}{2-a}$ $-\dfrac{2}{3}$

WRITTEN EXERCISES

Goal: To multiply rational expressions that involve binomials

Sample Problem: $\dfrac{x-2}{2} \cdot \dfrac{x+5}{2-x}$ **Answer:** $-\dfrac{x+5}{2}$

Multiply. Simplify where necessary. (Examples 1–4)

1. $\dfrac{11}{x-2} \cdot \dfrac{x-2}{12}$ $\dfrac{11}{12}$

2. $\dfrac{x+7}{9} \cdot \dfrac{5}{x+7}$ $\dfrac{5}{9}$

3. $-\dfrac{7}{x+y} \cdot \dfrac{2(x+y)}{15}$ $-\dfrac{14}{15}$

4. $-\dfrac{2(a+b)}{11} \cdot \dfrac{5}{a+b}$ $-\dfrac{10}{11}$

5. $\dfrac{x-3}{x+2} \cdot \dfrac{x+2}{x-1}$ $\dfrac{x-3}{x-1}$

6. $\dfrac{a+7}{a-3} \cdot \dfrac{a-1}{a+7}$ $\dfrac{a-1}{a-3}$

7. $\dfrac{x(x-3)}{y(x+5)} \cdot \dfrac{x+5}{2x}$ $\dfrac{x-3}{2y}$

8. $\dfrac{a+6}{3(a+2)} \cdot \dfrac{12a(a+2)}{4b}$ $\dfrac{a^2+6a}{b}$

9. $\dfrac{x-3}{x+2}(x+2)$ $x-3$

10. $\dfrac{a+7}{a-5}(a-5)$ $a+7$

11. $\dfrac{1}{x-3} \cdot \dfrac{2(x-3)}{x+2}$ $\dfrac{2}{x+2}$

12. $\dfrac{x(x-y)}{x+y} \cdot \dfrac{1}{x}$ $\dfrac{x-y}{x+y}$

13. $\dfrac{x-3}{x+2} \cdot \dfrac{x-2}{x+3}$ $\dfrac{x^2-5x+6}{x^2+5x+6}$

14. $\dfrac{2a+b}{a-b} \cdot \dfrac{a+b}{a+2b}$ $\dfrac{2a^2+3ab+b^2}{a^2+ab-2b^2}$

15. $\dfrac{x-5}{x} \cdot \dfrac{5}{5-x}$ $-\dfrac{5}{x}$

16. $\dfrac{a}{a-7} \cdot \dfrac{7-a}{5}$ $-\dfrac{a}{5}$

17. $\dfrac{a-3}{a+4} \cdot \dfrac{-a-4}{3-a}$ 1

18. $\dfrac{-x-2}{x-3} \cdot \dfrac{3-x}{x+2}$ 1

19. $\dfrac{2(x-y)}{2x-3y} \cdot \dfrac{2x-3y}{6(y-x)}$ $-\dfrac{1}{3}$

20. $\dfrac{-2a-3b}{15(a-b)} \cdot \dfrac{6(b-a)}{2a+3b}$ $\dfrac{2}{5}$

21. $\dfrac{x+2}{(x-1)(x+3)} \cdot \dfrac{x(x-1)}{x+2}$ $\dfrac{x}{x+3}$

22. $\dfrac{(x-5)(x-2)}{x+3} \cdot \dfrac{x+3}{2(x-2)}$ $\dfrac{x-5}{2}$

Products and Quotients of Rational Expressions / 279

Common Error
When multiplying rational expressions, students do not use the Properties of the Opposite of a Sum or a Difference to simplify.

Example

$$\dfrac{x-5}{x} \cdot \dfrac{5}{5-x} = \dfrac{5(x-5)}{x(5-x)}$$
$$= \dfrac{5x-25}{5x-5x^2}$$

Prescription
Review the Property of the Opposite of a Sum on page 277 and the Property of the Opposite of a Difference on page 278. Point out to students that they can use these properties in Written Exercises 15–20.

Additional Answers
Classroom Exercises
3. $\dfrac{x^2+4x+3}{x^2+6x+8}$
4. $\dfrac{x-3}{3}$
12. $\dfrac{2a+2b}{3a-3b}$

Assignment Guide
Basic p. 279: 1–23 odd
Average pp. 279–280: 1–25 odd
Above Average pp. 279–280: 1–23 odd, 24, 25

APPLICATIONS

A fraction such as the one shown at the right is called a **continued fraction.** You can use a calculator to approximate its value.

EXAMPLE   Find a decimal approximation of the above continued fraction.

SOLUTION   Start by finding the reciprocal of 5.

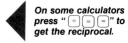

On some calculators press "÷ = =" to get the reciprocal.

0.6470588

Find a decimal approximation.

23. $\dfrac{1}{1+\dfrac{1}{1+\dfrac{1}{1+\frac{1}{7}}}}$ 0.6521739

24. $\dfrac{1}{1+\dfrac{1}{1+\dfrac{1}{1+\frac{1}{8}}}}$ 0.6538462

25. $\dfrac{1}{1+\dfrac{1}{1+\dfrac{1}{1+\frac{1}{11}}}}$ 0.6034483

MID-CHAPTER REVIEW

Evaluate each rational expression if x equals −2. (Section 10.1)

1. $\dfrac{x-5}{x+7}$ $-\dfrac{7}{5}$

2. $\dfrac{x^2+3}{x^2-2}$ $\dfrac{7}{2}$

3. $\dfrac{x^2-x+3}{2x^2+x-1}$ $\dfrac{9}{5}$

4. $\dfrac{-x^2+2x-5}{3x^2-x+2}$

Simplify. No divisor equals 0. (Section 10.2)

5. $\dfrac{3x(x+5)}{6y(x+5)}$ $\dfrac{x}{2y}$

6. $\dfrac{5x-15}{10x+20}$ $\dfrac{x-3}{2x+4}$

7. $\dfrac{x^2-36}{2x^2-12x}$ $\dfrac{x+6}{2x}$

8. $\dfrac{x^2-2x-8}{3x^2-12}$

Multiply. Simplify where necessary. (Section 10.3)

9. $\dfrac{6x}{35}\cdot\dfrac{14}{9x}$ $\dfrac{4}{15}$

10. $\dfrac{21a}{8bc}\cdot\dfrac{6c^2}{7a^2b}$ $\dfrac{9c}{4ab^2}$

11. $-\dfrac{5r^2}{4st^2}\cdot\dfrac{10s^2}{3r}$ $-\dfrac{25rs}{6t^2}$

12. $\dfrac{mn^2}{6}\cdot\dfrac{4}{-m^2}\cdot\dfrac{-3}{10n}$

Multiply. (Section 10.4)

13. $\dfrac{3(x-5)}{2x}\cdot\dfrac{4x}{5(x-5)}$ $\dfrac{6}{5}$

14. $\dfrac{n+5}{n-2}\cdot\dfrac{n+3}{n+5}$

15. $\dfrac{2r-3}{3r+5}\cdot(3r+5)$

16. $\dfrac{m-3}{5m}\cdot\dfrac{10}{3-m}$

REVIEW CAPSULE FOR SECTION 10.5

Factor. (Pages 246–248)

1. x^2-2x-3
$(x+1)(x-3)$

2. x^2-5x+6
$(x-2)(x-3)$

3. $2x^2+5x-3$
$(2x-1)(x+3)$

4. $4x^2+4x-3$
$(2x-1)(2x+3)$

280 / *Chapter 10*

10.5 Trinomial Numerators and Denominators

OBJECTIVE: To multiply rational expressions that involve trinomials, writing the product in simplest form

Teaching Suggestions p. M–36

Lesson Resources
Warm–Up: p. M–37
Maintenance: See below.
Practice Worksheet 10.5
Transparencies 24, 26, 27, 28

Note, in Example 1, that numerators and denominators are factored <u>before</u> multiplying the rational expressions.

Maintenance
1. Solve and check:
   $4(2x - 3) = 15x - 4x$
   Ans: x = −4 (Section 4.3)
2. Write $x^{-4} \cdot x^{-1}$ as one power.
   Ans: $\frac{1}{x^5}$ (Section 6.3)
3. Simplify: $\frac{\sqrt{x^3y^7}}{\sqrt{xy^3}}$
   Ans: xy^2 (Section 7.6)
4. Multiply:
   $(2x + 3)(3x^2 + 2x - 5)$
   Ans: $6x^3 + 13x^2 - 4x - 15$
   (Section 8.5)
5. The area of a square slab of concrete is 116 square feet. Find the length of each side of the square to the nearest tenth of a foot.
   Ans: 10.8 feet
   (Section 7.7)

EXAMPLE 1

Multiply and simplify: $\dfrac{x^2 - x - 6}{x^2 + 2x + 1} \cdot \dfrac{x + 1}{x + 2}$

Solution:

1. Factor. $\quad\dfrac{x^2 - x - 6}{x^2 + 2x + 1} \cdot \dfrac{x + 1}{x + 2} = \dfrac{(x - 3)(x + 2)}{(x + 1)(x + 1)} \cdot \dfrac{x + 1}{x + 2}$ **Factor first.**

2. Product Rule for Rational Expressions $\quad= \dfrac{(x - 3)(x + 2)(x + 1)}{(x + 1)(x + 1)(x + 2)}$

3. Identify common factors. $\quad= \dfrac{(x - 3)\overset{1}{\cancel{(x + 2)}}\overset{1}{\cancel{(x + 1)}}}{(x + 1)\cancel{(x + 1)}\cancel{(x + 2)}}$

4. Simplify. $\quad= \dfrac{x - 3}{x + 1}$

Students will need to be reminded again and again of the difference between <u>factors</u> and <u>terms</u>. You can divide numerators and denominators by common factors but not by common terms.

P–1 **What are the prime factors of $x^2 + 2x$? of $x^2 - 4$?** $x + 2, x - 2$
$x, x + 2$

EXAMPLE 2

Multiply and simplify: $\dfrac{x^2 + 2x}{x^2 - 4} \cdot \dfrac{x - 2}{x + 2}$

Solution:

$\dfrac{x^2 + 2x}{x^2 - 4} \cdot \dfrac{x - 2}{x + 2} = \dfrac{x(x + 2)}{(x + 2)(x - 2)} \cdot \dfrac{x - 2}{x + 2}$ ◀ **Prime factorization**

$= \dfrac{x\overset{1}{\cancel{(x + 2)}}}{(x + 2)\underset{1}{\cancel{(x - 2)}}} \cdot \dfrac{\overset{1}{\cancel{x - 2}}}{\underset{1}{\cancel{x + 2}}}$ ◀ **A name for 1 is identified.**

$= \dfrac{x}{x + 2}$ ◀ **Simplest form**

Additional Examples
Example 1
Multiply and simplify.
$\dfrac{x + 4}{x - 4} \cdot \dfrac{x^2 - 2x - 8}{x^2 + 5x + 4}$
Ans: $\dfrac{x + 2}{x + 1}$

Example 2
Multiply and simplify.
$\dfrac{x + 5}{x - 1} \cdot \dfrac{3x^2 - 3x}{x^2 - 25}$
Ans: $\dfrac{3x}{x - 5}$

Products and Quotients of Rational Expressions / 281

Write the prime factors of each polynomial below.

a. $2x^2 + x - 1$ **b.** $x^2 - 36$ **c.** $6x^3 - 4x^2$
    $2x - 1, x + 1$ $x - 6, x + 6$ $2x^2, 3x - 2$

It may be helpful to review each of these types of factoring.

$$2x^2 + x - 1 = 2x^2 + rx + sx - 1$$

     $r + s = 1$
     $r \cdot s = -2$
     Let $r = 2$
     and $s = -1$.

$$= 2x^2 + 2x - x - 1$$

$$= 2x(x + 1) - 1(x + 1)$$

$$= (2x - 1)(x + 1)$$

$$x^2 - 36 = (x + 6)(x - 6)$$ **Difference of two squares**

$$6x^3 - 4x^2 = 2x^2(3x - 2)$$ **Common monomial factor**

P–3   **What are the prime factors of $x^2 - 3x - 10$?**   **of $x^2 - 9$?** $x - 3, x + 3$
                                       $x - 5, x + 2$
   of $x^2 + 4x + 4$?
     $x + 2, x + 2$

EXAMPLE 3   Multiply and simplify: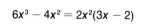

Solution: $$\frac{x^2 - 3x - 10}{x^2 - 9} \cdot \frac{x + 3}{x^2 + 4x + 4} = \frac{(x - 5)(x + 2)}{(x + 3)(x - 3)} \cdot \frac{x + 3}{(x + 2)(x + 2)}$$

$$= \frac{(x - 5)\cancel{(x + 2)}\cancel{(x + 3)}}{\cancel{(x + 3)}(x - 3)\cancel{(x + 2)}(x + 2)}$$

$$= \frac{x - 5}{(x - 3)(x + 2)}$$

Additional Examples
Example 3
Multiply and simplify.

1. $\dfrac{3x - 6}{6x + 6} \cdot \dfrac{x^2 + 3x + 2}{x^2 - 3x + 2}$
   Ans: $\dfrac{x + 2}{2x - 2}$

2. $\dfrac{x^2 + 5x + 6}{3x - 3} \cdot \dfrac{x^2 - x}{x + 2}$
   Ans: $\dfrac{x^2 + 3x}{3}$

3. $\dfrac{x^2 - 2x - 3}{x^2 - 9} \cdot \dfrac{x^2 + 5x + 6}{x^2 - 1}$
   Ans: $\dfrac{x + 2}{x - 1}$

> *Steps for multiplying rational expressions:*
>
> 1. Factor all numerators and denominators.
> 2. Apply the Product Rule for Rational Expressions.
> 3. Identify a name for 1.
> 4. Use the Multiplication Property of One and simplify the result.

Simplify. (Examples 1–3)

1. $\dfrac{(x-2)(x+3)}{(x+3)(x+4)}$ $\dfrac{x-2}{x+4}$

2. $\dfrac{(x+5)(x+2)}{(x+5)(x+5)(x+2)}$ $\dfrac{1}{x+5}$

3. $\dfrac{(x+5)(x-2)}{x^2-4}$ $\dfrac{x+5}{x+2}$

4. $\dfrac{x-3}{x+7}\cdot\dfrac{x+7}{x+3}$ $\dfrac{x-3}{x+3}$

5. $\dfrac{x^2-2x+1}{2x}\cdot\dfrac{x}{x-1}$ $\dfrac{x-1}{2}$

6. $\dfrac{x^2+2x}{x+3}\cdot\dfrac{x+3}{x(x-1)}$

7. $\dfrac{5x}{x^2-9}\cdot\dfrac{x(x-3)}{x^2-4x+3}$

8. $\dfrac{x^2-x}{2x+4}\cdot\dfrac{x^2-4}{x-1}$ $\dfrac{x^2-2x}{2}$

9. $\dfrac{x^2-x-2}{x-3}\cdot\dfrac{3x-9}{3x^2-6x}$

WRITTEN EXERCISES

Goal: To multiply rational expressions that involve trinomials

Sample Problem: $\dfrac{x^2-7x+12}{x^2-16}\cdot\dfrac{x+4}{3x-9}$ **Answer:** $\dfrac{1}{3}$

Multiply and simplify. (Examples 1–3)

1. $\dfrac{3x+6}{x^2+x}\cdot\dfrac{4x+4}{12}$ $\dfrac{x+2}{x}$

2. $\dfrac{x^2-2x}{15}\cdot\dfrac{3x-9}{2x-4}$ $\dfrac{x^2-3x}{10}$

3. $\dfrac{x^2-4}{x+2}\cdot\dfrac{5x+10}{3x-6}$ $\dfrac{5x+10}{3}$

4. $\dfrac{2x-6}{25}\cdot\dfrac{5x+15}{x^2-9}$ $\dfrac{2}{5}$

5. $\dfrac{x^3-2x^2}{x^2-25}\cdot\dfrac{5x+10}{x^2-2x}$ $\dfrac{5x^2+10x}{x^2-25}$

6. $\dfrac{9y^2+9}{y^2-1}\cdot\dfrac{2y^3+2y^2}{3y}$

7. $\dfrac{x^2-x-2}{10}\cdot\dfrac{2x+4}{x^2-4}$ $\dfrac{x+1}{5}$

8. $\dfrac{x^2-2x-3}{3x+3}\cdot\dfrac{15}{x^2-9}$ $\dfrac{5}{x+3}$

9. $\dfrac{x^2+8x+16}{12}\cdot\dfrac{3x+3}{x^2+5x+4}$ $\dfrac{x+4}{4}$

10. $\dfrac{x^2-5x+6}{5x-10}\cdot\dfrac{15}{x^2-6x+9}$ $\dfrac{3}{x-3}$

11. $\dfrac{x^2-16}{2x}\cdot\dfrac{3x^2-6x}{16-x^2}$ $-\dfrac{3x-6}{2}$ or $\dfrac{6-3x}{2}$

12. $\dfrac{5a}{a^2-9}\cdot\dfrac{9-a^2}{5a^2-a}$ $\dfrac{5}{1-5a}$

MORE CHALLENGING EXERCISES

13. $\dfrac{2x^2-5x-3}{4x^2-1}\cdot\dfrac{3}{x^2+2x-15}$ $\dfrac{3}{2x^2+9x-5}$

14. $\dfrac{3x^2-14x-5}{9x^2-1}\cdot\dfrac{1}{x^2-6x+5}$

15. $\dfrac{2x^2-4x+8}{6x^2+7x-3}(2x^2+x-3)$

16. $\dfrac{3x^2+15x-3}{3x^2-27}(3x^2-7x-6)$

17. If $\dfrac{x}{6}\cdot\dfrac{3}{y}=1$, how does the value of x compare to the value of y? $x=2y$

18. If $\dfrac{ax^2+ax}{6x}\cdot\dfrac{2x-6}{x^2-2x-3}=3$, what is the value of a? 9

19. If $\dfrac{a}{b}\cdot\dfrac{3}{5}=1$, can you conclude that the value of a is 5 and the value of b is 3? Explain.

20. Write a rational expression that, when multiplied by $\dfrac{x^2+6x+5}{x+1}$, gives a product of 3. $\dfrac{3}{x+5}$

Products and Quotients of Rational Expressions / 283

Common Error
When multiplying rational expressions, students do not use all the remaining factors after simplifying.

Example
$$\dfrac{x^2-x}{2x+4}\cdot\dfrac{x+2}{x-1}=\dfrac{x(x-1)}{2(x+2)}\cdot\dfrac{x+2}{x-1}$$
$$=\dfrac{1}{2}$$

Prescription
Emphasize that students must be careful to use all the remaining factors. Encourage students to do their work neatly and to check that they have used all the factors.

Additional Answers
Classroom Exercises

6. $\dfrac{x+2}{x-1}$

7. $\dfrac{5x^2}{x^3-x^2-9x+9}$

9. $\dfrac{x+1}{x}$

Assignment Guide
Basic p. 283: 1–11 odd
Average p. 283: 1–19 odd
Above Average pp. 283:
1–11 odd, 13–20

Problem–Solving Skills
Using logical reasoning
(Ex. 17–20)

Critical Thinking
See Exercises 17–20.

Additional Answers
Written Exercises

6. $\dfrac{6y^3+6y}{y-1}$

14. $\dfrac{1}{3x^2-4x+1}$

15. $\dfrac{2x^3-6x^2+12x-8}{3x-1}$

16. $\dfrac{3x^3+17x^2+7x-2}{x+3}$

19. No, a and b can have any values such that the ratio $\dfrac{a}{b}$ is equivalent to $\dfrac{5}{3}$.

Maintenance
1. Solve and check:
   $7(2x - 2) - 5x = 4x + 2$
   Ans: $x = 3\frac{1}{5}$
   (Section 4.3)
2. Simplify $\left(\dfrac{3}{p}\right)^{-2}$
   Ans: $\dfrac{p^2}{9}$ (Section 6.4)
3. Evaluate $x^2y^2 + 2x - 3y$
   when $x = -1$ and $y = -2$.
   Ans: 8 (Section 8.3)
4. Factor: $y^2 - t^2$
   Ans: $(y - t)(y + t)$
   (Section 9.5)
5. A rectangular water trough
   has a length of $(x + 5)$ units
   and a width of $(x - 3)$ units.
   Find the area of the water
   trough.
   **Ans: $x^2 + 2x - 15$ square
   units**
   (Section 8.5)

Additional Examples
Example 1
Divide and simplify.
1. $\dfrac{5}{8x} \div \dfrac{3}{4x}$
   Ans: $\dfrac{5}{6}$
2. $\dfrac{2x}{15} \div \dfrac{6x}{5}$
   Ans: $\dfrac{1}{9}$
3. $\dfrac{21x^2}{12xz} \div \dfrac{7x}{2z^2}$

   Ans: $\dfrac{z}{2}$

OBJECTIVES: To divide two rational expressions involving only monomials,
writing the quotient in simplest form

10.6 Dividing Rational Expressions

To divide two rational expressions involving monomials and
binomials, writing the quotient in simplest form

The Quotient Rule for Fractions can be applied to rational
expressions.

Quotient Rule for Rational Expressions

If $\dfrac{a}{b}$ and $\dfrac{c}{d}$ are rational expressions,

then $\dfrac{a}{b} \div \dfrac{c}{d} = \dfrac{a}{b} \cdot \dfrac{d}{c}$. ($b$, c, and d are not zero.)

Dividing by a rational expression is the
same as multiplying by its reciprocal.

$$\dfrac{x}{x + 2} \div \dfrac{x - 1}{x}$$

$$\dfrac{x}{x + 2} \cdot \dfrac{x}{x - 1}$$

$$\dfrac{x^2}{(x + 2)(x - 1)}$$

Expressions such as $\dfrac{x}{x + 1}$ and $\dfrac{x + 1}{x}$ are reciprocals. The product of a
rational expression and its reciprocal is 1.

P–1 Write each quotient below as a product.

a. $\dfrac{x - 3}{x + 2} \div \dfrac{x}{x - 2}$ b. $\dfrac{3x}{5} \div \dfrac{4}{9x}$ c. $\dfrac{2x^2}{5} \div (x + 1)$

$\dfrac{x - 3}{x + 2} \cdot \dfrac{x - 2}{x}$ $\dfrac{3x}{5} \cdot \dfrac{9x}{4}$ $\dfrac{2x^2}{5} \cdot \dfrac{1}{x + 1}$

EXAMPLE 1   Divide and simplify: $\dfrac{2x}{3} \div \dfrac{5x}{6}$

Solution:

$$\dfrac{2x}{3} \div \dfrac{5x}{6} = \dfrac{2x}{3} \cdot \dfrac{6}{5x}$$

Remind students that
this shortcut is a way of
identifying common
factors (or a name for 1).

$\dfrac{3x}{3x} \cdot \dfrac{2 \cdot 2}{5} = 1 \cdot \dfrac{4}{5}$

$= \dfrac{4}{5}$

$= \dfrac{2x}{3} \cdot \dfrac{6}{5x}$

$= \dfrac{4}{5}$ ◀ **Simplest form**

P-2 Write $\dfrac{x-2}{x} \div \dfrac{x-2}{5}$ as a product. $\dfrac{x-2}{x} \cdot \dfrac{5}{x-2}$

EXAMPLE 2   Divide and simplify: $\dfrac{x-2}{x} \div \dfrac{x-2}{5}$

Solution:

$$\frac{x-2}{x} \div \frac{x-2}{5} = \frac{x-2}{x} \cdot \frac{5}{x-2}$$

$$= \frac{5(x-2)^{\,1}}{x(x-2)_{\,1}}$$

$$= \frac{5}{x}$$

P-3 What is the reciprocal of $\dfrac{9pq}{4(r-3)}$? $\dfrac{4(r-3)}{9pq}$

EXAMPLE 3   Divide and simplify: $\dfrac{3p^2q}{2r(r-3)} \div \dfrac{9pq}{4(r-3)}$

Solution:

$$\frac{3p^2q}{2r(r-3)} \div \frac{9pq}{4(r-3)} = \frac{3p^2q}{2r(r-3)} \cdot \frac{4(r-3)}{9pq}$$

$$= \frac{\overset{1 \cdot p \cdot 1}{3p^2q} \cdot \overset{2}{4(r-3)}^{\,1}}{\underset{1}{2r(r-3)} \cdot \underset{3 \cdot 1 \cdot 1}{9pq}}$$

$$= \frac{2p}{3r}$$

Additional Examples
Example 2
Divide and simplify.
1. $\dfrac{x+1}{15} \div \dfrac{x+1}{3}$
Ans: $\dfrac{1}{5}$
2. $\dfrac{20x}{x-6} \div \dfrac{5}{x-6}$
Ans: 4x
3. $\dfrac{6}{x+5} \div \dfrac{x-3}{x+5}$
Ans: $\dfrac{6}{x-3}$

Example 3
Divide and simplify.
1. $\dfrac{2(x-8)}{9x^2y} \div \dfrac{5(x-8)}{6xy}$
Ans: $\dfrac{4}{15x}$
2. $\dfrac{6ab^3}{5x(x-6)} \div \dfrac{15a^2b}{x(x-6)}$
Ans: $\dfrac{2b^2}{25a}$
3. $6h(x-6) \div \dfrac{8h^2}{x+6}$
Ans: $\dfrac{3x^2-108}{4h}$

Common Error
Students do not multiply by the reciprocal when dividing rational expressions.

Example
$\dfrac{2}{x-2} \div \dfrac{x}{5} = \dfrac{2}{x-2} \cdot \dfrac{x}{5}$
$= \dfrac{2x}{5x-10}$

Prescription
Review the Quotient Rule for Rational Expressions on page 284. For Written Exercises 1–20, have students name the reciprocal that they will use when they write the quotient as a product.

▩▩▩ **CLASSROOM EXERCISES** ▩▩▩

Write each quotient as an equivalent product. (P-1, P-2)

1. $\dfrac{2}{3} \div \dfrac{5}{6}$ $\dfrac{2}{3} \cdot \dfrac{6}{5}$

2. $\dfrac{x}{5} \div \dfrac{3}{x}$ $\dfrac{x}{5} \cdot \dfrac{x}{3}$

3. $\dfrac{2x}{6} \div \dfrac{x}{3}$ $\dfrac{2x}{6} \cdot \dfrac{3}{x}$

4. $\dfrac{x}{3y} \div \dfrac{3x}{y}$ $\dfrac{x}{3y} \cdot \dfrac{y}{3x}$

Products and Quotients of Rational Expressions / **285**

5. $\dfrac{2}{x-2} \div \dfrac{x}{5}$ $\dfrac{2}{x-2} \cdot \dfrac{5}{x}$

6. $\dfrac{x+1}{2} \div \dfrac{3}{5}$ $\dfrac{x+1}{2} \cdot \dfrac{5}{3}$

7. $\dfrac{x-7}{5x} \div \dfrac{x+1}{4}$

8. $\dfrac{x+3}{x} \div \dfrac{x-5}{2x}$ $\dfrac{x+3}{x} \cdot \dfrac{2x}{x-5}$

9. $\dfrac{1}{x+2} \div (x-3)$ $\dfrac{1}{x+2} \cdot \dfrac{1}{x-3}$

10. $(x+3) \div \dfrac{1}{x+5}$

11. $\dfrac{x+1}{x} \div \dfrac{x-10}{2x}$ $\dfrac{x+1}{x} \cdot \dfrac{2x}{x-10}$

12. $\dfrac{x+3}{x-4} \div \dfrac{x+7}{x-5}$ $\dfrac{x+3}{x-4} \cdot \dfrac{x-5}{x+7}$

13. $\dfrac{x+8}{x-5} \div \dfrac{x+2}{x-10}$

WRITTEN EXERCISES

Goal: To divide rational expressions

Sample Problem: $\dfrac{5a^2b^2}{a(1-b)} \div \dfrac{25ab^2}{b(1-b)}$ **Answer:** $\dfrac{b}{5}$

Divide and simplify. (Examples 1–3)

1. $\dfrac{2}{5} \div \dfrac{3}{5}$ $\dfrac{2}{3}$

2. $\dfrac{5}{8} \div \dfrac{3}{16}$ $\dfrac{10}{3}$

3. $\dfrac{3}{x} \div \dfrac{1}{2x}$ 6

4. $\dfrac{x}{5} \div \dfrac{3x}{10}$ $\dfrac{2}{3}$

5. $\dfrac{a^2}{3} \div \dfrac{5a}{9}$ $\dfrac{3a}{5}$

6. $\dfrac{2}{b^2} \div \dfrac{4}{3b}$ $\dfrac{3}{2b}$

7. $\dfrac{3}{x-5} \div \dfrac{6}{x-5}$ $\dfrac{1}{2}$

8. $\dfrac{x+2}{5} \div \dfrac{x+2}{3}$

9. $\dfrac{x^2y}{7} \div \dfrac{3x}{14}$ $\dfrac{2xy}{3}$

10. $\dfrac{5}{2a^2b} \div \dfrac{15}{ab}$ $\dfrac{1}{6a}$

11. $\dfrac{2}{x^2y^2} \div \dfrac{4}{5xy}$ $\dfrac{5}{2xy}$

12. $\dfrac{5xy}{3z} \div \dfrac{10x^2}{9z^2}$

13. $\dfrac{x-5}{x+2} \div \dfrac{5}{x+2}$ $\dfrac{x-5}{5}$

14. $\dfrac{3}{a+b} \div \dfrac{a+3}{a+b}$ $\dfrac{3}{a+3}$

15. $\dfrac{x^2+2}{5} \div \dfrac{x^2}{10}$ $\dfrac{2x^2+4}{x^2}$

16. $\dfrac{3}{x+6} \div \dfrac{9}{x}$

17. $\dfrac{x-6}{x+5} \div 2(x-6)$ $\dfrac{1}{2x+10}$

18. $\dfrac{x+3}{x-10} \div 5(x+3)$ $\dfrac{1}{5x-50}$

19. $x(x^2+3) \div \dfrac{3x}{x^2+3}$ $\dfrac{x^4+6x^2+9}{3}$

20. $3a(a^2+2) \div \dfrac{a^2}{a-2}$ $\dfrac{3a^3-6a^2+6a-12}{a}$

MORE CHALLENGING EXERCISES

21. $\dfrac{5}{x} \div \dfrac{3}{5x} \div \dfrac{10}{3x}$ $\dfrac{5x}{2}$

22. $\dfrac{a}{3} \div \dfrac{a^2}{2} \div \dfrac{7}{3a}$ $\dfrac{2}{7}$

23. $\dfrac{x+1}{x+3} \div \dfrac{x-1}{2(x+3)} \div \dfrac{3(x-1)}{x+1}$

NON-ROUTINE PROBLEM

24. The figure at the right shows three different views of a cube.

   a. What shape is shown on that side of the cube which is opposite the side showing a circle?

   b. What shape is shown on that side of the cube which is opposite the side showing a triangle?

OBJECTIVES: To divide two rational expressions involving monomials, binomials, or trinomials as numerators, or denominators, writing the quotient in simplest form

10.7 Quotients and Complex Fractions

To simplify a complex fraction having a variable in one or more of the numerators

P–1 **What is the reciprocal of $(x^2 - 25)$?** $\quad \dfrac{1}{x^2 - 25}$

EXAMPLE 1 Divide and simplify: $\dfrac{x^2 + 4x - 5}{3x - 3} \div (x^2 - 25)$

Solution:

1 Factor. $\longrightarrow \dfrac{x^2 + 4x - 5}{3x - 3} \div (x^2 - 25) = \dfrac{(x + 5)(x - 1)}{3(x - 1)} \div (x + 5)(x - 5)$

2 Quotient Property of Rational Expressions $\longrightarrow = \dfrac{(x + 5)(x - 1)}{3(x - 1)} \cdot \dfrac{1}{(x + 5)(x - 5)}$

3 Identify common factors. $\longrightarrow = \dfrac{\overset{1}{\cancel{(x + 5)}}\overset{1}{\cancel{(x - 1)}}}{3\underset{1}{\cancel{(x - 1)}}} \cdot \dfrac{1}{\underset{1}{\cancel{(x + 5)}}(x - 5)}$

4 Simplify. $\longrightarrow = \dfrac{1}{3(x - 5)}$

EXAMPLE 2 Divide and simplify: $\dfrac{x^2 - 3x - 4}{6} \div \dfrac{x^2 - 1}{2x - 2}$

Solution: $\dfrac{x^2 - 3x - 4}{6} \div \dfrac{x^2 - 1}{2x - 2} = \dfrac{(x - 4)(x + 1)}{6} \div \dfrac{(x + 1)(x - 1)}{2(x - 1)}$

$= \dfrac{(x - 4)\overset{1}{\cancel{(x + 1)}}}{\underset{3}{\cancel{6}}} \cdot \dfrac{\overset{1}{\cancel{2}}\cdot\overset{1}{\cancel{(x - 1)}}}{\underset{1}{\cancel{(x + 1)}}\underset{1}{\cancel{(x - 1)}}}$

$= \dfrac{x - 4}{3}$

A quotient such as $\left(\dfrac{x^2}{2} + \dfrac{x}{3}\right) \div \dfrac{x}{6}$ is often written as a fraction.

$\left(\dfrac{x^2}{2} + \dfrac{x}{3}\right) \div \dfrac{x}{6} = \dfrac{\dfrac{x^2}{2} + \dfrac{x}{3}}{\dfrac{x}{6}}$ 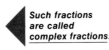 *Such fractions are called complex fractions.*

Ask students to name the numerator and the denominator.

Products and Quotients of Rational Expressions / **287**

Teaching Suggestions p. M–36

Lesson Resources
Warm–Up: p. M–37
Maintenance: See below.
Practice Worksheet 10.7
Transparencies 24, 26, 27, 28

Maintenance
1. Solve: $-2y + 9 > 2y - 2$
   Ans: {all real numbers less than $2\frac{3}{4}$} (Section 5.6)
2. Simplify: $\left(\dfrac{4a^{-3}}{b^4}\right)^{-2}$
   Ans: $\dfrac{a^6 b^8}{16}$ (Section 6.5)
3. Divide:
   $(4m^2 - 25) \div (2m - 5)$
   Ans: $2m + 5$
   (Section 8.6)
4. Factor: $2x^2 + 2x - 24$
   Ans: $2(x + 4)(x - 3)$
   (Section 9.7)
5. A rectangular field has a length of $(x + 7)$ units and a width of $(x + 5)$ units. Find the area of the field.
   Ans: $x^2 + 12x + 35$ square units (Section 8.5)

Additional Examples
Example 1
Divide and simplify.
1. $\dfrac{x^2 - x - 6}{x + 4} \div (x - 3)$
   Ans: $\dfrac{x + 2}{x + 4}$
2. $\dfrac{x^2 - 9x + 20}{x^2 - 5x} \div (x^2 - 16)$
   Ans: $\dfrac{1}{x^2 + 4x}$

Example 2
Divide and simplify.
1. $\dfrac{x^2 - 4}{6x^2} \div \dfrac{x^2 - 4x + 4}{9x}$
   Ans: $\dfrac{3x + 6}{2x^2 - 4x}$
2. $\dfrac{x^2 - 36}{x^2 + 12x + 36} \div \dfrac{5x - 30}{x^2 + 6x}$
   Ans: $\dfrac{x}{5}$

1. $\dfrac{\dfrac{x}{4} + \dfrac{x^2}{6}}{\dfrac{x^2}{12}}$

    Ans: $\dfrac{3 + 2x}{x}$

2. $\dfrac{\dfrac{x}{10} + \dfrac{x}{2}}{\dfrac{3}{5} + \dfrac{x}{4}}$

    Ans: $\dfrac{12x}{12 + 5x}$

3. $\dfrac{\dfrac{a}{3} + \dfrac{b}{4}}{\dfrac{a}{6} + \dfrac{b}{2}}$

    Ans: $\dfrac{4a + 3b}{2a + 6b}$

EXAMPLE 3   Simplify $\dfrac{\dfrac{x^2}{2} + \dfrac{x}{3}}{\dfrac{x}{6}}$ ◀ *First find a common multiple.*

Solution: You may want to compare this method with the method used in Examples 1 and 2.

☐1 Multiplication Property of One ⟶ $\dfrac{\dfrac{x^2}{2} + \dfrac{x}{3}}{\dfrac{x}{6}} = \dfrac{6}{6} \cdot \dfrac{\dfrac{x^2}{2} + \dfrac{x}{3}}{\dfrac{x}{6}}$ ◀ *6 is a common multiple of 2, 3, and 6.*

☐2 Product Property of Rational Expressions ⟶ $= \dfrac{6\left(\dfrac{x^2}{2} + \dfrac{x}{3}\right)}{6\left(\dfrac{x}{6}\right)}$

☐3 Distributive Property ⟶ $= \dfrac{6\left(\dfrac{x^2}{2}\right) + 6\left(\dfrac{x}{3}\right)}{6\left(\dfrac{x}{6}\right)}$

☐4 Multiply. ⟶ $= \dfrac{3x^2 + 2x}{x}$

☐5 Factor. ⟶ $= \dfrac{x(3x + 2)}{x}$

☐6 Simplify. ⟶ $= 3x + 2$

Steps for simplifying a complex fraction:

1. Determine m, a common multiple of all the denominators.

2. Multiply the complex fraction by $\dfrac{m}{m}$.

3. Simplify the result.

▓▓▓▓ CLASSROOM EXERCISES ▓▓▓▓

Write the prime factors of each polynomial. (Step 1 of Examples 1 and 2)

1. $2x - 6$ 2, $x - 3$ **2.** $x^2 - 9$ $x + 3, x - 3$ **3.** $3x^2 - x$ $x, 3x - 1$ **4.** $5x + 10$ 5, $x + 2$ **5.** $x^2 - 4$ $x + 2, x - 2$

6. $x^2 - 81$ **7.** $4x - 16$ **8.** $x^2 + 4x$ **9.** $3x - 21$ 3, $x - 7$ **10.** $x^2 - 49$
    $x + 9, x - 9$ 4, $x - 4$ $x, x + 4$ $x + 7, x - 7$

11. $x^2 - 7x$ $x, x - 7$

12. $x^2 + x$ $x, x + 1$

13. $2x^2 - 10x$ $2x, x - 5$

14. $x^2 - 10x + 25$ $x - 5, x - 5$

15. $x^2 - 16$ $x + 4, x - 4$

16. $5x - 20$

17. $3x^2 - 15x$ $3x, x - 5$

18. $x^2 + 2x - 15$ $x + 5, x - 3$

19. $5x^2 + 15x$ $5x, x + 3$

Additional Answers
Classroom Exercises
16. $5, x - 4$

▨ WRITTEN EXERCISES ▨

Goal: To simplify complex fractions

Sample Problem: $\dfrac{\dfrac{a}{2} - \dfrac{a}{4}}{\dfrac{a}{6}}$ **Answer:** $\dfrac{3}{2}$

Divide and simplify. (Examples 1 and 2)

1. $\dfrac{2x - 6}{5x^2} \div \dfrac{x^2 - 9}{x}$ $\dfrac{2}{5x^2 + 15x}$

2. $\dfrac{3x^2 - x}{5x + 10} \div \dfrac{3x}{x^2 - 4}$ $\dfrac{3x^2 - 7x + 2}{15}$

3. $\dfrac{x^2 - 16}{4x - 16} \div \dfrac{x^2 + 4x}{2x}$ $\dfrac{1}{2}$

4. $\dfrac{3x - 21}{x^2 - 49} \div \dfrac{3x}{x^2 - 7x}$ $\dfrac{x - 7}{x + 7}$

5. $\dfrac{x^2 - 10x + 25}{x + 1} \div \dfrac{2x^2 - 10x}{x^2 + x}$ $\dfrac{x - 5}{2}$

6. $\dfrac{x^2 - 16}{6x} \div \dfrac{5x - 20}{3x^2 - 15x}$ $\dfrac{x^2 - x - 20}{10}$

7. $\dfrac{x^2 + 2x - 15}{x + 3} \div \dfrac{x^2 + 7x + 10}{x - 2}$ $\dfrac{x^2 - 5x + 6}{x^2 + 5x + 6}$

8. $\dfrac{x - 5}{x^2 + 3x - 10} \div \dfrac{9x^2}{3x^2 - 6x}$

9. $\dfrac{x^2 + 10x + 24}{3x^2 - 12x} \div (x^2 - 3x - 18)$

10. $\dfrac{x^2 - 2x - 15}{5x^2 + 15x} \div (x^2 - 6x + 5)$

11. $\dfrac{x^2 - y^2}{6x} \div \dfrac{x^2y + xy^2}{3x^2y^2}$ $\dfrac{xy - y^2}{2}$

12. $\dfrac{4x^2 - 25y^2}{2x^2y + 5xy^2} \div \dfrac{6x^2 - 15xy}{9x^2y^2}$ $3y$

Simplify. (Example 3)

13. $\dfrac{\dfrac{x}{5}}{\dfrac{x}{2} + \dfrac{x}{5}}$ $\dfrac{2}{7}$

14. $\dfrac{\dfrac{y}{3} + \dfrac{y}{5}}{\dfrac{y}{5}}$ $\dfrac{8}{3}$

15. $\dfrac{\dfrac{a}{6} + \dfrac{3}{5}}{\dfrac{2}{5}}$

16. $\dfrac{\dfrac{b}{8} + \dfrac{1}{3}}{\dfrac{5}{3}}$

17. $\dfrac{\dfrac{a}{10} + \dfrac{b}{5}}{\dfrac{a}{3} + \dfrac{b}{6}}$

18. $\dfrac{\dfrac{x}{6} + \dfrac{y}{8}}{\dfrac{x}{2} + \dfrac{y}{4}}$

MORE CHALLENGING EXERCISES

Divide and simplify.

19. $\dfrac{6x^2 - 7x - 3}{10x - 15} \div \dfrac{3x^2 - 14x - 5}{5x - 25}$ 1

20. $\dfrac{10x^2 - 13x - 3}{15x + 3} \div \dfrac{2x^2 + 7x - 15}{3x^2 + 15x}$ x

21. $\dfrac{10x + 15}{2x^2 - 3x - 9} \div \dfrac{x^2 + 6x + 9}{9 - x^2}$ $-\dfrac{5}{x + 3}$

22. $\dfrac{3x^2 - 15x}{2x^2 + x} \div \dfrac{25 - x^2}{2x^2 + 7x + 5}$

Products and Quotients of Rational Expressions / **289**

Assignment Guide
Basic p. 289: 1, 5, 9, 13, 17
Average p. 289: 1–21 odd
Above Average p. 289:
1–17 odd, 19–22

Additional Answers
Written Exercises
8. $\dfrac{x - 5}{3x^2 + 15x}$

9. $\dfrac{x^2 + 10x + 24}{3x^4 - 21x^3 - 18x^2 + 216x}$

10. $\dfrac{1}{5x^2 - 5x}$

15. $\dfrac{5a + 18}{12}$

16. $\dfrac{3b + 8}{40}$

17. $\dfrac{3a + 6b}{10a + 5b}$

18. $\dfrac{4x + 3y}{12x + 6y}$

22. $\dfrac{-6x^2 - 21x - 15}{2x^2 + 11x + 5}$

Quiz Sections 10.5–10.7
After completing Sections 10.5–10.7, you may want to administer a quiz covering the same sections. See the quiz provided on page 212 in the *Teacher's ResourceBank*.

This special topic includes applications of both direct and indirect proportions as they are used to determine the output voltage and current of an electrical transformer.

Using Proportions

Electricity

Electricians keep electric equipment such as **transformers** in working order. The main parts of a transformer are the core, the primary coil to which the input power is supplied, and the secondary coil from which output power is delivered. The following **direct proportion** relates the input (V_P) and output (V_S) <u>voltages</u> and the number of turns on the primary (N_P) and the secondary (N_S) coils.

$$\frac{V_S}{V_P} = \frac{N_S}{N_P}$$

◀ The ratio $\frac{N_S}{N_P}$ is called the <u>turns ratio</u>.

Power Input

The following **indirect proportion** relates the input (I_P) and output (I_S) <u>amperages</u> and the number of turns on the coils.

$$\frac{I_S}{I_P} = \frac{N_P}{N_S}$$

◀ Note that the <u>turns ratio</u> is inverted.

Primary coil

Core   Secondary coil

Power Output

EXAMPLE: The number of turns on the primary coil of a transformer is 200. The number of turns on the secondary coil is 800. The input voltage is 110 volts. The input current is 1.2 amperes. Find **a.** the output voltage in volts and **b.** the output current in amperes.

SOLUTION: **a.** $\dfrac{V_S}{V_P} = \dfrac{N_S}{N_P}$ $\dfrac{V_s}{110} = \dfrac{800}{200}$ **b.** $\dfrac{I_S}{I_P} = \dfrac{N_P}{N_S}$ $\dfrac{I_S}{1.2} = \dfrac{200}{800}$

$V_S = \dfrac{(110)(800)}{200}$ $I_S = \dfrac{(1.2)(200)}{800}$

$V_S =$ **440 volts** $I_S =$ **0.3 amperes**

EXERCISES

Find the output voltage (V_S) in volts of each transformer.

1. Turns Ratio: $\frac{20}{1}$ 2300 volts
   Input voltage: 115 volts

2. Turns Ratio: $\frac{1}{3}$ 230 volts
   Input voltage: 690 volts

Find the output current (I_S) in amperes for each transformer.

3. Turns Ratio: $\frac{4}{5}$ 1.625 amperes
   Input current: 1.3 amperes

4. Turns Ratio: $\frac{25}{3}$ 1.44 amperes
   Input current: 12 amperes

CHAPTER SUMMARY

IMPORTANT TERMS	Rational expression *(p. 268)* Complex fraction *(p. 287)*

IMPORTANT IDEAS

1. Any polynomial is also a rational expression.

2. Any number that will result in a divisor of 0 must be restricted from the domain of the variable in a rational expression.

3. Product Rule of Rational Expressions: If $\frac{a}{b}$ and $\frac{c}{d}$ are rational expressions, then $\frac{a}{b} \cdot \frac{c}{d} = \frac{ac}{bd}$. (*b* and *d* are not zero.)

4. Property of the Opposite of a Difference: If *a* and *b* are any real numbers, then $-(a - b) = b - a$.

5. Quotient Rule for Rational Expressions: If $\frac{a}{b}$ and $\frac{c}{d}$ are rational expressions, then $\frac{a}{b} \div \frac{c}{d} = \frac{a}{b} \cdot \frac{d}{c}$. (*b, c,* and *d* are not zero.)

CHAPTER REVIEW

SECTION 10.1

Write Yes or No to show whether each expression is a rational expression.

1. $\frac{2x^2 - 5x + 3}{x^2 + 7}$ Yes **2.** $-5x^3 - 2x$ Yes **3.** $3x^{-2} + 4x^{-1} + 10$ No **4.** $\frac{\sqrt{x + 1}}{|x - 3|}$ No

Write the number or numbers that cannot be in the domain of each variable.

5. $\frac{x + 1}{x + 3}$ -3 **6.** $\frac{x - 3}{x - 2}$ 2 **7.** $\frac{x + 3}{2x - 1}$ $\frac{1}{2}$ **8.** $\frac{x + 6}{3x + 2}$ $-\frac{2}{3}$

SECTION 10.2

Simplify. No divisor equals 0.

9. $\frac{10x^2y}{12xy}$ $\frac{5x}{6}$ **10.** $\frac{8ab^2c}{20a^2bc}$ $\frac{2b}{5a}$ **11.** $\frac{x^2 - 100}{3x^2 - 300}$ $\frac{1}{3}$ **12.** $\frac{-3x^2 - 36x}{x^2 - 144}$ $\frac{-3x}{x - 12}$

Products and Quotients of Rational Expressions / **291**

Chapter Test
Two Chapter Tests (Form A and Form B) are provided on pages 213–216 in the *Teacher's ResourceBank.*

SECTION 10.3

Multiply and simplify.

13. $\dfrac{12x}{5} \cdot \dfrac{15}{18x^2}$ $\dfrac{2}{x}$

14. $\dfrac{9}{14x} \cdot \dfrac{8x^3}{30}$ $\dfrac{6x^2}{35}$

15. $\dfrac{3xy^2}{5a^2b} \cdot \dfrac{10a}{9by}$ $\dfrac{2xy}{3ab^2}$

16. $\dfrac{5ab}{18xy} \cdot \dfrac{8x^2y}{15a^2b^2}$ $\dfrac{4x}{27ab}$

SECTION 10.4

Multiply and simplify.

17. $\dfrac{ax}{x^2 - a^2} \cdot \dfrac{x^2 + xa}{a^2x^2}$ $\dfrac{1}{ax - a^2}$

18. $\dfrac{5x}{x^2 - 25} \cdot \dfrac{x^2 + 5x}{2x^2y}$ $\dfrac{5}{2xy - 10y}$

19. $\dfrac{x^2 - 4}{14} \cdot \dfrac{6x^2}{3x^2 + 6x}$

20. $\dfrac{5r^2 - 15r}{22} \cdot \dfrac{2r + 6}{r^2 - 9}$ $\dfrac{5r}{11}$

21. $\dfrac{3 - x}{6} \cdot \dfrac{15}{3x - 9}$ $-\dfrac{5}{6}$

22. $\dfrac{2y - 4}{21} \cdot \dfrac{7y}{6 - 3y}$

SECTION 10.5

Multiply and simplify.

23. $\dfrac{x^2 + 2x - 15}{x^2 - 6x + 9} \cdot \dfrac{x^2 - 2x - 3}{x^2 + 7x + 10}$ $\dfrac{x + 1}{x + 2}$

24. $\dfrac{x^2 + 6x + 8}{x^2 + 8x + 16} \cdot \dfrac{x^2 - x - 20}{x^2 + 4x + 4}$ $\dfrac{x - 5}{x + 2}$

25. $\dfrac{x^2 + 4x + 4}{x^2 - x - 6} \cdot \dfrac{x^2 - 5x + 6}{x^2 - 4}$ 1

26. $\dfrac{x^2 - 1}{x^2 - 6x + 5} \cdot \dfrac{x^2 - 10x + 25}{x^2 - 4x - 5}$ 1

SECTION 10.6

Divide and simplify.

27. $\dfrac{-10abc}{21xy} \div \dfrac{-2a^2bc}{-14x^2y}$ $-\dfrac{10x}{3a}$

28. $\dfrac{49xy^2}{24c^2d^2} \div \dfrac{21x^2y}{-8cd}$ $-\dfrac{7y}{9cdx}$

29. $\dfrac{x - 2}{x(x + 3)} \div \dfrac{x + 2}{2(x + 3)}$ $\dfrac{2x - 4}{x^2 + 2x}$

30. $\dfrac{2x + 1}{x + 7} \div \dfrac{5(2x - 1)}{x(x + 7)}$ $\dfrac{2x^2 + x}{10x - 5}$

SECTION 10.7

Divide and simplify.

31. $\dfrac{2x^3 - 6x^2 + 4x}{x - 2} \div \dfrac{3x^3 - 3x^2}{x^2 + 4}$ $\dfrac{2x^2 + 8}{3x}$

32. $\dfrac{x^3 + 2x^2 - 3x}{x^2 - 5} \div \dfrac{x^2 - 4x - 12}{2x^2}$

Simplify.

33. $\dfrac{\frac{a}{10} + \frac{a}{3}}{\frac{5}{6}}$ $\dfrac{13a}{25}$

34. $\dfrac{\frac{x}{12}}{\frac{3}{4} - \frac{x}{8}}$ $\dfrac{2x}{18 - 3x}$

35. $\dfrac{\frac{a}{6} - \frac{b}{5}}{\frac{a}{12} + \frac{b}{20}}$ $\dfrac{10a - 12b}{5a + 3b}$

36. $\dfrac{\frac{x}{7} + \frac{x}{5}}{\frac{y}{2} - \frac{y}{3}}$

Additional Answers

19. $\dfrac{x^2 - 2x}{7}$

22. $-\dfrac{2y}{9}$

32. $\dfrac{2x^5 + 4x^4 - 6x^3}{x^4 - 4x^3 - 17x^2 + 20x + 60}$

36. $\dfrac{72x}{35y}$

Sums and Differences of Rational Expressions

A silicon chip smaller than the size of a human fingernail can contain as many as 256,000 separate circuits. Each circuit can be as small as $\frac{1}{2000}$ millimeter, or $\frac{1}{500,000}$ inch.

Background

The chips shown above are tiny pieces of silicon packed with thousands of electronic elements called **transistors.** These transistors actually carry on the control and logic operations in a computer. A combination of transistors and circuits on a chip results in an **integrated circuit** as shown in the center photo above. Since the chip is so small, it must be attached to a carrier device that has wire prongs that plug into the computer's main circuit board.

Lesson Resources
Warm–Up: p. M–38
Maintenance: See below.
Practice Worksheet 11.1
Exploration 11
Transparency 20

Maintenance
1. Solve and check:
   $2y - 8 - 5y = -44$
   Ans: $y = 12$
   (Section 3.5)
2. Write $\dfrac{p^3 p^{-4}}{p}$ as one power.
   Simplify.
   Ans: $\dfrac{1}{p^2}$ (Section 6.3)
3. Multiply:
   $(4x - 3)(2x + 5)$
   Ans: $8x^2 + 14x - 15$
   (Section 8.5)
4. Evaluate $\dfrac{x^2 - 3x}{x + 2}$ when
   $x = -3$.
   Ans: -18 (Section 10.1)
5. At 9:00 P.M., the wind chill
   temperature was three times
   the thermometer reading of
   $-8°C$. What was the wind
   chill temperature?
   Ans: $-24°C$ (Section 1.4)

Additional Examples
Example 1
Add or subtract.
1. $\dfrac{7}{x + 8} + \dfrac{10}{x + 8}$
   Ans: $\dfrac{17}{x + 8}$
2. $\dfrac{9}{x - 1} - \dfrac{11}{x - 1}$
   Ans: $-\dfrac{2}{x - 1}$

Example 2
Add or subtract.
1. $\dfrac{3x}{2x + y} + \dfrac{y}{2x + y}$
   Ans: $\dfrac{3x + y}{2x + y}$
2. $\dfrac{x}{x - 3y} - \dfrac{y}{x - 3y}$
   Ans: $\dfrac{x - y}{x - 3y}$

294

11.1 Sums/Differences: Like Denominators

OBJECTIVE: To add or subtract with two or more rational expressions having
monomial numerators and like monomial or binomial denominators

P–1 Add or subtract: **a.** $\dfrac{3}{7} + \dfrac{2}{7}$ $\dfrac{5}{7}$ **b.** $\dfrac{5}{12} - \dfrac{11}{12}$ $-\dfrac{1}{2}$

The Sum and Difference Rules for Fractions also apply to rational
expressions.

> **Sum and Difference Rules for Rational Expressions**
>
> If $\dfrac{a}{c}$ and $\dfrac{b}{c}$ are rational expressions, then $\dfrac{3}{x + 1} + \dfrac{2}{x + 1} = \dfrac{5}{x + 1}$
>
> 1. $\dfrac{a}{c} + \dfrac{b}{c} = \dfrac{a + b}{c}$ 2. $\dfrac{a}{c} - \dfrac{b}{c} = \dfrac{a - b}{c}$. $\dfrac{3x + 2}{11x} - \dfrac{4}{11x} = \dfrac{3x - 2}{11x}$
> (c is not zero.)

P–2 Add or subtract.

a. $\dfrac{4}{x} + \dfrac{5}{x}$ $\dfrac{9}{x}$ **b.** $\dfrac{10}{a} - \dfrac{3}{a}$ $\dfrac{7}{a}$ **c.** $\dfrac{x}{3y} + \dfrac{2x}{3y}$ $\dfrac{x}{y}$

P–3 In $\dfrac{2}{x - 3}$, what number must be restricted from the domain of x? 3

In Examples 1 and 2 you
might ask students to
identify what a, b, and c
of the Sum and Difference
Rules represent.

EXAMPLE 1   Add: $\dfrac{2}{x - 3} + \dfrac{12}{x - 3}$

Solution:

1 Sum Rule for Rational Expressions ⟶ $\dfrac{2}{x - 3} + \dfrac{12}{x - 3} = \dfrac{2 + 12}{x - 3}$

2 Simplify. ⟶ $= \dfrac{14}{x - 3}$

EXAMPLE 2   Subtract: $\dfrac{x}{x + y} - \dfrac{y}{x + y}$

Solution: $\dfrac{x}{x + y} - \dfrac{y}{x + y} = \dfrac{x - y}{x + y}$ **Simplest form**

EXAMPLE 3   Subtract: $\dfrac{3x}{x+5} - \dfrac{10x}{x+5}$

Solution:

1. Difference Rule for Rational Expressions $\longrightarrow$ $\dfrac{3x}{x+5} - \dfrac{10x}{x+5} = \dfrac{3x - 10x}{x+5}$

2. Combine like terms. $\longrightarrow$ $= \dfrac{-7x}{x+5}$

3. Simplify. $\longrightarrow$ $= -\dfrac{7x}{x+5}$

EXAMPLE 4   Subtract: $\dfrac{3x}{5y} - \dfrac{x}{5y} - \dfrac{4x}{5y}$

Solution: Remember that addition and subtraction in an expression are performed from left to right.

$\dfrac{3x}{5y} - \dfrac{x}{5y} - \dfrac{4x}{5y} = \dfrac{3x - x - 4x}{5y}$ ◀ **Difference rule**

$= \dfrac{-2x}{5y}$ ◀ **Like terms are combined.**

$= -\dfrac{2x}{5y}$

▊ **CLASSROOM EXERCISES** ▊

Add or subtract. Simplify where necessary. (Examples 1–3)

1. $\dfrac{1}{5} + \dfrac{2}{5}$ $\dfrac{3}{5}$

2. $\dfrac{2}{9} + \dfrac{5}{9}$ $\dfrac{7}{9}$

3. $\dfrac{5}{12} - \dfrac{3}{12}$ $\dfrac{1}{6}$

4. $\dfrac{8}{15} - \dfrac{11}{15}$ $-\dfrac{1}{5}$

5. $\dfrac{2}{x} + \dfrac{1}{x}$ $\dfrac{3}{x}$

6. $\dfrac{8}{a} - \dfrac{5}{a}$ $\dfrac{3}{a}$

7. $\dfrac{5}{3x} + \dfrac{2}{3x}$ $\dfrac{7}{3x}$

8. $\dfrac{7}{5a} - \dfrac{4}{5a}$ $\dfrac{3}{5a}$

9. $\dfrac{5}{x+1} + \dfrac{6}{x+1}$

10. $\dfrac{10}{x-2} - \dfrac{3}{x-2}$

11. $\dfrac{x}{x+3} + \dfrac{2x}{x+3}$

12. $\dfrac{3y}{y-2} - \dfrac{y}{y-2}$

13. $\dfrac{a}{a+1} - \dfrac{5a}{a+1}$

14. $\dfrac{3x}{x-5} - \dfrac{3x}{x-5}$ 0

15. $\dfrac{2x}{x+y} + \dfrac{3y}{x+y}$

16. $\dfrac{2a}{2a+3b} - \dfrac{3b}{2a+3b}$

Sums and Differences of Rational Expressions / **295**

Additional Examples
Example 3
Add or subtract.

1. $\dfrac{4x}{x-4} - \dfrac{11x}{x-4}$
   Ans: $-\dfrac{7x}{x-4}$

2. $\dfrac{y}{x+7} + \dfrac{8y}{x+7}$
   Ans: $\dfrac{9y}{x+7}$

Example 4
Add or subtract.

1. $\dfrac{9}{3x} + \dfrac{8}{3x} - \dfrac{10}{3x}$
   Ans: $\dfrac{7}{3x}$

2. $\dfrac{5y}{9x} - \dfrac{y}{9x} + \dfrac{4y}{9x}$
   Ans: $\dfrac{8y}{9x}$

Common Error
Students add or subtract the denominators when finding the sums or differences of rational expressions.

Example
$\dfrac{x}{x+3} + \dfrac{2x}{x+3} = \dfrac{3x}{2x+6}$

Prescription
Remind students that fraction sums such as $\frac{1}{3} + \frac{1}{3} = \frac{2}{3}$, not $\frac{2}{6}$. Then go over the Sum and Difference Rules for Rational Expressions on page 294.

Additional Answers

9. $\dfrac{11}{x+1}$

10. $\dfrac{7}{x-2}$

11. $\dfrac{3x}{x+3}$

12. $\dfrac{2y}{y-2}$

13. $-\dfrac{4a}{a+1}$

15. $\dfrac{2x+3y}{x+y}$

16. $\dfrac{2a-3b}{2a+3b}$

Add or subtract. (Example 4)

17. $\dfrac{x}{11} - \dfrac{3x}{11} + \dfrac{5x}{11}$ $\dfrac{3x}{11}$

18. $\dfrac{4}{a} + \dfrac{9}{a} - \dfrac{6}{a}$ $\dfrac{7}{a}$

19. $\dfrac{3x}{y+1} + \dfrac{5}{y+1} - \dfrac{x}{y+1}$ $\dfrac{2x+5}{y+1}$

WRITTEN EXERCISES

Goal: To add or subtract rational expressions with equal denominators

Sample Problems: a. $\dfrac{6}{x-4} + \dfrac{x}{x-4}$ **b.** $\dfrac{6}{x-4} - \dfrac{x}{x-4}$

Answers: a. $\dfrac{6+x}{x-4}$ **b.** $\dfrac{6-x}{x-4}$

Add or subtract. (Examples 1–4)

1. $\dfrac{5}{13} + \dfrac{6}{13}$ $\dfrac{11}{13}$

2. $\dfrac{2}{11} + \dfrac{8}{11}$ $\dfrac{10}{11}$

3. $\dfrac{3}{17} - \dfrac{8}{17} - \dfrac{5}{17}$

4. $\dfrac{3}{7} - \dfrac{5}{7}$

5. $\dfrac{5}{a} + \dfrac{11}{a}$ $\dfrac{16}{a}$

6. $\dfrac{2}{b} + \dfrac{13}{b}$ $\dfrac{15}{b}$

7. $\dfrac{8}{y} - \dfrac{13}{y} - \dfrac{5}{y}$

8. $\dfrac{12}{x} - \dfrac{27}{x}$

9. $\dfrac{5}{x+3} + \dfrac{7}{x+3}$ $\dfrac{12}{x+3}$

10. $\dfrac{3}{y+2} + \dfrac{7}{y+2}$ $\dfrac{10}{y+2}$

11. $\dfrac{3}{a-3} - \dfrac{4}{a-3}$

12. $\dfrac{10}{b+7} - \dfrac{11}{b+7}$ $-\dfrac{1}{b+7}$

13. $\dfrac{2x}{x+2} + \dfrac{5x}{x+2}$ $\dfrac{7x}{x+2}$

14. $\dfrac{y}{y-1} + \dfrac{4y}{y-1}$

15. $\dfrac{a}{a+1} - \dfrac{5a}{a+1}$ $-\dfrac{4a}{a+1}$

16. $\dfrac{2r}{r-6} - \dfrac{3r}{r-6} - \dfrac{r}{r-6}$

17. $\dfrac{a}{2a+b} + \dfrac{b}{2a+b}$

18. $\dfrac{x}{x-3y} + \dfrac{y}{x-3y}$ $\dfrac{x+y}{x-3y}$

19. $\dfrac{3r}{r-2s} - \dfrac{5s}{r-2s}$ $\dfrac{3r-5s}{r-2s}$

20. $\dfrac{c}{c-d} - \dfrac{2d}{c-d}$

21. $\dfrac{3y}{19} - \dfrac{8y}{19} + \dfrac{y}{19}$ $-\dfrac{4y}{19}$

22. $\dfrac{t}{8} + \dfrac{4t}{8} - \dfrac{10t}{8}$ $-\dfrac{5t}{8}$

23. $\dfrac{4x}{3y} - \dfrac{x}{3y} - \dfrac{5x}{3y}$

24. $\dfrac{3p}{10q} - \dfrac{2p}{10q} - \dfrac{8p}{10q}$ $-\dfrac{7p}{10q}$

25. $\dfrac{5}{9t} - \dfrac{12}{9t} + \dfrac{3}{9t}$ $-\dfrac{4}{9t}$

26. $\dfrac{4}{7y} - \dfrac{9}{7y} + \dfrac{1}{7y}$

REVIEW CAPSULE FOR SECTION 11.2

Subtract. Then combine like terms. (Pages 217–219)

1. $(4x - 2) - (3x + 1)$ $x - 3$

2. $(5t + 1) - (6t - 5)$ $-t + 6$

3. $(3m - 6) - (-m + 10)$ $4m - 16$

4. $(r^2 - r) - (3r + 2)$ $r^2 - 4r - 2$

5. $(3n^2 + n) - (12 - n)$ $3n^2 + 2n - 12$

6. $(s^2 - 3s) - (3s^2 - 5s + 3)$ $-2s^2 + 2s - 3$

11.2 More Sums and Differences

OBJECTIVE: To add or subtract with two rational expressions having monomial or binomial numerators and like binomial denominators

In many of the examples and exercises of this section, it will be necessary to simplify the result after applying the Sum or Difference Rule.

P–1 In $\dfrac{a}{a-b}$, what values cannot be assigned to *a* and *b*?

Their values cannot be the same.

EXAMPLE 1 Subtract and simplify: $\dfrac{a}{a-b} - \dfrac{b}{a-b}$

Solution:

1. Difference Rule for Rational Expressions ⟶ $\dfrac{a}{a-b} - \dfrac{b}{a-b} = \dfrac{a-b}{a-b}$

2. Simplify. ⟶ $= 1$

EXAMPLE 2 Add: $\dfrac{2x+3}{x} + \dfrac{x-5}{x}$

Solution:

1. Sum Rule for Rational Expressions ⟶ $\dfrac{2x+3}{x} + \dfrac{x-5}{x} = \dfrac{(2x+3)+(x-5)}{x}$

2. Add binomials in the numerator. ⟶ $= \dfrac{3x-2}{x}$

P–2 How do you know that $\dfrac{3x-2}{x}$ in Example 2 cannot be simplified?

The numerator and denominator have no common factors.

Recall how to subtract binomials.

Vertical Method

$$4x - 3$$
$$(-)\ \underline{\quad x + 1\quad}$$
$$3x - 4$$

Horizontal Method

$(4x - 3) - (x + 1) = (4x - 3) + -(x + 1)$

$= (4x - 3) + (-x - 1)$

$= 3x - 4$

◄ The binomial *x* + 1 is subtracted by adding its opposite (−*x* − 1).

Sums and Differences of Rational Expressions / **297**

Teaching Suggestions p. M–38

Lesson Resources
Warm–Up: p. M–39
Maintenance: See below.
Practice Worksheet 11.2
Transparency 20

Maintenance
1. Solve and check:
$\frac{3}{4}a - \frac{1}{4} = \frac{1}{4}$
Ans: $a = \frac{2}{3}$
(Section 3.3)
2. Simplify: $(2^{-3}x^4yz^{-1})^3$
Ans: $\dfrac{x^{12}y^3}{512z^3}$
(Section 6.5)
3. Divide:
$(5x^2 - 15x) \div 5x$
Ans: $x - 3$ (Section 8.6)
4. Write the number or numbers that cannot be in the domain of *x* in $\dfrac{x-4}{3x+5}$.
Ans: $-\frac{5}{3}$ (Section 10.1)
5. The perimeter of a rectangle is 15 meters. The width is 4 meters. Find the length.
Ans: 3.5 meters
(Section 3.4)

Additional Examples
Example 1
Add or subtract. Then simplify.
1. $\dfrac{2x}{2x+7} + \dfrac{7}{2x+7}$
Ans: 1
2. $\dfrac{3x}{x+8} - \dfrac{2x-8}{x+8}$
Ans: 1

Example 2
Add or subtract.
1. $\dfrac{3x+1}{x} + \dfrac{x-6}{x}$
Ans: $\dfrac{4x-5}{x}$
2. $\dfrac{x-6}{4x} - \dfrac{4x+1}{4x}$
Ans: $\dfrac{-3x-7}{4x}$

EXAMPLE 3 $\qquad$ Subtract: $\dfrac{5x - 2}{x - 3} - \dfrac{2x + 3}{x - 3}$

Special attention should be given to step ② in both Examples 3 and 4. This concept causes much difficulty.

Solution:

① Difference Rule → $\dfrac{5x - 2}{x - 3} - \dfrac{2x + 3}{x - 3} = \dfrac{(5x - 2) - (2x + 3)}{x - 3}$

② Meaning of subtraction → $= \dfrac{(5x - 2) + (-2x - 3)}{x - 3}$

◄ **$2x + 3$ is subtracted by adding its opposite.**

③ Combine like terms. → $= \dfrac{3x - 5}{x - 3}$

EXAMPLE 4 $\qquad$ Subtract and simplify: $\dfrac{3x + 2}{x + 2} - \dfrac{x - 2}{x + 2}$

Solution:

① Difference Rule → $\dfrac{3x + 2}{x + 2} - \dfrac{x - 2}{x + 2} = \dfrac{(3x + 2) - (x - 2)}{x + 2}$

② Meaning of subtraction → $= \dfrac{(3x + 2) + (-x + 2)}{x + 2}$

③ Combine like terms. → $= \dfrac{2x + 4}{x + 2}$

④ Factor the numerator. → $= \dfrac{2(x + 2)}{x + 2}$

⑤ Identify a name for 1. → $= \dfrac{(x + 2)}{(x + 2)} \cdot \dfrac{2}{1}$

⑥ Simplify. → $= 2$

◄ **Simplifying a fraction is also called reducing the fraction.**

CLASSROOM EXERCISES

Add or subtract. Simplify where necessary. (Examples 1–4)

1. $\dfrac{x + 3}{x + 2} + \dfrac{2}{x + 2}$ $\dfrac{x + 5}{x + 2}$

2. $\dfrac{3a}{a - 5} + \dfrac{2a + 3}{a - 5}$

3. $\dfrac{3y + 5}{y + 3} - \dfrac{2}{y + 3}$

4. $\dfrac{2x - 3}{x + 1} - \dfrac{1}{x + 1}$ $\dfrac{2x - 4}{x + 1}$

5. $\dfrac{3b + 1}{b + 2} + \dfrac{b}{b + 2}$

6. $\dfrac{5r - 2}{r - 4} - \dfrac{2r}{r - 4}$

7. $\dfrac{2a + b}{a + b} - \dfrac{a}{a + b}$ 1

8. $\dfrac{x + y}{2x + y} + \dfrac{x}{2x + y}$ 1

9. $\dfrac{x + 2}{x - 3} + \dfrac{x + 3}{x - 3}$

10. $\dfrac{x + 5}{x + 1} + \dfrac{3 - x}{x + 1}$

11. $\dfrac{3x + 3}{x + y} - \dfrac{x + 1}{x + y}$ $\dfrac{2x + 2}{x + y}$

12. $\dfrac{3x + 1}{x + 4} - \dfrac{x + 4}{x + 4}$ $\dfrac{2x - 3}{x + 4}$

WRITTEN EXERCISES

Goal: To add or subtract rational expressions

Sample Problem: $\dfrac{4x + 13}{x + 4} - \dfrac{x + 1}{x + 4}$

Answer: 3

Add or subtract. Simplify where necessary. (Examples 1–4)

1. $\dfrac{2x + 1}{x + y} + \dfrac{4}{x + y}$ $\dfrac{2x + 5}{x + y}$

2. $\dfrac{2}{a - b} + \dfrac{3a + 5}{a - b}$ $\dfrac{3a + 7}{a - b}$

3. $\dfrac{6r + 3}{r + s} - \dfrac{4}{r + s}$

4. $\dfrac{5d + 1}{c + d} - \dfrac{3}{c + d}$ $\dfrac{5d - 2}{c + d}$

5. $\dfrac{3x - 5}{x + 2} + \dfrac{x - 2}{x + 2}$ $\dfrac{4x - 7}{x + 2}$

6. $\dfrac{5y - 1}{y - 3} + \dfrac{2y - 7}{y - 3}$

7. $\dfrac{5a - 2}{a + b} - \dfrac{a + 3}{a + b}$ $\dfrac{4a - 5}{a + b}$

8. $\dfrac{7c - 1}{c - d} - \dfrac{2c + 3}{c - d}$ $\dfrac{5c - 4}{c - d}$

9. $\dfrac{5x - 7}{2x - 3} + \dfrac{4 - 3x}{2x - 3}$

10. $\dfrac{10 - 3y}{5y + 1} + \dfrac{8y - 9}{5y + 1}$ 1

11. $\dfrac{5a + 1}{a + 2} - \dfrac{3a - 3}{a + 2}$ 2

12. $\dfrac{4b - 9}{b - 1} - \dfrac{2b - 7}{b - 1}$

13. $\dfrac{x^2 + 3x}{4x + 3} + \dfrac{3x^2}{4x + 3}$ x

14. $\dfrac{4y^2}{y - 1} + \dfrac{y^2 - 5y}{y - 1}$ $5y$

15. $\dfrac{9a - 2a^2}{a + 2} - \dfrac{3a - 5a^2}{a + 2}$ $3a$

16. $\dfrac{b^2 + 5b}{b + 2} - \dfrac{-5b - 4b^2}{b + 2}$ $5b$

17. $\dfrac{x^2 + 3x}{x + 4} - \dfrac{16 + 3x}{x + 4}$ $x - 4$

18. $\dfrac{2a^2 + 2x}{a + 3} - \dfrac{a^2 + 2x + 9}{a + 3}$ $a - 3$

19. $\dfrac{y^2 - 3y}{y - 4} + \dfrac{16 - 5y}{y - 4}$ $y - 4$

20. $\dfrac{25 - 8b}{b - 5} + \dfrac{b^2 - 2b}{b - 5}$ $b - 5$

MORE CHALLENGING EXERCISES

21. $\dfrac{3x + 2}{x - 3} - \dfrac{x + 7}{x - 3} + \dfrac{2x - 7}{x - 3}$ 4

22. $\dfrac{x + 3}{2x - 1} + \dfrac{3x - 5}{2x - 1} - \dfrac{1 - 2x}{2x - 1}$ 3

23. $\dfrac{x - 5}{x + 5} - \dfrac{2x}{x + 5} - \dfrac{1 - x}{x + 5}$ $-\dfrac{6}{x + 5}$

24. $\dfrac{a^2 + a}{a + 4} + \dfrac{a}{a + 4} - \dfrac{a^2 + 2a}{a + 4}$ 0

REVIEW CAPSULE FOR SECTION 11.3

Write the prime factors of each polynomial. (Pages 258–260)

1. $24x^3y^2$

2. $56r^2s^4t^3$

3. $4m^2 - 36$

4. $x^2 - 6x + 5$ $(x - 1), (x - 5)$

5. $6x^2 + 6x - 72$

6. $(a^2 - 4)(4a^2 - 4a - 8)$

*Sums and Differences of Rational Expressions / **299***

Assignment Guide
Basic p. 299: 1–19 odd
Average p. 299: 3, 6, 9, ⋯, 24;
p. 296: 3, 6, 9 ⋯ 24
Above Average p. 299:
1–19 odd, 21–24; p.296: 3, 6,
9 ⋯ 24

Additional Answers
Written Exercises
3. $\dfrac{6r - 1}{r + s}$
6. $\dfrac{7y - 8}{y - 3}$
9. 1
12. 2

Additional Answers
Review Capsule
1. 2, 2, 2, 3, x, x, x, y, y
2. 2, 2, 2, 7, r, r, s, s, s, s, t, t, t
3. 2, 2, $(m + 3)$, $(m - 3)$
5. 2, 3, $(x + 4)$, $(x - 3)$
6. 2, 2, $(a + 2)$, $(a - 2)$,
  $(a + 1)$, $(a - 2)$

Maintenance

1. Simplify: $3t - t - 7t$
   Ans: −5t (Section 2.6)
2. Write 38,400 in scientific notation.
   Ans: 3.84 × 10⁴
   (Section 6.6)
3. Factor: $2x^4 + 16x^2 + 12x$
   Ans: 2x(x³ + 8x + 6)
   (Section 9.1)
4. Simplify: $\frac{5x + 15}{x^2 - 9}$
   Ans: $\frac{5}{x - 3}$ (Section 10.2)
5. The cost of the hotel was $108 less than the cost of the airfare (Condition 1). The total amount for the hotel and the airfare was $288 (Condition 2). Find the cost of the hotel and the airfare.
   **Ans: airfare: $198;
   hotel: $90** (Section 4.1)

Additional Examples
Example 1
Find each LCM.
1. 4; 10; 12
   Ans: 60
2. 9; 15; 6
   Ans: 90
3. 7; 4; 3
   Ans: 84

Example 2
Find each LCM.
1. $8xy$; $10x^2$; $5y$
   Ans: $40x^2y$
2. $6a^2$; $15ab^2$; $10a^3$
   Ans: $30a^3b^2$
3. $5x$; $12y$; $9x^2$
   Ans: $180x^2y$

11.3 Least Common Multiple of Polynomials

OBJECTIVE: To write the least common multiple of two or more polynomials in factored form

Definition

It is important to have students follow the steps as shown in Example 1 and summarized on page 302.

> The **least common multiple** (*LCM*) of two or more counting numbers is the smallest counting number that is divisible by the given numbers.
>
> The LCM of 4 and 6 is 12.

EXAMPLE 1   Find the *LCM* of 12, 30, and 45.

Solution:

1. Write the prime factorization of each number. ⟶ $12 = 2^2 \cdot 3$

   $30 = 2 \cdot 3 \cdot 5$

   $45 = 3^2 \cdot 5$

2. Write a product using each prime factor only once. ⟶ $2 \cdot 3 \cdot 5$

3. For each factor, write the greatest exponent used in any of the prime factorizations. ⟶ $2^2 \cdot 3^2 \cdot 5$

4. Multiply. ⟶ $4 \cdot 9 \cdot 5 = 180$

The method of Example 1 can also be used with polynomials.

EXAMPLE 2   Find the *LCM* of $10x^2$, $4xy$, and $15x^2y$.

Solution:

1. $10x^2 = 2 \cdot 5 \cdot x^2$

   $4xy = 2^2 \cdot x \cdot y$

   $15x^2y = 3 \cdot 5 \cdot x^2 \cdot y$

2. $2 \cdot 3 \cdot 5 \cdot x \cdot y$

3. $2^2 \cdot 3 \cdot 5 \cdot x^2 \cdot y$

4. $60x^2y$

300 / *Chapter 11*

You can write the *LCM* in factored form if one or more of the factors is a binomial.

EXAMPLE 3   Find the *LCM* of $(2x - 4)$ and $(x^2 - 4)$.

Solution:

1   Factor. ──────────────────────▶ $2x - 4 = 2(x - 2)$

                                             $x^2 - 4 = (x + 2)(x - 2)$

2   Write a product using each factor only once. ──▶ $2(x + 2)(x - 2)$

3   Write the greatest exponent for each factor. ──▶ $2(x + 2)(x - 2)$ ◀ *Factored form*

EXAMPLE 4   Find the *LCM* of $(x - 3)$ and $(x + 5)$.

Solution: Both of the binomials are prime.

$$LCM = (x - 3)(x + 5)$$ ◀ *Since the factors are prime, the LCM is their product.*

> The least common multiple of prime polynomials is their product.

P–1   **What is the *LCM* of the polynomials below?** $(x - y)(x + y)3xy$

       $(x - y)$ $(x + y)$ $3xy$

EXAMPLE 5   Find the *LCM* of $x^2 - 9$, $5x^2 + 5x - 30$, and $x^2 - 4x + 4$.

Solution:

The LCM of two or more polynomials must contain each prime factor the greatest number of times it appears in any one of the given polynomials.

1   Factor. ────────▶

$$x^2 - 9 = (x + 3)(x - 3)$$
$$5x^2 + 5x - 30 = 5(x^2 + x - 6)$$
$$= 5(x + 3)(x - 2)$$
$$x^2 - 4x + 4 = (x - 2)(x - 2)$$

◀ *Prime factorizations*

2   Write a product using each factor only once. ──▶ $5(x + 3)(x - 3)(x - 2)$

3   Write the greatest exponent for each factor. ──▶ $5(x + 3)(x - 3)(x - 2)^2$

Additional Examples
Example 3
Find each LCM.
1. $x^2 - 6x$; $x^2 - 36$
   Ans: $x(x - 6)(x + 6)$
2. $7x + 35$; $x^2 + 6x + 5$
   Ans: $7(x + 5)(x + 1)$

Example 4
Find each LCM.
1. $x - 6$; $x + 4$
   Ans: $(x - 6)(x + 4)$
2. $5x + 5$; $x - 1$
   Ans: $5(x + 1)(x - 1)$

Example 5
Find each LCM.
1. $4x^2 - 8x$; $x^2 - 4$;
   $x^2 + 5x + 6$
   **Ans: $4x(x - 2)(x + 2) \cdot$
   $(x + 3)$**
2. $5x^2 - 5$; $10x - 60$;
   $x^2 - 3x - 18$
   **Ans: $10(x - 1)(x + 1) \cdot$
   $(x - 6)(x + 3)$**

Additional Answers
Classroom Exercises

3. $2 \cdot 3 \cdot x \cdot x \cdot y$
5. $2 \cdot 3 \cdot 5 \cdot a \cdot b \cdot c$
6. $2 \cdot 5 \cdot 7 \cdot x \cdot y \cdot z$
8. $(2x + 3)(3x - 5)$
9. $(a + b)(a - b)$
10. $2 \cdot 2 \cdot 3 \cdot x \cdot x \cdot y$
13. $(2x - y)(x + 2y)$
14. $(5x - 3)(x + 2)(x - 4)$

Assignment Guide
Basic p. 302: 1–19 odd
Average p. 302: 1–21 odd
Above Average p. 302: 1–21 odd, 22

Problem–Solving Skills
Using logical reasoning (Ex. 21–22)

Critical Thinking
See Exercises 21 and 22.

Non–Routine Problems
For a description of non–routine problems, see page M–13.

Additional Answers
Written Exercises

6. $60x^2y^2$
7. $(x - 3)(x + 3)$
8. $(a + 5)(a - 5)$
9. $3(x - y)(x + y)$
10. $5(a + b)(a - b)$
11. $6b(b + 5)(b - 6)$
12. $10r(r + 3)(r - 3)$
14. $2a^2(a + 6)(a - 6)$
18. $(x - 3)(x + 6)^2$
20. $(x + 2)(x - 2)(x - 4)$
21. No. Other possible monomials are $6c$, $12c$, and $24c$.
22. $6(x - 4)(x + 4)$ is the LCM if $a = 3$ or $a = 6$.

Steps in finding the LCM of two or more polynomials:

1. Write the prime factorization of each polynomial.
2. Write a product using each prime factor only once.
3. For each factor, write the greatest exponent used in any prime factorization.
4. Multiply where necessary.

CLASSROOM EXERCISES

Write the factored form of each LCM. (Examples 1–4)

1. 3; 7 $3 \cdot 7$ **2.** $2x$; y $2xy$ **3.** $3x^2$; $2y$ **4.** 2; 3; 5 $2 \cdot 3 \cdot 5$ **5.** $2a$; $3b$; $5c$
6. $5x$; $7y$; $2z$ **7.** $(x - 3)$; $(x - 5)$ $(x-3)(x-5)$ **8.** $(2x + 3)$; $(3x - 5)$
9. $(a + b)$; $(a - b)$ **10.** $12x^2y$; $3xy$ **11.** $3(a - b)$; $(a - b)$ $3(a-b)$
12. $2x$; $3x$ $2 \cdot 3 \cdot x$ **13.** $(2x - y)$; $(x + 2y)$ **14.** $(5x - 3)$; $(x + 2)$; $(x - 4)$
15. $5xy$; $3xy$ $3 \cdot 5 \cdot x \cdot y$ **16.** $2(x + y)$; $3(x + y)$ $2 \cdot 3(x + y)$ **17.** $5(a + b)$; $3(a - b)$ $3 \cdot 5(a + b)(a - b)$

WRITTEN EXERCISES

Goal: To find the least common multiple of two or more polynomials
Sample Problem: $2x + 2y$; $x^2 - y^2$; $x^2 + xy$ **Answer:** $2x(x + y)(x - y)$

Find each LCM. (Examples 1–5)

1. 10; 6 30 **2.** 12; 30 60 **3.** $4x^2$; $6xy$ $12x^2y$
4. $2ab^2$; $8a^2b$ $8a^2b^2$ **5.** $12a^2b$; $3ab$; $10a^2b$ $60a^2b$ **6.** $4xy$; $6xy^2$; $15x^2$
7. $(x^2 - 9)$; $(x + 3)$ **8.** $(a - 5)$; $(a^2 - 25)$ **9.** $x^2 - y^2$; $3x - 3y$
10. $5a + 5b$; $a^2 - b^2$ **11.** $3b^2 + 15b$; $2b^2 - 12b$ **12.** $5r^2 - 15r$; $2r^2 + 6r$
13. $x^2 - 5x$; $x^2 - 25$; $x^3 + 5x^2$ $x^2(x+5)(x-5)$ **14.** $a^2 - 36$; $2a^2 + 12a$; $a^3 - 6a^2$
15. $x^2 + 4x + 4$; $5x + 10$ $5(x+2)^2$ **16.** $x^2 - 6x + 9$; $4x - 12$ $4(x-3)^2$
17. $x^2 - 8x + 16$; $x^2 - 2x - 8$ $(x+2)(x-4)^2$ **18.** $x^2 + 12x + 36$; $x^2 + 3x - 18$
19. $x^2 - 5x + 6$; $x^2 + x - 6$ $(x+3)(x-2)(x-3)$ **20.** $x^2 - 2x - 8$; $x^2 - 6x + 8$

MORE CHALLENGING EXERCISES

21. The least common multiple of two monomials is $24cd$. One of the monomials is $8d$. Must the other monomial be $3c$? Explain.

22. If the least common multiple of the polynomials $2x^2 - 32$ and $ax - 4a$ is $6(x - 4)(x + 4)$, find two possible values for a. 3, 6

11.4 Unlike Monomial Denominators

OBJECTIVE: To add or subtract with rational expressions having monomial numerators and unlike monomial denominators

EXAMPLE 1   Add: $\dfrac{2}{a} + \dfrac{3}{b}$

Solution:

1. Find the *LCM* of a and b. ────────▶ ab

2. Multiplication Property of One ────────▶ $\dfrac{2}{a} = \dfrac{2}{a}\left(\dfrac{b}{b}\right)$ and $\dfrac{3}{b} = \dfrac{3}{b}\left(\dfrac{a}{a}\right)$

    You must get ab as the common denominator.

3. Substitute. ────────▶ $\dfrac{2}{a} + \dfrac{3}{b} = \dfrac{2}{a} \cdot \dfrac{b}{b} + \dfrac{3}{b} \cdot \dfrac{a}{a}$

4. Product Rule ────────▶ $= \dfrac{2b}{ab} + \dfrac{3a}{ab}$

5. Sum Rule ────────▶ $= \dfrac{2b + 3a}{ab}$ Simplest form

EXAMPLE 2   Subtract: $\dfrac{5}{2x} - \dfrac{2}{3x}$

Solution:

1. Find the *LCM* of $2x$ and $3x$. ────────▶ $6x$

2. Multiplication Property of One ────────▶ $\dfrac{5}{2x} = \dfrac{5}{2x}\left(\dfrac{3}{3}\right)$ and $\dfrac{2}{3x} = \dfrac{2}{3x}\left(\dfrac{2}{2}\right)$

3. Substitute. ────────▶ $\dfrac{5}{2x} - \dfrac{2}{3x} = \dfrac{5}{2x} \cdot \dfrac{3}{3} - \dfrac{2}{3x} \cdot \dfrac{2}{2}$

   Students will probably need some special help in choosing the names for 1 in step 2 that will convert each denominator to the LCM. (See P-1 below.)

4. Product Rule ────────▶ $= \dfrac{15}{6x} - \dfrac{4}{6x}$

5. Difference Rule ────────▶ $= \dfrac{15 - 4}{6x}$

6. Simplify. ────────▶ $= \dfrac{11}{6x}$

Multiplying by these numbers gives the LCM, $6x$, as the denominator for both fractions.

P–1 In Step 2 of Example 2, how were $\dfrac{3}{3}$ and $\dfrac{2}{2}$ chosen as names for 1?

Multiplying by these numbers gives the LCM, $6x$, as the denominator for both fractions.

Sums and Differences of Rational Expressions / 303

Teaching Suggestions p. M-38

Lesson Resources
Warm–Up: p. M–39
Maintenance: See below.
Practice Worksheet 11.4
Transparency 29

Maintenance
1. Simplify: $y - 12 + z + 6$
   Ans: $y + z - 6$
   (Section 2.2)
2. Solve and check:
   $2(4x + 7) = 6(x - 4)$
   Ans: $x = -19$
   (Section 4.3)
3. Factor: $x^2 - 11x + 24$
   Ans: $(x - 8)(x - 3)$
   (Section 9.3)
4. Multiply and simplify:
   $\dfrac{3ab}{7b} \cdot \dfrac{14}{a}$
   Ans: 6 (Section 10.3)
5. Mary's monthly income is $40 more than Deanna's monthly income (Condition 1). Mary deposits 60% of her monthly income in a checking account. Deanna deposits 65% of her monthly income in a checking account. Both deposit the same amount (Condition 2). Find both monthly incomes.
   **Ans: Mary: $520;
   Deanna: $480** (Section 4.4)

Additional Examples
Example 1
Add.
1. $\dfrac{3}{x} + \dfrac{7}{y}$ Ans: $\dfrac{3y + 7x}{xy}$
2. $\dfrac{9}{2a} + \dfrac{1}{5b}$ Ans: $\dfrac{45b + 2a}{10ab}$

Example 2
Add or subtract.
1. $\dfrac{2}{5y} + \dfrac{9}{4y}$ Ans: $\dfrac{53}{20y}$
2. $\dfrac{1}{7x} - \dfrac{3}{2x}$ Ans: $-\dfrac{19}{14x}$

What is the *LCM* of $10x^2y$ and $8xy$? $40x^2y$

EXAMPLE 3   Add: $\dfrac{3}{10x^2y} + \dfrac{5}{8xy}$

Solution: $\dfrac{3}{10x^2y} + \dfrac{5}{8xy} = \dfrac{3}{10x^2y}\left(\dfrac{4}{4}\right) + \dfrac{5}{8xy}\left(\dfrac{5x}{5x}\right)$

◄ $10x^2y = 2 \cdot 5 \cdot x^2 \cdot y$
$8xy = 2^3 \cdot x \cdot y$
$LCM = 2^3 \cdot 5 \cdot x^2 \cdot y$ or $40x^2y$

$= \dfrac{12}{40x^2y} + \dfrac{25x}{40x^2y}$

$= \dfrac{12 + 25x}{40x^2y}$

P–3   How do you know that $\dfrac{12 + 25x}{40x^2y}$ in Example 3 is in simplest form?

There is no common factor.

P–4   How can **5r** be written as a fraction? $\dfrac{5r}{1}$

EXAMPLE 4   Subtract: $5r - \dfrac{3r}{8t}$

Solution: $\dfrac{5r}{1} - \dfrac{3r}{8t} = \dfrac{5r}{1}\left(\dfrac{8t}{8t}\right) - \dfrac{3r}{8t}$

$= \dfrac{40rt}{8t} - \dfrac{3r}{8t}$

$= \dfrac{40rt - 3r}{8t}$

CLASSROOM EXERCISES

Write the missing name for 1.

1. $\dfrac{2}{x}\left(\dfrac{?}{?}\right) = \dfrac{2x}{x^2}$ $\dfrac{x}{x}$

2. $\dfrac{3}{ab}\left(\dfrac{?}{?}\right) = \dfrac{15}{5ab}$ $\dfrac{5}{5}$

3. $\dfrac{2}{5y}\left(\dfrac{?}{?}\right) = \dfrac{4}{10y}$ $\dfrac{2}{2}$

4. $\dfrac{3}{a^2b}\left(\dfrac{?}{?}\right) = \dfrac{3b}{a^2b^2}$ $\dfrac{b}{b}$

5. $\dfrac{5}{6xy}\left(\dfrac{?}{?}\right) = \dfrac{10x}{12x^2y}$ $\dfrac{2x}{2x}$

6. $\dfrac{3a}{2b}\left(\dfrac{?}{?}\right) = \dfrac{18a^2b}{12ab^2}$ $\dfrac{6ab}{6ab}$

7. $\dfrac{2}{5r}\left(\dfrac{?}{?}\right) = \dfrac{6s}{15rs}$ $\dfrac{3s}{3s}$

8. $\dfrac{5}{12xy}\left(\dfrac{?}{?}\right) = \dfrac{15xy}{36x^2y^2}$

$\dfrac{3xy}{3xy}$

Additional Examples
Example 3
Add or subtract.

1. $\dfrac{1}{6x^2} - \dfrac{3}{4xy}$
   Ans: $\dfrac{2y - 9x}{12x^2y}$

2. $\dfrac{3}{5ab} + \dfrac{7}{15b^3}$
   Ans: $\dfrac{9b^2 + 7a}{15ab^3}$

3. $\dfrac{1}{12xy} - \dfrac{5}{18x^2y}$
   Ans: $\dfrac{3x - 10}{36x^2y}$

Example 4
Add or subtract.

1. $4x + \dfrac{3x}{2y}$
   Ans: $\dfrac{8xy + 3x}{2y}$

2. $5a - \dfrac{1}{7b}$
   Ans: $\dfrac{35ab - 1}{7b}$

3. $x + \dfrac{3}{y}$
   Ans: $\dfrac{xy + 3}{y}$

Common Error
When finding the sums or differences of rational expressions with unlike monomial denominators, students forget to change the numerator after finding the common denominator.

Example
$\dfrac{3}{2x} - \dfrac{5}{x} = \dfrac{3}{2x} - \dfrac{5}{2x}$
$= -\dfrac{2}{2x}$

Prescription
Require students to follow the steps as shown in Examples 1–4. Remind students that when the denominator is changed, the numerator must also be changed.

Write the missing polynomials in each exercise. (Step 2 of Examples 1 and 2)

9. $\dfrac{2}{3ab}\left(\dfrac{?}{?}\right)=\dfrac{?}{9a^2b^2}$ $\quad \dfrac{3ab}{3ab}; 6ab$

10. $\dfrac{9s}{5r^2}\left(\dfrac{?}{?}\right)=\dfrac{?}{15r^2s}$ $\quad \dfrac{3s}{3s}; 27s^2$

11. $\dfrac{?}{5y}\left(\dfrac{2x}{2x}\right)=\dfrac{6ax}{?}$

12. $\dfrac{3b}{?}\left(\dfrac{10b}{10b}\right)=\dfrac{?}{40a^2b}$ $\quad 4a^2; 30b^2$

13. $\dfrac{15x}{22y}\left(\dfrac{?}{?}\right)=\dfrac{?}{66xy}$ $\quad \dfrac{3x}{3x}; 45x^2$

14. $\dfrac{7s}{5r}\left(\dfrac{3rs}{3rs}\right)=\dfrac{?}{?}$

15. $\dfrac{2}{3x}+\dfrac{5}{2y}=\dfrac{2}{3x}\left(\dfrac{?}{?}\right)+\dfrac{5}{2y}\left(\dfrac{?}{?}\right)$ $\quad \dfrac{2y}{2y}; \dfrac{3x}{3x}$

$=\dfrac{?}{6xy}+\dfrac{15x}{6xy}$ $\quad 4y$

$=\dfrac{?}{6xy}$ $\quad 4y+15x$

16. $\dfrac{3}{a^2b}-\dfrac{5}{2ab}=\dfrac{3}{a^2b}\left(\dfrac{2}{2}\right)-\dfrac{5}{2ab}\left(\dfrac{?}{?}\right)$ $\quad \dfrac{a}{a}$

$=\dfrac{?}{?}-\dfrac{5a}{2a^2b}$ $\quad \dfrac{6}{2a^2b}$

$=\dfrac{?}{2a^2b}$ $\quad 6-5a$

WRITTEN EXERCISES

Goal: To add or subtract rational expressions that have unlike monomial denominators

Sample Problem: $\dfrac{7}{4x^2}+\dfrac{5}{6xy}$ **Answer:** $\dfrac{21y+10x}{12x^2y}$

Add or subtract. (Examples 1–4)

1. $\dfrac{3}{7}+\dfrac{2}{5}$ $\quad \dfrac{29}{35}$

2. $\dfrac{2}{3}+\dfrac{5}{11}$ $\quad \dfrac{37}{33}$

3. $\dfrac{5}{a}+\dfrac{7}{b}$ $\quad \dfrac{5b+7a}{ab}$

4. $\dfrac{1}{x}+\dfrac{4}{y}$

5. $\dfrac{3}{2x}-\dfrac{5}{x}$ $\quad -\dfrac{7}{2x}$

6. $\dfrac{3}{y}-\dfrac{4}{3y}$ $\quad \dfrac{5}{3y}$

7. $\dfrac{2}{5x}+\dfrac{3}{10x}$ $\quad \dfrac{7}{10x}$

8. $\dfrac{5}{3a}+\dfrac{2}{15a}$

9. $\dfrac{5}{4a}-\dfrac{5}{12b}$ $\quad \dfrac{15b-5a}{12ab}$

10. $\dfrac{5}{6x}-\dfrac{3}{24y}$ $\quad \dfrac{20y-3x}{24xy}$

11. $\dfrac{2}{a^2b}+\dfrac{3}{ab^2}$ $\quad \dfrac{2b+3a}{a^2b^2}$

12. $\dfrac{5}{xy}+\dfrac{3}{xy^2}$

13. $\dfrac{5}{2a^2}-\dfrac{1}{6b^2}$ $\quad \dfrac{15b^2-a^2}{6a^2b^2}$

14. $\dfrac{2}{5xy}-\dfrac{7}{10y^2}$ $\quad \dfrac{4y-7x}{10xy^2}$

15. $\dfrac{2a}{bc}+\dfrac{3b}{ac}$ $\quad \dfrac{2a^2+3b^2}{abc}$

16. $\dfrac{t}{2rs}+\dfrac{3s}{rt}$

17. $\dfrac{2}{3a^2b}+5a$ $\quad \dfrac{2+15a^3b}{3a^2b}$

18. $\dfrac{5}{7xy}+4y$ $\quad \dfrac{5+28xy^2}{7xy}$

19. $5cd-\dfrac{3d}{8c}$

20. $10rt-\dfrac{4r}{9t}$

MORE CHALLENGING EXERCISES

21. If $\dfrac{1}{3c}-\dfrac{1}{6d}=0$, how does the value of c compare with the value of d? $\quad c=2d$

22. If $x^a-\dfrac{1}{x}=\dfrac{x^2-1}{x}$, what is the value of a? $\quad 1$

23. Write a rational expression that, when subtracted from $\dfrac{3x-7}{2x+3}$, gives a difference of $\dfrac{x-5}{2x+3}$. $\quad \dfrac{2x-2}{2x+3}$

24. Write a rational expression that, when added to $\dfrac{3}{x+4}$, gives a sum of $\dfrac{7x+5}{x^2+4x}$. $\quad \dfrac{4x+5}{x^2+4x}$

Sums and Differences of Rational Expressions / **305**

Additional Answers
Classroom Exercises
11. $3a; 10xy$
14. $\dfrac{21rs^2}{15r^2s}$

Assignment Guide
Basic p. 305: 1–19 odd
Average p. 305: 1–23 odd
Above Average p. 305:
1–19 odd, 21–23

Additional Answers
Written Exercises
4. $\dfrac{y+4x}{xy}$
8. $\dfrac{27}{15a}$
12. $\dfrac{5y+3}{xy^2}$
16. $\dfrac{t^2+6s^2}{2rst}$
19. $\dfrac{40c^2d-3d}{8c}$
20. $\dfrac{90rt^2-4r}{9t}$

Problem–Solving Skills
Using logical reasoning
(Ex. 21–28)

Critical Thinking
See Exercises 21–24.

Problem–Solving Skills
Drawing a diagram (Ex. 25–27)
Using guess and check
(Ex. 25–26)
Solving a simpler problem
(Ex. 27)
Making a list (Ex. 28)

Non–Routine Problems
For a description of non–routine problems, see page M–13.

Additional Answers
Written Exercises
25. One solution is shown in the diagram below.

There are 4 pens and 9 pigs. Each pen holds an odd number of pigs.
26. A pattern of 9 small squares within one large square can be drawn with 8 straight lines. The pattern has 1 size 3 x 3, 4 size 2 x 2, and 9 size 1 x 1 squares, or exactly 14 squares.
27. Since Jim runs 8 meters for each of Todd's 7 meters, he catches up one meter. Jim must do this 50 times, so he runs the 8 meters 50 times. The track is 8 x 50, or 400 meters long.
28. Student #1 can be paired with #2–#10, or with 9 other students. Student #2 can be paired with #3–#10, or with 8 other students. Thus, the pattern is 9 + 8 + 7 + 6 + 5 + 4 + 3 + 2 + 1, or a total of 45 pairs.

Quiz Sections 11.1–11.4
After completing this Mid-Chapter Review, you may want to administer a quiz covering the same sections. See the quiz provided on page 229 in the *Teacher's ResourceBank*.

306

NON-ROUTINE PROBLEMS

25. How is it possible to put nine pigs in four pens so that there is an odd number of pigs in each pen?

26. What is the least number of straight lines you can draw to make exactly 14 squares?

27. Jim and Todd each ran two laps around a track. When Jim completed the first lap, he noted that he was 50 meters behind Todd. Jim then began to run a distance of eight meters for every seven meters that Todd ran. The two boys were tied after completing the second lap. How long was the track?

28. In a shop class of ten students, exactly two students are given the responsibility of cleaning the shop each day after class. A schedule is made pairing each student with every other student in the class exactly once. How many pairs of students are listed on the schedule?

■ MID-CHAPTER REVIEW ■

Add or subtract. (Section 11.1)

1. $\dfrac{5}{t} - \dfrac{9}{t}$ $-\dfrac{4}{t}$

2. $\dfrac{3}{5t} + \dfrac{1}{5t}$ $\dfrac{4}{5t}$

3. $\dfrac{3m}{m+8} + \dfrac{5m}{m+8}$ $\dfrac{8m}{m+8}$

4. $\dfrac{6r}{r-3} - \dfrac{r}{r-3}$ $\dfrac{5r}{r-3}$

Add or subtract. Simplify where necessary. (Section 11.2)

5. $\dfrac{3a-1}{a+1} - \dfrac{2}{a+1}$ $\dfrac{3a-3}{a+1}$

6. $\dfrac{4r+8}{r+5} + \dfrac{r+7}{r+5}$ $\dfrac{5r+15}{r+5}$

7. $\dfrac{m^2-3}{m+4} - \dfrac{5-2m}{m+4}$ $m-2$

Find each LCM. (Section 11.3)

8. $10a^2b$; $12abc$ $60a^2bc$

9. $2t^2 - 18$; $3t^2 - 9t$ $6t(t+3)(t-3)$

10. $x^2 + x - 2$; $x^2 - x - 6$ $(x+2)(x-1)(x-3)$

Add or subtract. (Section 11.4)

11. $\dfrac{2}{5a} - \dfrac{3}{b}$ $\dfrac{2b-15a}{5ab}$

12. $\dfrac{4}{3t} + \dfrac{7}{12t}$ $\dfrac{23}{12t}$

13. $\dfrac{c}{6d} + \dfrac{d}{3c}$ $\dfrac{c^2+2d^2}{6cd}$

14. $\dfrac{2m}{6np} - \dfrac{5n}{8mp}$ $\dfrac{8m^2-15n^2}{24mnp}$

REVIEW CAPSULE FOR SECTION 11.5

Simplify. (Pages 271–273)

1. $\dfrac{x^2+3x-10}{(x+5)(x-1)}$ $\dfrac{x-2}{x-1}$

2. $\dfrac{5x^2-15x}{(x-3)(x-3)}$ $\dfrac{5x}{x-3}$

3. $\dfrac{2x^2-2x-4}{(x-4)(x-2)}$ $\dfrac{2(x+1)}{x-4}$

4. $\dfrac{4x^3-2x^2}{3x(2x-1)}$ $\dfrac{2x}{3}$

306 / *Chapter 11*

OBJECTIVE: To add or subtract with rational expressions having monomial or binomial numerators and unlike monomial or binomial denominators

11.5 Sums/Differences: Unlike Denominators

Teaching Suggestions p. M–38

Lesson Resources
Warm–Up: p. M–39
Maintenance: See below.
Practice Worksheet 11.5
Transparencies 20, 23, 29

EXAMPLE 1   Add: $\dfrac{3}{x} + \dfrac{2}{x-2}$

Solution:

1. Find the *LCM* of x and $x - 2$. ⟶ $x(x-2)$

2. Multiplication Property of One ⟶ $\dfrac{3}{x} = \dfrac{3}{x}\left(\dfrac{x-2}{x-2}\right)$ and $\dfrac{2}{x-2} = \dfrac{2}{x-2}\left(\dfrac{x}{x}\right)$

3. Substitute. ⟶ $\dfrac{3}{x} + \dfrac{2}{x-2} = \dfrac{3}{x}\left(\dfrac{x-2}{x-2}\right) + \dfrac{2}{x-2}\left(\dfrac{x}{x}\right)$

4. Product Rule ⟶ $= \dfrac{3(x-2)}{x(x-2)} + \dfrac{2x}{x(x-2)}$

5. Sum Rule ⟶ $= \dfrac{3(x-2)+2x}{x(x-2)}$

6. Distributive Property ⟶ $= \dfrac{3x-6+2x}{x(x-2)}$ Explain that the denominators are usually left in factored form.

7. Combine like terms. ⟶ $= \dfrac{5x-6}{x(x-2)}$

Maintenance
1. Evaluate:
   $(-6)(8) \div (-2)(-3)$
   Ans: −72   (Section 1.6)
2. Multiply and simplify:
   $(\sqrt{2} \cdot \sqrt{5})^2$
   Ans: 10   (Section 7.2)
3. Factor: $x^2 - 8x - 48$
   Ans: (x − 12) (x + 4)
   (Section 9.3)
4. Multiply and simplify:
   $\left(\dfrac{a^2}{2}\right)\left(\dfrac{-b^3}{ac}\right)\left(\dfrac{-6c}{b^2}\right)$
   Ans: 3ab   (Section 10.3)
5. At a certain concert, the number of orchestra seat tickets sold was 10 less than twice the number of balcony tickets sold (Condition 1). More than 230 tickets were sold in all (Condition 2). Is it possible that only 80 balcony tickets were sold?
   Ans: No   (Section 5.7)

EXAMPLE 2   Subtract: $\dfrac{5}{x+3} - \dfrac{4}{x-1}$

Solution:

1. Find the *LCM* of $x+3$ and $x-1$. ⟶ $(x+3)(x-1)$

2. Multiplication Property of One ⟶ $\dfrac{5}{x+3} = \dfrac{5}{x+3}\left(\dfrac{x-1}{x-1}\right)$ and $\dfrac{4}{x-1} = \dfrac{4}{x-1}\left(\dfrac{x+3}{x+3}\right)$

3. Substitute. ⟶ $\dfrac{5}{x+3} - \dfrac{4}{x-1} = \dfrac{5}{x+3}\left(\dfrac{x-1}{x-1}\right) - \dfrac{4}{x-1}\left(\dfrac{x+3}{x+3}\right)$

4. Product Rule ⟶ $= \dfrac{5(x-1)}{(x+3)(x-1)} - \dfrac{4(x+3)}{(x+3)(x-1)}$

5. Difference Rule ⟶ $= \dfrac{5(x-1)-4(x+3)}{(x+3)(x-1)}$

6. Distributive Property ⟶ $= \dfrac{5x-5-4x-12}{(x+3)(x+1)}$

7. Combine like terms. ⟶ $= \dfrac{x-17}{(x+3)(x+1)}$

Additional Examples
Example 1
Add or subtract.
1. $\dfrac{5}{x+4} + \dfrac{8}{x}$
   Ans: $\dfrac{13x + 32}{x(x + 4)}$
2. $\dfrac{1}{x} - \dfrac{4}{x-1}$
   Ans: $\dfrac{-3x - 1}{x(x - 1)}$

Example 2
Add or subtract.
1. $\dfrac{2}{x+4} + \dfrac{5}{x+3}$
   Ans: $\dfrac{7x + 26}{(x + 4)(x + 3)}$
2. $\dfrac{9}{x+6} - \dfrac{4}{x-2}$
   Ans: $\dfrac{5x - 42}{(x + 6)(x - 2)}$

Sums and Differences of Rational Expressions / **307**

P–1　**What names for 1 are needed as suggested below?**

$$\dfrac{x+4}{x+4}\quad \dfrac{x+1}{x-1}\left(\dfrac{?}{?}\right) = \dfrac{?}{(x-1)(x+4)} \qquad \dfrac{x+3}{x+4}\left(\dfrac{?}{?}\right) = \dfrac{?}{(x-1)(x+4)}$$

$$\dfrac{x-1}{x-1}$$

EXAMPLE 3　　Add: $\dfrac{x+1}{x-1} + \dfrac{x+3}{x+4}$

Solution:

$\boxed{1}$ Multiplication Property of 1 $\longrightarrow$ $\dfrac{x+1}{x-1} + \dfrac{x+3}{x+4} = \dfrac{x+1}{x-1}\left(\dfrac{x+4}{x+4}\right) + \dfrac{x+3}{x+4}\left(\dfrac{x-1}{x-1}\right)$

$\boxed{2}$ Product Rule $\longrightarrow$ $= \dfrac{(x+1)(x+4)}{(x-1)(x+4)} + \dfrac{(x+3)(x-1)}{(x+4)(x-1)}$

$\boxed{3}$ Sum Rule $\longrightarrow$ $= \dfrac{(x+1)(x+4) + (x+3)(x-1)}{(x-1)(x+4)}$

$\boxed{4}$ Distributive Property $\longrightarrow$ $= \dfrac{x^2 + 5x + 4 + x^2 + 2x - 3}{(x-1)(x+4)}$

$\boxed{5}$ Add like terms. $\longrightarrow$ $= \dfrac{2x^2 + 7x + 1}{(x-1)(x+4)}$

Test to see whether the result in Step $\boxed{5}$ is in simplest form.

Try to factor $2x^2 + 7x + 1$.

$$2x^2 + rx + sx + 1$$
$$r + s = 7 \qquad r \cdot s = 2$$

There are no integers for r and s that meet both of these conditions. The polynomial $2x^2 + 7x + 1$ is prime.

$\dfrac{2x^2 + 7x + 1}{(x-1)(x+4)}$ **Simplest form**　There are no common *factors* of numerator and denominator.

CLASSROOM EXERCISES

Write the LCM. (Step 1 of Examples 1 and 2)

1. x and $(x+1)$　$x(x+1)$

2. b and $(b-1)$　$b(b-1)$

3. $(x+5)$ and $(x-3)$　$(x+5)(x-3)$

4. $(y-2)$ and $(y+2)$

5. r and $(r+4)$　$r(r+4)$

6. $(a+b)$ and $(a-b)$

7. $(2x-1)$ and $(x+3)$

8. $(3x+2)$ and $(2x+3)$

9. $x(x-1)$ and $(x-1)$　$x(x-1)$

Write the missing name for 1. Then write the numerator of the expression on the right. (Step 2 of Examples 1 and 2)

10. $\dfrac{5}{x-2}\left(\dfrac{?}{?}\right) = \dfrac{?}{x(x-2)}$ $\dfrac{x}{x}$; $5x$

11. $\dfrac{3}{a}\left(\dfrac{?}{?}\right) = \dfrac{?}{a(a+3)}$ $\dfrac{a+3}{a+3}$; $3a+9$

12. $\dfrac{x}{x+5}\left(\dfrac{?}{?}\right) = \dfrac{?}{x(x+5)}$ $\dfrac{x}{x}$; x^2

13. $\dfrac{2}{y+2}\left(\dfrac{?}{?}\right) = \dfrac{?}{(y+2)(y-2)}$

14. $\dfrac{a}{a+b}\left(\dfrac{?}{?}\right) = \dfrac{?}{(a+b)(a-b)}$ $\dfrac{a-b}{a-b}$; a^2-ab

15. $\dfrac{x-1}{x+3}\left(\dfrac{?}{?}\right) = \dfrac{?}{(x-2)(x+3)}$

16. $\dfrac{b+5}{b-7}\left(\dfrac{?}{?}\right) = \dfrac{?}{(b-7)(b+1)}$ $\dfrac{b+1}{b+1}$; b^2+6b+5

17. $\dfrac{x+5}{2(x+3)}\left(\dfrac{?}{?}\right) = \dfrac{?}{2(x+3)(x-2)}$

WRITTEN EXERCISES

Goal: To add or subtract rational expressions that have binomial denominators

Sample Problem: $\dfrac{7}{x-4} - \dfrac{3}{x+1}$ **Answer:** $\dfrac{4x+19}{(x-4)(x+1)}$

Add or subtract. (Examples 1–3)

1. $\dfrac{1}{x+1} + \dfrac{2}{x}$ $\dfrac{3x+2}{x(x+1)}$

2. $\dfrac{3}{x-1} + \dfrac{5}{x}$ $\dfrac{8x-5}{x(x-1)}$

3. $\dfrac{4}{a} - \dfrac{2}{a+2}$ $\dfrac{2a+8}{a(a+2)}$

4. $\dfrac{2}{b} - \dfrac{3}{b-1}$

5. $\dfrac{x}{x+5} + \dfrac{5}{x-3}$

6. $\dfrac{3}{y-2} + \dfrac{y}{y+2}$

7. $\dfrac{a}{a-5} - \dfrac{2}{a}$

8. $\dfrac{r}{r+4} - \dfrac{1}{r}$

9. $\dfrac{a}{a+b} + \dfrac{b}{a-b}$

10. $\dfrac{r}{r-s} - \dfrac{s}{r+s}$

11. $\dfrac{a+1}{a-3} + \dfrac{3}{a+2}$

12. $\dfrac{y-2}{y+1} + \dfrac{5}{y+3}$

13. $\dfrac{3}{2x-1} - \dfrac{5}{x+3}$

14. $\dfrac{2}{3x+2} - \dfrac{1}{2x+3}$

15. $\dfrac{x+3}{x-1} + \dfrac{x+4}{x-2}$

16. $\dfrac{y+5}{y+2} + \dfrac{y-3}{y-1}$

MORE CHALLENGING EXERCISES

17. $\dfrac{b+2}{b-1} - \dfrac{b+5}{b-2}$ $\dfrac{-4b+1}{(b-1)(b-2)}$

18. $\dfrac{t+2}{t-4} - \dfrac{t+1}{t+2}$ $\dfrac{7t+8}{(t-4)(t+2)}$

19. $\dfrac{x+1}{x(x-1)} + \dfrac{5}{x-1}$ $\dfrac{6x+1}{x(x-1)}$

REVIEW CAPSULE FOR SECTION 11.6

Find the LCM. (Pages 300–302)

1. $4x-2$ and $6x-3$
   $6(2x-1)$

2. $5x+15$ and x^2-9
   $5(x+3)(x-3)$

3. x^2-2x and $x^2+3x-10$
   $x(x-2)(x+5)$

Sums and Differences of Rational Expressions / **309**

309

Lesson Resources
Warm–Up: p. M–39
Maintenance: See below.
Practice Worksheet 11.6
Transparencies 20, 24, 26, 27, 29

Maintenance
1. Solve: $-8 > -\frac{4}{7}x$
   Ans: {all real numbers greater than 14}
   (Section 5.5)
2. Add: $-3\sqrt{6} + 2\sqrt{24} + \sqrt{2}$
   Ans: $\sqrt{6} + \sqrt{2}$
   (Section 7.5)
3. Factor: $9x^2 + 30x + 25$
   Ans: $(3x + 5)^2$
   (Section 9.6)
4. Multiply and simplify:
   $\frac{x^2 - 4y^2}{7x} \cdot \frac{21y}{4y - 2x}$
   Ans: $\frac{-3y(x + 2y)}{2x}$
   (Section 10.5)
5. The area of a square is 576 square units. Find the length of each side of the square to the nearest tenth of a unit.
   Ans: 24.0 units
   (Section 7.7)

Additional Examples
Example 1
Add.
1. $\frac{6}{5x - 10} + \frac{3}{8x - 16}$
   Ans: $\frac{63}{40(x - 2)}$
2. $\frac{7}{3x + 3} + \frac{5}{4x + 4}$
   Ans: $\frac{43}{12(x + 1)}$

Example 2
Add or subtract.
1. $\frac{6}{x^2 - 49} + \frac{2}{3x + 21}$
   Ans: $\frac{2x + 4}{3(x - 7)(x + 7)}$
2. $\frac{10}{6x - 30} - \frac{11}{x^2 - 3x - 10}$
   Ans: $\frac{10x - 46}{6(x - 5)(x + 2)}$

11.6 More Sums and Differences

OBJECTIVE: To add or subtract with rational expressions having monomial or binomial numerators and unlike binomial or trinomial denominators

P–1 **What are the factors of $2x + 4$? of $3x + 6$?** 3, $x + 2$
2, $x + 2$

EXAMPLE 1 Add: $\dfrac{3}{2x + 4} + \dfrac{2}{3x + 6}$

In step 3 below, ask "By what must $2(x + 2)$ be multiplied in order to get the LCM?" Ask a similar question about the denominator $3(x + 2)$.

Solution:

1. Factor the denominators. ⟶ $2x + 4 = 2(x + 2)$ and $3x + 6 = 3(x + 2)$

2. Use the factors to write the LCM. ⟶ $2 \cdot 3(x + 2)$ or $6(x + 2)$

3. Multiplication Property of 1 ⟶ $\dfrac{3}{2(x + 2)} + \dfrac{2}{3(x + 2)} = \dfrac{3}{2(x + 2)}\left(\dfrac{3}{3}\right) + \dfrac{2}{3(x + 2)}\left(\dfrac{2}{2}\right)$

4. Product Rule for Rational Expressions ⟶ $= \dfrac{9}{6(x + 2)} + \dfrac{4}{6(x + 2)}$

5. Sum Rule for Rational Expressions ⟶ $= \dfrac{13}{6(x + 2)}$ ◀ **Simplest form**

EXAMPLE 2 Subtract: $\dfrac{3}{5x - 10} - \dfrac{2}{x^2 - 4}$

Solution:

1. Factor the denominators. ⟶ $5x - 10 = 5(x - 2)$ and $x^2 - 4 = (x - 2)(x + 2)$

2. Use the factors to write the LCM. ⟶ $5 \cdot (x - 2)(x + 2)$

3. Multiplication Property of 1 ⟶ $\dfrac{3}{5x - 10} - \dfrac{2}{x^2 - 4} = \dfrac{3}{5(x - 2)}\left(\dfrac{x + 2}{x + 2}\right) - \dfrac{2}{(x - 2)(x + 2)}\left(\dfrac{5}{5}\right)$

4. Product Rule ⟶ $= \dfrac{3(x + 2)}{5(x - 2)(x + 2)} - \dfrac{2(5)}{5(x - 2)(x + 2)}$

5. Distributive Property and the Difference Rule ⟶ $= \dfrac{3x + 6 - 10}{5(x - 2)(x + 2)}$

6. Combine like terms. ⟶ $= \dfrac{3x - 4}{5(x - 2)(x + 2)}$

P-2 How do you know that $\dfrac{3x - 4}{5(x - 2)(x + 2)}$ in Example 2 is in simplest form?   There are no common factors.

P-3 What are the prime factors of $x^2 + x$? $x, x + 1$

P-4 What are the prime factors of $x^2 + 3x + 2$? $x + 1, x + 2$

P-5 What is the *LCM* of $x(x + 1)$ and $(x + 1)(x + 2)$? $x(x + 1)(x + 2)$

EXAMPLE 3   Add: $\dfrac{4}{x^2 + x} + \dfrac{1}{x^2 + 3x + 2}$ Ask students how the names for 1 in step ③ are obtained.

Solution:

① Factor the denominators. ⟶ $x^2 + x = x(x + 1);\ (x^2 + 3x + 2) = (x + 2)(x + 1)$

② Use the factors to write the *LCM*. ⟶ $x(x + 1)(x + 2)$

③ Multiplication Property of 1 ⟶ $\dfrac{4}{x^2 + x} + \dfrac{1}{x^2 + 3x + 2} = \dfrac{4}{x(x + 1)}\left(\dfrac{x + 2}{x + 2}\right) + \dfrac{1}{(x + 1)(x + 2)}\left(\dfrac{x}{x}\right)$

④ Product Rule ⟶ $= \dfrac{4(x + 2)}{x(x + 1)(x + 2)} + \dfrac{x}{(x + 1)(x + 2)x}$

⑤ Sum Rule ⟶ $= \dfrac{4(x + 2) + x}{x(x + 1)(x + 2)}$

⑥ Distributive Property ⟶ $= \dfrac{4x + 8 + x}{x(x + 1)(x + 2)}$

⑦ Combine like terms. ⟶ $= \dfrac{5x + 8}{x(x + 1)(x + 2)}$

The steps of Examples 1–3 are summarized.

> **Steps in adding or subtracting rational expressions:**
>
> 1. Factor the denominators.
> 2. Determine the *LCM*.
> 3. Change the rational expressions to equivalent expressions with the *LCM* as the denominator of each.
> 4. Add or subtract.
> 5. Simplify.

Sums and Differences of Rational Expressions / **311**

Additional Examples
Example 3
Add or subtract.

1. $\dfrac{5}{x^2 - 8x + 15} + \dfrac{4}{x^2 - 3x}$

   Ans: $\dfrac{9x - 20}{x(x - 5)(x - 3)}$

2. $\dfrac{3}{x^2 - 36} - \dfrac{8}{x^2 + 9x + 18}$

   Ans: $\dfrac{-5x + 57}{(x - 6)(x + 6)(x + 3)}$

Write the LCM. (Step 2 of Examples 1–3)

1. $x(x - 1)$; $(x - 1)(x + 1)$ $x(x-1)(x+1)$ **2.** $2(x + 2)$; $3(x + 2)$ $6(x + 2)$

3. $(x + 3)(x + 3)$; $(x - 2)(x + 3)$ $(x + 3)^2(x - 2)$ **4.** $a(a - 1)$; $a(a + 1)$ $a(a - 1)(a + 1)$

5. $(y + 2)(y - 3)$; $(y + 2)(y + 1)$ $(y + 2)(y - 3)(y + 1)$ **6.** $10(x - 5)$; $2(x - 2)$ $10(x - 5)(x - 2)$

7. $2(x + 1)(x - 2)$; $(x - 2)(x - 2)$ $2(x + 1)(x - 2)^2$ **8.** $3(a + 2)$; $(a - 2)(a - 2)$ $3(a + 2)(a - 2)^2$

Write the prime factors. (P-3, P-4)

9. $2x - 6$ $2, x - 3$ **10.** $6x + 6$ $6, x + 1$ **11.** $4x + 12$ $4, x + 3$ **12.** $8x - 8$ $8, x - 1$

13. $x^2 + x - 2$ $x + 2, x - 1$ **14.** $x^2 - x - 6$ $x - 3, x + 2$ **15.** $a^2 + 3a - 10$ $a + 5, a - 2$ **16.** $a^2 + 2a - 15$ $a + 5, a - 3$

Goal: To add or subtract rational expressions that have binomial or trinomial denominators

Sample Problem: $\dfrac{5}{4x - 12} + \dfrac{1}{x^2 - 6x + 9}$ **Answer:** $\dfrac{5x - 11}{4(x - 3)(x - 3)}$

Copy each exercise. Replace each question mark with the correct polynomial. (Examples 1 and 2)

1. $\dfrac{3}{2x + 2} + \dfrac{1}{x^2 - 1} = \dfrac{3}{2(?)} + \dfrac{1}{(x - 1)(?)}$ $x + 1; x + 1$

$= \dfrac{3}{2(?)}\left(\dfrac{?}{?}\right) + \dfrac{1}{(x - 1)(?)}\left(\dfrac{?}{?}\right)$ $x + 1; \dfrac{x - 1}{x - 1}; x + 1; \dfrac{2}{2}$

$= \dfrac{3(?)}{2(x + 1)(x - 1)} + \dfrac{1(?)}{2(x + 1)(x - 1)} = \dfrac{?}{2(x + 1)(x - 1)}$ $x - 1; 2; 3x - 1$

2. $\dfrac{2}{x^2 - 25} + \dfrac{4}{3x - 15} = \dfrac{2}{(x - 5)(?)} + \dfrac{4}{3(?)}$ $x + 5; x - 5$

$= \dfrac{2}{(x - 5)(?)}\left(\dfrac{?}{?}\right) + \dfrac{4}{3(?)}\left(\dfrac{?}{?}\right)$ $x + 5; \dfrac{3}{3}; x - 5; \dfrac{x + 5}{x + 5}$

$= \dfrac{2(?)\ 3}{3(x - 5)(x + 5)} + \dfrac{4(?)\ x + 5}{3(x - 5)(x + 5)} = \dfrac{?\ 4x + 26}{3(x - 5)(x + 5)}$

Assignment Guide
Basic pp. 312–313: 1–17 odd
Average pp. 312–313: 1–17 odd
Above Average pp. 312–313:
1–17 odd

Copy each exercise. Replace each question mark with the correct polynomial.
(Example 3)

3. $\dfrac{1}{x^2 + x - 6} - \dfrac{2}{x^2 - 2x} = \dfrac{1}{(x - 2)(?)} - \dfrac{2}{x(?)}$ $x + 3; x - 2$

$$= \dfrac{1}{(x - 2)(?)}\left(\dfrac{?}{?}\right) - \dfrac{2}{x(?)}\left(\dfrac{?}{?}\right)$$ $x + 3; \dfrac{x}{x}; x - 2; \dfrac{x + 3}{x + 3}$

$$= \dfrac{1(?) \quad x}{x(x + 3)(x - 2)} - \dfrac{2(?) \quad x + 3}{x(x + 3)(x - 2)} = \dfrac{? \quad -x - 6}{x(x + 3)(x - 2)}$$

4. $\dfrac{2}{x^2 + 4x} - \dfrac{3}{x^2 + 3x - 4} = \dfrac{2}{x(?)} - \dfrac{3}{(x - 1)(?)}$ $x + 4; x + 4$

$$= \dfrac{2}{x(?)}\left(\dfrac{?}{?}\right) - \dfrac{3}{(x - 1)(?)}\left(\dfrac{?}{?}\right)$$ $x + 4; \dfrac{x - 1}{x - 1}; x + 4; \dfrac{x}{x}$

$$= \dfrac{2(?) \quad x - 1}{x(x + 4)(x - 1)} - \dfrac{3(?) \quad x}{x(x + 4)(x - 1)} = \dfrac{? \quad -x - 2}{x(x + 4)(x - 1)}$$

MIXED PRACTICE   *Add or subtract.*

5. $\dfrac{1}{2x - 6} + \dfrac{1}{6x - 6}$

6. $\dfrac{1}{4x + 12} + \dfrac{1}{8x - 8}$

7. $\dfrac{3}{x^2 + x - 2} - \dfrac{2}{x^2 - x - 6}$

8. $\dfrac{2}{x^2 - 9} + \dfrac{1}{2x + 6}$

9. $\dfrac{7}{10x - 50} - \dfrac{3}{x^2 - 25}$

10. $\dfrac{4}{a^2 + 3a - 10} - \dfrac{5}{a^2 + 2a - 15}$

11. $\dfrac{x}{2x - 4} + \dfrac{x - 1}{x^2 - 4}$ $\dfrac{x^2 + 4x - 2}{2(x - 2)(x + 2)}$

12. $\dfrac{5a}{a^2 + 2a - 15} + \dfrac{a + 3}{a^2 + 6a + 5}$

13. $\dfrac{a + 7}{a^2 - a - 6} - \dfrac{a}{a^2 - 10a + 21}$

14. $\dfrac{2x + 1}{x^2 - 81} - \dfrac{x}{x^2 - 10x + 9}$

15. $\dfrac{x - 2}{x^2 - 1} + \dfrac{x}{x^2 + 2x - 3}$ $\dfrac{2(x^2 + x - 3)}{(x + 1)(x - 1)(x + 3)}$

16. $\dfrac{y}{y^2 + 2y - 8} + \dfrac{y + 3}{y^2 - 16}$

17. $\dfrac{2a - 1}{2a^2 - 5a - 3} - \dfrac{3a + 1}{2a^2 - a - 1}$

18. $\dfrac{2x + 1}{3x^2 + 8x - 3} - \dfrac{x - 5}{3x^2 + 2x - 1}$

REVIEW CAPSULE FOR SECTION 11.7

Multiply. (Pages 274–279)

1. $6r^2s \cdot \dfrac{3}{2rs}$ $9r$

2. $(t + 1)(t - 1) \cdot \dfrac{t}{t + 1}$ $t(t - 1)$

3. $3q(q + 2)\left(\dfrac{3}{q} - \dfrac{5}{3q}\right)$ $4(q + 2)$

Sums and Differences of Rational Expressions / **313**

Maintenance
1. Write $t^5 \cdot t \cdot t^2$ as one power.
   Ans: t^8 (Section 6.1)
2. Evaluate:
   $-x^2y + y^2x - x + y$ when
   $x = -3$ and $y = -2$
   Ans: 7 (Section 8.3)
3. Divide:
   $(6x^3 + 17x^2 + 10x - 5) \div (2x + 3)$
   Ans: $3x^2 + 4x - 1 - \dfrac{2}{2x + 3}$
   (Section 8.6)
4. Divide and simplify:
   $\dfrac{x^2 - y^2}{8a} \div \dfrac{3x - 3y}{2}$
   Ans: $\dfrac{x + y}{12a}$ (Section 10.6)
5. A rectangular patio has a
   length of $(3x + 2)$ units and a
   width of $(2x - 7)$ units. Find
   the area of the patio.
   **Ans: $6x^2 - 17x - 14$
   square units**
   (Section 8.5)

Additional Examples
Example 1
Solve.
1. $\dfrac{8}{x + 6} = 2$

   Ans: $x = -2$
2. $-7 = \dfrac{7}{x - 3}$
   Ans: $x = 2$

Example 2
Solve.
1. $\dfrac{6}{x + 2} = \dfrac{7}{x - 5}$
   Ans: $x = -44$
2. $\dfrac{9}{x} = \dfrac{2}{x - 7}$
   Ans: $x = 9$

OBJECTIVE: To solve and check an equation containing one or more
rational expressions

11.7 Equations with Rational Expressions

P–1 In $\dfrac{3}{x - 3}$, what real number must be excluded from the domain of $\overset{3}{x}$?

EXAMPLE 1   Solve $\dfrac{3}{x - 3} = 5$.

Solution:

$\boxed{1}$ Multiply each side by $x - 3$. $\longrightarrow$ $\dfrac{3}{x - 3}(x - 3) = 5(x - 3)$

$\boxed{2}$ Simplify the left side. $\longrightarrow$ $3 = 5(x - 3)$

$\boxed{3}$ Distributive Property $\longrightarrow$ $3 = 5x - 15$

$\boxed{4}$ Addition Property of Equality $\longrightarrow$ $18 = 5x$

$\boxed{5}$ Division Property of Equality $\longrightarrow$ $\dfrac{18}{5} = x$

To check the solution set of Example 1, replace x in $\dfrac{3}{x - 3} = 5$ with $\dfrac{18}{5}$.

EXAMPLE 2   Solve $\dfrac{2}{x - 1} = \dfrac{3}{x + 1}$.   In step $\boxed{1}$, each side is multiplied
by the LCM of the denominators.

Solution:

$\boxed{1}$ Multiply each side by $(x - 1)(x + 1)$. $\rightarrow$ $\dfrac{2}{x - 1}(x - 1)(x + 1) = \dfrac{3}{x + 1}(x - 1)(x + 1)$

$\boxed{2}$ Simplify each side. $\longrightarrow$ $\dfrac{2}{x - 1} \cdot \dfrac{(x - 1)(x + 1)}{1} = \dfrac{3}{x + 1} \cdot \dfrac{(x - 1)(x + 1)}{1}$

$2(x + 1) = 3(x - 1)$

$\boxed{3}$ Distributive Property $\longrightarrow$ $2x + 2 = 3x - 3$

$\boxed{4}$ Subtract $2x$ from each side. $\longrightarrow$ $2 = x - 3$

$\boxed{5}$ Add 3 to each side. $\longrightarrow$ $5 = x$

Check: $\dfrac{2}{x - 1} \overset{=}{\vert} \dfrac{3}{x + 1}$

$\dfrac{2}{5 - 1} \ \vert \ \dfrac{3}{5 + 1}$

$\dfrac{2}{4} \ \vert \ \dfrac{3}{6}$

$\dfrac{1}{2} \ \vert \ \dfrac{1}{2}$

P–2 **What is the *LCM* of x, 3x, and x + 2?** $3x(x + 2)$

EXAMPLE 3   Solve $\dfrac{1}{x} + \dfrac{1}{3x} = \dfrac{2}{x + 2}$.

Solution:

1. Multiply each side by $3x(x + 2)$. ⟶ $3x(x + 2)\left(\dfrac{1}{x} + \dfrac{1}{3x}\right) = 3x(x + 2) \cdot \dfrac{2}{x + 2}$

2. Distributive Property ⟶ $3x(x + 2) \cdot \dfrac{1}{x} + 3x(x + 2) \cdot \dfrac{1}{3x} = 3x(x + 2) \cdot \dfrac{2}{x + 2}$

3. Simplify each side. ⟶ $3x + 6 + x + 2 = 6x$

4. Combine like terms. ⟶ $4x + 8 = 6x$

5. Subtract 4x from each side. ⟶ $8 = 2x$

6. Divide each side by 2. ⟶ $4 = x$

Check: $\dfrac{1}{x} + \dfrac{1}{3x} = \dfrac{2}{x + 2}$

$\dfrac{1}{4} + \dfrac{1}{12} \,\bigg|\, \dfrac{2}{4 + 2}$

$\dfrac{3}{12} + \dfrac{1}{12} \,\bigg|\, \dfrac{2}{6}$

$\dfrac{4}{12} \,\bigg|\, \dfrac{1}{3}$

$\dfrac{1}{3}$

Additional Examples
Example 3
Solve.
1. $\dfrac{4}{x} + \dfrac{3}{2x} = \dfrac{4}{x + 6}$
   Ans: $x = -22$
2. $\dfrac{5}{x + 3} + \dfrac{1}{x} = \dfrac{12}{4x}$
   Ans: $x = 2$
3. $\dfrac{3}{2x} + \dfrac{1}{x + 1} = \dfrac{5}{6x}$
   Ans: $x = -\dfrac{2}{5}$

The steps of Examples 1–3 are summarized.

> ***Steps in solving equations with rational expressions:***
>
> 1. Exclude numbers that will make zero divisors.
> 2. Multiply each side by the *LCM* of the denominators.
> 3. Solve the resulting equation.

Sums and Differences of Rational Expressions / **315**

Common Error

When solving equations with rational expressions, students do not multiply binominal factors correctly.

Example

$$\frac{\frac{4}{x + 3}}{} = 2$$

$$\frac{4}{x + 3} \cdot x + 3 = 2 \cdot x + 3$$

$$4 = 2x + 3$$

Prescription

Encourage students to use parentheses whenever they multiply by a binomial factor. Review the Distributive Property shown in Examples 1–3.

Assignment Guide

Basic p. 316: Omit
Average p. 316: 1–15 odd, 17
Above Average p. 316:
1–15 odd, 17

Problem–Solving Skills

Drawing a diagram (Ex. 17)
Using guess and check (Ex. 17)
Using logical reasoning (Ex. 17)

Non–Routine Problems

For a description of non–routine problems, see page M–13.

Additional Answers
Written Exercises

4. $x = -3$
8. No solution
12. $x = -18$
16. $x = -8$
See page 319 for the answer to Exercise 17.

Quiz Sections 11.5–11.7

After completing Sections 11.5–11.7, you may want to administer a quiz covering the same sections. See the quiz provided on page 230 in the *Teacher's ResourceBank*.

CLASSROOM EXERCISES

Which real number(s) must be excluded from each domain? (P-1)

1. $\frac{2}{x} = \frac{5}{x - 3}$ 0, 3

2. $\frac{3}{x + 2} = \frac{1}{x - 5}$ −2, 5

3. $\frac{4}{3x} = \frac{2}{x(x + 1)}$ 0, −1

4. $\frac{x}{x + 7} = 12$ −7

5. $\frac{3x}{2x + 1} = 5$ $-\frac{1}{2}$

6. $\frac{5}{3x + 1} = \frac{4}{2x - 1}$ $-\frac{1}{3}, \frac{1}{2}$

Write the LCM of the denominators. (P-1)

7. $\frac{x}{x - 3} = \frac{3}{4}$ $4(x - 3)$

8. $\frac{6}{x} = \frac{8}{x - 5}$ $x(x - 5)$

9. $\frac{3}{2x} = \frac{6}{5x - 1}$ $2x(5x - 1)$

10. $\frac{5}{3} = \frac{2}{6(x - 1)} + 3$ $6(x - 1)$

11. $\frac{5}{2x} = \frac{1}{x + 3}$ $2x(x + 3)$

12. $\frac{10}{7x} = \frac{3}{7(2x + 1)}$ $7x(2x + 1)$

WRITTEN EXERCISES

Goal: To solve equations that have rational expressions

Sample Problem: $\frac{1}{2x} + \frac{1}{x - 1} = \frac{1}{x}$

Answers: $x = -1$

Solve and check each equation. (Examples 1–3)

1. $\frac{12}{x} = 3$ $x = 4$

2. $\frac{18}{x} = 6$ $x = 3$

3. $\frac{6}{x + 2} = 3$ $x = 0$

4. $\frac{4}{x + 5} = 2$

5. $\frac{3}{4} = \frac{12}{x}$ $x = 16$

6. $\frac{9}{x} = \frac{3}{8}$ $x = 24$

7. $\frac{2}{x - 1} = \frac{3}{x - 1}$ No solution

8. $\frac{6}{x + 3} = \frac{5}{x + 3}$

9. $\frac{5}{x} = \frac{4}{x - 3}$ $x = 15$

10. $\frac{2}{x - 7} = \frac{3}{x}$ $x = 21$

11. $\frac{7}{x + 1} = \frac{3}{x + 5}$ $x = -8$

12. $\frac{2}{x + 4} = \frac{3}{x - 3}$

13. $\frac{2}{3} + \frac{3}{x} = \frac{5}{2x}$ $x = -\frac{3}{4}$

14. $\frac{1}{3x} + \frac{2}{5} = \frac{5}{x}$ $x = \frac{35}{3}$

15. $\frac{1}{x} + \frac{2}{x - 2} = \frac{1}{2x}$ $x = \frac{2}{5}$

16. $\frac{1}{x - 4} + \frac{1}{3x} = \frac{1}{x}$

NON-ROUTINE PROBLEM

17. A large block of cheese is 16 centimeters long, 8 centimeters wide, and $7\frac{1}{2}$ centimeters high. What is the greatest number of blocks 5 centimeters long, 3 centimeters wide, and $2\frac{1}{2}$ centimeters high that can be cut from the large block?

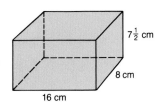

Using Statistics

Misleading Graphs

Statistical data is often presented in bar graphs. **Graphs** are an efficient, easily–read way to display information. However, graphs can be **misleading** in the way in which they display data. Special attention should be paid to the horizontal and vertical scales used in the graphs.

In the graph at the right, are the profits really going up as indicated by the rising line?

This special topic focuses on examples of graphs of data that lead to incorrect conclusions. Sometimes such graphs are intended to be misleading. Students are challenged to answer questions about several misleading graphs in the exercises.

EXERCISES

The two bar graphs below show the number of stereos sold by two electronics stores for each of the past five years.

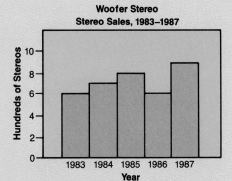

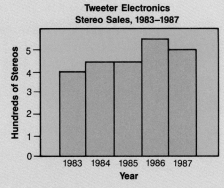

1. Which company appears to have sold the greater number of stereos each year?   Tweeter Electronics

2. What is the value of each vertical unit on the graph for Woofer Stereo? for Tweeter Electronics?

3. How many stereos did Woofer Stereo sell in 1983? in 1984? in 1987?   600; 700; 900

4. How many stereos did Tweeter Electronics sell in 1983? in 1984? in 1987?   400; 450; 500

5. Did Tweeter Electronics sell more stereos than Woofer Stereo in <u>any</u> year from 1983 to 1987?   No

6. The graphs are the same size and are placed side by side. Do you think this encourages the consumer to compare the height of the bars without examining the scales? Explain.

7. Suppose the graphs were <u>not</u> side by side. Do you think Tweeter Electronics would still appear to have sold more stereos? Explain.

Additional Answers
2. 200; 100
6. Yes; Placed side by side, the bars give the impression that Tweeter Electronics sold more stereos.
7. No; The bars would not be as easily compared and the reader would be more likely to use the vertical scales.

The owner of Hacker's Computer Outlet drew the two line graphs below.

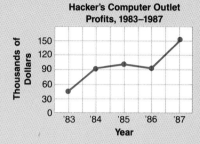

Hacker's Computer Outlet Profits, 1983–1987

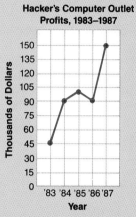

Hacker's Computer Outlet Profits, 1983–1987

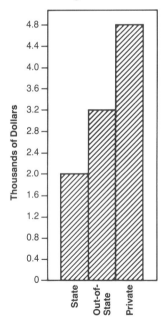

8. Do the two graphs show the same amount of profit for 1983? for 1986? for 1987? Yes; Yes; Yes

9. In which graph does the amount of profit seem to be increasing faster in a shorter period of time?

10. Compare the scales of the two graphs. How are they different?

11. The owner of Hacker's Computer Outlet wants to borrow some money to expand the business. Which graph do you think the owner should show to the loan officer at the bank? Explain.

The graph at the right is misleading.

12. Measure the length of each bar in the graph.

13. What is the yearly tuition cost at each college?

14. How many times as long as the bar for State College is the bar for Out-of-State College? 4 times

15. Is the tuition at Out-of-State College actually 4 times the tuition at State College? No

16. How many times as long as the bar for State College is the bar for Private College? 8 times

17. Is the tuition at Private College actually 8 times the tuition at State College? No

18. Redraw the graph using a scale that begins with 0.

19. On your graph, is the bar for Out-of-State College 4 times as long as the bar for State College? No

20. On your graph, is the bar for Private College 8 times as long as the bar for State College? No

21. Is your graph misleading? Explain.

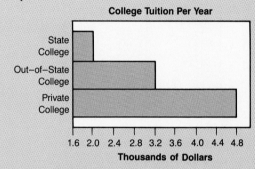

College Tuition Per Year

CHAPTER SUMMARY

IMPORTANT TERMS	Least common multiple of counting numbers *(p. 300)* Least common multiple of polynomials *(p. 301)*

IMPORTANT IDEAS

1. Sum and Difference Rules for Rational Expressions:

If $\frac{a}{c}$ and $\frac{b}{c}$ are rational expressions, then

$\frac{a}{c} + \frac{b}{c} = \frac{a+b}{c}$ and $\frac{a}{c} - \frac{b}{c} = \frac{a-b}{c}$. $(c \neq 0)$

2. The least common multiple of prime polynomials is their product.

3. Steps in finding the *LCM* of two or more polynomials:
a. Write the prime factorization of each polynomial.
b. Write a product using each prime factorization only once.
c. For each factor, write the greatest exponent used in any prime factorization.
d. Multiply.

4. Steps in adding or subtracting rational expressions:
a. Factor the denominators.
b. Determine the *LCM*.
c. Change the rational expressions to equivalent expressions with the *LCM* as the denominator of each.
d. Add or subtract.
e. Simplify.

5. Steps in solving an equation with rational expressions:
a. Exclude numbers from the domain that will make zero divisors.
b. Multiply each side by the *LCM* of the denominators.
c. Solve the resulting equation.

CHAPTER REVIEW

SECTION 11.1

Add or subtract.

1. $\frac{13}{3x} - \frac{5}{3x}$ $\frac{8}{3x}$ **2.** $\frac{3}{5a} + \frac{1}{5a}$ $\frac{4}{5a}$ **3.** $\frac{2a}{a+5} - \frac{a}{a+5}$ $\frac{a}{a+5}$ **4.** $\frac{b}{b+2} + \frac{-3b}{b+2}$ $\frac{-2b}{b+2}$

SECTION 11.2

Add or subtract. Simplify where necessary.

5. $\frac{x+1}{2x+3} + \frac{x+2}{2x+3}$ 1   **6.** $\frac{3x+1}{x-1} - \frac{x+3}{x-1}$ 2   **7.** $\frac{3a+1}{5a-1} - \frac{a-3}{5a-1}$ $\frac{2(a+2)}{5a-1}$

Sums and Differences of Rational Expressions / **319**

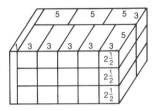

8. $\dfrac{a-5}{3a+2}+\dfrac{5a+2}{3a+2}$ $\dfrac{3(2a-1)}{3a+2}$ 9. $\dfrac{5x-4}{2x-3}-\dfrac{x+2}{2x-3}$ 2   10. $\dfrac{8x+3}{5x-1}-\dfrac{-2x^2+5}{5x-1}$

SECTION 11.3

Find each LCM.

11. $2x;\ 3x\ ;\ x-2$ $6x(x-2)$ 12. $5a;\ 10a^2;\ a+5$ $10a^2(a+5)$ 13. $y-2;\ y-3;\ y^2$ $y^2(y-2)(y-3)$

14. $b+2;\ b+5;\ 3b$ 15. $y^2-4;\ y+2;\ 2$ $2(y+2)(y-2)$ 16. $a^2-9;\ a-3;\ 3a$ $3a(a+3)(a-3)$

SECTION 11.4

Add or subtract.

17. $\dfrac{1}{2a}+\dfrac{1}{3a}$ $\dfrac{5}{6a}$ 18. $\dfrac{5}{4x}+\dfrac{3}{10x}$ $\dfrac{31}{20x}$ 19. $\dfrac{3}{4x}-\dfrac{5}{6y}$ $\dfrac{9y-10x}{12xy}$ 20. $\dfrac{5}{8a}-\dfrac{7}{12b}$ $\dfrac{15b-14a}{24ab}$

SECTION 11.5

Add or subtract.

21. $\dfrac{1}{a+b}+\dfrac{a}{a-b}$ 22. $\dfrac{1}{2x+3}+\dfrac{3}{x-1}$ $\dfrac{7x+8}{(2x+3)(x-1)}$ 23. $\dfrac{5a-1}{a+7}-\dfrac{a}{a-3}$

24. $\dfrac{y+1}{y-1}-\dfrac{1}{y+3}$ 25. $\dfrac{x-2}{x+1}+\dfrac{x-3}{x+2}$ $\dfrac{2x^2-2x-7}{(x+1)(x+2)}$ 26. $\dfrac{x+4}{x+2}+\dfrac{x-5}{x-4}$

SECTION 11.6

Add or subtract.

27. $\dfrac{1}{3x+6}+\dfrac{x-1}{x+2}$ $\dfrac{3x-2}{3(x+2)}$ 28. $\dfrac{2}{x-3}+\dfrac{x+1}{x^2-3x}$ $\dfrac{3x+1}{x(x-3)}$ 29. $\dfrac{x+1}{4x^2-9}-\dfrac{5}{2x+3}$

30. $\dfrac{3}{x+y}-\dfrac{2x+y}{x^2-y^2}$ 31. $\dfrac{2}{x^2-2x+1}+\dfrac{3}{x^2+x}$ 32. $\dfrac{4}{x^2-x-2}+\dfrac{1}{x^2-2x}$ $\dfrac{5x+1}{x(x-2)(x+1)}$

SECTION 11.7

Solve and check each equation.

33. $\dfrac{2}{x-3}=3$ $x=\dfrac{11}{3}$ 34. $\dfrac{10}{2x-1}=2$ $x=3$ 35. $\dfrac{3}{x}=\dfrac{1}{x+5}$ $x=-\dfrac{15}{2}$

36. $\dfrac{4}{3x}=\dfrac{2}{x+1}$ $x=2$ 37. $\dfrac{5}{x+2}=\dfrac{6}{x+2}$ No solution   38. $\dfrac{1}{x-5}=\dfrac{2}{x-5}$ No solution

---

Additional Answers

14. $3b(b+2)(b+5)$

21. $\dfrac{a^2+ab+a-b}{(a+b)(a-b)}$

23. $\dfrac{4a^2-23a+3}{(a+7)(a-3)}$

24. $\dfrac{y^2+3y+4}{(y-1)(y+3)}$

26. $\dfrac{2x^2-3x-26}{(x+2)(x-4)}$

29. $\dfrac{-9x+16}{(2x-3)(2x+3)}$

30. $\dfrac{x-4y}{(x+y)(x-y)}$

31. $\dfrac{5x^2-4x+3}{x(x+1)(x-1)^2}$

Review of Basic Algebra Skills
Before teaching Chapters 12–14, you may wish to use Survey Test 4 on pages 9–10 in the *Teacher's ResourceBank* to assess student competency with the basic algebra skills related to functions and linear systems covered in *Holt Introductory Algebra 1*.

Additional practice with these skills can be found on pages 17–18 in the *Teacher's ResourceBank*.

Relations and Functions

Locating points in space requires a method different from that used to locate points on earth. To locate points on earth, you use *latitude* and *longitude*. Early Polynesian sailors constructed a navigational chart out of shells, which represent islands, and sticks.

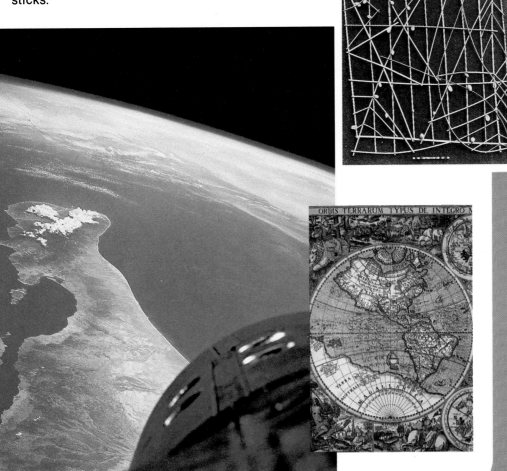

Background
The center photo above was taken by Astronaut L. Gordon Cooper from the Gemini 5 spacecraft on August 5, 1965. The photo shows Mexico and Baja California.
The top right photo is the Polynesian navigational chart of the Wojte Atoll in the Marshall Islands.
The bottom right photo shows part of Pieter van den Keere's 1607 map of the world.

Maintenance
1. Simplify:
$x^2 + 5 - 3x - 2x^2 + 4x$
Ans: $-x^2 + x + 5$
(Section 2.6)
2. Simplify: $\left(\dfrac{10^{-3}}{m^{-2}}\right)^{-4}$
Ans: $\dfrac{10^{12}}{m^8}$ (Section 6.5)
3. Factor: $5x^2 - 3x - 2$
Ans: $(5x + 2)(x - 1)$
(Section 9.4)
4. Multiply and simplify:
$\dfrac{4x}{x^2 - 4} \cdot \dfrac{6x + 12}{18x^2}$
Ans: $\dfrac{4}{3x(x - 2)}$
(Section 10.5)
5. Add: $\dfrac{7x}{x - 2} + \dfrac{x - 1}{x + 2}$
Ans: $\dfrac{8x^2 + 11x + 2}{(x + 2)(x - 2)}$
(Section 11.5)

Additional Examples
Example 1
Write the two moves needed to
locate the point corresponding to
each ordered pair.
1. $A(2, 5)$
Ans: right 2; up 5
2. $B(3, -4)$
Ans: right 3; down 4
3. $C(1, -5)$
Ans: right 1; down 5

OBJECTIVES: To graph a point having integral coordinates

12.1 Graphs of Ordered Pairs

To write the integral coordinates of a point from a graph of the point

Figure 1 shows two intersecting real-number lines and two points,
P and Q. The horizontal line is the **x axis,** and the vertical line is the
y axis. The point where the two lines meet is called the **origin.**

P–1 On the x axis, what direction is positive? ~~Right~~ negative? ~~Left~~

P–2 On the y axis, what direction is positive? ~~Up~~ negative? ~~Down~~

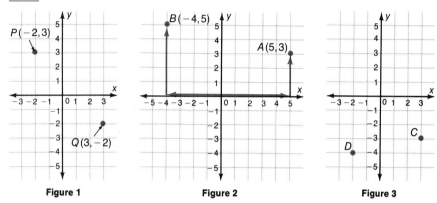

Figure 1   Figure 2   Figure 3

The pair of real numbers $(-2, 3)$ are the **coordinates** of point P in
Figure 1. Their order is important. For that reason they are called
an **ordered pair.**

P–3 What is the first coordinate of point Q in Figure 1? 3   the second
coordinate of point Q? -2

P–4 How does Figure 1 show that $(-2, 3)$ and $(3, -2)$ are not the same?
They do not have the same location.

EXAMPLE 1   In Figure 2, locate $A(5, 3)$.

> $A(5, 3)$ is the Point A
> which corresponds to
> the ordered pair (5, 3).

Solution:
1. Start at 0, and move five units to the <u>right</u> along the x axis.
2. Next, move <u>up</u> three units in a direction parallel to the y axis.

P–5 Why is the move of five units made along the x axis?
The first number is x.

P–6 Why do you move five units to the **right**? The first coordinate, 5, is positive.

P–7 Why is the second move of three units **up** and not **down**?
The second coordinate, 3, is positive.

EXAMPLE 2   Locate B(−4, 5) in Figure 2.

Solution: [1] Start at 0, and move four units to the **left**
along the *x* axis.

Left because the first coordinate is negative.

[2] Next, move five units **up** in a direction
parallel to the *y* axis.

EXAMPLE 3   Write the ordered pair that corresponds to point *C* in
Figure 3. The ordered pair is (3, −3).

Solution:
[1] Start at the origin.

[2] Move to the right, directly over point *C*.

Right means a positive coordinate.

[3] Write the first coordinate, 3.

[4] Move down, counting the units.

Down means a negative coordinate.

[5] Write the second coordinate, −3.   The ordered pair is (3, −3).

P–8 To name the ordered pair that corresponds to point *D*, what move do
you make first? What is the first coordinate?
Start at 0, and move left 2 units along the *x* axis; −2

P–9 What is the second move?   the second coordinate?
Down 4 units parallel to the *y* axis; −4

P–10 What is the ordered pair for point *D*? (−2, −4)

You may also want to ask what ordered pair corresponds to the origin.

━━━ **CLASSROOM EXERCISES** ━━━

For Exercises 1-10, the moves described locate the desired points. Always start at the origin.

*Write the two moves needed to locate the point corresponding to each
ordered pair.* (Examples 1 and 2) Left 3, Up 7   Right 5, Up 5   Right 7, Down 1

1. (3, −2)   **2.** (−10, −3)   **3.** (−3, 7)   **4.** (5, 5)   **5.** (7, −1)

6. (10, −10)   **7.** (8, 3)   **8.** (1, −1)   **9.** (−3, −3)   **10.** (−5, 5)

Relations and Functions   /   **323**

Additional Examples
Example 2
Write the two moves needed to
locate the point corresponding to
each ordered pair.
1. D(−7, 2)
   Ans: left 7; up 2
2. E(−5, −5)
   Ans: left 5; down 5

Example 3
Write the ordered pair that corresponds to each point.

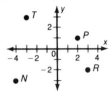

1. *P* **Ans: (2, 1)**
2. *R* **Ans: (3, −2)**
3. *N* **Ans: (−4, −3)**

Common Error
Students plot the *x* values vertically and the *y* values horizontally.

Example
Graph (2,1).

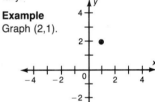

Prescription
This is a matter of remembering
that the *x* value or first coordinate must be counted along the
x axis. Have students take turns
naming the two moves needed in
Classroom Exercises 1–10.

Additional Answers
1. Right 3, Down 2
2. Left 10, Down 3
6. Right 10, Down 10
7. Right 8, Up 3
8. Right 1, Down 1
9. Left 3, Down 3
10. Left 5, Up 5

Write the ordered pair that corresponds to each point. (Example 3, P-8, P-9)

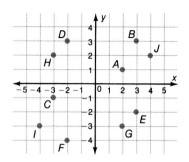

11. A (2, 1)   **12.** B (3, 3)

13. C (−3, −1)   **14.** D (−2, 3)

15. E (3, −2)   **16.** F (−2, −4)

17. G (2, −3)   **18.** H (−3, 2)

19. I (−4, −3)   **20.** J (4, 2)

Assignment Guide
Basic p. 324: 1–24
Average p. 324: 1–23 odd;
p. 327: 1–21 odd
Above Average p. 324:
1–23 odd; p. 327: 1–21 odd

Additional Answers
Written Exercises
4. Left 2, Up 4
8. Left 2, Down 5
12. Right 2, Down 2

WRITTEN EXERCISES

Goals: To locate the point that corresponds to a given ordered pair of integers, and to name the ordered pair that corresponds to a given point

Sample Problems: a. Locate G(4, −1)   **b.** Name the ordered pair for Z in Figure 4 below.

Answers: a. Move four units right, one unit down, and draw the point. Label it G(4, −1).   **b.** (−5, 4)

For Ex. 1-12, the moves described locate the desired points. Always start at the origin.

Draw an x axis and y axis on graph paper. Locate the point corresponding to each of the following. Name it with the capital letter. (Examples 1 and 2)

1. A(1, 4) Right 1, Up 4   **2.** B(5, 2) Right 5, Up 2   **3.** C(−5, 3) Left 5, Up 3   **4.** D(−2, 4)

5. E(6, −1) Right 6, Down 1 **6.** F(3, −5) Right 3, Down 5 **7.** G(−4, −1) Left 4, Down 1 **8.** H(−2, −5)

9. I(−4, −4) Left 4, Down 4 **10.** J(−5, 5) Left 5, Up 5   **11.** K(3, 3) Right 3, Up 3   **12.** L(2, −2)

Write the ordered pair that corresponds to each point. (Example 3)

(−5, −1)   (2, 1)   (−2, 3)   (5, 2)   (−4, −2)   (5, −1)
13. P   **14.** Q   **15.** R   **16.** S   **17.** T   **18.** U

19. V   **20.** W   **21.** M   **22.** N   **23.** K   **24.** L
(4, 3)   (−4, 1)   (−1, 1)   (3, −2)   (−2, −1)   (−5, 2)

Figure 4

REVIEW CAPSULE FOR SECTION 12.2

Name the two consecutive integers that each number lies between.

EXAMPLE: −1.6   **SOLUTION:** −2 < −1.6;   −1.6 < −1;   The integers are −2 and −1.

1. $-3\frac{1}{4}$ **2.** $5\frac{7}{8}$ **3.** −0.9   **4.** −3.08   **5.** −π   **6.** −√3

−3 and −4   5 and 6   −1 and 0   −3 and −4   −3 and −4   −2 and −1

324 / Chapter 12

12.2 Points and Ordered Pairs

OBJECTIVE: To graph a point having real-number coordinates

The plane of Figure 1 is called a **coordinate plane**. Any point of the coordinate plane may be represented as an ordered pair of real numbers (x, y).

The four regions in Figure 1 are called **quadrants**. Both x and y are positive real numbers for each point in quadrant I. In quadrant II, x is negative, and y is positive.

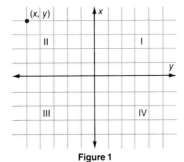

Figure 1

P–1 How can you describe the ordered pair of coordinates corresponding to each point in quadrant III? in quadrant IV? *x and y are negative.*
x positive, y negative

EXAMPLE 1   Graph $A(3\frac{1}{2}, 5)$.

Solution: ☐1 Start at the origin, and move three and one-half units to the <u>right</u> on the x axis.

☐2 Move <u>up</u> five units parallel to the y axis. ◄ **See Figure 2.**

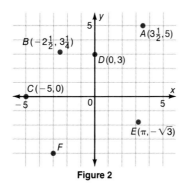

Figure 2

P–2 To locate $B(-2\frac{1}{2}, 3\frac{1}{4})$ in Figure 2 what move is made first?
Start at 0 and move $2\frac{1}{2}$ units left.

P–3 What move is made to represent the y value?
Move $3\frac{1}{4}$ units up, parallel to the y axis.

Relations and Functions / 325

Teaching Suggestions p. M–40

Lesson Resources
Warm–Up: p. M–41
Maintenance: See below.
Practice Worksheet 12.2
Transparency 30

Maintenance
1. Solve and check:
   $-y = 0$
   Ans: $y = 0$ (Section 3.2)
2. Write 0.00738 in scientific notation.
   Ans: 7.38×10^{-3}
   (Section 6.6)
3. Factor: $9y^2 - 16$
   Ans: $(3y + 4)(3y - 4)$
   (Section 9.5)
4. Divide and simplify:
   $\frac{3a + 9}{5b + 10} \div \frac{6}{25}$
   Ans: $\frac{5(a + 3)}{2(b + 2)}$
   (Section 10.6)
5. Subtract:
   $\frac{5}{x^2 - 2x} - \frac{6}{x^2 - 4}$
   Ans: $\frac{-x + 10}{x(x + 2)(x - 2)}$
   (Section 11.6)

Additional Examples
Example 1
Graph each point on the same set of axes.

a. $B(3, 2\frac{1}{2})$ **b.** $C(-1\frac{1}{2}, 2)$
c. $D(3\frac{1}{3}, -1)$ **d.** $E(-3, -1\frac{3}{4})$
Ans:

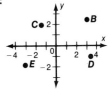

325

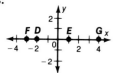
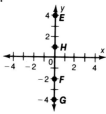
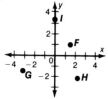
EXAMPLE 2   Graph $C(-5, 0)$.   **A coordinate of 0 indicates no motion.**

Solution: ☐1 Start at 0 and move five units to the left.

☐2 Do not move up or down.

Point C is on the x axis.   **C is between quadrants II and III. (Figure 2, p. 325)**

Give special attention to Examples 2 and 3. Students tend to make mistakes when 0 is one of the coordinates.

EXAMPLE 3   Graph $D(0, 3)$.

Solution: ☐1 Start at 0 and do not move to the left or right.

☐2 Move up three units.

Point D is on the y axis.   **D is between quadrants I and II. (Figure 2, p. 325)**

EXAMPLE 4   Graph $E(\pi, -\sqrt{3})$.
Use approximate values of π and $-\sqrt{3}$.

$$\pi \approx 3.1 \qquad -\sqrt{3} \approx -1.7$$ **The symbol "≈" means "is approximately equal to."**

Solution: ☐1 Start at 0 and move right 3.1 units.

☐2 Move down 1.7 units.   **See Figure 2.**

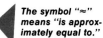

$(0, 0)$ $(-3, -4)$

P-4   **What are the coordinates of the origin?   of point F on page 325?**

Every ordered pair of real numbers has a corresponding point in the coordinate plane. Also, every point in the coordinate plane has an ordered pair of real numbers corresponding to it.

> There is a one-to-one correspondence between points of a coordinate plane and ordered pairs of real numbers.

326   /   Chapter 12

Write the quadrant in which each point lies. (Figure 1)

1. $A(-2, 7)$ II

2. $B(-5, -6)$ III

3. $C(1.7, 2.9)$ I

4. $D(3\frac{1}{2}, -5)$ IV

Locate each point. (Examples 2 and 3)

5. $E(0, -5)$ On the y axis, 5 units down from the origin

6. $F(10, 0)$ On the x axis, 10 units to the right of the origin

7. $G(-5\frac{1}{2}, 0)$ On the x axis, $5\frac{1}{2}$ units to the left of the origin

8. $H(0, 2.7)$

WRITTEN EXERCISES

Goal: To graph points with real-number coordinates

Sample Problem: Graph $P(2\frac{1}{4}, -1\frac{1}{2})$.   **Answer:**

For Ex. 1-20, each point may be located by the directions following each coordinate pair. Always start at the origin. Note that in Ex. 5, 6, 9, 10, and 15, there is only one move to locate each point.

Represent a coordinate plane on graph paper. Graph the point that corresponds to each ordered pair. Label each point. (Examples 1–4)

1. $N(4\frac{1}{2}, 3)$

2. $Z(2\frac{1}{2}, 5)$

3. $P(-2, -4\frac{1}{2})$

4. $W(4, -3\frac{1}{2})$

5. $K(0, 4)$ Up 4

6. $U(5, 0)$ Right 5

7. $S(-3.5, 2.2)$

8. $T(1.7, -2.5)$

9. $Q(-3\frac{1}{2}, 0)$ Left $3\frac{1}{2}$

10. $R(0, -5\frac{1}{2})$ Down $5\frac{1}{2}$

11. $L(-3, \sqrt{5})$

12. $M(4.5, \sqrt{3})$

Graph these points in the same coordinate plane. (Examples 1–4)

13. $A(-6, -2)$

14. $E(9, 3)$ Right 9, Up 3

15. $B(0, -2)$ Down 2

16. $F(3, 3)$

17. $C(-1, 2)$

18. $G(3, -3)$

19. $D(-7, 2)$ Left 7, Up 2

20. $H(9, -3)$

21. Connect points A and B, B and C, C and D, and D and A with line segments. What name is given to this geometric figure?
parallelogram

22. Connect points E and F, F and G, G and H, and H and E with line segments. What kind of figure do you get?   square

REVIEW CAPSULE FOR SECTION 12.3

Evaluate each expression. (Pages 21–23)

1. $-x + \frac{3}{4}$ when $x = -\frac{1}{2}$ $\frac{5}{4}$

2. $x^2 - x + 3$ when $x = \frac{1}{2}$ $\frac{11}{4}$

3. $-x^2 - 5$ when $x = -5$ -30

4. $|x - 1.2|$ when $x = -0.9$ 2.1

Assignment Guide
Basic p. 327: 1–22
Average p. 327: 1–21 odd; p. 324: 1–23 odd
Above Average p. 327: 1–21 odd; p. 324: 1–23 odd

Additional Answers
Written Exercises
  1. Right $4\frac{1}{2}$, Up 3
  2. Right $2\frac{1}{2}$, Up 5
  3. Left 2, Down $4\frac{1}{2}$
  4. Right 4, Down $3\frac{1}{2}$
  7. Left 3.5, Up 2.2
  8. Right 1.7, Down 2.5
  11. Left 3, Up 2.2
  12. Right 4.5, Up 1.7
  13. Left 6, Down 2
  16. Right 3, Up 3
  17. Left 1, Up 2
  18. Right 3, Down 3
  20. Right 9, Down 3

Problem-Solving Skills
Making a graph (Ex. 21–22)

328

OBJECTIVES: To write a relation as a set of ordered pairs when given an
equation and a set of x values
To graph a relation when its set of ordered pairs is given

12.3 Relations

To graph a relation when its equation and a set of x values is given

 **What are the coordinates of the six points
graphed at the right?** $(-2, 1)$, $(0, 1)$, $(2, 1)$
$(2, -1)$, $(0, -1)$, $(-2, -1)$

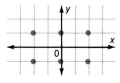

Coordinates of these points are elements of
set A below.

$A = \{(2, 1), (2, -1), (0, 1), (0, -1), (-2, 1), (-2, -1)\}$

Elements of
set A are
ordered pairs.

Set A is a <u>relation</u>.

Definition | A ***relation*** is a set of ordered pairs.

The equation $y = 2x - 1$ describes a relation.

Here are some ways to describe a relation.

Table

Kind of Description	Description
1. Equation	1. $y = 2x - 1$ when the x values are $-2, -1, 0, 1, 2, 3$, and 4.
2. Table of ordered pairs	2.
3. Graph	3.

x	-2	-1	0	1	2	3	4
y	-5	-3	-1	1	3	5	7

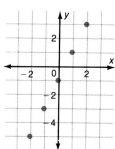

Some relations such as set A in P-1 cannot be described
by an equation.

 EXAMPLE 1   Write the relation T described by the equation
$y = x^2 + 1$ as a set of ordered pairs when the x values
are $-2, -1, 0, 1$, and 2.

Solution:

x	y
−2	?
−1	?
0	?
1	?
2	?

$y = x^2 + 1$
$y = (-2)^2 + 1 = 5$
$y = (-1)^2 + 1 = 2$
$y = (0)^2 + 1 = 1$
$y = (1)^2 + 1 = 2$
$y = (2)^2 + 1 = 5$

x	y
−2	5
−1	2
0	1
1	2
2	5

$T = \{(-2, 5), (-1, 2), (0, 1), (1, 2), (2, 5)\}$

EXAMPLE 2   Draw a graph of relation K described by the equation
$y = |x| + 1$ when the x values are −2, −1, 0, 1, and 2.

Solution:

x	y
−2	?
−1	?
0	?
1	?
2	?

$y = |x| + 1$
$y = |-2| + 1 = 3$
$y = |-1| + 1 = 2$
$y = |0| + 1 = 1$
$y = |1| + 1 = 2$
$y = |2| + 1 = 3$

x	y
−2	3
−1	2
0	1
1	2
2	3

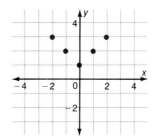

Additional Examples
Example 2
Draw a graph of the relation in a coordinate plane. The x values are −2, −1, 0, 1, and 2.
$y = x^2 - 2$
Ans:

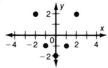

Common Error
Students forget to include the constant when computing the y values of a relation.

Example
Find the y values.
$y = x^2 - 1$

x	−2	−1	0	1
y	4	1	0	1

Prescription
Require students to write the relation each time they use an x value. As they substitute the given value for x, have them solve for y step by step.

▨▨▨ **CLASSROOM EXERCISES** ▨▨▨

Name three ordered pairs of each relation below. (P-1)   Answers will vary.

1. (−3, 3), (0, 0), (3, 3)
2. (−3, 1), (−2, 3), (−1, 3)
3. (−2, 3), (0, 1), (3, −2)
4. (1, 2), (1, 0), (1, −1)

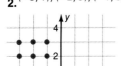
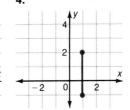

Relations and Functions / **329**

5.
x	y
-2	-2
-1	-1
0	0
1	1
2	2

6.
x	y
-2	2
-1	1
0	0
1	-1
2	-2

7.
x	y
-2	4
-1	1
0	0
1	1
2	4

8.
x	y
-2	-3
-1	-2
0	-1
1	0
2	1

9.
x	y
-2	1
-1	0
0	1
1	4
2	9

10.
x	y
-2	2
-1	1
0	0
1	1
2	2

Write a table of ordered pairs for each of the following relations if the x values are −2, −1, 0, 1, and 2. (Table, Examples 1 and 2)

5. $y = x$ **6.** $y = -x$ **7.** $y = x^2$

8. $y = x - 1$ **9.** $y = x^2 + 2x + 1$ **10.** $y = |x|$

Write two ordered pairs of each relation if the x values are 1 and −1. (Example 1)

11. $y = x$ $(1, 1), (-1, -1)$ **12.** $y = -x$ $(1, -1), (-1, 1)$ **13.** $y = x^2$ $(1, 1), (-1, 1)$

14. $y = x - 1$ $(1, 0)(-1, -2)$ **15.** $y = x^2 + 2x + 1$ $(1, 4), (-1, 0)$ **16.** $y = |x|$ $(1, 1), (-1, 1)$

WRITTEN EXERCISES

Goals: To write a relation as a set of ordered pairs and to graph a relation

Sample Problem: Draw a graph of the relation described by the equation $y = x$ if the x values are 0, 1, and 2.

Set: $\{(0, 0), (1, 1), (2, 2)\}$

Graph: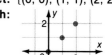

Write each relation as a set of ordered pairs if the x values are 1, 2, 3, and 4. (Example 1)

1. $y = x - 1$ $\{(1, 0), (2, 1), (3, 2), (4, 3)\}$ **2.** $y = x - 3$ $\{(1, -2), (2, -1), (3, 0), (4, 1)\}$

3. $y = x^2$ $\{(1, 1), (2, 4), (3, 9), (4, 16)\}$ **4.** $y = x^2 - 1$ $\{(1, 0), (2, 3), (3, 8), (4, 15)\}$

5. $y = |x|$ $\{(1, 1), (2, 2), (3, 3), (4, 4)\}$ **6.** $y = \sqrt{x^2}$ $\{(1, 1), (2, 2), (3, 3), (4, 4)\}$

7. $y = \sqrt{x}$ $\{(1, 1), (2, \sqrt{2}), (3, \sqrt{3}), (4, 2)\}$ **8.** $y = -\sqrt{x}$ $\{(1, -1), (2, -\sqrt{2}), (3, -\sqrt{3}), (4, -2)\}$

9. $y = 0 \cdot x + 1$ $\{(1, 1), (2, 1), (3, 1), (4, 1)\}$ **10.** $y = 0 \cdot x - 2$ $\{(1, -2), (2, -2), (3, -2), (4, -2)\}$

11. $y = x^2 - x + 1$ $\{(1, 1), (2, 3), (3, 7), (4, 13)\}$ **12.** $y = x^2 + x - 1$ $\{(1, 1), (2, 5), (3, 11), (4, 19)\}$

Draw a graph of each relation in a separate coordinate plane. (Example 2)

13. $\{(1, 2), (1, 3), (-3, 5), (-3, -4\frac{1}{2})\}$ **14.** $\{(-3, 4), (-3, -2), (4, 1), (4, 6\frac{1}{2})\}$

15. $\{(0, 0), (0, \pi), (0, -\pi), (\pi, 0)\}$ **16.** $\{(\pi, 3), (\pi, -3), (\pi, 5\frac{1}{2}), (\pi, 0)\}$

In Ex. 17-28, each group consists of 4 pairs. Ordered pairs are given.

Draw a graph of each relation in a separate coordinate plane. The x values are −1, 0, 1 and 2. (Example 2)

17. $y = x + 1$ $(-1, 0), (0, 1), (1, 2), (2, 3)$ **18.** $y = x - 3$ $(-1, -4), (0, -3), (1, -2), (2, -1)$ **19.** $y = x^2 - 7$

20. $y = x^2 - 10$ $(-1, -9), (0, -10), (1, -9), (2, -6)$ **21.** $y = |x|$ $(-1, 1), (0, 0), (1, 1), (2, 2)$ **22.** $y = \sqrt{x^2}$

23. $y = 0 \cdot x + 1$ $(-1, 1), (0, 1)(1, 1), (2, 1)$ **24.** $y = 0 \cdot x - 2$ **25.** $y = -\sqrt{x^2}$

26. $y = x^2 - x + 1$ $(-1, 3), (0, 1), (1, 1), (2, 3)$ **27.** $y = x^2 + x - 1$ **28.** $y = 2x + 1$

NON-ROUTINE PROBLEMS

29. The owner of the Centerville Bike Shop sold two different bicycles for $99 each. The owner made a 10% profit on the sale of one bike, but took a 10% loss on the sale of the other bike. On the two sales, did the owner make a profit, take a loss, or break even? Explain.

31. Arrange eight 8's in an addition problem so that the sum is 1000.

30. Copy the figure shown below. Then arrange the digits 2 through 9 in the empty boxes so that the sum of the number formed by the first row and the number formed by the second row equals the number formed by the third row.

Problem-Solving Skills
Using logical reasoning (Ex. 29–31)
Making a comparison (Ex. 29)
Solving a multi–step problem (Ex. 29)
Using guess and check (Ex. 30–31)

Additional Answers
Written Exercises
29. For 10% profit, cost = $90.
$99 − $90 = $9 (profit)
For 10% loss, cost = $110.
$110 − $99 = $11 (loss)
The shop owner had a net loss of $2.
30. Use guess and check. Other solutions are possible.

3	1	7
6	2	8
9	4	5

31. Use guess and check.
888 + 88 + 8 + 8 + 8

━━━━━ **MID-CHAPTER REVIEW** ━━━━━

Write the ordered pair that corresponds to each point of the graph. (Section 12.1)

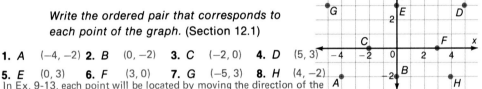

1. A $(-4, -2)$ **2.** B $(0, -2)$ **3.** C $(-2, 0)$ **4.** D $(5, 3)$

5. E $(0, 3)$ **6.** F $(3, 0)$ **7.** G $(-5, 3)$ **8.** H $(4, -2)$

In Ex. 9-13, each point will be located by moving the direction of the stated number of units.

Graph these points in the same coordinate plane. (Section 12.2)

9. $A(2, -1)$
Right 2, Down 1
10. $B(-4, -3)$
Left 4, Down 3
11. $C(-2.5, -3.5)$
12. $D(4, 3.5)$
Right 4, Up 3.5
13. $E(4, -3\frac{1}{4})$

Write each relation as a set of ordered pairs and then draw its graph. The x values are −2, −1, 0, 1, and 2. (Section 12.3) Each graph consists of the given points.

14. $y = -2x + 3$
Line: $\{(-2,7),(-1,5),(0,3),(1,1),(2,-1)\}$
15. $y = |-x| + 2$
16. $y = -x^2 - 1$

Quiz Sections 12.1–12.3
After completing this Mid-Chapter Review, you may want to administer a quiz covering the same sections. See the quiz provided on page 247 in the *Teacher's ResourceBank.*

Additional Answers
Mid-Chapter Review
11. Left 2.5, Down 3.5
13. Right 4, Down $3\frac{1}{4}$
15. Two lines (V-shaped):
$\{(-2,4), (-1,3), (0,2), (1,3), (2,4)\}$
16. Curve: $\{(-2,-5), (-1,-2), (0,-1), (1,-2), (2,-5)\}$

REVIEW CAPSULE FOR SECTION 12.4

Write the ordered pairs for the points A, B, C, and D. (Pages 322–324)

1.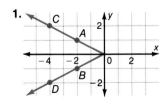

2.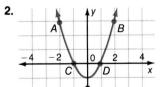

3.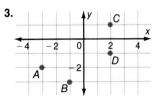

Additional Answers
Review Capsule
1. A $(-2, 1)$; B $(-2, -1)$; C $(-4, 2)$; D $(-4, -2)$
2. A$(-2, 3)$; B$(2, 3)$; C$(-1, 0)$; D$(1, 0)$
3. A$(-3, -2)$; B$(-1, -3)$; C$(2, 1)$; D$(2, -1)$

*Relations and Functions / **331***

Using Graphs

Air Conditioning

One skill of an **air conditioning technician** is the ability to determine whether an air conditioner is of the right <u>cooling capacity</u> for a particular room. The **cooling capacity** of an air conditioner is measured in **British thermal units (Btu's)**.

The graph shown at the right can be used to estimate how much cooling capacity is needed for a given room of a house or apartment. (The graph can be used for any room that is about eight feet high, provided that it is not a kitchen and not directly below an attic floor.) The exposure of the room's exterior wall determines which part of the shaded band you should use.

Monitoring a large air conditioning unit

EXAMPLE: A room of a house has a northern exposure. Its floor area is 300 square feet. What must be the size (cooling capacity) of the air conditioner for this room?

SOLUTION: 1. On the vertical scale, find the room's floor area, 300 square feet.

2. Since the room has a northern exposure, find the point on the <u>left</u> portion of the band directly to the right of the 300 reading.

3. From this point on the band read directly <u>down</u> on the horizontal scale to find the correct cooling capacity, **5500 Btu's per hour.**

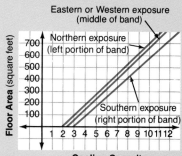

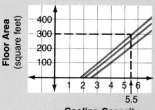

EXERCISES

Estimate the size of an air conditioner needed to cool each room described. Round your answer to the nearest 500 Btu's.

1. Southern exposure; 400 square feet
   8000 Btu's

2. Northern exposure; 500 square feet
   8000 Btu's

3. Eastern exposure; 300 square feet
   6000 Btu's

4. Southern exposure; 250 square feet
   6000 Btu's

5. Western exposure; 100 square feet
   3500 Btu's

6. Northern exposure; 700 square feet
   10,000 Btu's

OBJECTIVES: To write the domain and range of a relation when either its set of ordered pairs or a table of ordered pairs is given

12.4 Functions

To identify whether or not a relation is a function when either its set of ordered pairs or its graph is given

Teaching Suggestions p. M–40

Lesson Resources
Warm–Up: p. M–41
Maintenance: See below.
Practice Worksheet 12.4
Exploration 12

 P–1 **What are the x values of relation T below?   the y values?**

−1, 0, 4, −3   3, −2, 0

$$T = \{(-1, 3), (0, -2), (4, 0), (-3, -2)\}$$

Remember, no element of a set is written more than once. Thus, D and R do not have the same number of elements.

Set D below is the <u>domain</u> of relation T. Set R is the <u>range</u>.

$$D = \{-1, 0, 4, -3\} \qquad R = \{3, -2, 0\}$$

Definitions

The **domain** of a relation is the set of x values of its ordered pairs. The **range** of a relation is the set of y values of its ordered pairs.

EXAMPLE 1   Write the domain D and range R of relation W.

$$W = \{(-\tfrac{1}{2}, 0), (-2, 3), (-\tfrac{1}{4}, 2), (-\tfrac{1}{2}, -1), (2, -1)\}$$

Solution: $D = \{-\tfrac{1}{2}, -2, -\tfrac{1}{4}, 2\} \qquad R = \{0, 3, 2, -1\}$

Two x values are equal in A; no two x values are equal in B.

 P–2 **What is different about the domains of relations A and B?**

$$A = \{(-2, 5), (4, 8), (-2, 3)\}$$
$$B = \{(-2, 5), (4, 8), (9, 3)\}$$

 In A, two x values are equal.
In B, no two x values are equal.

Relation B is an example of a special relation called a <u>function</u>.

Definition

A **function** is a relation in which no two ordered pairs have the same x value. In a function, each different x value has only one y value.

 P–3 **How can you show that relation T following P-1 is a function?**

Show that no two x values of relation T are equal.

P–4 **Which of the following relations are also functions?**

P and S

$$P = \{(-1, 3), (2, 7), (0, 0)\}$$
$$Q = \{(1.3, 2), (-0.5, 1), (1.3, 0)\}$$
$$R = \{(\tfrac{1}{2}, \tfrac{1}{4}), (-\tfrac{1}{4}, 0), (\tfrac{2}{3}, 5), (0.5, 0.7)\}$$
$$S = \{(\pi, -2), (3, \sqrt{5}), (-1, -\sqrt{5})\}$$

Relations and Functions   /   333

Maintenance
1. Solve:
   $2(x - 4) + 3 > 2x - x$
   Ans: {all real numbers greater than 5}
   (Section 5.4)
2. Add: $4\sqrt{3} + \sqrt{12}$
   Ans: $6\sqrt{3}$ (Section 7.5)
3. Subtract:
   $(c^2 - 2c - 7) -$
   $(c^2 - 6c - 7)$
   Ans: 4c
   (Section 8.4)
4. Add and simplify:
   $\dfrac{4x + 1}{2x - 5} + \dfrac{-2x + 4}{2x - 5}$
   Ans: $\dfrac{2x + 5}{2x - 5}$
   (Section 11.2)
5. The width and length of a carpet runner are in the ratio 2:13 (Condition 1). The perimeter of the runner is 63 feet (Condition 2). Find the width and the length of the carpet runner.
   Ans: width: 4.2 feet; length: 27.3 feet   (Section 4.2)

Additional Examples
Example 1
Write the domain, D, and range, R, of each relation.
1. $A = \{(-1, 2), (-2, 5), (0, 7), (-2, 9), (3, 2)\}$
   Ans: $D = \{-1, -2, 0, 3\}$
   $R = \{2, 5, 7, 9\}$
2. $B = \{(6, \sqrt{6}), (3, \sqrt{3}), (0, 0), (9, 3)\}$
   Ans: $D = \{6, 3, 0, 9\}$
   $R = \{\sqrt{6}, \sqrt{3}, 0, 3\}$

The graph of a certain function is shown in Figure 1. The "hollow dot" means that $(-1, -2)$ is not in the function. The "heavy dot" emphasizes that $(3, 1)$ is in the function.

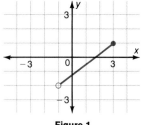
Figure 1

The graph of this function consists of all points on the portion of the line as shown.

There is a simple test to tell whether or not a graph represents a function.

> If any vertical line has more than one point in common with a graph, then the graph does not represent a function. Otherwise, it is the graph of a function.

P-5 **Does the graph in Figure 1 represent a function?** Yes **Why?**
A vertical line will have only one point in common with the graph.

EXAMPLE 2   Write <u>Yes</u> or <u>No</u> to show whether the graph in Figure 2 represents a function.

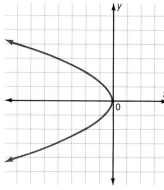
Figure 2

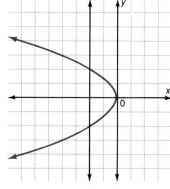
Figure 3

Solution: A vertical line is shown in Figure 3. It has two points in common with the graph. <u>No</u>, the graph of Figure 2 does not represent a function.

334

Additional Examples
Example 2
Write <u>Yes</u> or <u>No</u> to show whether the graph represents a function.

1.

Ans: Yes

2.

Ans: No

3.

Ans: No

CLASSROOM EXERCISES

See the Answers to Selected Exercises for the answers to Ex. 1-7 odd.
Write the domain and range of each relation. (Example 1)

1. $\{(3, 7), (-5, 2), (4, -1), (0, 0)\}$

2. $\{(\frac{1}{2}, 3), (\frac{1}{4}, 5), (\frac{1}{5}, 7), (0, 0)\}$

3. $\{(3, 8), (\sqrt{2}, -1), (\pi, -2), (\frac{1}{3}, -1)\}$

4. $\{(10, 1), (11, 2), (12, 3), (10, \frac{1}{2})\}$

5.

x	2	3	4	5
y	3	4	5	6

6.

x	$\frac{1}{2}$	$\frac{1}{3}$	$\frac{1}{4}$
y	$-\frac{1}{2}$	$-\frac{1}{3}$	$-\frac{1}{4}$

7.

x	1	-1	0	$\frac{1}{2}$	$-\frac{1}{2}$
y	1	1	0	$\frac{1}{2}$	$\frac{1}{2}$

8.

x	1	1	4	4	9
y	1	-1	2	-2	3

Write Yes or No to indicate if each set is a function. (P-2, P-3, P-4)

9. $\{(-1, 2), (2, -1)\}$ Yes

10. $\{(3, -1), (3, -2), (3, -3)\}$ No

11. $\{(\frac{1}{2}, 1), (\frac{1}{3}, 1), (\frac{1}{4}, 1)\}$ Yes

12. $\{(1.6, 2), (0.5, -3), (\frac{1}{2}, 0.7)\}$ No

13. $\{(1, 2), (3, 4), (5, 6)\}$ Yes

14. $\{(1, 1), (1, 2), (1, 3), (1, 4)\}$ No

15.

x	0	-3	2	-1
y	2	5	7	2

Yes

16.

x	$\frac{1}{2}$	$-\frac{1}{2}$	$\frac{3}{4}$	$-\frac{3}{4}$
y	5	6	7	8

Yes

WRITTEN EXERCISES

Goals: To determine whether a relation is a function by examining its set of ordered pairs or by applying the vertical line test to its graph

Sample Problems: a. $\{(0, 1), (1, 1)\}$ **b.** See Example 2.

Answers: a. Yes.   **b.** No.

See the Answers to Selected Exercises for the answers to Ex. 1-9 odd.
Write the domain, D, and range, R, of each relation. (Example 1)

1. $\{(5, 10), (-3, 2), (-1, 0), (1, -3)\}$

2. $\{(6, 5), (2, 3), (0, 0), (-3, -1)\}$

3. $\{(\frac{1}{2}, 2), (-1, \frac{1}{2}), (-\frac{1}{2}, -1), (2, -1)\}$

4. $\{(\frac{2}{3}, -1), (0, \frac{1}{2}), (-\frac{1}{2}, 0), (\frac{1}{4}, \frac{1}{4})\}$

5. $\{(-1, \sqrt{2}), (-2, \sqrt{3}), (\pi, -1), (0, -\pi)\}$

6. $\{(\sqrt{3}, -1), (0, \sqrt{2}), (0, \pi), (-1, \sqrt{2})\}$

7.

x	0	1	1.5	-1	$-\pi$
y	0	1	1	-1	-1

8.

x	0	1	2	1.5	-1	-1.5
y	0	1	2	2	0	-1

9.

x	0	$\frac{1}{2}$	1	$1\frac{1}{2}$	$-\frac{1}{2}$	-1
y	0	0	1	1	-1	-1

10.

x	0	1	2	$\frac{1}{2}$	-1	$-\frac{1}{2}$
y	0	1	4	$\frac{1}{4}$	1	$\frac{1}{4}$

Relations and Functions / **335**

Additional Answers
Classroom Exercises
2. $D = \{\frac{1}{2}, \frac{1}{4}, \frac{1}{5}, 0\}$
    $R = \{3, 5, 7, 0\}$
4. $D = \{10, 11, 12\}$
    $R = \{1, 2, 3, \frac{1}{2}\}$
6. $D = \{\frac{1}{2}, \frac{1}{3}, \frac{1}{4}\}$
    $R = \{-\frac{1}{2}, -\frac{1}{3}, -\frac{1}{4}\}$
8. $D = \{1, 4, 9\}$
    $R = \{1, -1, 2, -2, 3\}$

Additional Answers
Written Exercises
2. $D = \{6, 2, 0, -3\}$
    $R = \{5, 3, 0, -1\}$
4. $D = \{\frac{2}{3}, 0, -\frac{1}{2}, \frac{1}{4}\}$
    $R = \{-1, \frac{1}{2}, 0, \frac{1}{4}\}$
6. $D = \{\sqrt{3}, 0, -1\}$
    $R = \{-1, \sqrt{2}, \pi\}$
8. $D = \{0, 1, 2, 1.5, -1, -1.5\}$
    $R = \{0, 1, 2, -1\}$
10. $D = \{0, 1, 2, \frac{1}{2}, -1, -\frac{1}{2}\}$
    $R = \{0, 1, 4, \frac{1}{4}\}$

335

Write Yes or No to show whether each relation is a function. (P-2, P-3, P-4)

11. $\{(5, 7), (9, 11)\}$ Yes

12. $\{(2, 4), (6, 8)\}$ Yes

13. $\{(-2, 3), (-3, 4), (-5, 3)\}$ Yes

14. $\{(-\frac{1}{2}, 1), (-\frac{3}{2}, 2), (\frac{5}{2}, 1)\}$ Yes

15. $\{(0, 0), (2, 10), (-1, 3), (0, 5)\}$ No

16. $\{(-2, -1), (0, -3), (-2, 0), (-3, 0)\}$ No

17. $\{(\pi, 3), (\frac{1}{2}, \frac{1}{3}), (0.5, 5)\}$ No

18. $\{(\sqrt{2}, 3), (-\pi, 2), (-\sqrt{2}, 1)\}$ Yes

Write Yes or No to show whether each graph represents a function. Explain your answer. (Example 2)

19. Yes

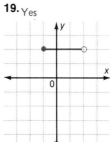

20. Yes

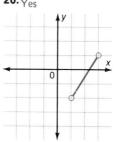

21. No

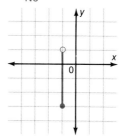

22. No

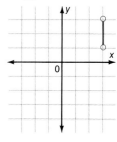

23. Yes

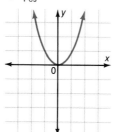

24. No

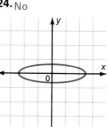

25. No

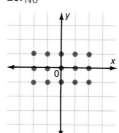

26. No

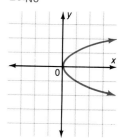

REVIEW CAPSULE FOR SECTION 12.5

Find which of these ordered pairs are in each relation. (Pages 328–330)

$(-3, 3)$ $(-2, 2)$ $(1, 4)$ $(-2, -2)$ $(1, 1)$ $(0, 0)$ $(0, 2)$ $(7, -1)$ $(-4, -2)$

1. $y = -\frac{1}{3}x + \frac{4}{3}$ $(-2,2), (1,1), (7,-1)$

2. $y = -x^2 + 2$ $(-2,-2), (1,1), (0,2)$

3. $y = |x|$

4. $y = -\sqrt{x^2}$ $(-2, -2), (0, 0)$

5. $y = -|x|$ $(-2, -2), (0, 0)$

6. $y = x^2 - x$ $(0, 0)$

12.5 Graphs of Functions

OBJECTIVE: To graph a function when (a) its domain is not more than five real numbers and (b) its domain is {real numbers}

EXAMPLE 1   Using {real numbers} as the domain, draw a graph of $y = 2x + 1$.

Solution: Choose a few x values that are small in absolute value. The corresponding y values are obtained from the formula to complete the table.   Discuss why x values small in absolute value should be chosen.

Values Chosen for x

x	y
−1	?
0	?
1	?
2	?

Values for y Obtained from the Formula
$$y = 2x + 1$$

For $x = -1$, $y = 2(-1) + 1 = -1$.
For $x = 0$, $y = 2(0) + 1 = 1$.
For $x = 1$, $y = 2(1) + 1 = 3$.
For $x = 2$, $y = 2(2) + 1 = 5$.

x	y
−1	−1
0	1
1	3
2	5

These ordered pairs are graphed in Figure 1. You can see that the points lie on a line.

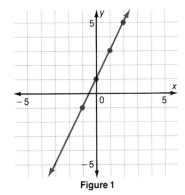

Figure 1

 P–1   If $x = -2$, what is the corresponding value for y from the formula $y = 2x + 1$?   −3

> The point (−2, −3) is on the graph of Figure 1.

Relations and Functions / 337

Teaching Suggestions p. M–40

Lesson Resources
Warm–Up: p. M–41
Maintenance: See below.
Practice Worksheet 12.5
Transparency 31

Maintenance
1. Simplify: t^{-1}
   Ans: $\frac{1}{t}$ (Section 6.3)
2. Factor: $x^2 - 3x - 28$
   Ans: $(x - 7)(x + 4)$
   (Section 9.3)
3. Multiply and simplify:
   $\frac{x + 3}{5} \cdot \frac{10}{x - 3}$
   Ans: $\frac{2(x + 3)}{x - 3}$
   (Section 10.4)
4. Solve and check:
   $\frac{y}{4y + 3} = \frac{2}{5}$
   Ans: $y = -2$
   (Section 11.7)
5. A boat traveled 9 miles due east into the ocean from the marina. The boat then traveled 12 miles due south to a reef. Find the direct distance from the marina to the reef.
   Ans: **15 miles**
   (Section 7.8)

Additional Example
Example 1
Using {real numbers} as the domain, draw a graph of $y = 3 - 2x$.
Ans:

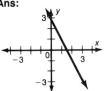

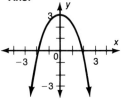
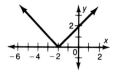
338

EXAMPLE 2   Using {real numbers} as the domain, draw a graph of
$y = x^2 - 1$.

Solution: Choose values for x and compute the corresponding values for y.

x	y
-2	?
-1	?
0	?
1	?
2	?

Emphasize that only a few values are selected from the domain in order to determine the location and shape of the graph.

For $x = -2$, $y = (-2)^2 - 1 = 3$.
For $x = -1$, $y = (-1)^2 - 1 = 0$.
For $x = 0$, $y = (0)^2 - 1 = -1$.
For $x = 1$, $y = (1)^2 - 1 = 0$.
For $x = 2$, $y = (2)^2 - 1 = 3$.

x	y
-2	3
-1	0
0	-1
1	0
2	3

These ordered pairs are graphed in Figure 2, and a curved dashed line is drawn through the points. The complete graph is shown in Figure 3.

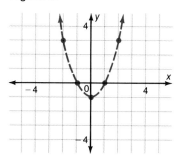

Figure 2

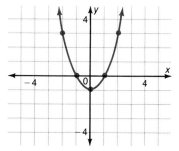

Figure 3

P-2 What is the range of the function shown in Figure 3?
{All real numbers greater than or equal to -1}

P-3 If $x = \frac{3}{2}$, what is the corresponding y value from the formula $y = x^2 - 1$? $\frac{5}{4}$

 The point $(\frac{3}{2}, \frac{5}{4})$ is on the graph of Figure 3.

CLASSROOM EXERCISES

Write the y values that correspond to the given x values. (Examples 1 and 2)

1. $y = 2x$

x	0	-1	1	-2	2
y	?	?	?	?	?
	0	-2	2	-4	4

2. $y = 3x + 1$

x	-2	-1	0	1	2
y	?	?	?	?	?
	-5	-2	1	4	7

Refer to Figure 4 for Exercises 3–7. (P-1, P-2, P-3)

3. Why does the graph represent a function?

4. What is the domain? {real numbers}

5. What is the range? {real numbers}

6. What y value corresponds to $x = -1$? $_{-1}$

7. What x value corresponds to $y = 1$? $_0$

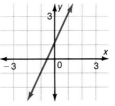

Figure 4

Refer to Figure 5 for Exercises 8–12. (P-1, P-2, P-3)

8. Why does the graph represent a function?

9. What is the domain?

10. What is the range?

11. What y value corresponds to $x = 1$? $1\frac{1}{2}$

12. What x value corresponds to $y = 1\frac{3}{4}$? $_0$

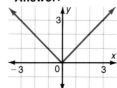

Figure 5

WRITTEN EXERCISES

Goal: To draw the graph of a function given the set description and domain

Sample Problem: Using {real numbers} as the domain, draw a graph of $y = |x|$.

Answer:

In Exercises 1–15, the graph is one of the following: a straight line, a "v", an inverted "v", or a curve (parabola). Points on the graph are given.

Draw a graph of each function in a separate coordinate plane. First graph five ordered pairs with x values −2, −1, 0, 1, and 2. The domain is {real numbers}. (Examples 1 and 2)

1. $y = -2x$ Line: (0, 0), (2, −4)

2. $y = -3x$ Line: (0, 0), (2, −6)

3. $y = -x + 1$

4. $y = -2x + 1$ Line: (0, 1), (2, −3)

5. $y = -|x|$

6. $y = -\sqrt{x^2}$

7. $y = |x + 1|$

8. $y = |x - 1|$

9. $y = 0 \cdot x - 3$

10. $y = 0 \cdot x + 4$ Line parallel to x axis: (0, 4), (2, 4)

11. $y = -x^2 + 2$ Curve: (−2, −2), (−1, 1), (0, 2)

12. $y = x^2 - 3$

MORE CHALLENGING EXERCISES

13. $y = \frac{1}{2}x^2 - x$

14. $y = x^2 - x + 1$

15. $y = -x^2 + x$

NON-ROUTINE PROBLEM

16. Copy the figure at the right. Then place 8 dots on the figure so that two dots are on each of the three lines and two dots are on each of the four circles.

Relations and Functions / **339**

Teaching Suggestions p. M–40

Lesson Resources
Warm–Up: p. M–41
Maintenance: See below.
Practice Worksheet 12.6

Maintenance
1. Simplify: $(-3rk)^4$
   Ans: $81r^4k^4$ (Section 6.4)
2. Factor: $2x^2 - 3x - 9$
   Ans: $(2x + 3)(x - 3)$
   (Section 9.4)
3. Multiply and simplify:
   $$\frac{x^2 - x - 12}{x^2 - 6x + 8} \cdot \frac{x^2 - 16}{x^2 + 7x + 12}$$
   Ans: $\dfrac{x - 4}{x - 2}$ (Section 10.5)
4. Solve and check:
   $$\frac{x}{10} - \frac{1}{5} = 3$$
   Ans: $x = 32$ (Section 11.7)
5. A gym floor has a length of
   $(3x + 4)$ units and a width of
   $(4x - 3)$ units. Find the area
   of the gym floor.
   Ans: $12x^2 + 7x - 12$
   (Section 8.5)

Additional Examples
Example 1
A projectile is fired into the sky
and returns to the ground. The
graph below describes the relationship between the distance
down range and the height of the
projectile during flight. Use the
graph to answer each of the following.

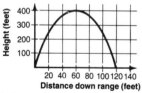

1. How high is the projectile
   when it has traveled 80 feet
   down range?
   Ans: About 350 feet
2. How far down range does the
   projectile hit the ground?
   Ans: 120 feet

PROBLEM SOLVING AND APPLICATIONS

12.6 Using Graphs: Prediction

Roger is learning to parachute. His
instructor told him that he will be
falling faster and faster each
second after leaving the airplane
until he opens his parachute. The
distance that he falls is a function
of time.

EXAMPLE 1

The graph below describes the relationship between
the distance in meters that an object falls and the time
in seconds.

Find the approximate distance, in meters, for each of
the following times.

a. 1 second

b. 3 seconds

Solutions: **a.** Locate the point on the graph that
corresponds to $t = 1$ on the t axis.
Read the approximate corresponding
distance in meters on the d axis.

Ask students
whether or
not this is a
linear relation.

When $t = 1$, d is approximately 5 meters.

b. Locate the point on the graph that
corresponds to $t = 3$ on the t axis.
Read the approximate corresponding
distance in meters on the d axis.

When $t = 3$, d is approximately 45 meters.

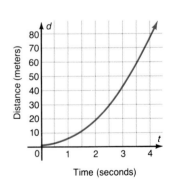

Emphasize
the information
in this
paragraph.

By examining the change in values at regular intervals on a graph,
predictions can be made for points not shown on the graph. This
procedure is illustrated in Example 2. When making predictions
based on a graph, you assume the graph extends without limit
according to the pattern shown.

EXAMPLE 2

The graph below describes the relationship between the distance, *d*, that sound travels in water and the traveling time, *t*.

a. Use the graph to find *d* for *t* = 2, 4, 6, and 8.

b. By how much does *d* increase when *t* increases by 2 seconds?

c. Use your answer to *b* to predict the approximate distance traveled by sound in 10 seconds.

Solutions:

a.

t	2	4	6	8
d	3	6	9	12

b. When *t* increases by 2 seconds, *d* increases by 3 kilometers.

c. Thus, sound will travel about 15 kilometers in 10 seconds.

Additional Examples
Example 2
The graph below describes the relationship between the amount of sales and the commission paid. Use the graph to predict the approximate commission paid for each amount of sales.

1. $2000   **Ans: $100**
2. $5000   **Ans: $250**
3. $8000   **Ans: $400**

CLASSROOM EXERCISES

Use the graph in Example 1 to find the approximate distance that an object falls for each of the following times.
(Example 1)

1. 2 seconds   20 m
2. $3\frac{1}{2}$ seconds   55 m
3. 4 seconds   75 m

4. How far would an object fall between the first and second seconds?   15 m

5. Does an object fall twice as far in the first 4 seconds as it does in the first 2 seconds?   No

6. How does the distance fallen between the second and third seconds compare with the distance fallen between the third and fourth seconds?   25 m; 30 m

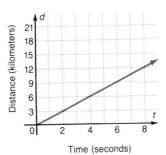

Galileo is reported to have dropped two iron balls having different weights from the Leaning Tower of Pisa. Both hit the ground at the same time, supporting Galileo's theory that gravity pulls all bodies to earth with the same acceleration, regardless of their weight.

*Relations and Functions / **341***

341

Use the graph in Example 2 to predict the distance that sound travels in water for each of the following times.
(Example 2)

7. 12 seconds
18 km

8. 16 seconds
24 km

9. 20 seconds
30 km

10. 14 seconds
21 km

WRITTEN EXERCISES

Goal: To use the graph of a function to solve word problems
Sample Problem: From the graph in Example 1, find the approximate distance that an object falls in $2\frac{1}{2}$ seconds.
Answer: 31 meters

The graph at the right shows the average stopping distance for a car traveling at various speeds. Write the stopping distance for each of the following speeds.
(Example 1)

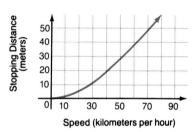

1. 20 km/hr 7 m

2. 40 km/hr 19 m

3. 30 km/hr 10 m

4. 50 km/hr 30 m

5. 60 km/hr 38 m

6. 70 km/hr 48 m

Assume that the graph below represents the cost of sending first-class mail. The domain of the function is the set of weights in ounces. The range is the corresponding set of costs in cents. Use the graph to predict the approximate cost of sending first-class mail having the following weights.
(Example 2)

7. $5\frac{1}{2}$ ounces $1.07

8. $6\frac{3}{4}$ ounces $1.24

9. 8 ounces $1.41

10. $9\frac{1}{4}$ ounces $1.75

11. $7\frac{1}{2}$ ounces $1.41

12. 10 ounces $1.75

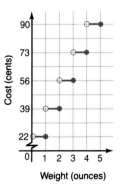

342 / Chapter 12

Assignment Guide
Basic p. 342: 1–12
Average p. 342: 1–12
Above Average p. 342: 1–12

Problem-Solving Skills
Reading a graph (Ex. 1–12)
Using patterns (Ex. 7–12)

Quiz Sections 12.4–12.6
After completing Sections 12.4–12.6, you may want to administer a quiz covering the same sections. See the quiz provided on page 248 in the *Teacher's ResourceBank*.

Additional Answers
Written Exercises, p. 339
16. Answers may vary. One answer is given.

Focus on Reasoning:
Too Little/Too Much Information

This special topic contains topics that are important in problem solving but are often neglected. In solving practical problems, a person must be able to sort out the pertinent information from all the information available about a problem. Notice that **guess and check** and **making a table** are the two strategies that are applied in the examples.

When solving a problem, it is important to identify what information is useful and whether enough information is given.

If enough information is given to solve the problem, find the solution. Then state whether you used <u>Clue 1 only</u>, <u>Clue 2 only</u>, or <u>both Clue 1 and Clue 2</u>. If not enough information is given, write "Can't tell."

EXAMPLE 1 **CLUE 1** $5 = 2 \cdot x\%$ **CLUE 2** $x > 100$ What is the value of x?

Solution: Since $5 = 2 \cdot x\%$, $x\% = \dfrac{5}{2}$, or 2.5.

Since $x\% = 2.5$, $\dfrac{x}{100} = 2.5$. $\quad$ $x\% = \frac{x}{100}$

$x = 2.5(100)$

$x = \mathbf{250}$ **Clue 1 only** was used.

EXAMPLE 2 **CLUE 1** Ariel has 13 nickels and dimes.   **CLUE 2** The coins have a total value of 95¢.

How many dimes does Ariel have?

Solution: Use guess and check.

Think: Since the total value is 95¢, Ariel has an odd number of nickels.

Guess 1: 5 nickels and 8 dimes   **Check 1:** $25¢ + 80¢ = \$1.05$ ⟵ *Too high*
Guess 2: 11 nickels and 2 dimes   **Check 2:** $55¢ + 20¢ = 75¢$ ⟵ *Too low*
Guess 3: 7 nickels and 6 dimes   **Check 3:** $35¢ + 60¢ = 95¢$ ✔

Ariel has **6 dimes.**   **Both Clue 1 and Clue 2** were used.

EXAMPLE 3 **CLUE 1** One number is 3 more than a second number.
CLUE 2 Both numbers are integers.   Numbers: __?__ and __?__

Solution: Use a table to make a list of possible pairs of numbers.

First integer, n	-7	-4	0	3	8	13	$\cdots$
Second integer, $n + 3$	-4	-1	3	6	11	16	$\cdots$

You <u>**can't tell**</u> what the two numbers are because not enough information is given.

EXERCISES

1. CLUE 1 Janet and her roommate divide their rent evenly.

    CLUE 2 Janet's rent is 20% more than the $200 she paid last month.

How much is Janet's rent this month? $240; Clue 2

2. CLUE 1 Video tape sales were 30% of Monday's total sales.

    CLUE 2 Video tapes were on sale for $4.99 each.

What was the total of Monday's video tape sales?   Can't tell

3. CLUE 1 The ratio of two numbers is 2:3.

    CLUE 2 The smaller number is $\frac{2}{3}$ of the larger number.

What is the smaller number?   Can't tell

4. CLUE 1 The square of an integer is between 40 and 50.

    CLUE 2 The integer is odd.

What is the integer?   7 and −7; Clue 1

5. CLUE 1 A trophy weighs 20 ounces.

    CLUE 2 The trophy is made of pewter. Pewter is 60% tin and 20% lead.

How much tin was used to make the trophy?   12 ounces; Both clues

6. CLUE 1 David's first four algebra test scores were 72, 72, 73, and 83.

    CLUE 2 His average after the fifth test is 77.

What was David's score on his fifth algebra test?

7. CLUE 1 In a 10–mile race, Beth ran the first 3 miles in 25 minutes.

    CLUE 2 She ran the final 7 miles in 1 hour 35 minutes.

What was Beth's average speed for the race, in miles per hour?

8. CLUE 1 The ages of two sisters are both prime numbers.

    CLUE 2 Both sisters are between 20 and 30 years old.

What are the ages of the two sisters?   23 and 29; Both clues

9. CLUE 1 $s + t = 12$ and $s + w = 5$

    CLUE 2 $r + s = 15$ and $t + w = 5$

What is the value of $r + s + t + w$?

10. CLUE 1 $10^{3k} \cdot 10^2 = 10^8$

    CLUE 2 k is an integer.

What is the value of k?

Additional Answers
6. 85; Both clues
7. 5 miles per hour; Both clues
9. 20; Clue 2
10. 2; Clue 1

CHAPTER SUMMARY

IMPORTANT TERMS		
	x axis *(p. 322)*	Quadrants *(p. 325)*
	y axis *(p. 322)*	Relation *(p. 328)*
	Origin *(p. 322)*	Domain *(p. 333)*
	Coordinates *(p. 322)*	Range *(p. 333)*
	Ordered pair *(p. 322)*	Function *(p. 333)*
	Coordinate plane *(p. 325)*	

IMPORTANT IDEAS

1. There is a one-to-one correspondence between points of a coordinate plane and ordered pairs of real numbers.

2. A relation can be described by an equation, by a table of ordered pairs, by a graph in a coordinate plane, or by a set of ordered pairs.

3. If any vertical line has more than one point in common with a graph, then the graph does not represent a function.

CHAPTER REVIEW

For Ex. 1-16, each point is located by moving in the stated direction, the stated number of units.

SECTION 12.1

Draw an x axis and y axis. Then locate these points.

1. $A(-3, 2)$ Left 3, Up 2 **2.** $B(4, -3)$ Right 4, Down 3 **3.** $C(-3, -4)$ Left 3, Down 4 **4.** $D(2, 5)$

5. $E(3, -3)$ Right 3, Down 3 **6.** $F(-4, -2)$ Left 4, Down 2 **7.** $G(3, 1)$ Right 3, Up 1 **8.** $H(0, 0)$ Origin

SECTION 12.2

Graph these points in the same coordinate plane.

9. $P(2, -3\frac{1}{2})$ **10.** $Q(-3, 2\frac{1}{2})$ **11.** $R(2.25, 1.75)$ **12.** $S(-4.5, -2.5)$

13. $T(0, 3)$ Up 3 **14.** $W(-3, 0)$ Left 3 **15.** $U(-2\frac{1}{2}, -3\frac{1}{2})$ **16.** $Z(-3\frac{1}{2}, -2\frac{1}{2})$

SECTION 12.3

Write each relation as a set of ordered pairs if the x values are −1, 0, 2, and 3. Then draw its graph.

17. $y = 3x - 5$ $\{(-1, -8), (0, -5), (2, 1), (3, 4)\}$ **18.** $y = 4 - x$ $\{(-1, 5), (0, 4), (2, 2), (3, 1)\}$

19. $y = x^2 - 1$ $\{(-1, 0), (0, -1), (2, 3), (3, 8)\}$ **20.** $y = 3 - x^2$ $\{(-1, 2), (0, 3), (2, -1), (3, -6)\}$

21. $y = |x - 1|$ $\{(-1, 2), (0, 1), (2, 1), (3, 2)\}$ **22.** $y = 2 - |x|$ $\{(-1, 1), (0, 2), (2, 0), (3, -1)\}$

Relations and Functions / **345**

Chapter Test
Two Chapter Tests (Form A and Form B) are provided on pages 249–252 in the *Teacher's ResourceBank.*

Additional Answers
4. Right 2, Up 5
9. Right 2, Down $3\frac{1}{2}$
10. Left 3, Up $2\frac{1}{2}$
11. Right 2.25, Up 1.75
12. Left 4.5, Down 2.5
15. Left $2\frac{1}{2}$, Down $3\frac{1}{2}$
16. Left $3\frac{1}{2}$, Down $2\frac{1}{2}$

SECTION 12.4

Write the domain D and the range R of each relation.

23. $\{(-\frac{1}{2}, 0), (2, 6), (4, -3)\}$

24. $\{(1.7, -2), (-2.3, -0.9), (0, -6)\}$

25. $\{(-3, \frac{1}{2}), (0, -5), (-3, 0), (-2, -4)\}$

26. $\{(3, 2), (-1.6, 3), (3.5, -2.7), (-1, 2)\}$

Write Yes or No to show whether each relation is a function.

27. $\{(12, -10), (-25, 5), (-4, -10), (17, 5)\}$ Yes

28. $\{(1, 6), (0, 5), (-4, -2.7), (1.8, -5.3)\}$ Yes

29. No

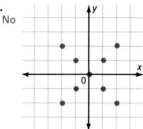

30. Yes

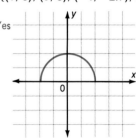

SECTION 12.5 For Ex. 31-32, each graph is a line through the points located by the stated ordered pairs.

Draw a graph of each function in a separate coordinate plane. The domain is {real numbers}.

31. $y = x - 3$ Line: $\{(0, -3), (1, -2), (2, -1), (3, 0)\}$

32. $y = -2x + 1$

33. $y = |x| - 2$ Two lines: $\{(-2, 0), (0, -2), (2, 0)\}$

34. $y = 3 - \sqrt{x^2}$

SECTION 12.6

The graph at the right describes the relationship between the distance, d, a car travels in one second and the speed, s, at which it is traveling. Use the graph to determine each of the following.

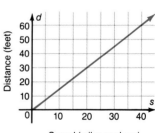

Speed (miles per hour)

35. The approximate distance traveled in one second by a car traveling 20 mi/hr 30 ft

36. The approximate distance traveled in one second by a car traveling 30 mi/hr 45 ft

37. The approximate distance traveled in one second by a car traveling 40 mi/hr 60 ft

38. Predict the approximate distance traveled in one second by a car traveling 50 mi/hr. 75 ft

39. Predict the approximate distance traveled in one second by a car traveling 55 mi/hr. 83 ft

346 / *Chapter 12*

346

Use Condition 1 to represent the unknowns. Use Condition 2 to write an equation for the problem. Solve. (Sections 4.1 and 4.2)

1. The parts used to repair a car at a local body shop cost $500 less than the labor. The total bill was $2100. Find the cost of the parts.

2. A sailboat race is along a triangular route. The ratio of the lengths of the three legs of the race is 7:8:10. The race has a length of 75 miles. Find the length of each leg.

Solve. Check your answer. (Section 4.3)

3. $5 - 3y = -16 + 2y$ $y = \frac{21}{5}$

4. $2x - 3(x - 9) = 2(x + 3)$ $x = 7$

Solve each inequality. (Sections 5.3 through 5.6)

5. $2d - 5 < d + 11$

6. $\frac{a}{-6} > -1.5$

7. $-3(y + 8) < 2y - 1$

Simplify. (Sections 6.4 and 6.5)

8. $(7ad)^2$ $49a^2d^2$

9. $(-5m)^{-3}$ $-\frac{1}{125m^3}$

10. $(3a^2c^3)^2$ $9a^4c^6$

11. $\left(\frac{-2x^3}{y^2}\right)^4$ $\frac{16x^{12}}{y^8}$

Simplify. (Sections 7.1 through 7.4)

12. $-\sqrt{\frac{25}{36}}$ $-\frac{5}{6}$

13. $\sqrt{2} \cdot \sqrt{32}$ 8

14. $\sqrt{150}$ $5\sqrt{6}$

15. $\sqrt{12a^6n^9}$ $2a^3n^4\sqrt{3n}$

Add or subtract. Simplify where necessary. (Section 7.5)

16. $7\sqrt{5} + 9\sqrt{5}$ $16\sqrt{5}$

17. $4\sqrt{6} - \sqrt{24}$ $2\sqrt{6}$

18. $\sqrt{12} + \sqrt{3} - \sqrt{75}$ $-2\sqrt{3}$

19. $\sqrt{72g} + \sqrt{32g}$ $10\sqrt{2g}$

Evaluate each expression. (Sections 8.1 and 8.3)

20. $-6x^3$ when $x = 2$ -48

21. $-x^2 + 7x + 9$ when $x = -5$ -51

Multiply. (Sections 8.2 and 8.5)

22. $(-4c^6)(-3c^2)$ $12c^8$

23. $(-6c^2d^3e)(\frac{1}{2}cd^4)$ $-3c^3d^7e$

24. $(6x + 5)(2x - 3)$ $12x^2 - 8x - 15$

Divide. (Section 8.6)

25. $x - 3\overline{)5x^2 - 17x + 6}$ $5x - 2$

26. $x + 2\overline{)4x^3 + 6x^2 + 9}$ $4x^2 - 2x + 4 + \frac{1}{x + 2}$

Factor. (Sections 9.1 through 9.7)

27. $3x^2 - 12x$ $3x(x - 4)$

28. $n^2 + 6n + 8$ $(n + 2)(n + 4)$

29. $g^2 + 4g - 45$

30. $5m^2 - 18m - 8$ $(5m + 2)(m - 4)$

31. $6x^2 - 17x + 5$ $(3x - 1)(2x - 5)$

32. $9y^2 - 1$ $(3y + 1)(3y - 1)$

33. $4h^2 - 20h + 25$ $(2h - 5)(2h - 5)$

34. $48y^2 - 27$ $3(4y + 3)(4y - 3)$

35. $2n^2 + 6n - 36$ $2(n - 3)(n + 6)$

Cumulative Review / **347**

Cumulative Test

A cumulative test covering Chapters 10–12 is provided on pages 253–256 in the *Teacher's ResourceBank*.

Additional Answers

1. Let x = cost of labor. Then $x - 500$ = cost of parts.
   $x + x - 500 = 2100$
   Parts: $800

2. Let lengths of legs be $7x$, $8x$, and $10x$.
   $7x + 8x + 10x = 75$
   Length of legs: 21 mi, 24 mi, and 30 mi

5. {all real numbers less than 16}

6. {all real numbers less than 9}

7. {all real numbers greater than $-4\frac{3}{5}$}

29. $(g - 5)(g + 9)$

Write the number or numbers that cannot be in the domain of each variable. (Section 10.1)

36. $\dfrac{6}{7x}$ 0

37. $\dfrac{x-6}{2x+1}$ $-\dfrac{1}{2}$

38. $\dfrac{9}{x^2-16}$ 4, −4

39. $\dfrac{8x+1}{(x-2)(x+5)}$ 2, −5

Simplify. No divisor equals 0. (Section 10.2)

40. $\dfrac{18y}{24y^3}$ $\dfrac{3}{4y^2}$

41. $\dfrac{3x-9}{x^2-5x+6}$ $\dfrac{3}{x-2}$

42. $\dfrac{x^2-10x+25}{x^2-25}$

43. $\dfrac{12b^2-6}{4ab^3-2ab}$ $\dfrac{3}{ab}$

Multiply and simplify. (Sections 10.3 through 10.5)

44. $\dfrac{-10ab}{2}\cdot\dfrac{12a}{9a^2b}$ $-\dfrac{20}{3}$

45. $\dfrac{3x^2}{8}\cdot\dfrac{4x}{6y^2}\cdot\dfrac{16y^3}{3x}$ $\dfrac{4x^2y}{3}$

46. $\dfrac{4x}{x+1}\cdot\dfrac{x+1}{2x(x-3)}$

47. $\dfrac{x-9}{x+9}\cdot\dfrac{x-1}{9-x}$ $-\dfrac{x-1}{x+9}$

48. $\dfrac{a^2-4}{a^2-1}\cdot\dfrac{3a-3}{a-2}$ $\dfrac{3(a+2)}{a+1}$

49. $\dfrac{x^2-6x-7}{x^2+x}\cdot\dfrac{x^2-x}{3x-21}$

Divide and simplify. (Sections 10.6 and 10.7)

50. $\dfrac{x+3}{5x}\div\dfrac{2(x+3)}{5}$ $\dfrac{1}{2x}$

51. $\dfrac{x^2-x-6}{x^2-4}\div(x-3)$ $\dfrac{1}{x-2}$

52. $\dfrac{h^2-9}{h^2-6h+9}\div\dfrac{3h+9}{7h-21}$ $\dfrac{7}{3}$

Add or subtract. Simplify where necessary. (Sections 11.1, 11.2, and 11.4 through 11.6)

53. $\dfrac{9}{x+2}+\dfrac{8}{x+2}$ $\dfrac{17}{x+2}$

54. $\dfrac{3x-1}{x-4}-\dfrac{x+7}{x-4}$ 2

55. $\dfrac{7}{5y}-\dfrac{3}{2y}$ $-\dfrac{1}{10y}$

56. $\dfrac{9}{x}+\dfrac{1}{x-1}$ $\dfrac{10x-9}{x(x-1)}$

57. $\dfrac{3}{2x-6}-\dfrac{1}{3x-9}$ $\dfrac{7}{6(x-3)}$

58. $\dfrac{7x}{x^2-1}+\dfrac{8}{3x-3}$

Solve and check each equation. (Section 11.7)

59. $\dfrac{15}{2x-1}=-3$ −2

60. $\dfrac{7}{x-6}=\dfrac{9}{x+2}$ 34

61. $\dfrac{1}{x-2}=\dfrac{2}{x}+\dfrac{3}{x-2}$ 1

For Ex. 62-65, the moves described locate the desired points.

Graph these points in the same coordinate plane. (Sections 12.1 and 12.2)

62. $P(-3, 4)$

63. $W(5, -6)$ Right 5, Down 6

64. $R(4, 0)$ Right 4

65. $T(-1\frac{1}{2}, -2)$

Write Yes or No to show whether each relation is a function. (Section 12.4)

66. $\{(1,2), (3,6), (1,5)\}$ No

67. $\{(6,1), (4,-3), (-2,1)\}$ Yes

Draw a graph of each function. The domain is {real numbers}. (Section 12.5)

68. $y = 3x - 2$

69. $y = \frac{1}{2}|x|$

70. $y = x^2 - 3$

Linear Functions

The two towers of the Verrazano-Narrows Bridge in New York City are perpendicular to the earth but are not parallel to each other. They are about four centimeters farther apart at the top than at their bases.

Background

The top right photo shows a laser experiment. Lasers produce very narrow and intense beams of light that are used in industry to cut diamonds and steel, in medicine to perform delicate operations, and in communications to transmit voice messages and television signals.

The bottom right photo shows a beam of light passing through a prism, separating the light into the colors of the spectrum (red, orange, yellow, green, blue, indigo, and violet).

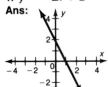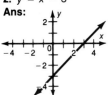
13.1 Graphs of Linear Functions

To identify from the rule of a function whether or not the function is linear

The steps for graphing a function are reviewed.

| Select values for x. | → | Compute the y values. | → | Graph the ordered pairs. | → | Draw a line containing the points. |

EXAMPLE 1   Draw a graph of $y = 2x - 1$. The domain is {real numbers}.

Solution: $y = 2x - 1$ ◀ **Rule for the function**

Point out again that values small in absolute value are chosen for x.

If $x = -2$, $y = 2(-2) - 1$
   $= -5$.

If $x = -1$, $y = 2(-1) - 1$
   $= -3$.

If $x = 0$, $y = 2(0) - 1$
   $= -1$.

If $x = 1$, $y = 2(1) - 1$
   $= 1$.

If $x = 2$, $y = 2(2) - 1$
   $= 3$.

x	-2	-1	0	1	2
y	-5	-3	-1	1	3

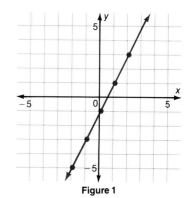

Figure 1

These ordered pairs are graphed in Figure 1. The points lie on a line.

P–1   **What are the coordinates of some other points on the line?**
   $(3, 5)$, $(0, -1)$, $(-2, -5)$, etc.

P–2   **How do you know whether (3, 5) belongs to this function?**
   Substitute 3 for x, 5 for y in $y = 2x - 1$

   ☐1 Replace x with 3. ⟶ $y = 2(3) - 1$

   ☐2 Solve for y. ⟶ $y = 5$

Thus, $(3, 5)$ belongs to the function. The point $(3, 5)$ lies on the graph of Figure 1.

350 / *Chapter 13*

EXAMPLE 2   Draw a graph of $y = -\frac{1}{2}x + 2$.

Solution: When the domain is not given, assume that it is {real numbers}.

$$y = -\frac{1}{2}x + 2$$
$$y = -\frac{1}{2}(-4) + 2 = 4$$
$$y = -\frac{1}{2}(-2) + 2 = 3$$
$$y = -\frac{1}{2}(0) + 2 = 2$$
$$y = -\frac{1}{2}(2) + 2 = 1$$
$$y = -\frac{1}{2}(4) + 2 = 0$$

x	−4	−2	0	2	4
y	4	3	2	1	0

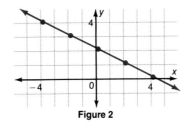

The graph of the function is shown in Figure 2.

Figure 2

Definition
You will want to explain that a vertical line is the graph of a relation but not of a function.

A function that has the points of its graph lying in a straight line is a *linear function.*

The rule for a linear function is
$$y = mx + b.$$
(m and b are real numbers.)

$$y = 3x + 5$$
$$y = -\frac{5}{2}x + \sqrt{3}$$

$m: -1;\ b: \frac{1}{2}$

P–3 What are the values of m and b in the rule $y = -x + \frac{1}{2}$? in the rule $y = \frac{2}{3}x - \frac{1}{4}$? in the rule $y = 10 - 3x$?

$m: \frac{2}{3};\ b: -\frac{1}{4}$ $m: -3;\ b: 10$

P–4 Which functions below are linear?   a and b

a. $y = 2x + 5$ b. $y = x - 3$ c. $y = x^2 + 1$ d. $y = \sqrt{x}$

CLASSROOM EXERCISES

Write the y values that are missing from each table. (Examples 1 and 2)

1. $y = 3x$

x	−2	−1	0	2
y	?	?	?	?
	−6	−3	0	6

2. $y = x + 3$

x	−5	−2	0	4
y	?	?	?	?
	−2	1	3	7

3. $y = 10x + 1$

x	−5	−2	0	3
y	?	?	?	?
	−49	−19	1	31

Linear Functions / 351

Draw a graph of each function.
1. $y = 3x - 2$
Ans:

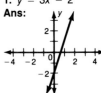

2. $y = 4 - \frac{1}{2}x$
Ans:

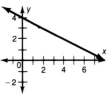

Common Error
Students incorrectly solve for y with one or more of the x values and then draw a graph which is not linear.

Example
Graph $y = -x + 3$.

x	y
−1	2
0	3
1	2
2	1

Prescription
Remind students that these are called linear functions because all of the points of each function lie in a straight line. If the points do not lie in a straight line, insist that students check their calculations to find the mistake.

Compare each rule with $y = mx + b$. Then write the values of m and b for each function. (P-3)

4. $y = -10x + 1$ m: -10; b: 1

5. $y = -3x + 7$ m: -3; b: 7

6. $y = 5 - x$

7. $y = -x - 3$ m: -1; b: -3

8. $y = 10x$ m: 10; b: 0

9. $y = -7x$

10. $y = -\frac{3}{5}$ m: 0; b: $-\frac{3}{5}$

11. $y = -0.5x + 3.8$ m: -0.5; b: 3.8

12. $y = -\frac{2}{3}x - \frac{1}{3}$

WRITTEN EXERCISES

Goals: To identify and to graph linear functions
Sample Problems: **a.** $y = 3x^2$ **b.** $y = -x$
Answers: **a.** Not linear   **b.** Linear; the graph
is at the right.

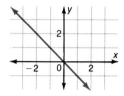

See the Answers to Selected Exercises for the answers to Ex. 5-11 odd.

Graph at least four ordered pairs of each function. Then draw the graph. The domain is {real numbers}. (Examples 1 and 2)

1. $y = -x + 3$

2. $y = -x - 2$

3. $y = 3x - 1$

4. $y = 2x - 3$

5. $y = 2x$

6. $y = -2x$

7. $y = 0 \cdot x - 2$

8. $y = 0 \cdot x + 3$

9. $y = -4x + 3$

10. $y = \frac{1}{2}x - \frac{3}{2}$

11. $y = -\frac{1}{2}x + \frac{1}{2}$

12. $y = -2x + 1$

Write Yes or No to show whether each function is linear. (P-4)

13. $y = -4x + 10$ Yes

14. $y = x^2 + 1$ No

15. $y = -\frac{2}{3}x - 5$ Yes

16. $y = x(x - 1)$ No

17. $y = 0 \cdot x + 0$ Yes

18. $y = 0 \cdot x + \frac{1}{3}$ Yes

19. $y = -2x + 0$ Yes

20. $y = 6x + 0$ Yes

21. $y = (x + 1)(x - 1)$ No

22. $y = x^2 - 2x + 5$ No

23. $y = -4$ Yes

24. $y = 3$ Yes

NON-ROUTINE PROBLEM

25. The figure at the right shows three squares inside one another. The small square has sides which are 7 centimeters in length, the middle square has sides which are 9 centi-meters in length, and the large square has sides which are 12 centimeters in length. Find the area of the shaded portion of the figure.

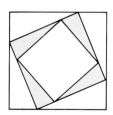

OBJECTIVES: To write the Y intercept of a linear function from its rule
To graph a linear function from its given rule using the y intercept

13.2 Y Intercept

Teaching Suggestions p. M–42

Lesson Resources
Warm–Up: p. M–43
Maintenance: See below.
Practice Worksheet 13.2
Exploration 13

P–1 **How are the equations of these functions alike?** In each, the value
of b is 2.

a. $y = \frac{1}{2}x + 2$ **b.** $y = x + 2$ **c.** $y = 2x + 2$

d. $y = -2x + 2$ **e.** $y = -x + 2$ **f.** $y = -\frac{1}{2}x + 2$

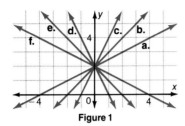

Figure 1

Point out that
the equations all
have the number
2 in common.

Maintenance
1. Solve and check:
   $95 = 8c - 3c + 15$
   Ans: c = 16
   (Section 3.5)
2. Simplify: $\sqrt{144a^4b^3}$
   Ans: $12a^2b\sqrt{b}$
   (Section 7.4)
3. Divide:
   $2x + 3 \overline{)6x^2 + 5x - 15}$
   Ans: $3x - 2 - \dfrac{9}{2x + 3}$
   (Section 8.7)
4. Factor: $18m^2 - 8$
   Ans: 2(3m − 2)(3m + 2)
   (Section 9.7)
5. Subtract: $\dfrac{4}{x - 2} - \dfrac{3}{x + 2}$
   Ans: $\dfrac{x+14}{(x-2)(x+2)}$
   (Section 11.5)

P–2 **What are the coordinates of the common point of the graphs in
Figure 1?** $(0, 2)$

Graphs of the following functions
are shown in Figure 2.

1. $y = 3x + 1$
2. $y = 2x - 3$
3. $y = -\frac{1}{2}x + 3$
4. $y = -x - 1$

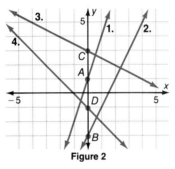

Figure 2

P–3 **What is the x value of each of
the points A, B, C, and D? the
y value of each point?** x value: 0
y value: (A) 1; (B) −3; (C) 3; (D) −1

Definition   The y value of the point that a line has in common with the
y axis is the **y intercept.**

EXAMPLE 1   Write the y intercept of the graph of $y = 3x - 2$.

Solution: $y = 3x - 2$

☐1 Replace x with 0. ———————→ $y = 3(0) - 2$ For any point
on the y axis,
x is zero.
☐2 Solve for y. ———————————→ $y = -2$ y intercept

Additional Examples
Example 1
Write the y intercept of the graph
of each function.
1. $y = -2x + 5$
   Ans: 5
2. $y = 6 - x$
   Ans: 6
3. $y = 4x$
   Ans: 0

In the rule for the linear function
$y = mx + b$, b represents the y intercept
of its graph.

$$y = -x + \frac{1}{2}$$
$$y \text{ intercept} = \frac{1}{2}$$

P–4 **What is the y intercept of the graph of each function below?**

a. $y = \frac{3}{2}x + \frac{5}{2}$ $\frac{5}{2}$, or $2\frac{1}{2}$ **b.** $y = 4 - x$ 4

Additional Examples
Example 2
Graph by using at least three ordered pairs, including one with
the y intercept.
1. $y = -x + 1$
   Ans:

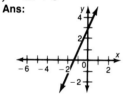

2. $y = 2x + 3$
   Ans:

EXAMPLE 2   Graph $y = -2x - 3$. Graph at least three ordered pairs,
including one with the y intercept.

Solution: $y = -2x - 3$

$y = -2(0) - 3 = -3$

$y = -2(-3) - 3 = 3$

$y = -2(2) - 3 = -7$

Point out that
an x value of
0 yields the y
intercept as a
value for y.

x	y
0	−3
−3	3
2	−7

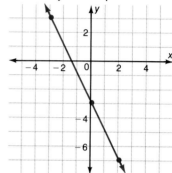

CLASSROOM EXERCISES

Estimate the y intercept of each graph. (P-3, second part)

1. 1

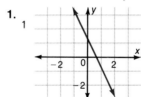

2. −2

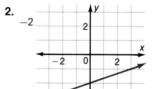

3. $\frac{2}{3}$

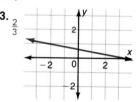

4. 0

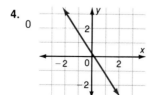

5. $1\frac{1}{2}$

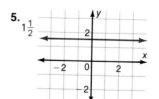

6. $-\frac{3}{4}$

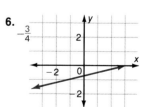

354 / *Chapter 13*

Write each y intercept. (P-4)

7. $y = 3x + \frac{1}{2}$ $\frac{1}{2}$

8. $y = -2x + 5$ 5

9. $y = \frac{2}{3}x - \frac{1}{3}$

10. $y = -3x$ 0

11. $y = -x - 1$ −1

12. $y = 3x + \frac{15}{2}$

13. $y = x$ 0

14. $y = 5 - 3x$ 5

15. $y = 2(x + 3)$

Additional Answers
Classroom Exercises
 9. $-\frac{1}{3}$
12. $\frac{15}{2}$
15. 6

Assignment Guide
Basic p. 355: 1–21 odd
Average p. 355: 3, 6, 9, ···, 21;
p. 358: 3, 6, 9, ···, 30
Above Average p. 355:
3, 6, 9, ···, 21; p. 358: 3, 6, 9
··· 30

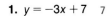

WRITTEN EXERCISES

Goal: To identify the y intercept of a linear function and to graph the function
Sample Problem: $y = 3(2 - \frac{1}{3}x)$
Answer: 6 The graph is at the right.

Write each y intercept. (Example 1)

1. $y = -3x + 7$ 7

2. $y = 5x - 2$ −2

3. $y = -\frac{1}{2}x + \frac{7}{2}$ $\frac{7}{2}$

4. $y = 1.73x + 3.85$ 3.85

5. $y = \frac{2}{3}x - \frac{5}{3}$ $-\frac{5}{3}$

6. $y = -1.17x + 3.14$ 3.14

7. $y = 0.3 - 5x$ 0.3

8. $y = 14 - \frac{1}{2}x$ 14

9. $y = 3(2 - x)$ 6

10. $y = -5(1 - 2x)$ −5

11. $y = -x$ 0

12. $y = \frac{7}{5} + 2x$ $\frac{7}{5}$

Graph each function. Graph at least three ordered pairs including one with the y intercept. (Example 2) In Ex. 13-22 each graph is a straight line. The coordinates of two points of each line are given.

13. $y = x + 1$ (0, 1), (−1, 0)

14. $y = -x + 2$ (0, 2), (2, 0)

15. $y = -2x - 3$ (0, −3), (−2, 1)

16. $y = 3x - 2$ (0, −2), (2, 4)

17. $y = \frac{1}{2}x + 5$ (0, 5), (4, 7)

18. $y = -\frac{1}{2}x + 4$ (0, 4), (2, 3)

19. $y = -3x$ (0, 0), (2, −6)

20. $y = 2x$ (0, 0), (2, 4)

21. $y = \frac{1}{2}x + \frac{3}{2}$ (0, $\frac{3}{2}$), (5, 4)

22. $y = -2x - \frac{3}{2}$ (0, $-\frac{3}{2}$), (2, $-\frac{11}{2}$)

REVIEW CAPSULE FOR SECTION 13.3

Each graph is a straight line. The coordinates of two points of each line are given.

Graph these functions in the same coordinate plane. (Pages 350–352)

1. $y = \frac{1}{2}x + 5$ (0, 5), (2, 6)

2. $y = \frac{1}{2}x + 3$ (0, 3), (2, 4)

3. $y = \frac{1}{2}x + 1$ (0, 1), (2, 2)

4. $y = \frac{1}{2}x$ (0, 0), (2, 1)

5. $y = \frac{1}{2}x - 3$ (0, −3), (2, −2)

6. $y = \frac{1}{2}x - 5$ (0, −5), (2, −4)

Maintenance
1. Solve and check:
   $3a + 2a - 5 = 13 - 2(a+2)$
   Ans: a = 2
   (Section 4.3)
2. Add: $\sqrt{80} - \sqrt{5}$
   Ans: $3\sqrt{5}$ (Section 7.5)
3. Simplify: $\dfrac{4y^2 - 8y}{12}$
   Ans: $\dfrac{y^2 - 2y}{3}$
   (Section 10.2)
4. Add: $\dfrac{3}{x + 5} + \dfrac{2}{x + 5}$
   Ans: $\dfrac{5}{x + 5}$
   (Section 11.1)
5. Write the domain, D, and the range, R, of the relation, $\{(\frac{1}{2}, \frac{1}{4}),$ (2,1), (0,0), (4,2)\}.
   Ans: D: $\{\frac{1}{2}, 2, 0, 4\}$
   R: $\{\frac{1}{4}, 1, 0, 2\}$
   (Section 12.4)

13.3 Slope of a Line

To name the quadrants in which the graph of a linear function will lie given the equation $y = ax$

Compare the graph of 1 in Fig. 1 with the graph of 1 in Fig. 2. Also compare the graphs of 2, 3, and 4 in Fig. 1 with the graphs of 2, 3, and 4 in Fig. 2, respectively.

P–1 **What is the y intercept for each graph in Figure 1?** 2

The rules for these linear functions are listed below.

1. $y = \frac{1}{2}x + 2$ 2. $y = x + 2$
3. $y = \frac{3}{2}x + 2$ 4. $y = 2x + 2$

The lines get "steeper" with respect to the x axis in the order from **1** to **4**.

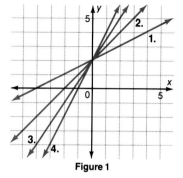
Figure 1

P–2 **What number in the rule $y = mx + b$ affects the "steepness" of a line?** m

The graphs of four more linear functions are shown in Figure 2.

1. $y = -\frac{1}{2}x - 1$ 2. $y = -x - 1$
3. $y = -\frac{3}{2}x - 1$ 4. $y = -4x - 1$

P–3 **In what special way are the rules for these functions related?**
In all four, the value of b is −1.
The "steepness" of a line is described by a number called **slope**.

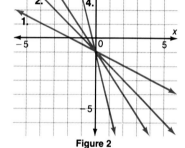
Figure 2

In the rule for the linear function $y = mx + b$, m represents the **slope** of its graph.	$y = -2x + 5$ Slope $= -2$

The rule for a linear function, $y = mx + b$, is called the **slope-intercept form**.

P–4 **What is the slope of the graph of the function having each rule below? the y intercept?**

a. $y = -\frac{3}{4}x - 2$ $-\frac{3}{4}; -2$ **b.** $y = 5x - \frac{1}{2}$ $5; -\frac{1}{2}$ **c.** $y = \frac{5}{2} - \frac{3}{2}x$ $-\frac{3}{2}; \frac{5}{2}$

356 / Chapter 13

EXAMPLE 1   Write the slope-intercept
form of the rule for
the line in the figure.
The slope is −2.

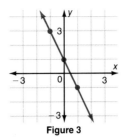
Figure 3

Solution: $y = mx + b$ $m = -2$
$y = -2x + 1$ $b = 1$

P–5 **In what quadrants does the graph of Figure 3 lie?** I, II, and IV

EXAMPLE 2   Graph these functions.

1. $y = 2x$ **2.** $y = -\frac{1}{3}x$

Solutions: $y = 2x$ $y = -\frac{1}{3}x$

x	y
−2	−4
0	0
2	4

x	y
−3	1
0	0
3	−1

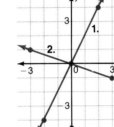

P–6 **In what quadrants does the graph of $y = 2x$ lie? the graph of**
$y = -\frac{1}{3}x$? II and IV   I and III

It is important
that students have
an understanding
of positive and
negative slopes.

A line such as the graph of $y = 2x$ has a positive slope. It "slopes"
upward and to the right. The graph of $y = -\frac{1}{3}x$ has a negative slope.
It "slopes" upward and to the left.

CLASSROOM EXERCISES

Find the slope and the y intercept of the graph of each function. (P-4)

1. $y = 3x + 2$ 3; 2   **2.** $y = -5x + 1$ −5; 1   **3.** $y = x + 5$ 1; 5

4. $y = -x - 1$ −1; −1   **5.** $y = \frac{1}{3}x - 6$ $\frac{1}{3}$; −6   **6.** $y = -\frac{1}{2}x + 1$

Write the slope-intercept form of the rule for each function. (Example 1) $-\frac{1}{2}$; 1

7. The slope is 5 and the y intercept is 3.   **8.** The slope is $\frac{1}{4}$ and the y intercept is $\frac{1}{2}$.

$y = 5x + 3$

$y = \frac{1}{4}x + \frac{1}{2}$

*Linear Functions / **357***

Additional Examples
Example 1
Write the slope–intercept form of
the rule for each function.
1. The slope is 4 and the y intercept is 7.
Ans: $y = 4x + 7$
2. The slope is $\frac{1}{2}$ and the y intercept is −3.
Ans: $y = \frac{1}{2}x - 3$
3. The slope is $\frac{2}{3}$ and the y intercept is 0.
Ans: $y = \frac{2}{3}x + 0$ or $y = \frac{2}{3}x$

Example 2
Graph each function.
1. $y = 3x$
Ans:
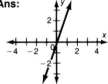

2. $y = -\frac{5}{3}x$
Ans:
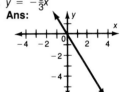

Write the slope-intercept form of the rule for each function.
(Example 1)

9. The slope is −1 and the y intercept is 0. $y = -x$ or $y = -1x + 0$

10. The slope is $\frac{-2}{3}$ and the y intercept is $\frac{3}{5}$. $y = -\frac{2}{3}x + \frac{3}{5}$

11. The line crosses the y axis at (0, 10) and has slope 3. $y = 3x + 10$

12. The line crosses the y axis at (0, −6) and has slope −1. $y = -x - 6$

Assignment Guide
Basic p. 358: 1–29 odd
Average p. 358: 3, 6, 9, ···, 30;
p. 355: 3, 6, 9, ···, 21
Above Average p. 358:
3, 6, 9, ···, 30; p. 355: 3, 6, 9,
···, 21

WRITTEN EXERCISES

Goal: To write the slope-intercept form of the rule for a linear function
Sample Problem: Slope is $\frac{2}{3}$. See graph below. **Answer:** $y = \frac{2}{3}x + 2$

Write the slope of the graph of each function. (P-4)

1. $y = x - 5$ 1
2. $y = -x + 2$ −1
3. $y = -\frac{1}{2}x + 7$ $-\frac{1}{2}$
4. $y = \frac{1}{2}x - 6$ $\frac{1}{2}$

5. $y = 2(x - 3)$ 2
6. $y = 3(x + 1)$ 3
7. $y = 2 - 3x$ −3
8. $y = 5 - 7x$ −7

9. $y = \frac{1}{2}(3 - x)$ $-\frac{1}{2}$
10. $y = -\frac{1}{3}(5 - x)$ $\frac{1}{3}$
11. $y = -\frac{1}{4}(8 + 16x)$ −4
12. $y = \frac{1}{5}(10 - 10x)$
−2

Write the slope-intercept form of the rule for each function graphed at the right. The slopes are given below. (Example 1)

13. Slope is −1.
14. Slope is $\frac{1}{6}$.
15. Slope is $-\frac{1}{3}$.
16. Slope is $-\frac{4}{3}$.
17. Slope is −1.
18. Slope is 1.

Name the quadrants in which each graph lies. (Example 2, P-5, P-6)

19. $y = 5x$ I, III
20. $y = 10x$ I, III
21. $y = -2x$ II, IV
22. $y = -6x$

23. $y = -\frac{1}{2}x$ II, IV
24. $y = \frac{2}{3}x$ I, III
25. $y = \frac{5}{6}x$ I, III
26. $y = -\frac{3}{4}x$

27. $y = -0.7x$ II, IV
28. $y = 0.56x$ I, III
29. $y = x$ I, III
30. $y = -x$

Additional Answers
Written Exercises
13. $y = -x + 5$
14. $y = \frac{1}{6}x + 3$
15. $y = -\frac{1}{3}x - 3$
16. $y = -\frac{4}{3}x - 2$
17. $y = -x$
18. $y = x$
22. II, IV
26. II, IV
30. II, IV

REVIEW CAPSULE FOR SECTION 13.4

Simplify each fraction. (Pages 8–10 and 16–18)

1. $\dfrac{2 - 5}{7 - 2}$ $-\frac{3}{5}$
2. $\dfrac{0 - (-3)}{2 - 4}$ $-\frac{3}{2}$
3. $\dfrac{(-2) - 3}{4 - 5}$ 5
4. $\dfrac{7 - 2}{5 - (-3)}$ $\frac{5}{8}$
5. $\dfrac{5 - (-3)}{7 - (-5)}$
$\frac{2}{3}$

OBJECTIVES: To write the slope of a line containing two points with their coordinates given

13.4 Slope Formula

To write the slope-intercept form of a line containing two points with their coordinates given

Teaching Suggestions p. M–42

P–1 **What is the slope of the graph of $y = \frac{3}{2}x + 2$ shown in the figure at the right?** $\frac{3}{2}$

Suppose that you move from B to C in Figure 1 by one vertical and one horizontal motion.

P–2 **How many units and in what direction is the vertical motion?** 3 units **the horizontal motion?** 2 units

$$\frac{5-2}{2-0} = \frac{3}{2} \blacktriangleleft \frac{\text{Vertical motion: 3}}{\text{Horizontal motion: 2}}$$

Now suppose that you move from A to C by one vertical motion and one horizontal motion.

P–3 **How many units and in what direction is the vertical motion?** the **horizontal motion?** 4 units 6 units

$$\frac{\text{Difference in y coordinates}}{\text{Difference in x coordinates}} \blacktriangleright \frac{5-(-1)}{2-(-2)} = \frac{5+1}{2+2}$$
$$= \frac{6}{4} \text{ or } \frac{3}{2} \blacktriangleleft \frac{\text{Vertical motion}}{\text{Horizontal motion}}$$

A quotient of two numbers such as $\frac{3}{2}$ is called a <u>ratio</u>. The <u>slope</u> of a line is the ratio of the vertical motion to the horizontal motion from one point to another on the line.

You may want to warn students of some of the incorrect ways of substituting in the Slope Formula.

Slope Formula

If the coordinates of two points on a nonvertical line are (x_1, y_1) and (x_2, y_2), then m, the **slope** of the line, is given by

$$m = \frac{y_2 - y_1}{x_2 - x_1}.$$

 1 and 2 are subscripts. Read x sub 1, x sub 2, etc.

$(2, -3), (-5, 1)$

$$m = \frac{1 - (-3)}{-5 - 2}$$

$$m = -\frac{4}{7}$$

Linear Functions / **359**

Lesson Resources
Warm–Up: p. M–43
Maintenance: See below.
Practice Worksheet 13.4
Transparency 32

Maintenance
1. Write 34,200 in scientific notation.
   Ans: 3.42 × 10⁴
   (Section 6.6)
2. Multiply:
   $(3x - 4)(2x + 1)$
   Ans: 6x² − 5x − 4
   (Section 8.5)
3. Factor:
   $2x^2 + 11x + 5$
   Ans: (2x + 1)(x + 5)
   (Section 9.4)
4. Solve and check:
   $$\frac{4}{a - 1} = \frac{7}{a - 5}$$
   Ans: $a = -\frac{13}{3}$
   (Section 11.7)
5. Name two points on the graph of $y = -3x + 2$.
   **Ans: sample points:
   (0,2), (−1,5), (2,−4)
   Answers will vary.**
   (Section 12.5)

Additional Examples

Example 1

Write the slope of each line.

1.

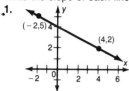

Ans: $m = -\dfrac{1}{2}$

2.

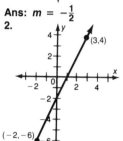

Ans: $m = 2$

Example 2

Write the slope of the line that contains each pair of points.

1. $(-2, 5);\ (4, 3)$
   Ans: $m = -\dfrac{1}{3}$
2. $(-1, -6);\ (7, 3)$
   Ans: $m = \dfrac{9}{8}$
3. $(4, -2);\ (3, 5)$
   Ans: $m = -7$

Example 3

Write the slope–intercept form of the line that contains each pair of points.

1. $(0, 2);\ (-2, 6)$
   Ans: $y = -2x + 2$
2. $(0, -5);\ (2, 1)$
   Ans: $y = 3x - 5$
3. $(0, 4);\ (6, 7)$
   Ans: $y = \dfrac{1}{2}x + 4$

EXAMPLE 1

Write the slope of the line in Figure 2 by using the following ratio:

$$\frac{\text{vertical motion from } A \text{ to } B}{\text{horizontal motion from } A \text{ to } B}$$

Solution:

$$\frac{\text{vertical motion from } A \text{ to } B}{\text{horizontal motion from } A \text{ to } B} = \frac{-5 - 1}{1 - (-3)}$$

The vertical motion is *down* 6 units. The horizontal motion is to the *right* 4 units.

$$\text{slope} = \frac{-6}{4} \quad \text{or} \quad -\frac{3}{2}$$

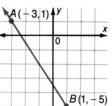

Figure 2

EXAMPLE 2

Write the slope of the line that contains the points $(-2, 5)$ and $(3, -1)$.

Solution:

Method 1

$x_1 = -2,\ x_2 = 3$
$y_1 = 5,\ y_2 = -1$

$$m = \frac{y_2 - y_1}{x_2 - x_1}$$

$$= \frac{-1 - 5}{3 - (-2)}$$

$$= -\frac{6}{5} \quad \textit{Slope of the line}$$

Method 2

$$m = \frac{y_2 - y_1}{x_2 - x_1}$$

$x_1 = 3,\ x_2 = -2$
$y_1 = -1,\ y_2 = 5$

$$= \frac{5 - (-1)}{-2 - 3}$$

$$= -\frac{6}{5} \quad \textit{Slope of the line}$$

EXAMPLE 3

Write the slope-intercept form of the line that contains the points $(0, 3)$ and $(5, 2)$.

Solution:

1. Slope formula

$$\frac{y_2 - y_1}{x_2 - x_1} = \frac{2 - 3}{5 - 0}$$

$$= -\frac{1}{5}$$

2. y intercept

$b = 3$

If $x = 0$, the point is on the y axis.

3. Slope-intercept form

$y = mx + b$

4. Substitute.

$y = -\frac{1}{5}x + 3$

$m = -\frac{1}{5}$
$b = 3$

CLASSROOM EXERCISES

Find the value of each ratio. (P-2, P-3)

1. $\dfrac{8-5}{7-2}$ $\dfrac{3}{5}$

2. $\dfrac{3-7}{9-4}$ $-\dfrac{4}{5}$

3. $\dfrac{-1-2}{7-9}$ $\dfrac{\frac{3}{2}}{-\frac{1}{8}}$

4. $\dfrac{5-(-3)}{2-3}$ -8

5. $\dfrac{0-(-1)}{-1-2}$ $-\dfrac{1}{3}$

6. $\dfrac{-3-(-2)}{7-(-1)}$

7. $\dfrac{-3-0}{0-(-3)}$ -1

8. $\dfrac{(-3)-(-3)}{2-7}$ 0

9. $\dfrac{1-2}{4-6}$ $\dfrac{1}{2}$

Write the slope of each line. (Example 1)

10.   1

11. -1

12. 2

13. 2

14. -3

15. 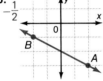 $-\dfrac{1}{2}$

16. 0

17. 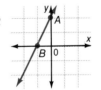 2

WRITTEN EXERCISES

Goal: To write the slope of a line given coordinates of two points on the line
Sample Problem: (4, 5) and (−2, −1)   **Answer:** $m = 1$

Write the slope of each line. (Example 1)

1. 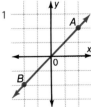 1

2. 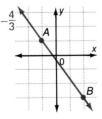 $-\dfrac{4}{3}$

3. 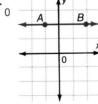 0

4. 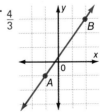 $\dfrac{4}{3}$

Linear Functions / 361

Common Error
Students use the *x* coordinates in the numerator and the *y* coordinates in the denominator of the slope formula.

Example
Write the slope of the line that contains (−1,5) and (3,2).
$$\dfrac{-1-3}{5-2} = -\dfrac{4}{3}$$

Prescription
Review the slope formula with the students. Remind them that the vertical axis is the *y* axis and the horizontal axis is the *x* axis.

Assignment Guide
Basic pp. 361–362: 1–27 odd
Average pp. 361–362: 1–37 odd
Above Average pp. 361–362: 1–27 odd, 29–37

Write the slope of the line that contains the two points. (Example 2)

5. $(-1, 5)$; $(3, 2)$ $-\frac{3}{4}$
6. $(4, 6)$; $(-2, 3)$ $\frac{1}{2}$
7. $(5, -3)$; $(-1, 4)$

8. $(2, 3)$; $(-1, -2)$ $\frac{5}{3}$
9. $(-3, -4)$; $(-2, -5)$ -1
10. $(-1, -5)$; $(-2, -7)$

11. $(3, -5)$; $(-1, -5)$ 0
12. $(4, -3)$; $(-1, -3)$ 0
13. $(10, -5)$; $(6, -2)$

14. $(12, 8)$; $(-4, -10)$ $\frac{9}{8}$
15. $(3\frac{1}{2}, -5\frac{3}{4})$; $(-\frac{1}{2}, \frac{1}{4})$ $-\frac{3}{2}$
16. $(-\frac{3}{2}, -\frac{5}{2})$; $(\frac{1}{2}, \frac{1}{2})$

Write the slope-intercept form of the line that contains the two points.
(Example 3)

17. $(0, 5)$; $(-3, 2)$ $y = x + 5$
18. $(0, 1)$; $(5, -3)$ $y = -\frac{4}{5}x + 1$
19. $(0, -3)$; $(4, 6)$

20. $(0, -5)$; $(-1, -1)$ $y = -4x - 5$
21. $(0, 0)$; $(5, 8)$ $y = \frac{8}{5}x$
22. $(0, 0)$; $(-4, 7)$

23. $(0, -6)$; $(2, -6)$ $y = -6$
24. $(0, 10)$; $(-1, 10)$ $y = 10$
25. $(0, 5)$; $(5, 0)$

26. $(0, -2)$; $(-2, 0)$ $y = -x - 2$
27. $(0, 6)$; $(3, 8)$ $y = \frac{2}{3}x + 6$
28. $(0, 2)$; $(4, 10)$

MORE CHALLENGING EXERCISES

Draw the line with the given slope and that contains the given point.

EXAMPLE: $m = -\frac{1}{2}$; $(5, -3)$

SOLUTION: Locate the point as shown in the figure. Call it P.

Use the slope as $\dfrac{\text{vertical motion from } P \text{ to } Q}{\text{horizontal motion from } P \text{ to } Q}$

to locate a second point Q.

$$-\frac{1}{2} = \frac{1 \text{ unit upward } (+)}{2 \text{ units left } (-)}$$

Draw the line containing P and Q.

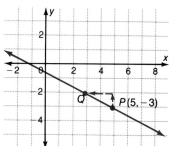

In Ex. 29-34 a second point is given that can be found using the slope.
A line can be drawn using the 2 points.

29. $m = 3$; $(0, 0)$ $(1, 3)$
30. $m = \frac{1}{2}$; $(1, -1)$ $(3, 0)$
31. $m = -\frac{3}{4}$; $(2, -3)$ $(6, -6)$

32. $m = -5$; $(-2, -3)$ $(-3, 2)$
33. $m = 0$; $(-5, 3)$ $(0, 3)$
34. $m = \frac{5}{3}$; $(1, -5)$ $(4, 0)$

APPLICATIONS

Suppose a sheet of paper has a thickness of 0.0015 inch. What would be the thickness of the paper if it could be folded in half several times? A calculator can be used to find the answer.

EXAMPLE Find the thickness of the folded paper when the number of folds is 15.

SOLUTION

Number of Folds	Number of Thicknesses	Thickness in Inches
1	$2^1 = 2$	$2 \times 0.0015 = 0.0030$ in
2	$2^2 = 4$	$4 \times 0.0015 = 0.0060$ in
3	$2^3 = 8$	$8 \times 0.0015 = 0.012$ in
⋮	⋮	⋮
15	$2^{15} = $ __?__	__?__ $\times 0.0015 = $ __?__ in

2 × = × . 0 0 1 5 =   `49.152`

▲ Press " = " 14 times.   ▲ About 49 inches thick

Find the following thicknesses for the sheet of paper in the Example.

35. 16 folds, in inches   **36.** 25 folds, in feet   **37.** 50 folds, in miles (HINT: $2^{50} = 2^{25} \times 2^{25}$)
About 98 inches   About 4194 feet   About 27 million miles
                             (See "NOTE" in margin.)   (See "NOTE" in margin.)

NOTE: To avoid calculator over-load, change 0.0015 inches to 0.000125 feet in Exercise 36. In Exercise 37, work in this order:
$2^{25} \times 0.000125 \div 5280 \times 2^{25}$

■■■ **MID-CHAPTER REVIEW** ■■■

Graph at least four ordered pairs of each function. Then draw the graph. The domain is {real numbers}. (Section 13.1) In Ex. 1-6 each graph is a straight line. The coordinates of two points of each line are given.

1. $y = x - 3$ (0, −3), (1, −2)   **2.** $y = -x + 2$ (0, 2), (−1, 3)   **3.** $y = -2x + 1$ (0, 1), (−2, 5)

4. $y = 3x + 2$ (0, 2)(1, 5)   **5.** $y = \frac{1}{2}x - 2$ (0, −2), (2, −1)   **6.** $y = -\frac{1}{2}x + 3$ (0, 3)(−2, 4)

Use the y intercept and two ordered pairs to graph each function. (Section 13.2) In Ex. 7-9 each graph is a straight line. Two points are given. The first is the *y* intercept.

7. $y = -x + 3$ (0, 3), (1, 2)   **8.** $y = 2x - 1$ (0, −1), (2, 3)   **9.** $y = -\frac{1}{2}x + \frac{5}{2}$ (0, $\frac{5}{2}$), (1, 2)

Write the slope-intercept form of the rule for each function. (Section 13.3)

10. Slope: −2; *y* intercept: 5   **11.** Slope: −$\frac{1}{4}$; *y* intercept: −1   **12.** Slope: −1; *y* intercept: $\frac{1}{2}$
$y = -2x + 5$ $y = -\frac{1}{4}x - 1$ $y = -x + \frac{1}{2}$

Write the slope of the line that contains the two points. (Section 13.4)

13. (2, 3);(4, 1)   **14.** (−4, 0);(3, −2)   **15.** (−2, −1);(3, −1)   **16.** (−7, −4);(−5, −1)
−1   −$\frac{2}{7}$ 0 $\frac{3}{2}$

*Linear Functions / **363***

Quiz Sections 13.1–13.4
After completing this Mid-Chapter Review, you may want to administer a quiz covering the same sections. See the quiz provided on page 271 in the *Teacher's ResourceBank.*

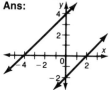
OBJECTIVES: Given the rules of two linear functions with equal slopes, to graph the functions in the same coordinate plane
Given the rules of three linear functions with equal slopes, to graph one function using three ordered pairs and to graph the other two functions from the first graph using y intercepts and the meaning of slope
To graph a linear function given its rule in the form $y = b$

13.5 Parallel Lines

EXAMPLE 1 Graph in the same coordinate plane.

1. $y = 2x - 3$ **2.** $y = 2x + 1$

Solutions: $y = 2x - 3$ $y = 2x + 1$

x	y
−1	−5
0	−3
2	1

x	y
−2	−3
0	1
2	5

Figure 1

P–1   **What appears to be true about the two lines of Figure 1?**   They appear to be parallel.

Lines **1** and **2** of Figure 1 are <u>parallel</u>.   Explain that you can't tell just by looking at the graphs of two lines that they are parallel.

Definition   **Parallel lines** are two lines in the same plane that do not have a common point.

Lines **1** and **2** also have the same slope.

If two lines have the same slope, they are parallel.

P–2   **Which functions described below have parallel graphs?** a and c

a. $y = -\frac{1}{2}x - 5$ **b.** $y = 2x - 5$ **c.** $y = 4 - \frac{1}{2}x$

EXAMPLE 2   Write the slope-intercept form of the function that has its graph parallel to the graph of $y = 2x - 5$ and that contains $(0, 3)$.

Solution:

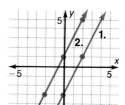

$m = 2$
$b = 3$ $y = mx + b$

$y = 2x + 3$ ◀ **Slope-intercept form**

EXAMPLE 3   Graph the three functions. Use three points for the
first line. Graph the other two lines using y intercepts
and the meaning of slope.

  1. $y = -\frac{1}{2}x + 1$ **2.** $y = -\frac{1}{2}x - 3$ **3.** $y = -\frac{1}{2}x$

Solution:

1. Draw the graph of $y = -\frac{1}{2}x + 1$.

x	-4	0	4
y	3	1	-1

2. Locate $(0, -3)$. Move 1 unit up
and 2 units left to locate $(-2, -2)$.
Draw the graph of $y = -\frac{1}{2}x - 3$.

3. Locate $(0, 0)$. Move 1 unit up and
2 units left to locate $(-2, 1)$.
Draw the graph of $y = -\frac{1}{2}x$.

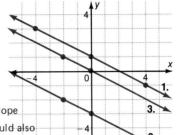

Think of the slope
as $\frac{+1}{-2}$. You could also
think of it as $\frac{-1}{+2}$.

P–3   **What is the value of $0 \cdot x$ for any value of x?**   0

EXAMPLE 4   Graph the three functions.

1. $y = 4$
2. $y = 2$
3. $y = -1$

  1. $y = 0 \cdot x + 4$
  2. $y = 0 \cdot x + 2$
  3. $y = 0 \cdot x - 1$

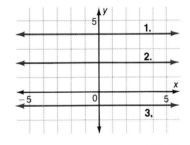

Solution:

All the rules are of the form
$y = 0 \cdot x + b$ or simply $y = b$. The
three graphs are parallel to
the x axis.

CLASSROOM EXERCISES

Write _Yes_ or _No_ to tell whether the graphs of each pair of functions are
parallel. (P-1, P-2)

1. $y = \frac{1}{2}x + 2$ **2.** $y = 5 - 2x$ **3.** $y = 1$ **4.** $y = \frac{3}{4}x$ No
   $y = 0.5x + 3$ Yes $y = 5x - 2$ No $y = 3$ Yes $y = -\frac{3}{4}x$

Linear Functions   /   365

Additional Examples
Example 3
Graph the three functions in the
same coordinate plane. Use
three points for the first line.
Graph the other two lines using
the y intercepts and the meaning
of slope.
$y = -x + 2; y = -x - 4;$
$y = -x$
Ans:

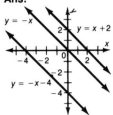

Example 4
Graph the functions in the same
coordinate plane.
$y = 0 \cdot x - 3; y = 0 \cdot x + 2$
Ans:

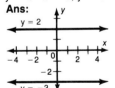

Write the slope-intercept form of each function. (Example 2)

5. Its graph is parallel to the graph of $y = 4x - 3$ and contains (0, 3). $y = 4x + 3$

6. Its graph is parallel to the graph of $y = -5x + 2$ and contains (0, -3). $y = -5x - 3$

7. Its graph is parallel to the graph of $y = \frac{3}{2}x - 1$ and contains (0, 10). $y = \frac{3}{2}x + 10$

8. Its graph is parallel to the graph of $y = -\frac{1}{4}x + \frac{1}{2}$ and contains $(0, \frac{5}{2})$. $y = -\frac{1}{4}x + \frac{5}{2}$

9. Its graph is parallel to the graph of $y = -5.3x + 2.5$ and contains (0, 0.6).

$$y = -5.3x + 0.6$$

▓▓▓ WRITTEN EXERCISES ▓▓▓

In Ex. 1-6 each graph is a pair of parallel lines. Two points are given for each line.

Goal: To use slopes to identify and graph parallel lines

Sample Problem: $y = -x$, $y = \frac{1}{2} - x$, $y = -x + 7$

Answer: Each slope: -1; lines are parallel.   Graphs:

Graph each pair of functions in the same coordinate plane (Example 1)

1. $y = -2x + 3$ (0,3), (1,1)   (0, 4), (-3, 1)
$y = -2x - 1$ (0,-1), (-1,1)

2. $y = x + 4$
$y = x - 3$ (0,-3), (2,-1)

(0, -2), (4, 4)

3. $y = \frac{3}{2}x - 2$
$y = \frac{3}{2}x + 1$
(0, 1), (-2, -2)

4. $y = -\frac{1}{2}x$ (0,0), (-2, 1)
$y = -\frac{1}{2}x - 4$ (0,-4), (-2,-3)

5. $y = \frac{5}{4}x + 3$(0,3,), (4,8)
$y = \frac{5}{4}x$ (0, 0), (4, 5)

6. $y = -\frac{3}{2}x + 2$ (0, 2), (2, -1)
$y = -\frac{3}{2}x - 5$ (0, -5), (2, -8)

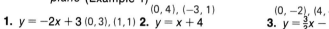

Graph the three functions in the same coordinate plane. Use three points for the first line. Graph the other two lines using the y intercepts and the meaning of slope. (Example 3)

7. $y = 2x - 3$
$y = 2x + 3$
$y = 2x$

8. $y = 5x + 1$
$y = 5x - 5$
$y = 5x$

9. $y = x + 3$
$y = x - 3$
$y = x$

10. $y = -x + 2$
$y = -x - 3$
$y = -x$

The graphs of Ex. 11-14 are horizontal lines. One point is given.

Graph the functions in the same coordinate plane. (Example 4)

11. $y = 5$ (0, 5)

12. $y = -2$ (0, -2)

13. $y = \frac{3}{2}$ $(0, \frac{3}{2})$

14. $y = -\frac{3}{2}$ $(0, -\frac{3}{2})$

MORE CHALLENGING EXERCISES

15. If $a = b$, are the graphs of the functions described by $ax + by = c$ and $bx + ay = d$ parallel? Explain.

16. What value of k would make the graph of $6x + 2y = 5$ and the graph of $4x + ky = 3$ parallel? $\frac{4}{3}$

REVIEW CAPSULE FOR SECTION 13.6

Solve for x. (Pages 64–66)

1. $0 = 3x + 6$ $x = -2$

2. $0 = \frac{1}{2}x - 1$ $x = 2$

3. $0 = -\frac{3}{4}x + 9$ $x = 12$

4. $0 = -3x + \frac{9}{4}$
$x = \frac{3}{4}$

OBJECTIVES: To solve for the x intercept of a line using the slope-intercept form of the rule for its corresponding linear function
To graph a linear function given the slope-intercept form of its rule and using only the x and y intercepts
To graph a linear function given its rule in the form $x = k$

13.6 **X Intercept**

P–1 **What are the coordinates of the point common to the graph of Figure 1 and the x axis?** $(-4, 0)$

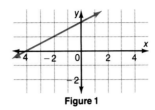

Figure 1

Any point on the x axis has a y value equal to 0.

Definition

The x value of a point common to the x axis and a straight line is called the **x intercept** of the line.

In Figure 1, the x intercept $= -4$.

EXAMPLE 1   Write the x intercept of the graph of $y = -\frac{1}{2}x + 1$.

Solution:

$y = -\frac{1}{2}x + 1$ ◄ *Replace y with 0.*

$0 = -\frac{1}{2}x + 1$

$-1 = -\frac{1}{2}x$ ◄ *Solve for x.*

$-2(-1) = -2(-\frac{1}{2}x)$

$2 = x$ ◄ *The x intercept is 2.*

P–2 **What is the x intercept of the graph of each function below?**

a. $y = x - 5$ 5   **b.** $y = x + \frac{1}{2}$ $-\frac{1}{2}$ **c.** $y = 2x + 8$ -4

EXAMPLE 2   Graph $y = \frac{3}{2}x - 3$ by using the x and y intercepts.

Solution:

$y = \frac{3}{2}x - 3$

$0 = \frac{3}{2}x - 3$ ◄ *Let y = 0.*

$3 = \frac{3}{2}x$

$\frac{2}{3}(3) = \frac{2}{3}(\frac{3}{2}x)$

$2 = x$ if $y = 0$.

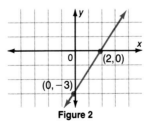

Figure 2

The y intercept is -3.

The line is drawn through $(2, 0)$ and $(0, -3)$.

Linear Functions / **367**

Teaching Suggestions p. M–42

Lesson Resources
Warm–Up: p. M–43
Maintenance: See below.
Practice Worksheet 13.6
Transparency 32

Maintenance
1. Solve and check:
$\frac{m}{12} - 4 = 18$
Ans: 264   (Section 3.3)
2. Write as one power: $\frac{x^{-3} \cdot x^4}{x^{-2}}$
Ans: x^3   (Section 6.3)
3. Subtract:
$(3y - 6) - (8 - 9y)$
Ans: 12y − 14
(Section 8.4)
4. Multiply and simplify:
$\frac{3x}{x^2 - 4} \cdot \frac{x^2 + x - 2}{6x^2}$
Ans: $\frac{x - 1}{2x^2 - 4x}$
(Section 10.6)

5. Write the relation described by $y = 3x + 4$ as a set of ordered pairs if the x values are -2, 0, and 1.
Ans: {(−2, −2), (0,4), (1, 7)}   (Section 12.3)

Additional Examples
Example 1
Write the x intercept of the graph of each function.
1. $y = \frac{1}{3}x - 5$
Ans: 15 = x
2. $y = 3x + 12$
Ans: −4 = x

Example 2
Graph $y = -\frac{1}{2}x + 1$ by using the x and y intercepts.
Ans:

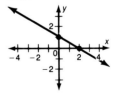

Figure 3 shows the graph of a linear relation that is not a function.

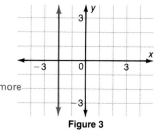

Figure 3

P-3 **What is the domain of the relation?** $\{-2\}$

P-4 **Why is the relation not a function?** There is more than one y value for the x value, -2.

Let (x, y) represent the coordinates of any point on the vertical line of Figure 3. Then x equals -2 for any value of y.

$$x = -2 \blacktriangleleft \quad \textbf{\textit{Rule of}}\\ \textbf{\textit{the relation}}$$

> The rule of a relation having a vertical line as its graph is $x = k$ in which k is the x intercept of the line.

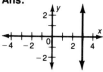
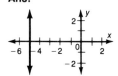
EXAMPLE 3   Graph $x = 2.7$.

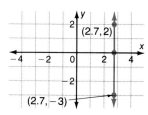

Figure 4

Solution: The rule is of the form $x = k$. The graph is a vertical line 2.7 units to the right of the origin.

The ordered pairs (2.7, 2) and (2.7, −3) are coordinates of two points on the vertical line in Figure 4. Use these points to determine the slope of the line.

$$\frac{y_2 - y_1}{x_2 - x_1} = \frac{-3 - 2}{2.7 - 2.7}$$

$$= \frac{-5}{0}$$

Since you cannot divide by 0, $\frac{-5}{0}$ is not a number.

This is a difficult concept for students. Remind them that slope is a number.

> A vertical line does not have a slope.

368 / Chapter 13

CLASSROOM EXERCISES

Write each x intercept. (Example 1)

1. $y = x + 3$ −3
2. $y = x - 1$ 1
3. $y = 2x - 2$ 1
4. $y = 3x - 5$ $\frac{5}{3}$

5. $y = 2x + 4$ −2
6. $y = x - \frac{1}{2}$ $\frac{1}{2}$
7. $y = 2x - 5$ $\frac{5}{2}$
8. $y = 8x$ 0

9–14. *Write the x intercept of each line in Figure 5.* (P-1)

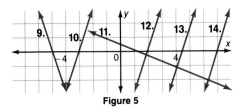

Figure 5

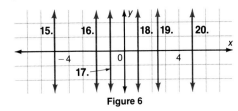

Figure 6

15–20. *Give a rule for each relation graphed in Figure 6.* (Example 3)

WRITTEN EXERCISES

Goal: To graph a linear function using x and y intercepts
Sample Problem: $y = \frac{3}{2}x - 3$ **Answer:** See Example 2.

Write each x intercept. (Example 1)

1. $y = x + 6$ −6
2. $y = x + 10$ −10
3. $y = x - 3$ 3
4. $y = x - 7$

5. $y = 3x + 6$ −2
6. $y = 2x + 10$ −5
7. $y = 4x - 12$ 3
8. $y = 3x - 15$

9. $y = \frac{1}{2}x - 2$ 4
10. $y = \frac{1}{2}x - 5$ 10
11. $y = \frac{1}{4}x + 1$ −4
12. $y = \frac{1}{4}x + 2$

In Ex. 13-20 each graph is a straight line. Two ordered pairs are given for each line, one using the x intercept, the other using the y intercept.

Graph each function by using only the x and y intercepts. (Example 2)

13. $y = 2x - 4$ (2,0),(0,−4)
14. $y = 3x + 6$ (−2,0)(0,6)
15. $y = \frac{1}{2}x - 3$ (6,0),(0,−3)
16. $y = \frac{1}{3}x - 2$

17. $y = 2x + 6$ (−3,0),(0,6)
18. $y = 3x - 3$ (1,0),(0,−3)
19. $y = 2x - 5$ ($\frac{5}{2}$,0),(0,−5)
20. $y = 2x + 5$

In Ex. 21-24 the graphs are vertical lines. One point is given for each line.

Graph each linear relation. (Example 3)

21. $x = 5$ (5, 0)
22. $x = -4$ (−4, 0)
23. $x = 2\frac{1}{2}$ ($2\frac{1}{2}$, 0)
24. $x = -3\frac{1}{2}$ ($-3\frac{1}{2}$, 0)

REVIEW CAPSULE FOR SECTION 13.7

For each formula, write the slope and y intercept. (Pages 356–358)

1. $y = 2x + 3$
2. $F = 1.8C + 32$
3. $c = \pi d$
4. $K = C + 273$
5. $s = 3 + 2.5t$

Linear Functions / **369**

Common Error
Students incorrectly assume that the point where x is zero is the x intercept and the point where y is zero is the y intercept.

Example
$y = -2x + 4$
x intercept: (0,4)
y intercept: (2,0)

Prescription
Remind students that the x intercept is on the x axis and the y intercept is on the y axis. Have students graph some of the functions in Classroom Exercises 1–8. Identifying the coordinates of the points where the line intersects the axes should help clarify how to find the intercepts.

Additional Answers
Classroom Exercises
9. −5
10. −3
11. 1
12. 2
13. $4\frac{1}{2}$
14. 7
15. $x = -5$
16. $x = -2$
17. $x = -1$
18. $x = 1$
19. $x = 2\frac{1}{2}$
20. $x = 5$

Assignment Guide
Basic p. 369: 1–23 odd
Average p. 369: 1–23 odd
Above Average p. 369:
1–23 odd

Additional Answers
Written Exercises
4. 7
8. 5
12. −8
16. (6, 0), (0, −2)
20. ($-\frac{5}{2}$, 0), (0, 5)

Review Capsule
1. slope: 2; y intercept: 3
2. slope: 1.8; y intercept: 32
3. slope: π; y intercept: 0
4. slope: 1; y intercept: 273
5. slope: 2.5; y intercept: 3

OBJECTIVES: To graph a linear function representing a practical application given a description of the application and a rule for the function

PROBLEM SOLVING AND APPLICATIONS

13.7 Using Graphs: Estimation

To write an approximate value for one coordinate of a point given a graph of a linear function and the value of the other coordinate

Temperature in Fahrenheit degrees is a function of temperature in Celsius degrees.

$$F = 1.8C + 32$$

◀ **Compare $F = 1.8C + 32$ with $y = 1.8x + 32$. The function is linear.**

EXAMPLE 1

a. Graph $F = 1.8C + 32$.

b. Use the graph to estimate the value of F when $C = 10$.

Solutions:

a. The ordered pairs of this function are represented by (C, F).

C	F
-20	-4
0	32
20	68

b. From the graph, the value of F is approximately 50 when $C = 10$.

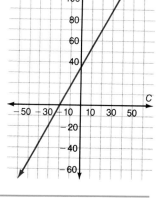

Figure 1

The distance, d, in meters that a car traveling at r kilometers per hour will cover in a certain fraction of a second is given by the function rule $d = 0.28r$.

EXAMPLE 2

a. Graph $d = 0.28r$.

b. Use the graph to estimate the value of d when $r = 60$.

◀ **Compare with $y = mx + b$. m is 0.28; b is 0.**

Solutions:

a. The ordered pairs of this function are represented by (r, d).

r	d
10	2.8
30	8.4
50	14

b. From the graph, the value of d is approximately 17 when $r = 60$.

Check the estimate by substituting in the equation.

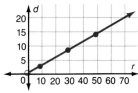

Figure 2

The circumference of a circle is a function of the length of a diameter.

$$c = \pi d$$

Compare with
$$y = mx + b.$$

EXAMPLE 3

a. Graph $c = \pi d$. The domain is {positive real numbers}. ($\pi \approx 3.14$)

b. Use the graph to estimate the value of c when $d = 0.5$.

Solutions: **a.** The ordered pairs of this function are represented by (d, c).

d	c
1	3.14
2	6.28
3	9.42

b. From the graph, the value of c is approximately 1.6 when $d = 0.5$.

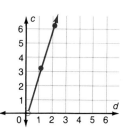

Figure 3

Additional Examples
Example 3
Refer to the graph of Figure 3. Estimate the diameter corresponding to each circumference.
1. 2.5 cm
    Ans: approximately 0.9 cm
2. 4.5 cm
    Ans: approximately 1.5 cm

The simple interest on money in savings for one year at 5% is a function of the principal.

Rule: $i = 0.05p$

$i = interest$
$p = principal$

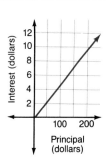

P–1 **Use the graph to estimate the simple interest for one year on a principal of \$175.** \$9

▒▒▒▒ CLASSROOM EXERCISES ▒▒▒▒

Refer to the graph of Figure 1. Estimate the Celsius temperature for each Fahrenheit temperature. (Example 1)

1. 100°F
    38°C
2. 60°F
    16°C
3. 20°F
    −7°C
4. 0°F
    −18°C
5. −20°F
    −29°C
6. −60°F
    −50°C

Refer to the graph of Figure 2. Estimate the distance traveled at each speed. (Example 2)

7. 20 km/hr
    6 meters
8. 35 km/hr
    10 meters
9. 25 km/hr
    7 meters
10. 40 km/hr
    11 meters

Linear Functions / **371**

Assignment Guide
Basic p. 372: 1–17 odd
Average p. 372: 1–18
Above Average p. 372: 1–18

Problem–Solving Skills
Using a formula (Ex. 1–4)
Reading a graph (Ex. 5–18)
Using estimation (Ex. 5–18)

Additional Answers
1. See Example 2 on page 357 ($y = 2x$). This graph is similar except that only the portion in the first quadrant is drawn.
2. The graph is a portion of a line beginning at (0, 0) and through (10, 5). The graph is in Quadrant I only.
3. The graph is a portion of a line and contains the points (−273, 0) and (0, 273). The graph is in Quadrants I and II only.
4. The graph is a portion of a line and contains the points (0, 3) and (6, 18). The graph is in Quadrant I only.

Quiz Sections 13.5–13.7
After completing Sections 13.5–13.7, you may want to administer a quiz covering the same sections. See the quiz provided on page 272 in the *Teacher's ResourceBank*.

▬▬ WRITTEN EXERCISES ▬▬

Goal: To graph linear functions that represent practical applications and to estimate values from the graphs

Sample Problem: Use the graph of Figure 2 to estimate the distance traveled at a speed of 55 km/hr.

Answer: 15 meters

Graph each function as described. Unless otherwise stated the domain is {nonnegative real numbers}. (Examples 1–3)

1. The distance in millimeters that a coil spring stretches when w grams are attached is a function with this rule.
   Rule: $d = 2w$

2. The simple interest on money in savings for ten years at 5% is a function of the principal.
   Rule: $i = 10(0.05)p$ or $i = 0.5p$

3. The temperature in Kelvin degrees is a function of temperature in Celsius degrees.
   Rule: $K = C + 273$
   The domain is {real numbers greater than −273}.

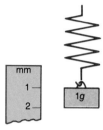

4. In an auto race one racer has a 3 kilometer head start. His speed is 2.5 kilometers per minute. His total distance is a function of the number of minutes he drives.
   Rule: $s = 3 + 2.5t$

Refer to the graph of Figure 1. Estimate the Fahrenheit temperature for each Celsius temperature. (Example 1)

5. 20°C	6. 30°C	7. 40°C	8. −10°C	9. −30°C	10. −50°C
68°F	86°F	104°F	14°F	−22°F	−58°F

Refer to the graph of Figure 2. Estimate the speed for each distance traveled. (Example 2)

11. 5 m	12. 10 m	13. 15 m	14. 20 m
18 km/hr	36 km/hr	54 km/hr	71 km/hr

Refer to the graph of Figure 3. Estimate the diameter corresponding to each circumference. (Example 3)

15. 2 cm	16. 3 cm	17. 4 cm	18. 5 cm
0.6 cm	1 cm	1.3 cm	1.6 cm

Using Formulas

Gravity

Gravity is the force that causes an object to fall when it is released above the ground.

The speed of a falling object starting from rest increases as the object falls. The table at the right shows the object's speed in meters per second after each of the first five seconds. This increase in speed, or <u>acceleration</u>, is due to gravity. The **acceleration due to gravity**, represented by "g," is constant on and near the earth. It has a value of 9.8 meters per second per second.

Time (seconds)	Speed (meters/second)
1	9.8
2	19.6
3	29.4
4	39.2
5	49.0

The following formula computes the distance s that a falling object, starting from rest, travels in a length of time t, with acceleration g.

$$s = \tfrac{1}{2}gt^2$$

EXAMPLE: Find the approximate distance in meters a parachutist would fall after jumping from a plane and free-falling for 12 seconds before opening the chute.

SOLUTION: $s = \tfrac{1}{2}gt^2$

$s = \tfrac{1}{2}(9.8)(12)^2$ $g = 9.8$ $t = 12$

$= 4.9(144)$

$= 705.6$ or about **706 meters**

Skydivers fall freely at speeds of more than 160 kilometers per hour.

This is another special topic that features a formula as a problem-solving strategy. For the formula, $s = \tfrac{1}{2}gt^2$, the value of g is given in meters per second. Students are to compute the distance a freely falling object will travel in a given amount of time.

EXERCISES

1. A plane releases its auxiliary fuel tanks at a low altitude. The tanks take 15 seconds to reach the ground. At what altitude were they released? 1102.5 meters

2. A helicopter drops bales of hay to feed snowbound cattle. Each bale takes 5 seconds to reach the ground. At what height were the bales of hay released? 122.5 meters

3. A rock dropped from the Royal Gorge Bridge in Colorado takes about 8 seconds to reach the river below. Find the approximate height of the bridge above the river to the nearest meter.
314 meters

4. A ball dropped from the observation level of the Washington Monument reaches the ground in about $5\tfrac{1}{2}$ seconds. Find the height to the nearest meter. 148 meters

Linear Functions / **373**

CHAPTER SUMMARY

IMPORTANT TERMS	Linear function *(p. 351)* Ratio *(p. 359)* *y* intercept *(p. 353)* Parallel lines *(p. 364)* Slope *(p. 356)* *x* intercept *(p. 367)* Slope-intercept form *(p. 356)*

IMPORTANT IDEAS

1. The rule for a linear function is $y = mx + b$.

2. In the rule for the linear function $y = mx + b$, b represents the y intercept of the graph.

3. In the rule for the linear function $y = mx + b$, m represents the slope of its graph.

4. The slope of a line is the ratio of the vertical motion to the horizontal motion from one point to another on the line.

5. *Slope Formula:* If (x_1, y_1) and (x_2, y_2) are coordinates of two points on a nonvertical line, then the slope m equals $\frac{y_2 - y_1}{x_2 - x_1}$.

6. If two lines have the same slope, they are parallel.

7. The rule of a relation having a vertical line as its graph is $x = k$ in which k is the x intercept of the line.

8. A vertical line does not have a slope.

Chapter Test

Two Chapter Tests (Form A and Form B) are provided on pages 273–276 in the *Teacher's ResourceBank.*

CHAPTER REVIEW

SECTION 13.1

Write Yes or No to show whether each function is linear.

1. $y = x^2 + 2x$ No

2. $y = 5 - 7x$ Yes

3. $y = \frac{1}{2}x + \frac{3}{4}$ Yes

4. $y = |3x| - 2$ No

Graph at least four ordered pairs of each function. Then draw the graph with {real numbers} as the domain.   Each graph is a straight line.
The coordinates of four points are given for each.

5. $y = -2x + 1$
    (0, 1), (2, −3), (−1, 3), (1, −1)

6. $y = \frac{3}{2}x - 3$
    (0, −3), (2, 0), (−2, −6), (4, 3)

SECTION 13.2

Write each y intercept.

7. $y = 3x - 10$ −10

8. $y = -4x + 7$ 7

9. $y = -0.7x + 3.8$ 3.8   **10.** $y = 2.4x - 1.3$ -1.3

SECTION 13.3

Write the slope-intercept form of the rule for each function.

11. The slope is 4 and the *y* intercept is -1. $y = 4x - 1$

12. The slope is -2 and the *y* intercept is 5. $y = -2x + 5$

13. The line crosses the *y* axis at (0, 2.9) and has slope -2.8. $y = -2.8x + 2.9$

14. The line crosses the *y* axis at (0, -5.3) and has slope 4.9. $y = 4.9x - 5.3$

Name the quadrants in which each graph lies.

15. $y = -7x$ II, IV   **16.** $y = 4x$ I, III   **17.** $y = \frac{5}{2}x$ I, III   **18.** $y = -0.3x$
                                                                                   II, IV

Write the slope of the graph of each function.

19. $y = -\frac{1}{2}x + 2$ $-\frac{1}{2}$ **20.** $y = \frac{5}{4}x - 3$ $\frac{5}{4}$

21. $y = 0.8x - 1.2$ 0.8   **22.** $y = -0.3x$ -0.3

SECTION 13.4

Write the slope of the line that contains the two points.

23. (4, -3); (-1, 7) -2 **24.** (-2, 4); (-1, -2) -6

25. $(-\frac{1}{2}, \frac{5}{2})$; $(\frac{3}{2}, -\frac{7}{2})$ -3 **26.** $(-\frac{3}{4}, -\frac{1}{4})$; $(\frac{5}{4}, -\frac{3}{4})$ $-\frac{1}{4}$

Write the slope-intercept form of the line that contains the two points.

27. (0, -2); (3, 4) $y = 2x - 2$ **28.** (0, 3); (-4, -2) $y = \frac{5}{4}x + 3$

29. $(0, -\frac{5}{2})$; $(4, \frac{1}{2})$ $y = \frac{3}{4}x - \frac{5}{2}$ **30.** $(0, \frac{3}{2})$; $(-2, \frac{1}{2})$ $y = \frac{1}{2}x + \frac{3}{2}$

In Ex. 31 and 32, each graph is a set of three parallel lines.
       SECTION 13.5   Two points are given for each line.

Graph the three functions in the same coordinate plane.

31. $y = \frac{1}{2}x - 2$ (0, -2), (4, 0)   **32.** $y = x - 5$ (0, -5), (2, -3)
      $y = \frac{1}{2}x + 3$ (0, 3), (4, 5) $y = x + 2$ (0, 2), (2, 4)
      $y = \frac{1}{2}x$ (0, 0), (4, 2) $y = x$ (0, 0), (2, 2)

Linear Functions   /   **375**

In Ex. 33-36 each graph is a horizontal line. One point is given.
Graph the functions in the same coordinate plane.

33. $y = -3$ (0, -3)   **34.** $y = 4$ (0, 4)   **35.** $y = 3.5$ (0, 3.5)   **36.** $y = -\frac{5}{2}$

(0, $-\frac{5}{2}$)

SECTION 13.6

Write each x intercept.

37. $y = -2x - 5$ $-2\frac{1}{2}$ **38.** $y = 3x + 8$ $-2\frac{2}{3}$

39. $y = -\frac{1}{2}x + 4$ 8   **40.** $y = \frac{1}{4}x - 3$ 12

Graph each function by using only the x and y intercepts.
Each graph is a straight line. Two points are given, one using the x intercept, the other the y intercept.
41. $y = x + 1$ (-1, 0), (0, 1)   **42.** $y = -x + 1$ (1, 0), (0, 1)

43. $y = 3x - 6$ (2, 0), (0, -6)   **44.** $y = -2x + 4$ (2, 0), (0, 4)

Graph each linear relation.
Each graph is a vertical line. One point is given for each line
45. $x = -1$ **46.** $x = 2$ **47.** $x = \frac{7}{2}$ **48.** $x = -\frac{3}{2}$

(-1, 0)   (2, 0)   ($\frac{7}{2}$, 0)   ($-\frac{3}{2}$, 0)

SECTION 13.7

Use the graph of Figure 2 on page 370 to estimate the distance in meters that a car will travel at each given speed. Compare your estimate with the answer you obtain using the function rule $d = 0.28r$.

49. 70 km/hr   **50.** 15 km/hr   **51.** 45 km/hr   **52.** 65 km/hr

19.6 meters   4.2 meters   12.6 meters   18.2 meters

Systems of Sentences

The use of systems of sentences is a valuable problem-solving tool for solving a variety of problems that deal with such things as money, mixture, distance/rate/time, and digit problems.

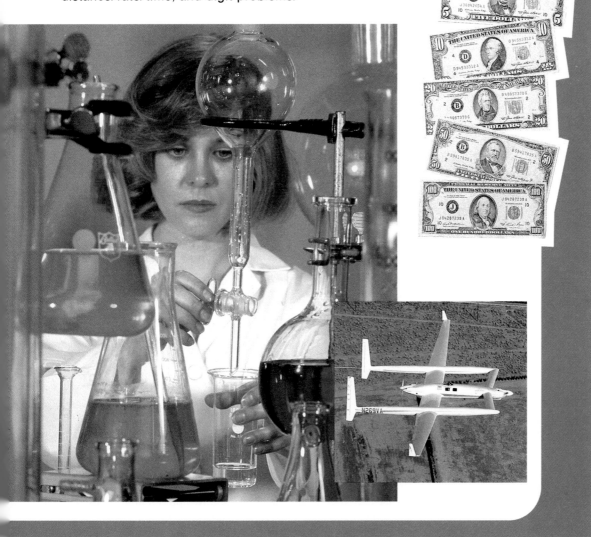

Background
The bottom right photo shows the airplane Voyager during its flight around the world. In December, 1986, Dick Rutan and Jeanna Yeager flew the Voyager approximately 26,000 miles around the world without landing or refueling.

Teaching Suggestions p. M–44

Lesson Resources
Warm–Up: p. M–45
Maintenance: See below.
Practice Worksheet 14.1

Maintenance
1. Simplify:
   $16k - 4t + 12k + 3t$
   Ans: $28k - t$
   (Section 2.6)
2. Solve: $4(z - 3) > 3z + 6$
   Ans: $z > 18$ (Section 5.4)
3. Simplify: $\sqrt{81r^4t^5}$
   Ans: $9r^2t^2\sqrt{t}$
   (Section 7.4)
4. Factor: $4x^2y - 64y$
   Ans: $4y(x - 4)(x + 4)$
   (Section 9.7)
5. Solve: $\dfrac{5}{y} - \dfrac{19}{2y} = 3$

   Ans: $y = -\dfrac{3}{2}$
   (Section 11.7)

Additional Examples
Example 1
Graph each function.
1. $x + y = 4$
Ans:

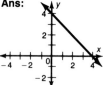

2. $y - x = 3$
Ans:

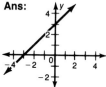

14.1 Graphs

OBJECTIVE: To graph the solution set of a linear equation

An equation such as $y = -2x + 3$ is called a ***linear equation*** in two variables. When you graph a linear equation, you graph its solution set. The ***solution set of a linear equation*** is the set of ordered pairs (x, y) that make the equation true.

EXAMPLE 1   Graph $x - y = -2$

"Graph" means "draw the graph of."

| Determine at least three ordered pairs. | → | Graph the ordered pairs. | → | Draw the line. |

Solution: It is usually easy to find the value of y when x equals 0. Similarly, it is usually easy to find the value of x when y equals 0.

Let $x = 0.$ Let $y = 0.$ Let $x = 2.$
$0 - y = -2$ $x - 0 = -2$ $2 - y = -2$
    $y = 2$ $x = -2$ $y = 4$

x	y
0	2
-2	0
2	4

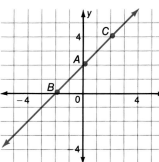

The solution set of $x - y = -2$ is the set of coordinates of points on the line containing points A, B and C. Thus the solution set is infinite.

P–1   **What is the y value of point A called?   the x value of point B?**
                    The y intercept   The x intercept

The intercepts are often used to locate two points on a graph. Other x or y values with small absolute values are then selected.

378 / *Chapter 14*

EXAMPLE 2   Graph $2x + 2y = 5$.

Solution:

$2(0) + 2y = 5$ **Let** $x = 0$. $2x + 2(0) = 5$ 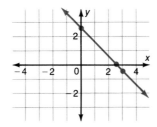 **Let** $y = 0$. $2(3) + 2y = 5$ **Let** $x = 3$.

$\quad\quad 2y = 5$ $\quad\quad 2x = 5$ $\quad\quad 6 + 2y = 5$

$\quad\quad y = 2\frac{1}{2}$ $\quad\quad x = 2\frac{1}{2}$ $\quad\quad\quad 2y = -1$

$\quad\quad\quad\quad\quad\quad\quad\quad\quad\quad\quad\quad\quad\quad\quad\quad\quad\quad\quad\quad y = -\frac{1}{2}$

x	y
0	$2\frac{1}{2}$
$2\frac{1}{2}$	0
3	$-\frac{1}{2}$

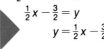

P–2 **Refer to the graph to determine what values of x will result in negative values for y.** $x > 2\frac{1}{2}$
What values of x will give positive values for y? $x \le 2\frac{1}{2}$

EXAMPLE 3   Graph $x - 2y = 3$. Use the slope-intercept form.

Solution:

$x - 2y = 3$

$\quad x = 3 + 2y$

$\quad x - 3 = 2y$

$\quad \frac{1}{2}(x - 3) = y$

$\quad \frac{1}{2}x - \frac{3}{2} = y$ **Slope-intercept form**

$\quad y = \frac{1}{2}x - \frac{3}{2}$

Let $x = 0$. $y = \frac{1}{2}(0) - \frac{3}{2}$

$\quad\quad\quad\quad\quad\quad\quad y = -\frac{3}{2}$

Let $y = 0$. 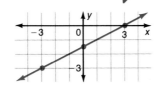 $0 = \frac{1}{2}x - \frac{3}{2}$

$\quad\quad\quad\quad\quad\quad\quad 3 = x$

Let $x = -3$. $y = \frac{1}{2}(-3) - \frac{3}{2}$

$\quad\quad\quad\quad\quad\quad\quad y = -3$

x	y
0	$-\frac{3}{2}$
3	0
-3	-3

Additional Examples
Example 2
Graph each function.
1. $2x - 5y = 10$
Ans:
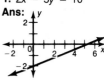

2. $x + 2y = 1$
Ans:
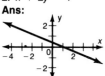

Example 3
Graph by using the slope-inter-cept form.
1. $2x + y = 4$
Ans:
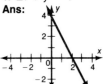

2. $5x - 3y = -9$
Ans:
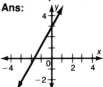

379

Common Error

Students select only two values for x when graphing a linear function and incorrectly solve for y using one or both values of x.

Example

Graph this function.

$y = x - 3$

x	y
0	3
4	1

Prescription

Insist that students plot at least three points. Show them why just two points can contain an incorrect graph.

Assignment Guide

Basic p. 380: 1–17 odd
Average p. 380: 1, 7, 9–17, odd; p. 384: 3, 6, 9, ···, 18, 19, 22
Above Average p. 380: 1, 7, 9–17 odd; p. 384: 3, 6, 9, ···, 18, 19–22

Additional Answers
Written Exercises

In Ex. 1–8, each graph is a straight line. Three ordered pairs are given from Ex. 1–8 of the Classroom Exercises.

1. (0, 2), (2, 0), (4, −2)
2. (0, 1), (−1, 0), (2, 3)
3. (−2, −1), (0, 0), (2, 1)
4. (0, 3), ($\frac{3}{2}$, 0), (−1, 5)
5. (0, −5), (5, 0), (2, −3)
6. (0, 2), (3, 0), (6, −2)
7. (0, −$\frac{5}{3}$), (5, 0), (−4, −3)
8. (0, 3), (−$\frac{3}{2}$, 0), (−4, −5)
11. (−2, −3), (2, −2)
14. (0, 0), (3, 5)
17. (0, −6), (2, 0)

CLASSROOM EXERCISES

Write the unknown values in each table. (Examples 1–3)

1. $x + y = 2$

		2	
x	0	?	4
y	?	0	?
	2		−2

2. $x - y = -1$

		−1	2
x	0	?	?
y	?	0	3
	1		

3. $x - 2y = 0$

		0	2
x	−2	?	?
y	?	0	1
	−1		

4. $2x + y = 3$

		$\frac{3}{2}$	
x	0	?	−1
y	?	0	?
	3		5

5. $x - y = 5$

		5	
x	0	?	2
y	?	0	?
	−5		−3

6. $2x + 3y = 6$

		3	6
x	0	?	?
y	?	0	−2
	2		

7. $x - 3y = 5$

		5	
x	0	?	−4
y	?	0	?
	−$\frac{5}{3}$		−3

8. $y - 3 = 2x$

		−$\frac{3}{2}$	
x	0	?	−4
y	?	0	?
	3		−5

WRITTEN EXERCISES

Goal: To graph the solution sets of linear equations
Sample Problem: $x + 3y = -3$
Answer: See the graph at the right.

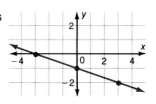

Graph each of the following. Use the tables in the Classroom Exercises. (Examples 1–3)

1. $x + y = 2$ **2.** $x - y = -1$ **3.** $x - 2y = 0$ **4.** $2x + y = 3$

5. $x - y = 5$ **6.** $2x + 3y = 6$ **7.** $x - 3y = 5$ **8.** $y - 3 = 2x$

Graph each of the following. (Examples 1–3) In Ex. 9–17, each graph is a straight line. Two points are given for each line.

9. $3x - 2y = 5$ (1, −1), (3, 2)   **10.** $4x - 2y = -3$ (0, $\frac{3}{2}$)(−$\frac{3}{4}$, 0)   **11.** $4y - x = -10$

12. $5y - 2x = 3$ (−4, −1), (1, 1)   **13.** $4x - 5y = 0$ (0, 0), (5, 4)   **14.** $3y - 5x = 0$

15. $5x + 5y = -5$ (0, −1), (−1, 0)   **16.** $3x - 3y = 3$ (0, −1), (1, 0)   **17.** $y - 3x = -6$

REVIEW CAPSULE FOR SECTION 14.2

Substitute to find one of the following ordered pairs that is in the solution set of both equations. (Pages 337–339)

(−3, 14)   (−3, −2)   (−1, 2)   (3, −10)   (−1, 1)

1. $\begin{cases} 2x - y = -4 \\ x + y = 1 \end{cases}$ (−1, 2)   **2.** $\begin{cases} x + 2y = 1 \\ x - 2y = -3 \end{cases}$ (−1, 1)   **3.** $\begin{cases} 4x + y = 2 \\ -3x - y = -5 \end{cases}$

(−3, 14)

OBJECTIVE: To solve a system of two linear equations by graphing

14.2 Solving Systems by Graphing

Teaching Suggestions p. M–44

Lesson Resources
Warm–Up: p. M–45
Maintenance: See below.
Practice Worksheet 14.2
Exploration 14
Transparencies 33, 34

EXAMPLE 1   Graph both linear equations in the same coordinate plane.

$$\begin{cases} \textbf{1. } x - 3y = 3 \\ \textbf{2. } 2x - y = -4 \end{cases}$$

Solution: Make a table of ordered pairs for each equation.

Emphasize the importance of selecting numbers that are small in absolute value as replacements for the variables.

1.

x	y
0	−1
3	0
6	1

2.

x	y
0	4
−2	0
−1	2

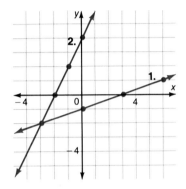

(−3, −2)

Maintenance
1. Write an algebraic expression for this word expression. Use the variable n.
   Fifteen less than twice the number of tennis balls
   Ans: $2t - 15$
   (Section 3.6)
2. Solve: $-\frac{3}{5}x > -18$
   Ans: $x < 30$
   (Section 5.5)
3. Divide: $\frac{8m^2p}{-2mp}$
   Ans: $-4m$ (Section 8.2)
4. Simplify: $\frac{2x + 4}{x^2 + 3x + 2}$
   Ans: $\frac{2}{x + 1}$
   (Section 10.2)
5. Write the relation described by $y = \frac{1}{2}x - 4$ as a set of ordered pairs when the x values are −2, 0, and 1.
   Ans: $\{(-2, -5), (0, -4),$
   $(1, -3\frac{1}{2})\}$ (Section 12.3)

P–1   **What are the coordinates of the common point of the two graphs?**

Two equations such as **1** and **2** in Example 1 form a *system of equations.*

Point out that the solution set contains just one ordered pair.

The solution set of the system of Example 1 is $\{(-3, -2)\}$. This is checked below.

Check: 1. $x - 3y = 3$
              $-3 - 3(-2)$
              $-3 + 6$
              3

2. $2x - y = -4$
         $2(-3) - (-2)$
         $-6 + 2$
         -4

◄ (−3, −2) *makes both equations true.*

Definition   The *solution set of a system of two equations* is the set of ordered pairs that make both equations true.

Additional Example
Example 1
Graph both linear equations in the same coordinate plane.
$$\begin{cases} x + y = 3 \\ x - y = 1 \end{cases}$$
Ans:

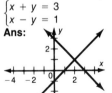

Systems of Sentences   /   381

Additional Examples
Example 2
Solve the system by graphing.
$\begin{cases} \textbf{1.}\ 3x - 2y = 6 \\ \textbf{2.}\ x + y = 2 \end{cases}$
Ans: $\{(2, 0)\}$

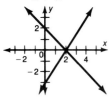

Example 3
Solve the system by graphing.
$\begin{cases} \textbf{1.}\ x - y = 3 \\ \textbf{2.}\ y - x = 2 \end{cases}$
Ans: ϕ

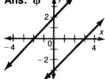

EXAMPLE 2   Solve the system below by graphing. Check.

$$\begin{cases} \textbf{1.}\ x - 2y = 0 \\ \textbf{2.}\ x + y = 3 \end{cases}$$

Solution: Make a table of ordered pairs for each equation.

1.

x	y
0	0
2	1
−4	−2

2.

x	y
0	3
3	0
−1	4

Check the point of intersection, (2, 1).

1. $x - 2y = 0$
      $2 - 2(1)$
      0

2. $x + y = 3$
    $2 + 1$
    3

Solution set: $\{(2, 1)\}$

EXAMPLE 3   Solve the system below by graphing.

$$\begin{cases} \textbf{1.}\ x - 2y = -2 \\ \textbf{2.}\ x - 2y = 4 \end{cases}$$

Solution:

1.

x	y
0	1
−2	0
2	2

2.

x	y
0	−2
4	0
2	−1

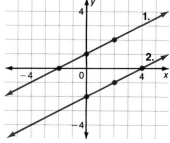

The graphs appear to be parallel. If this is true, the solution set of the system is the empty set, ϕ. To check this, write the equations in slope-intercept form.

1. $x - 2y = -2$

    $x = 2y - 2$

    $x + 2 = 2y$

    $\frac{1}{2}(x + 2) = y$

    $\frac{1}{2}x + 1 = y$

    $y = \frac{1}{2}x + 1$

2. $x - 2y = 4$

    $x = 4 + 2y$

    $x - 4 = 2y$

    $\frac{1}{2}(x - 4) = y$

    $\frac{1}{2}x - 2 = y$

    $y = \frac{1}{2}x - 2$ ◀ **The slope of each line is $\frac{1}{2}$.**

Since the slopes of the lines are equal, the lines are parallel. The solution set of the system is the empty set, ϕ.

CLASSROOM EXERCISES

Write $\underline{Yes}$ or $\underline{No}$ to tell whether each given set is the solution set of the system following it. (Checks of Examples 1 and 2)

1. $\{(1, 1)\};$ $\begin{cases} x + y = 2 \\ x - y = 3 \end{cases}$ No

2. $\{(3, -2)\};$ $\begin{cases} x + y = 1 \\ x - y = -1 \end{cases}$ No

3. $\{(0, 0)\};$ $\begin{cases} x + 2y = 0 \\ 2x - y = 0 \end{cases}$ Yes

4. $\{(0, 1)\};$ $\begin{cases} 2x - y = -1 \\ x + y = 1 \end{cases}$ Yes

5. $\{(0, 1)\};$ $\begin{cases} x - y = 2 \\ 3x - y = -1 \end{cases}$ No

6. $\{(2, -1)\};$ $\begin{cases} x - 2 = 0 \\ y + 1 = 0 \end{cases}$ Yes

7. $\phi;$ $\begin{cases} y = 2x - 1 \\ y = 2x + 3 \end{cases}$ Yes

8. $\{(-1, 5)\};$ $\begin{cases} x + 1 = 0 \\ y = 5 \end{cases}$ Yes

9. $\{(0, \frac{2}{3})\};$ $\begin{cases} 2x + 3y = 2 \\ 2x - 3y = -2 \end{cases}$ Yes

10. $\{(50, 50)\};$ $\begin{cases} x + y = 100 \\ x - y = 0 \end{cases}$ Yes

Write the solution set of each system from its graph. (P-1)

11.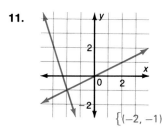

$\{(-2, -1)\}$

12.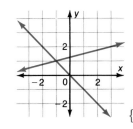

$\{(-1, 1)\}$

13.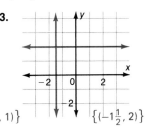

$\{(-1\frac{1}{2}, 2)\}$

Systems of Sentences / 383

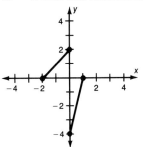

Goal: To solve systems of linear equations by graphing

Sample Problem: Solve by graphing and

check: $\begin{cases} \textbf{1. } 2x - y = 7 \\ \textbf{2. } 5x + 2y = 4 \end{cases}$

Answer: $\{(2, -3)\}$; see the graph at the right.

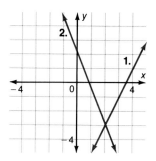

Solve each system by graphing. Check.
(Examples 1–3)

1. $\begin{cases} x + y = 4 \\ x - y = -2 \end{cases}$ $\{(1, 3)\}$

2. $\begin{cases} x + y = -3 \\ x - y = 1 \end{cases}$ $\{(-1, -2)\}$

3. $\begin{cases} 2x + y = 2 \\ x - y = -5 \end{cases}$

4. $\begin{cases} x - 2y = -3 \\ 2x + 2y = 6 \end{cases}$ $\{(1, 2)\}$

5. $\begin{cases} 2x + y = -2 \\ 2x + 3y = 6 \end{cases}$ $\{(-3, 4)\}$

6. $\begin{cases} 2x + y = 1 \\ 4x - 2y = 6 \end{cases}$

7. $\begin{cases} x - 2y = -3 \\ 2x - 4y = 1 \end{cases}$ ϕ

8. $\begin{cases} 2x + y = 1 \\ 4x + 2y = -1 \end{cases}$ ϕ

9. $\begin{cases} x = 5 \\ y = -2 \end{cases}$

10. $\begin{cases} x = -3 \\ y = 4 \end{cases}$ $\{(-3, 4)\}$

11. $\begin{cases} 2x - 3y = 0 \\ 5x + 2y = 0 \end{cases}$ $\{(0, 0)\}$

12. $\begin{cases} y - 3x = 0 \\ -2x - 3y = 0 \end{cases}$

13. $\begin{cases} x - y = 0 \\ x + y = 0 \end{cases}$ $\{(0, 0)\}$

14. $\begin{cases} 2y - 3x = 0 \\ 3y + 2x = 0 \end{cases}$ $\{(0, 0)\}$

15. $\begin{cases} 2x - y = 5 \\ x - 3y = 10 \end{cases}$

16. $\begin{cases} 3x - y = 5 \\ x + y = -1 \end{cases}$ $\{(1, -2)\}$

17. $\begin{cases} y = -x + 6 \\ y = 5 \end{cases}$ $\{(1, 5)\}$

18. $\begin{cases} 2x - y = 1 \\ x = 4 \end{cases}$

MORE CHALLENGING EXERCISES

Find the value for k that will make the lines in each system parallel.

19. $\begin{cases} 2x + y = 3 \\ kx + y = -1 \end{cases}$ $k = 2$

20. $\begin{cases} 2x - y = -1 \\ kx - 2y = 6 \end{cases}$ $k = 4$

$k = -1$
21. $\begin{cases} 4y + 2x = -4 \\ kx - 2y = -2 \end{cases}$

NON-ROUTINE PROBLEM

22. The cost of a watch and a bracelet is $77. The cost of the watch and a necklace is $95. The cost of the necklace and the bracelet is $108. Find the cost of each item.

Basic p. 384: 1–17 odd
Average p. 384: 3, 6, 9, ···, 18, 19, 22; p. 380: 1, 7, 9–17 odd
Above Average p. 384: 3, 6, 9, ···, 18, 19–22; p. 380: 1, 7, 9–17 odd

Problem-Solving Skills
Making a comparison
(Ex. 19–21)
Solving a multi-step problem
(Ex. 19–21)
Using logical reasoning
(Ex. 19–21)
Writing and solving an equation
(Ex. 22)

Critical Thinking
See Exercises 19–21.

Additional Answers
3. $(-1, 4)$
6. $(1, -1)$
9. $(5, -2)$
12. $(0, 0)$
15. $(1, -3)$
18. $(4, 7)$
22. Write an equation for each sentence.
$w + b = 77$, or $b = 77 - w$
$w + n = 95$, or $n = 95 - w$
$b + n = (77 - w) + (95 - w) = 108$;
$w = 32$; $b = 45$; $n = 63$
Watch: $32; Bracelet: $45; Necklace: $63

OBJECTIVE: To solve a system of two linear equations by the addition method

14.3 **Addition Method**

Sometimes you can solve a system of equations by adding the corresponding sides of the two equations to obtain a new equation with only one variable. This method is called the **addition method.**

EXAMPLE 1   Solve: $\begin{cases} \textbf{1. } x - y = -1 \\ \textbf{2. } x + y = 5 \end{cases}$

Solution:

1. Add the corresponding sides of both equations to eliminate y. Then solve for x.

$$\begin{array}{r} x - y = -1 \\ x + y = 5 \\ \hline 2x + 0 = 4 \\ x = 2 \end{array}$$

Note: The y terms, y and −y, are opposites.
$-y + y = 0$

2. Substitute 2 for x in either equation **1** or **2**.

$$\begin{array}{r} x + y = 5 \\ 2 + y = 5 \\ y = 3 \end{array}$$

Equation 2 is used.

3. **Check:** Substitute $x = 2$ and $y = 3$ in the original equations.

$$\begin{array}{c} x - y = -1 \\ 2 - 3 \\ -1 \end{array}$$

$$\begin{array}{c} x + y = 5 \\ 2 + 3 \\ 5 \end{array}$$

The solution set is $\{(2,3)\}$.

P–1   **Add the corresponding sides of both equations.**

a. $\begin{matrix} x + y = 3 \\ x - y = 6 \end{matrix}$ $2x = 9$

b. $\begin{matrix} -x + y = 5 \\ x + 2y = 7 \end{matrix}$ $3y = 12$

c. $\begin{matrix} 3x - y = 1 \\ 2x + y = 9 \end{matrix}$ $5x = 10$

EXAMPLE 2   Solve: $\begin{cases} \textbf{1. } -2x - y = 2 \\ \textbf{2. } 2x + 3y = 6 \end{cases}$

Solution:

1. Add the corresponding sides of both equations to eliminate the x terms. Then solve for y.

$$\begin{array}{r} -2x - y = 2 \\ 2x + 3y = 6 \\ \hline 2y = 8 \\ y = 4 \end{array}$$

2. Substitute 4 for y in either equation **1** or **2**.

$$\begin{array}{r} -2x - y = 2 \\ -2x - 4 = 2 \\ -2x = 6 \\ x = -3 \end{array}$$

Equation 1 is used.

3. The check is left for you.   The solution set is $\{(-3,4)\}$.

---

*Systems of Sentences / **385***

Teaching Suggestions p. M–44

Lesson Resources
Warm–Up: p. M–45
Maintenance: See below.
Practice Worksheet 14.3

Maintenance
1. A sprinter ran 72 yards in 4.8 seconds. Find the sprinter's speed in yards per second.
   Ans: 15 yds. per second
   (Section 3.4)
2. Write $(-8)^3(-8)^4(-8)^2$ as one power.
   Ans: $(-8)^9$
   (Section 6.1)
3. Subtract:
   $(3x^2 - 1) - (-4x^2 + 2)$
   Ans: $7x^2 - 3$
   (Section 8.4)
4. Divide and simplify:
   $$\frac{x+y}{x^2 - xy} \div \frac{1}{x^2 - y^2}$$
   Ans: $\dfrac{x^2 + 2xy + y^2}{x}$
   (Section 10.7)
5. Write the y intercept of $y = 3x + 5$.
   Ans: 5   (Section 13.2)

Additional Examples
Example 1
Solve.
1. $\begin{cases} x + y = 2 \\ x - y = -4 \end{cases}$
   Ans: $\{(-1,3)\}$
2. $\begin{cases} -x + y = 5 \\ x + 2y = 4 \end{cases}$
   Ans: $\{(-2,3)\}$

Example 2
Solve.
1. $\begin{cases} x + 2y = 7 \\ 2x - 2y = -4 \end{cases}$
   Ans: $\{(1,3)\}$
2. $\begin{cases} 3x + y = 1 \\ -3x + y = 7 \end{cases}$
   Ans: $\{(-1,4)\}$

Add the corresponding sides of both equations.

$$2x = -6$$

a. $2x + y = 0$
$-2x + y = 8$ $\quad 2y = 8$

b. $3x + 2y = 7$
$x - 2y = 5$ $\quad 4x = 12$

c. $x + 3y = -6$
$x - 3y = 0$

Explain that use
of the Addition
Method works
only if one of
the variables is
eliminated when
the two equations
are added.

Steps for the Addition Method

1. Add to eliminate one of the variables. Solve the resulting equation.
2. Substitute the known value of one variable in one of the original equations of the system. Solve for the other variable.
3. Check the solution in both equations of the system.

Common Error
Students get an incorrect equation when adding two equations of a system.

Example
$2x - 3y = 1$
$-2x + 5y = -2$
$\quad 8y = -1$

Prescription
If necessary, have students write any subtraction as addition. For example,
$2x + (-3y) = 1$
$-2x + 5y = -2$

Assignment Guide
Basic pp. 386–387: 1–17 odd
Average pp. 386–387: 1–17 odd
Above Average pp. 386–387:
1–17 odd

CLASSROOM EXERCISES

Write the equation that results if you add corresponding sides of the two equations in each system. (P-1 and P-2)

1. $\begin{cases} x - y = 4 \\ x + y = 2 \end{cases}$ $2x = 6$

2. $\begin{cases} -x + y = 6 \\ x + y = 2 \end{cases}$ $2y = 8$

3. $\begin{cases} x + 3y = 9 \\ -x + 2y = 6 \end{cases}$ $5y = 15$

4. $\begin{cases} 4x + y = 8 \\ -4x + y = 0 \end{cases}$ $2y = 8$

5. $\begin{cases} x - 3y = 7 \\ 2x + 3y = 5 \end{cases}$ $3x = 12$

6. $\begin{cases} x + 2y = 1 \\ x - 2y = -5 \end{cases}$ $2x = -4$

Add the corresponding sides of the equations. Then solve for x or y.
(Step 1 of Examples 1 and 2)

7. $\begin{cases} x - y = 10 \\ x + y = -2 \end{cases}$ $x = 4$

8. $\begin{cases} -x + 2y = 5 \\ x + y = 4 \end{cases}$ $y = 3$

9. $\begin{cases} 4x - 5y = -2 \\ 2x + 5y = 8 \end{cases}$ $x = 1$

WRITTEN EXERCISES

Goal: To solve a system of equations by the addition method

Sample Problem: Solve: $\begin{cases} 5x - 2y = 12 \\ 4x + 2y = 6 \end{cases}$ **Answer:** $\{(2, -1)\}$

For Exercises 1–18, solve each system by the addition method. Check.
(Examples 1 and 2)

1. $\begin{cases} x + y = 6 \\ x - y = -2 \end{cases}$ $\{(2, 4)\}$

2. $\begin{cases} -x + y = 5 \\ x + y = 9 \end{cases}$ $\{(2, 7)\}$

3. $\begin{cases} -x + 2y = 8 \\ x + 3y = 2 \end{cases}$
$\{(-4, 2)\}$

4. $\begin{cases} 3x - y = 5 \\ x + y = 7 \end{cases}$ $\{(3, 4)\}$ **5.** $\begin{cases} x + 3y = 1 \\ x - 3y = -5 \end{cases}$ $\{(-2, 1)\}$ **6.** $\begin{cases} x - 5y = -2 \\ 2x + 5y = -4 \end{cases}$ $\{(-2, 0)\}$

7. $\begin{cases} -x - 4y = -10 \\ -3x + 4y = -6 \end{cases}$ $\{(4, \frac{3}{2})\}$ **8.** $\begin{cases} 3x + y = 11 \\ -3x + 5y = -17 \end{cases}$ $\{(4, -1)\}$ **9.** $\begin{cases} x - 6y = 6 \\ -x + 2y = -6 \end{cases}$ $\{(6, 0)\}$

10. $\begin{cases} -3x + 7y = -1 \\ 2x - 7y = 3 \end{cases}$ $\{(-2, -1)\}$ **11.** $\begin{cases} 7x + 2y = -4 \\ 2x - 2y = -5 \end{cases}$ $\{(-1, \frac{3}{2})\}$ **12.** $\begin{cases} x + y = -8 \\ x - y = -3 \end{cases}$ $\{(-\frac{11}{2}, -\frac{5}{2})\}$

13. $\begin{cases} x + 3y = 1 \\ x - 3y = -5 \end{cases}$ $\{(-2, 1)\}$ **14.** $\begin{cases} -6x + 3y = 12 \\ 6x - 5y = 4 \end{cases}$ $\{(-6, -8)\}$ **15.** $\begin{cases} -3x + y = 1 \\ 5x - y = 3 \end{cases}$ $\{(2, 7)\}$

16. $\begin{cases} 2x - 3y = -2 \\ -x + 3y = -1 \end{cases}$ $\{(-3, -\frac{4}{3})\}$ **17.** $\begin{cases} -4x - y = 2 \\ 4x - y = 3 \end{cases}$ $\{(\frac{1}{8}, -\frac{5}{2})\}$ **18.** $\begin{cases} 7x + 3y = 19 \\ 4x - 3y = 3 \end{cases}$ $\{(2, \frac{5}{3})\}$

MID-CHAPTER REVIEW

In Ex. 1-6, each graph is a line. Two points on each line are given.

Graph each of the following equations. (Section 14.1)

1. $x + y = 5$ $(0, 5), (5, 0)$ **2.** $x - 2y = 3$ $(1, -1), (3, 0)$ **3.** $4x + y = 1$ $(0, 1), (1, -3)$

4. $3x - y = 4$ $(0, -4), (1, -1)$ **5.** $3x - 4y = -8$ $(0, 2), (4, 5)$ **6.** $y - 4 = 2x$ $(0, 4), (-2, 0)$

Solve each system by graphing. Check. (Section 14.2)

7. $\begin{cases} 5x - 6y = 8 \\ 2x - y = -1 \end{cases}$ $\{(-2, -3)\}$ **8.** $\begin{cases} x - 3y = -6 \\ 4x + 3y = -9 \end{cases}$ $\{(-3, 1)\}$ **9.** $\begin{cases} x + 2y = -2 \\ 5x + 2y = 6 \end{cases}$ $\{(2, -2)\}$

Solve each system by the addition method. Check. (Section 14.3)

10. $\begin{cases} x - 2y = 13 \\ -x - 3y = 2 \end{cases}$ $\{(7, -3)\}$ **11.** $\begin{cases} 3x + 3y = 7 \\ 4x - 3y = 7 \end{cases}$ $\{(2, \frac{1}{3})\}$ **12.** $\begin{cases} 2x - y = 12 \\ -2x + 3y = 8 \end{cases}$ $\{(11, 10)\}$

REVIEW CAPSULE FOR SECTION 14.4

Write the coefficient of each monomial. (Pages 208–209)

1. $3x$ 3 **2.** $\frac{1}{2}x^2$ $\frac{1}{2}$ **3.** $-4y$ -4 **4.** xy^2 1 **5.** $-y^2$ -1

Multiply. (Pages 47–49)

6. $2(x - y)$ $2x - 2y$ **7.** $3(x + y)$ $3x + 3y$ **8.** $3(x + 2y)$ $3x + 6y$ **9.** $4(2x - 3y)$ $8x - 12y$

10. $3(-3x + 5y)$ $-9x + 15y$ **11.** $-2(x - y)$ $-2x + 2y$ **12.** $-2(2x - y - 3)$ $-4x + 2y + 6$ **13.** $3(-5x + y + 5)$ $-15x + 3y + 15$

Systems of Sentences / **387**

Quiz Sections 14.1–14.3
After completing this Mid-Chapter Review, you may want to administer a quiz covering the same sections. See the quiz provided on page 289 in the *Teacher's ResourceBank*.

Lesson Resources
Warm–Up: p. M–45
Maintenance: See below.
Practice Worksheet 14.4

Maintenance
1. A racing bicycle is on sale for $254.83. The discount rate is 15%. Find the list price of the bicycle.
   Ans: $299.80
   (Section 4.4)
2. Simplify: $(3r^{-3}t^4)^{-2}$
   Ans: $\dfrac{r^6}{9t^8}$ (Section 6.5)
3. Multiply:
   $(x + 2)(3x^2 - 2x + 4)$
   Ans: $3x^3 + 4x^2 + 8$
   (Section 8.5)
4. Subtract: $\dfrac{8a}{b + c} - \dfrac{5a}{b + c}$
   Ans: $\dfrac{3a}{b + c}$ (Section 11.1)
5. Write the slope of the graph of $y = \frac{1}{2}(-2 + 4x)$.
   Ans: 2 (Section 13.3)

Additional Examples
Example 1
Solve.
1. $\begin{cases} 4x - y = 7 \\ 5x + 3y = 13 \end{cases}$
   Ans: $\{(2,1)\}$
2. $\begin{cases} 4x - 3y = -1 \\ 2x + 2y = 3 \end{cases}$
   Ans: $\{(\frac{1}{2},1)\}$

OBJECTIVE: To solve a system of two linear equations by the multiplication/addition method

14.4 Multiplication/Addition Method

When you add the corresponding sides of the equations of a system, you may get an equation that has two variables rather than only one. This occurs because the coefficients of the variables are not opposites.

To solve such a system of equations, you can multiply each side of one or both equations by a number or numbers that will make opposite coefficients for one variable. Then you add the corresponding sides of the two equations. This procedure is called the ***multiplication/addition method.***

EXAMPLE 1   Solve: $\begin{cases} \textbf{1. } x - y = -1 \\ \textbf{2. } 2x - 3y = -1 \end{cases}$

Solution:

1. Multiply each side of equation **1** by -2. ⟶ $-2x + 2y = 2$ ◀ **The coefficients of x are now opposites.**
2. Add equation **2**. Solve. ⟶ $\underline{2x - 3y = -1}$

   $-y = 1$

   $y = -1$

3. Substitute -1 for y in either equation **1** or **2**. Solve for x. ⟶
   $x - y = -1$
   $x - (-1) = -1$ ◀ **Equation 1 is used.**
   $x = -2$

4. **Check:** Substitute $x = -2$ and $y = -1$ in $x - y = -1$ and in $2x - 3y = -1$.

   The solution set is $\{(-2, -1)\}$.

P–1 **Multiply each side of equation 1 by a number that will form coefficients of x which are opposites.**

a. $\begin{cases} \textbf{1. } x + y = 7 \\ \textbf{2. } -3x + 2y = 4 \end{cases}$ b. $\begin{cases} \textbf{1. } x - 6y = -5 \\ \textbf{2. } x + 2y = 11 \end{cases}$ c. $\begin{cases} \textbf{1. } x + 2y = -2 \\ \textbf{2. } 7x - 5y = 43 \end{cases}$
Multiply by 3.   Multiply by -1.   Multiply by -7.

In Example 1, it was necessary to multiply the corresponding sides of only one of the equations to eliminate a variable. It is sometimes necessary to multiply corresponding sides of <u>both</u> equations by numbers that will make opposite coefficients for one variable.

388 / *Chapter 14*

EXAMPLE 2   Solve: $\begin{cases} \textbf{1. } 5x - 3y = -1 \\ \textbf{2. } -3x + 2y = 2 \end{cases}$ In step $\boxed{1}$, the absolute value of the new coefficient for y is the least common multiple of 2 and 3.

Solution:

$\boxed{1}$ Multiply each side of **1** by 2. ⟶ $10x - 6y = -2$ **The coefficients of y are now opposites.**

Multiply each side of **2** by 3. ⟶ $-9x + 6y = 6$

$\boxed{2}$ Add. $x \quad\quad = 4$

$\boxed{3}$ Substitute 4 for x in either **1** or **2**. ⟶ $5x - 3y = -1$ **Equation 1 is used.**

$5(4) - 3y = -1$

$20 - 3y = -1$

$-3y = -21$

$y = 7$

$\boxed{4}$ The check is left for you.   The solution set is $\{(4, 7)\}$.

P–2 **Multiply each side of both equations by numbers that will form coefficients of y which are opposites.**

Steps 1 and 2 produce an equation that can be used with either equation of the system to form a new system that is equivalent to the original system.

a. $\begin{cases} \textbf{1. } 3x + 8y = -16 \\ \textbf{2. } 4x - 5y = 10 \end{cases}$ b. $\begin{cases} \textbf{1. } 2x - 3y = -11 \\ \textbf{2. } 5x + 4y = 30 \end{cases}$ c. $\begin{cases} \textbf{1. } 7x + 2y = -36 \\ \textbf{2. } -8x + 3y = 57 \end{cases}$

> **Steps for the Multiplication/Addition Method**
>
> 1. Check for opposite coefficients of one variable. If necessary, multiply each side of either or both equations by numbers that will make opposite coefficients for one variable.
> 2. Add to eliminate one of the variables. Solve the resulting equation.
> 3. Substitute the known value of one variable in one of the original equations of the system. Solve for the other variable.
> 4. Check the solution in both equations of the system.

CLASSROOM EXERCISES

Multiply each side of one equation by a number to eliminate x or y.
(Steps 1 and 2 of Example 1)

1. $\begin{cases} x - 3y = -1 \\ -2x + y = -3 \end{cases}$ 2. $\begin{cases} 3x + 5y = 1 \\ 2x - y = -2 \end{cases}$ 3. $\begin{cases} -2x + 4y = 1 \\ 3x + 2y = -4 \end{cases}$
   Eq 1: Mult. by 2.   Eq 2: Mult. by 5.   Eq 2: Mult. by −2.

*Systems of Sentences / **389***

Additional Examples
Example 2
Solve.
1. $\begin{cases} 3x + 2y = 12 \\ 2x + 5y = 8 \end{cases}$
   Ans: $\{(4,0)\}$
2. $\begin{cases} 3x + 4y = -7 \\ 5x - 6y = -18 \end{cases}$
   Ans: $\{(-3,\frac{1}{2})\}$

Additional Answers
P–2 a. Eq. 1: Mult. by 5.
       Eq. 2: Mult. by 8.
   b. Eq. 1: Mult. by 4.
       Eq. 2: Mult. by 3.
   c. Eq. 1: Mult. by 3.
       Eq. 2: Mult. by −2.

Common Error
Students are not able to eliminate one of the variables when using the multiplication/addition method.

Example
$\begin{cases} \textbf{1. } \quad x - 3y = 2 \\ \textbf{2. } -3x + 2y = -3 \end{cases}$
$2x - 6y = 4$
$\underline{-6x + 4y = -6}$
$-4x - 2y = -2$

Prescription
First, instruct students to see if one variable has a coefficient of 1. Then multiply only this equation by the opposite of the corresponding coefficient in the other equation. Then explain how to use the least common multiple of the absolute values of the coefficients to determine what number to multiply by in each equation.

389

Multiply each side of both equations by numbers to eliminate x or y.
(Steps 1 and 2 of Example 2)

4. $\begin{cases} 2x - 4y = -1 \\ 3x - 3y = 2 \end{cases}$
Eq 1: Mult. by −3.
Eq 2: Mult. by 2.

5. $\begin{cases} 4x + 5y = 2 \\ -3x + 2y = 1 \end{cases}$
Eq 1: Mult. by 3.
Eq 2: Mult. by 4.

6. $\begin{cases} -6x - 2y = -3 \\ 4x + 3y = 1 \end{cases}$
Eq 1: Mult. by 3.
Eq 2: Mult. by 2.

WRITTEN EXERCISES

Goal: To solve systems of equations by the multiplication/addition method

Sample Problem: $\begin{cases} 5x - 4y = 2 \\ -x + y = 1 \end{cases}$ **Answer:** $\{(-2, -3)\}$

Solve each system by the multiplication/addition method. (Examples 1–3)

1. $\begin{cases} 2x + y = 10 \\ x - 2y = 0 \end{cases}$ $\{(4, 2)\}$

2. $\begin{cases} 2x - y = 5 \\ x + 2y = 25 \end{cases}$ $\{(7, 9)\}$

3. $\begin{cases} 4x + 3y = 6 \\ 2x - y = -2 \end{cases}$ $\{(0, 2)\}$

4. $\begin{cases} 5x + y = 4 \\ x - 2y = 3 \end{cases}$ $\{(1, -1)\}$

5. $\begin{cases} 2x - y = -3 \\ 4x - y = -2 \end{cases}$ $\{(\frac{1}{2}, 4)\}$

6. $\begin{cases} 4x - y = 1 \\ 2x - y = 3 \end{cases}$ $\{(-1, -5)\}$

7. $\begin{cases} 2x - 3y = 13 \\ 3x + y = 3 \end{cases}$ $\{(2, -3)\}$

8. $\begin{cases} 3x + 4y = 10 \\ 2x - y = -8 \end{cases}$ $\{(-2, 4)\}$

9. $\begin{cases} x + 3y = -3 \\ 2x - y = 4 \end{cases}$ $\{(\frac{9}{7}, -\frac{10}{7})\}$

10. $\begin{cases} 2x - 5y = -5 \\ x + 2y = -4 \end{cases}$ $\{(-\frac{10}{3}, -\frac{1}{3})\}$ **11.** $\begin{cases} -5x + 3y = 0 \\ x + 2y = 0 \end{cases}$ $\{(0, 0)\}$

12. $\begin{cases} 2x - y = 3 \\ -3x - y = -5 \end{cases}$ $\{(\frac{8}{5}, \frac{1}{5})\}$

13. $\begin{cases} 2x + 5y = 0 \\ 3x - y = 0 \end{cases}$ $\{(0, 0)\}$

14. $\begin{cases} 3x - y = -2 \\ -5x - y = 3 \end{cases}$ $\{(-\frac{5}{8}, \frac{1}{8})\}$

15. $\begin{cases} 3x + 2y = 2 \\ 2x + 3y = -2 \end{cases}$ $\{(2, -2)\}$

16. $\begin{cases} 5x - 2y = 11 \\ 3x + 5y = 19 \end{cases}$ $\{(3, 2)\}$

17. $\begin{cases} 4x - 3y = 5 \\ 2x + 9y = -1 \end{cases}$ $\{(1, -\frac{1}{3})\}$

18. $\begin{cases} 6x - 2y = -3 \\ 2x + 4y = 5 \end{cases}$ $\{(-\frac{1}{14}, \frac{9}{7})\}$

APPLICATIONS

You can use a calculator to check whether a given ordered pair is a solution of a system of equations.

EXAMPLE Check the solution set, $\{(4, 7)\}$, of Example 2 on page 389.

SOLUTION **1.** $5x - 3y = -1$

Store "5 · 4" in memory before adding it to "−3 · 7."

| 5 | × | 4 | = | M+ | 3 | +/− | × | 7 |

| + | MR | = |

$-1.$

2. $-3x + 2y = z$ This equation is left for you to check.

19–24. Use a calculator to do Exercises 1–6 on page 383.

19. No   20. No   21. Yes   22. Yes   23. No   24. Yes

Assignment Guide
Basic
Day 1 p. 390: 1–17 odd
Day 2 p. 390: 2–18 even
Average p. 390: 1–21 odd
Above Average p. 390:
1–17 odd, 19–24

Suppose x is the smaller and y is the greater of two unknown numbers.
Write an equation for each sentence.

1. The sum of the two numbers is -12. $x + y = -12$

2. The difference of the two numbers is 7. $y - x = 7$

3. The smaller of the two numbers is 6 less than the greater number. $x = y - 6$

4. The greater of the two numbers is 13 more than the smaller number. $y = x + 13$

■■■■ PROBLEM SOLVING AND APPLICATIONS ■■■■

OBJECTIVE: To solve a word problem involving two unknowns by solving

14.5 Using Systems of Equations

a system of two linear equations

The addition method and multiplication/addition method of solving a system of equations can be used to solve word problems. First, you use Condition 1 and two variables to represent the unknowns. You also use Condition 1 to write one equation of the system. You use Condition 2 to write the second equation for the system.

EXAMPLE 1   The sum of two numbers is -16 (Condition 1). The difference of the two numbers is 30 (Condition 2). Find the two numbers.

Solution:

1 Use Condition 1 to represent the unknowns. ⟶ Let $x =$ the greater number.
Let $y =$ the smaller number.

2 Write the equations.

Think: The sum of two numbers is -16. ◀ *Condition 1*

Translate: $x + y = -16$

Discuss how the equations would be written if x represents the smaller number and y represents the larger number.

Think: The difference of the two numbers is 30. ◀ *Condition 2*

Translate: $x - y = 30$

3 Solve. Use the ⟶ addition method.

$$x + y = -16$$
$$\underline{x - y = 30}$$
$$2x = 14$$
$$x = 7$$

◀ *Don't forget to find y.*

$$x + y = -16$$
$$7 + y = -16$$
$$y = -23$$

Systems of Sentences / **391**

Teaching Suggestions p. M–44

Lesson Resources
Warm–Up: p. M–45
Maintenance: See below.
Practice Worksheet 14.5

Maintenance
1. The width of a mirror is 28 inches less than the length (Condition 1). The perimeter is 192 inches (Condition 2). Find the length and width.
   Ans: length: 62 inches; width: 34 inches
   (Section 4.2)
2. Add: $3\sqrt{5} + \sqrt{20}$
   Ans: $5\sqrt{5}$ (Section 7.5)
3. Factor: $3x^2 - x - 10$
   Ans: $(3x + 5)(x - 2)$
   (Section 9.4)
4. Add:
   $$\frac{3}{y^2 - 3y + 2} + \frac{2}{y^2 - y - 2}$$
   Ans: $\dfrac{5y + 1}{(y - 2)(y + 1)(y - 1)}$
   (Section 11.6)
5. Write the slope of the line that contains the points $(0, -3)$ and $(1, 6)$.
   Ans: 9 (Section 13.4)

Additional Examples
Example 1
1. The sum of two numbers is 8 (Condition 1). The difference of the two numbers is 4 (Condition 2). Find the two numbers.
   Ans: 6; 2
2. The difference of two numbers is 21 (Condition 1). The sum of the numbers is -5 (Condition 2). Find the two numbers.
   Ans: 8; -13

☐4 **Check:** Condition 1   Is the sum of the two numbers −16?
                          Does $7 + (−23) = −16$?   Yes ✔

              Condition 2   Is the difference of the two numbers 30?
                          Does $7 − (−23) = 30$?   Yes ✔

                          The two numbers are 7 and −23.

Additional Examples
Example 2
1. The length of a rectangle is 8 inches more than the width (Condition 1). The perimeter is 96 inches (Condition 2). Find the length and width.
Ans: length: 28 in; width: 20 in
2. Betty's age is 3 years less than three times Carl's age (Condition 1). The sum of their ages is 57 years (Condition 2). Find their ages.
Ans: Carl: 15; Betty: 42

Additional Answers
1. First number: x; second number: y
   $x + y = 25$
2. First number: x; second number: y
   $x − y = −7$
3. Jerry's age: x; Mary's age: y
   $x + y = 25$
4. Greater number: x; smaller number: y
   $y + 12 = x$
5. Width: x; length: y
   $x + y = 51$
6. Length: x; width: y
   $2x + 2y = 93$
7. First number: x; second number: y
   $2x + 3y = 19$
8. Smaller number: x; greater number: y
   $y = x + 24$

EXAMPLE 2   The width of a rectangle is 5 centimeters less than the length (Condition 1). The perimeter is 38 centimeters (Condition 2). Find the length and width.

Solution:

Consider how to write the equations if x represents the width and y represents the length.

☐1 Let x = the length and let y = the width. Then $2x + 2y$ = the perimeter.

☐2 Write the equations. ⟶ $\begin{cases} \textbf{1.}\ y = x − 5,\ \text{or}\ −x + y = −5 \\ \textbf{2.}\ 2x + 2y = 38 \end{cases}$ 1. Condition 1
                                                                            2. Condition 2

☐3 To solve the system, multiply equation **1** by 2. ⟶

$$−2x + 2y = −10 \quad \textbf{(Equation 1)}$$
$$\underline{2x + 2y = 38} \quad \textbf{(Equation 2)}$$
$$4y = 28$$
$$y = 7$$

Don't forget to find x.

$$y = x − 5$$
$$7 = x − 5$$
$$12 = x$$

The length is 12 centimeters.
The width is 7 centimeters.

The check is left for you.

CLASSROOM EXERCISES

Write an equation in two variables that represents each sentence. Write what each variable represents. (Steps 1 and 2 of Examples 1 and 2)

1. The sum of two numbers is 25.

2. The difference of two numbers is −7.

3. The sum of the ages of Jerry and Mary is 25 years.

4. One number is 12 more than another number.

5. Half the perimeter of a rectangle is 51.

6. The perimeter of a rectangle is 93 millimeters.

7. Twice one number increased by three times another number is 19.

8. One number exceeds another number by 24.

▨▨▨ **WRITTEN EXERCISES** ▨▨▨

Goal: To solve number problems by systems of linear equations

Sample Problem: The greater of two numbers is 1 more than 3 times the smaller (Condition 1). If 8 times the smaller is decreased by 2 times the greater, the result is 10 (Condition 2). Find the numbers.

Answer: Greater number: 19   Smaller number: 6

For some problems, Condition 1 is <u>underscored once</u>. *Condition 2 is* <u>underscored twice</u>.

a. *Use Condition 1 and two variables to represent the unknowns.*

b. *Use Conditions 1 and 2 to write a system of equations for the problem.*

c. *Solve.* (Examples 1 and 2)

1. <u>The sum of two numbers is 28.</u> <u><u>The difference of the two numbers is 3.</u></u> Find the two numbers.

2. <u>The sum of two numbers is 73.</u> <u><u>The difference of the two numbers is 11.</u></u> Find the two numbers.

3. <u>The sum of two numbers is −12.</u> <u><u>One number exceeds the other by 42.</u></u> Find the two numbers.

4. <u>The sum of two numbers is −65.</u> <u><u>One number exceeds the other by 27.</u></u> Find the two numbers.

5. <u>Twice the smaller of two numbers is increased by 2. The result equals the greater number.</u> <u><u>Three times the smaller number is increased by twice the greater number. The result equals 39.</u></u> Find the two numbers.

6. <u>The greater of two numbers is 4 less than twice the smaller number.</u> <u><u>From twice the greater number, three times the smaller is subtracted. The result equals 7.</u></u> Find the two numbers.

7. The length of a rectangle is 8 meters more than the width. The perimeter is 34 meters. Find the length and width.

8. The width of a rectangle is 9 decimeters less than the length. The perimeter is 43 decimeters. Find the length and width.

9. John is seven years younger than his sister Ann. The sum of their ages is 45. Find their ages.

10. Kathy is 31 years younger than her father. The sum of their ages is 47. Find their ages.

11. Sam bought 6 fewer crates of corn than Tina. Twice the number of crates Tina bought is 5 more than 3 times the number of crates Sam bought. How many crates did each buy?

12. Brian played 4 fewer innings in a softball game than Denise. Twice the number of innings Brian played plus 2 equals the number of innings Denise played. How many innings did each play?

Systems of Sentences   /   **393**

Assignment Guide
Basic p. 393: 1–11 odd
Average p. 393: 1–11 odd
Above Average p. 393: 1–11 odd

Problem-Solving Skills
Writing and solving an equation (Ex. 1–12)

Additional Answers

2. **a.** Greater: x; smaller: y
   b. $\begin{cases} x + y = 73 \\ x - y = 11 \end{cases}$
   c. 42, 31

4. **a.** First: x; second: y
   b. $\begin{cases} x + y = -65 \\ x - y = 27 \end{cases}$
   c. $-19, -46$

6. **a.** Smaller: x; greater: y
   b. $\begin{cases} y = 2x - 4 \\ 2y - 3x = 7 \end{cases}$
   c. 15, 26

8. **a.** Length: x; width: y
   b. $\begin{cases} y + 9 = x \\ 2x + 2y = 43 \end{cases}$
   c. Width: $6\frac{1}{4}$ decimeters; length: $15\frac{1}{4}$ decimeters

10. **a.** Kathy's age: x; father's age: y
    b. $\begin{cases} x + 31 = y \\ x + y = 47 \end{cases}$
    c. Kathy: 8; father: 39

12. **a.** Innings Brian played: x; innings Denise played: y
    b. $\begin{cases} x = y - 4 \\ 2x + 2 = y \end{cases}$
    c. Brian: 2; Denise: 6

Quiz Sections 14.4–14.5
After completing Sections 14.4–14.5, you may want to administer a quiz covering the same sections. See the quiz provided on page 290 in the *Teacher's ResourceBank*.

This special topic illustrates how proportions can be applied to solving problems about chemical solutions. The molecular weights of some chemicals are displayed in a table for use in the example and exercises.

Using Tables

Chemistry

Pharmacists use their knowledge of **chemistry** to prepare various solutions. The table below shows the **molecular weights** of some chemicals.

Chemical	Molecular Weight
Boric acid	157.30
Calcium chloride	147.02
Magnesium sulfate	246.47
Potassium chloride	74.56

A **mole** in chemistry is the mass of a substance that equals its molecular weight. A **one molar solution** of boric acid and water would contain 157.3 grams of boric acid in 1 liter of solution.

EXAMPLE: Find the number of grams of boric acid to make 1500 milliliters of a solution with 5% molarity.

SOLUTION: **a.** Find the number of grams of boric acid to make 1 liter (1000 milliliters) of a 5% molarity solution.

◀ Find 5% of 157.3.

[1] [5] [7] [·] [3] [×] [5] [%] 7.865

One liter of a 5% molarity solution requires about 7.9 grams.

b. Find the number of grams of boric acid for 1500 milliliters. Solve the following proportion.

$$\frac{7.9}{1000} = \frac{n}{1500}$$ ◀ $n = \frac{(7.9)(1500)}{1000}$

$n = 11.85$ or about **11.9 grams** of boric acid

EXERCISES

Find the number of grams needed to make a solution of the given number of milliliters and molarity. Round the answer to the nearest tenth of a gram.

1. Magnesium sulfate
   1200 milliliters
   10% molarity   29.6 grams

2. Potassium chloride
   750 milliliters
   0.3% molarity   0.2 grams

3. Calcium chloride
   1750 milliliters
   8.5% molarity
   21.9 grams

CHAPTER SUMMARY

IMPORTANT TERMS	Linear equation *(p. 378)* Addition method *(p. 385)*
	Solution set Multiplication/addition
	of a linear equation *(p. 378)* method *(p. 388)*
	System of equations *(p. 381)*
	Solution set of a system
	of two equations *(p. 381)*

IMPORTANT IDEAS

1. To solve a system of equations by the *multiplication/addition method:*
  a. Check for opposite coefficients of one variable. If necessary, multiply each side of either or both equations by numbers that will make opposite coefficients for one variable.
  b. Add to eliminate one of the variables. Solve the resulting equation.
  c. Substitute the known value of one variable in one of the original equations of the system. Solve for the other variable.
  d. Check the solution in both equations of the system.

2. To solve word problems using a system of equations in two variables:
  a. Use Condition 1 and two variables to represent the unknowns.
  b. Use Condition 1 to write one equation of the system. Use Condition 2 to write the second equation for the system.
  c. Solve the system of equations.
  d. Check your solution in the original conditions of the problem. Answer the question.

CHAPTER REVIEW

SECTION 14.1

Each graph is a straight line. The coordinates of two points are given for each line.

Graph each of the following.

1. $2x + y = 3$ $(0, 3), (2, -1)$ **2.** $-3x + y = -2$ $(0, -2), (1, 1)$ **3.** $x - 5y = -1$ $(-1, 0), (4, 1)$

4. $-x - 3y = 1$ $(-1, 0), (2, -1)$ **5.** $-2x + 3y = -6$ $(0, -2), (3, 0)$ **6.** $4x - 2y = -8$ $(0, 4), (-2, 0)$

SECTION 14.2

Solve each system by graphing. Check.

7. $\begin{cases} x + 3y = 3 \\ 2x + y = -4 \end{cases}$ $\{(-3, 2)\}$ **8.** $\begin{cases} x - 2y = 2 \\ x + y = 2 \end{cases}$ $\{(2, 0)\}$

9. $\begin{cases} 2x - y = 0 \\ 2x - 3y = -4 \end{cases}$ $\{(1, 2)\}$ **10.** $\begin{cases} x - 2y = 4 \\ x + 3y = -1 \end{cases}$ $\{(2, -1)\}$

Systems of Sentences / **395**

Chapter Test
Two Chapter Tests (Form A and Form B) are provided on pages 291–294 in the *Teacher's ResourceBank.*

SECTION 14.3

Solve each system by the addition method. Check.

11. $\begin{cases} x - y = 3 \\ x + y = 1 \end{cases}$ $\{(2, -1)\}$

12. $\begin{cases} 2x + y = 1 \\ 2x - y = 7 \end{cases}$ $\{(2, -3)\}$

13. $\begin{cases} 3x + 2y = 6 \\ -3x - y = 3 \end{cases}$
$\{(-4, 9)\}$

SECTION 14.4

Solve each system by the multiplication/addition method. Check.

14. $\begin{cases} x - y = 1 \\ 2x + 3y = 7 \end{cases}$
$\{(2, 1)\}$

15. $\begin{cases} 2x + 3y = 6 \\ x + 2y = 1 \end{cases}$
$\{(9, -4)\}$

16. $\begin{cases} -2x + y = 1 \\ 4x + 3y = 23 \end{cases}$
$\{(2, 5)\}$

17. $\begin{cases} 3x - 4y = 5 \\ 6x - 8y = 3 \end{cases}$
ϕ

SECTION 14.5

For Exercises 18–21, Condition 1 is <u>underscored once</u>. Condition 2 is <u>underscored twice</u>.

Use Condition 1 to represent the unknowns. Use Condition 1 and Condition 2 to write a system of equations for the problem. Solve.

18. The sum of two numbers is 17. The smaller number is 33 less than the greater number. Find the two numbers.

19. The sum of two numbers is 8. Four times the greater number is two more than four times the smaller number. Find the two numbers.

20. The length of a rectangle is three meters less than twice the width. The perimeter is 24 meters. Find the length and width.

21. Three times the width of a rectangle increased by twice the length equals 40. The perimeter is 34 units. Find the number of units in the length and width.

Additional Answers

18. Smaller: x; greater: y
$\begin{cases} x + y = 17 \\ x = y - 33 \end{cases}$
smaller number: -8;
greater number: 25

19. Greater: x; smaller: y
$\begin{cases} x + y = 8 \\ 4x = 4y + 2 \end{cases}$
greater number: $4\frac{1}{4}$
smaller number: $3\frac{3}{4}$

20. Length: x; width: y
$\begin{cases} x = 2y - 3 \\ 2x + 2y = 24 \end{cases}$
length: 7 m;
width: 5 m

21. Width: x; length: y
$\begin{cases} 3x + 2y = 40 \\ 2x + 2y = 34 \end{cases}$
width: 6 units;
length: 11 units

More on Systems of Sentences

In mathematics, the formula $d = rt$ is used to find distance. Various instruments such as the telescope, the transit, and sonar equipment are used to measure distance in space, on the earth's surface, and in the ocean, respectively.

Background

The center photo above shows the Palomar Observatory in California which houses the 200-inch Hale reflector telescope.

The top right photo shows a transit which is a surveying instrument used to measure horizontal angles.

The photo at the bottom right shows sonar equipment used on a supertanker to determine the depth of the ocean.

OBJECTIVE: To solve a system of two linear equations by the substitution method

15.1 Substitution Method

Teaching Suggestions p. M–46

Lesson Resources
Warm–Up: p. M–46
Maintenance: See below.
Practice Worksheet 15.1

Maintenance
1. One season, a football team won 5 more games than it lost (Condition 1). The total number of games played was 13 (Condition 2). Find the number of wins and losses.
Ans: 9 wins, 4 losses
(Section 4.1)
2. Simplify: $\sqrt{250}$
Ans: $5\sqrt{10}$ (Section 7.3)
3. Factor:
$2x^2 - 12x + 16$
Ans: $2(x - 4)(x - 2)$
(Section 9.7)
4. Subtract:
$$\frac{3}{5x - 15} - \frac{1}{4x - 12}$$
Ans: $\frac{7}{20(x - 3)}$
(Section 11.6)
5. Write the x intercept of the graph of $y = \frac{1}{2}x - 3$.
Ans: 6 (Section 13.6)

Additional Examples
Example 1
Solve each system by the substitution method.
1. $\begin{cases} y = 3x - 13 \\ 3x + 4y = -7 \end{cases}$
Ans: $\{(3, -4)\}$
2. $\begin{cases} x = -3y - 10 \\ 2x - 3y = 7 \end{cases}$
Ans: $\{(-1, -3)\}$
3. $\begin{cases} 5x - 2y = -8 \\ y = 4 - 2x \end{cases}$
Ans: $\{(0,4)\}$

You have often replaced a variable by a number or expression. This process is called <u>substitution</u>. You can solve systems of linear equations by the **substitution method.**

EXAMPLE 1 Solve the system below by the substitution method.

$$\begin{cases} \textbf{1.}\ y = 2x - 3 \\ \textbf{2.}\ 2x + 3y = 7 \end{cases}$$

Solution: $\qquad\qquad\qquad\qquad\qquad\qquad 2x + 3y = 7$

1. Substitute $(2x - 3)$ for y in Equation **2.** $\longrightarrow 2x + 3(2x - 3) = 7$
2. Distributive Property $\longrightarrow 2x + 6x - 9 = 7$
3. Combine like terms. $\longrightarrow 8x - 9 = 7$
4. Addition Property of Equality $\longrightarrow 8x = 16$
5. Division Property of Equality $\longrightarrow x = 2$

Substitute 2 for x in either equation. Then solve for y.

$$\textbf{1.}\ y = 2x - 3$$
$$y = 2(2) - 3$$
$$y = 1$$

Check: 1. $y = 2x - 3$ $\qquad$ **2.** $2x + 3y = 7$ $\qquad$ Solution set: $\{(2, 1)\}$
$1 \mid 2(2) - 3$ $\qquad\qquad 2(2) + 3(1) \mid 7$
$\qquad\quad\rightarrow 1$ $\qquad\qquad\qquad 4 + 3$
$\qquad\qquad\qquad\qquad\qquad\qquad 7$

Suggest to students that they look for a variable whose coefficient is 1. Then solve for that variable.

EXAMPLE 2 Solve the system by the substitution method.

$$\begin{cases} \textbf{1.}\ 2x + 3y = 1 \\ \textbf{2.}\ x - 2y = -3 \end{cases}$$

Solution: $\qquad\qquad\qquad\qquad\qquad x - 2y = -3$
1. Solve Equation **2** for x. $\longrightarrow x = 2y - 3$

398 / *Chapter 15*

② Substitute $(2y - 3)$ for x in Equation 1. Solve for y.

The first substitution must always be in the other equation. You may demonstrate what happens when you substitute in the same equation.

$$2x + 3y = 1$$
$$2(2y - 3) + 3y = 1$$
$$4y - 6 + 3y = 1$$
$$7y - 6 = 1$$
$$7y = 7$$
$$y = 1$$

③ Substitute 1 for y in either equation. Solve for x.

$$x - 2y = -3$$
$$x - 2(1) = -3$$
$$x - 2 = -3$$
$$x = -1$$

④ Check:

1.
$$2x + 3y = 1$$
$$2(-1) + 3(1)$$
$$-2 + 3$$
$$1$$

2.
$$x - 2y = -3$$
$$-1 - 2(1)$$
$$-1 - 2$$
$$-3$$

Solution set: $\{(-1, 1)\}$

P-1 **What expression from $y = 2x - 3$ could be substituted for y in $x = \frac{1}{2}y + 2$?** $2x - 3$

P-2 **What expression from $x = \frac{1}{2}y + 2$ could be substituted for x in $2x - y = 3$?** $\frac{1}{2}y + 2$

EXAMPLE 3   Solve the system by the substitution method.

$$\begin{cases} 1.\ 2x - y = 3 \\ 2.\ x = \frac{1}{2}y + 2 \end{cases}$$

Solution:

① Substitute $(\frac{1}{2}y + 2)$ for x in Equation 1.

② Solve for y.

$$2x - y = 3$$
$$2(\frac{1}{2}y + 2) - y = 3$$
$$y + 4 - y = 3$$
$$4 = 3$$ ◄ **False sentence**

The resulting sentence is false. Therefore, the solution set of the system is ϕ.

More on Systems of Sentences / **399**

Additional Examples
Example 2
Solve each system by the substitution method.
1. $\begin{cases} 2x + y = 5 \\ 5x + 3y = 14 \end{cases}$
   Ans: $\{(1,3)\}$
2. $\begin{cases} x - 4y = -12 \\ 3x - 5y = -9 \end{cases}$
   Ans: $\{(-3,0)\}$
3. $\begin{cases} 2y - 4y = 14 \\ 3x - y = 11 \end{cases}$
   Ans: $\{(3,-2)\}$

Example 3
Solve each system by the substitution method.
1. $\begin{cases} y = 3x + 5 \\ 6x - 2y = 7 \end{cases}$
   Ans: ϕ
2. $\begin{cases} x = 5y - 1 \\ y = \frac{1}{5}x + 3 \end{cases}$
   Ans: ϕ
3. $\begin{cases} 2x - y = 0 \\ 6x = -3y + 1 \end{cases}$
   Ans: ϕ

> **Steps of the Substitution Method**
>
> 1. Solve one equation for one of the variables.
> 2. Substitute the resulting expression in the other equation. Solve the equation.
> 3. Substitute the value of the variable from Step **2** in either equation. Solve the resulting equation.
> 4. Check by substituting both values in both equations.

CLASSROOM EXERCISES

Solve the equation for one of the variables.
(Step 1 of Example 2)

EXAMPLE: $2x + y = 2$ **ANSWER:** $y = 2 - 2x$ or $y = -2x + 2$

1. $x - 2y = 3$ $x = 3 + 2y$
2. $3y + x = -5$ $x = -5 - 3y$
3. $x + 3y = 1$ $x = 1 - 3y$
4. $y + 3 = 5x$ $y = 5x - 3$
5. $x - 3 = -3y$ $x = 3 - 3y$
6. $y - x = 1$ $y = 1 + x$

Write the equation that results after making a substitution for one of the variables. (Step 1 of Example 1, Step 2 of Example 2)

7. $\begin{cases} y = 2x - 3 \\ x + y = 5 \end{cases}$ $x + (2x - 3) = 5$
8. $\begin{cases} x = -3y + 4 \\ x - 2y = -1 \end{cases}$ $(-3y + 4) - 2y = -1$
9. $\begin{cases} -2x + 5y = 2 \\ 3y - x = -1 \end{cases}$ $-2(3y + 1) + 5y = 2$

WRITTEN EXERCISES

Goal: To solve systems of linear equations by the substitution method

Sample Problem: $\begin{cases} x + y = 5 \\ 2x - y = -2 \end{cases}$ **Answer:** $\{(1, 4)\}$

Solve each system by the substitution method. Check. (Examples 1–3)

1. $\begin{cases} y = x - 4 \\ 2x - 5y = 2 \end{cases}$ $\{(6, 2)\}$
2. $\begin{cases} y = -x + 2 \\ 2x - y = 1 \end{cases}$ $\{(1, 1)\}$
3. $\begin{cases} x - y = 4 \\ 2x - 3y = -2 \end{cases}$
4. $\begin{cases} x - y = 3 \\ 5x + y = -15 \end{cases}$ $\{(-2, -5)\}$
5. $\begin{cases} x - 2y = -2 \\ 2x - 3y = 2 \end{cases}$ $\{(10, 6)\}$
6. $\begin{cases} 2x + y = 5 \\ 8x - y = 45 \end{cases}$
7. $\begin{cases} y = -3x + 4 \\ x = -\frac{1}{3}y + 2 \end{cases}$ ϕ
8. $\begin{cases} x = 2y - 1 \\ y = \frac{1}{2}x - 3 \end{cases}$ ϕ
9. $\begin{cases} 4x - 8 = -y \\ 5x - 3 = -3y \end{cases}$
10. $\begin{cases} 3x - 16 = -2y \\ 7x - 19 = -y \end{cases}$ $\{(2, 5)\}$
11. $\begin{cases} y = \frac{5}{2}x - 2 \\ 11x - 4y = 8 \end{cases}$ $\{(0, -2)\}$
12. $\begin{cases} x = -\frac{3}{2}y + 2 \\ 2x + 4y = 4 \end{cases}$

400 / Chapter 15

Common Error
Students incorrectly substitute the expression for a variable in the other equation.

Example
$\begin{cases} 1. \quad x + 2y = 5 \\ 2. \ -3x - 2y = -3 \end{cases}$
1. $x = 5 - 2y$
2. $-3 \cdot 5 - 2y - 2y = -3$
$-15 - 4y = -3$

Prescription
Help students develop a habit of writing an expression for one variable within parentheses before making the substitution.

Assignment Guide
Basic
Day 1 p. 400: 1–11 odd
Day 2 p. 400: 2–12 even
Average p. 400: 1–11 odd
Above Average p. 400: 1–11 odd

Additional Answers
Written Exercises
3. $\{(14,10)\}$
6. $\{(5,-5)\}$
9. $\{(3,-4)\}$
12. $\{(2,0)\}$

████ **PROBLEM SOLVING AND APPLICATIONS** ████

15.2 **Digit Problems**

Teaching Suggestions p. M–46

Lesson Resources
Warm–Up: p. M–46
Maintenance: See below.
Practice Worksheet 15.2
Transparency 35

The ten numerals below are *digits*.

0, 1, 2, 3, 4, 5, 6, 7, 8, 9

Each place that a digit occupies in a whole number has a name. In this section you will work with two-digit numbers.

TEN'S   ONE'S
DIGIT   DIGIT

└──→ 78 ←──┘

Any two-digit number can be represented as shown below.

Let x = the ten's digit of a two-digit number.
Let y = the one's digit.

Two-digit number: $10x + y$

> It may be helpful to show that a number like 35 can be represented as $10 \cdot 3 + 5$.

P–1 **Why is it incorrect to represent such a two-digit number as *xy*?**
Because *xy* represents the product of the digits, not the number

 EXAMPLE 1

The one's digit of a two-digit number is 4 more than the ten's digit (Condition 1). The sum of the digits is 14 (Condition 2). Find the one's and the ten's digits. Write the number.

Solution:

1 Use Condition 1 to represent the unknowns. ──────→ Let x = the ten's digit.
Let y = the one's digit.
Then $10x + y$ = the number.

2 Use Condition 1 and Condition 2 to write the equations.

Think: The one's digit is 4 more than the ten's digit.

Translate: y = $x + 4$ **Condition 1**

Think: The sum of the digits is 14.
 Condition 2
Translate: $x + y$ = 14

$$\begin{cases} 1.\ y = x + 4 \\ 2.\ x + y = 14 \end{cases}$$ *System of linear equations*

Maintenance
1. Multiply: $(t - 3)(-7)$
 Ans: $-7t + 21$
 (Section 2.5)
2. Multiply:
 $(x - 3)(2x^2 - 3x + 1)$
 Ans: $2x^3 - 9x^2 + 10x - 3$
 (Section 8.5)
3. Factor: $16x^2 - 1$
 Ans: $(4x - 1)(4x + 1)$
 (Section 9.5)
4. Write the slope-intercept form of the line that contains the points (0,6) and $(-3,4)$.
 Ans: $y = \frac{2}{3}x + 6$
 (Section 13.4)
5. Solve the system below by the addition method.
 $$\begin{cases} 2x + y = 1 \\ 3x - y = -11 \end{cases}$$
 Ans: $\{(-2,5)\}$
 (Section 14.3)

Additional Examples
Example 1
1. The one's digit of a two-digit number is 1 more than the ten's digit (Condition 1). The sum of the digits is 15 (Condition 2). Find the one's and the ten's digits. Write the number.
 Ans: 78
2. The one's digit of a two-digit number is 3 more than the ten's digit (Condition 1). The sum of the digits is 9 (Condition 2). Find the one's digit and the ten's digit. Write the number.
 Ans: 36

More on Systems of Sentences / **401**

The substitution method is used.

$\boxed{3}$ Substitute $(x + 4)$ for y in
Equation **2**. Solve for x. ⟶

$x + y = 14$

$x + (x + 4) = 14$

$2x + 4 = 14$

$2x = 10$

$x = 5$ ◀ *Ten's digit*

$\boxed{4}$ Substitute 5 for x in Equation **1**.
Solve for y. ⟶

$y = x + 4$

$y = 5 + 4$

$y = 9$ ◀ *One's digit*

$\boxed{5}$ Write the two-digit number. ⟶ $10x + y = 10\,(5) + 9$

$= 50 + 9$

$= 59$

Check: Does $5 + 9 = 14$? Yes ✔   Does $9 - 5 = 4$? Yes ✔

The two-digit number is 59.

EXAMPLE 2   In a two-digit number, the one's digit is 4 more than the ten's digit (Condition 1). The two-digit number is 2 more than five times the one's digit (Condition 2). Find the ten's and one's digits. Write the number.

Solution:

$\boxed{1}$ Use Condition 1 to represent the unknowns. ⟶

Let $x =$ the ten's digit.
Let $y =$ the one's digit.
Then $10x + y =$ the number.

$\boxed{2}$ Use Condition 1 and Condition 2 to write the equations.

Think: The one's digit is 4 more than the ten's digit.

Translate: y $=$ $x + 4$ ◀ *Condition 1*

Think: The number is 2 more than five times the one's digit.

Translate: $10x + y$ $=$ $5y + 2$ ◀ *Condition 2*

$\begin{cases} \textbf{1.}\ y = x + 4 \\ \textbf{2.}\ 10x + y = 5y + 2 \end{cases}$

402 / *Chapter 15*

The substitution method is used.

|3| Substitute $(x + 4)$ for y in
Equation **2**. Solve for x. ⟶

$$10x + y = 5y + 2$$
$$10x + (x + 4) = 5(x + 4) + 2$$
$$11x + 4 = 5x + 20 + 2$$
$$6x + 4 = 22$$
$$6x = 18$$
$$x = 3 \quad \blacktriangleleft \quad \textit{Ten's digit}$$

|4| Replace x by 3 in Equation **1**.
Solve for y. ⟶

$$y = x + 4$$
$$y = 3 + 4$$
$$y = 7 \quad \blacktriangleleft \quad \textit{One's digit}$$

Check: Does $7 = 3 + 4$?   Yes ✔   Does $37 = 5(7) + 2$?   Yes ✔

The number is 37.

■■■■■ **CLASSROOM EXERCISES** ■■■■■

*Write an equation for each sentence. Represent each ten's digit by x and
each one's digit by y. (Steps 1 and 2 of Examples 1 and 2)*

1. The sum of the digits of a two-digit number is 12. $x + y = 12$

2. The ten's digit is 3 greater than the one's digit. $x = y + 3$

3. The one's digit is 5 less than the ten's digit. $y = x - 5$

4. The one's digit equals the ten's digit less 5. $y = x - 5$

5. Five times the ten's digit equals seven times the one's digit. $5x = 7y$

6. A two-digit number is 9 more than the sum of its digits. $10x + y = (x + y) + 9$

7. The sum of the digits of a two-digit number is 1 less than twice the
ten's digit. $x + y = 2x - 1$

8. A two-digit number is five times its one's digit. $10x + y = 5y$

9. The one's digit of a two-digit number diminished by the ten's digit
equals -8. $y - x = -8$

10. Twice the one's digit of a two-digit number exceeds the ten's digit by 3. $2y = x + 3$

*More on Systems of Sentences / **403***

Assignment Guide
Basic
Day 1 p. 404: 1–9 odd
Day 2 p. 404: 2–10 even
Average p. 404: 1–9 odd
Above Average p. 404:
1–9 odd

Problem-Solving Skills
Writing and solving an equation
(Ex. 1–10)

Additional Answers
Written Exercises

1. $\begin{cases} y = x - 2 \\ x + y = 12 \end{cases}$
  Ans: 75

2. $\begin{cases} y = x + 7 \\ x + y = 11 \end{cases}$
  Ans: 29

3. $\begin{cases} x = 2y \\ x = y + 4 \end{cases}$
  Ans: 84

4. $\begin{cases} y = 3x \\ y = x + 6 \end{cases}$
  Ans: 39

5. $\begin{cases} x + y = 15 \\ 2x = 3y - 5 \end{cases}$
  Ans: 87

6. $\begin{cases} x = y - 2 \\ y = 2x - 4 \end{cases}$
  Ans: 68

7. $\begin{cases} y = x + 1 \\ 10x + y = 8y + 2 \end{cases}$
  Ans: 34

8. $\begin{cases} 3y = x + 2 \\ 10x + y = 3 + 10x \end{cases}$
  Ans: 73

9. $\begin{cases} x - y = -4 \\ 10x + y = 2\,(x + y) + 3 \end{cases}$
  Ans: 15

10. $\begin{cases} y - x = -5 \\ 10x + y = 8(x + y) \end{cases}$
  Ans: 72

▮ WRITTEN EXERCISES ▮

Goal: To solve digit problems by systems of linear equations
Sample Problem: The ten's digit of a two-digit number is 1 less than 5 times the unit's digit (Condition 1). The sum of the digits is 11 (Condition 2). Write the two-digit number.
Answer: The number is 92.

For Exercises 1–10, use Condition 1 and two variables to represent the unknowns. Use Condition 1 (underscored once) and Condition 2 (underscored twice) to write a system of equations for the problem. Solve. (Examples 1 and 2)

1. The one's digit of a two-digit number is 2 less than the ten's digit. The sum of the digits is 12. Write the two-digit number.

2. The one's digit of a two-digit number is 7 more than the ten's digit. The sum of the digits is 11. Write the two-digit number.

3. The ten's digit of a two-digit number is twice the one's digit. The ten's digit exceeds the one's digit by 4. Write the two-digit number.

4. The one's digit of a two-digit number is three times the ten's digit. The one's digit exceeds the ten's digit by 6. Write the number.

5. The sum of the digits of a two-digit number is 15. Twice the ten's digit is 5 less than three times the one's digit. Write the two-digit number.

6. The ten's digit of a two-digit number is 2 less than the one's digit. The one's digit is 4 less than twice the ten's digit. Write the number.

7. The one's digit of a two-digit number exceeds the ten's digit by 1. The number is two more than 8 times the one's digit. Write the two-digit number.

8. Three times the one's digit of a two-digit number is 2 more than the ten's digit. The number is 3 more than 10 times the ten's digit. Write the two-digit number.

9. The ten's digit of a two-digit number decreased by the one's digit equals −4. The number is 3 more than twice the sum of its digits. Write the two-digit number.

10. The one's digit of a two-digit number decreased by the ten's digit equals −5. The number is 8 times the sum of its digits. Write the two-digit number.

REVIEW CAPSULE FOR SECTION 15.3

Solve and check. (Pages 71–73 and 100–102)

1. $0.15x - 3.8 = 3.4$ $x = 48$

2. $0.08x + 0.12(200 - x) = 0.10(200)$ $x = 100$

3. $0.18x + 0.15(450 - x) = 0.16(450)$ $x = 150$

4. $0.06(x - 300) + 0.015(300) = 0.03x$ $x = 450$

OBJECTIVE: To solve a word problem involving mixtures by using a system
of two linear equations

■■■■ **PROBLEM SOLVING AND APPLICATIONS** ■■■■

15.3 Mixture Problems

Per cents are often used to express the strength of a solution or the relative amount of a specific ingredient in a solution or mixture.

a. 18 grams (g) of salt in 100 grams of a solution of salt and water is called an 18% solution $\left(\frac{18}{100} = 18\%\right)$.

b. The amount of salt in 320 grams of a 15% solution of salt and water is 15% of 320 = 0.15 (320), or 48 grams.

P–1 **How much sugar is in 80 grams of a 25% solution of sugar and water?** 20 grams

Making a table will help you to organize the information and to identify Condition 1 and Condition 2.

EXAMPLE   A 10% salt solution and an 18% salt solution are mixed to obtain 320 grams of a 15% salt solution (Condition 1). How many grams of the 10% solution and of the 18% solution are needed (Condition 2)?

Solution:

1 Make a table.

Ask students why the number of grams of salt is 0.10*x* rather than 10*x*.

Solution	Grams of Solution (Condition 1)	Grams of Salt (Condition 2)
10%	x	0.10x
18%	y	0.18y
15%	320	0.15(320)

2 Use Condition 1 and Condition 2 to write the equations.

Think: Grams of 10% solution plus grams of 18% solution equal 320 grams.

Translate: x + y =   320   ◀ Condition 1

Think: Salt (g) in 10% solution plus Salt (g) in 18% solution equals Salt (g) in 15% solution

Translate: 0.10x +   0.18y =   0.15(320)   ◀ Condition 2

$\begin{cases} \textbf{1. } x + y = 320 \\ \textbf{2. } 0.10x + 0.18y = 0.15(320) \end{cases}$ ◀ System of linear equations

More on Systems of Sentences / **405**

Teaching Suggestions p. M–46

Lesson Resources
Warm–Up: p. M–47
Maintenance: See below.
Practice Worksheet 15.3
Exploration 15
Transparency 36

Maintenance
1. Solve: $6 - 5r = 3r + 30$
   Ans: $r = -3$
   (Section 4.3)
2. Simplify: $\sqrt{72x^2y^5}$
   Ans: $6xy^2\sqrt{2y}$
   (Section 7.4)
3. Factor: $2x^2 + 13x + 18$
   Ans: $(2x + 9)(x + 2)$
   (Section 9.4)
4. Write the two moves needed to locate the point $(-5, 3)$ in a coordinate plane.
   Ans: Left 5, Up 3
   (Section 12.1)
5. Solve the system below by the multiplication/addition method.
   $\begin{cases} 4x + 5y = 23 \\ 2x - 3y = -5 \end{cases}$
   Ans: $\{(2,3)\}$ (Section 14.4)

Additional Examples
1. A 20% acid solution and an 80% acid solution are mixed to obtain 90 grams of a 60% acid solution (Conditon 1). How many grams of the 20% solution and of the 80% solution are needed (Condition 2)?
   Ans: 20% solution: 30 grams; 80% solution: 60 grams
2. A 10% chlorine solution is mixed with 20 grams of a 15% chlorine solution to obtain a 12% chlorine solution (Condition 1). Find the number of grams of the 10% solution and of the 12% solution. (Condition 2).
   Ans: 10% solution: 30 grams; 12% soluiton: 50 grams

405

The substitution method is used.

③ Solve for y in Equation **1**. ⟶ $y = 320 - x$

④ Substitute $(320 - x)$ for y in $0.10x + 0.18y = 0.15(320)$
Equation **2**. Solve for x. ⟶ $0.10x + 0.18(320 - x) = 0.15(320)$

$$0.10x + 57.6 - 0.18x = 48$$
$$57.6 - 0.08x = 48$$
$$-0.08x = -9.6$$
$$\frac{-0.08x}{-0.08} = \frac{-9.6}{-0.08}$$
$$x = 120$$ ◀ **Number of grams of 10% solution**

⑤ Substitute 120 for x in Equation **1**. $x + y = 320$
Solve for y. ⟶ $120 + y = 320$
$$y = 200$$ ◀ **Number of grams of 18% solution**

Check: Does $120 + 200 = 320$?   Yes ✓
Does $0.10(120) + 0.18(200) = 0.15(320)$?   Yes ✓ Both conditions must always be checked.

Thus, 120 grams of the 10% solution and 200 grams of the 18% solution are needed.

CLASSROOM EXERCISES

Write the numeral missing from each row of the table for salt solutions.
(Example)

	Per Cent Solution	Amount of Mixture	Amount of Salt
1.	8%	100 grams	_?_ 8
2.	15%	40 grams	_?_ 6
3.	4%	350 milligrams	_?_ 14
4.	12%	x milligrams	_?_ $0.12x$
5.	6%	y milligrams	_?_ $0.06y$
6.	18%	$(x + 20)$ decigrams	_?_ $0.18(x + 20)$
7.	10%	_?_ 120	12 grams
8.	20%	_?_ 40	8 grams

406 / *Chapter 15*

Common Error
Students are perplexed by the mixture problems and cannot get started with the solutions.

Example
Written Exercises 5–8

Prescription
The use of the table for organizing the information is important to success in solving mixture problems. Extra time should also be spent examining the solution to the Example.

Write the numeral or numerals missing from each row of the table for acid solutions. (Example)

	Per Cent Solution	Amount of Mixture	Amount of Acid	Amount of Water
9.	3%	500 grams	_?_ 15	_?_ 485
10.	10%	_?_ 800	80 milligrams	_?_ 720
11.	25%	_?_ 40	10 grams	_?_ 30
12.	20%	x decigrams	_?_ 0.1x	_?_ 0.9x

▒▒▒▒▒ **WRITTEN EXERCISES** ▒▒▒▒▒

Goal: To solve mixture problems by systems of linear equations

Sample Problem: A 30% salt solution and a 60% salt solution are mixed to obtain 60 grams of a 50% solution (Condition 1). How many grams of the 30% solution and of the 60% solution are needed (Condition 2)?

Answer: 20 grams of the 30% solution and 40 grams of the 60% solution

In Exercises 1–8, Condition 1 is <u>underscored once</u>. *Condition 2 is* <u>underscored twice</u>. (Example)

Use Condition 1 and Condition 2 to complete each table. Use Condition 1 and Condition 2 to write a system of equations for each problem. Solve.

1. <u>A 4% acid solution and an 8% acid solution are mixed to obtain 400 grams of a 5% acid solution.</u> <u><u>How many grams of the 4% solution and of the 8% solution are needed?</u></u>

$$\begin{cases} x + y = 400 \\ 0.04x + 0.08y = 0.05\,(400) \end{cases}$$

Solution	Amount of Solution (Condition 1)	Amount of Pure Acid (Condition 2)	
4%	x	0.04x	Ans: 4%: 300g
8%	y	_?_ 0.08y	8%: 100g
5%	400	_?_ 0.05 (400)	

2. <u>A 12% salt solution and a 20% salt solution are mixed to obtain 24 kilograms of a 15% salt solution.</u> <u><u>How many kilograms of the 12% solution and of the 20% solution are needed?</u></u> (See the table at the top of page 408.)

*More on Systems of Sentences / **407***

Assignment Guide
Basic
Day 1 pp. 407–408: 1–4
Day 2 p. 408: 5–8
Average
Day 1 pp. 407–409: 1–4, 9
Day 2 pp. 408–409: 5–8, 11
Above Average pp. 407–409: 1–12

Problem–Solving Skills
Writing and solving an equation (Ex. 1–2)
Completing a table (Ex. 1–2)

Additional Answer
Written Exercises
2. $\begin{cases} x + y = 24 \\ 0.12x + 0.20y = 0.15(24) \end{cases}$
Ans: 12% solution: 15 kg
20% solution: 9 kg

Solution	Amount of Solution (Condition 1)	Amount of Salt (Condition 2)
12%	x	$0.12x$
20%	y	_?_ $0.20y$
15%	24	_?_ $0.15\,(24)$

3. A 12% salt solution is mixed with 48 grams of a 20% salt solution to obtain a 15% salt solution. Find the number of grams of the 12% solution and of the 15% solution.

Solution	Amount of Solution (Condition 1)	Amount of Salt (Condition 2)
12%	x	_?_ $0.12x$
20%	48	_?_ $0.20\,(48)$
_?_ 15%	y	_?_ $0.15y$

4. An 8% acid solution is mixed with 200 grams of a 4% acid solution to obtain a $5\frac{1}{2}$% acid solution. Find the number of grams of the 8% solution and of the $5\frac{1}{2}$% solution.

Solution	Amount of Solution (Condition 1)	Amount of Pure Acid (Condition 2)
8%	x	_?_ $0.08x$
4%	200	_?_ $0.04\,(200)$
_?_ $5\frac{1}{2}$%	y	_?_ $0.055y$

Use Condition 1, Condition 2, and two variables to make a table for each problem. Use Condition 1 and Condition 2 to write a system of equations for each problem. Solve.

5. A 15% vinegar solution is mixed with a 30% vinegar solution to obtain 150 grams of an 18% vinegar solution. How many grams of the 15% solution and of the 30% solution are needed?

6. A 4% iodine solution is mixed with a 12% iodine solution to obtain 300 grams of a 6% iodine solution. How many grams of the 4% solution and of the 12% solution are needed?

7. A $2\frac{1}{2}$% acid solution is mixed with 250 grams of a 10% acid solution to obtain a 4% acid solution. Find the number of grams of the $2\frac{1}{2}$% solution and of the 4% solution.

8. An 18% antifreeze solution is mixed with 200 grams of a 30% antifreeze solution to obtain a 22% antifreeze solution. Find the number of grams of the 18% solution and of the 22% solution.

NON-ROUTINE PROBLEMS

9. A woman was born in the first half of the nineteenth century. She was x years old in the year x^2. In what year was the woman born?

10. In a group of cows and chickens, the number of legs was 14 more than twice the number of heads. Find the number of cows.

11. Four friends, Luis, Maria, Carol, and Glenn, are sitting in a row. Neither Maria nor Carol are sitting next to Luis. Maria is not sitting next to Glenn. Carol is sitting just to the right of Maria. Write the seating arrangement from left to right.

12. Look at the figure below. Name two different things that you see.

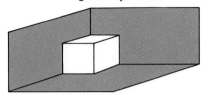

MID-CHAPTER REVIEW

Solve each system by the substitution method. (Section 15.1)

1. $\begin{cases} y = 2x - 3 \\ x - 2y = 9 \end{cases}$ $\{(-1, -5)\}$

2. $\begin{cases} x = -3y + 1 \\ 2x + 3y = 5 \end{cases}$ $\{(4, -1)\}$

3. $\begin{cases} 4x - y = 3 \\ 2x - y = -3 \end{cases}$ $\{(3, 9)\}$

For Exercises 4–7, Condition 1 is <u>underscored once</u>. Condition 2 is <u>underscored twice</u>.

Use Condition 1 and two variables to represent the unknowns for each problem. Use Condition 1 and Condition 2 to write a system of equations for each problem. Solve. (Section 15.2)

4. A two-digit number is 7 less than 4 times its one's digit. The one's digit is 1 more than 4 times the ten's digit. Write the two-digit number.

5. The ten's digit of a two-digit number is 2 more than twice the one's digit. The two-digit number is 11 more than 9 times its ten's digit. Write the two-digit number.

Use Condition 1, Condition 2, and two variables to set up a table for each problem. Use Condition 1 and Condition 2 to write a system of equations for each problem. Solve. (Section 15.3)

6. A 7% salt solution is mixed with a 15% salt solution to obtain 400 grams of a 12% salt solution. How many grams of the 7% solution and of the 15% solution are needed?

7. A 2% acid solution is mixed with a 3% acid solution to obtain 800 grams of a $2\frac{1}{2}$% acid solution. How many grams of the 2% solution and of the 3% solution are needed?

More on Systems of Sentences / **409**

Making Tables

Rate of Work

Rate of work on a job is the fraction of the job that can be completed in one unit of time.

$\frac{1}{5}$ is the rate of work on a job requiring 5 hours.

$\frac{1}{5}$ means "one-fifth of the job per hour."

In two hours, $2(\frac{1}{5})$ or $\frac{2}{5}$ of the job can be completed.

In three hours, $3(\frac{1}{5})$ or $\frac{3}{5}$ of the job can be completed.

$$\frac{2}{5} + \frac{3}{5} = 1$$ ◄ **The entire job is represented by 1.**

EXAMPLE 1: Two architects are working on the plans for a building. Jane estimates she could prepare the plans by herself in 100 hours. Melanie estimates she would need 120 hours for the entire job working alone. How many hours would the two need to prepare the plans working together? Round to the nearest whole hour.

SOLUTION:

Architect	Number of Hours Together on Job	Rate of Work	Part of Job Done
Jane	x	$\frac{1}{100}$	$x(\frac{1}{100})$, or $\frac{x}{100}$
Melanie	x	$\frac{1}{120}$	$x(\frac{1}{120})$, or $\frac{x}{120}$

Write an equation. ⟶ $\frac{x}{100} + \frac{x}{120} = 1$ ◄ **The entire job is represented by 1.**

Solve. ⟶ $600(\frac{x}{100} + \frac{x}{120}) = 600(1)$

$$6x + 5x = 600$$
$$11x = 600$$
$$x = 54\frac{6}{11}, \text{ or about } \textbf{55 hours}$$

EXAMPLE 2: A diesel–powered pump can drain a lake in 24 hours. After operating for 10 hours, the pump broke down. A smaller electric–powered pump, capable of doing the whole job in 36 hours, was substituted. How long did it take the smaller pump to complete the job?

SOLUTION:

Pump	Number of Hours on Job	Rate of Work	Part of Job Done
Diesel	10	$\frac{1}{24}$	$10\left(\frac{1}{24}\right)$, or $\frac{10}{24}$
Electric	x	$\frac{1}{36}$	$x\left(\frac{1}{36}\right)$, or $\frac{x}{36}$

Write an equation. $\longrightarrow$ $\frac{10}{24} + \frac{x}{36} = 1$

Solve. $\longrightarrow$ $72\left(\frac{10}{24} + \frac{x}{36}\right) = 72(1)$

$$30 + 2x = 72$$
$$2x = 42$$
$$x = 21 \text{ hours}$$

EXERCISES

1. An experienced roofer can roof a building alone in 20 hours. An apprentice roofer would need 30 hours working alone. How long would it take the two working together to do the job? 12 hours

2. Two printers are used to print mailing labels. The faster printer could do the job alone in 15 hours. The slower printer would require 35 hours working alone. How long would it take for both printers together to do the job? 10.5 hours

3. It takes 1000 hours for a worker to enter names and addresses onto a computer disk. After 450 hours this worker was replaced by a second worker who could do the entire job in 800 hours. How long did the second worker take to finish the job? 440 hours

4. A gravity drain can drain the city pool in 5 hours. A pump could drain the pool in 3 hours. Both methods were used together for one hour until the pump broke. How many hours were needed for the gravity drain to complete the job? $2\frac{1}{3}$ hours

More on Systems of Sentences / **411** ($3\frac{1}{3}$ hours in all)

Maintenance
1. The perimeter of a rectangular window is 168 inches. The length of the window is 48 inches. Find the width.
   Ans: 36 inches
   (Section 3.4)
2. Divide:
   $(12x^3 - 6x^2) \div 2x$
   Ans: $6x^2 - 3x$
   (Section 8.6)
3. Multiply and simplify:
   $\dfrac{3a - 6}{4} \cdot \dfrac{2a + 4}{a^2 - 4}$
   Ans: $\dfrac{3}{2}$ (Section 10.5)
4. Find the least common multiple of $(x^2 + 6x + 9)$ and $(x^2 - 9)$.
   Ans: $(x - 3)(x + 3)^2$
   (Section 11.3)
5. Write the slope and y intercept of the graph of $y = 2(3x - 1)$.
   Ans: slope: 6; y intercept: -2
   (Section 13.3)

Additional Examples
Example 1
Compute the value of each determinant.
1. $\begin{vmatrix} 3 & 6 \\ 9 & -2 \end{vmatrix}$
   Ans: -60
2. $\begin{vmatrix} -3 & -5 \\ 0 & 4 \end{vmatrix}$
   Ans: -12
3. $\begin{vmatrix} 7 & -11 \\ -2 & 4 \end{vmatrix}$
   Ans: 6

15.4 2 × 2 Determinants

OBJECTIVES: To compute the value of a 2 x 2 determinant
To write the determinant of a system of two linear equations and compute its value

Another method for solving systems of equations uses <u>determinants</u>.

Definition
Explain that this section is preparatory to Section 15.5 where determinants will be applied.

A **determinant** is a square array of the form $\begin{vmatrix} a & b \\ c & d \end{vmatrix}$ in which a, b, c, and d are real numbers.

$\begin{vmatrix} -3 & \frac{1}{2} \\ 5 & \sqrt{2} \end{vmatrix}$

Only 2 × 2 (two-by-two) determinants are used in this book. They have <u>two rows</u> and <u>two columns</u>.

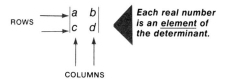

ROWS $\rightarrow \begin{vmatrix} a & b \\ c & d \end{vmatrix}$

Each real number is an <u>element</u> of the determinant.

COLUMNS

P–1 **What are the elements of the first row of the determinant at the right? the elements of the second column?** First row: π, -2
Second column: -2, $-\sqrt{3}$

$\begin{vmatrix} \pi & -2 \\ 0.6 & -\sqrt{3} \end{vmatrix}$

Each 2 × 2 determinant represents a real number.
$\begin{vmatrix} a & b \\ c & d \end{vmatrix} = ad - bc$

$\begin{vmatrix} 2 & 4 \\ 1 & 5 \end{vmatrix} = 2 \cdot 5 - 4 \cdot 1$
$= 6$

EXAMPLE 1   Compute the value of $\begin{vmatrix} 2 & 4 \\ -3 & 1 \end{vmatrix}$.

Solution: $\begin{vmatrix} 2 & 4 \\ -3 & 1 \end{vmatrix} = (2)(1) - (4)(-3)$

$= 2 - (-12)$

$= 2 + 12$

$= 14$

What is the value of each determinant below?

a. $\begin{vmatrix} 2 & 5 \\ 3 & 0 \end{vmatrix}$ -15 **b.** $\begin{vmatrix} 2 & 3 \\ 2 & 3 \end{vmatrix}$ 0 **c.** $\begin{vmatrix} 1 & 1 \\ -1 & 1 \end{vmatrix}$ 2

What expression represents the value of each determinant?

a. $\begin{vmatrix} p & q \\ 3 & 2 \end{vmatrix}$ **b.** $\begin{vmatrix} -x & x \\ 3 & 2 \end{vmatrix}$ **c.** $\begin{vmatrix} -s & t \\ -t & -s \end{vmatrix}$ $s^2 + t^2$

$2p - 3q$ $-5x$

The equations of the system must be written in **standard form.**

$$\begin{cases} 2x - 3y = -1 \\ 4x + 3y = 16 \end{cases}$$ ◄ **Standard form**

The **determinant of a system of equations** is formed with the coefficients of the variables as elements of the determinant.

Standard form ►
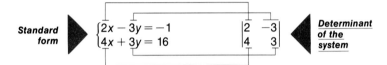
$$\begin{cases} 2x - 3y = -1 \\ 4x + 3y = 16 \end{cases} \qquad \begin{vmatrix} 2 & -3 \\ 4 & 3 \end{vmatrix}$$ ◄ **Determinant of the system**

EXAMPLE 2   Write the determinant of the system. Then compute its value.

$$\begin{cases} y - 5 = -3x \\ -2x + y = -3 \end{cases}$$

Solution: Write the equations in standard form. ⟶ $\begin{cases} 3x + y = 5 \\ -2x + y = -3 \end{cases}$

$\begin{vmatrix} 3 & 1 \\ -2 & 1 \end{vmatrix} = (3)(1) - (1)(-2)$ ◄ **Determinant of the system**

$= 3 - (-2)$

$= 3 + 2$

$= 5$ ◄ **Value of the determinant**

CLASSROOM EXERCISES

Write the value of each determinant. (Example 1)

1. $\begin{vmatrix} 1 & 2 \\ 1 & 4 \end{vmatrix}$ 2
2. $\begin{vmatrix} -2 & 0 \\ 1 & 3 \end{vmatrix}$ -6
3. $\begin{vmatrix} 2 & 3 \\ -1 & 5 \end{vmatrix}$ 13
4. $\begin{vmatrix} -1 & 1 \\ 0 & 1 \end{vmatrix}$

5. $\begin{vmatrix} 0 & 3 \\ 1 & 2 \end{vmatrix}$ -3
6. $\begin{vmatrix} 5 & -1 \\ 2 & -3 \end{vmatrix}$ -13
7. $\begin{vmatrix} 1 & -3 \\ 5 & 2 \end{vmatrix}$ 17
8. $\begin{vmatrix} 1 & -3 \\ 0 & 2 \end{vmatrix}$

9. $\begin{vmatrix} 2 & -3 \\ -1 & 5 \end{vmatrix}$ 7
10. $\begin{vmatrix} -1 & -4 \\ -3 & -2 \end{vmatrix}$ -10
11. $\begin{vmatrix} -5 & 1 \\ 3 & 2 \end{vmatrix}$ -13
12. $\begin{vmatrix} 3 & 6 \\ 2 & 4 \end{vmatrix}$

13. $\begin{vmatrix} -3 & -4 \\ -6 & 8 \end{vmatrix}$ -48
14. $\begin{vmatrix} 2 & -3 \\ 2 & -3 \end{vmatrix}$ 0
15. $\begin{vmatrix} 1 & -5 \\ -1 & 5 \end{vmatrix}$ 0
16. $\begin{vmatrix} \frac{1}{2} & -3 \\ 2 & 4 \end{vmatrix}$

17. $\begin{vmatrix} 1 & 6 \\ -\frac{1}{2} & 3 \end{vmatrix}$ 6
18. $\begin{vmatrix} 1 & 1 \\ -1 & 1 \end{vmatrix}$ 2
19. $\begin{vmatrix} \frac{1}{2} & \frac{1}{2} \\ 2 & -4 \end{vmatrix}$ -3
20. $\begin{vmatrix} -5 & -3 \\ -1 & -2 \end{vmatrix}$

WRITTEN EXERCISES

Goals: To write the determinant of a linear system and to compute the value of the determinant

Sample Problem: $\begin{cases} 2x - 3y = -6 \\ -x + 5y = 1 \end{cases}$ **Answer:** $\begin{vmatrix} 2 & -3 \\ -1 & 5 \end{vmatrix}$: 7

Compute the value of each determinant. (Example 1)

1. $\begin{vmatrix} -6 & 5 \\ 2 & 3 \end{vmatrix}$ -28
2. $\begin{vmatrix} 5 & 6 \\ 2 & -7 \end{vmatrix}$ -47
3. $\begin{vmatrix} 6 & -12 \\ -3 & 6 \end{vmatrix}$ 0
4. $\begin{vmatrix} -4 & 16 \\ 2 & -8 \end{vmatrix}$

5. $\begin{vmatrix} 1 & 5 \\ 0 & -2 \end{vmatrix}$ -2
6. $\begin{vmatrix} 2 & 0 \\ -10 & -3 \end{vmatrix}$ -6
7. $\begin{vmatrix} \frac{1}{3} & -4 \\ \frac{1}{2} & 6 \end{vmatrix}$ 4
8. $\begin{vmatrix} -\frac{1}{2} & -\frac{1}{3} \\ 9 & 8 \end{vmatrix}$

9. $\begin{vmatrix} -1 & 1 \\ 1 & -1 \end{vmatrix}$ 0
10. $\begin{vmatrix} 5 & 3 \\ 9 & 3 \end{vmatrix}$ -12
11. $\begin{vmatrix} 1 & -15 \\ -15 & 2 \end{vmatrix}$ -223
12. $\begin{vmatrix} -4 & -3 \\ 3 & -4 \end{vmatrix}$

Write the determinant of the system. Then compute its value. (Example 2)

13. $\begin{cases} 2x - y = -3 \\ x - 3y = 5 \end{cases}$ $\begin{vmatrix} 2 & -1 \\ 1 & -3 \end{vmatrix}$; -5
14. $\begin{cases} x - 3y = -5 \\ 5x + 2y = 3 \end{cases}$ $\begin{vmatrix} 1 & -3 \\ 5 & 2 \end{vmatrix}$; 17
15. $\begin{cases} -2x + 5y = 1 \\ x - 3y = -8 \end{cases}$

16. $\begin{cases} 4x - 3y = -2 \\ -3x - y = 5 \end{cases}$ $\begin{vmatrix} 4 & -3 \\ -3 & -1 \end{vmatrix}$; -13
17. $\begin{cases} y = 2x - 5 \\ y = 5x - 2 \end{cases}$ $\begin{vmatrix} 2 & -1 \\ 5 & -1 \end{vmatrix}$; 3
18. $\begin{cases} y = -3x + 7 \\ y = x - 13 \end{cases}$

19. $\begin{cases} 5x - 2 = -3y \\ y - 3x = -2 \end{cases}$ $\begin{vmatrix} 5 & 3 \\ -3 & 1 \end{vmatrix}$; 14
20. $\begin{cases} -3y + 10 = 5x \\ 2x - 8 = -3y \end{cases}$ $\begin{vmatrix} -5 & -3 \\ 2 & 3 \end{vmatrix}$; -9
21. $\begin{cases} 5y - 7 = -4x \\ 5x + 6y = 8 \end{cases}$

15.5 Solving Systems by Determinants

OBJECTIVES: To solve a system of two linear equations by determinants
To tell by use of determinants whether or not the graph of a system of two linear equations is a pair of parallel lines

Teaching Suggestions p. M–46

Lesson Resources
Warm–Up: p. M–47
Maintenance: See below.
Practice Worksheet 15.5
Transparency 37

Systems of equations can be solved by the method of determinants.

Remind students to make sure the equations are in this standard determinant form.

$$\begin{cases} 2x - y = 5 \\ x + 3y = -2 \end{cases}$$

$$x = \frac{\begin{vmatrix} 5 & -1 \\ -2 & 3 \end{vmatrix}}{\begin{vmatrix} 2 & -1 \\ 1 & 3 \end{vmatrix}}$$

◄ **First column has constant terms.**
Second column has y coefficients.

◄ **Determinant of the system**

$$y = \frac{\begin{vmatrix} 2 & 5 \\ 1 & -2 \end{vmatrix}}{\begin{vmatrix} 2 & -1 \\ 1 & 3 \end{vmatrix}}$$

◄ **First column has x coefficients.**
Second column has constant terms.

◄ **Determinant of the system**

P–1 What is the value of the numerator determinant for *x*? 13 the numerator determinant for *y*? −9 the determinant of the system? 7

The values of *x* and *y* are computed.

$$x = \frac{13}{7} \qquad y = \frac{-9}{7}$$

The solution set of the system is $\{(\frac{13}{7}, -\frac{9}{7})\}$.

EXAMPLE 1 Solve the following system by the method of determinants.

$$\begin{cases} x + 2 = 3y \\ 5y - 2x = 10 \end{cases}$$

Solution:

1. Write the equations in standard form. $\longrightarrow$ $\begin{cases} x - 3y = -2 \\ -2x + 5y = 10 \end{cases}$

2. Write *x* and *y* as quotients of determinants.

$$x = \frac{\begin{vmatrix} -2 & -3 \\ 10 & 5 \end{vmatrix}}{\begin{vmatrix} 1 & -3 \\ -2 & 5 \end{vmatrix}} \qquad y = \frac{\begin{vmatrix} 1 & -2 \\ -2 & 10 \end{vmatrix}}{\begin{vmatrix} 1 & -3 \\ -2 & 5 \end{vmatrix}}$$

More on Systems of Sentences / **415**

Maintenance
1. On a winter day, the wind chill temperature was twice the thermometer reading of −9°F. What was the wind chill temperature?
Ans: −18°F (Section 1.4)
2. Simplify: $(3t^{-2}p^4)^2$
Ans: $\dfrac{9p^8}{t^4}$ (Section 6.5)
3. Factor:
$x^2 + 3x + 2$
Ans: (x + 2)(x + 1)
(Section 9.3)
4. Add: $\dfrac{6}{x} + \dfrac{3}{xy}$
Ans: $\dfrac{6y + 3}{xy}$
(Section 11.4)
5. The sum of two numbers is 14 (Condition 1). The difference of the two numbers is 22 (Condition 2). Find the two numbers.
Ans: 18 and −4
(Section 14.5)

Additional Examples
Example 1
Solve each system by the method of determinants.
1. $\begin{cases} 5x + 2y = 7 \\ 5 - 4y = x \end{cases}$
Ans: {(1,1)}
2. $\begin{cases} 2x - 7y = 3 \\ 5x - 4y = -6 \end{cases}$
Ans: {(−2,−1)}
3. $\begin{cases} -2y = -3x + 6 \\ 5x - 41 = -7y \end{cases}$
Ans: {(4,3)}

$$\boxed{3}\ \text{Solve.} \longrightarrow x = \frac{(-2)(5) - (-3)(10)}{(1)(5) - (-3)(-2)} \qquad y = \frac{(1)(10) - (-2)(-2)}{(1)(5) - (-3)(-2)}$$

$$x = \frac{20}{-1} \qquad\qquad y = \frac{6}{-1}$$

$$x = -20 \qquad\qquad y = -6$$

The solution set of the system is $\{(-20, -6)\}$. If graphs of the equations are drawn, the lines intersect at the point $(-20, -6)$.

Additional Examples
Example 2
Solve each system by the method of determinants.

1. $\begin{cases} 4x + 6y = 11 \\ 6x + 9y = 11 \end{cases}$

Ans: ϕ

2. $\begin{cases} y = 5 - 3x \\ 9x + 3y = 4 \end{cases}$

Ans: ϕ

EXAMPLE 2   Solve the following system by the method of determinants.

$$\begin{cases} x - 2y = 5 \\ 2x - 4y = -3 \end{cases} \qquad \blacktriangleleft \text{ Standard form}$$

Solution:

$$x = \frac{\begin{vmatrix} 5 & -2 \\ -3 & -4 \end{vmatrix}}{\begin{vmatrix} 1 & -2 \\ 2 & -4 \end{vmatrix}} \qquad\qquad y = \frac{\begin{vmatrix} 1 & 5 \\ 2 & -3 \end{vmatrix}}{\begin{vmatrix} 1 & -2 \\ 2 & -4 \end{vmatrix}}$$

$$x = \frac{(5)(-4) - (-2)(-3)}{(1)(-4) - (-2)(2)} \qquad y = \frac{(1)(-3) - (5)(2)}{(1)(-4) - (-2)(2)}$$

$$x = \frac{-26}{0} \qquad\qquad y = \frac{-13}{0} \qquad \blacktriangleleft \text{ Division by zero is not possible.}$$

The two equations are changed to slope-intercept form below.

$$x - 2y = 5 \qquad\qquad 2x - 4y = -3$$

$$y = \tfrac{1}{2}x - \tfrac{5}{2} \qquad\qquad y = \tfrac{2}{4}x + \tfrac{3}{4} \qquad \blacktriangleleft \text{ The slopes are equal.}$$

The graphs of the two equations are parallel lines.
The solution set of the system is ϕ.

Example 2 suggests the following.

> If only the determinant of a system of equations equals 0, then the solution set of the system is empty.

CLASSROOM EXERCISES

Write the missing columns of elements. (Step 2 of Example 1)

1. $\begin{cases} 3x - y = 2 \\ x + 2y = -1 \end{cases}$

$$x = \dfrac{\begin{vmatrix} 2 & \!\!\begin{smallmatrix}? & -1\\? & 2\end{smallmatrix} \\ -1 & \end{vmatrix}}{\begin{vmatrix} 3 & -1 \\ 1 & 2 \end{vmatrix}}, \quad y = \dfrac{\begin{vmatrix} 3 & \!\!\begin{smallmatrix}?\\?\end{smallmatrix} \\ 1 & \end{vmatrix}\begin{smallmatrix}2\\-1\end{smallmatrix}}{\begin{vmatrix} 3 & -1 \\ 1 & 2 \end{vmatrix}}$$

2. $\begin{cases} -x + 4y = 6 \\ 2x - 3y = -1 \end{cases}$

$$x = \dfrac{\begin{vmatrix} 6 & \!\!\begin{smallmatrix}? & 4\\? & -3\end{smallmatrix} \\ -1 & \end{vmatrix}}{\begin{vmatrix} -1 & ? \\ 2 & ? \end{vmatrix}}, \quad y = \dfrac{\begin{vmatrix} -1 & \!\!\begin{smallmatrix}?\\?\end{smallmatrix} \\ 2 & \end{vmatrix}\begin{smallmatrix}6\\-1\end{smallmatrix}}{\begin{vmatrix} 4 & \\ -3 & \end{vmatrix}\begin{smallmatrix}1?\\2?\end{smallmatrix}\begin{smallmatrix}4\\-3\end{smallmatrix}}$$

For each system, write x and y as quotients of determinants.
(Step 2 of Example 1)

3. $\begin{cases} 0.5x - 0.5y = 1.5 \\ -1.5x + 1.5y = -2.5 \end{cases}$

4. $\begin{cases} x - 2y = -1 \\ 2x - y = 2 \end{cases}$

WRITTEN EXERCISES

Goal: To solve systems of linear equations by the method of determinants

Sample Problem: $\begin{cases} 3x + 7y = -4 \\ 2x + 5y = -3 \end{cases}$ **Answer:** $\{(1, -1)\}$

Solve each system by the method of determinants. (Examples 1 and 2)

1. $\begin{cases} x + y = 6 \\ x - 2y = 3 \end{cases}$ $\{(5, 1)\}$

2. $\begin{cases} 2x - y = 1 \\ x + y = -2 \end{cases}$ $\{(-\frac{1}{3}, -\frac{5}{3})\}$

3. $\begin{cases} x - 3y = -5 \\ 2x + 5y = 1 \end{cases}$

4. $\begin{cases} 2x - y = 3 \\ x + 3y = 5 \end{cases}$ $\{(2, 1)\}$

5. $\begin{cases} 4x + 3y = -6 \\ 5x - 2y = 3 \end{cases}$ $\{(-\frac{3}{23}, -\frac{42}{23})\}$

6. $\begin{cases} 2x - 5y = 10 \\ x - 6y = 8 \end{cases}$

7. $\begin{cases} x + 5 = 3y \\ 5x = 3y + 2 \end{cases}$ $\{(\frac{7}{4}, \frac{9}{4})\}$

8. $\begin{cases} -2y + 3x = 5 \\ 5y - x = -7 \end{cases}$ $\{(\frac{11}{13}, -\frac{16}{13})\}$

9. $\begin{cases} \frac{1}{3}x - 8y = -4 \\ \frac{3}{4}x + 6y = 3 \end{cases}$

10. $\begin{cases} x - 3y = -5 \\ -2x + 6y = 6 \end{cases}$ ϕ

11. $\begin{cases} 2x - y = -3 \\ -6x + 3y = 6 \end{cases}$ ϕ

12. $\begin{cases} -\frac{1}{4}x + 6y = -6 \\ \frac{1}{2}x - 4y = 12 \end{cases}$

REVIEW CAPSULE FOR SECTION 15.6

Write an algebraic expression. (Pages 74–77)

1. The value of n nickels in cents $5n$

2. The value of t dollars in cents $100t$

3. The cost of 5 records at w cents each $5w$

4. The cost of 12 pencils at y cents each $12y$

5. The cost in cents of k tickets at $8.50 each $850k$

6. The cost in cents of n cards at 75 cents each $75n$

More on Systems of Sentences / **417**

Additional Answers
Classroom Exercises

3. $$x = \dfrac{\begin{vmatrix} 1.5 & -0.5 \\ -2.5 & 1.5 \end{vmatrix}}{\begin{vmatrix} 0.5 & -0.5 \\ -1.5 & 1.5 \end{vmatrix}},$$

$$y = \dfrac{\begin{vmatrix} 0.5 & 1.5 \\ -1.5 & -2.5 \end{vmatrix}}{\begin{vmatrix} 0.5 & -0.5 \\ -1.5 & 1.5 \end{vmatrix}}$$

4. $$x = \dfrac{\begin{vmatrix} -1 & -2 \\ 2 & -1 \end{vmatrix}}{\begin{vmatrix} 1 & -2 \\ 2 & -1 \end{vmatrix}}, y = \dfrac{\begin{vmatrix} 1 & -1 \\ 2 & 2 \end{vmatrix}}{\begin{vmatrix} 1 & -2 \\ 2 & -1 \end{vmatrix}}$$

Assignment Guide
Basic Omit
Average p. 417: 1–11 odd
Above Average p. 417:
1–11 odd; p. 414: 1–21 odd

Additional Answers
Written Exercises

3. $\{(-2,1)\}$
6. $\{(\frac{20}{7}, \frac{-6}{7})\}$
9. $\{(0, \frac{1}{2})\}$
12. $\{(24,0)\}$

417

Maintenance

1. Solve: $-3x - 12 < 15$
   Ans: $x > -9$
   (Section 5.6)
2. Subtract:
   $(2x^2 - 3x + 2) - (3x^2 - 1)$
   Ans: $-x^2 - 3x + 3$
   (Section 8.4)
3. Evaluate:
   $\dfrac{x^2 - 5x}{x - 2}$ if $x = -1$
   Ans: -2 (Section 10.1)
4. Is the following relation a function?
   $\{(2,4), (-1,-1),(5,4)\}$
   Ans: Yes (Section 12.4)
5. Write the slope of the line that contains the points $(5, -1)$ and $(-3,2)$.
   Ans: $-\dfrac{3}{8}$ (Section 13.4)

Additional Example
Example 1
Roger bought a total of 7 T-shirts and pairs of socks (Condition 1). The T-shirts cost $4 each and the pairs of socks cost $3 each. He paid a total of $25 for the T-shirts and socks (Condition 2). How many of each did he buy?
Ans: T-shirts: 4
　　　pairs of socks: 3

OBJECTIVE: To solve a word problem involving costs by using a system of two linear equations

PROBLEM SOLVING AND APPLICATIONS

15.6 Money Problems

Word problems that involve the cost of two items can often be solved by systems of two linear equations. One equation usually represents the total number of items (Condition 1). The second equation represents the total cost of the items (Condition 2).

EXAMPLE 1　　Jane bought a total of 11 pencils and pens (Condition 1). The pencils cost 8 cents each and the pens cost 39 cents each. She paid a total of $1.50 for the pencils and pens (Condition 2). How many of each did she buy?

Solution:　1　Make a table to help you write the equations. Write all costs in cents.

Emphasize that total costs must be expressed either in dollars or in cents.

	Number of Items (Condition 1)	Price Per Item	Total Cost (Condition 2)
Pencil	x	8¢	$8x$
Pen	y	39¢	$39x$
Total	11		150

2　Equations: $\begin{cases} \textbf{1.}\ x + y = 11 \\ \textbf{2.}\ 8x + 39y = 150 \end{cases}$　　**1. Condition 1**
　　　　　　　　　　　　　　　　　　　　2. Condition 2

3　Solve. The method of determinants is shown. Other methods could also be used.

$$x = \frac{\begin{vmatrix} 11 & 1 \\ 150 & 39 \end{vmatrix}}{\begin{vmatrix} 1 & 1 \\ 8 & 39 \end{vmatrix}} = \frac{429 - 150}{39 - 8} \qquad y = \frac{\begin{vmatrix} 1 & 11 \\ 8 & 150 \end{vmatrix}}{31} = \frac{150 - 88}{31}$$

$$= \frac{279}{31} \qquad\qquad\qquad = \frac{62}{31}$$

$$= 9 \qquad\qquad\qquad\quad = 2$$

4　**Check:**　Condition 1　Did Jane buy a total of 11 pencils and pens?
　　　　　　　　　　　　Does $9 + 2 = 11$?　Yes ✔

　　　　　　　Condition 2　Was the total cost $1.50, or 150 cents?
　　　　　　　　　　　　Does $9(8) + 2(39) = 150$?　Yes ✔

Jane bought 9 pencils and 2 pens.

EXAMPLE 2   A school's Ski Club is planning a trip. A small bus
holds 6 fewer passengers than a large bus (Condition
1). A large bus will cost $18 per student if it is filled
and a small bus will cost $20 per student. The
difference between the cost of the large bus and the
small bus is $30 (Condition 2). Find the number of
passengers that each bus can carry.

Solution: 1

	Number of Passengers (Condition 1)	Cost Per Passenger	Total Cost (Condition 2)
Large bus	x	$18	$18x$
Small bus	y	$20	$20y$

2 Write the equations.

Think: The small bus carries 6 fewer than the large bus.

Translate: y = $x - 6$

Think: Cost of large bus minus cost of small bus is $30.

Translate: $18x$ — $20y$ = 30

$\begin{cases} \textbf{1. } y = x - 6 \\ \textbf{2. } 18x - 20y = 30 \end{cases}$ ◀   **1. Condition 1**
2. Condition 2

The substitution method is used.

3 Substitute $(x - 6)$ for y in Equation **2**. Solve for x.

$$18x - 20\textbf{y} = 30$$
$$18x - 20(\textbf{x} - \textbf{6}) = 30$$
$$18x - 20x + 120 = 30$$
$$-2x + 120 = 30$$
$$-2x = -90$$
$$x = 45$$ ◀ *Capacity of large bus*

4 Substitute 45 for x in Equation **1**. Solve for y.

$$y = \textbf{x} - 6$$
$$y = \textbf{45} - 6$$
$$y = 39$$ ◀ *Capacity of small bus*

The check is left for you.
Be sure to check both conditions.

Common Error
Students do not understand why the decimal point is not used in certain money problems.

Example
In Classroom Exercise 1, the second equation is written as $5x + 10y = 6.50$.

Prescription
Explain that $6.50 is the same as 650¢. Show students that they can use the equation $0.05x + 0.10y = 6.50$ but that $5x + 10y = 650$ is probably more convenient.

CLASSROOM EXERCISES

Complete the table. Then use the table to write two equations for the problem. (Examples 1 and 2)

1. Mike takes 60 coins consisting of *x* nickels and *y* dimes to the bank. The total deposit is $6.50. $x + y = 60; 5x + 10y = 650$

	Number of Coins	Value of Each	Total Value
Nickels	x	5¢	? 5x
Dimes	y	10¢	? 10y
Total	60		? 650

2. The Junior Class sold *x* boxes of pecans at $6 each and *y* boxes of peanuts at $5 each. They sold a total of 82 boxes for $440.
$x + y = 82; 6x + 5y = 440$

	Number of Boxes	Cost Per Box	Total Cost
Pecans	x	$6	? 6x
Peanuts	y	? $5	? 5y
Total	? 82		440

3. During a special sale, an appliance store sold *x* color TV sets for $325 each and *y* black and white TV sets for $135 each. The total receipts from the sale were $11,415 for selling 55 sets. $x + y = 55; 325x + 135y = 11,415$

	Number of Sets	Price Per Set	Total Receipts
Color	? x	$325	? 325x
Black and White	? y	$135	? 135y
Total	? 55		11,415

4. Gina bought *x* cans of peas for 85 cents each and *y* cans of corn for 78 cents each. She bought 2 more cans of peas than cans of corn. The total cost was $6.59. $x = y + 2; 85x + 78y = 659$

	Number of Cans	Price Per Can	Total Cost
Peas	? x	? 85¢	? 85x
Corn	? y	78¢	? 78y

5. At a recent bake sale, the Booster Club sold *x* cookies at 15 cents each and *y* cupcakes at 25 cents each. There were 20 more cookies sold than cupcakes. The total receipts from the sale were $15.00.

	Number of Item	Cost Per Item	Total Cost
Cookies	? x	15¢	? 15x
Cupcakes	? y	? 25¢	? 25y

$x = y + 20; 15x + 25y = 1500$

WRITTEN EXERCISES

Goal: To solve problems involving costs by systems of linear equations

Sample Problem: Allyson has a total of 20 dimes and quarters (Condition 1). The total value of the coins is $3.95 (Condition 2). How many of each coin does she have?

Answer: She has 7 dimes and 13 quarters.

For Exercises 1–8, Condition 1 is <u>underscored once</u>. Condition 2 is <u>underscored twice</u>.
   a. *Use Condition 1, Condition 2, and two variables to make a table.*
   b. *Use Condition 1 and Condition 2 to write a system of equations for the problem.*
   c. *Solve. (Examples 1 and 2)*

1. Brad Whitefoot bought a total of 9 records and cassette tapes. The records cost $6 each, and the cassette tapes cost $7 each. He paid a total of $58. How many of each did he buy?

2. A school lunch stand sold 230 hamburgers and hot dogs in one day for total receipts of $176.50. The hamburgers sold for 80 cents each and the hot dogs for 65 cents each. How many of each were sold?

3. Marc's scout troop bought 50 cans of beans and tomatoes for a camping trip at a total cost of $20.30. The tomatoes cost 35 cents per can and the beans 45 cents per can. How many cans of each kind were purchased?

4. Michelle sold two types of calendars for her club. She sold a total of 64 calendars for $140.50. One type of calendar sold for $2.50 each, and the other type sold for $1.75 each. How many calendars of each type did she sell?

5. Raphael deposits a total of 86 five- and ten-dollar bills. The amount of the deposit is $675. How many of each kind of bill are there?

6. Betty has 53 nickels and dimes in her bank. The total value of the coins is $3.75. How many of each type of coin does she have?

7. Susie won the ticket selling contest for the school carnival. She sold 82 tickets for total sales of $32.45. The student tickets sold for 35 cents and the adult tickets for 50 cents. How many tickets of each kind did she sell?

8. The Northeast Riders 4H Club sold orange juice and cocoa at their horse show. The orange juice cost 50 cents per glass, and the cocoa cost 75 cents per cup. They sold 20 more cups of cocoa than glasses of orange juice. How much of each did they sell if the total receipts were $45.00?

Assignment Guide
Basic
Day 1 p. 421: 1–4
Day 2 p. 421: 5–8
Average p. 421: 1–8
Above Average p. 421: 1–8

Problem–Solving Skills
Making a table (Ex. 1–8)
Writing and solving an equation (Ex. 1–8)

Additional Answers
1. a. Records: x; tapes: y
   b. $\begin{cases} x + y = 9 \\ 6x + 7y = 58 \end{cases}$
   c. 5 records, 4 tapes
2. a. Hamburgers: x; hot dogs: y
   b. $\begin{cases} x + y = 230 \\ 80x + 65y = 17,650 \end{cases}$
   c. 180 hamburgers, 50 hot dogs
3. a. Tomatoes: x; beans: y
   b. $\begin{cases} x + y = 50 \\ 35x + 45y = 2030 \end{cases}$
   c. 22 cans of tomatoes, 28 cans of beans
4. a. $2.50 calendars: x
      $1.75 calandars: y
   b. $\begin{cases} x + y = 64 \\ 250x + 175y = 14,050 \end{cases}$
   c. 38 calendars at $2.50; 26 caleandars at $1.75
5. a. Fives: x; tens: y
   b. $\begin{cases} x + y = 86 \\ 5x + 10y = 675 \end{cases}$
   c. 37 fives, 49 tens
6. a. Nickels: x; dimes: y
   b. $\begin{cases} x + y = 53 \\ 5x + 10y = 375 \end{cases}$
   c. 31 nickels, 22 dimes
7. a. Student tickets: x; adult tickets: y
   b. $\begin{cases} x + y = 82 \\ 35x + 50y = 3245 \end{cases}$
   c. 57 student tickets; 25 adult tickets
8. a. Juice: x; cocoa: y
   b. $\begin{cases} y = x + 20 \\ 50x + 75y = 4500 \end{cases}$
   c. 24 glasses of juice; 44 cups of cocoa

PROBLEM SOLVING AND APPLICATIONS

15.7 Distance/Rate/Time

The basic formula $d = rt$ relates distance, rate, and time.

EXAMPLE 1   Alexander travels a distance of 360 kilometers. The speed of the first part of his trip by plane is 240 kilometers per hour. He completes the journey by car at a speed of 80 kilometers per hour (Condition 2). The total time for the trip is 3 hours (Condition 1). How long does he travel by plane and how long by car?

Solution: ①

	Time (Condition 1)	Rate	Distance (Condition 2)
By plane	x hrs	240 km/hr	$240x$ km
By car	y hrs	80 km/hr	$80y$ km
Total	3 hrs		360 km

② $\begin{cases} \textbf{1.}\ x + y = 3 \\ \textbf{2.}\ 240x + 80y = 360 \end{cases}$ ◀ **Use the substitution method to solve.**

③ Solve for y in Equation **1.** ⟶ $y = -x + 3$

④ Substitute $(-x + 3)$ for y in Equation **2.** Solve for x.

$$240x + 80y = 360$$
$$240x + 80(-x + 3) = 360$$
$$240x - 80x + 240 = 360$$
$$160x = 120$$
$$x = \frac{120}{160}$$
$$x = \frac{3}{4}$$ ◀ **Number of hours by plane**

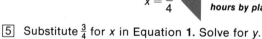

⑤ Substitute $\frac{3}{4}$ for x in Equation **1.** Solve for y.

$$x + y = 3$$
$$\frac{3}{4} + y = 3$$
$$y = 2\frac{1}{4}$$ ◀ **Number of hours by car**

The check is left for you.

EXAMPLE 2

Hisako swims 200 meters against a current in 6 minutes (Condition 1). She returns in 3 minutes swimming with the current (Condition 2). What is the speed of the current and the swimmer's speed in still water?

Solution:

1 Let x = the speed of the swimmer in still water (meters per minute).
Let y = the speed of the current.

	Time	Rate	Distance
Against current	6 minutes	$x - y$	200 meters
With current	3 minutes	$x + y$	200 meters

2 $\begin{cases} \textbf{1. } 6(x - y) = 200 \\ \textbf{2. } 3(x + y) = 200 \end{cases}$ ◀ **1. Condition 1**
                                                              2. Condition 2

3 Distributive Property   1. $6(x - y) = 200 \longrightarrow 6x - 6y = 200$ Discuss why $x - y$ and $x + y$
                         2. $3(x + y) = 200 \longrightarrow 3x + 3y = 200$ represent the speeds against the current and with the current.

4 Solve. The method of determinants is used.

$$x = \frac{\begin{vmatrix} 200 & -6 \\ 200 & 3 \end{vmatrix}}{\begin{vmatrix} 6 & -6 \\ 3 & 3 \end{vmatrix}} = \frac{600 - (-1200)}{18 - (-18)} \qquad y = \frac{\begin{vmatrix} 6 & 200 \\ 3 & 200 \end{vmatrix}}{36} = \frac{1200 - 600}{36}$$

$$= \frac{1800}{36} \qquad\qquad\qquad = \frac{600}{36}$$

$$= 50 \qquad\qquad\qquad\qquad = 16\tfrac{2}{3}$$

The speed of the swimmer in still water is 50 meters per minute. The speed of the current is approximately 16.7 meters per minute.

Additional Examples
Example 2

1. Sheila rows her boat 8 kilometers against a current in 2 hours (Condition 1). She returns in 1 hour rowing with the current (Condition 2). What is the speed of the current and the rower's speed in still water?
   Ans: speed in still water: 6 km/hr; speed of current: 2 km/hr

2. A small airplane travels 600 miles with a wind in 4 hours (Condition 2). It travels the same distance against the wind in 5 hours (Condition 1). Find the speed of the plane is still air and the speed of the wind.
   Ans: speed in still air: 135 mi/hr; wind speed: 15 mi/hr

▨▨▨ CLASSROOM EXERCISES ▨▨▨

Write an algebraic expression for each word expression. (Examples 1 and 2)

1. The distance traveled in x seconds by an object at a speed of 20 centimeters per second $20x$

2. The distance traveled in 5 minutes by an object at a speed of y meters per minute $5y$

3. The distance traveled in 10 hours by a ship whose speed with a current is $(x + y)$ miles per hour $10(x + y)$

4. The distance traveled in 6 hours by a ship whose speed against a current is $(x - y)$ miles per hour $6(x - y)$

More on Systems of Sentences / **423**

5. The distance between two cars 3 hours after they drive off in opposite directions, one at a speed of x kilometers per hour and the other at y kilometers per hour

$$3x + 3y$$

6. The distance between two planes 2 hours after they take off in opposite directions, one at a speed x kilometers per hour and the other at y kilometers per hour

$$2x + 2y$$

Assignment Guide
Basic
Day 1 pp. 424–425: 1–3
Day 2 p. 425: 4–7
Average pp. 424–425: 1–7
Above Average pp. 424–425: 1–7

Problem–Solving Skills
Completing a table (Ex. 1–7)
Using a formula (Ex. 1–7)
Writing and solving an equation (Ex. 1–7)

WRITTEN EXERCISES

Goal: To solve motion problems by systems of linear equations

Sample Problem: During a recent trip, Julia drove a total of 680 kilometers. She drove at a speed of 70 kilometers per hour on the first day and at a speed of 80 kilometers per hour on the second day (Condition 2). The total time for the trip was 9 hours (Condition 1). How long did she drive each day?

Answer: She drove 4 hours on the first day and 5 hours on the second day.

For Exercises 1–7, Condition 1 is <u>underscored once</u>. Condition 2 is <u>underscored twice</u>.

a. Use Condition 1 and Condition 2 to complete the table.
b. Use Condition 1 and Condition 2 to write a system of equations for the problem.
c. Solve. (Example 1)

1. Some scouts traveled by bus at a speed of 70 kilometers per hour and then hiked at a speed of 7 kilometers per hour. The distance traveled 224 kilometers. The total time for the trip was 5 hours. How long did they travel by bus and how long did they hike?

c. bus: 3 hr; hike: 2 hr

	Time	Rate	Distance
By bus	x hrs	70 km/hr	$70x$ km
Hiking	y hrs	? 7	? $7y$ km
Total	5 hrs		224 km

b. $\begin{cases} x + y = 5 \\ 70x + 7y = 224 \end{cases}$

2. Two trains leave the same city in opposite directions. The eastbound train travels 10 hours and the westbound train travels 5 hours. They are 1300 kilometers apart at their destinations. The westbound train is 20 kilometers per hour faster. What is the speed of each train?

	Rate	Time	Distance
Eastbound train	x km/hr	10 hrs	? $10x$ km
Westbound train	y km/hr	5 hrs	? $5y$ km
Total			? 1300 km

b. $\begin{cases} y = x + 20 \\ 10x + 5y = 1300 \end{cases}$

c. Eastbound: 80 km/hr
Westbound: 100 km/hr

See the margin for the answers to parts b. and c. of Ex. 3-7.

3. A plane and a carrier 3720 kilometers apart move toward each other and meet in 3 hours. The speed of the plane is 30 times the speed of the carrier in kilometers per hour. Find the speed of the plane and the speed of the carrier.

	Rate	Time	Distance
Plane	x km/hr	3 hrs	? $3x$ km
Carrier	y km/hr	? 3	? $3y$ km
Total			? 3720 km

(Example 2)

4. Linda rowed her boat 24 miles with a current in 3 hours. Rowing against the current, she returned to the point where she started in 4 hours. Find Linda's speed in still water and the speed of the current.

	Time	Rate	Distance
Against current	4 hrs	$(x - y)$ m/hr	24 m
With current	? 3	$(x + y)$ m/hr	? 24 m

5. A cross country runner runs a distance of 4500 meters against a wind in 18 minutes. He runs the same distance with the wind in 15 minutes. What is the speed of the runner and the speed of the wind?

	Time	Rate	Distance
Against wind	18 min	$(x - y)$ m/min	? 4500 m
With wind	? 15	$(x + y)$ m/min	? 4500 m

6. A plane travels 2240 kilometers with a wind in 7 hours. It travels the same distance against the wind in 8 hours. Find the speed of the plane in still air and the speed of the wind.

	Time	Rate	Distance
Against wind	? 8	$(x - y)$ km/hr	? 2240 km
With wind	7 hrs	$(x + y)$ km/hr	? 2240 km

7. Blake paddles a canoe 48 kilometers against a current in 6 hours. He paddles the same distance with the current in 4 hours. Find Blake's speed in still water and the speed of the current.

	Time	Rate	Distance
Against current	? 6	$(x - y)$ km/hr	? 48 km
With current	? 4	$(x + y)$ km/hr	? 48 km

More on Systems of Sentences / **425**

Quiz Sections 15.4–15.7
After completing Sections 15.4–15.7, you may want to administer a quiz covering the same sections. See the quiz provided on page 312 in the *Teacher's ResourceBank*.

Focus on Reasoning: Comparisons

Many problems involving the **comparison of two quantities** can
be solved by logical reasoning. Little or no computation with
paper and pencil may be necessary.

Refer to these instructions for the Examples and Exercises.

Each problem consists of two quantities, one in Column I and
one in Column II. Compare the two quantities.

- Write **A** if the quantity in Column I is greater.
- Write **B** if the quantity in Column II is greater.
- Write **C** if the two quantities are equal.
- Write **D** if there is not enough information to determine how
  the two quantities are related.

EXAMPLE 1

Column I	Column II
$1800 \times 9 \times 32$	$18 \times 32 \times 900$

Solution: Since 32 appears in both products, compare (1800×9) and
(18×900).
Since $(1800 \times 9) = 18 \times 100 \times 9$ and $(18 \times 900) = 18 \times 9 \times 100$,
the products are equal.

Answer: **C**

EXERCISES

	Column I	Column II	
1.	$5 \times 600 \times 3$	$500 \times 7 \times 3$	B
2.	$0.25 \times 800 \times 9$	$9 \times 25 \times 8$	C
3.	6% of 1010	1% of 606	A
4.	$\dfrac{(19)(7)(6)}{(32)(4)}$	$\dfrac{(2)(10)(42)}{(16)(8)}$	B
5.	$\dfrac{8}{8-7.9}$	$\dfrac{8}{8-7.99}$	B
6.	$(4)(6)(8)(2^5)$	$(2)(6)(16)(5^2)$	B

Sometimes additional information is given about the quantities being compared.

EXAMPLE 2	Column I	Column II

$$x^2 = 3, (x > 0)$$

	$2x\sqrt{3}$	$(2x)^2$

Solution: Since $x^2 = 3$ $(x > 0)$, $x = \sqrt{3}$.

Then $2x\sqrt{3} = 2(\sqrt{3})(\sqrt{3}) = 6$.

Also $(2x)^2 = (2\sqrt{3})^2 = 12$.

Answer: **B**

EXERCISES

	Column I	Column II	
7.	The average of 2, 6, and x is 5.		
	$\dfrac{6}{x}$	$\dfrac{x}{6}$	B
8.	$8x = 1$		
	$\sqrt{2x}$	$\sqrt[3]{x}$	C
9.	$3x + 2 = 20$		
	$\dfrac{5}{6}x + 1$	$\dfrac{1}{6}x + 5$	C
10.	$y = 2x, (x > 0)$		
	$\dfrac{y - x}{2}$	$\dfrac{y}{2} - x$	A
11.	$3x = 4y, (y \neq 0)$		
	$\dfrac{x}{y}$	$\dfrac{3}{4}$	A
12.	$x^2 = 8, (x > 0)$		
	$x\sqrt{2}$	4	C
13.	$x < 1$ and $x > -\dfrac{1}{4}$		
	$4x^2 + x$	$4x + 1$	B

CHAPTER SUMMARY

IMPORTANT TERMS	Substitution method *(p. 398)* Determinant Determinant *(p. 412)* of a system *(p. 413)* Standard form *(p. 413)*

IMPORTANT IDEAS

1. To solve a system of linear equations by the *substitution method:*
   a. Solve one equation for one of the variables.
   b. Substitute the resulting expression in the other equation. Solve the equation.
   c. Substitute the value of the variable from Step **2** in either equation. Solve the resulting equation.
   d. Check by substituting both values in both equations.

2. Any two-digit number can be represented by $10x + y$, in which x represents the ten's digit and y represents the one's digit.

3. The strength of a solution or the relative amount of a specific ingredient in a mixture is often expressed by per cents. Thus, ten grams of salt in 100 grams of a mixture of salt and water is called a 10% salt solution.

4. Each 2×2 determinant represents a real number.
$$\begin{vmatrix} a & b \\ c & d \end{vmatrix} = ad - bc$$

5. The determinant of a system of equations is formed with the coefficients of the variables as elements of the determinant.

6. If only the determinant of a system of equations equals 0, then the solution set of the system is empty.

7. The basic formula $d = rt$ relating distance, rate, and time is often used in motion problems.

8. Sometimes speeds can be added. Thus, if you let x represent the speed of an object in still water and y the speed of the current, then $(x + y)$ will represent the speed of the object with the current and $(x - y)$ the speed against the current.

Chapter Test

Two Chapter Tests (Form A and Form B) are provided on pages 313–316 in the *Teacher's ResourceBank*.

CHAPTER REVIEW

SECTION 15.1

Solve each system by the substitution method. Check.

1. $\begin{cases} y = -3x - 6 \\ x + 2y = 3 \end{cases} \{(-3, 3)\}$

2. $\begin{cases} y = \frac{1}{2}x - \frac{3}{2} \\ 3x - 2y = -3 \end{cases} \{(-3, -3)\}$

3. $\begin{cases} x - 3y = 5 \\ 2y + 3x = 4 \end{cases} \{(2, -1)\}$

4. $\begin{cases} 6 - x = 5y \\ -3x + 2y = 16 \end{cases} \{(-4, 2)\}$

*For Exercises 5–10, Condition 1 is <u>underscored once</u>. Condition 2 is
<u><u>underscored twice</u></u>.*

SECTION 15.2

*For each problem, use Condition 1 and two variables to represent the
unknowns. Use Condition 1 and Condition 2 to write a system of equations.
Solve.* In Ex. 5-8, x represents the ten's digit and y represents the one's digit.

5. <u>The ten's digit of a two-digit number is three times the one's digit.</u> $x = 3y$
<u><u>The sum of the digits is 12.</u></u> Write the two-digit number. 93 $x + y = 12$

6. <u>The one's digit of a two-digit number exceeds the ten's digit by 5.</u>
<u><u>Nine times the ten's digit increased by three times the one's digit</u></u> $y = x + 5$
<u><u>equals 39.</u></u> Write the two-digit number. 27 $9x + 3y = 39$

7. <u>A two-digit number equals 9 times its one's digit.</u> <u><u>The sum of the</u></u> $10x + y = 9y$
<u><u>digits is 9.</u></u> Write the two-digit number. 45 $x + y = 9$

8. <u>A two-digit number is 2 less than 5 times its one's digit.</u> <u><u>The one's digit</u></u>
<u><u>exceeds the ten's digit by 5.</u></u> Write the two-digit number. 38 $10x + y = 5y - 2$
   $y = x + 5$

SECTION 15.3

*For each problem, use Condition 1, Condition 2, and two variables to make a
table. Use Condition 1 and Condition 2 to write a system of equations. Solve.*

9. <u>A 12% salt solution is mixed with a 20% salt solution to obtain 800
grams of a 17% salt solution.</u> <u><u>How many grams of the 12% solution
and of the 20% are needed?</u></u>

10. <u>A 7% acid solution is mixed with 360 grams of a $2\frac{1}{2}$% acid solution to
obtain a 5% acid solution.</u> <u><u>Find the number of grams of the 7%
solution and of the 5% solution.</u></u>

SECTION 15.4

Write the value of each determinant.

11. $\begin{vmatrix} 5 & -2 \\ 3 & 7 \end{vmatrix}$ 41   12. $\begin{vmatrix} \frac{1}{2} & 0 \\ -6 & -4 \end{vmatrix}$ -2 13. $\begin{vmatrix} 0 & 1 \\ 1 & 1 \end{vmatrix}$ -1 14. $\begin{vmatrix} 0 & 1 \\ 1 & 0 \end{vmatrix}$ -1

SECTION 15.5

Solve each system by the method of determinants.

15. $\begin{cases} 3x - y = 4 \\ x + 2y = -1 \end{cases}$ 16. $\begin{cases} 3x + 2y = 13 \\ 2x + 3y = 12 \end{cases}$ 17. $\begin{cases} x - 3y = 10 \\ 2x + 5y = 1 \end{cases}$ 18. $\begin{cases} -2x + 5y = 6 \\ 3x - 4y = -2 \end{cases}$

More on Systems of Sentences / **429**

Additional Answers

9. Grams of 12%: x; grams of
20%: y
$\begin{cases} x + y = 800 \\ 0.12x + 0.20y = 0.17(800) \end{cases}$
Ans: 300 g of 12%;
     500 g of 20%

10. Grams of 7%: x; grams of
5%: y
$\begin{cases} x + 360 = y \\ 0.07x + 0.025(360) = 0.05y \end{cases}$
Ans: 450 g of 7%;
     810 g of 5%

15. $\{(1, -1)\}$
16. $\{(3,2)\}$
17. $\{(\frac{53}{11}, \frac{-19}{11})\}$
18. $\{(2,2)\}$

For Exercises 19–22, Condition 1 is underscored once. Condition 2 is underscored twice.

SECTION 15.6

For each problem, use Condition 1, Condition 2, and two variables to make a table. Use Condition 1 and Condition 2 to write a system of equations. Solve.

19. The Gymnastics Club sold 204 tickets for its show. The total receipts were $215. The adult tickets sold for $1.25 and the student tickets for $0.75. Find the number of each kind of ticket that was sold.

20. Mrs. Bell purchased 96 greeting cards on sale. The total cost was $10.20. One kind cost 15 cents each and another kind cost 8 cents each. Find the number of each kind she purchased.

SECTION 15.7

For each problem, use Condition 1 and Condition 2 to complete the table. Use Condition 1 and Condition 2 to write a system of equations. Solve.

See the margin for the system of equations and the solution.

21. A fishing group traveled by plane for 3 hours and then by jeep for 2 hours. The total distance traveled was 1000 kilometers. The speed of the plane in kilometers per hour was six times the speed of the jeep. Find the speed of the plane and the speed of the jeep.

	Rate	Time	Distance
By plane	x km/hr	? 3 hrs	? 3x km
By jeep	y km/hr	? 2 hrs	? 2y km
Total			? 1000 km

22. A motorboat travels 240 kilometers in 6 hours when it is traveling against a current. The same motorboat travels 240 kilometers in 4 hours when it is traveling with the current. Find the speed of the boat in still water and the speed of the current.

	Time	Rate	Distance
Against current	? 6 hrs	$(x - y)$ km/hr	? 240 km
With current	? 4 hrs	$(x + y)$ km/hr	? 240 km

Solve each inequality. (Sections 5.4 through 5.6)

1. $5x + 10. < 4x + 6$　　　**2.** $-6x < 18$　　　　　**3.** $4(x - 2) > 6x + 2$

Simplify. (Sections 6.4 and 6.5)

4. $(-6gh)^2$　$36g^2h^2$　　**5.** $(ab)^{-5}$　$\dfrac{1}{a^5b^5}$　　　**6.** $(2a^2b^3)^4$　$16a^8b^{12}$　　**7.** $\left(\dfrac{-3xy^4}{5}\right)^3$
$\dfrac{-27x^3y^{12}}{125}$

Simplify. (Sections 7.1 through 7.4)

8. $-\sqrt{144}$　-12　　**9.** $\sqrt{5} \cdot \sqrt{3}$　$\sqrt{15}$　　**10.** $\sqrt{50}$　$5\sqrt{2}$　　　**11.** $\sqrt{12x^6y^9}$
$2x^3y^4\sqrt{3y}$

Multiply. (Sections 8.2 and 8.5)

12. $(2d^3)(-6d)$　$-12d^4$　　**13.** $(-4cd^2e^3)(-2cd^4e^2)$　　**14.** $(3x - 1)(x + 7)$
$8c^2d^6e^5$　　　　　　$3x^2 + 20x - 7$

Factor. (Sections 9.4 through 9.7)

15. $12n^2 - 17n - 5$　　**16.** $16m^2 - 1$　　　　　**17.** $3x^2 - 18x + 27$
$(4n + 1)(3n - 5)$　　$(4m - 1)(4m + 1)$　　$3(x - 3)(x - 3)$

Multiply or divide. Then simplify. (Sections 10.3 through 10.7)

18. $\dfrac{7ab^2}{8a^2} \cdot \dfrac{16a^3c}{14bc}$　a^2b　　**19.** $\dfrac{a^2 + 5a}{a - 4} \cdot \dfrac{a^2 - 4a}{a^2 - 25}$　$\dfrac{a^2}{a - 5}$　　**20.** $\dfrac{x^2 - 16}{x^2 + x - 12} \div \dfrac{x^2 - 6x + 8}{5x - 15}$
$\dfrac{5}{x - 2}$

Add or subtract. Simplify where necessary.
(Sections 11.2 and 11.4 through 11.6)

21. $\dfrac{7x}{x + 2} - \dfrac{x - 12}{x + 2}$　6　　**22.** $\dfrac{5}{4x} + \dfrac{2}{3x}$　$\dfrac{23}{12x}$　　**23.** $\dfrac{x + 3}{x^2 - 4} + \dfrac{x - 5}{x + 2}$
$\dfrac{x^2 - 6x + 13}{x^2 - 4}$

Solve and check each equation. (Section 11.7)

24. $\dfrac{18}{2x + 3} = 3$　$\dfrac{3}{2}$　　**25.** $\dfrac{10}{x - 3} = \dfrac{9}{x - 5}$　23　　**26.** $\dfrac{6}{x - 3} - \dfrac{4}{x - 3} = \dfrac{5}{x}$　5

Graph at least four ordered pairs of each function. Then draw the graph.
The domain is {real numbers}. (Section 13.1)　Each graph is a line.
Two points are given.

27. $y = -2x + 3$　　　**28.** $y = \frac{1}{2}x - 4$　　　**29.** $y = 4x$
$(0, 3), (1, 1)$　　　$(0, -4), (8, 0)$　　　$(0, 0), (2, 8)$

Find the slope and Y-intercept of the graph of each function.
(Sections 13.2 and 13.3)

30. $y = \frac{2}{3}x - 6$　$\frac{2}{3}; -6$　　**31.** $y = 7 - x$　$-1; 7$　　**32.** $y = 3(x + 2)$　$3; 6$

Write the slope of the line that contains the two points. (Section 13.4)

33. $(2, 7); (-3, 5)$　$\frac{2}{5}$　　**34.** $(-2, 0); (-5, 9)$　-3　　**35.** $(3, -1); (-1, 8)$　$-\frac{9}{4}$

Cumulative Review / **431**

Cumulative Test
A cumulative test covering Chapters 13–15 is provided on pages 317–320 in the *Teacher's ResourceBank.*

Additional Answers
1. {all real numbers less than -4}
2. {all real numbers greater than -3}
3. {all real numbers less than -5}

Write _Yes_ or _No_ to tell whether the graphs of each pair of functions are parallel. (Section 13.5)

36. $y = 2x - 4$; $y = 2x + 3$ Yes

37. $y = \frac{1}{2}x - 4$; $y = 4x - \frac{1}{2}$ No

Solve each system by graphing. Check. (Section 14.2)

38. $\begin{cases} x - y = 9 \\ 2x + y = 6 \end{cases}$ $\{(5, -4)\}$

39. $\begin{cases} 3x + y = 6 \\ x - 2y = 2 \end{cases}$ $\{(2, 0)\}$

40. $\begin{cases} 2x - y = -3 \\ 4x - 2y = 2 \end{cases}$ ϕ

Solve each system using the addition method or the multiplication/addition method. (Sections 14.3 and 14.4)

41. $\begin{cases} 4x - y = -5 \\ -4x + 5y = -7 \end{cases}$ $\{(-2, -3)\}$

42. $\begin{cases} 2x + 3y = 8 \\ 3x + y = 5 \end{cases}$ $\{(1, 2)\}$

43. $\begin{cases} 3x - 2y = -6 \\ 4x - 5y = -1 \end{cases}$ $\{(-4, -3)\}$

Solve each system by the substitution method. Check. (Section 15.1)

44. $\begin{cases} y = 3 - x \\ 5x + 3y = -1 \end{cases}$ $\{(-5, 8)\}$

45. $\begin{cases} x - 3y = 0 \\ 2x - 5y = 4 \end{cases}$ $\{(12, 4)\}$

46. $\begin{cases} 2x - y = 6 \\ 2x + 3y = -2 \end{cases}$ $\{(2, -2)\}$

Use Condition 1 and two variables to represent the unknowns. Use Conditions 1 and 2 to write a system of equations for the problem. Solve. (Sections 14.5 and 15.2)

47. The width of a rectangle is 7 meters less than twice the length. The perimeter is 19 meters. Find the length and width.

48. The ten's digit of a two-digit number is 3 more than 5 times the one's digit. The sum of the digits is 9. Write the two-digit number.

Solve each system by the method of determinants. (Section 15.5)

49. $\begin{cases} 2x + 3y = 8 \\ 3x + y = 5 \end{cases}$ $\{(1, 2)\}$

50. $\begin{cases} 6x - 2y = -4 \\ 3x + y = -10 \end{cases}$ $\{(-2, -4)\}$

51. $\begin{cases} 4x - 3y = 5 \\ 2x - 9y = 1 \end{cases}$ $\{(\frac{7}{5}, \frac{1}{5})\}$

Use Condition 1, Condition 2, and two variables to make a table. Use Conditions 1 and 2 to write a system of equations for the problem. Solve. (Sections 15.3 and 15.6)

52. A 10% salt solution is mixed with 20 grams of a 15% salt solution to obtain a 12% salt solution. Find the number of grams of the 10% solution and of the 12% solution.

53. The junior class sold 520 tickets for a talent show. The total receipts were $1500. The adult tickets sold for $3, and the student tickets sold for $2. Find the number of each kind of ticket that was sold.

Systems of Inequalities

Given a choice of any geometric solid that you could use to enclose a given volume, you would choose the *sphere* because the sphere has the property of enclosing a given volume with a minimum amount of material.

Background

In addition to enclosing a given volume with a minimum of surface area, the tanks shown in the center photo above are structurally the strongest tanks that can be devised. The simplest way to form a sphere is to blow soap bubbles like those shown in the photo at the top right.

Maintenance
1. Solve and check:
   $6t - 11t + 24 = 4$
   Ans: $t = 4$
   (Section 3.5)
2. Simplify: $\sqrt[3]{8r^6t^9}$
   Ans: $2r^2t^3$
   (Section 7.4)
3. Multiply and simplify:
   $\dfrac{x^2 - x - 2}{x - 2} \cdot \dfrac{x - 4}{x^2 + 4x + 3}$
   Ans: $\dfrac{x - 4}{x + 3}$
   (Section 10.5)
4. Write the relation described by $y = 2x + 1$ as a set of ordered pairs when the x values are -2, 1, and 2.
   Ans: $\{(-2, -3), (1, 3), (2, 5)\}$
   (Section 12.3)
5. In a two-digit number, the sum of the digits is 10 (Condition 1). The ten's digit is 4 more than the one's digit (Condition 2). Write the two-digit number.
   Ans: 73 (Section 15.2)

Additional Example
Example 1
Write three ordered pairs having different x values that are in the solution set of the inequality.
$3x + y - 2 \geq 0$
Ans: Answers will vary. Three possible answers are given.
(1, −1), (2, −4), (3,3)

16.1 Inequalities in Two Variables

OBJECTIVES: To write ordered pairs that are in the solution set of a given inequality
To write the slope-intercept form of a given inequality

These three inequalities are examples of *linear inequalities*.

a. $y < 2x - 5$ **b.** $x - 3y + 2 > 0$ **c.** $-x + 2y + 1 \leq 0$

You can see that if you replace each inequality symbol by "=", you obtain three linear equations.

> The *solution set of a linear inequality* is the set of ordered pairs (x, y) that make the inequality true.

A linear inequality is the rule for a relation. ◄ *A relation is a set of ordered pairs.*

P–1 **What is one ordered pair in the solution set of $y < 2x - 5$?**
Many answers are possible. $(2, -1\frac{1}{2})$ is one answer.

EXAMPLE 1 Write three ordered pairs having different x values that are in the solution set of $-2x + y - 1 \leq 0$.

Solution:

1. Let $x = 0$. ⟶ $-2(0) + y - 1 \leq 0$
   $y - 1 \leq 0$
   $y \leq 1$ ◄ *Infinitely many y values*

   Choose a value for y that is less than or equal to 1. ◄ *The first ordered pair can be (0, 1), (0, 0.3), (0, −2), etc.*

2. Let $x = 2.7$. ⟶ $-2(2.7) + y - 1 \leq 0$
   $-5.4 + y - 1 \leq 0$
   $y - 6.4 \leq 0$
   $y \leq 6.4$

   Choose a value for y that is less than or equal to 6.4. ◄ *The second ordered pair can be (2.7, 6.2), (2.7, 0), (2.7, −1), etc.*

3. Let $x = -3\frac{1}{2}$. ⟶ $-2(-3\frac{1}{2}) + y - 1 \leq 0$
   $7 + y - 1 \leq 0$
   $y \leq -6$

   Choose a value for y that is less than or equal to -6. ◄ *The third ordered pair can be $(-3\frac{1}{2}, -6)$, $(-3\frac{1}{2}, -7)$, etc.*

A linear inequality can be written in <u>slope-intercept form</u>.

> The **slope-intercept form** of a linear
> inequality is one of the following:
>
> $y \le -2x + 7$
>
> $y < mx + b$ $\qquad$ $y > mx + b$
>
> $y > \frac{1}{2}x - \frac{5}{2}$
>
> $y \le mx + b$ $\qquad$ $y \ge mx + b$

EXAMPLE 2

Write the slope-intercept form of $2x + y < 5$.

Solution:

$$2x + y < 5$$

1. Addition Property of Order $\longrightarrow$ $2x + y + (-2x) < 5 + (-2x)$

2. Simplify. $\longrightarrow$ $y < 5 + (-2x)$

3. Commutative Property of Addition $\longrightarrow$ $y < -2x + 5$ ◀ **Slope-intercept form**

EXAMPLE 3

Write the slope-intercept form of $-3x - 5y + 2 \ge 0$.

Solution:

$$-3x - 5y + 2 \ge 0$$

1. Add $3x$ to each side. $\longrightarrow$ $-3x - 5y + 2 + 3x \ge 0 + 3x$

2. Simplify. $\longrightarrow$ $-5y + 2 \ge 3x$

3. Subtract 2 from each side. $\longrightarrow$ $-5y + 2 - 2 \ge 3x - 2$

4. Simplify. $\longrightarrow$ $-5y \ge 3x - 2$

5. Multiply each side by $-\frac{1}{5}$. $\longrightarrow$ $-\frac{1}{5}(-5y) \le -\frac{1}{5}(3x - 2)$ ◀ **The direction is reversed.**

6. Simplify. $\longrightarrow$ $y \le -\frac{3}{5}x + \frac{2}{5}$

CLASSROOM EXERCISES

Write three ordered pairs having different x values that are in the solution set of each inequality. (Example 1)

1. $y < x + 1$
2. $y < 2x - 1$
3. $y < -x + 2$
4. $y < x$
5. $y < -x$
6. $y > 3x + 5$
7. $y > \frac{1}{2}x + 3$
8. $y \le \frac{1}{3}x + 1$

Systems of Inequalities / **435**

Write the slope–intercept form of each inequality. (Examples 2 and 3)

9. $-2x + y - 1 < 0$ $\quad y < 2x + 1$ **10.** $3x + y > 2$ $\quad y > -3x + 2$ **11.** $-3x + y - 2 \leq 0$

12. $y + 5 \geq \frac{1}{2}x$ $\quad y \geq \frac{1}{2}x - 5$ **13.** $-y - x > 10$ $\quad y < -x - 10$ **14.** $2y < 2x + 4$
$\qquad\qquad\qquad\qquad\qquad\qquad\qquad\qquad\qquad\qquad\qquad\qquad\qquad y < x + 2$

WRITTEN EXERCISES

Goals: To write ordered pairs in the solution set of a given inequality and to write the slope-intercept form of the inequality

Sample Problems: **a.** Write three ordered pairs in the solution set of $-2x + 3y + 5 > 0$.

b. Write the slope-intercept form of $-2x + 3y + 5 > 0$.

Answers: **a.** $(0, 0), (\frac{5}{2}, 3), (-1, -2)$, etc.   **b.** $y > \frac{2}{3}x - \frac{5}{3}$

Write five ordered pairs having different x values that are in the solution set of each inequality. (Example 1)

1. $y < 2x$ **2.** $y < -5x$ **3.** $y \geq -3x + 1$

4. $y \geq 2x - 5$ **5.** $y < \frac{1}{2}x - 3$ **6.** $y > \frac{1}{3}x + 3$

7. $x + y \geq -2$ **8.** $x - y \leq 4$ **9.** $y \geq 10x + 2$

10. $x + y \leq 3$ **11.** $-2x + 3y - 5 > 0$ **12.** $4x - y + 3 < 0$

Write the slope-intercept form of each inequality. (Examples 2 and 3)

13. $-3x + y - 5 > 0$ $\quad y > 3x + 5$ **14.** $2x + y + 3 < 0$ $\quad y < -2x - 3$ **15.** $4x + y \leq 6$

16. $-5x + y \geq -2$ $\quad y \geq 5x - 2$ **17.** $x - y < -3$ $\quad y > x + 3$ **18.** $3x - y > 7$

19. $\frac{1}{3}y > x - 5$ $\quad y > 3x - 15$ **20.** $-\frac{1}{5}y \leq -x + 2$ $\quad y \geq 5x - 10$ **21.** $2x + 3y - 6 \geq 0$

22. $-5x + 2y + 3 \leq 0$ $\quad y \leq \frac{5}{2}x - \frac{3}{2}$ **23.** $4x - 2y < 5$ $\quad y > 2x - \frac{5}{2}$ **24.** $6x - 3y + 2 > 0$

25. $-7x + 14y - 3 \geq 0$
$\qquad y \geq \frac{1}{2}x + \frac{3}{14}$ **26.** $-9x + 27y > 3$ $\quad y > \frac{1}{3}x + \frac{1}{9}$ **27.** $3y + 5x - 1 < 0$
$\qquad\qquad\qquad\qquad\qquad\qquad\qquad\qquad\qquad\qquad\qquad\qquad y < -\frac{5}{3}x + \frac{1}{3}$

REVIEW CAPSULE FOR SECTION 16.2

In Ex. 1-8, each graph is a straight line. Two points of each line are given.

Graph each equation. (Pages 378–380)

$\qquad\qquad\qquad\qquad\qquad\qquad\qquad\qquad\qquad\qquad\qquad\qquad\qquad\qquad\qquad (0, 2), (4, 7)$

1. $y = -x + 2$ $(0,2), (2,0)$ **2.** $y = 2x - 1$ $(0, -1), (1, 1)$ **3.** $y = \frac{3}{2}x - 3$ $(0, -3), (2,0)$ **4.** $y = \frac{5}{4}x + 2$

5. $y = \frac{1}{2}x + \frac{3}{2}$ $(0, \frac{3}{2}), (1, 2)$ **6.** $y = -3x - 1$ **7.** $y = -2$ $(0, -2), (5, -2)$ **8.** $x = -3$
$\qquad\qquad\qquad\qquad\qquad\qquad\qquad (0, -1), (1, -4)$ $(-3, 0), (-3, -2)$

16.2 Graphs of Inequalities with $<$ and $>$

OBJECTIVES: To graph an inequality involving $<$
To graph an inequality involving $>$

Teaching Suggestions p. M–48

Lesson Resources
Warm–Up: p. M–49
Maintenance: See below.
Practice Worksheet 16.2
Exploration 16

Every linear inequality has a related linear equation.

Inequality: $y < 2x - 1$ **Related Equation:** $y = 2x - 1$

The graph of $y = 2x - 1$ is shown in Figure 1. The line separates the coordinate plane into three sets of points or *regions*.

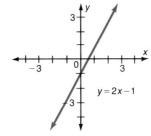

Figure 1

a. One region is one side of the line.
b. One region is the other side of the line.
c. One region is the line itself.

The region where $y > 2x - 1$

P–1 **In which region is the point with coordinates (0, 0)?   the point with coordinates (2, 3)?   the point with coordinates (5, −3)?**

The line $y = 2x - 1$ The region where $y < 2x - 1$

(2, 3) is in the solution set of $y = 2x - 1$.
(0, 0) is in the solution set of $y > 2x - 1$.
(5, −3) is in the solution set of $y < 2x - 1$.

In Figure 1, the region above, and to the left of, the line is the graph of $y > 2x - 1$. The region below, and to the right of, the line is the graph of $y < 2x - 1$.

> In graphing inequalities that contain $>$ or $<$, you must remember that the line is <u>not</u> part of the graph. Therefore, the graph of the related equation is drawn as a <u>dashed line</u>.

EXAMPLE 1   Graph $y < 2x - 1$.

◄ *"To graph" means "to graph the solution set of."*

Solution:

1. Graph the related equation $y = 2x - 1$. Use a dashed line.

2. Shade the region of the plane to the right of the dashed line.

The shaded region is the graph of $y < 2x - 1$.

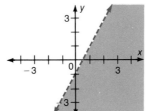

The shaded region contains all points and only those points with ordered pairs that are in the solueion set of the inequality.

Maintenance
1. Multiply:
   $(-2)(3)(h)(-p)(\frac{1}{2})$
   Ans: 3hp   (Section 2.3)
2. Evaluate:
   $r^3 - 2r^2 + r - 6$
   if $r = 2$
   Ans: −4   (Section 8.3)
3. Simplify:
   $3x^2 - 5 - 2x - 7x^2 + 4x - 7$
   Ans: −4x² + 2x − 12
   (Section 8.3)
4. Subtract:
   $$\frac{2}{2x + 8} - \frac{1}{x^2 - 16}$$
   Ans: $\dfrac{2x - 10}{2(x + 4)(x - 4)}$
   (Section 11.6)
5. Rosco has 35 dimes and quarters in a bank (Condition 1). The total value of the coins is $6.50 (Condition 2). How many of each coin does he have?
   Ans: 15 dimes; 20 quarters   (Section 15.6)

Additional Example
Example 1
Graph $y > x - 2$
Ans:

*Systems of Inequalities / **437***

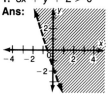
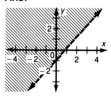
EXAMPLE 2   Graph $x + y - 1 > 0$.

Solution:

1. Write the slope-intercept form. ⟶ $y > -x + 1$

2. Graph the related equation $y = -x + 1$ as a dashed line as in Figure 2.

3. Choose a point not on the line. Check whether its coordinates make the inequality true or false.

$y > -x + 1$ **Try the origin, (0, 0).**

$0 > -0 + 1$

$0 > 1$ **False sentence**

The origin is not in the graph.

4. Shade the region on the side of the line not containing the origin as in Figure 3. The shaded region is the graph of $x + y - 1 > 0$.

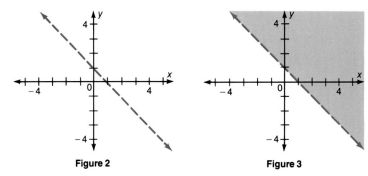

Figure 2   Figure 3

Steps 3 and 4 tell you which side of the line to shade.

Steps for graphing the relations $y < mx + b$ and $y > mx + b$:

1. Draw the graph of $y = mx + b$.
2. Select a point not on the line and substitute its coordinates in the inequality.
3. Shade the region of the plane containing this point if its coordinates make the inequality true.
4. Shade the region on the other side of the line from this point if its coordinates make the inequality false.

438   /   *Chapter 16*

CLASSROOM EXERCISES

Write <u>Yes</u> or <u>No</u> to tell whether (0, 0) is in each solution set. (P-1)

1. $y < 2x + 5$ Yes   2. $y > x - 1$ Yes   3. $y < -2x - 3$ No   4. $y < 3x$

5. $y = 5x$ Yes   6. $y > -\frac{1}{2}x - \frac{2}{3}$ Yes   7. $y > 3$ No   8. $x < -2$
                                                                                    No

No (above ex. 8)

Write an inequality for each graph. The equation of each dashed line is given.

9.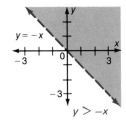

$y = -x$

$y > -x$

10.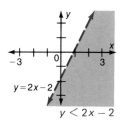

$y = 2x - 2$

$y < 2x - 2$

11.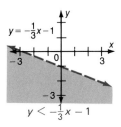

$y = -\frac{1}{3}x - 1$

$y < -\frac{1}{3}x - 1$

WRITTEN EXERCISES

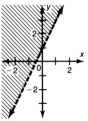

Goal: To graph linear inequalities that use $>$ or $<$
Sample Problem: $x + y - 3 > 0$
Answer: See the graph at the right.
In Ex. 1-18, each graph is a region that does not include the linear boundary.
Graph each inequality. (Examples 1 and 2)   The location of the region with respect to the boundary is indicated.

1. $y < 3x - 1$ right of the boundary 2. $y < -2x + 2$ left   3. $y > -x + 3$

4. $y > x - 3$ left   5. $y < \frac{1}{2}x - 3$ below   6. $y < -\frac{1}{2}x - 1$

7. $y < x$ right   8. $y > -x$ right   9. $y > -3$ above

10. $y < -2$ below   11. $3x + y - 2 < 0$ left   12. $-2x + y + 3 < 0$

13. $x < -2$ left   14. $x > -3$ right   15. $-3x + 2y - 2 > 0$

16. $4x + 2y - 3 > 0$ right   17. $3x - 4y - 12 < 0$ above   18. $2x - y + 4 > 0$
                                                                          right

REVIEW CAPSULE FOR SECTION 16.3

Write as one inequality using $\leq$ or $\geq$. (Pages 116–119)

1. $y < 2$ <u>or</u> $y = 2$ $y \leq 2$ 2. $y < x - 5$ <u>or</u> $y = x - 5$ $y \leq x - 5$

3. $y > 2x$ <u>or</u> $y = 2x$ $y \geq 2x$ 4. $y = \frac{1}{2} - 3x$ <u>or</u> $y > \frac{1}{2} - 3x$ $y \geq \frac{1}{2} - 3x$

5. $y = -x + 1$ <u>or</u> $y < -x + 1$ $y \leq -x + 1$ 6. $y > 2 - x$ <u>or</u> $y = 2 - x$ $y \geq 2 - x$

*Systems of Inequalities / **439***

Maintenance
1. Solve and check:
   $6 - 7r = 2 (9r - 22)$
   Ans: $r = 2$
   (Section 4.3)
2. Simplify: $(-2x)^3$
   Ans: $-8x^3$
   (Section 6.4)
3. Divide and simplify:
   $$\frac{9}{x^2 - 25} \div \frac{3}{2x - 10}$$
   Ans: $\dfrac{6}{x + 5}$
   (Section 10.7)
4. Name three points on the
   graph of $y = x^2 + 2$.
   Ans: Sample points:
   $(-1,3),(0,2),(3,11)$
   Answers will vary.
   (Section 12.5)
5. Solve the system below by
   graphing.
   $\begin{cases} y = x + 4 \\ y = 2x + 5 \end{cases}$
   Ans: $\{(-1,3)\}$
   (Section 14.2)

16.3 Graphs of Inequalities with ≤ and ≥

OBJECTIVES: To graph an inequality involving $\geq$ or $\leq$
To write the slope-intercept form of an inequality given its
graph and the slope of the boundary

Mathematical sentences can be classified as <u>simple</u> or <u>compound</u>.
An important example of a compound sentence is the <u>or</u> sentence.

Simple Sentences	Compound <u>Or</u> Sentences
$x > 9$	$x > 1$ <u>or</u> $x = -5$
$x = -2$	$x < 4$ <u>or</u> $x = 6$
$x < 6$	$x > -3$ <u>or</u> $x = -3$

◀ This can
also be
shown as
$x \geq -3$.

P–1 **What are four numbers in the solution set of $x > -3$?**
$-2.9, 0, 5, 17$

P–2 **Find the only number in the solution set of $x = -3$.**
-3

The **union of two sets** is the set of elements in either one of the two
sets or in both of them.

$$\{1, 3, 5\} \cup \{2, 3, 4\} = \{1, 2, 3, 4, 5\}$$

◀ **A ∪ B means
"the union
of A and B."**

The **solution set of a compound <u>or</u> sentence** is the union of the
solution sets of its two simple sentences.

P–3 **What are four numbers in the solution set of the compound <u>or</u>
sentence "$x > -3$ <u>or</u> $x = -3$"?** $-3, -2.9, 5, 17$

When a compound <u>or</u> sentence has *two* variables its graph is drawn
in a coordinate plane.

$(1, 1), (0, 2), (2, 0), (7, -5)$

P–4 **What are four ordered pairs in the solution set of $x + y = 2$?
of $x + y < 2$?**
$(0, 1), (1, 0), (2, -1), (-2, 2)$

Emphasize this
distinction between
graphs of inequalities
containing $<$ or $>$.

> In graphing inequalities that contain $\geq$ or $\leq$, you must remember
> that the line <u>is</u> part of the graph. Therefore, the graph of the
> related equation is drawn as a <u>solid line</u>.

EXAMPLE 1   Graph $x + y \leq 2$.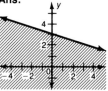

Solution:

[1] Write the slope-intercept form. ⟶ $y \leq -x + 2$

[2] Graph $y = -x + 2$.

[3] Choose a point not on the line
such as the origin, (0, 0). Check
whether its coordinates make
$y < -x + 2$ true or false.

$y < -x + 2$

$0 < -0 + 2$

$0 < 2$ ◀ **True sentence**

[4] Shade the region on the side of the line containing the origin.

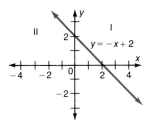   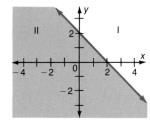

◀ **The graph is
the union of the
solid line
and Region II.**

The coordinates of any point
on the line makes the sentence
true. Also, the coordinates of
any point in the shaded region

make it true.

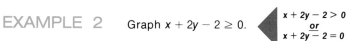

EXAMPLE 2   Graph $x + 2y - 2 \geq 0$.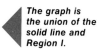

Solution:

[1] Write the slope-intercept form. ⟶ $y \geq -\frac{1}{2}x + 1$

[2] Graph $y = -\frac{1}{2}x + 1$.

[3] Choose a point not on the line
such as the origin, (0, 0). Check
whether its coordinates make
$y > -\frac{1}{2}x + 1$ true or false.

$y > -\frac{1}{2}x + 1$

$0 > -\frac{1}{2}(0) + 1$

$0 > 1$ ◀ **False sentence**

[4] Shade Region I above the line. It does not contain the origin.

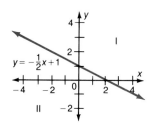   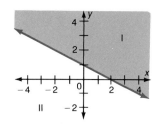

◀ **The graph is
the union of the
solid line and
Region I.**

Graph $x + 3y - 9 \leq 0$.
Ans:

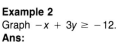

Example 2
Graph $-x + 3y \geq -12$.
Ans:

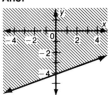

Additional Examples
Example 3
Write the slope-intercept form of an inequality for each graph. The slope, *m,* of each line is given.

1. $m = -2$

Ans: $y \le -2x - 4$

2. $m = \frac{1}{3}$

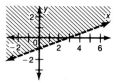

Ans: $y > \frac{1}{3}x - 1$

Common Error
Students do not include the graph of the related equation in the solution set of the inequality.

Example
The graph of $y \le x + 1$ is drawn using a dashed line for the graph of the related equation $y = x + 1$.

Prescription
Remind students that the graph of the equation is part of the solution set when using $\ge$ or $\le$. Review Examples 1 and 2.

EXAMPLE 3

Write the slope-intercept form of the inequality that is graphed below. The slope *m* of the line is $\frac{1}{2}$.

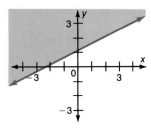

Solution:

The Y intercept of the line is 1 as shown on the graph.

Equation of the line: $\quad y = mx + b$

$$y = \tfrac{1}{2}x + 1$$

Inequality: $\quad y \ge \tfrac{1}{2}x + 1$ ◀ **The region above the line is shaded.**

Note, as a check, that the point with coordinates (0, 2) makes the inequality true. It is clear from the graph that the point with coordinates (0, 2) is in the graph.

CLASSROOM EXERCISES

Write Yes or No to indicate whether (0, 0) is in each solution set. (Step 3 of Example 1)

1. $y \le x + 1$ Yes

2. $y \le -x - 1$ No

3. $y \le 2x + 3$ Yes

4. $y \le -5x + 1$ Yes

5. $y \ge x$ Yes

6. $y \le -x$ Yes

7. $x + 2y - 5 \le 0$ Yes

8. $-2x + 3y - 1 \ge 0$ No

9. $x + y \ge -2$ Yes

Write the slope-intercept form for each graph. The slope, m, of each line is given. (Example 3)

10. $m = 1$

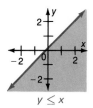

$y \le x$

11. $m = -1$

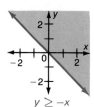

$y \ge -x$

12. $m = -\frac{1}{2}$

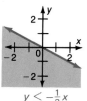

$y \le -\frac{1}{2}x$

13. $m = \frac{1}{2}$

$y \le \frac{1}{2}x + 1$

442 / Chapter 16

WRITTEN EXERCISES

Goal: To graph inequalities in two variables that use ≥ or ≤

Sample Problem: $x + y \leq 3$

Answer: See the graph at the right.

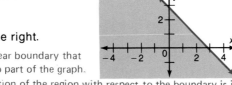

In Ex. 1-18, each graph is a region with a linear boundary that *Graph each inequality.* is also part of the graph. **(Examples 1 and 2)** The location of the region with respect to the boundary is indicated.

1. $y \leq x + 1$ right of the boundary
2. $y \leq -x + 1$ left
3. $y \leq 2x - 3$

4. $y \leq 3x - 2$ right
5. $y \geq -2x + 1$ right
6. $y \geq -3x - 1$

7. $-\frac{1}{2}x + y - 1 \leq 0$ below
8. $\frac{1}{2}x + y + 2 \leq 0$ below
9. $2x + 2y - 3 \geq 0$

10. $3x + 3y - 6 \geq 0$ above
11. $-\frac{1}{3}x + y \leq 1$ below
12. $-\frac{1}{4}x + y \leq -2$

13. $y \geq -1$ above
14. $y \leq -2$ below
15. $x \leq 3$ left

16. $x \geq -3$ right
17. $x \geq -y + 2$ right
18. $x \leq -y - 3$ below

Write the slope-intercept form of an inequality for each graph. The slope, m, of each line is given. (Example 3)

19. $m = 1$ $y \leq x + 1$

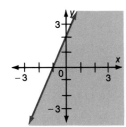

20. $m = -1$ $y \geq -x - 1$

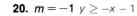

21. $m = -2$ $y \geq -2x - 2$

22. $m = 2$ $y \leq 2x + 2$

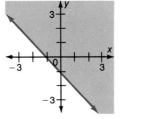

23. $m = -\frac{1}{3}$ $y \leq -\frac{1}{3}x + 1$

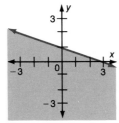

24. $m = \frac{1}{2}$ $y \geq \frac{1}{2}x + 1$

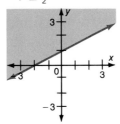

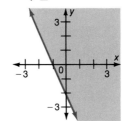

Assignment Guide
Basic Omit
Average pp. 443–444: 1–27 odd
Above Average pp. 443–444: 3, 6, 9, ···, 24, 25–27; p. 439: 3, 6, 9, ···, 18

Additional Answers
3. Right of the boundary
6. Right of the boundary
9. Right of the boundary
12. Below the boundary

Systems of Inequalities / **443**

443

APPLICATIONS

You can use a calculator to check whether an ordered pair is in the solution set of an inequality. First, substitute the ordered pair in the inequality. Then compare the two sides of the inequality.

EXAMPLE Is $(-2, 5)$ in the solution set of $6x - y > 1 + 3y$?

SOLUTION Left side:

[6] [×] [2] [+/−] [−] [5] [=] | -17. |

Right side:

[3] [×] [5] [+] [1] [=] | 16. |

▲ **Note: Multiply before adding. See page 19.**

Since "$-17 > 16$" is <u>false</u>, $(-2, 5)$ is not in the solution set.

Check whether the given ordered pair is in the solution set of the inequality.

25. $3x - 5 < 5y + 2; (3, 2)$ **26.** $4x + y > 2x - 4; (3, -7)$ **27.** $(x + 2)5 \leq x + 2y; (-5, 6)$
Yes   Yes   Yes

▮ MID-CHAPTER REVIEW ▮

Write the slope-intercept form of each inequality. (Section 16.1)

1. $5x + y - 2 < 0$ $y < -5x + 2$ **2.** $-2x + y + 6 > 0$ $y > 2x - 6$ **3.** $3x - 2y > -5$

4. $-5x + 3y < 8$ $y < \frac{5}{3}x + \frac{8}{3}$ **5.** $-3 > -8x + 6y$ $y < \frac{4}{3}x - \frac{1}{2}$ **6.** $12 < 20x - 8y$

Graph each inequality. (Section 16.2)

7. $y > 2x + 3$ **8.** $y < 3x - 2$ **9.** $y < \frac{1}{2}x - 2$

10. $y > -\frac{1}{3}x + 1$ **11.** $2x - y + 5 < 0$ **12.** $-3x + y - 3 > 0$

Graph each inequality. (Section 16.3)

13. $y \geq \frac{3}{2}x - 3$ **14.** $y \leq -\frac{7}{4}x + 2$ **15.** $y \leq -x + \frac{5}{2}$

16. $y \geq -x - \frac{9}{4}$ **17.** $3x - 2y + 6 \geq 0$ **18.** $-4x + 2y - 5 \leq 0$

REVIEW CAPSULE FOR SECTION 16.4

Which of the following are in the solution sets of <u>both</u> inequalities? $\{(-2, 3), (0, -1), (3, 5), (2, -4), (-4, 3)\}$ (Pages 434–436)

1. $\begin{cases} y > 2x - 1 & (-2, 3) \\ y > \frac{1}{2}x + 3 & (-4, 3) \end{cases}$ **2.** $\begin{cases} y < 4x + 3 & (0, -1) \\ y > x - 5 & (3, 5) \end{cases}$ **3.** $\begin{cases} y < -x + 3 & (2, -4) \\ y > -3x - 1 \end{cases}$ **4.** $\begin{cases} y > x + 5 & (-4, 3) \\ y < -2x - 4 \end{cases}$

444 / *Chapter 16*

Quiz Sections 16.1–16.3
After completing this Mid-Chapter Review, you may want to administer a quiz covering the same sections. See the quiz provided on page 331 in the *Teacher's ResourceBank*.

Additional Answers
Mid-Chapter Review

3. $y < \frac{3}{2}x + \frac{5}{2}$
6. $y < \frac{5}{2}x - \frac{3}{2}$

In Ex. 7–12, each graph is a region that does not include its linear boundary. The location of the region with respect to the boundary is indicated.
7. Left of the boundary
8. Right of the boundary
9. Below the boundary
10. Above the boundary
11. Left of the boundary
12. Left of the boundary

In Ex. 13–18, each graph is a region that does include its linear boundary. The location of the region with respect to the boundary is indicated.
13. Left of the boundary
14. Left of the boundary
15. Left of the boundary
16. Right of the boundary
17. Right of the boundary
18. Right of the boundary

444

16.4 Graphs of Systems with $<$ and $>$

OBJECTIVE: To graph a system of two linear inequalities involving $<$ or $>$

Teaching Suggestions p. M–48

Lesson Resources
Warm–Up: p. M–49
Maintenance: See below.
Practice Worksheet 16.4
Transparencies 40A, 40B

P–1 **What are three numbers in the solution set of $x < 4$?**

3.9, 1, −11

```
◄──┼──┼──┼──┼──┼──┼──┼──●──┼──┼►
  −3 −2 −1  0  1  2  3  4  5
```

P–2 **What are three numbers in the solution set of $x > -2$?**

−1.9, 0, 17

```
◄──┼──○──┼──┼──┼──┼──┼──┼──┼►
  −3 −2 −1  0  1  2  3  4  5
```

The *intersection of two sets* is the set of elements they have in common.

$$\{1, 5, 8\} \cap \{0, 5, 10\} = \{5\}$$

◄ *A* ∩ *B* means
"the intersection
of A and B."

The *solution set of a compound **and** sentence* is the intersection of the solution sets of its simple sentences.

P–3 **What are four numbers in the solution set of the compound *and* sentence "$x < 4$ *and* $x > -2$"?** −1.9, 0, 2, 3.8

```
◄──┼──○──┼──┼──┼──┼──○──┼──┼►
  −3 −2 −1  0  1  2  3  4  5
```

This transition from inequalities in one variable to two variables may require special explanation.

When a compound *and* sentence has two variables, its graph is drawn in a coordinate plane. A compound *and* sentence in two variables is often written as a *system of linear sentences* or, simply, as a <u>linear system</u>.

$$\begin{cases} \textbf{1. } y < -x + 3 \\ \textbf{2. } y > 3x - 1 \end{cases}$$

$y < -x + 3$
and
$y > 3x - 1$

The *solution set of a system of linear sentences* is the intersection of the solution sets of the sentences.

> The ***graph of the solution set of a system of sentences*** is the intersection of the graphs of the sentences.

Maintenance
1. Solve:
   $3x - 6 - 2x > 18$
   Ans: $x > 24$
   (Section 5.4)
2. Factor:
   $2x^2 - 3x + 1$
   Ans: $(2x - 1)(x - 1)$
   (Section 9.4)
3. Add: $\sqrt{28} + \sqrt{63}$
   Ans: $5\sqrt{7}$ (Section 7.5)
4. Write the x intercept of the graph of $y = -7x - 14$.
   Ans: -2 (Section 13.6)
5. Solve the system below by the multiplication/addition method.
   $$\begin{cases} 3x + y = -9 \\ x + 3y = -11 \end{cases}$$
   Ans: $\{(-2, -3)\}$
   (Section 14.4)

Systems of Inequalities / **445**

Additional Examples
Example 1
Graph each linear system.

1. $\begin{cases} y > -2x + 4 \\ y > x + 1 \end{cases}$
   Ans:
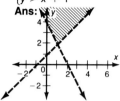

2. $\begin{cases} y > -\frac{1}{3}x - 3 \\ y > \frac{4}{3}x + 2 \end{cases}$
   Ans:
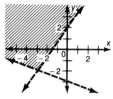

EXAMPLE 1   Graph the following linear system.

$\begin{cases} \textbf{1. } y < 2x + 1 \\ \textbf{2. } y < -\frac{1}{2}x - 2 \end{cases}$

$y < 2x + 1$
and
$y < -\frac{1}{2}x - 2$

Solution:

1. Graph $y = 2x + 1$ as the dashed line in Figure 1.

2. Shade the region below the dashed line to represent the solution set of $y < 2x + 1$.

3. Graph $y = -\frac{1}{2}x - 2$ as the dashed line in Figure 2.

4. Shade the region below the dashed line to represent the solution set of $y < -\frac{1}{2}x - 2$.

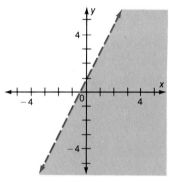

Figure 1

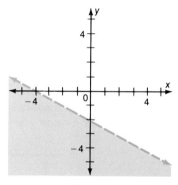

Figure 2

The graph of the solution set of the system of Example 1 is the intersection of the two graphs in Figures 1 and 2.

The intersection is shown by the darkest region in Figure 3.

The dashed lines are not included in the intersection.

Ask students to explain why the dashed lines are not in the intersection.

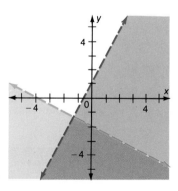

Figure 3

446 / *Chapter 16*

EXAMPLE 2 Graph the following linear system.

$$\begin{cases} \textbf{1.} \ y > 2x - 1 \\ \textbf{2.} \ y < -x + 3 \end{cases}$$

Solution:

☐1 Graph $y > 2x - 1$ in the coordinate plane of Figure 4.

☐2 Graph $y < -x + 3$.

The graph of the solution set of the system is the region showing the darkest shading in Figure 4.

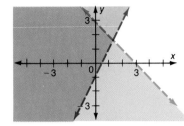

Figure 4

Additional Example
Example 2
Graph this linear system.
$$\begin{cases} \textbf{1.} \ y > x - 1 \\ \textbf{2.} \ y < -2x + 2 \end{cases}$$
Ans:

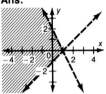

Assignment Guide
Basic Omit
Average p. 447: 1–11 odd
Above Average p. 447:
1–11 odd

Additional Answers
Written Exercises
In Ex. 1–12, each region has two straight lines as boundaries. Boundaries are not included in the graphs.
Abbreviations are:
Q: quadrants; L: left boundary; R: right boundary; T: top boundary; B: bottom boundary
1. Q: III, IV; L: $y = x + 1$; R: $y = -x - 1$
2. Q: I, II; L: $y = -x + 1$; R: $y = x - 1$
3. Q: I, II, III; B: $y = 2x + 1$; T: $y = -x + 2$
4. Q: I, III, IV; T: $y = \frac{1}{2}x + 2$; L: $y = 3x - 1$
5. Q: II, III; T: $x + 2y = 4$; B: $x - 2y = -4$
6. Q: I, IV; T: $x - 2y = 2$; B: $2x + y = -1$
7. Q: III, IV; L: $x - 3y = 3$; R: $2x + y = -1$
8. Q: I, II, III, IV; R: $2x + y = 4$; B: $x - y = 3$
9. Q: II, III; T: $y = 2$; R: $x = -1$
10. Q: II, III; B: $y = -2$; R: $x = -3$
11. Q: I, III, IV; L: $x = -1$; T: $y = 3x$
12. Q: I, II, IV; L: $y = -2x$; R: $x = 3$

CLASSROOM EXERCISES

Write the region that is the graph of each system.

1. $\begin{cases} y < -x + 1 \\ y < 2x \end{cases}$ iV

2. $\begin{cases} y > -x + 1 \\ y < 2x \end{cases}$ I

3. $\begin{cases} y > -x + 1 \\ y > 2x \end{cases}$ II

4. $\begin{cases} y < -x + 1 \\ y > 2x \end{cases}$ III

WRITTEN EXERCISES

Goal: To graph systems of linear sentences

Sample Problem: $\begin{cases} \textbf{1.} \ y > 2x + 5 \\ \textbf{2.} \ y < -x \end{cases}$

Answer: See the graph at the right.

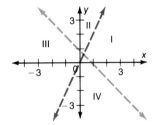

Graph each linear system. (Examples 1 and 2)

1. $\begin{cases} y < x + 1 \\ y < -x - 1 \end{cases}$

2. $\begin{cases} y > -x + 1 \\ y > x - 1 \end{cases}$

3. $\begin{cases} y > 2x + 1 \\ y < -x + 2 \end{cases}$

4. $\begin{cases} y < \frac{1}{2}x + 2 \\ y < 3x - 1 \end{cases}$

5. $\begin{cases} x + 2y < 4 \\ x - 2y < -4 \end{cases}$

6. $\begin{cases} 2x + y > -1 \\ x - 2y > 2 \end{cases}$

7. $\begin{cases} x - 3y > 3 \\ 2x + y < -1 \end{cases}$

8. $\begin{cases} 2x + y < 4 \\ x - y < 3 \end{cases}$

9. $\begin{cases} y < 2 \\ x < -1 \end{cases}$

10. $\begin{cases} y > -2 \\ x < -3 \end{cases}$

11. $\begin{cases} y < 3x \\ x > -1 \end{cases}$

12. $\begin{cases} y > -2x \\ x < 3 \end{cases}$

Systems of Inequalities / **447**

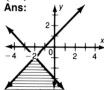
16.5 Graphs of Systems with $\le$ and $\ge$

OBJECTIVES: To graph a system of two linear inequalities involving $\le$, $\ge$, $<$ or $>$
To write a system of two linear inequalities that corresponds to a given graph

Recall that in graphing inequalities that contain $\le$ or $\ge$, the line is part of the graph. Therefore, the graph of the related equation is drawn as a solid line.

EXAMPLE 1   Graph the following linear system.

$$\begin{cases} 1.\ y \le 3x - 1 \\ 2.\ y \le -2x + 3 \end{cases}$$

Solution:

☐1 Graph $y = 3x - 1$ as the solid line in Figure 1.

☐2 Shade the region below the solid line to represent the solution set of $y < 3x - 1$.

☐3 Graph $y = -2x + 3$ as the solid line in Figure 2.

☐4 Shade the region below the solid line to represent the solution set of $y < -2x + 3$.

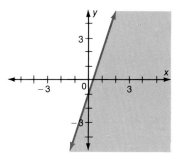

Figure 1

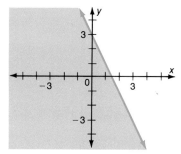

Figure 2

The graph of the solution set of the system of Example 1 is the intersection of the two graphs in Figures 1 and 2. The intersection is shown by the darkest region in Figure 3. The solid lines are included in the intersection.

Ask students if the point where the solid lines meet is in the intersection.

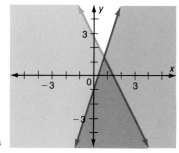

Figure 3

448 / Chapter 16

EXAMPLE 2

Graph the following linear system.

$$\begin{cases} \textbf{1. } y > 5x - 2 \\ \textbf{2. } y \geq \frac{1}{3}x \end{cases}$$

Solution:

1. Graph $y > 5x - 2$ in the coordinate plane of Figure 4.

2. Graph $y \geq \frac{1}{3}x$.

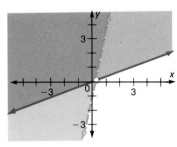

Figure 4

The graph of the solution set of the system is the region showing the darkest shading in Figure 4. A circle is drawn to emphasize that the point of intersection of the lines is not in the solution set of the system. See if students can explain why the point of intersection of the lines is not in the solution set.

EXAMPLE 3

Write a system of linear sentences that corresponds to the graph of Figure 5.

Solution:

Inequality for points on or above the graph of $y = x$: $y \geq x$

Inequality for points above the graph of $y = -x$: $y > -x$

System: $\begin{cases} \textbf{1. } y \geq x \\ \textbf{2. } y > -x \end{cases}$

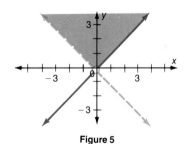

Figure 5

The circle drawn on the graph shows that the point of intersection of the lines is not in the solution set of the system.

*Systems of Inequalities / **449***

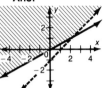
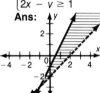

CLASSROOM EXERCISES

Match each graph in Exercises 1–4 with one of the systems given in a–d. (Examples 1 and 2)

1. 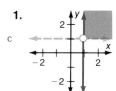   **2.** 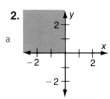   **3.** 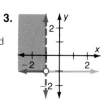   **4.**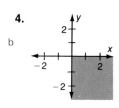

a. $\begin{cases} x \le 0 \\ y \ge 0 \end{cases}$ **b.** $\begin{cases} x \ge 0 \\ y \le 0 \end{cases}$ **c.** $\begin{cases} x \ge 1 \\ y > 1 \end{cases}$ **d.** $\begin{cases} x < -1 \\ y \ge -1 \end{cases}$

WRITTEN EXERCISES

See the Answers to Selected Exercises for the answers to Ex. 1-19 odd.

Goal: To graph systems of linear sentences

Sample Problem: $\begin{cases} 1.\ y \ge x + 4 \\ 2.\ y \ge -x \end{cases}$

Answer: See the graph at the right.

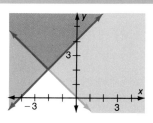

Graph each linear system. (Examples 1 and 2)

1. $\begin{cases} y \ge x + 1 \\ y \ge -x - 1 \end{cases}$ **2.** $\begin{cases} y \le 2x - 1 \\ y \le x \end{cases}$ **3.** $\begin{cases} y < -x + 1 \\ y \ge 2x \end{cases}$ **4.** $\begin{cases} y \ge \frac{1}{4}x + 1 \\ y > x - 1 \end{cases}$

5. $\begin{cases} x + y \le 4 \\ x - y \le -4 \end{cases}$ **6.** $\begin{cases} 2x + y \ge 1 \\ x - 2y \ge 1 \end{cases}$ **7.** $\begin{cases} x - 3y > 2 \\ 2x + y \le -2 \end{cases}$ **8.** $\begin{cases} 3x + y < -3 \\ x - y \ge 2 \end{cases}$

9. $\begin{cases} y > -x - 2 \\ y \ge x + 2 \end{cases}$ **10.** $\begin{cases} y \le -2x + 1 \\ y \le 2x - 3 \end{cases}$ **11.** $\begin{cases} y \ge 3x \\ y \le x \end{cases}$ **12.** $\begin{cases} y \ge \frac{1}{2}x - 2 \\ y \ge x - 3 \end{cases}$

13. $\begin{cases} y \le 3 \\ x < -2 \end{cases}$ **14.** $\begin{cases} y > -1 \\ x \le -2 \end{cases}$ **15.** $\begin{cases} y \le 3x \\ x > 0 \end{cases}$ **16.** $\begin{cases} y > -x \\ x \le 1 \end{cases}$

Write a system of linear sentences that corresponds to each graph. The equations of the lines are $y = x$ and $y = -x$. (Example 3)

17. 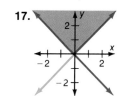   **18.** 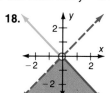   **19.** 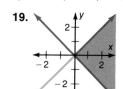   **20.**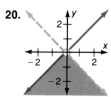

Using Tables and Graphs

Linear Programming

In large manufacturing companies there may be **production planners** or **industrial** or **production technicians** who plan the layout of machinery and movement of materials in order to obtain the most efficient way of operating.

A method called **linear programming** has been developed for solving some of these problems. It is based on the solution of systems of linear inequalities.

EXAMPLE: A cement mixing company has two plants. There are 5 trucks at Plant I and 10 trucks at Plant II. The company must supply mixed cement for two building sites A and B. Site A can handle 8 trucks, and Site B can handle 4. The company's production planner must decide how many trucks to send from each plant to each building site in order to make transportation costs as small as possible. How many trucks should be sent from each plant to each site?

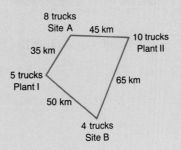

SOLUTION: Prepare a table showing the facts of the problem with the unknowns represented by two variables.

From Plant	To Site	Distance	Number of Trucks	Total Distance
I	A	35 km	x	$35x$
I	B	50 km	y	$50y$
II	A	45 km	$8 - x$	$45(8 - x)$
II	B	65 km	$4 - y$	$65(4 - y)$

Next, write a system of inequalities for the problem. Graph the system. In each case the number of trucks used must be greater than or equal to zero. Thus, six inequalities can be written.

1. $x \geq 0$
2. $y \geq 0$
3. $8 - x \geq 0$, or $x \leq 8$
4. $4 - y \geq 0$, or $y \leq 4$
5. $x + y \leq 5$
6. $(8 - x) + (4 - y) \leq 10$, or $x + y \geq 2$

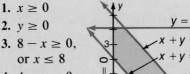

Systems of Inequalities / **451**

Additional Answers
Written Exercises, p. 450
6. Q: I, IV; T: $x - 2y = 1$; L: $2x + y = 1$; Boundaries included
8. Q: III, IV; L: $x - y = 2$; R: $3x + y = -3$; Left boundary included
10. Q: III, IV; R: $y = -2x + 1$; L: $y = 2x - 3$; Boundaries included
12. Q: all; R: $y = x - 3$; B: $y = \frac{1}{2}x - 2$; Boundaries included
14. Q: II, III; B: $y = -1$; R: $x = -2$; Right boundary included
16. Q: I, II, IV; R: $x = 1$; B: $y = -x$; Right boundary included
18. $y \leq -x; y < x$
20. $y \leq x; y < -x$

Represent the total distance by an algebraic expression.

Total Distance $= 35x + 50y + 45(8 - x) + 65(4 - y)$
$$= 620 - 10x - 15y$$

This distance must have a minimum value. Find the minimum value by substituting the x and y values of the five vertices of the polygon shown in the graph.

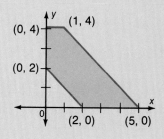

Vertex	Total Distance
(2, 0)	$620 - 10(2) - 15(0) = 600$ kilometers
(5, 0)	$620 - 10(5) - 15(0) = 605$ kilometers
(1, 4)	$620 - 10(1) - 15(4) = 550$ kilometers
(0, 4)	$620 - 10(0) - 15(4) = 560$ kilometers
(0, 2)	$620 - 10(0) - 15(2) = 590$ kilometers

The least value for the total distance is **550 kilometers**. This occurs when $x = 1$ and $y = 4$. Now the problem can be solved.

Since $x = 1$,   send **1** truck from **Plant I** to **Site A.**
Since $y = 4$,   send **4** trucks from **Plant I** to **Site B.**
Since $8 - x = 7$, send **7** trucks from **Plant II** to **Site A.**
Since $4 - y = 0$, send **0** trucks from **Plant II** to **Site B.**

EXERCISES

An electronics company has two plants that manufacture calculators. A special order of 10,000 calculators requires delivery within two weeks. To meet that deadline, the calculators must be produced in 5 or fewer working days (120 hours). Plant A can produce the calculators at a cost of $1.30 each and at a rate of 60 units per hour. Plant B can produce the calculators at a cost of $1.35 each and at a rate of 40 units per hour. On an order of 10,000 units, the quality control engineer recommends that 200 extra units be produced to assure at least 10,000 units of acceptable quality. Find how many calculators should be produced by each plant if the production cost is to be kept to a minimum. Complete the following table as the first step.

Plant	Number of Units	Hours Needed	Production Cost
A	x	?	?
B	y	?	?

$\frac{x}{60}$; $1.30x$

$\frac{y}{40}$; $1.35y$

Plant A should produce 7200 calculators and Plant B should produce 3000 calculators. The cost would be $13,410.00.

CHAPTER SUMMARY

IMPORTANT TERMS

Linear inequality *(p. 434)*
Solution set of a linear inequality *(p. 434)*
Slope-intercept form of a linear inequality *(p. 435)*
Region *(p. 437)*
Compound sentence *(p. 440)*
Union of two sets *(p. 440)*
Solution set of a compound *or* sentence *(p. 440)*
Intersection of two sets *(p. 445)*
Solution set of a compound *and* sentence *(p. 445)*
System of linear sentences *(p. 445)*
Solution set of a system of linear sentences *(p. 445)*

IMPORTANT IDEAS

1. The graph of a linear equation separates a coordinate plane into three regions: the points on the line and the points of the plane on each side of the line.

2. Steps for drawing the graph of the relations $y < mx + b$ and $y > mx + b$:
   a. Draw the graph of $y = mx + b$.
   b. Select a point, not on the line, and substitute its coordinates in the inequality.
   c. Shade the region of the plane containing this point if its coordinates make the inequality true.
   d. Shade the region on the other side of the line from this point if its coordinates make the inequality false.

3. The graph of the solution set of a system of sentences is the intersection of the graphs of the sentences.

CHAPTER REVIEW

In Ex. 1-4, many other answers are possible.

SECTION 16.1

Write five ordered pairs having different x values that are in the solution set of each inequality.

1. $y > -2x - 3$ (0, −2), (1, 0), (2, −6), (−1, 0) (−2, 2)

2. $y < 4x - 6$ (0, −7), (1, −4), (2, 0), (−1, −11), (−2, −20)

3. $y \leq \frac{5}{2}x + 1$ (0, 1), (2, 5), (−2, −4), (4, 2), (−4, −10)

4. $y \geq -2.3x + 1.9$ (0, 1.9), (1, 0), (2, −2), (−1, 5), (−2, 6.5)

Write the slope-intercept form of each inequality.

5. $-3x + y + 2 \geq 0$ $y \geq 3x - 2$

6. $5x - y + 4 > 0$ $y < 5x + 4$

7. $x - 3y + 5 < 0$ $y > \frac{1}{3}x + \frac{5}{3}$

8. $2x + 3y - 1 \leq 0$ $y \leq -\frac{2}{3}x + \frac{1}{3}$

Systems of Inequalities / 453

Chapter Test
Two Chapter Tests (Form A and Form B) are provided on pages 333–336 in the *Teacher's ResourceBank*.

SECTION 16.2

Graph each inequality.

In Ex. 9-14, each graph is a region that does not include the linear boundary. The location of each region with respect to the boundary is indicated.

9. $y > \frac{1}{2}x + 3$ above

10. $y < -x + 2$ below

11. $y < -2x - 1$ left

12. $y > 2x - 3$ left

13. $y < -x$ below

14. $y > x$ above

SECTION 16.3

Graph each inequality.

In Ex. 15-20, each graph is a region with a linear boundary that is also part of the graph. The location of the region with respect to the boundary is indicated.

15. $y \geq -x + 3$ above

16. $y \leq -\frac{1}{2}x + 1$ below

17. $x - 2y + 4 \leq 0$ above

18. $3x - 4y + 8 \geq 0$ below

19. $y \geq -4.5$ above

20. $x \leq -1.5$ left

SECTION 16.4

Graph each linear system.

21. $\begin{cases} y < \frac{1}{2}x + 2 \\ y < -x - 1 \end{cases}$

22. $\begin{cases} y < -2x + 3 \\ y > \frac{1}{3}x - 2 \end{cases}$

23. $\begin{cases} 2x + y < -3 \\ x + y > 2 \end{cases}$

SECTION 16.5

Graph each linear system.

In Ex. 24-26, each graph is a region with two boundaries that are straight lines. The quadrants that contain the region are indicated, as are the equations of the boundaries.

24. $\begin{cases} y \geq 3x - 1 \\ y < x + 2 \end{cases}$

Quadrants: All; Boundaries: Top: $y = x + 2$; Bottom: $y = 3x - 1$; Bottom boundary included.

25. $\begin{cases} -3x + y \leq -1 \\ x + 2y \geq 4 \end{cases}$

Quadrants: I, IV; Boundaries: Bottom: $x + 2y = 4$; Left: $-3x + y = -1$; Boundaries included

26. $\begin{cases} y < 2 \\ x \geq -4 \end{cases}$

Quadrants: All Boundaries: Top: $y = 2$; Left: $x = -4$; Left boundary included.

Quadratic Functions

The photographs below are real-world models of quadratic functions that you may study in a later high school mathematics course: the *circle*, the *ellipse*, and the *parabola*.

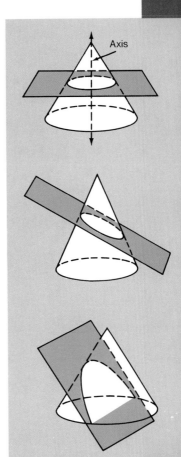

17.1 Quadratic Functions

OBJECTIVES: To identify whether or not a given rule describes a quadratic function
To write ordered pairs of a quadratic function by using the rule for the function

Quadratic trinomials such as $ax^2 + bx + c$ are used to form rules for quadratic functions.

Definition

You may want to give some examples of $y = ax^2 + bx + c$ when b is 0, when c is 0, and when both b and c are 0.

> A **quadratic function** is a function described by the rule $y = ax^2 + bx + c$, in which a, b, and c are rational numbers with $a \ne 0$.
>
> $y = 2x^2 - 3x + 5$
> $y = 0.5x^2 + 1.2x - 4.8$

P–1 **Which functions below are quadratic?** a, c

a. $y = 5x^2$ b. $y = 2x + 5$ c. $y = 3 - 2x^2$

EXAMPLE   Write three ordered pairs of $y = -3x^2 + x - 2$.

Solution: Let the x values equal -2, 0, and 2.

$y = -3(-2)^2 + (-2) - 2$ ◀ *Let $x = -2$.*
$y = -3(4) - 2 - 2$
$y = -12 - 2 - 2$
$y = -16$ The first ordered pair is $(-2, -16)$.

$y = -3(0)^2 + 0 - 2$ ◀ *Let $x = 0$.*
$y = -3(0) + 0 - 2$
$y = 0 + 0 - 2$
$y = -2$ The second ordered pair is $(0, -2)$.

$y = -3(2)^2 + 2 - 2$ ◀ *Let $x = 2$.*
$y = -3(4) + 2 - 2$
$y = -12 + 2 - 2$
$y = -12$ The third ordered pair is $(2, -12)$.

As with other functions, you can limit the domain of a quadratic function. Unless indicated otherwise, however, the domain will be {real numbers}.

456 / *Chapter 17*

▧ CLASSROOM EXERCISES ▧

Write Yes or No to indicate if each rule describes a quadratic function. (P-1)

1. $y = 5x^2 - 10x + 2$ Yes

2. $y = 3x - 10$ No

3. $y = x^3 + x^2 - 2x + 5$ No

4. $y = -3x^2$ Yes

5. $y = \dfrac{1}{x^2} + \dfrac{2}{x} + 3$ No

6. $y = x^2 - 3x$ Yes

7. $y = -2x + 1$ No

8. $y = \frac{3}{4}x^2 - \frac{1}{2}$ Yes

9. $y = \sqrt{x^2}$ No

10. $y = \sqrt{3}x^2 + \sqrt{2}x + \sqrt{5}$ No

11. $y = 2 - 3x^2 - 5x$ Yes

12. $y = (x + 1)^2$ Yes

13. $y = (x - 2)(x + 3)$ Yes

14. $y = x^{-2} + 3x^{-1} + 5$ No

15. $y = 3^2$ No

▧ WRITTEN EXERCISES ▧

Goal: To write ordered pairs of a quadratic equation

Sample Problem: Write three ordered pairs of $y = -x^2 + x + 1$.

Let $x = -2, 0,$ and 1.　　**Answer:** $(-2, -5), (0, 1), (1, 1)$

Write Yes or No to show whether each rule describes a quadratic function. (P-1)

1. $y = 2x - 5$ No

2. $y = -x + 3$ No

3. $y = -3x^2$ Yes

4. $y = 5x^2$ Yes

5. $y = x^3 + 2x - 6$ No

6. $y = x^4 - 2x - 3$ No

7. $y = (x - 2)^2$ Yes

8. $y = (2x - 3)(x + 1)$ Yes

9. $y = \dfrac{1}{x^2 + x - 5}$ No

10. $y = \sqrt{x^2 - 2x - 3}$ No

11. $y = 0.5x^2 + 0.7x - 0.3$ Yes

12. $y = -1.7x^2 - 3.1x + 2.7$ Yes

Write three ordered pairs of each function. Let the x values equal $-3, -1,$ and 2. (Example 1)

13. $y = -x^2$ $(-3, -9), (-1, -1), (2, -4)$

14. $y = -x^2 + 1$ $(-3, -8), (-1, 0), (2, -3)$

15. $y = x^2 + 2x + 1$ $(-3, 4), (-1, 0), (2, 9)$

16. $y = x^2 - x + 1$ $(-3, 13), (-1, 3), (2, 3)$

17. $y = x^2 - 3$ $(-3, 6), (-1, -2), (2, 1)$

18. $y = x^2 - 4$ $(-3, 5), (-1, -3), (2, 0)$

19. $y = -2x^2 - 3x$ $(-3, -9), (-1, 1), (2, -14)$

20. $y = -3x^2 - 1$
$(-3, -28), (-1, -4), (2, -13)$

Quadratic Functions / **457**

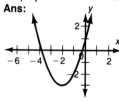
OBJECTIVES: To graph a quadratic function with {real numbers} as the domain, given a table of ordered pairs of the function

17.2 **Graphs of Quadratic Functions**

To graph a quadratic function with {real numbers} as the domain, given a rule for the function

You have already seen the graph of a quadratic function. It is a smooth curved line.

EXAMPLE 1 Graph $y = x^2 - 2x - 3$.

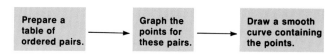

Solution:

$y = x^2 - 2x - 3$
$y = (-2)^2 - 2(-2) - 3 = 5$
$y = (-1)^2 - 2(-1) - 3 = 0$
$y = (0)^2 - 2(0) - 3 = -3$
$y = (1)^2 - 2(1) - 3 = -4$
$y = (2)^2 - 2(2) - 3 = \underline{\quad?\quad}$
$y = (3)^2 - 2(3) - 3 = \underline{\quad?\quad}$
$y = (4)^2 - 2(4) - 3 = \underline{\quad?\quad}$

x	y
-2	5
-1	0
0	-3
1	-4
2	?
3	?
4	?

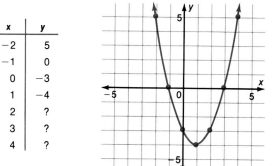

Figure 1

Ask questions such as these:
As x increases from -2 to 1, how does y change?
As x increases from 1 to 4, how does y change?

The points corresponding to these ordered pairs are graphed in Figure 1. A smooth curve containing the points is drawn.

The curve of Example 1 is called a *parabola.*

P–1 **Based on the graph, what is an estimate of each unknown y value?**

  a. $(2\frac{1}{2}, \underline{\quad?\quad}) -1\frac{3}{4}$ **b.** $(-1\frac{1}{2}, \underline{\quad?\quad}) 2\frac{1}{4}$ **c.** $(\frac{1}{2}, \underline{\quad?\quad}) -3\frac{3}{4}$

The two arrows on the graph of Figure 1 indicate that the parabola extends upward infinitely. The curve widens as it extends upward.

                                                 {all real numbers greater than or equal to -4}
P–2 **Based on the graph, what is the range of the function?**

There is no y value less than -4. The range is {all real numbers greater than or equal to -4}.

458 / *Chapter 17*

EXAMPLE 2   Graph $y = -x^2 + 3x$.

Solution:

$y = -x^2 + 3x$
$y = -(-1)^2 + 3(-1) = -4$
$y = -(0)^2 + 3(0) = 0$
$y = -(1)^2 + 3(1) = 2$
$y = -(2)^2 + 3(2) = 2$
$y = -(3)^2 + 3(3) = 0$
$y = -(4)^2 + 3(4) = -4$

x	y
−1	−4
0	0
1	2
2	2
3	0
4	−4

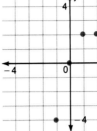

It is also helpful to locate the point at which the parabola "turns" or changes direction.

$y = -x^2 + 3x$ **Let** $x = \frac{3}{2}$.

$y = -\left(\frac{3}{2}\right)^2 + 3\left(\frac{3}{2}\right)$

$y = -\frac{9}{4} + \frac{9}{2}$ **Solve for y.**

$y = -\frac{9}{4} + \frac{18}{4}$

$y = \frac{9}{4}$

For this parabola, you can "see" that the x value of the turning point is $\frac{3}{2}$.

Discuss the range of this function based on its graph. Compare the parabola in Example 2 with the parabola in Example 1.

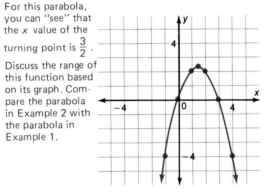

The point, $(\frac{3}{2}, \frac{9}{4})$, is located.

CLASSROOM EXERCISES

Write the unknown y values in each table. (Example 1)

1. $y = x^2$

x	y	
−2	?	4
−1	?	1
0	?	0
1	?	1
2	?	4

2. $y = -x^2$

x	y	
−2	?	−4
−1	?	−1
0	?	0
1	?	−1
2	?	−4

3. $y = x^2 + 2$

x	y	
−2	?	6
−1	?	3
0	?	2
1	?	3
2	?	6

Quadratic Functions / **459**

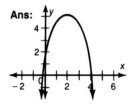

4. $y = x^2 + x + 1$

x	−2	−1	0	1	2
y	?	?	?	?	?

    3   1   1 3 7

5. $y = 2x^2 + x$

x	−2	−1	0	1	2
y	?	?	?	?	?

    6   1   0 3 10

6. $y = -x^2 + 2x - 3$

x	−2	−1	0	1	2
y	?	?	?	?	?

  −11 −6 −3 −2 −3

Refer to the graph at the right. Estimate one value for y and two values for x. (P-1)

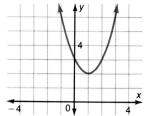

7. $(1, y)$ 2

8. $(2, y)$ 3

9. $(\frac{1}{2}, y)$ $2\frac{1}{4}$

10. $(-\frac{1}{2}, y)$ $4\frac{1}{4}$

11. $(x, 3)$ 0, 2

12. $(x, 6)$ −1, 3

13. $(x, 2\frac{1}{4})$ $\frac{1}{2}, 1\frac{1}{2}$

14. $(x, 5)$ $-\frac{3}{4}, 2\frac{3}{4}$

WRITTEN EXERCISES

Goal: To graph quadratic functions

Sample Problem: Graph $y = -x^2 + 3x$.   **Answer:** See Example 2.

In Ex. 1-6, each parabola turns at a high point (maximum), or a low point (minimum).

Graph the ordered pairs of each table. Draw a parabola that contains the points. (Examples 1 and 2) The point is given for each exercise.

1. Turns at (0, 1); low point

x	−2	−1	0	1	2
y	5	2	1	2	5

2. Turns at (0, −1); low point

x	−2	−1	0	1	2
y	3	0	−1	0	3

3. Turns at (1, 4); high point

x	−2	−1	0	1	2	3
y	−5	0	3	4	3	0

4. Turns at (−1, 1); high point

x	−3	−2	−1	0	1
y	−3	0	1	0	−3

5. Turns at (1, 2); low point

x	−1	0	1	2	3
y	6	3	2	3	6

6.

x	−2	−1	0	1	2	3
y	4	0	−2	−2	0	4

Turns at $(\frac{1}{2}, -2\frac{1}{4})$; low point

Graph each function. (Examples 1 and 2)

7. $y = x^2 - 2$

8. $y = -x^2 + 3$

9. $y = -x^2 - x$

10. $y = 2x^2 + x$

11. $y = x^2 - 6x + 10$

12. $y = -x^2 - 2x + 1$

NON-ROUTINE PROBLEM

13. Find the missing digits that make the three statements at the right true. The digits 0 through 9 can each appear exactly once in the three statements.

$7 + \blacksquare = \blacksquare$

$\blacksquare - 6 = \blacksquare$

$\blacksquare \times \blacksquare = 20$

460 / *Chapter 17*

OBJECTIVES: To graph two functions of the form $y = ax^2$ in the same coordinate plane

17.3 Functions Defined by $y = ax^2$

To graph a relation of the form $x = ay^2$

You have learned that any quadratic function has a rule of the form $y = ax^2 + bx + c$.

P–1 **Find the values of _a_, _b_, and _c_ in each rule below.**

1. $y = -\frac{5}{2}x^2$ **2.** $y = 3x^2 - 4x - 1$ **3.** $y = x^2 + 3x$

$a = -\frac{5}{2}; b = 0; c = 0$ $a = 3; b = -4; c = -1$ $a = 1; b = 3; c = 0$

EXAMPLE 1 Graph in the same coordinate plane.

1. $y = 2x^2$ **2.** $y = x^2$ **3.** $y = \frac{1}{2}x^2$

Solutions:

1.

x	y
−2	8
−1	2
0	0
1	2
2	8

2.

x	y
−2	4
−1	1
0	0
1	1
2	4

3.

x	y
−4	8
−2	2
0	0
2	2
4	8

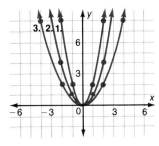

If a > 0, the parabola opens upward.

P–2 **Which function in Example 1 has the "widest" graph? the next widest? the least wide?**
$y = x^2$ $y = 2x^2$ $y = \frac{1}{2}x^2$

EXAMPLE 2 Graph in the same coordinate plane.

1. $y = -2x^2$ **2.** $y = -x^2$ **3.** $y = -\frac{1}{2}x^2$

Solutions:

1.

x	y
−2	−8
−1	−2
0	0
1	−2
2	−8

2.

x	y
−2	−4
−1	−1
0	0
1	−1
2	−4

3.

x	y
−4	−8
−2	−2
0	0
2	−2
4	−8

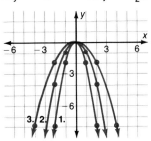

If a < 0, the parabola opens downward.

Quadratic Functions / 461

Teaching Suggestions p. M–50

Lesson Resources
Warm–Up: p. M–51
Maintenance: See below.
Practice Worksheet 17.3
Exploration 17
Transparency 41

Maintenance
1. The cost of Jack's lunch was $3 more than twice the cost of Kim's lunch (Condition 1). The total cost of the two lunches was $15 (Condition 2). Find the cost of each lunch.
   Ans: Jack's lunch: $11; Kim's lunch: $4
   (Section 4.1)
2. Multiply:
   $(2x + 1)(3x^2 + 2x - 1)$
   Ans: $6x^3 + 7x^2 - 1$
   (Section 8.5)
3. Factor: $3x^2 - 8x + 5$
   Ans: $(3x - 5)(x - 1)$
   (Section 9.4)
4. Add and simplify:
   $\dfrac{4}{x^2 - 9} + \dfrac{2}{x^2 + 4x + 3}$
   Ans: $\dfrac{6x - 2}{(x - 3)(x + 3)(x + 1)}$
5. Solve the system below by the multiplication/addition method.
   $\begin{cases} 3x - y = 13 \\ 2x + 3y = 16 \end{cases}$
   Ans: $\{(5,2)\}$ (Section 14.4)

Additional Examples
Example 1
Graph in the same coordinate plane.
a. $y = 3x^2$
b. $y = x^2$
c. $y = \frac{1}{3}x^2$
Ans:

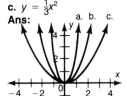

461

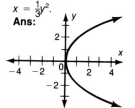
462

$y = -\frac{1}{2}x^2$

P–3 **Which function in Example 2 has the widest graph?** the next **widest?** **the least wide?** $y = -2x^2$

$y = -x^2$

Examples 1 and 2 show that as a gets smaller in absolute value, the corresponding parabola gets wider.

> **Effect of a on the graph of $y = ax^2 + bx + c$:**
>
> 1. If a is positive, the parabola opens upward.
> 2. If a is negative, the parabola opens downward.
> 3. As $|a|$ decreases, the parabola becomes wider.

A relation defined by a rule of the form $x = ay^2$ also has a parabola for its graph.

EXAMPLE 3 $\quad$ Graph the parabola defined by $x = -\frac{1}{2}y^2$.

Solution: Choose values for y and compute corresponding values for x.

$x = -\frac{1}{2}y^2$
$x = -\frac{1}{2}(-2)^2 = -2$
$x = -\frac{1}{2}(-1)^2 = -\frac{1}{2}$
$x = -\frac{1}{2}(0)^2 = 0$
$x = -\frac{1}{2}(1)^2 = -\frac{1}{2}$
$x = -\frac{1}{2}(2)^2 = -2$

x	y
-2	-2
-\frac{1}{2}	-1
0	0
-\frac{1}{2}	1
-2	2

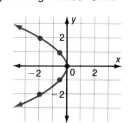

You can see that $x = -\frac{1}{2}y^2$ does not define a function. There are many ordered pairs that have the same x values.

CLASSROOM EXERCISES

Refer to the graphs of Examples 1 and 2.

1. What point is in all six graphs? (0, 0)

2. How do the graphs of $y = x^2$ and $y = -x^2$ compare? They are mirror images in the x axis.

3. Does the graph of $y = -2x^2$ open upward or downward? Downward

4. Which parabola is wider, $y = -x^2$ or $y = 2x^2$? $y = -x^2$

5. What is the range of the function defined by $y = 2x^2$?

6. What is the range of the function defined by $y = -2x^2$?

Indicate if the graph of each function opens upward or downward.
(Examples 1 and 2)

7. $y = -10x^2$ **8.** $y = 0.3x^2$ **9.** $y = 4 - \frac{5}{2}x^2$ **10.** $y = (2x - 1)^2$
     Downward   Upward   Downward   Upward

Indicate which of the two functions in each exercise has the wider graph.
(P-2, P-3)

11. $y = 5x^2$ **12.** $y = \frac{1}{2}x^2$ **13.** $y = 0.15x^2$ **14.** $y = x^2 - 5x + 1$
$\ y = 3x^2$ $\ y = \frac{3}{2}x^2$ $\ y = 0.09x^2$ $\ y = -2x^2 + x - 3$
$\ y = 3x^2$ $\ y = \frac{1}{2}x^2$ $\ y = 0.09x^2$ $\ y = x^2 - 5x + 1$

WRITTEN EXERCISES

Goal: To use a in the rule $y = ax^2 + bx + c$ to compare the graphs of quadratic functions

Sample Problem: Graph in the same coordinate plane: $y = 2x^2$, $y = \frac{1}{2}x^2$, $y = x^2$, $y = -2x^2$, $y = -\frac{1}{2}x^2$, $y = -x^2$.

Answer: See Examples 1 and 2.

In Ex. 1-10, the graphs are all parabolas that open upward or downward.

Complete each table. Graph the two functions in the same coordinate plane.
(Examples 1 and 2)

1. $y = x^2$ Opens upward at (0, 0) $y = \frac{1}{4}x^2$ Opens upward at (0, 0)

x	y		x	y	
-2	?	4	-4	?	4
-1	?	1	-2	?	1
0	?	0	0	?	0
1	?	1	2	?	1
2	?	4	4	?	4

2. $y = -x^2$ Opens downward at (0, 0) $y = -\frac{1}{4}x^2$ Opens downward at (0, 0)

x	y		x	y	
-2	?	-4	-4	?	-4
-1	?	-1	-2	?	-1
0	?	0	0	?	0
1	?	-1	2	?	-1
2	?	-4	4	?	-4

3. $y = -\frac{1}{2}x^2$ $y = -\frac{3}{2}x^2$ Opens downward at (0, 0)

x	y		x	y	
-3	?	$-\frac{9}{2}$	-2	?	-6
-2	?	-2	-1	?	$-\frac{3}{2}$
0	?	0	0	?	0
2	?	-2	1	?	$-\frac{3}{2}$
3	?	$-\frac{9}{2}$	2	?	-6

Opens downward at (0, 0)

4. $y = \frac{1}{2}x^2$ Opens upward at (0, 0) $y = \frac{3}{2}x^2$ Opens upward at (0, 0)

x	y		x	y	
-3	?	$\frac{9}{2}$	-2	?	6
-2	?	2	-1	?	$\frac{3}{2}$
0	?	0	0	?	0
2	?	2	1	?	$\frac{3}{2}$
3	?	$\frac{9}{2}$	2	?	6

*Quadratic Functions / **463***

5. $y = -2x^2$ Opens downward at (0, 0)

x	y
-2	?-8
-1	?-2
0	?0
1	?-2
2	?-8

$y = \frac{3}{2}x^2$ Opens upward at (0, 0)

x	y
-2	?6
-1	?$\frac{3}{2}$
0	?0
1	?$\frac{3}{2}$
2	?6

6. $y = 2x^2$ Opens upward at (0, 0)

x	y
-2	?8
-1	?2
0	?0
1	?2
2	?8

$y = -\frac{3}{2}x^2$

x	y
-2	?-6
-1	?$-\frac{3}{2}$
0	?0
1	?$-\frac{3}{2}$
2	?-6

7. $y = \frac{1}{2}x^2$ Opens upward at (0, 0)

x	y
-2	?2
-1	?$\frac{1}{2}$
0	?0
1	?$\frac{1}{2}$
2	?2

$y = \frac{1}{2}x^2 - 2$ Opens upward at (0, -2)

x	y
-2	?0
-1	?$-\frac{3}{2}$
0	?-2
1	?$-\frac{3}{2}$
2	?0

8. $y = -\frac{1}{4}x^2$

x	y
-4	?-4
-2	?-1
0	?0
2	?-1
4	?-4

$y = -\frac{1}{4}x^2 + 3$

x	y
-4	?-1
-2	?2
0	?3
2	?2
4	?-1

9. $y = -\frac{3}{4}x^2$ Opens downward at (0, 0)

x	y
-4	?-12
-2	?-3
0	?0
2	?-3
4	?-12

$y = -\frac{3}{4}x^2 + 1$ Opens downward at (0, 1)

x	y
-4	?-11
-2	?-2
0	?1
2	?-2
4	?-11

10. $y = \frac{3}{8}x^2$ Opens upward at (0, 0)

x	y
-4	?6
-2	?$\frac{3}{2}$
0	?0
2	?$\frac{3}{2}$
4	?6

$y = \frac{3}{8}x^2 - 2$

x	y
-4	?4
-2	?$-\frac{1}{2}$
0	?-2
2	?$-\frac{1}{2}$
4	?4

In Ex. 11-14, the parabolas open at the origin to the right or to the left of the y axis.

Complete each table. Then graph each parabola. (Example 3)

11. $x = y^2$ Opens to the right

x	y
1 ?	1
1 ?	-1
0 ?	0
4 ?	2
4 ?	-2

12. $x = -y^2$ Opens to the left

x	y
-1 ?	1
-1 ?	-1
0 ?	0
-4 ?	2
-4 ?	-2

13. $x = 2y^2$ Opens to the right

x	y
2 ?	1
2 ?	-1
0 ?	0
8 ?	2
8 ?	-2

14. $x = -2y^2$

x	y
-2 ?	1
-2 ?	-1
0 ?	0
-8 ?	2
-8 ?	-2

Opens to the left

REVIEW CAPSULE FOR SECTION 17.4

Multiply and simplify. (Pages 220–223)

1. $(x - 2)(x - 2)$ $x^2 - 4x + 4$
2. $(x + 3)(x + 3)$ $x^2 + 6x + 9$
3. $(x - 5)(x - 5)$ $x^2 - 10x + 25$
4. $(x + 6)(x + 6)$ $x^2 + 12x + 36$
5. $(x + 2.1)(x + 2.1)$ $x^2 + 4.2x + 4.41$
6. $(x - 2.7)(x - 2.7)$ $x^2 - 5.4x + 7.29$

OBJECTIVES: To write the number of units and the direction which the graph of a function
$y = a(x - h)^2$ is from the graph of a function $y = ax^2$ in the same coordinate plane

17.4 Functions Defined by $y = a(x - h)^2$

To graph a function $y = ax^2$ by using several ordered pairs and then to graph $y = a(x - h)^2$ by using the effect that h has on the graph

Explain that in a rule, such as $y = 3x^2 - 5x + 4$ it is not at once apparent what the value of h is. This will be learned in Section 17.6.

The equation $y = \frac{1}{2}(x - 2)^2$ is a rule for a quadratic function.

$\frac{1}{2}(x - 2)^2 = \frac{1}{2}(x - 2)(x - 2)$

$= \frac{1}{2}(x^2 - 4x + 4)$

$= \frac{1}{2}x^2 - 2x + 2$ ◀ In the form $ax^2 + bx + c$

 What are the values of a, b, and c? $a = \frac{1}{2}$; $b = -2$; $c = 2$

Replace $(x - 2)$ by A in $y = \frac{1}{2}(x - 2)^2$. ⟶ $y = \frac{1}{2}A^2$

This form suggests that the graph of $y = \frac{1}{2}(x - 2)^2$ is related to the graph of $y = \frac{1}{2}x^2$.

EXAMPLE 1 Graph the parabolas $y = \frac{1}{2}x^2$ and $y = \frac{1}{2}(x - 2)^2$ in the same coordinate plane.

Solutions:

$y = \frac{1}{2}x^2$

x	y
−3	$4\frac{1}{2}$
−2	2
−1	$\frac{1}{2}$
0	0
1	$\frac{1}{2}$
2	2
3	$4\frac{1}{2}$

$y = \frac{1}{2}(x - 2)^2$

x	y
−1	$4\frac{1}{2}$
0	2
1	$\frac{1}{2}$
2	0
3	$\frac{1}{2}$
4	2
5	$4\frac{1}{2}$

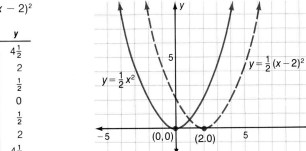

▲ The parabolas have the same shape.

Note that the graph of $y = \frac{1}{2}x^2$ can be moved **two** units to the **right** to obtain the graph of $y = \frac{1}{2}(x - 2)^2$.

EXAMPLE 2 Graph in the same coordinate plane.

1. $y = -\frac{1}{2}x^2$ **2.** $y = -\frac{1}{2}(x + 3)^2$

Lesson Resources
Warm–Up: p. M–51
Maintenance: See below.
Practice Worksheet 17.4
Transparency 41

Maintenance
1. Factor: $-18x^2 - 9x$
   Ans: $-9x(2x + 1)$
   (Section 2.5)
2. Subtract:
   $(-3x + 1) - (-4x + 2)$
   Ans: $x - 1$ (Section 8.4)
3. Simplify: $\frac{6a^2b}{18ab^2}$
   Ans: $\frac{a}{3b}$ (Section 10.2)
4. Write the domain, D, and the range, R, of the relation $\{(-2, -3), (0,4), (2,0)\}$.
   Ans: D = $\{-2, 0, 2\}$;
   R = $\{-3, 4, 0\}$
   (Section 12.4)
5. Write three ordered pairs having different x values that are in the solution set of $y > 3x + 2$.
   Ans: Answers will vary.
   Sample points: $(-2,0)$,
   $(0,3)$, $(1,7)$ (Section 16.1)

Additional Example
Example 1
Graph $y = x^2$ and $y = (x - 3)^2$ in the same coordinate plane.
Ans:

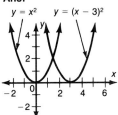

Additional Example
Example 2
Graph $y = -x^2$ and $y = -(x + 4)^2$ in the same co-ordinate plane.
Ans: $y = -x^2$

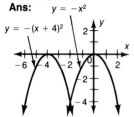

Explain that $y = -\frac{1}{2}(x + 3)^2$ can be expressed as $y = -\frac{1}{2}(x - (-3))^2$. Thus, h has the value -3.

Solutions:

$y = -\frac{1}{2}x^2$

x	y
-3	$-4\frac{1}{2}$
-2	-2
-1	$-\frac{1}{2}$
0	0
1	$-\frac{1}{2}$
2	-2
3	$-4\frac{1}{2}$

$y = -\frac{1}{2}(x + 3)^2$

x	y
-6	$-4\frac{1}{2}$
-5	-2
-4	$-\frac{1}{2}$
-3	0
-2	$-\frac{1}{2}$
-1	-2
0	$-4\frac{1}{2}$

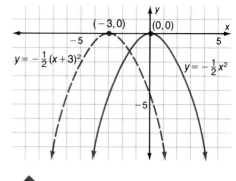

▲ *The parabolas have the same shape.*

Note that the graph of $y = -\frac{1}{2}x^2$ can be moved **three** units to the **left** to obtain the graph of $y = -\frac{1}{2}(x + 3)^2$.

Graphs of $y = ax^2$ and $y = a(x - h)^2$

The graph of $y = ax^2$ can be moved horizontally $|h|$ units to obtain the graph of $y = a(x - h)^2$. If h is negative, the motion is to the left. If h is positive, the motion is to the right.

Example 3
Graph the first parabola using x values $-2, -1, 0, 1,$ and 2. Then graph the second parabola. $y = x^2; y = (x - 3)^2$

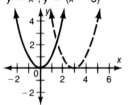

EXAMPLE 3   Graph the parabola $y = (x + 4)^2$ in the same plane as the parabola $y = x^2$ shown in Figure 1. Locate points of $y = (x + 4)^2$ that correspond to points A, B, C, D, and E.

Solution:

1. Locate points A', B', C', D', and E' in Figure 2 that are 4 units to the left of the corresponding points of Figure 1.

2. Draw the parabola.

Ask what value h has in $y = (x + 4)^2$ as it is compared with $y = a(x - h)^2$. Also ask what the value of a is.

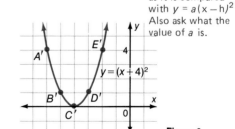

Figure 1

Figure 2

For each exercise, indicate how to move the graph of the first function to obtain the graph of the second function.

1. $y = x^2$; $y = (x - 2)^2$ 2 units right

2. $y = x^2$; $y = (x + 2)^2$ 2 units left

3. $y = 5x^2$; $y = 5(x - 1)^2$ 1 unit right

4. $y = 5x^2$; $y = 5(x + 4)^2$ 4 units left

5. $y = -3x^2$; $y = -3(x + 2)^2$ 2 units left

6. $y = -3x^2$; $y = -3(x - 1)^2$ 1 unit right

7. $y = \frac{1}{3}x^2$; $y = \frac{1}{3}(x + 10)^2$ 10 units left

8. $y = \frac{1}{3}x^2$; $y = \frac{1}{3}(x - 5)^2$ 5 units right

9. $y = \frac{3}{4}x^2$; $y = \frac{3}{4}(x - 7)^2$ 7 units right

10. $y = \frac{3}{4}x^2$; $y = \frac{3}{4}(x + 6)^2$ 6 units left

11. $y = 2(x - 3)^2$; $y = 2(x + 3)^2$ 6 units left

12. $y = 2(x - 1)^2$; $y = 2(x + 2)^2$
   3 units left

Goal: To use a graph of $y = ax^2$ to draw a graph of the form $y = a(x - h)^2$
Sample Problem: Graph $y = (x - 4)^2$ and $y = (x + 4)^2$.
Answer: See Example 3.

Write the number of units and the direction the graph of the first function can be moved to obtain the graph of the second function.

EXAMPLE: $y = 5x^2$; $y = 5(x + 6)^2$ **ANSWER:** 6 units left

1. $y = -3x^2$; $y = -3(x - 5)^2$ 5 units right

2. $y = 4x^2$; $y = 4(x - 10)^2$ 10 units right

3. $y = \frac{5}{2}x^2$; $y = \frac{5}{2}(x + 8)^2$ 8 units left

4. $y = -\frac{7}{2}x^2$; $y = -\frac{7}{2}(x + 6)^2$ 6 units left

5. $y = -x^2$; $y = -(x + 3\frac{1}{2})^2$ $3\frac{1}{2}$ units left

6. $y = x^2$; $y = (x - 2\frac{1}{4})^2$ $2\frac{1}{4}$ units right

7. $y = 1.4x^2$; $y = 1.4(x - 2.7)^2$ 2.7 units right

8. $y = -3.6x^2$; $y = -3.6(x + 2.1)^2$
   2.1 units left

Graph the first parabola in each exercise using the x values −2, −1, 0, 1, and 2. Then graph the second parabola. (Examples 1–3)

9. $y = 2x^2$; $y = 2(x + 2)^2$

10. $y = -2x^2$; $y = -2(x - 2)^2$

11. $y = -\frac{1}{2}x^2$; $y = -\frac{1}{2}(x - 3)^2$

12. $y = \frac{1}{2}x^2$; $y = \frac{1}{2}(x + 3)^2$

13. $y = -x^2$; $y = -(x + \frac{3}{2})^2$

14. $y = x^2$; $y = (x - \frac{5}{2})^2$

15. $y = \frac{3}{2}x^2$; $y = \frac{3}{2}(x - 2)^2$

16. $y = -\frac{3}{2}x^2$; $y = -\frac{3}{2}(x + 2)^2$

17. $y = -3x^2$; $y = -3(x + 1)^2$

18. $y = 3x^2$; $y = 3(x - 1)^2$

Quadratic Functions / 467

Common Error
When determining the direction the graph of a function can be moved to obtain the graph of a second function, students confuse the subtraction and addition signs with the value of h.

Example
Write the number of units and the direction the graph of $y = 3x^2$ can be moved to obtain the graph of $y = 3(x - 4)^2$.
Answer: Since h is -4, the graph can be moved left 4 units.

Prescription
Emphasize that the function must be in the form $y = a(x - h)^2$. If h is 3, then the function is $y = a(x - 3)^2$. If h is -3, then the function is $y = a(x - (-3))^2$ or $y = a(x + 3)^2$.

Assignment Guide
Basic Omit
Average Omit
Above Average pp. 467–468:
1–17 odd, 19–21

Additional Answers
Written Exercises
9. Second parabola is 2 units to left of the first.
10. Second parabola is 2 units to right of the first.
11. Second parabola is 3 units to right of the first.
12. Second parabola is 3 units to left of the first.
13. Second parabola is $\frac{3}{2}$ units to left of the first.
14. Second parabola is $\frac{5}{2}$ units to right of the first.
15. Second parabola is 2 units to right of the first.
16. Second parabola is 2 units to left of the first.
17. Second parabola is 1 unit to left of the first.
18. Second parabola is 1 unit to right of the first.

The expression $c - \dfrac{b^2}{4a}$ represents the _least_ value of the quadratic function $y = ax^2 + bx + c$, when $a > 0$.

EXAMPLE Find the approximate least value of $y = 21x^2 + 9x + 52$.

SOLUTION $a = 21$, $b = 9$, $c = 52$ $c - \dfrac{b^2}{4a} = 52 - \dfrac{9^2}{(4)(21)}$

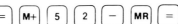

◄ **Least value**

Find the least value.

19. $y = 34x^2 + 68x + 109$
75

20. $y = 176x^2 + 465x + 239$
−68.13778

21. $y = 1084x^2 + 1296x + 238$
−149.36531

▌ MID-CHAPTER REVIEW ▌

Write four ordered pairs for each function. Let the x values equal −2, −1, 0, and 1. (Section 17.1)

1. $y = -\frac{1}{2}x^2 - 3x$
(−2, 4), (−1, $\frac{5}{2}$)
(0, 0), (1, $-\frac{7}{2}$)

2. $y = 5x^2 - 2$
(−2, 18), (−1, 3),
(0, −2), (1, 3)

3. $y = x^2 + x - 3$
(−2, −1), (−1, −3),
(0, −3), (1, −1)

Graph each function. (Section 17.2)

4. $y = \frac{1}{2}x^2 - 3$

5. $y = -\frac{1}{2}x^2 + 2$

6. $y = x^2 - 4x + 2$

Graph the two functions in the same coordinate plane. (Section 17.3)

7. $y = \frac{3}{4}x^2$
$y = -\frac{3}{4}x^2$

8. $y = \frac{5}{2}x^2$
$y = -\frac{5}{2}x^2$

9. $y = 2x^2$
$y = \frac{1}{2}x^2$

10. $y = -\frac{3}{2}x^2$
$y = -\frac{3}{4}x^2$

11. $y = -2x^2 - 2$
$y = \frac{1}{2}x^2 + 1$

Write the number of units and the direction the graph of the first function can be moved to obtain the graph of the second function. (Section 17.4)

12. $y = \frac{1}{4}x^2$ 3 units right
$y = \frac{1}{4}(x - 3)^2$

13. $y = -3x^2$ 5 units right
$y = -3(x - 5)^2$

14. $y = -\frac{5}{2}x^2$ $\frac{3}{2}$ units left
$y = -\frac{5}{2}(x + \frac{3}{2})^2$

15. $y = 5x^2$ 4 units left
$y = 5(x + 4)^2$

REVIEW CAPSULE FOR SECTION 17.5

Write three ordered pairs for each function below. Let the x values equal −2, 0, and 2. (Pages 456–457)

1. $y = \frac{1}{2}(x - 2)^2 + 1$
(−2, 9), (0, 3), (2, 1)

2. $y = 2x^2 - x - 3$ (−2, 7), (0, −3)
(2, 3)

3. $y = -2(x + 1)^2 - 3$
(−2, −5), (0, −5), (2, −21)

Additional Answers
Mid–Chapter Review
In Ex. 4–6, the high or low point and three points of the graph (parabola) are given.

4. Low point: (0, −3);
(−1, −2$\frac{1}{2}$), (0, 3), (1, −2$\frac{1}{2}$)

5. High point: (0, 2); (−1, 1$\frac{1}{2}$),
(0, 2), (1, 1$\frac{1}{2}$)

6. Low point: (2, −2); (1, −1),
(3, −1), (2, −2)

In Ex. 7–11, three points are given for each curve (parabola).

7. $y = \frac{3}{4}x^2$: (0, 0), (2, 3), (−2, 3);
$y = -\frac{3}{4}x^2$: (0, 0), (2, −3),
(−2, −3)

8. $y = \frac{5}{2}x^2$: (0, 0), (1, $\frac{5}{2}$), (−1, $\frac{5}{2}$);
$y = -\frac{5}{2}x^2$: (0, 0), (1, $-\frac{5}{2}$),
(−1, $-\frac{5}{2}$)

9. $y = 2x^2$: (0, 0), (1, 2),
(−1, 2); $y = \frac{1}{2}x^2$: (0, 0), (2, 2),
(−2, 2)

10. $y = -\frac{3}{2}x^2$: (0, 0), (1, $-\frac{3}{2}$),
(−1, $-\frac{3}{2}$); $y = -\frac{3}{4}x^2$: (0, 0),
(1, $-\frac{3}{4}$), (−1, $-\frac{3}{4}$)

11. $y = \frac{1}{2}x^2 + 1$: (0, 1), (2, 3),
(−2, 3); $y = -2x^2 - 2$:
(0, −2), (1, −4), (−1, −4)

17.5 Functions Defined by $y = a(x - h)^2 + k$

OBJECTIVES: To graph a function $y = ax^2$ using several ordered pairs and then to graph
$y = a(x - h)^2 + k$ by using the effect that h and k have on the first graph
To write an equation for a function obtained by moving the graph of $y = ax^2$
according to stated horizontal and vertical motions

The expression $\frac{1}{2}(x - 3)^2 + 2$ can be changed to the form $ax^2 + bx + c$.

$$\frac{1}{2}(x - 3)^2 + 2 = \frac{1}{2}(x^2 - 6x + 9) + 2$$

$$= \frac{1}{2}x^2 - 3x + \frac{9}{2} + 2$$

$$= \frac{1}{2}x^2 - 3x + \frac{13}{2} \quad \blacktriangleleft \quad \text{In the form } ax^2 + bx + c$$

P–1 What kind of function does $y = \frac{1}{2}(x - 3)^2 + 2$ represent? Quadratic

EXAMPLE 1 Graph the following functions in the same coordinate plane.

1. $y = \frac{1}{2}x^2$ **2.** $y = \frac{1}{2}(x - 3)^2$ **3.** $y = \frac{1}{2}(x - 3)^2 + 2$

Solutions:

1.

x	y
−3	$4\frac{1}{2}$
−2	2
−1	$\frac{1}{2}$
0	0
1	$\frac{1}{2}$
2	2
3	$4\frac{1}{2}$

2.

x	y
0	$4\frac{1}{2}$
1	2
2	$\frac{1}{2}$
3	0
4	$\frac{1}{2}$
5	2
6	$4\frac{1}{2}$

3.

x	y
0	$6\frac{1}{2}$
1	4
2	$2\frac{1}{2}$
3	2
4	$2\frac{1}{2}$
5	4
6	$6\frac{1}{2}$

You may want to demonstrate how
to obtain the ordered pairs
especially in **3**.

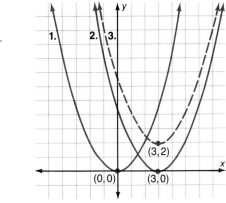

$\blacktriangle$ *The parabolas have same shape.*

P–2 How could the graph of $y = \frac{1}{2}(x - 3)^2$ be moved to make it coincide with the graph of $y = \frac{1}{2}(x - 3)^2 + 2$? 2 units up

P–3 How could the graph of $y = \frac{1}{2}x^2$ be moved in two steps to make it coincide with the graph of $y = \frac{1}{2}(x - 3)^2 + 2$? 3 units right and 2 units up

Quadratic Functions / **469**

Teaching Suggestions p. M–50

Lesson Resources
Warm–Up: p. M–51
Maintenance: See below.
Practice Worksheet 17.5
Transparency 41

Maintenance

1. Write $\dfrac{r^6 r^{-1}}{r^3}$ as one power.
   Ans: r^2 (Section 6.3)

2. Simplify: $\dfrac{\sqrt{x^4}}{\sqrt{9x^2}}$
   Ans: $\dfrac{x}{3}$ (Section 7.6)

3. Divide and simplify:
   $\dfrac{16xy}{9x} \div \dfrac{4xy}{3x^2z}$
   Ans: $\dfrac{4xz}{3}$ (Section 10.6)

4. Is $y = x^2 + 2$ a linear function?
   Ans: No (Section 13.1)

5. Tom deposited a total of 21 ten- and twenty-dollar bills (Condition 1). The amount of the deposit was $260 (Condition 2). How many of each kind of bill was deposited?
   Ans: tens: 16; twenties: 5 (Section 15.6)

Additional Example
Example 1
Graph the following functions in the same coordinate plane.
a. $y = x^2$
b. $y = (x + 2)^2$
c. $y = (x + 2)^2 + 3$
Ans:

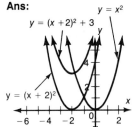

Graph the following functions in
the same coordinate plane.
a. $y = -x^2$
b. $y = -(x - 2)^2$
c. $y = -(x - 2)^2 + 3$
Ans:

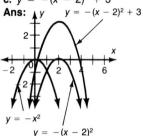

EXAMPLE 2 Graph the following functions in the same coordinate plane.

Remind students that rules 2
and 3 can be written as
$y = -2(x - (-4)^2$ and
$y = -2(x - (-4))^2 -3$.

1. $y = -2x^2$ **2.** $y = -2(x + 4)^2$

3. $y = -2(x + 4)^2 - 3$

Solutions:

1.

x	y
-2	-8
-1	-2
0	0
1	-2
2	-8

2.

x	y
-6	-8
-5	-2
-4	0
-3	-2
-2	-8

3.

x	y
-6	-11
-5	-5
-4	-3
-3	-5
-2	-11

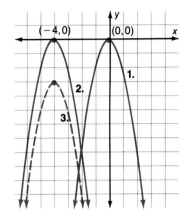

▲ *The parabolas have same shape.*

P–4 **How could the graph of $y = -2(x + 4)^2$ be moved to make it coincide with the graph of $y = -2(x + 4)^2 - 3$?** 3 units down

P–5 **How could the graph of $y = -2x^2$ be moved in two steps to make it coincide with the graph of $y = -2(x + 4)^2 - 3$?** 4 units left and
3 units down

Make sure
students clearly
understand this
summary.

Graphs of $y = ax^2$ and $y = a(x - h)^2 + k$

1. The graph of $y = a(x - h)^2 + k$ has the same shape as the graph of $y = ax^2$.
2. The graph of $y = a(x - h)^2 + k$ is $|h|$ units to the right or to the left of the graph of $y = ax^2$.
3. The graph of $y = a(x - h)^2 + k$ is $|k|$ units above or below the graph of $y = ax^2$.

P–6 **How do you know whether the graph of $y = a(x - h)^2 + k$ is to the right or to the left of the graph of $y = ax^2$?**

If h is positive, it is to the right. If h is negative, it is to the left.

How do you know whether the graph of $y = a(x - h)^2 + k$ is above or below the graph of $y = ax^2$?
If k is positive, it is above. If k is negative, it is below.

EXAMPLE 3   Write an equation of a function for which the graph is obtained as described below.

"The graph of $y = -\frac{3}{2}x^2$ is moved 4 units to the left and 3 units upward."

Solution: Let $h = -4$, $k = 3$, $a = -\frac{3}{2}$.

◀ *The graph is 4 units to the left, 3 units upward, and has the same shape as $y = -\frac{3}{2}x^2$.*

$y = a(x - h)^2 + k$

$y = -\frac{3}{2}(x - (-4))^2 + 3$

$y = -\frac{3}{2}(x + 4)^2 + 3$ ◀ *Equation of the function*

CLASSROOM EXERCISES

Write Yes or No to indicate if the graph of each function below has the same shape as the graph of $y = \frac{1}{2}(x - 3)^2 + 5$.

1. $y = \frac{1}{2}x^2$ Yes

2. $y = 2x^2$ No

3. $y = 3(x - 3)^2 + 5$ No

4. $y = \frac{1}{2}(x + 3)^2 - 5$ Yes

5. $y = -\frac{1}{2}x^2$ Yes

6. $y = \frac{1}{2}(x - 4)^2 + 3$ Yes

7. What variable in $a(x - h)^2 + k$ represents a number that determines the shape of the parabola? a

8. What variable in $a(x - h)^2 + k$ represents a number that determines the horizontal position of the parabola? h

9. What variable in $a(x - h)^2 + k$ represents a number that determines the vertical position of the parabola? k

Write the two motions needed to make the graph of the first function coincide with the graph of the second function. (P-3, P-5)

10. $y = x^2$ 2 units right, 3 units up
    $y = (x - 2)^2 + 3$

11. $y = 3x^2$ 5 units right, 2 units down
    $y = 3(x - 5)^2 - 2$

12. $y = \frac{1}{3}x^2$ 3 units left, 5 units up
    $y = \frac{1}{3}(x + 3)^2 + 5$

13. $y = -\frac{1}{2}x^2$ 2 units left, 3 units down
    $y = -\frac{1}{2}(x + 2)^2 - 3$

14. $y = -x^2$ 2 units right, 1 unit up
    $y = -(x - 2)^2 + 1$

15. $y = \frac{3}{4}x^2$
    $y = \frac{3}{4}(x + 1)^2 - 1$ 1 unit left, 1 unit down

Quadratic Functions / 471

■ WRITTEN EXERCISES ■

Goal: To use a graph of $y = ax^2$ to graph a function of the form
$$y = a(x - h)^2 + k$$
Sample Problem: Graph in the same coordinate plane: $y = -2x^2$,
$$y = -2(x + 1)^2 + 5, \quad y = -2(x + 1)^2 - 5.$$
Answer: See Examples 1 and 2.

*Graph the first function using the x values −2, −1, 0, 1, and 2. Then graph
the second function.* (Examples 1 and 2)

1. $y = \frac{1}{2}x^2$; $y = \frac{1}{2}(x - 2)^2 + 3$

2. $y = -\frac{1}{2}x^2$; $y = -\frac{1}{2}(x - 5)^2 + 2$

3. $y = -x^2$; $y = -(x + 3)^2 + 2$

4. $y = x^2$; $y = (x + 2)^2 + 4$

5. $y = 2x^2$; $y = 2(x + 1)^2 - 3$

6. $y = -2x^2$; $y = -2(x - 3)^2 - 3$

7. $y = -\frac{3}{2}x^2$; $y = -\frac{3}{2}(x - 3\frac{1}{2})^2 + 2\frac{1}{2}$

8. $y = \frac{3}{2}x^2$; $y = \frac{3}{2}(x + 2\frac{1}{2})^2 - 3\frac{1}{2}$

*Write an equation of a function for which the graph is obtained as
described in each exercise.* (Example 3)

9. The graph of $y = 2x^2$ is moved
three units to the right. $y = 2(x - 3)^2$

10. The graph of $y = -3x^2$ is moved
five units to the right. $y = -3(x - 5)^2$

11. The graph of $y = \frac{1}{2}x^2$ is moved
two units to the left and five units
upward. $y = \frac{1}{2}(x + 2)^2 + 5$

12. The graph of $y = 5x^2$ is moved
three units to the left and two units
upward. $y = 5(x + 3)^2 + 2$

13. The graph of $y = \frac{1}{3}(x + 1)^2$ is moved
five units downward. $y = \frac{1}{3}(x + 1)^2 - 5$

14. The graph of $y = -3(x - 5)^2$ is
moved four units downward.
$y = -3(x - 5)^2 - 4$

15. The graph of $y = -x^2$ is moved
three units to the right and two
units downward. $y = -(x - 3)^2 - 2$

16. The graph of $y = \frac{1}{2}x^2$ is moved five
units to the right and six units
downward. $y = \frac{1}{2}(x - 5)^2 - 6$

17. The graph of $y = 3x^2$ is moved five
units downward. $y = 3x^2 - 5$

18. The graph of $y = -\frac{1}{2}x^2$ is moved
three units downward. $y = \frac{1}{2}x^2 - 3$

REVIEW CAPSULE FOR SECTION 17.6

*Find the missing term to form a trinomial square (take $\frac{1}{2}$ the coefficient of x
and square it). Then factor the trinomial.* (Pages 255–257)

1. $x^2 - 4x +$ __?__

2. $x^2 + 6x +$ __?__

3. $x^2 + x +$ __?__

4. $x^2 - 3x +$ __?__

5. $x^2 + 10x +$ __?__

6. $x^2 + 18x +$ __?__

7. $x^2 - 5x +$ __?__

8. $x^2 - 7x +$ __?__

17.6 Standard Form of Quadratic Polynomials

Teaching Suggestions p. M–50

Lesson Resources
Warm–Up: p. M–51
Maintenance: See below.
Practice Worksheet 17.6

P–1   **What are the values of a, h, and k in each expression below?**

a. $-(x + 3)^2 - \frac{3}{4}$ **b.** $\frac{3}{2}(x - \frac{1}{4})^2 + \frac{5}{4}$ $a = \frac{3}{2}, h = \frac{1}{4}, k = \frac{5}{4}$

$a = -1, h = -3, k = -\frac{3}{4}$

EXAMPLE 1   Write $2(x - 3)^2 + 5$ in the form $ax^2 + bx + c$.

Solution: $2(x - 3)^2 + 5 = 2(x^2 - 6x + 9) + 5$ ◀ $x^2 - 6x + 9$ is a *trinomial square.*

$= 2x^2 - 12x + 18 + 5$

$= 2x^2 - 12x + 23$

Every quadratic polynomial can be changed from the form
$ax^2 + bx + c$ to the **standard form** $a(x - h)^2 + k$.

EXAMPLE 2   Write $x^2 + 2x + 5$ in standard form.

Solution:

1. Group terms to form a trinomial square. $\longrightarrow$ $(x^2 + 2x + \underline{\ ?\ }) + 5 + \underline{\ ?\ }$

2. Multiply the coefficient of x by $\frac{1}{2}$. Square. $\longrightarrow$ $\frac{1}{2}(2) = 1$; $(1)^2 = 1$

3. Add 1 to make a trinomial square.
Then add -1. $\longrightarrow$ $(x^2 + 2x + 1) + 5 + (-1)$

4. Factor and simplify. $\longrightarrow$ $(x + 1)^2 + 4$

5. Write the standard form. $\longrightarrow$ $(x - (-1))^2 + 4$ ◀ $a = 1$ $h = -1$ $k = 4$

EXAMPLE 3   Write $x^2 - 6x - 2$ in standard form.

Solution:

1. Group terms. $\longrightarrow$ $(x^2 - 6x + \underline{\ ?\ }) - 2 + \underline{\ ?\ }$

2. Multiply the coefficient of x by $\frac{1}{2}$. Square. $\longrightarrow$ $\frac{1}{2}(6) = 3$; $3^2 = 9$

3. Add 9 to make a trinomial square.
Then add -9. $\longrightarrow$ $(x^2 - 6x + 9) - 2 + (-9)$ ◀ $9 + (-9) = 0$

4. Factor and simplify. $\longrightarrow$ $(x - 3)^2 - 11$

Quadratic Functions / **473**

Maintenance

1. Alan's scores on three tests were 87, 93, and 68. He wants his mean score on four tests to be greater than 80 (Condition 2). Find the lowest score he can get on the fourth test (Condition 1).
Ans: 73   (Section 5.7)

2. Simplify: $\sqrt{288}$
Ans: $12\sqrt{2}$ (Section 7.3)

3. Factor: $3x^2 - 48$
Ans: $3(x - 4)(x + 4)$
(Section 9.7)

4. Write the y intercept of the graph of $y = 2x - 4$.
Ans: -4   (Section 13.2)

5. Solve the system below by the substitution method.
$\begin{cases} x - 2y = 8 \\ -3x + 2y = -16 \end{cases}$
Ans: $\{(4, -2)\}$ (Section 15.1)

Additional Examples
Example 1
Write each expression in the form $ax^2 + bx + c$.
1. $(x + 2)^2 + 6$
Ans: $x^2 + 4x + 10$
2. $3(x - 5)^2 + 2$
Ans: $3x^2 - 30x + 77$

Example 2
Write each polynomial in standard form.
1. $x^2 + 8x + 10$
Ans: $(x - (-4))^2 - 6$
2. $x^2 + 12x - 3$
Ans: $(x - (-6))^2 - 39$

Example 3
Write each polynomial in standard form.
1. $x^2 - 2x - 5$
Ans: $(x - 1)^2 - 6$
2. $x^2 - 6x + 11$
Ans: $(x - 3)^2 + 2$

Additional Examples
Example 4
Write each polynomial in standard form.
1. $x^2 + 7x - 2$
   Ans: $(x - (-\frac{7}{2}))^2 - \frac{57}{4}$
2. $x^2 - 3x - 5$
   Ans: $(x - \frac{3}{2})^2 - \frac{29}{4}$

Common Error
When adding a number to make a trinomial square, students forget to also add the opposite of the number.

Example
$x^2 - 4x + 5 =$
$(x^2 - 4x + 4) =$
$(x - 2)^2 + 5$

Prescription
Remind students that they do not want to change the value of the polynomial. When they add a number, they must also add its opposite so that the amount they are actually adding is 0. Have students check their answers by changing from standard form to the form $ax^2 + bx + c$.

Assignment Guide
Basic Omit
Average Omit
Above Average p. 474:
1–19 odd

Additional Answers
Written Exercises
1. $x^2 - 6x + 11$
2. $x^2 - 4x + 9$
3. $2x^2 - 4x - 5$
4. $-2x^2 - 4x - 7$
9. $(x - (-1))^2 + 2$
10. $(x - 1)^2 - 6$
11. $(x - 2)^2 - 3$
12. $(x - (-2))^2 - 1$
13. $(x - (-3))^2 - 14$
14. $(x - 3)^2 + 1$
15. $(x - 5)^2 - 26$
16. $(x - (-6))^2 - 33$

EXAMPLE 4 Write $x^2 + 5x + 1$ in standard form.
Solution:

① Group terms. ⟶ $(x^2 + 5x + \underline{\ ?\ }) + 1 + \underline{\ ?\ }$

② Multiply the coefficient of x by $\frac{1}{2}$. Square. ⟶ $\frac{1}{2}(5) = \frac{5}{2}; (\frac{5}{2})^2 = \frac{25}{4}$

③ Add $\frac{25}{4}$. Then add its opposite. ⟶ $(x^2 + 5x + \frac{25}{4}) + 1 + (-\frac{25}{4})$ ◀ $\frac{25}{4} + (-\frac{25}{4}) = 0$

④ Factor and simplify. ⟶ $(x + \frac{5}{2})^2 - \frac{21}{4}$

⑤ Write the standard form. ⟶ $(x - (-\frac{5}{2}))^2 - \frac{21}{4}$

CLASSROOM EXERCISES

Write each trinomial in the form $(x - h)^2$.

1. $x^2 + 2x + 1$ $(x - (-1))^2$
2. $x^2 + 4x + 4$ $(x - (-2))^2$
3. $x^2 - 6x + 9$ $(x - 3)^2$
4. $x^2 - 8x + 16$ $(x - 4)^2$
5. $x^2 + x + \frac{1}{4}$ $(x - (-\frac{1}{2}))^2$
6. $x^2 - x + \frac{1}{4}$ $(x - \frac{1}{2})^2$
7. $x^2 - 5x + \frac{25}{4}$ $(x - \frac{5}{2})^2$
8. $x^2 - 3x + \frac{9}{4}$ $(x - \frac{3}{2})^2$

WRITTEN EXERCISES

Goal: To write a quadratic polynomial in standard form
Sample Problem: $x^2 - 7x + 10$ **Answer:** $(x - \frac{7}{2})^2 - \frac{9}{4}$

Write each expression in the form $ax^2 + bx + c$. (Example 1)

1. $(x - 3)^2 + 2$
2. $(x - 2)^2 + 5$
3. $2(x - 1)^2 - 7$
4. $-2(x + 1)^2 - 5$
5. $3(x + 2)^2 + 3$ $3x^2 + 12x + 15$
6. $5(x - 5)^2 + 3$ $5x^2 - 50x + 128$
7. $\frac{1}{2}(x - 2)^2 + 6$ $\frac{1}{2}x^2 - 2x + 8$
8. $-\frac{1}{2}(x + 2)^2 - 3$ $-\frac{1}{2}x^2 - 2x - 5$

Write each polynomial in standard form. (Examples 2–4)

9. $x^2 + 2x + 3$
10. $x^2 - 2x - 5$
11. $x^2 - 4x + 1$
12. $x^2 + 4x + 3$
13. $x^2 + 6x - 5$
14. $x^2 - 6x + 10$
15. $x^2 - 10x - 1$
16. $x^2 + 12x + 3$
17. $x^2 - x + 5$ $(x - \frac{1}{2})^2 + \frac{19}{4}$
18. $x^2 + x - 2$ $(x - (-\frac{1}{2}))^2 - \frac{9}{4}$
19. $x^2 + 3x - 4$ $(x - (-\frac{3}{2}))^2 - \frac{25}{4}$
20. $x^2 - 5x - 3$ $(x - \frac{5}{2})^2 - \frac{37}{4}$

REVIEW CAPSULE FOR SECTION 17.7

Identify the values of a, h, and k in each function. (Pages 469–472)

1. $y = 3(x - 2)^2 + 1$
   $a = 3; h = 2; k = 1$
2. $y = 4(x - 5)^2 - 2$
   $a = 4; h = 5; k = -2$
3. $y = \frac{1}{2}(x + \frac{3}{2})^2 - 5$
   $a = \frac{1}{2}; h = -\frac{3}{2}; k = -5$

OBJECTIVES: To write the coordinates of the vertex of the parabola that is the
graph of a given function having the rule $y = a(x - h)^2 + k$

17.7 Vertex and Axis of a Parabola

To write an equation for the axis of the parabola that is the graph
of a given function having the rule $y = a(x - h)^2 + k$

Teaching Suggestions p. M–50

Lesson Resources
Warm–Up: p. M–51
Maintenance: See below.
Practice Worksheet 17.7

You have seen how the values of a, h, and k in $y = a(x - h)^2 + k$
affect the shape and position of the graph. There are other facts
about the graph that the standard form can provide.

EXAMPLE 1   The graph of $y = \frac{1}{2}(x - 2)^2 + 3$ is drawn. Write
the coordinates of its lowest point.

Solution: $y = \frac{1}{2}(x - 2)^2 + 3$

Explain that a
parabola opening
upward has a
lowest point (the
vertex). A parabola
opening downward
has a highest
point (also the
vertex).

x	y
0	5
1	$3\frac{1}{2}$
2	3
3	$3\frac{1}{2}$
4	5

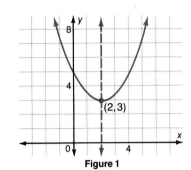

Figure 1

The coordinates of the lowest
point of the graph are (2, 3).

Maintenance
1. Solve: $-8x < 20$
   Ans: $x > -\frac{5}{2}$ (Section 5.5)
2. Divide:
   $(8x^2 + 6x - 9) \div (2x + 3)$
   Ans: $4x - 3$
   (Section 8.6)
3. Subtract and simplify:
   $\frac{9}{y + 1} - \frac{3}{y}$
   Ans: $\frac{6y - 3}{y(y + 1)}$
   (Section 11.5)
4. Write the slope of the line that
   contains the points $(-4, -1)$
   and $(-1, -4)$.
   Ans: -1 (Section 13.4)
5. The length of a rectangle is
   twice the width. The perimeter
   is 42 feet. Find the length and
   the width.
   **Ans: length: 14 feet; width:
   7 feet** (Section 14.5)

The point with coordinates (2, 3) is called the ***vertex of the parabola.***
The vertex of a parabola that is a graph of a quadratic function is
either the lowest point or the highest point of the graph.

P–1   **Is the vertex in Figure 1 the highest point or the lowest point of
the graph?**   Lowest point

Compare $\frac{1}{2}(x - 2)^2 + 3$ with the standard form $y = a(x - h)^2 + k$.

P–2   **What is the value of a?** $\frac{1}{2}$ **of h?** 2 **of k?** 3

The coordinates of the vertex of the parabola in Figure 1 are (2, 3).
They correspond to (h, k).

Additional Example
Example 1
Write the coordinates of the low-
est point of the parabola
$y = 2(x + 3)^2 + 1$.

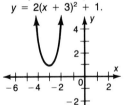

Ans: $(-3,1)$

> The vertex of a parabola with a
> rule of the form $y = a(x - h)^2 + k$
> has coordinates (h, k).
>
> $y = -2(x - (-5))^2 - 1$
>
> Vertex: $(-5, -1)$

Quadratic Functions / **475**

P–3 **What are the coordinates of the vertex of the graph of each function below?**

    a. $y = (x - \frac{1}{2})^2 + \frac{5}{2}$ $(\frac{1}{2}, \frac{5}{2})$ **b.** $y = (x + 3)^2 - 7$ $(-3, -7)$

A vertical line that contains the vertex of a parabola is called the *axis* of the parabola. The dashed line in Figure 1 is the axis.

P–4 **What is an equation for the axis in Figure 1?** $x = 2$

If a parabola has the rule $y = a(x - h)^2 + k$, its axis has the equation $x = h$.	$y = 5(x - 2)^2 + 3$ Equation of axis: $x = 2$

P–5 **What is the value of h if you compare each of the following with $y = a(x - h)^2 + k$?**

    a. $y = -3(x - 5)^2 + 2$ $h = 5$ **b.** $y = -(x + 3)^2 - 2$ $h = -3$

    c. $y = \frac{1}{2}(x - \frac{1}{4})^2 + 1$ $h = \frac{1}{4}$ **d.** $y = -2(x + \frac{4}{3})^2 - 5$ $h = -\frac{4}{3}$

Additional Examples
Example 2
Write an equation for the axis of the graph of each parabola.
1. $y = 2(x - 4)^2 - 6$
   Ans: $x = 4$
2. $y = -3(x + 5)^2 - 2$
   Ans: $x = -5$
3. $y = 5(x - 3)^2 + 4$
   Ans: $x = 3$

EXAMPLE 2   Write an equation for the axis of the graph of $y = -(x + 3)^2 - 2$.

Solution: Since $h = -3$,   ◀ **Equation**
                       $x = -3$.   **of the axis**

P–6 **What are the coordinates of the vertex of the parabola in Example 2?** $(-3, -2)$

CLASSROOM EXERCISES

Write the coordinates of the vertex of each graph. (P-3)

1. $y = 2(x - 5)^2 + 2$ $(5, 2)$ **2.** $y = 3(x - 1)^2 - 4$ $(1, -4)$

3. $y = -\frac{1}{2}(x + 2)^2 + 6$ $(-2, 6)$ **4.** $y = (x + 1)^2 - 6$ $(-1, -6)$

5. $y = x^2$ (0, 0)

6. $y = x^2 + 1$ (0, 1)

7. $y = -(x - \frac{1}{2})^2 + \frac{3}{4}$ $(\frac{1}{2}, \frac{3}{4})$

8. $y = 2.5(x + 0.3)^2 - 1.2$ (−0.3, −1.2)

9. $y = (x - 5)^2$ (5, 0)

10. $y = -(x + 4)^2$ (−4, 0)

Write an equation for the axis of each graph. (Example 2)

11. $y = \frac{1}{2}(x + 1)^2 - 2$ $x = -1$

12. $y = -2(x - 3)^2 + 4$ $x = 3$

13. $y = -3(x - \frac{1}{2})^2 + \frac{3}{2}$ $x = \frac{1}{2}$

14. $y = (x + 4.5)^2 - 2.5$ $x = -4.5$

15. $y = -\frac{1}{2}x^2$ $x = 0$

16. $y = \frac{4}{3}x^2$ $x = 0$

17. $y = 2x^2 + 3$ $x = 0$

18. $y = -3(x + 5)^2$ $x = -5$

▓▓▓ **WRITTEN EXERCISES** ▓▓▓

Goals: To write the coordinates of the vertex and the equation of the axis of the graph of a parabola

Sample Problem: $y = -3(x + 5)^2 - 2$

Answer: Vertex: (−5, −2);   Equation of axis: $x = -5$

Write the coordinates of the vertex of each graph. (Example 1, P-3)

1. $y = 3(x - 1)^2 + 10$ (1, 10)

2. $y = \frac{1}{2}(x - 5)^2 + 2$ (5, 2)

3. $y = -5(x - 7)^2 - 2$ (7, −2)

4. $y = -3(x - 3)^2 - 5$ (3, −5)

5. $y = -(x + 2)^2 + 3$ (−2, 3)

6. $y = 2(x + 5)^2 + 7$ (−5, 7)

7. $y = \frac{2}{3}(x + 1)^2 - 8$ (−1, −8)

8. $y = -\frac{1}{2}(x + 4)^2 - 3$ (−4, −3)

9. $y = \frac{3}{4}(x - \frac{1}{2})^2 - \frac{2}{3}$ $(\frac{1}{2}, -\frac{2}{3})$

10. $y = \frac{2}{3}(x - \frac{1}{3})^2 - \frac{3}{4}$ $(\frac{1}{3}, -\frac{3}{4})$

Write an equation for the axis of each graph. (Example 2)

11. $y = 3(x - 5)^2 + 2$ $x = 5$

12. $y = -3(x - 2)^2 + 1$ $x = 2$

13. $y = -2(x - 3)^2 + 5$ $x = 3$

14. $y = \frac{1}{2}(x - 1)^2 + 5$ $x = 1$

15. $y = -(x + 1)^2 - 5$ $x = -1$

16. $y = (x + 2)^2 - 6$ $x = -2$

17. $y = 3x^2 - 2x + 3$ $x = \frac{1}{3}$

18. $y = 2x^2 + 3x - 5$ $x = -\frac{3}{4}$

19. $y = -x^2 - 7x + 2$ $x = -\frac{7}{2}$

20. $y = -3x^2 - x + 7$ $x = -\frac{1}{6}$

21. $y = -5x^2 - x + 3$ $x = -\frac{1}{10}$

22. $y = -x^2 - 9x + 1$ $x = -\frac{9}{2}$

*Quadratic Functions   /   **477***

Assignment Guide
Basic Omit
Average Omit
Above Average p. 477:
1–21 odd

Quiz Sections 17.5–17.7
After completing Sections 17.5–17.7, you may want to administer a quiz covering the same sections. See the quiz provided on page 352 in the *Teacher's ResourceBank*.

The basis for making predictions is a straight line graph which is the median line of fit drawn on a scatter plot of data. Students will learn the technique of drawing a median line of fit and making predictions based on it.

Using Statistics

Predicting

Jeanine Curren is the manager of an automobile parts supply house. She must be able to predict the arrival time for a shipment of parts sent to a customer by train.

Jeanine used records of past deliveries to construct the table below. In the table, x is the delivery distance in miles and y is the delivery time in days.

x	210	290	350	480	490	730	780	850	920	1010
y	5	7	6	11	8	11	12	8	15	12

EXAMPLE: Find the predicted delivery time for a shipment to a customer located 600 rail miles away.

SOLUTION:

1. Graph the ordered pairs (x, y) in the table. This is a **scatter plot.**

2. Separate the data into three nearly equal groups with an odd number of ordered pairs in the first and third groups.

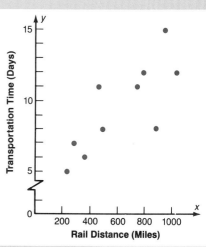

First Group		Second Group		Third Group	
x	y	x	y	x	y
210	5	480	11	850	8
290	7	490	8	920	15
350	6	730	11	1010	12
		780	12		

3. Find the median for the x and y values in the first and third groups. (For a review of median, see pages 52–53.) Since there is an odd number of values in each group, the **median** is the middle value.

Medians:		x	y
	First group	290	6
	Third group	920	12

478 / Chapter 17

478

4 Graph the median points (290, 6) and (920, 12). Draw the line passing through the two points. This is the **median line of fit.**

5 To predict the delivery time, locate 600 on the *x* axis. Draw a vertical line to the median line of fit. Then draw a horizontal line to the *y* axis. Read the value on the *y* axis.

The *y* value (delivery time) is about 9.

Thus, Jeanine can predict that the delivery time will be **at least 9 days.**

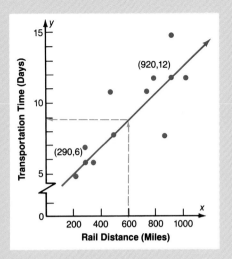

EXERCISES

For each exercise:
a. *Construct a scatter plot.*
b. *Determine the median points.*
c. *Graph the median line of fit.*
d. *Find the predicted value.*

1.

Pages	x	10	20	20	30	40	50	60	100	110	120
Hours to Type	y	6	8	12	16	18	21	32	46	60	58

Find the predicted number of hours needed to type 70 pages. d. About 36 hours

2.

Tons of Crunchola	x	100	150	170	180	220	240	300	350	400	500
Total Cost (thousands of dollars)	y	65	70	65	85	115	125	140	190	200	200

Find the predicted total cost for 200 tons of crunchola. d. About $90,000

3.

Height in Inches	x	62	64	65	66	67	67	69	71	72	74
Weight in Pounds	y	105	115	115	135	140	150	165	170	180	180

Find the predicted weight for a person whose height is 68 inches. d. About 148 pounds

4.

Number of Push-ups	x	15	20	22	25	27	27	32	35	40	42
Number of Chin-ups	y	3	5	4	6	6	9	8	9	11	10

Find the predicted number of chin-ups done by a person who can do 30 push-ups.
d. About 7 chin-ups

Quadratic Functions / **479**

1. a. c.
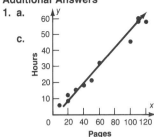
b. (20,8) and (110,58)

2. a. c.
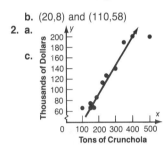
b. (150,65) and (400,200)

3. a. c.
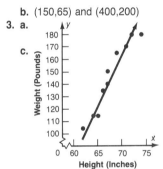
b. (64,115) and (72,180)

4. a. c.
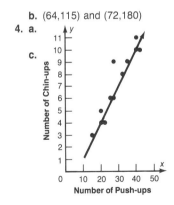
b. (20,4) and (40,10)

CHAPTER SUMMARY

IMPORTANT TERMS	Quadratic function *(p. 456)* Parabola *(p. 458)* Standard form of a quadratic polynomial *(p. 473)* Vertex of a parabola *(p. 475)* Axis of a parabola *(p. 476)*

IMPORTANT IDEAS

1. Effect of a on the graph of $y = ax^2 + bx + c$:

   a. If a is positive, the parabola opens upward.

   b. If a is negative, the parabola opens downward.

   c. If $|a|$ decreases, the parabola becomes wider.

2. The graph of $y = ax^2$ can be moved horizontally $|h|$ units to obtain the graph of $y = a(x - h)^2$. If h is negative, the motion is to the left. If h is positive, the motion is to the right.

3. The graph of $y = a(x - h)^2 + k$ has the same shape as the graph of $y = ax^2$.

4. The graph of $y = a(x - h)^2 + k$ is $|h|$ units to the right or to the left of the graph of $y = ax^2$.

5. The graph of $y = a(x - h)^2 + k$ is $|k|$ units above or below the graph of $y = ax^2$.

6. Every quadratic polynomial can be changed from the form $ax^2 + bx + c$ to the standard form $a(x - h)^2 + k$.

7. The vertex of a parabola with a rule of the form $y = a(x - h)^2 + k$ has coordinates (h, k).

8. If a parabola has the rule $y = a(x - h)^2 + k$, its axis has the equation $x = h$.

Chapter Test

Two Chapter Tests (Form A and Form B) are provided on pages 353–356 in the *Teacher's ResourceBank*.

CHAPTER REVIEW

SECTION 17.1

Write <u>Yes</u> or <u>No</u> to show whether each rule describes a quadratic function.

1. $y = -5x + 1$ No

2. $y = -x^2 + 3x - 5$ Yes

Write three ordered pairs of each function below. Let the x values equal −1, 0, and 2.

3. $y = -3x^2 + x - 1$ (−1, −5), (0, −1), (2, −11)

4. $y = (2x - 1)^2$ (−1, 9), (0, 1), (2, 9)

480 / Chapter 17

SECTION 17.2

Graph each function.

5. $y = x^2 - 1$ **6.** $y = -x^2 + 2$ **7.** $y = -\frac{1}{2}x^2 + x$

8. $y = \frac{1}{2}x^2 - 2x$ **9.** $y = x^2 + 2x - 3$ **10.** $y = -x^2 + 4x + 5$

SECTION 17.3

Graph the two functions in the same coordinate plane.

11. $y = -\frac{1}{2}x^2$; $y = -2x^2$ **12.** $y = -x^2$; $y = -\frac{3}{2}x^2$

13. $y = \frac{5}{2}x^2$; $y = -\frac{5}{2}x^2$ **14.** $y = -\frac{1}{2}x^2$; $y = \frac{1}{2}x^2$

Graph each parabola.

15. $x = \frac{1}{2}y^2$ **16.** $x = \frac{3}{2}y^2$ **17.** $x = -y^2$ **18.** $x = -2y^2$

SECTION 17.4

Write the number of units and the direction the graph of the first function can be moved to obtain the graph of the second function.

19. $y = 2x^2$; $y = 2(x + 3)^2$ 3 units left   **20.** $y = -\frac{1}{2}x^2$; $y = -\frac{1}{2}(x - 5)^2$ 5 units right

21. $y = -x^2$; $y = -(x - \frac{5}{2})^2$ $\frac{5}{2}$ units right   **22.** $y = x^2$; $y = (x + 12)^2$ 12 units left

SECTION 17.5   In Ex. 23-26, the relationship between the first and second parabola is given.

Graph the first function using the x values −2, −1, 0, 1, and 2. Then graph the second function.

23. $y = x^2$
$y = (x - 3)^2 - 2$ Moves 2 units down, 3 units right.

24. $y = -x^2$
$y = -(x + 2)^2 - 3$ Moves 3 units down, 2 units left.

25. $y = -\frac{1}{2}x^2$
$y = -\frac{1}{2}(x + 1)^2 + 3$ Moves 3 units up, 1 unit left.

26. $y = \frac{3}{2}x^2$
$y = \frac{3}{2}(x - 3)^2 + 1$ Moves 1 unit up, 3 units right.

Write an equation of a function for which the graph is obtained as described in each exercise.

27. The graph of $y = -10x^2$ is moved 5 units to the left. $y = -10(x + 5)^2$

28. The graph of $y = 4x^2$ is moved 8 units to the right. $y = 4(x - 8)^2$

Quadratic Functions / **481**

Write an equation of a function for which the graph is obtained as described in each exercise. $y = \frac{5}{2}(x - 2)^2 - 3$

29. The graph of $\frac{5}{2}x^2$ is moved 2 units to the right and 3 units downward.

30. The graph of $-3.7x^2$ is moved 4 units to the left and 2 units upward.
$$y = -3.7(x + 4)^2 + 2$$

SECTION 17.6

Write each expression in the form $ax^2 + bx + c$.

$-3x^2 + 12x - 11$

31. $(x + 1)^2 - 5$ $x^2 + 2x - 4$ **32.** $(x - 3)^2 + 4$ $x^2 - 6x + 13$ **33.** $-3(x - 2)^2 + 1$

34. $4(x + 1)^2 - 6$ $4x^2 + 8x - 2$ **35.** $\frac{1}{2}(x + 4)^2 - 8$ $\frac{1}{2}x^2 + 4x$ **36.** $-\frac{1}{2}(x - 2)^2 + 3$
$-\frac{1}{2}x^2 + 2x + 1$

Write each polynomial in the form $(x - h)^2 + k$.

$(x - (-\frac{3}{2}))^2 - \frac{5}{2}$

37. $x^2 - 6x + 2$ $(x - 3)^2 - 7$ **38.** $x^2 + 4x - 5$ $(x - (-2))^2 - 9$ **39.** $x^2 + 3x - \frac{1}{4}$

40. $x^2 - 5x + \frac{3}{4}$ $(x - \frac{5}{2})^2 - \frac{11}{2}$ **41.** $x^2 - 2x + \frac{1}{2}$ $(x - 1)^2 - \frac{1}{2}$ **42.** $x^2 + 7x + \frac{5}{4}$
$(x - (-\frac{7}{2}))^2 - 11$

SECTION 17.7

Write the coordinates of the vertex of each graph.

43. $y = \frac{1}{2}(x - 10)^2 + 5$ $(10, 5)$ **44.** $y = -2(x + 8)^2 - 3$ $(-8, -3)$

45. $y = 3(x - 6)^2$ $(6, 0)$ **46.** $y = -4(x + 5)^2$ $(-5, 0)$

Write an equation for the axis of each graph.

47. $y = \frac{3}{2}(x - 2)^2 + 5$ $x = 2$ **48.** $y = -5(x + 1)^2 - 6$ $x = -1$

49. $y = -(x + \frac{9}{2})^2 - \frac{7}{2}$ $x = -\frac{9}{2}$ **50.** $y = (x - \frac{4}{3})^2 + \frac{5}{3}$ $x = \frac{4}{3}$

Quadratic Equations

A golf ball being driven from the tee follows a path which is a curve of the type called a *parabola*. Quadratic functions whose graphs are parabolas have many interesting applications besides their purely mathematical uses.

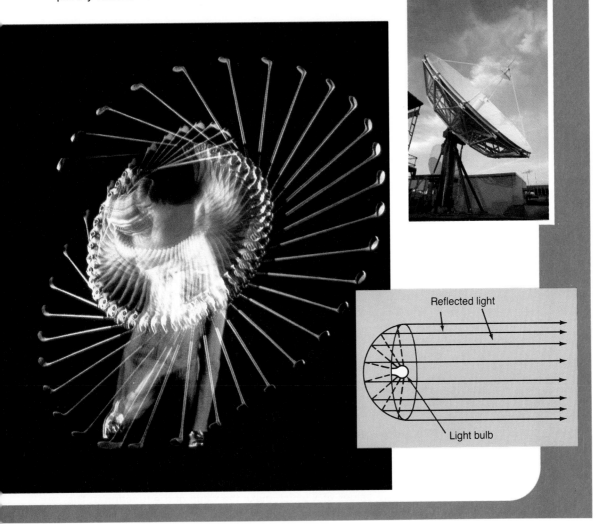

Reflected light

Light bulb

Background
The center photo above is a stroboscopic photograph of a golf ball being driven from a tee.
The top right photo shows a parabolic-shaped radar dish. Waves are reflected from a parabolic surface in parallel rays.

Lesson Resources
Warm–Up: p. M–52
Maintenance: See below.
Practice Worksheet 18.1
Exploration 18
Transparencies 24, 41

Maintenance
1. A special triangle has sides in the ratio 3:5:6 (Condition 1). Find the lengths of the sides if the perimeter is 42 centimeters (Condition 2).
   Ans: 9 cm, 15 cm, 18 cm
   (Section 4.2)
2. Divide:
   $(3x^3 - 6x^2 - 9x) \div (3x)$
   Ans: $x^2 - 2x - 3$
   (Section 8.6)
3. Factor: $2x^2 + 15x + 25$
   Ans: $(2x + 5)(x + 5)$
   (Section 9.4)
4. Write the relation described by $y = -2x + 3$ as a set of ordered pairs when the x values are -2, 0, and 3.
   Ans: $\{(-2,7), (0,3), (3,-3)\}$ (Section 12.3)
5. Write the coordinates of the vertex of the parabola.
   $y = 3(x - 1)^2 + 2$
   Ans: (1,2) (Section 17.7)

Additional Examples
Example 1
Change each equation to the form $ax^2 + bx + c = 0$.
1. $x^2 - 7x - 4 = 2$
   Ans: $x^2 - 7x - 6 = 0$
2. $-3x^2 = 9 - 11x$
   Ans: $3x^2 - 11x + 9 = 0$
3. $16x - x^2 = 25$
   Ans: $x^2 - 16x + 25 = 0$

OBJECTIVES: To write any quadratic equation with rational coefficients in the form $ax^2 + bx + c = 0$, where a, b, and c are integers

18.1 Quadratic Equations and Graphs

To graph a function defined by $y = ax^2 + bx + c$ and then write the approximate solutions of an equation $ax^2 + bx + c = k$ where a, b, and c have the same values as in $y = ax^2 + bx + c$ and where k is a rational number

The equation $ax^2 + bx + c = 0$ can be obtained from the rule of the function $y = ax^2 + bx + c$ if y is replaced with 0.

Definition

> A ***quadratic equation*** is an equation of the form $ax^2 + bx + c = 0$, ($a \neq 0$). (a, b, and c represent rational numbers in this course.)
>
> $2x^2 - 3x + 1 = 0$
>
> $-\frac{1}{2}x^2 + \frac{3}{2}x + \frac{1}{4} = 0$

EXAMPLE 1   Change $x^2 - 3x - 5 = 2$ to the form $ax^2 + bx + c = 0$.

Solution:

① Subtract 2 from each side. ⟶ $x^2 - 3x - 5 - 2 = 2 - 2$

② Simplify. ⟶ $x^2 - 3x - 7 = 0$ ◀ *In the form $ax^2 + bx + c = 0$*

A graph of the function $y = x^2 - x - 2$ is shown in Figure 1.

The standard form of $y = x^2 - x - 2$ is $y = (x - \frac{1}{2})^2 - \frac{9}{4}$. You can see that the coordinates of the vertex are $(\frac{1}{2}, -2\frac{1}{4})$.

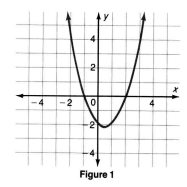
Figure 1

P–1   **What are the x values of the points which the parabola and the x axis have in common?** $-1, 2$

The numbers -1 and 2 are ***zeros of the function*** $y = x^2 - x - 2$. They are values of x which make the value of the function 0. They are also the ***solutions*** of $x^2 - x - 2 = 0$.

484 / *Chapter 18*

EXAMPLE 2   Use the graph of $y = x^2 - x - 2$ to estimate the
              solutions of $x^2 - x - 2 = 4$.

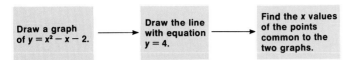

| Draw a graph of $y = x^2 - x - 2$. | → | Draw the line with equation $y = 4$. | → | Find the x values of the points common to the two graphs. |

Solution: Notice the x values of points
           A and B in Figure 2.

           The numbers -2 and 3 are the
           solutions of $x^2 - x - 2 = 4$.

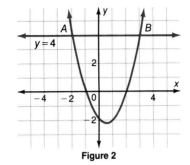

Figure 2

Additional Examples
Example 2
The graph $y = x^2 + 4x + 1$ is
given below. Refer to this graph
to estimate the solutions of each
equation.

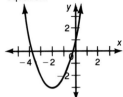

1. $x^2 + 4x + 1 = 0$
   Ans: -3.7; -0.3
2. $x^2 + 4x + 1 = -2$
   Ans: -3; -1
3. $x^2 + 4x + 1 = 2$
   Ans: -4.1; 0.2

You can estimate the solutions
of other equations from
the same graph. Note
the horizontal lines and
their equations.

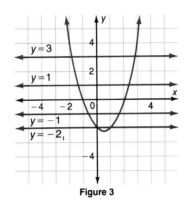

Figure 3

Equation	Estimated Solutions
$x^2 - x - 2 = 3$	$-1.8, 2.8$
$x^2 - x - 2 = -1$	$-0.6, 1.6$
$x^2 - x - 2 = 1$	$-1.3, 2.3$
$x^2 - x - 2 = -2$	$0, 1$

The line with equation $y = -3$ does not intersect the parabola. There
are no real solutions for $x^2 - x - 2 = -3$.

CLASSROOM EXERCISES

Change each equation to the form $ax^2 + bx + c = 0$. (Example 1)

1. $x^2 - 2x - 1 = 2$ $\quad x^2 - 2x - 3 = 0$
2. $2x^2 - 3x + 5 = 1$ $\quad 2x^2 - 3x + 4 = 0$
3. $-3x^2 + 5x = 6$ $\quad 3x^2 - 5x + 6 = 0$
4. $5x - x^2 + 3 = 0$ $\quad x^2 - 5x - 3 = 0$
5. $4 + x^2 + 2x = 0$ $\quad x^2 + 2x + 4 = 0$
6. $x^2 = 2x$ $\quad x^2 - 2x = 0$

*Quadratic Equations / **485***

The graph of $y = x^2 - 4x + 3$ is shown at the right. Estimate the solutions of each equation from the graph. (Example 2)

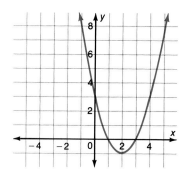

7. $x^2 - 4x + 3 = 0$ 1, 3

8. $x^2 - 4x + 3 = 1$ 0.5, 3.5

9. $x^2 - 4x + 3 = 3$ 0, 4

10. $x^2 - 4x + 3 = 5$ −0.25, 4.25

11. $x^2 - 4x + 3 = -1$ 2

12. $x^2 - 4x + 3 = -2$ No real solutions

Assignment Guide
Basic Omit
Average Omit
Above Average p. 486:
1–17 odd

Additional Answers
Written Exercises
3. $2x^2 - 5x + 10 = 0$

WRITTEN EXERCISES

Goal: To estimate the solutions of a quadratic equation from the graph of a related function

Sample Problem: $x^2 - x - 2 = 2$ (Refer to Figure 2.)

Answer: The solutions are approximately 2.6 and −1.6.

Change each equation to the form $ax^2 + bx + c = 0$. (Example 1)

1. $x^2 = 5x - 3$ $x^2 - 5x + 3 = 0$ **2.** $x^2 = 5 - 3x$ $x^2 + 3x - 5 = 0$ **3.** $5x - 2x^2 = 10$

4. $3x - x^2 = -5$ $x^2 - 3x - 5 = 0$ **5.** $x^2 = 9$ $x^2 - 9 = 0$ **6.** $x^2 = 3x$
                                                                 $x^2 - 3x = 0$

Graph $y = x^2 + 2x - 3$. Refer to this graph to estimate the solutions of each equation. (Example 2)

                                                                                        −3.4, 1.4

7. $x^2 + 2x - 3 = 0$ −3, 1   **8.** $x^2 + 2x - 3 = 1$ −3.2, 1.2   **9.** $x^2 + 2x - 3 = 2$

10. $x^2 + 2x - 3 = 4$ −3.8, 1.8   **11.** $x^2 + 2x - 3 = -5$ No real solutions   **12.** $x^2 + 2x - 3 = -6$ No real solutions

Graph $y = -x^2 + 4x - 5$. Estimate the solutions of each equation. (Example 2)

13. $-x^2 + 4x - 5 = 0$ No real solutions   **14.** $-x^2 + 4x - 5 = 1$ No real solutions

15. $-x^2 + 4x - 5 = -1$ 2   **16.** $-x^2 + 4x - 5 = -2$ 1, 3

17. $-x^2 + 4x - 5 = -3$ 0.6, 3.4   **18.** $-x^2 + 4x - 5 = -4$ 0.3, 3.7

REVIEW CAPSULE FOR SECTION 18.2

Factor each trinomial. (Pages 246–248)

                                                              $(x - 6)(x + 4)$ $(x + 8)(x - 7)$

1. $x^2 - 4x - 12$ **2.** $x^2 - 7x + 10$ **3.** $x^2 - 2x - 24$ **4.** $x^2 + x - 56$

5. $2x^2 + 7x - 4$ **6.** $3x^2 - 5x + 2$ **7.** $7 - x^2 - 6x$ **8.** $9x^2 + 14 - 65x$
                                                                         $(7 + x)(1 - x)$ $(9x - 2)(x - 7)$

Review Capsule
1. $(x - 6)(x + 2)$
2. $(x - 5)(x - 2)$
5. $(2x - 1)(x + 4)$
6. $(3x - 2)(x - 1)$

18.2 Solving by Factoring

OBJECTIVE: To solve a quadratic equation by using the Factors of Zero Property

In this course, polynomials represent real numbers. Therefore, the **Factors of Zero Property** applies to polynomials.

> **Factors of Zero Property**
>
> If a and b are real numbers and $ab = 0$, then $a = 0$ or $b = 0$.

P–1 What is the value of x in the equation $x + 5 = 0$? -5
in the equation $x - 2 = 0$? 2

EXAMPLE 1   Solve $(x + 5)(x - 2) = 0$.

Solution: $(x + 5)(x - 2) = 0$

[1] Factors of Zero Property ⟶ $x + 5 = 0$ *or* $x - 2 = 0$

[2] Solve both equations. ⟶ $x = -5$ *or* $x = 2$

You may want to demonstrate checks of the solution in all examples.

The solution set is $\{-5, 2\}$.

EXAMPLE 2   Solve $x^2 - 2x - 3 = 0$.

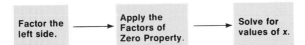

Solution: $x^2 - 2x - 3 = 0$

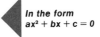
In the form $ax^2 + bx + c = 0$

[1] Factor. ⟶ $(x - 3)(x + 1) = 0$

[2] Factors of Zero Property ⟶ $x - 3 = 0$ *or* $x + 1 = 0$

[3] Solve both equations. ⟶ $x = 3$ *or* $x = -1$

The solution set is $\{3, -1\}$.

Teaching Suggestions p. M–52

Lesson Resources
Warm–Up: p. M–53
Maintenance: See below.
Practice Worksheet 18.2
Transparency 25

Maintenance
1. Celia spent more than $20 for an album and a compact disc (Condition 2). The cost of the compact disc was $4 more than the cost of the album (Condition 1). Is it possible that the album cost $7?
   Ans: No (Section 5.7)
2. Factor: $2y^2 + 2y - 12$
   Ans: 2(y + 3)(y − 2)
   (Section 9.7)
3. Add and simplify:
   $$\frac{1}{x^2 - 16} + \frac{1}{2x + 8}$$
   Ans: $\dfrac{x - 2}{2(x - 4)(x + 4)}$
   (Section 11.6)
4. Solve the system below by graphing.
   $$\begin{cases} y + 2x = 4 \\ y + 2x = 6 \end{cases}$$
   Ans: ϕ (Section 14.2)
5. Write $(x^2 + 4x + 3)$ in the form $(x - h)^2 + k$.
   Ans: (x + 2)² − 1
   (Section 17.6)

Additional Examples
Example 1
Solve.
1. $(x + 4)(x - 6) = 0$
   Ans: {−4, 6}
2. $(x - 3)(x + 8) = 0$
   Ans: {3, −8}

Example 2
Solve.
1. $x^2 + 2x - 8 = 0$
   Ans: {2, −4}
2. $x^2 + 11x + 30 = 0$
   Ans: {−5, −6}

Additional Examples
Example 3
Solve.
1. $x^2 + 2x + 1 = 0$
   Ans: $\{-1\}$
2. $x^2 - 10x + 25 = 0$
   Ans: $\{5\}$
3. $x^2 + 8x + 16 = 0$
   Ans: $\{-4\}$

Example 4
Solve.
1. $x^2 = 8x$
   Ans: $\{0, 8\}$
2. $x^2 = 9x$
   Ans: $\{0, 9\}$
3. $3x^2 = 15x$
   Ans: $\{0, 5\}$

Example 5
Solve.
1. $2x^2 - 7x = -3$
   Ans: $\{\frac{1}{2}, 3\}$
2. $3x^2 - 2 = x$
   Ans: $\{-\frac{2}{3}, 1\}$
3. $x^2 + 21 = 10x$
   Ans: $\{7, 3\}$

EXAMPLE 3   Solve $x^2 - 6x + 9 = 0$.

Solution: $x^2 - 6x + 9 = 0$

$$(x - 3)(x - 3) = 0$$

$$x - 3 = 0 \quad \underline{or} \quad x - 3 = 0$$

Both equations have the same solution, 3.
Solution set: $\{3\}$

Before solving a quadratic equation, be sure that it is in the form $ax^2 + bx + c = 0$.

EXAMPLE 4   Solve $x^2 = 5x$.

Solution: $x^2 = 5x$

1. Subtract $5x$ from each side. $\longrightarrow$ $x^2 - 5x = 0$ ◀ *In the form* $ax^2 + bx + c = 0$

2. Factor. $\longrightarrow$ $x(x - 5) = 0$

3. Factors of Zero Property $\longrightarrow$ $x = 0 \quad \underline{or} \quad x - 5 = 0$

4. Solve both equations. $\longrightarrow$ $x = 0 \quad \underline{or} \quad x = 5$

Solution set: $\{0, 5\}$

EXAMPLE 5   Solve $2x^2 + x = 6$.

Solution: $2x^2 + x = 6$

1. Subtract 6 from each side. $\longrightarrow$ $2x^2 + x - 6 = 0$

2. Factor. $\longrightarrow$ $(2x - 3)(x + 2) = 0$

3. Factors of Zero Property $\longrightarrow$ $2x - 3 = 0 \quad \underline{or} \quad x + 2 = 0$

4. Solve both equations. $\longrightarrow$ $2x - 3 + 3 = 0 + 3 \quad \underline{or} \quad x + 2 - 2 = 0 - 2$

$$2x = 3 \qquad\qquad x = -2$$

$$\tfrac{1}{2}(2x) = \tfrac{1}{2}(3)$$

$$x = \tfrac{3}{2} \qquad \text{Solution set: } \{\tfrac{3}{2}, -2\}$$

CLASSROOM EXERCISES

Solve. (Example 1)

1. $(x - 3)(x + 5) = 0$ $\{3, -5\}$

2. $x(x - 1) = 0$ $\{0, 1\}$

3. $(x - 6)(x + 6) = 0$ $\{6, -6\}$

4. $(x - 2)(x - 7) = 0$ $\{2, 7\}$

5. $(x + 4)(x + 7) = 0$ $\{-4, -7\}$

6. $(x - 3)(x - 3) = 0$ $\{3\}$

7. $(2x - 1)(x + 5) = 0$ $\{\frac{1}{2}, -5\}$

8. $(x - 1)(2x + 1) = 0$ $\{1, -\frac{1}{2}\}$

9. $(2x + 3)(x + 1) = 0$ $\{-\frac{3}{2}, -1\}$

10. $(4x - 1)(3x + 2) = 0$ $\{\frac{1}{4}, -\frac{2}{3}\}$

WRITTEN EXERCISES

Goal: To use the Factors of Zero Property to solve quadratic equations

Sample Problem: $x^2 - 56 = -x$

Answer: $\{7, -8\}$

Solve. (Examples 2–4)

1. $x^2 + 3x = 0$ $\{0, -3\}$

2. $x^2 - 4x = 0$ $\{0, 4\}$

3. $x^2 - 25 = 0$

4. $x^2 - 49 = 0$ $\{7, -7\}$

5. $x^2 = 10x$ $\{0, 10\}$

6. $x^2 = -8x$

7. $x^2 = 100$ $\{10, -10\}$

8. $x^2 = 81$ $\{9, -9\}$

9. $x^2 + 4x - 5 = 0$

10. $x^2 - 5x + 6 = 0$ $\{2, 3\}$

11. $x^2 + 3x = 4$ $\{1, -4\}$

12. $x^2 - 3x = 10$

13. $x^2 + 8x + 16 = 0$ $\{-4\}$

14. $x^2 - 12x + 36 = 0$ $\{6\}$

15. $x^2 + 4 = 4x$

16. $x^2 + 9 = -6x$ $\{-3\}$

17. $8x - 12 = x^2$ $\{2, 6\}$

18. $2x + x^2 = 24$

19. $x^2 + 24 = 11x$ $\{3, 8\}$

20. $x^2 + 15 = 8x$ $\{3, 5\}$

21. $x^2 + 18x = -81$

22. $x^2 + 100 = 20x$ $\{10\}$

23. $x^2 = x + 72$ $\{9, -8\}$

24. $7 = x^2 + 6x$

Solve. (Example 5)

25. $3x^2 + 8x - 3 = 0$ $\{-3, \frac{1}{3}\}$

26. $2x^2 - 9x + 7 = 0$ $\{1, \frac{7}{2}\}$

27. $6x^2 + x - 2 = 0$ $\{-\frac{2}{3}, \frac{1}{2}\}$

28. $9x^2 - 9x + 2 = 0$ $\{\frac{1}{3}, \frac{2}{3}\}$

29. $4 - 12x^2 = 13x$ $\{-\frac{4}{3}, \frac{1}{4}\}$

30. $10 - 15x^2 = 19x$ $\{\frac{2}{5}, -\frac{5}{3}\}$

31. $8x^2 - 5 + 6x = 0$ $\{-\frac{5}{4}, \frac{1}{2}\}$

32. $6x^2 - 7x - 5 = 0$ $\{\frac{5}{3}, -\frac{1}{2}\}$

33. $2x^2 - 13x = 45$ $\{9, -\frac{5}{2}\}$

34. $9x^2 + 14 = 65x$ $\{7, \frac{2}{9}\}$

35. $4x^2 + x - 3 = 0$ $\{\frac{3}{4}, -1\}$

36. $25x^2 + 9 - 30x = 0$ $\{\frac{3}{5}\}$

*Quadratic Equations / **489***

Common Error

Students do not apply the Factors of Zero Property when solving quadratic equations.

Example

Solve $x^2 - x - 12 = 0$.

$x^2 - x - 12 = 0$

$(x - 4)(x + 3) = 0$

$x = -4$ or $x = 3$

Prescription

Require that students follow the steps show in Example 1–5. Have them show each factor equal to zero and then solve for x in both equations.

Assignment Guide

Basic Omit

Average Omit

Above Average p. 489: 1–35 odd

Additional Answers

Written Exercises

3. $\{5, -5\}$

6. $\{0, -8\}$

9. $\{-5, 1\}$

12. $\{-2, 5\}$

15. $\{2\}$

18. $\{4, -6\}$

21. $\{-9\}$

24. $\{-7, 1\}$

OBJECTIVES: To solve an equation of the form $x^2 = k$, where k is a nonnegative rational number
To write an equivalent compound *or* sentence for a quadratic equation of the form
$(x - a)^2 = k$ where a and k are rational numbers, $k \geq 0$

18.3 Special Quadratic Equations

To solve a quadratic equation of the form $(x - a)^2 = k$ where a and k are rational numbers, $k \geq 0$
To solve a quadratic equation of the form $x^2 + bx + c = k$ where $x^2 + bx + c$ is a trinomial square and $k \geq 0$

P–1 **Solve each equation below.**

a. $x^2 = 16$ **b.** $x^2 = 49$ **c.** $x^2 = 100$
$\{4, -4\}$ $\{7, -7\}$ $\{10, -10\}$

If the left side of an equation is the square of a binomial, the equation can be solved in the same way you solved P-1.

EXAMPLE 1   Solve $(x - 3)^2 = 25$.

Solution:

① Write the equivalent compound *or* sentence. ⟶ $x - 3 = 5$ *or* $x - 3 = -5$

② Solve both equations. ⟶ $x = 8$ *or* $x = -2$

Solution set: $\{8, -2\}$

Check:

$(x - 3)^2 = 25$	$(x - 3)^2 = 25$
$(8 - 3)^2$	$(-2 - 3)^2$
5^2	$(-5)^2$
25	25

This method is actually a short cut of the method using the Factors of Zero Property.

This rule will be useful in the sections that follow.

> Any quadratic equation of the form $(x + r)^2 = s$ in which r and s are rational numbers, $s \geq 0$, is equivalent to "$x + r = \sqrt{s}$ *or* $x + r = -\sqrt{s}$."
>
> "$(x + 1)^2 = 5$" is equivalent to "$x + 1 = \sqrt{5}$ *or* $x + 1 = -\sqrt{5}$."

P–2 **What is the value of r if you compare $(x - 5)^2 = 81$ with $(x + r)^2 = s$?**
−5

EXAMPLE 2   Solve $(x - 5)^2 = 81$.

Solution: $(x - 5)^2 = 81$

$x - 5 = 9$ *or* $x - 5 = -9$
$x = 14$ *or* $x = -4$

Solution set: $\{14, -4\}$

 The check is left for you.

You have learned to tell whether a quadratic trinomial of the form $ax^2 + bx + c$ is a trinomial square.

1. The leading coefficient a must be a perfect square.
2. The constant term c must be a perfect square.
3. The sentence $(\frac{1}{2}b)^2 = ac$ must be true.

P-3 **What is the value of a in $x^2 - 8x + 16$? of c? of $(\frac{1}{2}b)^2$? of ac?**
  1   16   16   16

EXAMPLE 3   Solve $x^2 - 8x + 16 = 36$.

Solution: $x^2 - 8x + 16 = 36$ ◄ $x^2 - 8x + 16$ is a trinomial square.

$(x - 4)^2 = 36$

$x - 4 = 6$ _or_ $x - 4 = -6$

$x = 10$ _or_ $x = -2$ ◄ The check is left for you.

Solution set: $\{10, -2\}$

EXAMPLE 4   Solve $x^2 + 3x + \frac{9}{4} = 5$.

Solution: $x^2 + 3x + \frac{9}{4}$ is a trinomial square over {rational numbers}.

$a = 1$ $b = 3$ $c = \frac{9}{4}$ ◄ a and c are perfect squares.

$(\frac{1}{2}b) = \frac{3}{2}$

$(\frac{1}{2}b)^2 = \frac{9}{4}$ $ac = \frac{9}{4}$ ◄ $(\frac{1}{2}b)^2 = ac$

$x^2 + 3x + \frac{9}{4} = 5$

$(x + \frac{3}{2})^2 = 5$

$x + \frac{3}{2} = \sqrt{5}$ _or_ $x + \frac{3}{2} = -\sqrt{5}$

$x = -\frac{3}{2} + \sqrt{5}$ _or_ $x = -\frac{3}{2} - \sqrt{5}$

Solution set: $\{-\frac{3}{2} + \sqrt{5}, -\frac{3}{2} - \sqrt{5}\}$

■■■ **CLASSROOM EXERCISES** ■■■

Solve. (P-1)

1. $x^2 = 1$ $\{1, -1\}$
2. $x^2 = 5$ $\{\sqrt{5}, -\sqrt{5}\}$
3. $x^2 = 121$ $\{11, -11\}$
4. $x^2 = 3$ $\{\sqrt{3}, -\sqrt{3}\}$

*Quadratic Equations / **491***

Write an equivalent compound or sentence. (Step 1 of Example 1)

5. $(x - 6)^2 = 36$ **6.** $(x + 3)^2 = 49$ **7.** $(x - \frac{1}{2})^2 = 16$

Write the values of $(\frac{1}{2}b)^2$ and ac for each trinomial. Write Yes or No to indicate if each trinomial is a trinomial square. (P-3)

8. $x^2 + 2x + 2$ 1; 2; No   **9.** $x^2 - 4x + 2$ 4; 2; No   **10.** $x^2 - 12x + 36$

11. $x^2 + 10x + 25$ 25; 25; Yes   **12.** $x^2 + x + \frac{1}{4}$ $\frac{1}{4}$; $\frac{1}{4}$; Yes   **13.** $x^2 - 3x + \frac{3}{4}$

Write two equal factors of each trinomial square over {rational numbers}.

14. $x^2 - 8x + 16$ $x - 4, x - 4$ **15.** $x^2 + 20x + 100$ $x + 10, x + 10$ **16.** $x^2 + x + \frac{1}{4}$

17. $x^2 - \frac{1}{2}x + \frac{1}{16}$ $x - \frac{1}{4}, x - \frac{1}{4}$ **18.** $x^2 + 5x + \frac{25}{4}$ $x + \frac{5}{2}, x + \frac{5}{2}$ **19.** $x^2 - 3x + \frac{9}{4}$

WRITTEN EXERCISES

Goal: To solve quadratic equations of the form $(x + r)^2 = s$
Sample Problem: Solve $x^2 + x + \frac{1}{4} = 9$.   **Answer:** $\{\frac{5}{2}, -\frac{7}{2}\}$

Solve each equation. (P-1)

                                                          $\{15, -15\}$

1. $x^2 = 9$ $\{3, -3\}$ **2.** $x^2 = 16$ $\{4, -4\}$ **3.** $x^2 = 169$ $\{13, -13\}$ **4.** $x^2 = 225$

5. $x^2 = 7$ $\{\sqrt{7}, -\sqrt{7}\}$ **6.** $x^2 = 10$ $\{\sqrt{10}, -\sqrt{10}\}$ **7.** $x^2 = 75$ $\{5\sqrt{3}, -5\sqrt{3}\}$ **8.** $x^2 = 48$

Write an equivalent compound or sentence. (Step 1 of Example 1)

9. $(x - 5)^2 = 36$ **10.** $(x - 4)^2 = 49$ **11.** $(x + 6)^2 = 25$

12. $(x - 1)^2 = 5$ **13.** $(x + \frac{1}{2})^2 = \frac{1}{4}$ **14.** $(x + \frac{2}{3})^2 = \frac{4}{9}$

Solve. (Examples 1–4)

15. $(x + 1)^2 = 121$ $\{10, -12\}$ **16.** $(x + 2)^2 = 169$ $\{11, -15\}$ **17.** $(x - 3)^2 = 5$

18. $(x - 5)^2 = 3$ $\{5 + \sqrt{3}, 5 - \sqrt{3}\}$ **19.** $(x - \frac{1}{2})^2 = \frac{1}{4}$ $\{1, 0\}$ **20.** $(x - \frac{3}{4})^2 = \frac{1}{16}$

21. $(x + \frac{2}{3})^2 = \frac{5}{9}$ $\left\{\dfrac{-2 + \sqrt{5}}{3}, \dfrac{-2 - \sqrt{5}}{3}\right\}$ **22.** $(x + \frac{1}{3})^2 = \frac{2}{9}$ **23.** $x^2 + 8x + 16 = 0$

24. $x^2 + 12x + 36 = 0$ $\{-6\}$ **25.** $x^2 - 10x + 25 = 0$ $\{5\}$ **26.** $x^2 - 6x + 9 = 0$

27. $x^2 - \frac{1}{2}x + \frac{1}{16} = 0$ $\{\frac{1}{4}\}$ **28.** $x^2 - 3x + \frac{9}{4} = 0$ $\{\frac{3}{2}\}$ **29.** $x^2 - \frac{4}{5}x + \frac{4}{25} = 0$

MORE CHALLENGING EXERCISES

30. If $r = 3$ and the solution set of $(x + r)^2 = s$ is $\{-1, -5\}$, what is the value of s? $s = 4$

31. If $r = -2$ and the solution set of $(x + r)^2 = s$ is $\{\frac{1}{2}, \frac{7}{2}\}$, what is the value of s? $s = \frac{9}{4}$

APPLICATIONS

You can use a calculator to evaluate an expression such as the following that involves several radicals.

$$\sqrt{x + \sqrt{x + \sqrt{x + \sqrt{x}}}}$$

Suggest that students extend the number of radicals and conjecture what happens to the approximate values obtained. Hint: In this Example the result approaches 2.

EXAMPLE   Evaluate the above expression for $x = 2$.

SOLUTION   Start with 2 at the extreme right. Then work to the left.

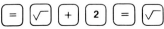

$$\boxed{1.9903694}$$

Evaluate the radical expression shown above for each given value of x.

32. $x = 4$ 2.5572611   **33.** $x = 3$ 2.2967226   **34.** $x = 10$ 3.7002176   **35.** $x = 16$

36. $x = 5.2$ 2.8314161   **37.** $x = 0.67$ 1.4299625   **38.** $x = 100$ 10.512437   **39.** $x = 359$

**Additional Answers
Written Exercises**
35. 4.53041
39. 19.453883

■■■■■ **MID-CHAPTER REVIEW** ■■■■■■■■■■

Graph $y = x^2 + 3x - 4$. Estimate the solutions in each equation.
(Section 18.1)

$\{1.3, -4.3\}$

1. $x^2 + 3x - 4 = 0$ $\{-4, 1\}$ **2.** $x^2 + 3x - 4 = 1$ $\{1.2, -4.1\}$ **3.** $x^2 + 3x - 4 = 2$

4. $x^2 + 3x - 4 = -1$ $\{0.8, -3.8\}$ **5.** $x^2 + 3x - 4 = -2$ $\{0.6, -3.6\}$ **6.** $x^2 + 3x - 4 = -3$
$\{0.3, -3.3\}$

Solve. (Section 18.2)

7. $2x^2 + 9x - 5 = 0$ $\{\frac{1}{2}, -5\}$ **8.** $2x^2 - 5x - 3 = 0$ $\{3, -\frac{1}{2}\}$ **9.** $3x^2 - 2x - 1 = 0$

10. $4x^2 - 5x + 1 = 0$ $\{1, \frac{1}{4}\}$ **11.** $6x^2 + 5x + 1 = 0$ $\{-\frac{1}{2}, -\frac{1}{3}\}$ **12.** $4x^2 + 4x - 3 = 0$
$\{\frac{1}{2}, -\frac{3}{2}\}$

Solve. (Section 18.3)

13. $(x - 3)^2 = 144$ $\{15, -9\}$ **14.** $(x + 5)^2 = 81$ $\{4, -14\}$ **15.** $x^2 - x + \frac{1}{4} = \frac{9}{4}$

16. $x^2 + x + \frac{1}{4} = \frac{9}{16}$ $\{\frac{1}{4}, -\frac{5}{4}\}$ **17.** $x^2 + 10x + 25 = 9$ $\{-2, -8\}$ **18.** $x^2 + 14x + 49 = 121$
$\{4, -18\}$

Quiz Sections 18.1–18.3
After completing this Mid-Chapter Review, you may want to administer a quiz covering the same sections. See the quiz provided on page 371 in the *Teacher's ResourceBank*.

**Additional Answers
Mid-Chapter Review**
9. $\{1, -\frac{1}{3}\}$
15. $\{2, -1\}$

REVIEW CAPSULE FOR SECTION 18.4

Write each equation in the form $x^2 + bx = -c$. (Pages 58–63)

$x^2 - 4x = -10$

1. $x^2 - 6x + 3 = 0$ $x^2 - 6x = -3$ **2.** $x^2 + 5x - 4 = 0$ $x^2 + 5x = 4$ **3.** $x^2 = 4x - 10$

4. $x^2 - \frac{1}{2}x + \frac{3}{2} = 0$ $x^2 - \frac{1}{2}x = -\frac{3}{2}$ **5.** $2x^2 - 4x + 5 = 0$ $2x^2 - 4x = -5$ **6.** $2x^2 + 3x - 10 = 0$
$2x^2 + 3x = 10$

*Quadratic Equations   /   **493***

Lesson Resources
Warm–Up: p. M–53
Maintenance: See below.
Practice Worksheet 18.4

Maintenance

1. Simplify: $(-3r^{-1}t^2)^3$

 Ans: $\dfrac{-27t^6}{r^3}$ (Section 6.5)

2. Evaluate $-4x^3$ if $x = -2$.

 Ans: 32 (Section 8.1)

3. Simplify: $\dfrac{x^2 - 2x}{x^2 - 4}$

 Ans: $\dfrac{x}{x + 2}$ (Section 10.2)

4. Find the least common multiple of $(x^2 + 5x + 4)$ and $(4x + 4)$.

 Ans: 4(x + 1)(x + 4)
 (Section 11.3)

5. A purse contains 40 dimes and quarters (Condition 1). The total value of the coins is $7.60 (Condition 2). How many coins of each type are in the purse? **Ans: dimes: 16; quarters: 24**
 (Section 15.6)

Additional Examples
Example 1
Solve by completing the square.
1. $x^2 + 6x - 7 = 0$
 Ans: {−7, 1}
2. $x^2 - 10x + 16 = 0$
 Ans: {8, 2}
3. $x^2 - 2x - 15 = 0$
 Ans: {5, −3}

Example 2
Solve.
1. $x^2 - 3x - 5 = 0$
 Ans: $\left\{ \dfrac{3 + \sqrt{29}}{2}, \dfrac{3 - \sqrt{29}}{2} \right\}$
2. $x^2 - 8x - 3 = 0$
 Ans: {4 + √19, 4 − √19}
3. $x^2 + 12x + 9 = 0$
 Ans: {−6 + 3√3, −6 − 3√3}

494

18.4 Completing the Square

OBJECTIVE: To solve a quadratic equation by completing the square

Lead students through the steps of these examples carefully.

It is possible to write any quadratic equation in a form that makes the left side a trinomial square. This is called **completing the square**.

EXAMPLE 1   Solve by completing the square: $x^2 - 8x - 9 = 0$

Solution:

1. Add 9 to each side. → $x^2 - 8x = 9$

2. Multiply the coefficient of x by $\frac{1}{2}$. → $\frac{1}{2}(-8) = -4$

3. Square (-4). → $(-4)^2 = 16$

4. Add 16 to each side. → $x^2 - 8x + 16 = 9 + 16$

 $x^2 - 8x + 16$ is a trinomial square.

5. Factor and simplify. → $(x - 4)^2 = 25$

6. Write the equivalent compound *or* sentence. → $x - 4 = 5$ *or* $x - 4 = -5$

7. Solve both equations. → $x = 9$ *or* $x = -1$

 Solution set: $\{9, -1\}$

EXAMPLE 2   Solve $x^2 - x - 1 = 0$.

Solution:

1. Add 1 to each side. → $x^2 - x = 1$

2. Multiply the coefficient of x by $\frac{1}{2}$. → $\frac{1}{2}(-1) = -\frac{1}{2}$

3. Square $(-\frac{1}{2})$. → $(-\frac{1}{2})^2 = \frac{1}{4}$

4. Add $\frac{1}{4}$ to each side. → $x^2 - x + \frac{1}{4} = 1 + \frac{1}{4}$

5. Factor and simplify. → $(x - \frac{1}{2})^2 = \frac{5}{4}$

6. Write the equivalent compound *or* sentence. → $x - \frac{1}{2} = \sqrt{\frac{5}{4}}$ *or* $x - \frac{1}{2} = -\sqrt{\frac{5}{4}}$

7. Simplify the radicals. → $x - \frac{1}{2} = \frac{\sqrt{5}}{2}$ *or* $x - \frac{1}{2} = -\frac{\sqrt{5}}{2}$

8. Solve both equations. → $x = \frac{1}{2} + \frac{\sqrt{5}}{2}$ *or* $x = \frac{1}{2} - \frac{\sqrt{5}}{2}$

 Solution set: $\left\{ \dfrac{1 + \sqrt{5}}{2}, \dfrac{1 - \sqrt{5}}{2} \right\}$ *Students may want to see these solutions as approximations. A calculator makes it easy!*

EXAMPLE 3   Solve $2x^2 + 3x - 5 = 0$.

Solution:

1 Add 5 to each side. ⟶ $2x^2 + 3x = 5$

2 Multiply each side by $\frac{1}{2}$. ⟶ $x^2 + \frac{3}{2}x = \frac{5}{2}$

3 Multiply the coefficient of x by $\frac{1}{2}$. ⟶ $\frac{1}{2}(\frac{3}{2}) = \frac{3}{4}$

4 Square $\frac{3}{4}$. ⟶ $(\frac{3}{4})^2 = \frac{9}{16}$

5 Add $\frac{9}{16}$ to each side. ⟶ $x^2 + \frac{3}{2}x + \frac{9}{16} = \frac{5}{2} + \frac{9}{16}$

6 Factor and simplify. ⟶ $(x + \frac{3}{4})^2 = \frac{49}{16}$

7 Write the equivalent compound
  _or_ sentence. ⟶ $x + \frac{3}{4} = \sqrt{\frac{49}{16}}$ _or_ $x + \frac{3}{4} = -\sqrt{\frac{49}{16}}$

8 Simplify the radicals. ⟶ $x + \frac{3}{4} = \frac{7}{4}$ _or_ $x + \frac{3}{4} = -\frac{7}{4}$

9 Solve both equations. ⟶ $x = -\frac{3}{4} + \frac{7}{4}$ _or_ $x = -\frac{3}{4} - \frac{7}{4}$

10 Simplify. ⟶ $x = 1$ _or_ $x = -2\frac{1}{2}$

Solution set: $\{1, -2\frac{1}{2}\}$

Additional Examples
Example 3
Solve.
1. $3x^2 - 11x - 4 = 0$
   Ans: $\{-\frac{1}{3}, 4\}$
2. $2x^2 - 11x + 14 = 0$
   Ans: $\{\frac{7}{2}, 2\}$
3. $2x^2 + 8x - 5 = 0$
   Ans: $\left\{-2 + \frac{\sqrt{26}}{2}, -2 - \frac{\sqrt{26}}{2}\right\}$

The solutions of the equations of Examples 1–3 can be checked by replacing x in the original equation with each number from the solution set.

Show how these generalized steps apply to the previous examples.

Steps for solving $ax^2 + bx + c = 0$ by completing the square:

1. Add $-c$ to each side.
2. Multiply each side by $\frac{1}{a}$.
3. Compute $\left(\frac{1}{2} \cdot \frac{b}{a}\right)$.
4. Compute $\left(\frac{1}{2} \cdot \frac{b}{a}\right)^2$.
5. Add $\left(\frac{1}{2} \cdot \frac{b}{a}\right)^2$ to each side.
6. Factor the left side.
7. Write the equivalent compound _or_ sentence.
8. Solve.

Quadratic Equations / **495**

CLASSROOM EXERCISES

Write the number that you should add to each side to form a trinomial square in the left side. (Examples 1 and 2)

1. $x^2 + 4x = 1$ 4
2. $x^2 - 6x = 5$ 9
3. $x^2 - 2x = 3$ 1
4. $x^2 + 10x = 12$ 25
5. $x^2 - x = 2$ $\dfrac{1}{4}$
6. $x^2 + 3x = 5$ $\dfrac{9}{4}$
7. $x^2 + \dfrac{1}{3}x = 3$ $\dfrac{1}{36}$
8. $x^2 - \dfrac{2}{3}x = 4$ $\dfrac{1}{9}$

Solve each equation.

9. $x^2 = \dfrac{3}{4}$ $\left\{ \dfrac{\sqrt{3}}{2}, -\dfrac{\sqrt{3}}{2} \right\}$
10. $x^2 = \dfrac{5}{9}$ $\left\{ \dfrac{\sqrt{5}}{3}, -\dfrac{\sqrt{5}}{3} \right\}$
11. $x^2 = \dfrac{1}{2} + \dfrac{3}{4}$
12. $x^2 = \dfrac{3}{2} + \dfrac{5}{16}$ $\left\{ \dfrac{\sqrt{29}}{4}, -\dfrac{\sqrt{29}}{4} \right\}$
13. $(x - \dfrac{1}{2})^2 = \dfrac{3}{4}$ $\left\{ \dfrac{1 + \sqrt{3}}{2}, \dfrac{1 - \sqrt{3}}{2} \right\}$
14. $(x - 3)^2 = \dfrac{13}{25}$

Write the factors of each trinomial square over {rational numbers}.

15. $x^2 + 18x + 81$ $x + 9, x + 9$
16. $x^2 - 14x + 49$ $x - 7, x - 7$
17. $x^2 + 3x + \dfrac{9}{4}$
18. $x^2 + 5x + \dfrac{25}{4}$ $x + \dfrac{5}{2}, x + \dfrac{5}{2}$
19. $x^2 - \dfrac{2}{5}x + \dfrac{1}{25}$ $x - \dfrac{1}{5}, x - \dfrac{1}{5}$
20. $x^2 + \dfrac{3}{2}x + \dfrac{9}{16}$

WRITTEN EXERCISES

Goal: To solve quadratic equations by completing the square
Sample Problem: Solve $2x^2 - 5x - 7 = 0$.   **Answer:** $\{\dfrac{7}{2}, -1\}$

Solve by completing the square. (Examples 1–3)

1. $x^2 - 2x - 15 = 0$ $\{-3, 5\}$
2. $x^2 + 2x - 8 = 0$ $\{-4, 2\}$
3. $x^2 + 4x - 5 = 0$
4. $x^2 - 4x - 12 = 0$ $\{-2, 6\}$
5. $x^2 + x - 6 = 0$ $\{-3, 2\}$
6. $x^2 - x - 20 = 0$
7. $x^2 + 2x - 5 = 0$
8. $x^2 + 4x - 1 = 0$
9. $x^2 - 4x + 7 = 0$
10. $x^2 + 10x + 2 = 0$
11. $x^2 - 6x - 5 = 0$
12. $x^2 - 4x - 10 = 0$
13. $x^2 + 8x + 13 = 0$
14. $x^2 - x - 3 = 0$
15. $x^2 - 3x - 1 = 0$
16. $x^2 - 5x - 3 = 0$
17. $x^2 + 3x - 5 = 0$
18. $x^2 + \dfrac{2}{3}x - \dfrac{2}{3} = 0$
19. $x^2 + \dfrac{2}{5}x - \dfrac{1}{5} = 0$
20. $2x^2 + 5x - 3 = 0$ $\{-3, \dfrac{1}{2}\}$
21. $2x^2 + x - 3 = 0$
22. $3x^2 + 8x - 3 = 0$
23. $3x^2 - 5x - 2 = 0$ $\{-\dfrac{1}{3}, 2\}$
24. $2x^2 - 4x - 3 = 0$
25. $3x^2 - 6x + 2 = 0$
26. $2x^2 + 4x - 5 = 0$
27. $-2x + 4x^2 - 10 = 0$ $\left\{ \dfrac{1 + \sqrt{41}}{4}, \dfrac{1 - \sqrt{41}}{4} \right\}$

REVIEW CAPSULE FOR SECTION 18.5

Simplify. (Pages 180–182 and 187–189)

1. $\dfrac{3 + \sqrt{49}}{4}; \dfrac{3 - \sqrt{49}}{4}$ $\dfrac{5}{2}; -1$
2. $\dfrac{-2 + \sqrt{36}}{2}; \dfrac{-2 - \sqrt{36}}{2}$ $2; -4$
3. $\dfrac{4 + \sqrt{24}}{8}; \dfrac{4 - \sqrt{24}}{8}$ $\dfrac{2 + \sqrt{6}}{4}, \dfrac{2 - \sqrt{6}}{4}$

18.5 The Quadratic Formula

OBJECTIVE: To solve a quadratic equation by using the quadratic formula

A special formula called the **quadratic formula** can be used to solve <u>any</u> quadratic equation.

This Example develops the Quadratic Formula. Go through the steps carefully with students.

─── EXAMPLE 1   Solve $ax^2 + bx + c = 0$ for x.

Solution:

1.   Add $-c$ to each side. ──────────────▶ $ax^2 + bx = -c$

2.   Multiply each side by $\dfrac{1}{a}$. ──────────▶ $x^2 + \dfrac{b}{a}x = \dfrac{-c}{a}$

3.   Add $\left(\dfrac{1}{2} \cdot \dfrac{b}{a}\right)^2$ or $\dfrac{b^2}{4a^2}$ to each side. ──▶ $x^2 + \dfrac{b}{a}x + \dfrac{b^2}{4a^2} = -\dfrac{c}{a} + \dfrac{b^2}{4a^2}$

4.   Factor the left side and write the simplest name for the right side. ──▶ $\left(x + \dfrac{b}{2a}\right)^2 = \dfrac{b^2 - 4ac}{4a^2}$

5.   Write the equivalent compound <u>or</u> sentence. ──────────▶ $x + \dfrac{b}{2a} = \dfrac{\sqrt{b^2 - 4ac}}{2a}$

         <u>or</u> $x + \dfrac{b}{2a} = \dfrac{-\sqrt{b^2 - 4ac}}{2a}$

6.   Solve both equations. ──────────▶ $x = \dfrac{-b}{2a} + \dfrac{\sqrt{b^2 - 4ac}}{2a}$

         <u>or</u> $x = \dfrac{-b}{2a} - \dfrac{\sqrt{b^2 - 4ac}}{2a}$

7.   Simplify. ──────────────▶ $x = \dfrac{-b + \sqrt{b^2 - 4ac}}{2a}$

         <u>or</u> $x = \dfrac{-b - \sqrt{b^2 - 4ac}}{2a}$

Solution set: $\left\{ \dfrac{-b + \sqrt{b^2 - 4ac}}{2a}, \dfrac{-b - \sqrt{b^2 - 4ac}}{2a} \right\}$

Explain to students that this formula is sometimes shown in this condensed form:
$x = \dfrac{-b \pm \sqrt{b^2 - 4ac}}{2a}$

> **Quadratic Formula**
>
> Any quadratic equation of the form $ax^2 + bx + c = 0$ is equivalent to the following compound sentence:
>
> $$x = \frac{-b + \sqrt{b^2 - 4ac}}{2a} \quad \underline{or} \quad x = \frac{-b - \sqrt{b^2 - 4ac}}{2a}$$

Quadratic Equations   /   **497**

Teaching Suggestions p. M–52

Lesson Resources
Warm–Up: p. M–53
Maintenance: See below.
Practice Worksheet 18.5
Transparency 18

Maintenance
1. Solve and check:
   $36 = 4(m - 3) - 7m$
   Ans: $m = -16$
   (Section 3.5)
2. Add: $\sqrt{32} + \sqrt{8} + \sqrt{18}$
   Ans: $9\sqrt{2}$ (Section 7.5)
3. Subtract:
   $\dfrac{2x - 1}{x + 3} - \dfrac{x - 3}{x + 3}$
   Ans: $\dfrac{x + 2}{x + 3}$ (Section 11.1)
4. The sum of two numbers is 10 (Condition 1). Three times one number decreased by twice the other is 15 (Condition 2). Find the two numbers.
   Ans: 7, 3   (Section 14.5)
5. Solve the system below by the substitution method.
   $\begin{cases} -3x + y = -1 \\ 7x + 2y = 37 \end{cases}$
   Ans: $\{(3, 8)\}$
   (Section 15.1)

Example 2
Solve by using the quadratic
formula.
1. $x^2 - 4x - 45 = 0$
   Ans: $\{9, -5\}$
2. $3x^2 + 17x + 10 = 0$
   Ans: $\{-5, -\frac{2}{3}\}$
3. $8x^2 + 10x - 3 = 0$
   Ans: $\{\frac{1}{4}, -\frac{3}{2}\}$

Example 3
Solve.
1. $3x^2 - 4x = 3$
   Ans: $\left\{\dfrac{2 + \sqrt{13}}{3}, \dfrac{2 - \sqrt{13}}{3}\right\}$
2. $4x^2 = 3 - 5x$
   Ans: $\left\{\dfrac{-5 + \sqrt{73}}{8}, \dfrac{-5 - \sqrt{73}}{8}\right\}$
3. $5 = 9x - 3x^2$
   Ans: $\left\{\dfrac{9 + \sqrt{21}}{6}, \dfrac{9 - \sqrt{21}}{6}\right\}$

EXAMPLE 2   Solve $6x^2 - x - 1 = 0$ by the quadratic formula.

Solution:

1 Write the values of a, b, and c. ⟶ $a = 6$ $b = -1$ $c = -1$

2 Substitute these values in the two equations of the quadratic formula. Then simplify.

$$x = \frac{-b + \sqrt{b^2 - 4ac}}{2a} \qquad \underline{or} \qquad x = \frac{-b - \sqrt{b^2 - 4ac}}{2a}$$

$$= \frac{-(-1) + \sqrt{(-1)^2 - 4(6)(-1)}}{2(6)} \qquad \underline{or} \qquad = \frac{1 - \sqrt{25}}{12}$$

$$= \frac{1 + \sqrt{1 + 24}}{12} \qquad \underline{or} \qquad = \frac{1 - 5}{12}$$

$$= \frac{1 + \sqrt{25}}{12} \qquad \underline{or} \qquad = \frac{-4}{12}$$

$$= \frac{1 + 5}{12} = \frac{1}{2} \qquad \underline{or} \qquad = -\frac{1}{3}$$

Solution set: $\{\frac{1}{2}, -\frac{1}{3}\}$

EXAMPLE 3   Solve $2x^2 - x = 5$.

Solution:

1 Subtract 5 from each side. ⟶ $2x^2 - x - 5 = 0$ ◀ *In the form* $ax^2 + bx + c = 0$

2 Write the values of a, b, and c. ⟶ $a = 2$ $b = -1$ $c = -5$

3 Substitute these values in the two equations of the quadratic formula. Then simplify.

$$x = \frac{-b + \sqrt{b^2 - 4ac}}{2a} \qquad \underline{or} \qquad x = \frac{-b - \sqrt{b^2 - 4ac}}{2a}$$

$$= \frac{-(-1) + \sqrt{(-1)^2 - 4(2)(-5)}}{2(2)} \qquad \underline{or} \qquad = \frac{-(-1) - \sqrt{(-1)^2 - 4(2)(-5)}}{2(2)}$$

$$= \frac{1 + \sqrt{1 + 40}}{4} \qquad \underline{or} \qquad = \frac{1 - \sqrt{1 + 40}}{4}$$

$$= \frac{1 + \sqrt{41}}{4} \qquad \underline{or} \qquad = \frac{1 - \sqrt{41}}{4}$$

Solution set: $\left\{\dfrac{1 + \sqrt{41}}{4}, \dfrac{1 - \sqrt{41}}{4}\right\}$

The Table of Square Roots on page 194 can be used to approximate the solutions of Example 3.

$$\frac{1}{4} + \frac{\sqrt{41}}{4} \approx 0.25 + \frac{6.4}{4} \quad or \quad \frac{1}{4} - \frac{\sqrt{41}}{4} \approx 0.25 - \frac{6.4}{4}$$

You can also write in the following forms in order to approximate by use of a calculator.
$(\sqrt{41} + 1) \div 4$
and
$(\sqrt{41} - 1) \div 4$

$$\approx 0.25 + 1.6 \quad or \quad \approx 0.25 - 1.6$$
$$\approx 1.9 \quad or \quad \approx -1.35 \text{ or } -1.4$$

The solutions of $2x^2 - x = 5$ are approximately 1.9 and −1.4.

CLASSROOM EXERCISES

Compare each equation to $ax^2 + bx + c = 0$. Then write the values of a, b, and c for each equation. (Examples 2 and 3)

1. $x^2 + 3x - 2 = 0$ $a = 1, b = 3, c = -2$
2. $-2x^2 - x + 5 = 0$ $a = -2, b = -1, c = 5$
3. $x^2 + 5x = 0$ $a = 1, b = 5, c = 0$
4. $4 - x^2 - x = 0$ $a = -1, b = -1, c = 4$
5. $x^2 = -3x - 5$ $a = 1, b = 3, c = 5$
6. $3x^2 - 1 = 5x$ $a = 3, b = -5, c = -1$

WRITTEN EXERCISES

Goal: To solve quadratic equations by the quadratic formula
Sample Problem: Solve $3x^2 - 8x + 5$.   **Answer:** $\{1, \frac{5}{3}\}$

Solve by the quadratic formula. (Examples 1–3)

1. $x^2 + 5x - 6 = 0$ $\{-6, 1\}$
2. $x^2 + 3x - 10 = 0$ $\{-5, 2\}$
3. $x^2 + 6x - 7 = 0$ $\{1, -7\}$
4. $x^2 + 2x - 8 = 0$ $\{-4, 2\}$
5. $x^2 - 7x + 10 = 0$ $\{2, 5\}$
6. $x^2 - 7x + 12 = 0$
7. $x^2 = 4x + 12$ $\{-2, 6\}$
8. $x^2 = 2x + 15$ $\{-3, 5\}$
9. $-x^2 + 6x - 9 = 0$
10. $-x^2 + 10x - 25 = 0$ $\{5\}$
11. $6x^2 + x - 2 = 0$ $\{-\frac{2}{3}, \frac{1}{2}\}$
12. $4x^2 + 12x + 5 = 0$
13. $x^2 - 3 = 0$ $\{\sqrt{3}, -\sqrt{3}\}$
14. $x^2 - 5 = 0$ $\{\sqrt{5}, -\sqrt{5}\}$
15. $x^2 - 6x = 0$
16. $x^2 + 9x = 0$ $\{-9, 0\}$
17. $3x - 5x^2 + 1 = 0$
18. $5x^2 + 7x - 10 = 0$

Solve by the quadratic formula. Approximate to one decimal place.

19. $x^2 + 2x - 1 = 0$ $\{0.4, -2.4\}$
20. $x^2 + 4x + 1 = 0$ $\{-0.3, -3.7\}$
21. $2x^2 - 3x - 1 = 0$
22. $-3x^2 + 6x - 2 = 0$ $\{1.6, 0.4\}$
23. $2x^2 - 6x + 3 = 0$ $\{2.4, 0.6\}$
24. $-5x^2 + x + 1 = 0$ $\{0.6, -0.4\}$

REVIEW CAPSULE FOR SECTION 18.6

Find the value of $b^2 - 4ac$. (Pages 74–77)

1. $a = -2; b = -5; c = 1$ 33
2. $a = 4; b = 3; c = 5$ −71
3. $a = 1; b = 6; c = -2$ 44

Quadratic Equations / 499

Common Error
Students often substitute values for a, b, and c in the quadratic formula before writing the equation in the form $ax^2 + bx + c = 0$.

Example
Solve $x^2 = 4x + 3$.
$a = 1, b = 4,$ and $c = 3$

Prescription
Remind students that they must write equations in the proper form before finding values. Have students tell which of Written Exercises 1–18 are not in the form $ax^2 + bx + c = 0$.

Assignment Guide
Basic Omit
Average Omit
Above Average p. 499: 1–23 odd

Additional Answers
Written Exercises
6. $\{3, 4\}$
9. $\{3\}$
12. $\{-\frac{1}{2}, -\frac{5}{2}\}$
15. $\{6, 0\}$
17. $\left\{\frac{3 + \sqrt{29}}{10}, \frac{3 - \sqrt{29}}{10}\right\}$
18. $\left\{\frac{-7 + \sqrt{249}}{10}, \frac{-7 - \sqrt{249}}{10}\right\}$
21. $\{1.8, -0.3\}$

499

OBJECTIVE: To compute the value of the discriminant $b^2 - 4ac$, and identify
the solutions of the quadratic equation as One Rational, Two Rational,
Two Irrational, or No Real

18.6 The Discriminant

In using the quadratic formula, you find the value of the expression $b^2 - 4ac$.

> The expression $b^2 - 4ac$ is called the ***discriminant.***

The value of the discriminant describes the kinds of numbers that belong to the solution set of a quadratic equation. It tells whether the numbers are <u>rational</u>, <u>irrational</u>, or <u>not real</u>.

The relationship between the solutions of quadratic equations and their discriminants is exhibited below for several cases.

Discuss all these cases in detail with your class.

Equation: $x^2 + x - 12 = 0$ **Solution Set:** $\{-4, 3\}$
$(x + 4)(x - 3) = 0$

Value of Discriminant: $b^2 - 4ac = 1^2 - 4(1)(-12) = 49$ *Perfect square*

The solutions are <u>rational numbers</u> because the value of the discriminant is a perfect square.

Equation: $x^2 - 5 = 0$ **Solution Set:** $\{\sqrt{5}, -\sqrt{5}\}$
$x^2 = 5$

Value of Discriminant: $b^2 - 4ac = 0^2 - 4(1)(-5) = 20$ *Not a perfect square*

The solutions are <u>irrational numbers</u> because the value of the discriminant is positive and not a perfect square.

Equation: $x^2 - 4x + 4 = 0$ **Solution Set:** $\{2\}$
$(x - 2)(x - 2) = 0$

Value of Discriminant: $b^2 - 4ac = (-4)^2 - 4(1)(4) = 0$ *Perfect square*

The solutions are rational <u>and</u> equal because the value of the discriminant is 0. The solution set contains one number.

Equation: $x^2 + 4 = 0$ **Solution Set:** ϕ
$x^2 = -4$ There is no real number whose square is −4.

Value of Discriminant: $b^2 - 4ac = 0^2 - 4(1)(4) = -16$ *The value of the discriminant is negative.*

There are <u>no real solutions</u> because the value of the discriminant is a negative number.

500 / *Chapter 18*

> **Value of the discriminant and the solution set of $ax^2 + bx + c = 0$**
>
> 1. If $b^2 - 4ac$ is 0 or a perfect-square counting number, the solutions are rational.
> 2. If $b^2 - 4ac$ is positive and not a perfect square, the solutions are irrational.
> 3. If $b^2 - 4ac = 0$, the solution set contains one number. The solution is rational.
> 4. If $b^2 - 4ac$ is negative, no real number solutions exist.

EXAMPLE 1   Compute the value of $b^2 - 4ac$ for each equation. Write <u>One rational number</u>, <u>Two rational numbers</u>, <u>Two irrational numbers</u>, or <u>No real numbers</u> to describe the solutions.

     a. $2x^2 - x + 3 = 0$ **b.** $2x^2 + 3x = 2$

Solutions:

a. $2x^2 - x + 3 = 0$

$a = 2,\ b = -1,\ c = 3$

$b^2 - 4ac = (-1)^2 - 4(2)(3)$

            $= 1 - 24$

            $= -23$ ◀ *Discriminant is negative.*

No real numbers

b. $2x^2 + 3x = 2$

      $2x^2 + 3x - 2 = 0$

      $a = 2,\ b = 3,\ c = -2$

      $b^2 - 4ac = 3^2 - 4(2)(-2)$

              $= 9 + 16 = 25$ ◀ *Discriminant is a non-zero perfect square.*

Two rational numbers

EXAMPLE 2   Compute the value of $b^2 - 4ac$ for each equation. Write <u>One rational number</u>, <u>Two rational numbers</u>, <u>Two irrational numbers</u>, or <u>No real numbers</u> to describe the solutions.

     a. $x^2 + 3x = 5$ **b.** $4x^2 - 4x + 1 = 0$

Solutions:

a. $x^2 + 3x = 5$

      $x^2 + 3x - 5 = 0$

      $a = 1,\ b = 3,\ c = -5$

      $b^2 - 4ac = 3^2 - 4(1)(-5)$

              $= 9 + 20 = 29$ ◀ *Discriminant is not a perfect square.*

Two irrational numbers

b. $4x^2 - 4x + 1 = 0$

      $a = 4,\ b = -4,\ c = 1$

      $b^2 - 4ac = (-4)^2 - 4(4)(1)$

              $= 16 - 16$

              $= 0$ ◀ *Discriminant is zero.*

One rational number

Quadratic Equations / **501**

Additional Examples
Example 1
Compute the value of $b^2 - 4ac$ for each equation. Then write <u>One rational number</u>, <u>Two rational numbers</u>, <u>Two irrational numbers</u>, or <u>No real numbers</u> for the solutions.
1. $2x^2 - 5x + 3 = 0$
    Ans: 1; Two rational numbers
2. $6x^2 - 7x = -3$
    Ans: -23; No real numbers

Example 2
Compute the value of $b^2 - 4ac$ for each equation. Write <u>One rational number</u>, <u>Two rational numbers</u>, <u>Two irrational numbers</u>, or <u>No real numbers</u> to describe the solutions.
1. $x^2 - 6x + 9 = 0$
    Ans: 0; One rational number
2. $3x^2 - 6x = 8$
    Ans: 132; Two irrational numbers

$$\boxed{\text{P-1}}$$ **Use One rational number, Two rational numbers, Two irrational numbers, or No real numbers to describe the solution of each equation.**

a. $8x^2 - 2x - 1 = 0$ $\overset{\text{Two rational}}{\text{numbers}}$ **b.** $x^2 - x - 1 = 0$ $\overset{\text{Two irrational}}{\text{numbers}}$

c. $x^2 - 2x + 4 = 0$ $\underset{\text{numbers}}{\text{No real}}$ **d.** $x^2 - 16x + 64 = 0$ $\underset{\text{number}}{\text{One rational}}$

CLASSROOM EXERCISES

Write One rational number, Two rational numbers, Two irrational numbers, or No real numbers to describe the solutions for each equation having the given value of its discriminant. (Examples 1–2)

1. 25 Two rat.   **2.** 36 Two rat.   **3.** 0 One rat.   **4.** 12 Two irrat.   **5.** 17 Two irrat.   **6.** 1

7. 100 Two rat.   **8.** 4 Two rat.   **9.** −5 No real   **10.** 22 Two irrat.   **11.** 9 Two rat.   **12.** −32

Write the value of $b^2 - 4ac$ for each equation. Then write One rational number, Two rational numbers, Two irrational numbers, or No real numbers to describe the solutions. (Examples 1–2)

13. $x^2 + x - 2 = 0$ 9; Two rat.   **14.** $x^2 + 2x + 1 = 0$ 0; One rat.   **15.** $x^2 - 3x + 2 = 0$

16. $x^2 - 5x + 5 = 0$ 5; Two irrat.   **17.** $x^2 + 16 = 0$ −64; No real   **18.** $x^2 - 7 = 0$

19. $2x^2 - 3x - 1 = 0$ 17; Two irrat.   **20.** $5x^2 + 5x + 1 = 0$ 5; Two irrat.   **21.** $x^2 - x + 1 = 0$
   −3; No real

WRITTEN EXERCISES

Goal: To describe the solutions of quadratic equations by use of the discriminant

Sample Problems: a. $2x^2 - 7x + 1 = 0$ **b.** $x^2 - 8x + 16 = 0$
Answers: a. Two irrational numbers   **b.** One rational number

Compute the value of $b^2 - 4ac$ for each equation. Then write One rational number, Two rational numbers, Two irrational numbers, or No real numbers for the solutions. (Examples 1–2)

1. $x^2 - x - 6 = 0$ 25; Two rat.   **2.** $x^2 + 4x - 5 = 0$ 36; Two rat.   **3.** $-x^2 + 2x + 8 = 0$

4. $-x^2 + 8x - 12 = 0$ 16; Two rat.   **5.** $2x^2 - x - 2 = 0$ 17; Two irrat.   **6.** $3x^2 + 3x - 1 = 0$

7. $-x^2 + 4x - 4 = 0$ 0; One rat.   **8.** $-x^2 + 10x - 25 = 0$ 0; One rat.   **9.** $x^2 - 7 = 0$

10. $x^2 - 13 = 0$ 52; Two irrat.   **11.** $x^2 + 25 = 0$ −100; No real   **12.** $2x^2 + x + 6 = 0$

13. $2x^2 = 3 - 5x$ 49; Two rat.   **14.** $3x^2 = 9x - 5$ 21; Two irrat.   **15.** $x^2 = 10x$

16. $3x^2 = 12x$ 144; Two rat.   **17.** $5x^2 - 2x = 0$ 4; Two rat.   **18.** $3x - 2x^2 = 0$

19. $x^2 - 2x + 7 = 0$ −24; No real   **20.** $3x^2 - 2x + 8 = 0$ −92; No real   **21.** $-8x^2 - 9x + 2 = 0$

22. $9x^2 = 5$ 180; Two irrat.   **23.** $3x^2 - 10 = 0$ 120; Two irrat.   **24.** $-x^2 - x + 1 = 0$

REVIEW CAPSULE FOR SECTION 18.7

Write each equation in the form $ax^2 + bx + c = 0$. (Pages 484–486)

1. $x(x - 4) = 78$ $x^2 - 4x - 78 = 0$ **2.** $x(2x + 5) = 102$

3. $5x + 1 = x(x - 2)$

4. $3x - 4 = x(x + 5)$
$x^2 + 2x + 4 = 0$

5. $x(3x - 1) - x^2 = 128$
$2x^2 - x - 128 = 0$

6. $2x(x + 3) - 5x = 10$
$2x^2 + x - 10 = 0$

███████ **PROBLEM SOLVING AND APPLICATIONS** ███████

18.7 Using Quadratic Equations

OBJECTIVE: To solve a word problem that can be described by a quadratic equation

EXAMPLE 1   The length of a rectangle is 3 centimeters more than its width (Condition 1). The area is 18 square centimeters (Condition 2). Find the length and width.

Solution:

1️⃣ Use Condition 1 to represent the unknowns. ➝ Let x = the width.
Then $x + 3$ = the length.

2️⃣ Use Condition 2 to write an equation.

$$\underset{\downarrow}{A} = \underset{\downarrow}{\ell} \cdot \underset{\downarrow}{w}$$

 Formula for the area of a rectangle

$$18 = (x + 3)\ x$$

3️⃣ Solve the equation. ➝ $18 = x^2 + 3x$

$$0 = x^2 + 3x - 18$$
$$0 = (x + 6)(x - 3)$$
$$x + 6 = 0 \quad \underline{or} \quad x - 3 = 0$$
$$x = -6 \quad \underline{or} \quad x = 3$$
$$\text{Length: } x + 3 = 6$$

 The measure must be positive.

The number -6 is a solution of the equation but is not part of the solution to the problem.

1️⃣ **Check:** Condition 1 Is the length 3 centimeters more than the width?
Does $6 = 3 + 3$?   Yes ✔

Condition 2 Is the area 18 square centimeters?
Does $(6)(3) = 18$?   Yes ✔

The width is 3 centimeters, and the length is 6 centimeters.

Quadratic Equations / **503**

In Examples 2 and 3, Condition 1 is <u>underscored once</u> and Condition 2 is <u>underscored twice</u>.

Addition Examples
Example 2
1. <u>The product of two consecutive integers is 156</u>. What are the integers?
   Ans: 12, 13; or −12, −13
2. <u>One integer is 6 less than another integer</u>. Their product is 55. Find two pairs of such integers.
   Ans: 5 and 11; −5 and −11

EXAMPLE 2   <u>The product of two consecutive integers</u> <u><u>is 72</u></u>. What are the integers?

Solution:

☐1 Let n = the smaller integer.
Then $n + 1$ = the greater integer. *Condition 1*

☐2 Write an equation.

Think: <u>The product of two consecutive integers</u> <u><u>is 72</u></u>. *Condition 2*

Translate: $n(n + 1)$ $= 72$

☐3 Solve the equation. ⟶ $n^2 + n = 72$

$$n^2 + n - 72 = 0$$
$$(n + 9)(n - 8) = 0$$
$$n + 9 = 0 \quad \underline{or} \quad n - 8 = 0$$
$$n = -9 \quad \underline{or} \quad n = 8$$

Smaller Integer	Greater Integer
−9	−8
8	9

◀ *Two answers are possible.*
1. −9 and −8
2. 8 and 9

☐4 **Check:** Condition 1   Are the two integers consecutive?
Does $-8 - (-9) = 1$?   Yes ✔   Does $9 - 8 = 1$?   Yes ✔
Condition 2   Is the product 72?   Does $(-9)(-8) = 72$?   Yes ✔
Does $(8)(9) = 72$?   Yes ✔

Example 3
A square hole is cut in a rectangular plate as shown below. <u>The area of the plate after the square hole is cut out is 335 square units</u>. <u>The width of the rectangle is three times the length of each side of the square hole</u>. <u><u>The length of the rectangle is 24 units</u></u>. Find the width of the rectangle.

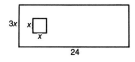

Ans: 15 units

EXAMPLE 3   <u>The area of a rectangle less the area of a square equals 15 square meters</u>. <u>The width of the rectangle equals the length of each side of the square</u>. <u><u>The length of the rectangle is 2 meters more than twice the width</u></u>. Find the width and length of the rectangle.

Solution:

☐1 Let x = the width of the rectangle.
Let x = the length of each side of the square.
Then $2x + 2$ = the length of the rectangle.

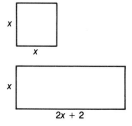

504 / *Chapter 18*

504

Think: The area of a rectangle less the area of a square equals 15.

⎣2⎦ **Translate:** $x(2x + 2)$ $-$ x^2 $=$ 15

⎣3⎦ $2x^2 + 2x - x^2 = 15$

   $x^2 + 2x - 15 = 0$

   $(x + 5)(x - 3) = 0$

   $x + 5 = 0$ *or* $x - 3 = 0$

   $x = -5$ *or* $x = 3$ **The width cannot be negative.**

   Length: $2x + 2 = 8$

Explain that two pairs of answers are possible in Example 2 because negative numbers are acceptable. In Examples 1 and 3 the numbers represent lengths and must be positive.

⎣4⎦ **Check:**   Condition 1   Is the length of the rectangle 2 meters more than twice the width?   Does $2(3) + 2 = 8$?   Yes ✔

   Condition 2   Is the area of the rectangle less the area of the square 15 square meters?   Does $2(3)^2 + 2(3) - 3^2 = 15$?   Yes ✔

   The width is 3 meters, and the length is 8 meters.

Common Error
When using a quadratic equation to solve an integer problem, students use the solutions to the equation as the answer.

Example
The product of two consecutive integers is 56. Find the integers.
   $n(n + 1) = 56$
   $n^2 + n - 56 = 0$
   $(n + 8)(n - 7) = 0$
   $n = -8$ or $n = 7$
The integers are -8 and 7.

Prescription
Make sure students check their answers with the two conditions as shown in Example 2. Remind them that they have solved for n, but must still solve for the second number.

CLASSROOM EXERCISES

For Exercises 1–8, Condition 1 is <u>underscored once</u>. Condition 2 is <u>underscored twice</u>.
a. Use Condition 1 to represent the unknowns.
b. Use Condition 2 to write an equation for the problem.
(Examples 1 and 2)

1. The product of two consecutive integers is 210.

2. The product of two consecutive even integers is 48.

3. One integer is three more than another integer. Their product is 108.

4. One number is 15 less than another number. Their product is -36.

5. The length of a rectangle is 5 units greater than the width. The area is 24 square units.

6. The length of a rectangle is three units more than twice the width. The area is 65 square units.

7. One number is three times as great as another number. The square of the smaller number is 18 less than the square of the greater number.

8. One number is three more than another number. The square of the smaller number is 51 more than the square of the greater number.

Additional Answers
1. **a.** n = smaller integer;
      $n + 1$ = greater integer
   b. $n(n + 1) = 210$
2. **a.** n = smaller integer;
      $n + 2$ = greater integer
   b. $n(n + 2) = 48$
3. **a.** n = smaller integer;
      $n + 3$ = greater integer
   b. $n(n + 3) = 108$
4. **a.** n = greater number;
      $n - 15$ = smaller number
   b. $n(n - 15) = -36$
5. **a.** w = width; $w + 5$ = length
   b. $w(w + 5) = 24$
6. **a.** w = width; $2w + 3$ = length
   b. $w(2w + 3) = 65$
7. **a.** n = smaller number;
      $3n$ = greater number
   b. $n^2 = (3n)^2 - 18$
8. **a.** n = smaller number;
      $n + 3$ = greater number
   b. $n^2 = (n + 3)^2 + 51$

WRITTEN EXERCISES

Goal: To use quadratic equations to solve problems

Sample Problem: One integer is 10 less than another integer (Condition 1). Their product is −16 (Condition 2). Find the integers.

Answer: Equation: $x(x - 10) = -16$; $x^2 - 10x + 16 = 0$
The integers are 8 and −2 *or* 2 and −8.

For Exercises 1–8, Condition 1 is underscored once. Condition 2 is underscored twice.
a. Use Condition 1 to represent the unknowns.
b. Use Condition 2 to write an equation for the problem.
c. Solve. (Examples 1–3)

1. The width of a rectangle is 2 units less than the length. If the area is 80 square units, what are the length and width?

2. The length of a rectangle is 5 units more than the width. If the area is 50 square units, what are the length and width?

3. The product of two consecutive integers is 30. Find two pairs of such integers.

4. The product of two consecutive integers is 56. Find two pairs of such integers.

5. One integer is 13 less than another integer. Their product is −40. Find all possible pairs of such integers.

6. One integer is 6 more than another integer. Their product is 55. Find all possible pairs of such integers.

7. A rectangular plate has a rectangular piece cut from one corner as shown in the figure. The area after the rectangular piece is cut out is 40 square units. Use the information given in the figure to find the length and width of the original rectangular plate.

8. A square hole is cut in a rectangular plate as shown in the figure. The area of the plate after the square hole is cut out is 105 square units. The width of the rectangle is twice the length of each side of the square hole. The length of the rectangle is 13 units. Find the width of the rectangle.

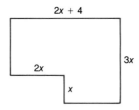

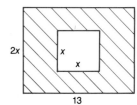

Using a Geometric Model   # Factoring

An algebraic method for solving a quadratic equation by **completing the square** was shown on pages 494–495. The following method of completing the square uses a **geometric model.** This method works for finding positive solutions only to quadratic equations of the form $ax^2 + bx + c = 0$ where $a > 0$, $b > 0$, and $c < 0$.

EXAMPLE: Solve $x^2 + 4x - 12 = 0$ by completing the square.

SOLUTION:

1. Write the equation in the form $x^2 + bx = -c$. ⟶ $x^2 + 4x = 12$

2. Represent x^2 as a square with sides of length x.

   Area: x^2

3. Represent $4x$ as two rectangles, each with an area of $\frac{1}{2}(4x)$, or $2x$.

   Position the rectangles as shown.

   Area of the L–shaped figure: $2x + x^2 + 2x$, or $x^2 + 4x$

   From step 1, $x^2 + 4x = 12$.
   Thus, the area of the L–shaped figure is 12.

   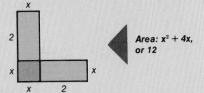

   Area: $x^2 + 4x$, or 12

4. Complete the square by adding the red region shown.

   Why must the lengths of the sides of the red region equal 2?

   What is the area of the red square?

   $$\frac{\text{Area of}}{\text{L–shaped Region}} + \frac{\text{Area of}}{\text{Red Square}} = \underline{}$$

   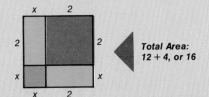

   Total Area: $12 + 4$, or 16

5. Since the large square has a total area of 16, the length of each side must equal 4. Why?

   Thus, $x + 2 = 4$, or $x = 2$.

6. **Check:** Does $(2)^2 + 4(2) - 12 = 0$?

   Does $4 + 8 - 12 = 0$?   Yes ✔

Quadratic Equations   /   507

This discussion covers the classical geometrical method of solving a quadratic equation by completing the square. This is a good way to show students a relationship between algebra and geometry and, in particular, to show the use of a geometric model in problem solving.

EXERCISES

For Exercise 1, replace each __?__ to solve $2x^2 + 5x - 3 = 0$ by completing the square.

1. ① Write the equation in the form $x^2 + \frac{b}{a}x = -\frac{c}{a}$ ⟶ $x^2 + \frac{5}{2}x =$ __?__ $\frac{3}{2}$

② Represent x^2.

③ Represent $\frac{5}{2}x$ as two rectangles, each with an area of $\frac{1}{2}(\frac{5}{2}x)$, or __?__ $\frac{5}{4}x$

Area of L-shaped figure:
$\frac{5}{4}x + x^2 + \frac{5}{4}x =$ __?__ $x^2 + \frac{5}{2}x$

From step 1, $x^2 + \frac{5}{2}x = \frac{3}{2}$.
Thus, the area of the L-shaped figure = __?__ . $\frac{3}{2}$

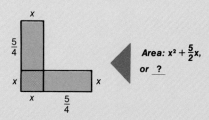

④ Complete the square.

The area of the red square equals __?__ . $\frac{25}{16}$

$$\text{Area of L-shaped Region} + \text{Area of Red Square} = \underline{\quad?\quad}$$

$$\frac{3}{2} + \frac{25}{16} = \frac{49}{16}$$

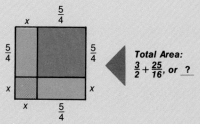

⑤ Since the large square has a total area of $\frac{49}{16}$, the length of each side must equal __?__ . $\frac{7}{4}$

Thus, $x + \frac{5}{4} = \frac{7}{4}$, or $x =$ __?__ . $\frac{2}{4}$, or $\frac{1}{2}$

2. Why does the method of Exercise 1 work only for finding a positive solution of a quadratic equation? Because the lengths of the sides of the square must be positive numbers.

Use the geometric method of completing the square to find a positive solution for each quadratic equation.

3. $x^2 + 2x - 3 = 0$ $x = 1$ **4.** $x^2 + 4x - 21 = 0$ $x = 3$

5. $x^2 + 6x - 16 = 0$ $x = 2$ **6.** $x^2 + 8x - 20 = 0$ $x = 2$

7. $2x^2 + 5x - 12 = 0$ $x = \frac{3}{2}$ **8.** $2x^2 + 3x - 9 = 0$ $x = \frac{3}{2}$

CHAPTER SUMMARY

IMPORTANT TERMS	Quadratic equation *(p. 484)* Quadratic formula *(p. 497)* Zeros of a function *(p. 484)* Discriminant *(p. 500)* Completing the square *(p. 494)*

IMPORTANT IDEAS

1. The solutions of a quadratic equation of the form $ax^2 + bx + c = 0$ are the same as the zeros of the function $y = ax^2 + bx + c$.

2. *Factors of Zero Property:* If a and b are real numbers and $ab = 0$, then $a = 0$ or $b = 0$.

3. Any quadratic equation of the form $(x + r)^2 = s$ in which r and s are rational numbers, $s \geq 0$, is equivalent to "$x + r = \sqrt{s}$ or $x + r = -\sqrt{s}$."

4. It is possible to write any quadratic equation in a form that makes the left side a trinomial square.

5. *Steps for solving $ax^2 + bx + c = 0$ by completing the square:*
   a. Add $-c$ to each side.
   b. Multiply each side by $\frac{1}{a}$.
   c. Compute $\left(\frac{1}{2} \cdot \frac{b}{a}\right)$.
   d. Compute $\left(\frac{1}{2} \cdot \frac{b}{a}\right)^2$.
   e. Add $\left(\frac{1}{2} \cdot \frac{b}{a}\right)^2$ to each side.
   f. Factor the left side.
   g. Write the equivalent compound *or* sentence.
   h. Solve.

6. *Quadratic Formula:* Any quadratic equation of the form $ax^2 + bx + c = 0$ is equivalent to the following compound sentence:
$$x = \frac{-b + \sqrt{b^2 - 4ac}}{2a} \text{ or } x = \frac{-b - \sqrt{b^2 - 4ac}}{2a}$$

7. *Value of the discriminant and the solution set of $ax^2 + bx + c = 0$*
See page 501.

CHAPTER REVIEW

SECTION 18.1

Change each equation to the form $ax^2 + bx + c = 0$.

1. $x^2 + 5x = 2$ $x^2 + 5x - 2 = 0$

2. $x^2 = 3 - 2x$ $x^2 + 2x - 3 = 0$

Graph $y = x^2 - 1$. Refer to this graph to estimate the solutions of each equation.

3. $x^2 - 1 = 2$ $\{1.7, -1.7\}$

4. $x^2 - 1 = 0$ $\{-1, 1\}$

*Quadratic Equations / **509***

Chapter Test
Two Chapter Tests (Form A and Form B) are provided on pages 373–376 in the *Teacher's ResourceBank*.

SECTION 18.2

Solve by using the Factors of Zero Property.

5. $x^2 - 9x - 10 = 0$ $\{10, -1\}$ **6.** $x^2 = -21 + 10x$ $\{3, 7\}$ **7.** $2x^2 + 10x = 0$
$$\{0, -5\}$$

SECTION 18.3

Write the solutions of each equation.

8. $x^2 = 256$ $\{16, -16\}$ **9.** $x^2 = 361$ $\{19, -19\}$ **10.** $(x - 6)^2 = 64$
$$\{14, -2\}$$

SECTION 18.4

Solve by completing the square.

11. $x^2 - 8x - 20 = 0$ $\{10, -2\}$ **12.** $x^2 + 4x - 32 = 0$ $\{4, -8\}$ **13.** $4x^2 - x - 3 = 0$
$$\{1, -\tfrac{3}{4}\}$$

SECTION 18.5

Solve by the quadratic formula.

14. $8x^2 - 6x + 1 = 0$ $\{\tfrac{1}{4}, \tfrac{1}{2}\}$ **15.** $6x^2 - x - 1 = 0$ $\{-\tfrac{1}{3}, \tfrac{1}{2}\}$ **16.** $x^2 + 5x + 7 = 0$
No real

SECTION 18.6

Compute the value of $b^2 - 4ac$ for each equation. Then write <u>One rational</u> <u>number</u>, <u>Two rational numbers</u>, <u>Two irrational numbers</u>, or <u>No real numbers</u> to describe the solutions.

17. $x^2 - 4x + 4 = 0$ 0; One rat.   **18.** $x^2 + 5x - 14 = 0$ **19.** $-3x^2 + 4x + 5 = 0$
81; Two rat.   76; Two irrat.

SECTION 18.7

For Exercises 20–21, Condition 1 is <u>underscored once</u>, and Condition 2 is <u>underscored twice</u>. Use Condition 1 to represent the unknowns. Use Condition 2 to write an equation for the problem. Solve.

Additional Answers
20. n = smaller number;
$n + 3$ = greater number;
$n(n + 3) = 10$; 2 and 5;
-5 and -2
21. w = width; $w + 4$ = length;
$w(w + 4) = 60$; length:
10 cm; width: 6 cm

20. <u>One number is 3 greater than</u> <u>another number.</u> Their product is 10. What are the two pairs of numbers that satisfy these conditions?

21. <u>The length of a rectangle is 4</u> <u>centimeters more than its width.</u> The area is 60 square centimeters. Find the length and width.

CHAPTER 19

Probability

When all of the steel balls have fallen in the probability machine below and in the device at the right, a *bell-shaped curve* is formed. As each ball falls, it is equally likely to go to the right or to the left.

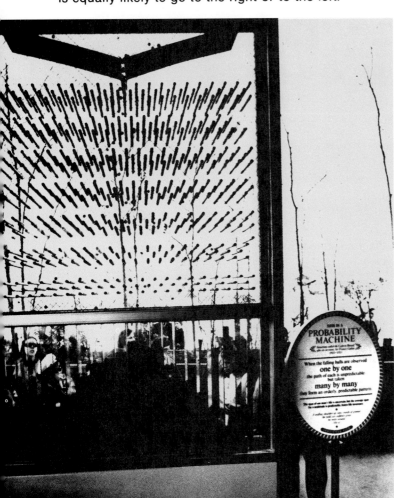

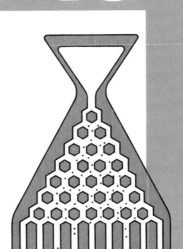

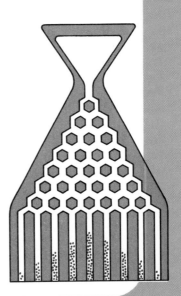

Background

The diagram shown at the right is a **hexstat,** a smaller version of the probability machine shown above. In conducting the probability experiment, 256 balls are generally used. Because it is equally likely for each ball to go to the left or to the right when it comes to a channel branch, the final figure would be as shown in the bottom drawing, a bell-shaped curve.

Maintenance
1. Simplify: $-6k^2 + 3k + 2k^2$
   Ans: $-4k^2 + 3k$
   (Section 2.6)
2. Factor: $2b^2 - 6b + 4$
   Ans: $2(b - 2)(b - 1)$
   (Section 9.7)
3. Name two points on the graph of $y = -2x + 3$.
   Ans: Sample points:
   $(-1, 5), (0, 3), (2, -1)$
   Answers will vary.
   (Section 12.5)
4. Beth is twice as old as Donna. The sum of their ages is 54. Find their ages.
   Ans: Beth: 36; Donna: 18
   (Section 14.5)
5. Write an equation for the axis of the graph of $y = 2(x - 1)^2 + 3$.
   Ans: $x = 1$ (Section 17.7)

OBJECTIVE: To count the number of ways a situation can occur by using the Fundamental Counting Principle

19.1 Fundamental Counting Principle

Karen and Sarah are trying out for the female lead in this year's school play. John, Mike, and Henry are auditioning for the male lead. The director asked Sarah to make a chart that shows how many pairings are possible for the two leading roles. A tree diagram can be used to show the pairings.

Table

Note that the interest here is in the pairings — not the order of the pairs. Karen-John is the same pairing as John-Karen.

Female Lead	Male Lead	Pairings	
Karen	John	Karen and John	1
	Mike	Karen and Mike	2
	Henry	Karen and Henry	3
Sarah	John	Sarah and John	4
	Mike	Sarah and Mike	5
	Henry	Sarah and Henry	6

The tree diagram shows that there are 6 different pairings for the two leading roles.

The total number of pairings can also be found by multiplying.

$$\left(\begin{array}{c}\textbf{Number of choices:}\\ \textbf{Female lead}\end{array}\right) \times \left(\begin{array}{c}\textbf{Number of choices:}\\ \textbf{Male lead}\end{array}\right) = \left(\begin{array}{c}\textbf{Number of}\\ \textbf{pairings}\end{array}\right)$$

$$2 \qquad \times \qquad 3 \qquad = \qquad 6$$

This method for multiplying the choices applies the Fundamental Counting Principle.

Fundamental Counting Principle

If there are p choices for the first decision and q choices for the second decision, then there are $p \times q$ ways that the two decisions can be made together.

The following example shows how to use the Fundamental Counting Principle for situations involving more than two decisions.

512 / *Chapter 19*

EXAMPLE

The members of the senior class are choosing class rings. They can decide on the material (white gold or yellow gold), style (traditional, contemporary, or signet), and the color of the stone (red, blue, green, or yellow). How many different choices does each student have?

Solution: Use the Fundamental Counting Principle.

$$\begin{pmatrix} \text{Number of} \\ \text{choices:} \\ \text{Metal type} \end{pmatrix} \times \begin{pmatrix} \text{Number of} \\ \text{choices:} \\ \text{Style} \end{pmatrix} \times \begin{pmatrix} \text{Number of} \\ \text{choices:} \\ \text{Stone color} \end{pmatrix} = \begin{pmatrix} \text{Total} \\ \text{choices} \end{pmatrix}$$

$$2 \quad \times \quad 3 \quad \times \quad 4 \quad = \quad 24$$

Each student has 24 different choices for a class ring.

CLASSROOM EXERCISES

Draw a tree diagram to illustrate each exercise. Then write the number of different choices that are possible. (Table)

1. The number of ways to choose the color of a new convertible sports car if the top can be white or black and the car body can be red, blue, brown, or black   8

2. The number of different outfits possible if a girl has four sweaters, three skirts, and two pairs of shoes that are all color coordinated   24

3. The number of ways to choose a president and vice-president for the senior class if there are six candidates for president and four for vice-president   24

4. The number of ways to choose a lunch in the cafeteria if there are two choices for an entree, three choices for vegetable, and two choices for dessert (only one choice is allowed per category)   12

5. The number of ways that a two-digit number can be written using the following digits:

    1, 2, 3, 4

    NOTE: The digits can be repeated.
    16

Probability / 513

Additional Examples

1. A salesperson wants to fly from city A to city B, change planes, and then fly to city C. Four airlines have flights from city A to city B. Five airlines have flights from city B to city C. In how many possible ways can the salesperson fly from city A to city B to city C?
   Ans: 20

2. The Quick Frame shop offers a choice of twelve styles of frames, eight types of mattes, and three types of glass. Choosing one of each, in how many possible ways can a picture be framed?
   Ans: 288

513

![WRITTEN EXERCISES]

Goal: To use the Fundamental Counting Principle
Sample Problem: A man has two sports coats, three pairs of slacks, and five ties that are color coordinated. In how many different ways can he choose an outfit from these items?
Answer: $2 \times 3 \times 5 = 30$

Use the Fundamental Counting Principle to solve each problem.
(Example)

1. A car manufacturer offers three new models in six different colors. How many cars will a car dealer need in order to display all the possible selections? 18

2. There are five candidates for junior class president and three candidates for vice-president. In how many possible ways can the two offices be filled? 15

3. An English class must choose from a list of twenty books to write a book report, and from ten topics to write a theme. In how many ways can a student choose a book and a theme topic? 200

4. Freight can be shipped from city A to city B either by plane or by boat. It can be shipped from city B to city C by plane, truck, or bus. In how many possible ways can freight be shipped from city A to city B to city C? 6

5. The identification code used by a computer operator can be any two letters. The two letters can be the same. How many different identification codes are possible under this system? 676

6. A dinner menu has a choice of four appetizers, nine entrees, three beverages, and five desserts. In how many ways can a dinner be ordered, choosing one item from each category? 540

7. A commuter going to work can walk, ride a bike, or share a ride to the station. At the station she can either take a train or a bus. After arriving in the city, she can walk, ride a shuttle bus, or take a subway. How many possible ways are there for her to get to work? 18

8. For a required reading assignment, a student must choose one novel from a list of ten, one book of nonfiction from a list of six, and one short story from a list of eight. In how many possible ways can the reading assignment be completed? 480

Write each fraction in lowest terms.

1. $\frac{5}{20}$ $\frac{1}{4}$

2. $\frac{6}{15}$ $\frac{2}{5}$

3. $\frac{12}{28}$ $\frac{3}{7}$

4. $\frac{10}{16}$ $\frac{5}{8}$

5. $\frac{6}{24}$ $\frac{1}{4}$

OBJECTIVE: To compute the probability of an outcome

19.2 Sample Space and Probabilities

Suppose you toss a coin and observe the way in which it lands. It can land "heads" up or "tails" up. Both results are equally likely to occur. Now you toss the coin twice. The tree diagram below shows the ways the coin can land after being tossed.

First toss	Second toss	Both tosses	
H	H	HH	1
	T	HT	2
T	H	TH	3
	T	TT	4

The equal likelihood of the occurrence of events is an important assumption in many probability experiments. Often, it is not explicitly stated.

Tossing the coin twice is an example of an **experiment.** Each result of an experiment is called an **outcome.** The set of all the different possible outcomes of an experiment is called the **sample space.** An **event** is a subset of a sample space. Thus, it is a set of one or more of the possible outcomes of an experiment. The sample space for tossing a coin twice consists of the four outcomes shown in the tree diagram. The sample space is written as a set.

$$\{HH, HT, TH, TT\}$$

Each of the four outcomes in this sample space is equally likely to occur. Therefore, the probability of each outcome is 1 out of 4, or $\frac{1}{4}$.

Definition

> The probability P of an event E is the ratio of the number of favorable outcomes of the event to the total number of possible outcomes in the sample space.
>
> $$P(E) = \frac{\text{number of favorable outcomes}}{\text{total number of possible outcomes}}$$

Probability / 515

Teaching Suggestions p. M–54

Lesson Resources
Warm–Up: p. M–55
Maintenance: See below.
Practice Worksheet 19.2
Exploration 19
Transparencies 44, 45A, 45B, 45C

Maintenance
1. Solve and check:
   $8t = 23 + 3(2t - 5)$
   Ans: t = 4 (Section 4.3)
2. The area of a square court-yard is 676 square yards. Find the length of each side of the courtyard.
   Ans: 26 yds (Section 7.7)
3. Is the following relation a function?
   $\{(1, 0), (2, 3), (4, 3), (-1, 0)\}$
   Ans: Yes (Section 12.4)
4. Write three ordered pairs having different x values that are in the solution set of
   $2x - y + 3 > 0$.
   Ans: Sample points:
   (-1, 0), (2, 4), (3, 8)
   Answers will vary.
   (Section 16.1)
5. Write $x^2 - 6x + 3$ in the form $(x - h)^2 + k$.
   Ans: $(x - 3)^2 - 6$
   (Section 17.6)

A pen is drawn from a drawer without looking. The drawer contains 5 red pens, 2 black pens, and 3 blue pens. Find the probability of each event.
1. Drawing a black pen
   Ans: $\frac{1}{5}$
2. Drawing a red pen
   Ans: $\frac{1}{2}$

EXAMPLE 1 A coin is tossed twice. What is the probability of tossing one head and one tail?

Solution: Possible outcomes (sample space): {HH, HT, TH, TT}

Favorable outcomes: {HT, TH}   *In this Example, the order is not important.*

1. Determine the total number of possible outcomes. ⟶ 4

2. Determine the number of favorable outcomes for the event. ⟶ 2

3. Write the probability ratio.

P(tossing one head and one tail) $= \frac{2}{4}$ ← *Number of favorable outcomes*
   ← *Number of possible outcomes*

$= \frac{1}{2}$ *Lowest terms*

The probability of tossing one head and one tail is $\frac{1}{2}$.

Always write probabilities in lowest terms.

Example 2
The arrow of the spinner shown on page 516 is equally likely to stop on any of the six regions. The arrow is spun once. Find each probability.
1. Spinning a number less than 3
   Ans: $\frac{1}{3}$
2. Spinning a number greater than 2
   Ans: $\frac{2}{3}$

EXAMPLE 2

The arrow of the spinner shown is equally likely to stop on any of the six regions. The arrow is spun once. Find each probability.

a. Spinning a number less than 2
b. Spinning a number greater than 1

The usual agreement is to ignore a spin that stops on the boundary between two regions.

Solutions: Possible outcomes: {1, 2, 3, 4, 5, 6}

	a.	**b.**
Favorable outcomes:	{1}	{2, 3, 4, 5, 6}

1. Determine the total number of possible outcomes. ⟶ 6   6

2. Determine the number of favorable outcomes for the event. ⟶ 1   5

3. Write the probability ratio. ⟶ P(number < 2) $= \frac{1}{6}$ P(number > 1) $= \frac{5}{6}$

The probability of spinning a number less than 2 is $\frac{1}{6}$.

The probability of spinning a number greater than 1 is $\frac{5}{6}$.

In Example 2, spinning a number less than 2 and spinning a number greater than 1 are complementary events. Since two **complementary events** include all of the possible outcomes of an experiment, the sum of their probabilities is always 1.

NOTE: So, when the probability of an event is known, the probability of its complement can be found.

EXAMPLE 3.   For the spinner in Example 2, find these probabilities.

   a. Spinning a whole number
   b. Not spinning a whole number

Solutions: Possible outcomes: {1, 2, 3, 4, 5, 6}

           a.   **b.**

Favorable outcomes: {1, 2, 3, 4, 5, 6} ϕ

① Determine the total number of possible outcomes. ⟶ 6   6

② Determine the number of favorable outcomes for the event. ⟶ 6 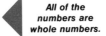 **All of the numbers are whole numbers.** 0

③ Write the probability ratio. ⟶ $P(\text{whole}) = \frac{6}{6}$, or 1 $P(\text{not whole}) = \frac{0}{6}$, or 0

The probability of spinning a whole number is 1.
The probability of <u>not spinning</u> a whole number is 0.

> The probability of an event that is certain to happen is 1.
> The probability of an event that cannot happen is 0.

Additional Examples
Example 3
For the spinner in Example 2 on page 516, find these probabilities.
1. Spinning a whole number less than 7
 Ans: 1
2. Spinning a negative integer
 Ans: 0

CLASSROOM EXERCISES

A coin is tossed twice. Find each probability.
(Examples 1 and 3)

1. Tossing two heads
2. Tossing two tails
3. Not tossing a head
4. Not tossing a tail
5. Tossing more than one head
6. Tossing more than two tails

Probability / **517**

Common Error
Students incorrectly write a probability ratio.

Example
Find the probability of getting a 5 on the spinner in Example 2.
Answer: $\frac{1}{5}$

Prescription
Emphasize that the second term in a probability ratio must be the <u>total</u> number of outcomes that are possible.

The arrow of the spinner at the right is equally likely to stop on any of the 4 regions. The arrow is spun once.

For Exercises 9–12, find each probability. (Examples 2 and 3)

7. Stopping on an even number $\frac{1}{2}$

8. Stopping on a prime number $\frac{1}{2}$

9. Not stopping on 4 $\frac{3}{4}$

10. Not stopping on a whole number 0

11. Stopping on a whole number less than 5 1

12. Stopping on a whole number greater than 3 $\frac{1}{4}$

13. Are the events in Exercises 7 and 8 complementary events? Explain.
No. The two events do not include all the possible outcomes.

14. Are the events in Exercises 9 and 12 complementary events? Explain.
Yes. The two events include all the possible outcomes.

Assignment Guide
Basic Omit
Average Omit
Above Average p. 518–519:
1–19 odd, 21–28; p. 514: 1–8

WRITTEN EXERCISES

Goal: To compute probabilities
Sample Problem: A container holds three yellow balls and four red balls. One ball is drawn without looking. What is the probability that a red ball is drawn?

Answer: $\frac{4}{7}$

A ball is drawn from a bag without looking. The bag contains 4 red balls, 3 white balls, 5 black balls, and 2 green balls. Find the probability of each event. (Example 1)

1. Drawing a green ball $\frac{1}{7}$

2. Drawing a red ball $\frac{2}{7}$

3. Drawing a black ball $\frac{5}{14}$

4. Drawing a white ball $\frac{3}{14}$

An experiment involves tossing a game marker which has four sides numbered 1, 2, 3, and 4. The marker is equally likely to land on any of the four sides. The marker is tossed once.

For Exercises 5–10, find each probability. (Examples 2 and 3)

5. Landing on 2 $\frac{1}{4}$

6. Landing on an odd number $\frac{1}{2}$

7. Landing on a prime number $\frac{1}{2}$

8. Landing on an even number $\frac{1}{2}$

9. Landing on a counting number   1

10. Landing on a number greater than $\overset{0}{4}$

11. Are the events in Exercises 5 and 7 complementary events?   No. The Explain.   two events do not include all the possible outcomes.

12. Are the events in Exercises 6 and 8 complementary events?   Yes. The Explain.   two events include all the possible outcomes.

The cards shown at the right are placed in a box. The cards are shuffled and one card is drawn without looking.

Find each probability.
(Examples 2 and 3)

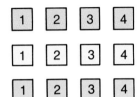

13. Drawing a green card $\frac{1}{3}$

14. Not drawing a blue card $\frac{2}{3}$

15. Drawing a 4 $\frac{1}{4}$

16. Drawing a blue 2 $\frac{1}{12}$

17. Drawing a number less than 3 $\frac{1}{2}$

18. Not drawing an odd number $\frac{1}{2}$

19. Drawing a red card   0

20. Drawing a number less than 5   1

MORE CHALLENGING EXERCISES

A coin is tossed three times. Draw a tree diagram to identify the sample space for this experiment. Then find each probability.

21. Tossing two heads and one tail $\frac{3}{8}$

22. Tossing one head and two tails $\frac{3}{8}$

23. Tossing two or more heads $\frac{1}{2}$

24. Tossing less than three heads $\frac{7}{8}$

25. Tossing more tails than heads $\frac{1}{2}$

26. Tossing no heads $\frac{1}{8}$

NON-ROUTINE PROBLEMS

27. Doris took some coins out of her bank and arranged the coins in four rows with four coins in each row. Each of the four rows, the four columns, and the two diagonals had exactly one penny, one nickel, one dime, and one quarter. Show how Doris could have arranged the coins.

28. Roberto needs to draw a line that is six inches long, but he does not have a ruler. He does have some sheets of notebook paper that are each $8\frac{1}{2}$ inches wide and 11 inches long. Describe how Roberto can use the notebook paper to measure six inches.

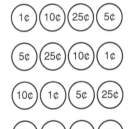

Maintenance
1. Solve and check:

$$41 = \frac{3x}{2} - 10$$

Ans: x = 34   (Section 3.3)
2. Use the Rule of Pythagoras to determine whether a triangle having sides of lengths 9 inches, 12 inches, and 15 inches is a right triangle.
Ans: Yes   (Section 7.8)
3. Multiply and simplify:

$$\frac{6}{2y + 6} \cdot \frac{y^2 - 9}{y^2 + y - 12}$$

Ans: $\frac{3}{y + 4}$ (Section 10.5)
4. Solve the system below by the multiplication/addition method.

$$\begin{cases} 2x + y = 2 \\ 3x + 4y = 18 \end{cases}$$

Ans: {(−2,6)}
(Section 14.4)
5. Solve $2x^2 = 5 - 9x$ by the quadratic formula.
Ans: $\{\frac{1}{2}, -5\}$ (Section 18.5)

Additional Examples
Example 1
A red number cube and a green number cube are tossed. Find each probability.
1. Tossing a 2 on the red cube and a 3 on the green cube
Ans: $\frac{1}{36}$
2. Tossing a number less than 4 on the red cube and an odd number on the green cube
Ans: $\frac{1}{4}$

19.3 Independent Events

OBJECTIVES: To determine the probability of independent events
To determine the probability of mutually exclusive events

When tossing two number cubes, the outcome of tossing the second cube is not affected by the outcome of tossing the first. The two outcomes are said to be independent. Independence of events is not always easy to establish.

> Two events are ***independent*** if the occurrence of one event is not influenced by the occurrence of the other.

When two events A and B are independent, the probability that both events will occur together is the product of their probabilities.

$$P(A \text{ and } B) = P(A) \cdot P(B)$$

EXAMPLE 1   A red number cube and a green number cube are tossed. Find the probability of tossing a number less than 3 on the red cube and an even number on the green cube.

Solution: Let event A be tossing a number less than 3 on the red cube.
Let event B be tossing an even number on the green cube.

1 Identify the elements of events A and B. ⟶ $A = \{1, 2\}$ $B = \{2, 4, 6\}$

2 Find the probability of each event. ⟶ $P(A) = \frac{2}{6},$ or $\frac{1}{3}$ $P(B) = \frac{3}{6},$ or $\frac{1}{2}$

3 Find the probability of both events. ⟶ $P(A \text{ and } B) = P(A) \cdot P(B)$

$$= \frac{1}{3} \cdot \frac{1}{2}$$

$$= \frac{1}{6}$$

The probability of tossing a number less than 3 on the red cube and an even number on the green cube is $\frac{1}{6}$.

P–1 **Find these probabilities for the number cube experiment in Example 1.**

   a. Tossing a number greater than 4 on the red cube <u>and</u> a 2 on the green cube $\frac{1}{18}$
   b. Tossing a prime number on the red cube <u>and</u> an odd number on the green cube $\frac{1}{4}$

We have to assume that Marian cannot identify the coins by touch before she draws one.

Marian has 3 pennies, 5 nickels, and 4 dimes in her pocket. Suppose she draws one coin. She wants to know the probability of drawing a penny <u>or</u> a dime. Since no coin is both a penny and a dime, the two events are said to be **mutually exclusive.** Mutually exclusive events have no outcomes in common.

An outcome is successful if either A or B occurs.

> The probability that one of two mutually exclusive events will occur is the sum of their probabilities.
>
> $$P(A \text{ \underline{or} } B) = P(A) + P(B)$$

EXAMPLE 2

Marian needs a coin for the parking meter. Without looking, she draws one coin from the 3 pennies, 5 nickels, and 4 dimes in her pocket. What is the probability that it will be a nickel <u>or</u> a dime?

Solution: Let event A be drawing a nickel.
Let event B be drawing a dime.

1 Find the probability of each event. ⟶ $P(A) = \frac{5}{12}$ $P(B) = \frac{4}{12}$

2 Find the probability of drawing a nickel or a dime. ⟶ $P(A \text{ \underline{or} } B) = P(A) + P(B)$

$$= \frac{5}{12} + \frac{4}{12}$$

$$= \frac{9}{12}, \text{ or } \frac{3}{4} \blacktriangleleft \text{ Lowest terms}$$

The probability that Marian will draw a nickel <u>or</u> a dime is $\frac{3}{4}$.

Common Error

Students always multiply the probabilities of the events when finding the probability of mutually exclusive events.

Example

Find the probability of drawing a penny or a dime in Example 2.

$$P(A) = \frac{3}{12} \qquad P(B) = \frac{4}{12}$$

$$P(A \text{ or } B) = \frac{3}{12} \cdot \frac{4}{12} = \frac{1}{12}$$

Prescription

Go over Classroom Exercises 1–8 with the students. Explain the difference of using and and or in finding probabilities. Emphasize that multiplication is used with and and addition is used with or.

CLASSROOM EXERCISES

Four cards are marked with the letters M, A, T, and H. They are shuffled and placed with the letters down. One card is drawn. Then the spinner is spun. Find each probability. (Example 1)

1. Drawing a consonant <u>and</u> spinning a number greater than 2 $\frac{9}{16}$

2. Drawing a vowel <u>and</u> spinning a number less than 5 $\frac{1}{8}$

3. Drawing the T <u>and</u> spinning a prime number $\frac{1}{8}$

4. Drawing a consonant <u>and</u> spinning an even number $\frac{3}{8}$

The seven cards shown at the right are shuffled and placed with the letters down. One card is drawn. Find each probability. (Example 2)

5. Drawing the G <u>or</u> a vowel $\frac{4}{7}$

6. Drawing the B <u>or</u> an A $\frac{3}{7}$

7. Drawing the R <u>or</u> an A $\frac{3}{7}$

8. Drawing the E <u>or</u> a consonant $\frac{5}{7}$

WRITTEN EXERCISES

Goal: To find the probability of independent events
Sample Problem: A coin is tossed twice. Find the probability of getting a head on the first toss <u>and</u> a tail on the second.
Answer: $\frac{1}{2} \cdot \frac{1}{2} = \frac{1}{4}$

A red number cube and a green number cube are tossed. Find each probability. (Example 1)

1. Tossing a 3 on the red cube <u>and</u> a 5 on the green cube $\frac{1}{36}$

2. Tossing a 1 on the red cube <u>and</u> an even number on the green cube $\frac{1}{12}$

3. Tossing a number less than 6 on the red cube <u>and</u> a number greater than 4 on the green cube $\frac{5}{18}$

4. Tossing an odd number on the red cube <u>and</u> an even number on the green cube $\frac{1}{4}$

5. Tossing a number less than 3 on the red cube <u>and</u> a number greater than 3 on the green cube $\frac{1}{6}$

6. Tossing a number greater than 2 on the red cube <u>and</u> a number less than 4 on the green cube $\frac{1}{3}$

7. Tossing a negative number on the red cube <u>and</u> a number greater than 0 on the green cube $\quad 0$

8. Tossing a counting number on the red cube <u>and</u> a whole number on the green cube $\quad 1$

522 / *Chapter 19*

522

A bag contains 6 white, 4 orange, 7 yellow, and 3 green balls. One ball is drawn. Find each probability. (Example 2)

9. Drawing a white ball <u>or</u> a yellow ball $\frac{13}{20}$

10. Drawing an orange ball <u>or</u> a green ball $\frac{7}{20}$

11. Drawing a white ball <u>or</u> a green ball $\frac{9}{20}$

12. Drawing a yellow ball <u>or</u> a green ball $\frac{1}{2}$

13. Drawing a white ball <u>or</u> an orange ball $\frac{1}{2}$

14. Drawing an orange ball <u>or</u> a yellow ball $\frac{11}{20}$

■■■■ MID-CHAPTER REVIEW ■■■■■■■■■

Use the Fundamental Counting Principle to write the number of different choices that are possible. (Section 19.1)

1. The number of ways to choose classes for the first two periods of the school day if there are three choices for the first period and six choices for the second 18

2. The number of ways to choose a computer system if there are four computer models, two different disc drives, and two different printers 16

The arrow of the spinner is equally likely to stop on any of the eight regions. The arrow is spun once. Find each probability. (Section 19.2)

3. Stopping on a number less than 3 $\frac{1}{4}$

4. Stopping on a prime number $\frac{1}{2}$

5. Stopping on a counting number 1

6. Not stopping on an integer 0

The arrow of the spinner is equally likely to stop on any of the four regions. The arrow is spun and then a number cube is tossed. Find each probability. (Section 19.3)

7. Spinning an even number <u>and</u> tossing a multiple of 3 $\frac{1}{6}$

8. Spinning a prime number <u>and</u> tossing a number greater than 4 $\frac{1}{6}$

A sack contains 4 red, 2 white, 6 green, and 8 blue marbles. One marble is drawn without looking. Find each probability. (Section 19.3)

9. Drawing a red marble <u>or</u> a blue marble $\frac{3}{5}$

10. Drawing a white marble <u>or</u> a green marble $\frac{2}{5}$

*Probability / **523***

Quiz Sections 19.1–19.3
After completing this Mid-Chapter Review, you may want to administer a quiz covering the same sections. See the quiz provided on page 389 in the *Teacher's ResourceBank*.

Using Statistics

Collecting Data

How do manufacturers determine
which colors to use on cereal boxes in
order to attract the most customers?
They do it by taking a **sample**. For example, they might ask people entering
supermarkets in several cities in different parts of the country which of
two or three cereal boxes they prefer.
The designs and colors for cereal boxes
are determined on the basis of these
choices.

A **sample** is a group chosen to represent a larger group called the
population. In order for the sample to represent the population, the
sample must be <u>sufficiently large</u> and it must be <u>chosen at random</u>.
The items in a random sample are chosen completely by chance.

EXERCISES

1. This advertisement is based on a
   sample of eight plumbers. Is the
   sample sufficiently large? Explain.

3. A book has 700 pages. To check
   for errors, an editor randomly
   chose 70 pages to be examined. Is
   this a good sample of the
   population? Explain.

2. This advertisement is based on a
   sample of Senator Win's friends.
   Is this a random sample? Explain.

4. Estimate the number of left-handed
   students in your school by asking
   25 students. Tell how you selected
   the students. Explain why you think
   the sample does, or does not, provide a good basis for the estimate.

5. A local television station includes a poll as part of its late night
   newscast. A "question of the day" is flashed on the screen and
   viewers are asked to call in their "yes" or "no" votes. Do you
   think the results represent the community's opinion? Explain.

19.4 Dependent Events

OBJECTIVE: To determine the probability of dependent events

Twenty students in a biology class are assigned to one of four lab stations by drawing cards from a box. The box contains twenty cards. Five cards are marked with a 1, five are marked with a 2, five are marked with a 3, and five are marked with a 4.

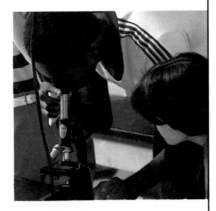

Each time a student draws a card, there is one fewer card for the next student to choose from. Thus, the outcome of each drawing is <u>dependent</u> on the previous drawing.

> Two events are ***dependent*** if the occurrence of one event is influenced by the occurrence of the other.

As with independent events, the probability that two dependent events will both occur is the product of their probabilities. The difference is that the number of possible outcomes for the second event has been affected by the occurrence of the first event.

EXAMPLE 1   Jane is the first and Linda is the second biology student to draw a card. Find the probability that Jane draws a 1 <u>and</u> Linda draws a 4.

Solution: Let event A be Jane's drawing a 1.
Let event B be Linda's drawing a 4, given that Jane draws a 1.

1. Determine the total number of possible outcomes for each event. $\longrightarrow$ A: 20 B: 19   **Jane has drawn one card.**

2. Determine the number of favorable outcomes for each event. $\longrightarrow$ A: 5 B: 5

3. Find the probability of each event. $\longrightarrow$ $P(A) = \frac{5}{20}$, or $\frac{1}{4}$ $P(B) = \frac{5}{19}$

*Probability / **525***

Lesson Resources
Warm–Up: p. M–55
Maintenance: See below.
Practice Worksheet 19.4

Maintenance
1. Solve: $10y - 24 > 3y - 3$
   Ans: $y > 3$ (Section 5.6)
2. Multiply: $(2x + 3)(x - 5)$
   Ans: $2x^2 - 7x - 15$
   (Section 8.5)
3. Write the slope of the line that contains the points $(1, -6)$ and $(3, -9)$.
   Ans: $-\frac{3}{2}$ (Section 13.4)
4. A plane travels 720 miles with a wind in 3 hours (Condition 2). It travels the same distance against the wind in 4 hours (Condition 1). Find the speed of the plane in still air and the speed of the wind.
   Ans: Speed in still air: 210 mi/hr; speed of wind: 30 mi/hr (Section 15.7)
5. Write the slope-intercept form of $-4x - 2y + 3 \leq 0$.
   Ans: $y \geq -2x + \frac{3}{2}$
   (Section 16.1)

Additional Example
Example 1
Jane has drawn a 1 and Linda has drawn a 4. Carl is the third and Sidney is the fourth biology student to draw a card. Find the probability that Carl draws a 4 and Sidney draws a 3.

Ans: $\frac{10}{153}$

4. Find the probability of
both events. —————————→ $P(A \text{ and } B) = P(A) \cdot P(B)$

$$= \frac{1}{4} \cdot \frac{5}{19}$$

$$= \frac{5}{76}$$

The probability that Jane draws a 1 and Linda draws a 4 is $\frac{5}{76}$.

In Example 1, the total number of possible outcomes for Linda's drawing was fewer because Jane had drawn before her. Example 2 shows a situation in which both the total number of possible outcomes and the number of favorable outcomes for Linda's drawing are fewer than Jane's.

EXAMPLE 2   As in Example 1, Jane is the first and Linda is the second student to draw a card. Find the probability that both Jane and Linda draw a 1.

Solution: Let event A be Jane's drawing a 1.
Let event B be Linda's drawing a 1, given that Jane draws a 1.

1. Determine the total number
of possible outcomes for
each event. —————————→ A: 20   B: 19   ◀ Jane has drawn one card.

2. Determine the number of
favorable outcomes for
each event. —————————→ A: 5   B: 4   ◀ Jane has drawn a 1.

3. Find the probability of
each event. —————————→ $P(A) = \frac{5}{20}$, or $\frac{1}{4}$ $P(B) = \frac{4}{19}$

4. Find the probability of
both events. —————————→ $P(A \text{ and } B) = P(A) \cdot P(B)$

$$= \frac{1}{\cancel{4}} \cdot \frac{\overset{1}{\cancel{4}}}{19}$$

$$= \frac{1}{19}$$

The probability that both Jane and Linda draw a 1 is $\frac{1}{19}$.

In Example 1, $P(B) = \frac{5}{19}$. In Example 2, $P(B) = \frac{4}{19}$. Both represent the probability that event B will occur, given that event A has already occurred. This kind of probability is called a conditional probability.

**Additional Example
Example 2**
The box now contains 16 cards. Three cards are marked with a 1, five are marked with a 2, four are marked with a 3, and four are marked with a 4. Bob draws a card and then Jenny draws a card. Find the probability that both Bob and Jenny draw a 3.

Ans: $\frac{1}{20}$

CLASSROOM EXERCISES

There are 5 seniors, 2 juniors, and 3 sophomores who are reporters for the school newspaper. Reporting assignments are handed out by drawing names from a box. If two names are chosen (without replacement), find each probability. (Examples 1 and 2).

1. Drawing a senior, <u>and</u> then a junior $\frac{1}{9}$

2. Drawing a sophomore, <u>and</u> then a junior $\frac{1}{15}$

3. Drawing a senior <u>and</u> then a sophomore $\frac{1}{6}$

4. Drawing a junior, <u>and</u> then a senior $\frac{1}{9}$

5. Drawing two seniors $\frac{2}{9}$

6. Drawing two juniors $\frac{1}{45}$

7. Drawing two sophomores $\frac{1}{15}$

Common Error
When finding the probability of dependent events, students do not change the number of possible outcomes after the first event.

Example
In Classroom Exercise 1, the probability is written as $\frac{5}{10} \cdot \frac{2}{10}$, or $\frac{1}{10}$.

Prescription
Explain to students that each time a dependent event occurs, the number of possible outcomes decreases by 1. Make sure students understand what is meant by "without replacement."

WRITTEN EXERCISES

Goal: To find the probability of dependent events

Sample Problem: Six cards are numbered 1 through 6 and then placed with the numbers down. A card is drawn. Without replacing the first card, a second card is drawn. What is the probability of drawing an even number first, <u>and</u> then a 5?

Answer: $\frac{1}{2} \cdot \frac{1}{5} = \frac{1}{10}$

The Homecoming Planning Committee consists of 4 seniors (3 girls and 1 boy), 4 juniors (2 girls and 2 boys), and 2 sophomores (both girls). The president and vice-president of the committee will be selected by drawing names from a box.

The name of the president is drawn first and then, without replacing the first name, the name of the vice-president is drawn. Find the probability of each event. (Examples 1 and 2)

1. Drawing a senior for president <u>and</u> a junior for vice-president $\frac{8}{45}$

2. Drawing a girl for president <u>and</u> a boy for vice-president $\frac{7}{30}$

Assignment Guide
Basic Omit
Average Omit
Above Average p. 527–528: 1–17

Probability / 527

3. Drawing a senior boy for president and a junior girl for vice-president $\frac{1}{45}$

4. Drawing a junior girl for president and a sophomore girl for vice-president $\frac{2}{45}$

5. Drawing a senior for president and another senior for vice-president $\frac{2}{15}$

6. Drawing a sophomore for president and another sophomore for vice-president $\frac{1}{45}$

7. Drawing a girl for president and another girl for vice-president $\frac{7}{15}$

8. Drawing a boy for president and another boy for vice-president $\frac{1}{15}$

9. Drawing a junior boy for president and another junior boy for vice-president $\frac{1}{45}$

10. Drawing a senior girl for president and another senior girl for vice-president $\frac{1}{15}$

MORE CHALLENGING EXERCISES

The Homecoming Planning Committee decided they need to choose a secretary/treasurer in addition to a president and a vice-president. Three names will be drawn from the box, without replacement. The first name drawn will be for president, the second will be for vice-president, and the third will be for secretary/treasurer. Find each probability.

11. Drawing the names of three girls $\frac{7}{24}$

12. Drawing the names of three boys $\frac{1}{120}$

13. Drawing a senior for president, a junior for vice-president, and a sophomore for secretary/treasurer $\frac{2}{45}$

14. Drawing a girl for president, a boy for vice-president, and a boy for secretary/treasurer $\frac{7}{120}$

15. Drawing a senior boy for president, a junior boy for vice-president, and a junior girl for secretary/treasurer $\frac{1}{180}$

16. Drawing a junior girl for president, a junior girl for vice-president, and a sophomore girl for secretary/treasurer $\frac{1}{180}$

NON-ROUTINE PROBLEM

17. Jerome and his sister Latasha walk to school every day. They start at the same time and walk at the same speed. Who gets to school first? Explain. Refer to the figure at the right.

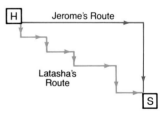

REVIEW CAPSULE FOR SECTION 19.5

Multiply.

1. $\frac{4}{5} \times 240$ 192

2. $\frac{2}{3} \times 123$ 82

3. $\frac{3}{8} \times 488$ 183

4. $\frac{5}{6} \times 234$ 195

5. $\frac{7}{12} \times 1512$ 882

6. $\frac{3}{4} \times 1064$ 798

Using a diagram (Ex. 17)
Using logical reasoning (Ex. 17)

Non–Routine Problems
For a description of non–routine problems, see page M–13.

Additional Answer Written Exercises
17. They arrive at the same time. The total length of Latasha's paths to the right in the figure is the same as the length of Jerome's path to the right. Likewise, the total length of Latasha's path down in the figure is the same as the length of Jerome's path down.

19.5 Experimental Probability

OBJECTIVES: To calculate the experimental probability given the data from an experiment
To use the experimental probability from an experiment to calculate the expected outcome

In many situations, the probability of an event is determined by performing an experiment a large number of times. Probabilities determined in this way are called <u>experimental probabilities</u>.

> If an experiment is carried out n times and a certain event A occurs a times, then the **experimental probability** for event A is $\frac{a}{n}$.
>
> $$P(A) = \frac{\text{Number of times event } A \text{ occurred}}{\text{Total number of times the experiment is tried}} = \frac{a}{n}$$

EXAMPLE 1   A survey was conducted at Pleasant Mountain High School. Students were asked to recite the words to the first verse of the school song. Of the 420 students surveyed, 240 students correctly recited the words. Find the probability that a student at Pleasant Mountain High School can recite the words to the first verse of the school song.

Solution: Let event A be correctly reciting the words.

$$P(A) = \frac{\text{Number of students who correctly recited the words}}{\text{Total number of students surveyed}} = \frac{\overset{4}{\cancel{240}}}{\underset{7}{\cancel{420}}} = \frac{4}{7}$$

The probability that a student at Pleasant Mountain High School can recite the words to the first verse of the school song is $\frac{4}{7}$. Discuss with students why you can't tell that the probability of a student correctly reciting the verse is exactly $\frac{4}{7}$

Once the probability of an event is known, that probability can be used to predict how many times the event will occur if the experiment is conducted again under the same circumstances.

> If an experiment is carried out n times, the number of **expected outcomes** E of an event A is the product of the probability of event A and n.
>
> $$E(A) = P(A) \cdot n$$

Probability / 529

Lesson Resources
Warm–Up: p. M–55
Maintenance: See below.
Practice Worksheet 19.5
Manipulative 15, p. 568
Teaching Aid 6

Maintenance

1. Simplify: $\left(\frac{3x^2}{y}\right)^3$
   Ans: $\frac{27x^6}{y^3}$ (Section 6.5)

2. Divide and simplify:
   $\frac{a^2 + 4a - 12}{a^2 + 8a + 12} \div \frac{2a - 4}{2}$
   Ans: $\frac{1}{a + 2}$ (Section 10.7)

3. Add: $\frac{3}{x + 1} + \frac{5}{x + 4}$
   Ans: $\frac{8x + 17}{(x + 1)(x + 4)}$
   (Section 11.5)

4. Write the slope-intercept form of a line with a slope of -3 and a y intercept of -2.
   Ans: $y = -3x - 2$
   (Section 13.3)

5. Solve $x^2 - 6x + 8 = 0$ by completing the square.
   Ans: $\{2, 4\}$ (Section 18.4)

Additional Examples
Example 1
A survey of 200 households in Baytown is conducted. The results of the survey showed that 90 households have no pets, 55 households have dogs, 40 households have cats, 10 households have birds, and 5 households have hamsters. Find each probability.

1. Of a household having no pets
   Ans: $\frac{9}{20}$
2. Of a household having dogs
   Ans: $\frac{11}{40}$
3. Of a household having cats
   Ans: $\frac{1}{5}$
4. Of a household having birds
   Ans: $\frac{1}{20}$

529

In Additional Example 1, the 200 households surveyed were from a total number of 8200 households in Baytown. According to the results of the survey, how many of the remaining households could be expected to have the following kinds of pets.

1. No pets   **Ans: 3690**
2. Dogs   **Ans: 2255**
3. Cats   **Ans: 1640**
4. Birds   **Ans: 410**

EXAMPLE 2

In Example 1, the 420 students surveyed were from a total of 1015 students at Pleasant Mountain High School. According to the results of the survey, how many of the remaining 595 students could be expected to correctly recite the words to the first verse of the school song?

Solution: Let event A be correctly reciting the words.

$$E(A) = P(A) \cdot n$$

$$= \frac{4}{\cancel{7}} \cdot \cancel{595}^{\,85}$$

From Example 1, $P(A) = \frac{4}{7}$.

$$= 340$$

Of the remaining 595 students at Pleasant Mountain High School, 340 could be expected to correctly recite the words to the first verse of the school song.

CLASSROOM EXERCISES

A survey of 200 households in Pleasant Valley is conducted. The results of the survey showed that 4 households have no automobiles, 48 households have 1 automobile, 132 households have 2 automobiles, and 16 households have 3 or more automobiles. Find each probability for the households surveyed.
(Example 1)

1. Of a household having no automobiles $\frac{1}{50}$

2. Of a household having 1 automobile $\frac{6}{25}$

3. Of a household having 2 automobiles $\frac{33}{50}$

4. Of a household having 3 or more automobiles $\frac{2}{25}$

There are 650 households in Pleasant Valley. Use the probabilities found in Exercises 1–4 to find the expected number of households having the following number of automobiles.
(Example 2)

5. 0 automobiles 13

6. 1 automobile 156

7. 2 automobiles 429

8. 3 or more automobiles 52

WRITTEN EXERCISES

Goal: To find the experimental probability of an event

Sample Problem: A radio station found that 36 out of 90 telephone requests taken from 7:00 A.M. to 5:00 P.M. were for the same song. What is the probability that the next request will be for this song?

Answer: $\frac{36}{90}$, or $\frac{2}{5}$

A survey of 450 households in Riverview is conducted. The results of the survey showed that 108 households have no children, 90 households have 1 child, 180 households have 2 children, 45 households have 3 children, 18 households have 4 children, and 9 households have 5 or more children. Find each probability. (Example 1)

1. Of a household having no children $\frac{6}{25}$

2. Of a household having 1 child $\frac{1}{5}$

3. Of a household having 2 children $\frac{2}{5}$

4. Of a household having 3 children $\frac{1}{10}$

5. Of a household having 4 children $\frac{1}{25}$

6. Of a household having 5 or more children $\frac{1}{50}$

There are 1150 households in Riverview. Use the probabilities found in Exercises 1–6 to find the expected number of households having the following number of children. (Example 2)

7. 0 children   276

8. 1 child   230

9. 2 children   460

10. 3 children   115

11. 4 children   46

12. 5 or more children   23

Solve. (Examples 1 and 2)

13. The results of a survey of 200 students at South High School show that 35 students ride a bike to school. There are 2400 students at the school. How many of the 2400 can be expected to ride a bike to school?   420

14. At the Timely Company clock factory, 150 clocks are tested. The test results show that 2 of the clocks are defective. In a shipment of 1200 clocks, how many can be expected to be defective?   16

Probability / **531**

Problem–Solving Skills
Solving a multi-step problem
(Ex. 13–14)

Quiz Sections 19.4–19.5
After completing Sections 19.4–19.5, you may want to administer a quiz covering the same sections. See the quiz provided on page 390 in the *Teacher's ResourceBank*.

Focus on Reasoning: Venn Diagrams

Venn diagrams can sometimes help you to organize information so that it is easier to see relationships.

EXAMPLE In a sample of 180 students, 119 said that they had seen the movie Outer Stars 1 and 110 said that they had seen Outer Stars 2. Thirteen students had not seen either movie.

a. How many students had seen <u>both</u> movies?
b. How many students had seen Outer Stars 1 <u>only</u>?
c. How many students had seen Outer Stars 2 <u>only</u>?

Solutions:

The interior of the rectangle represents the students sampled.
The interior of the circle on the left represents the students who saw Outer Stars 1.
The interior of the circle on the right represents the students who saw Outer Stars 2.
The overlap of the circles represents the students who saw both movies.

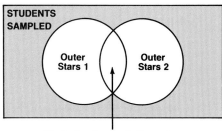

Students who saw both movies

Think: Since 13 students did not see either movie, there are 180 − 13, or **167 students** who saw <u>one or both</u> movies.

a. Since 119 students saw Outer Stars 1 and 110 students saw Outer Stars 2, the sum of these is 229. But there are only 167 students who saw <u>one or both</u> movies. Therefore, the number of students who saw <u>both</u> movies is 229 − 167, or **62 students.**

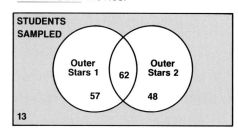

b. Since 62 students saw <u>both</u> movies, 119 − 62, or **57 students** saw Outer Stars 1 <u>only</u>.

c. Since 62 students saw <u>both</u> movies, 110 − 62, or **48 students** saw Outer Stars 2 only.

Check: Does 13 + 57 + 62 + 48 = 180?   Yes ✔

EXERCISES

For Exercises 1–8, refer to the information below and the Venn diagram at the right.

There are 153 students taking algebra.
There are 135 students taking biology.
There are 94 students taking Spanish.
There are 55 students taking biology <u>and</u> Spanish.
There are 15 students taking algebra, biology, <u>and</u> Spanish.

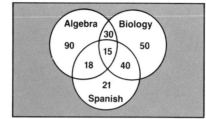

1. How many students are taking algebra <u>and</u> biology, but <u>not</u> Spanish? 30

2. How many students are taking algebra <u>and</u> Spanish, but <u>not</u> biology? 18

3. How many students are taking biology, but <u>not</u> algebra or Spanish? 50

4. How many students are taking algebra, but <u>not</u> biology or Spanish? 90

5. How many students are taking Spanish, but <u>not</u> algebra or biology? 21

6. How many students are taking <u>exactly</u> one of these three courses? 161

7. How many students are taking <u>at least</u> one of these three courses? 264

8. How many students are taking <u>exactly</u> two of these three courses? 88

For Exercises 9–10, use a Venn diagram to solve each problem.

9. In a class of 31 students, there are 17 who play softball and 22 who play tennis. There are 4 students who do not play <u>either</u> game. How many students play <u>both</u> softball and tennis? 12 students

10. At a party, 27 persons played the game "Mystery Clues" and 22 played the game "Skill Shot." Fourteen persons played <u>both</u> games. How many persons played <u>one</u> game, but <u>not both</u>?
21 persons

IMPORTANT TERMS	Tree diagram *(p. 512)*	Complementary events *(p. 517)*
	Fundamental Counting Principle *(p. 512)*	Independent events *(p. 520)*
		Mutually exclusive events *(p. 521)*
	Experiment *(p. 515)*	Dependent events *(p. 525)*
	Outcome *(p. 515)*	Conditional probability *(p. 526)*
	Sample space *(p. 515)*	Experimental probability *(p. 529)*
	Event *(p. 515)*	Expected outcomes *(p. 529)*
	Probability of an event *(p. 515)*	

IMPORTANT IDEAS

1. *Fundamental Counting Principle:* If there are p choices for the first decision and q choices for the second decision, then there are $p \times q$ ways that the two decisions can be made together.

2. The probability of an event that is certain to happen is 1.

3. The probability of an event that cannot happen is 0.

4. For two independent or two dependent events, A and B, $P(A \text{ and } B) = P(A) \cdot P(B)$.

5. For two mutually exclusive events A and B, $P(A \text{ or } B) = P(A) + P(B)$.

6. If an experiment is carried out n times, $E(A) = P(A) \cdot n$.

Chapter Test
Two Chapter Tests (Form A and Form B) are provided on pages 301–304 in the *Teacher's ResourceBank*.

CHAPTER REVIEW

SECTION 19.1

Use the Fundamental Counting Principle to solve each problem.

1. A student will choose three elective courses. There are three choices for the first elective, five choices for the second, and three choices for the third. In how many ways can the electives be chosen? 45

2. The Freshness Bakery bakes three types of bread. Loaves of bread are made in four sizes, and sliced two different ways. How many choices are there when buying a loaf of bread at the bakery? 8

3. A dinner menu has a choice of three appetizers, seven entrees, four beverages, and three desserts. In how many ways can a dinner be ordered, choosing one item from each category? 252

4. When using the Automatic Teller Machine at the American Trust Bank, a customer must enter a four-digit security code. If the digits can be repeated, how many different security codes are possible?
10,000

The arrow of the spinner is equally likely to stop on any of the ten regions. The arrow is spun once. Find each probability.

5. Stopping on a multiple of 3 $\frac{3}{10}$

6. Stopping on a number less than 7 $\frac{3}{5}$

7. Stopping on a number greater than 8 $\frac{1}{5}$

8. Not stopping on a prime number $\frac{3}{5}$

9. Not stopping on a whole number 0

10. Stopping on a counting number 1

SECTION 19.3

A golfer has 10 white, 6 yellow, and 4 orange golf balls. She selects one ball without looking. Find each probability.

11. Selecting a white <u>or</u> orange ball $\frac{7}{10}$

12. Selecting a yellow <u>or</u> orange ball $\frac{1}{2}$

The golfer has 8 red and 12 white tees. She selects one ball and one tee without looking. Find each probability.

13. Selecting a white ball <u>and</u> a white tee $\frac{3}{10}$

14. Selecting a yellow ball <u>and</u> a white tee $\frac{9}{50}$

15. Selecting a yellow ball <u>and</u> a red tee $\frac{3}{25}$

16. Selecting an orange ball <u>and</u> a red tee $\frac{2}{25}$

SECTION 19.4

Jim has 7 nickels, 5 dimes, and 3 quarters in his pocket. He selects two coins, one at a time. Find each probability.

17. Selecting a quarter first <u>and</u> then a dime $\frac{1}{14}$

18. Selecting a nickel first <u>and</u> then a dime $\frac{1}{6}$

19. Selecting two nickels $\frac{1}{5}$

20. Selecting two quarters $\frac{1}{35}$

SECTION 19.5

Two new products are being manufactured. The first 200 samples of each product were tested for defects. Thirty-six of the Product A samples were defective and twenty-eight of the Product B samples were defective.

21. If another Product A sample is tested, what is the probability that it is defective? $\frac{9}{50}$

22. If another Product B sample is tested, what is the probability that it is defective? $\frac{7}{50}$

23. In a shipment of 1500, how many of Product A can be expected to be defective? 270

24. In a shipment of 1200, how many of Product B can be expected to be defective? 168

*Probability / **535***

CUMULATIVE REVIEW: CHAPTERS 1–19

Solve. Check each answer. (Section 4.3)

1. $15 - 4a = 7a - 7$ $a = 2$

2. $4x - 5 = 2(x - 1) + 6$ $x = \frac{9}{2}$

Multiply. (Sections 8.2 and 8.5)

3. $(5e^6)(3e^3)$ $15e^9$

4. $(-6x^2y^4)(5x^7y)$ $-30x^9y^5$

5. $(4x - 5)(x - 11)$ $4x^2 - 49x + 55$

Perform the indicated operations. Simplify where necessary.
(Sections 10.3, 10.5, 10.7, 11.2, 11.4, and 11.6)

6. $\dfrac{3ac^2}{2a^2b^3} \cdot \dfrac{4ab}{9c}$ $\dfrac{2c}{3b^2}$

7. $\dfrac{2x + 1}{x - 3} \cdot \dfrac{x^2 - 9}{4x + 2}$ $\dfrac{x + 3}{2}$

8. $\dfrac{x^2 + 5x + 4}{x^2 - 16} \div \dfrac{x^2 - 1}{3x - 12}$ $\dfrac{3}{x - 1}$

9. $\dfrac{9x}{x + 6} - \dfrac{7x - 12}{x + 6}$ 2

10. $\dfrac{2}{3x} - \dfrac{3}{4x}$ $-\dfrac{1}{12x}$

11. $\dfrac{3}{2x - 6} + \dfrac{5}{x^2 - 9}$

Solve each system of equations. (Sections 14.4, 15.1, and 15.5)

12. $\begin{cases} 2x - 3y = -14 \\ 4x + 2y = -12 \end{cases}$ $\{(-4, 2)\}$

13. $\begin{cases} 5x + y = 7 \\ 8x + 2y = 10 \end{cases}$ $\{(2, -3)\}$

14. $\begin{cases} 5x + 7y = -23 \\ 8x - y = 12 \end{cases}$

Write the slope-intercept form of each inequality. (Section 16.1)

15. $3x + y - 1 > 0$ $y > -3x + 1$

16. $5x - y < -3$ $y > 5x + 3$

17. $-7x + 2y \leq 8$ $y \leq \frac{7}{2}x + 4$

Graph each linear system. (Sections 16.4 and 16.5)

18. $\begin{cases} y < -x + 1 \\ y > 2x - 3 \end{cases}$

19. $\begin{cases} x \leq 3 \\ x - y \leq 3 \end{cases}$

20. $\begin{cases} x + y < 3 \\ 2x - y \geq -1 \end{cases}$

Write three ordered pairs of each function. Let the x values equal -2, 0, and 3.
(Section 17.1)

21. $y = -x^2$ $(-2, -4), (0, 0), (3, -9)$

22. $y = x^2 - 2$ $(-2, 2), (0, -2), (3, 7)$

23. $y = x^2 - 2x + 1$ $(-2, 9), (0, 1), (3, 4)$

Graph each function. (Section 17.2)

24. $y = x^2 + 3$ Turns at $(0, 3)$ low point

25. $y = 2x^2 - x$ Turns at $(\frac{1}{4}, -\frac{1}{8})$ low point

26. $y = -x^2 + 4x - 4$ Turns at $(2, 0)$ high point

Graph the two functions in the same coordinate plane.
(Section 17.3)

27. $y = x^2$; $y = 2x^2$ Turns at $(0, 0)$ low point

28. $y = -x^2$; $y = -\frac{1}{2}x^2$ Turns at $(0, 0)$ high point

Write the motion(s) needed to make the graph of the first function coincide with the graph of the second function. (Sections 17.4 and 17.5)

29. $y = x^2$; $y = (x - 5)^2$
5 units right

30. $y = \frac{1}{2}x^2$; $y = \frac{1}{2}(x - 3)^2 + 4$
3 units right;
4 units up

Write each polynomial in the form $(x - h)^2 + k$. (Section 17.6)

31. $x^2 + 4x + 1$
$(x - (-2))^2 - 3$

32. $x^2 - 6x - 3$
$(x - 3)^2 - 12$

33. $x^2 + 3x - 2$
$(x - (-\frac{3}{2}))^2 - \frac{17}{4}$

Write an equation for the axis of each graph. (Section 17.7)

34. $y = 2(x - 3)^2 + 4$
$x = 3$

35. $y = \frac{1}{4}(x - 1)^2 - 3$
$x = 1$

36. $y = x^2 - 10x + 27$
$x = 5$

Solve. (Sections 18.2 through 18.4)

37. $x^2 - 7x + 10 = 0$ $\{2, 5\}$

38. $x^2 = 81$ $\{-9, 9\}$

39. $x^2 + 8x - 1 = 0$
$\{-4 + \sqrt{17}, -4 - \sqrt{17}\}$

Solve by the quadratic formula. (Section 18.5)

40. $x^2 + 7x + 2 = 0$
$\{\frac{-7 + \sqrt{41}}{2}, \frac{-7 - \sqrt{41}}{2}\}$

41. $2x^2 = 5x + 3$ $\{3, -\frac{1}{2}\}$

42. $x^2 - 5x = 0$
$\{0, 5\}$

Use Condition 1 to represent the unknowns. Use Condition 2 to write an equation for the problem. Solve. (Section 18.7)

43. The product of two consecutive integers is 56. Find two pairs of such integers.
$n, n + 1$;
$n(n + 1) = 56$; Ans: 7, 8 or −7, −8

44. The length of a rectangle is 3 feet more than the width. If the area is 40 square feet, what are the length and width?
$w, w + 3$; $w(w + 3) = 40$;
Ans: 8 and 5

A blue number cube and a red number cube are tossed. Find each probability. (Sections 19.2 and 19.3)

45. Tossing a 4 on the blue cube $\frac{1}{6}$

46. Tossing an odd number on the red cube $\frac{1}{2}$

47. Tossing a 5 on the blue cube and a 5 on the red cube $\frac{1}{36}$

48. Tossing a 4 on the blue cube or a number less than 4 on the red cube
$\frac{2}{3}$

Solve. (Section 19.5)

49. The results of a survey of 350 people at an airport showed that 84 people had eaten eggs for breakfast that morning. Find the probability that a person at the airport ate eggs for breakfast that morning. $\frac{6}{25}$

50. Professor Plank's Hardware store examined a shipment of 550 nails and found 11 to be defective. In a shipment of 1800 nails, how many can be expected to be defective?
36

*Cumulative Review   /   **537***

Using Manipulatives

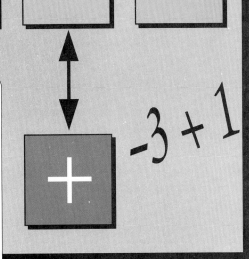

$-3 + 1 = -2$

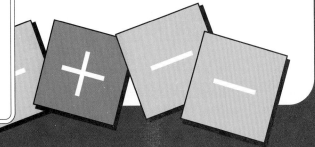

Manipulative 1 *(Use with Lesson 1.2.)*

You can use ⊞ and ⊟ tiles to model addition of integers. The value of a square is determined by the number and type of tiles in the square.

Value: +2, or 2

Value: −3

Value: 0

A ⊞ -tile and a ⊟ -tile have opposite signs. A pair of tiles having opposite signs forms a **neutral pair**. A neutral pair has a value of 0.

Value: −2

You can use algebra tiles to find a sum.

EXAMPLES   Use algebra tiles to find each sum.

a. −2 + (−4) = ___?___

Solution: **A.** Put in two ⊟ -tiles.

B. Add four ⊟ -tiles.

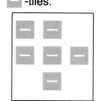

C. Count the tiles. Record the sum.

−2 + (−4) = **−6**

b. −5 + 2 = ___?___

Solution: **A.** Put in five ⊟ -tiles.

B. Add two ⊞ -tiles.

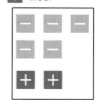

C. Remove neutral pairs. Record the sum.

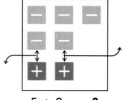

−5 + 2 = **−3**

540   /   *Using Manipulatives*

c. $6 + (-3) = \underline{\ ?\ }$

A. Put in six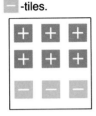
   +-tiles.

B. Add three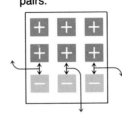
   -tiles.

C. Remove neutral
   pairs.

D. Record the sum.

$6 + (-3) = \mathbf{3}$

■■■ **EXERCISES** ■■■

Goal: To add positive and negative numbers

Give the value of each square.

1.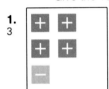
3

2.
0

3.
−6

Use algebra tiles to model each sum.

4. −3 + 6 3
5. −5 + (−4) −9
6. 4 + (−7) −3
7. −1 + 1 0
8. 7 + (−5) 2
9. −2 + (−2) −4
10. 3 + (−2) 1
11. 0 + (−3) −3
12. Use algebra tiles to show that 6 + (−2) = (−2) + 6. Check students' models.

Use algebra tiles to solve.

13. 8 + (−9) + 1 0
14. −3 + 2 + 5 4
15. 6 + (−2) + (−7) −3
16. −2 + (−2) + (−2) −6
17. −4 + 9 + (−3) 2
18. 5 + (−3) + (−7) −5

19. Tom gained 4 pounds in the month of March. He lost 6 pounds in April. What is his overall loss or gain in weight? −2 lb

20. At 4 A.M., the temperature was −3°C. At 10 A.M., the temperature had risen 7°. What was the temperature then? 4°C

Write true (T) or false (F) for each statement. When a statement is false, explain why.

21. The sum of two negative numbers is always a positive number.

22. The sum of a positive and a negative number is always a negative number.

Using Manipulatives / **541**

Manipulative 2 *(Use with Lesson 1.3.)*

You can use algebra tiles to show that

$4 - 3$ means the same as $4 + (-3)$.

EXAMPLE 1   Use algebra tiles to show that $4 - 3 = 4 + (-3)$.

a. $4 - 3 = \underline{\ ?\ }$

Solution:

A. Put in four ($+$)-tiles.

B. Remove three ($+$)-tiles.

C. Record the result.

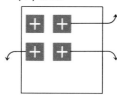

$4 - 3 = \mathbf{1}$

b. $4 + (-3) = \underline{\ ?\ }$

Solution:

A. Put in four ($+$)-tiles.

B. Add three ($-$)-tiles.

C. Remove neutral pairs. Record the result.

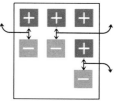

$4 + (-3) = \mathbf{1}$

Example 1 suggests the following.

To subtract an integer, add its opposite.

EXAMPLE 2   Use algebra tiles to subtract: $-1 - (-3)$

Solution: Rewrite $-1 - (-3)$ as $-1 + 3$.

A. Put in one ($-$)-tile.

B. Add three ($+$)-tiles.

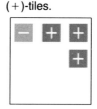

C. Remove neutral pairs. Record the result.

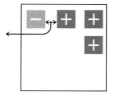

$-1 - (-3) = \mathbf{2}$

542   /   *Using Manipulatives*

EXERCISES

Goal: To use algebra tiles to subtract integers

Write an addition problem that will give the same answer as the subtraction problem.

1. $9 - 5$

2. $-9 - 5$

3. $-9 - (-5)$

4. $9 - (-5)$

5. $-13 - (-7)$

6. $13 - (-1)$

7. $1 - 8$

8. $1 - (-10)$

Use algebra tiles to subtract.

9. $5 - 9$ -4

10. $-4 - 6$ -10

11. $-3 - 6$ -9

12. $4 - (-5)$

13. $-6 - (-7)$ 1

14. $-8 - 7$ -15

15. $6 - (-9)$ 3

16. $-5 - (-8)$

17. $7 - (-3)$ 10

18. $1 - (-1)$ 2

19. $3 - 5$ -2

20. $-3 - 5$

21. Use algebra tiles to determine which is greater, $1 - (-8)$ or $8 - 1$.

Use algebra tiles to illustrate your answer to each question.

22. You start with 6 and subtract a negative integer having an absolute value less than 6. Will the result be a positive or a negative integer?

23. You start with 6 and subtract a positive integer greater than 6. Will the result be a positive or a negative integer?

24. You start with -6 and subtract a positive integer with a value less than $|-6|$. Will the result be a positive or a negative integer?

25. You start with -6 and subtract a positive integer with a value greater than $|-6|$. Will the result be a positive or a negative integer?

26. You start with -6 and subtract a negative integer having an absolute value greater than $|-6|$. Will the result be a positive or a negative integer?

27. You start with -6 and subtract a negative integer having an absolute value less than $|-6|$. Will the result be a positive or a negative integer?

Additional Answers
1. $9 + (-5)$
2. $-9 + (-5)$
3. $-9 + 5$
4. $9 + 5$
5. $-13 + 7$
6. $13 + 1$
7. $1 + (-8)$
8. $1 + 10$
12. 9
16. 3
20. -8
21. $1 - (-8) = 1 + 8 = 9$
and $8 - 1 = 7$.
So $1 - (-8) > 8 - 1$.
22. Positive
Example: $6 - (-4) = 6 + 4 = 10$
23. Negative
Example: $6 - 9 = -3$
24. Negative
Example: $-6 - 5 = -11$
25. Negative
Example: $-6 - 8 = -14$
26. Positive
Example:
$-6 - (-8) = -6 + 8$
$= 2$
27. Negative
Example:
$-6 - (-4) = -6 + 4$
$= -2$

Teaching Suggestions
After introducing the x-title and
the x^2-tile to students, point out
that the dimensions of a (1)-tile
are 1 by 1, the dimensions of an
x-tile are 1 by x, and the dimensions of an x^2-tile are x by x.

Have them use algebra tiles to
give examples of neutral pairs
and of tiles that represent like
terms. Then have them decide
whether each of these statements is true or false and ask a
student volunteer to use algebra
tiles to explain each answer.

1. Algebra tiles that represent
   neutral pairs will also represent like terms.   (T)

2. Algebra tiles that represent
   like terms will also represent
   neutral pairs.   (F)

As student work with the Example and Exercises, emphasize
that only like terms can be
added and subtracted and that
combining like terms means to
add and subtract like terms.

Manipulative 3 *(Use with Lesson 2.6.)*

You can use algebra tiles like these to model algebraic expressions.

1-tile   **x-tile**   **x²-tile**

For each positive tile, there is a corresponding negative tile.

(−1)-tile   **(−x)-tile**   **(−x²)-tile**

Just as with the ➕↔◻ pairs you used to model a neutral pair of integers, neutral
pairs can also be formed as shown below. The value of each neutral pair is 0.

Neutral Pairs

If two terms are **like terms**, the algebra tiles that model them will have the same
size, but may have different signs. The pairs of tiles below show like terms.

EXAMPLE   Simplify $2x^2 - x^2 + x - 1$.

Solution: Think of $2x^2 - x^2 + x - 1$ as $2x^2 + (-x^2) + x + (-1)$.

A. Model the expression.

$$2x^2 + (-x^2) + x + (-1)$$

B. Group like terms. Remove neutral pairs and record the result.

$$[2x^2 + (-x^2)] + x - 1$$
$$x^2 + x - 1$$

544   /   *Using Manipulatives*

EXERCISES

Goal: To combine like terms

Which pairs of tiles show like terms? Answers Yes or No.

1. No

2.  No

3.  Yes

4. No

5.  No

6. No

Use algebra tiles to simplify each expression.

7. $-x^2 - 5x^2$ $-6x^2$ **8.** $-3 + 2x - 2x$ -3 **9.** $3x^2 - 5x^2$

10. $x^2 + x + 1 - x$ $x^2 + 1$ **11.** $-5 + x^2 - 6x^2$ $-5 - 5x^2$ **12.** $-x - 2 - 8x$

13. $x^2 - 7x - 5x - 3$ **14.** $3 + 4x - x + 6$ **15.** $x + 3 - x^2 + 2x^2$

16. $-x^2 + 3x - 7x + 4$ **17.** $5x - x^2 + 2x$ **18.** $8 - 3x^2 + 7x + 2x^2$

Use algebra tiles to determine whether each statement is true (T) or false (F).

19. $-(-x) + x = 0$ F   **20.** $x^2 - (-x^2) = 2x^2$ T   **21.** $x^2 - x \neq x - x^2$ T

22. Use algebra tiles to show an algebraic expression of three terms that, when simplified, can be represented as $-x$.

23. Use algebra tiles to show an algebraic expression of four terms that, when simplified, can be represented as $x^2 - 1$.

Using Manipulatives / **545**

545

Critical Thinking
See Exercises 19–23.

Additional Answers
  9. $-2x^2$
12. $-9x - 2$
13. $x^2 - 12x - 3$
14. $3x + 9$
15. $x^2 + x + 3$
16. $-x^2 - 4x + 4$
17. $-x^2 + 7x$
18. $-x^2 + 7x + 8$
21. Point out that $x^2 - x = x - x^2$ only when $x = 0$ or $x = 1$.
22. Answers will vary. Sample answer: $-x - x^2 + x^2$
23. Answers will vary. Sample answer: $3x^2 + 6 - 7 - 2x^2$

Manipulative 4 *(Use with Lesson 3.1.)*

You can use algebra tiles to solve equations involving addition and subtraction.

EXAMPLE 1   Use algebra tiles to solve $x - 5 = -1$.

Solution: **A.** Model the equation. **Think:** $x - 5 = x + (-5)$

$$x + (-5) = -1$$

B. Think: There are five (-1)-tiles on the left side of the equation. To get x alone, add five (1)-tiles to each side.

$$x + (-5) + 5 = -1 + 5$$

C. Identify and remove any neutral pairs from each side.

$$x + 0 = 4$$

D. Show the result.

$$x = 4$$

EXAMPLE 2   Use algebra tiles to solve $x + 2 = 5$.

Solution: **A.** Model the equation.

$$x + 2 = 5$$

B. To get x alone, subtract two (1)-tiles from each side.

$$x + 2 - 2 = 5 - 2$$

C. Show the result.

$x = 3$

EXERCISES

Goal: To use algebra tiles to solve equations of the form $x \pm a = b$

1. Use algebra tiles to show that 9 is the solution of $x - 8 = 1$. Check students' models.

Use algebra tiles to solve each equation.

2. $x - 3 = -4$ $x = -1$ **3.** $x + 3 = 7$ $x = 4$ **4.** $x - 5 = 8$

5. $x + 4 = -6$ $x = -10$ **6.** $x - 7 = 1$ $x = 8$ **7.** $x + 4 = 7$

8. $x - (-3) = 7$ $x = 4$ **9.** $-2 = x - 4$ $x = 2$ **10.** $2 = x - 7$

Use algebra tiles to model an equation having the given solution.
Write the equation you have modeled.

11.   **12.**   **13.**

Use algebra tiles to model two equations having the given solution.
Write the equations you have modeled.

14. $x = -1$ **15.** $x = 5$ **16.** $x = 0$

17. Write a paragraph that summarizes how you solve equations of the form $x \pm a = b$. Answers will vary.

Using Manipulatives / **547**

Critical Thinking
See Exercises 14–16.

Additional Answers
 4. $x = 13$
 7. $x = 3$
10. $x = 9$
11. Sample answer: $x - 2 = 1$
12. Sample answer: $x + 5 = 3$
13. Sample answer: $x + 8 = 2$
14. Sample answer: $x + 1 = 0$;
     $x + 2 = 1$
15. Sample answer: $x + 2 = 7$;
     $x - 1 = 4$
16. Sample answer: $x + 9 = 9$;
     $x - 9 = -9$

Lesson Resources
Teaching Aid 1
Teaching Aid 3
Transparencies 50, 51, 53, 54

Teaching Suggestions
Have students leave their textbooks closed. Ask them to use algebra tiles to model $3x = -12$. Then ask for suggestions for finding the value of one x-tile. Guide the discussion to the procedure of dividing both sides of the equation into two equal groups. Then ask students to write an equation to represent the procedure. $\left(\dfrac{3x}{3} = \dfrac{-12}{3}\right)$

Have them complete the solution and verify that the solution is correct by using algebra tiles.

Then focus the students' attention on Example 2. Remind them that solving an equation means to find the value of x that makes the equation true. So in Example 2, the goal is to find the value of one positive x-tile. Have students work in pairs to complete Example 2 and the Exercises.

Manipulative 5 *(Use with Lesson 3.2.)*

You can also use algebra tiles to solve equations involving multiplication.

EXAMPLE 1   Use algebra tiles to solve: $3x = -12$

Solution: **A.** Model the equation.

$$3x = -12$$

B. Think: There are three x-tiles on the left of the equation. To find the value of one x-tile, form 3 equal groups on each side.

$$\frac{3x}{3} = \frac{-12}{3}$$

C. Show the value of one x-tile.

$$x = -4$$

Example 2 shows that you may sometimes need to add neutral pairs to solve for x.

EXAMPLE 2   Use algebra tiles to solve: $6 = -x$

Solution: **A.** Model the equation.

$$6 = -x$$

B. Think: There is a negative x-tile on the right side of the equation. Add a positive x-tile to each side.

$$x + 6 = x + (-x)$$

C. Identify and remove neutral pairs.

$$x + 6 = 0$$

D. To get x-alone, add six (-1)-tiles to each side.
Remove neutral pairs.

$$x + 6 + (-6) = 0 + (-6)$$
$$x + 0 = -6$$

E. Show the value of x.

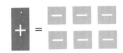

$$x = -6$$

EXERCISES

Goal: To use algebra tiles to solve equations of the form $ax = b$

Use algebra tiles to solve each equation.

1. $2x = 8$ $x = 4$ **2.** $4x = -8$ $x = -2$ **3.** $3x = 12$ $x = 4$ **4.** $4x = 12$

5. $4 = -x$ $x = -4$ **6.** $-x = 1$ $x = -1$ **7.** $-3 = -x$ $x = 3$ **8.** $-x = -1$

9. $4x = -4$ $x = -1$ **10.** $5x = -15$ $x = -3$ **11.** $-2x = -6$ $x = 3$ **12.** $-2x = 6$

13. $9 = -3x$ $x = -3$ **14.** $-6 = -6x$ $x = 1$ **15.** $-5x = -10$ $x = 2$ **16.** $-2x = 4$

Solve. Use algebra tiles if you need help.

17. If $x = -5$, what is the value of $3x$? -15

18. If $-x = 1$, what is the value of $-5x$? 5

19. If $-x = -3$, what is the value of $-4x$? -12

20. If $-1 = -x$, what is the value of $6x$? 6

21. If $-x = 2$, what is the value of $8x$? -16

22. List five equations of the form $ax = b$ that have a solution of $x = -2$.

Teaching Suggestions
Emphasize that in solving equations involving more than one operation, students should apply the Addition or Subtraction Properties for Equations before applying the Multiplication or Division Properties. Ask them how this procedure differs from the rules for order of operations used in simplifying arithmetic expressions. (When following the rules for order of operations, multiplication and division are performed before addition and subtraction.) Have students work in pairs, with one student manipulating the tiles and one student writing the equation that corresponds to the related step in the modeling procedure. Students should exchange roles for every other problem.

Manipulative 6 *(Use with Lesson 3.3.)*

The first step in solving equations that involve more than one operation is to get the term with the variable alone on one side of the equation. That is, you apply the Addition or Subtraction Property of Equations *before* applying the Multiplication or Division Property.

EXAMPLE

Use algebra tiles to solve: $2x - 3 = 7$

Solution: Think of $2x - 3$ as $2x + (-3)$.

A. Model the equation.

$$2x + (-3) = 7$$

B. **Think:** There are three (-1)-tiles on the left side of the equation. To get the two x-tiles alone, add three (1)-tiles to each side.

$$2x + (-3) + 3 = 7 + 3$$

C. Identify and remove any neutral pairs.

$$2x = 10$$

D. Since there are two x-tiles on the left side of the equation, form two equal groups on each side.

$$\frac{2x}{2} = \frac{10}{2}$$

E. Pair one x-tile with one group of (1)-tiles.

$$x = 5$$

550 / *Using Manipulatives*

550

EXERCISES

Goal: To use algebra tiles to solve equations involving two or more operations

Write the equation shown by each model.

1.

$3x + 1 = -4$

2.

$2x - 4 = 4$

Use algebra tiles to solve each equation.

3. $2x + 1 = 5$ $x = 2$ **4.** $3x - 2 = 7$ $x = 3$ **5.** $2x + 4 = -6$

6. $2x - 3 = 7$ $x = 5$ **7.** $9 = 2x + 5$ $x = 2$ **8.** $3 + 2x = -9$

9. $7 = 2x + 3$ $x = 2$ **10.** $8 = 3x - 1$ $x = 3$ **11.** $-10 = 3x + 5$

Write an equation having two or more operations and the solution shown by the model.

12.

13.

14.

Use algebra tiles to solve each problem.

15. When 3 is subtracted from four times a number, x, the result is 1. What is the number?

16. Four more than three times a number, x, is 16. What is the number?

Answers will vary.

17. Write a word problem that can be solved by using the equation $3x + 5 = 14$.

18. Write a word problem that can be solved by using the equation $2x - 7 = 11$.

Answers will vary.

Additional Answers

5. $x = -5$

8. $x = -6$

11. $x = -5$

12. Sample answer:
$2x + 5 = 3$

13. Sample answer:
$3x - 4 = 8$

14. Sample answer:
$5x + 2 = -13$

15. $4x - 3 = 1; x = 1$

16. $3x + 4 = 16; x = 4$

Manipulative 7 *(Use with Lesson 3.5.)*

To solve some equations, you may have to combine like terms first.

EXAMPLE   Use algebra tiles to solve: $3x + 5 + x = -3$

Solution:

A. Model the equation.

$$3x + 5 + x = -3$$

B. Group the like terms.

$$4x + 5 = -3$$

C. To get the four x-tiles alone, add five (-1)-tiles to each side of the equation.

$$4x + 5 + (-5) = -3 + (-5)$$

D. Identify and remove neutral pairs.

$$4x = -8$$

E. Since there are four x-tiles on the left side of the equation, form four equal groups on each side.

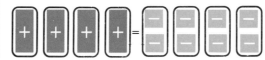

$$\frac{4x}{4} = \frac{-8}{4}$$

F. Write the value of x.

$$x = -2$$

EXERCISES

Goal: To use algebra tiles to solve equations having like terms.

Check students' models.

1. Use algebra tiles to show that $x = -2$ is the solution of $3x + 5 + x = -3$.

2. Use algebra tiles or drawings to illustrate two pairs of tiles that represent like terms and two pairs of tiles that represent unlike terms. Check students' work.

Use algebra tiles to solve each equation.

3. $2x + 3 + x = 9$ $x = 2$

4. $5x - 3 - x = 9$ $x = 3$

5. $3x + 6 - x = -8$ $x = -7$

6. $2x + 1 - 9 = 6$ $x = 7$

7. $2x + 3x - 7 = 3$ $x = 2$

8. $7 - 3x + x = 5$ $x = 1$

9. $8 = 3x + 2 - x$ $x = 3$

10. $5 = 2x + 1 - 4$ $x = 4$

11. Use algebra tiles to show that the solution of $3(x - 1) = 6$ is $x = 3$. Check students' models.

12. Use algebra tiles to solve and check: $2(x - 3) = 4$. $x = 5$

For Exercises 13–16, write an equation that satisfies the given condition. Exchange equations with a partner and solve the equations by using algebra tiles. Check each other's solutions.

13. Write an equation involving like terms and having a solution of 0.

14. Write an equation involving like terms and having a solution of -7.

15. Write an equation having a solution of 5 and all the like terms on the right side of the equation.

16. Write an equation having a solution of -1 and having like terms of x and $(-x)$.

Additional Answers

11. Solutions may vary. One way to solve the equation is to use algebra tiles to show three groups of $(x - 1)$ on the left side of the equation and six ones on the right side. Then have the students separate each side into two equal groups, model the equation showing the value of one group, and solve for x. This gives the result $(x - 1) = 2$ and $x = 3$.

13. Sample answer:
$2x + 1 = 1$

14. Sample answer:
$x + 3 + 2x = -18$

15. Sample answer:
$14 = 6x - 1 - 3x$

16. Sample answer:
$x - (-x) + 3 = 1$

Teaching Suggestions
Remind students that like terms cannot be added or subtracted until they are on the same side of the equation.

When working the exercises, students may not all follow the same order of steps in solving the equations. Have volunteers demonstrate alternate methods of solution and have the class discuss the advantages and disadvantages, if any, of each method.

Manipulative 8 *(Use with Lesson 4.3.)*

You can use algebra tiles to solve equations having the variable on both sides. The first step in solving these equations is to eliminate the variable from one side.

EXAMPLE

Use algebra tiles to find the solution of $3x + 1 = x - 5$.

Solution: **A.** Model the equation.

$$3x + 1 = x - 5$$

B. To eliminate the x-tile from the right side of the equation, add one $(-x)$-tile to each side.

$$3x + (-x) + 1 = x + (-x) - 5$$

C. Identify and remove the neutral pairs.

$$2x + 1 = -5$$

D. To eliminate the (1)-tile from the left side of the equation, add a (-1)-tile to each side.

$$2x + 1 + (-1) = -5 + (-1)$$

E. Identify and remove any neutral pairs.

$$2x = -6$$

F. Since there are two x-tiles, form 2 equal groups on each side.

$$\frac{2x}{2} = \frac{-6}{2}$$

554 / *Using Manipulatives*

G. Write the value of *x*.

$x = -3$

EXERCISES

Goal: To use algebra tiles to solve equations that have the variable on both sides

Find the value of one x-tile.

1.
     $x = 1$

2.

$x = -4$

3. Use algebra tiles to show that $x = -3$ is the solution of $3x + 1 = x - 5$.
Check students' models.

Use algebra tiles to solve each equation.

4. $2x + 1 = x + 5$ **5.** $x + 8 = -3x + 4$ **6.** $3x - 4 = x + 2$

7. $4x + 1 = 6 - x$ **8.** $-x - 1 = -2x + 4$ **9.** $-2x + 6 = 3x + 1$

10. $3x + 2 = -2x - 8$ **11.** $3x - 4 = 8 - x$ **12.** $3x - 8 = 2 - 2x$

Which equation has a solution of $x = -3$? Write <u>Yes</u> or <u>No</u>.

13.
   No

14.
  Yes

Which equation has a solution of $x = -1$? Write <u>Yes</u> or <u>No</u>.

15.
 No

16.
 Yes

17. Write a letter to a friend explaining how to solve the equation $3 - 2x = 17 - x$ by using algebra tiles. Explain also how you would check to be sure that your solution is correct. $x = -14$

Critical Thinking
See Exercise 17.

Additional Answers
  4. $x = 4$
  5. $x = -1$
  6. $x = 3$
  7. $x = 1$
  8. $x = 5$
  9. $x = 1$
  10. $x = -2$
  11. $x = 3$
  12. $x = 2$

Manipulative 9 *(Use with Lesson 8.3.)*

To model polynomials, you will use positive and negative (1)-tiles, positive and negative x-tiles, and positive and negative x^2-tiles.

x^2-tile   **$(-x^2)$-tile**

Recall that algebra tiles used to model like terms will have the same size and shape, but may have different signs.

EXAMPLE 1   Tell whether each pair of tiles represents like terms.
Give a reason for your answer.

Model   **Like terms?**

a.    Yes; the tiles have the same size and shape.

b.    No; the x^2-tile is wider than the x-tile.

c.   No; the x-tile is taller than the (1)-tile.

You can use algebra tiles to simplify polynomials.

EXAMPLE 2   Use algebra tiles to simplify $x^2 - x + 1 + x^2 + 2x - 3$

Solution: **A.** Model the polynomial. Think of $x^2 - x$ as $x^2 + (-x)$ and think of $2x - 3$ as $2x + (-3)$.

 $x^2 + (-x) + 1 + x^2 + 2x + (-3)$

B. Group the like terms.

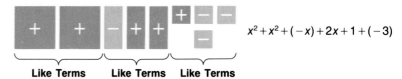 $x^2 + x^2 + (-x) + 2x + 1 + (-3)$

Like Terms   Like Terms   Like Terms

556 / *Using Manipulatives*

C. Identify and remove neutral pairs. Count the remaining tiles and record the results.

$2x^2 + x - 2$

Additional Answers
5. $2x^2 + 3x + 2$
6. $2x - 9$
7. $2x^2 - 7x + 7$
8. $2x^2 - 2x + 6$
9. $-2x - 1$
10. $x^2 - 8x + 13$
11. $-3x^2 + 4x + 6$
12. $2x^2 - x + 9$
13. $-x^2 - 4x + 7$
14. $x^2 - 3x - 2$

For Ex. 15–18, examples may vary.
15. F; Example: one x^2-tile
16. F; Example: one x-tile and one (-1)-tile
17. T; the trinomial is $x^2 + x + 1$.
18. T; the monomial is -3.

EXERCISES

Goal: To use algebra tiles to simplify polynomials

Which pair of algebra tiles shows like terms? Answer <u>Yes</u> or <u>No</u>.

1.
Yes

2.
No

3.
No

4.
Yes

Use algebra tiles to simplify each polynomial.

5. $x^2 - 3x + 4 + x^2 + 6x - 2$

6. $x^2 - 5x - 1 - x^2 + 7x - 8$

7. $x^2 + 2 - 3x + 5 + x^2 - 4x$

8. $-x^2 + 2x + 3x^2 - 4x + 6$

9. $x^2 - 7 + 2x - x^2 - 4x + 6$

10. $3 - x - x^2 + 2x^2 - 7x + 10$

11. $2 - x^2 + 7x - 2x^2 - 3x + 4$

12. $8 - x + x^2 + 3 + x^2 - 2$

13. $7 - 8x + x^2 + 4x - 2x^2$

14. $-2x^2 + 6 - 3x + 3x^2 - 8$

Write true (T) or false (F) for each statement. When a statement is false, give an example to show why it is false.

15. The only way to model a monomial is to use one x-tile.

16. The only way to model a binomial in simplest form is to use one x^2-tile and one x-tile.

17. To model a trinomial in simplest form, you can use one x^2-tile, one x-tile, and one (1)-tile. The tiles can be positive or negative.

18. Three (-1)-tiles can be used to model a monomial.

Teaching Suggestions
Be sure that students understand
that using algebra tiles to model
addition and subtraction of polynomials is similar to using the
tiles to model simplifying polynomials.

Encourage students to use a column format to model addition
and subtraction problems in order to emphasize the correspondence with the written form for
column addition and subtraction.
When subtracting polynomials,
remind students <u>often</u> that the
opposite of <u>every</u> term of the
polynomial being subtracted
must be found.

Manipulative 10 *(Use with Lesson 8.4.)*

You can use algebra tiles to model addition and subtraction of polynomials.

EXAMPLE 1   Use algebra tiles to add: $x^2 + 3x - 4$
$-x^2 - x + 1$

Solution: **A.** Model each polynomial. Align tiles representing like terms.

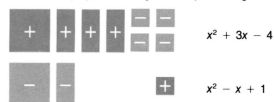

$x^2 + 3x - 4$

$x^2 - x + 1$

B. Remove neutral pairs and count the remaining tiles.

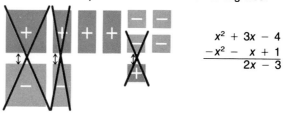

$x^2 + 3x - 4$
$\underline{-x^2 - x + 1}$
$2x - 3$

Recall that subtracting a number is the same as adding its opposite. That is,
$x^2 - (3x^2 - 7x) = x^2 + [-(3x^2 - 7x)] = x^2 - 3x^2 + 7x.$

EXAMPLE 2   Subtract: $x^2 - 3x + 2$
$(-)\ \underline{-x^2 - 2x + 1}$

Solution: **A.** Add the opposite of $(-x^2 - 2x + 1)$ to $x^2 - 3x + 2$.

$x^2 - 3x + 2$ $x^2 - 3x + 2$
$(-)\ \underline{-x^2 - 2x + 1}$ ⟶ $+\ \underline{x^2 + 2x - 1}$

B. Model the polynomials. Align like terms.

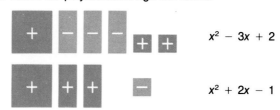

$x^2 - 3x + 2$

$x^2 + 2x - 1$

558 / *Using Manipulatives*

558

C. Remove neutral pairs and count the remaining like tiles.

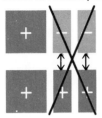

$$x^2 - 3x + 2$$
$$\underline{+\ x^2 + 2x - 1}$$
$$2x^2 - \quad x + 1$$

EXERCISES

Goal: To use algebra tiles to add and subtract polynomials

Use algebra tiles to add.

1. $x^2 - 2x + 1$
$\underline{x^2 + 4x - 3}$

2. $2x^2 + 7x - 4$
$\underline{x^2 - 5x + 8}$

3. $-x^2 + 2x - 6$
$\underline{2x^2 + 3x + 8}$

Use algebra tiles to find the opposite of each polynomial.

4. $x - 7 \quad -x + 7$

5. $-x^2 + x \quad x^2 - x$

6. $2x^2 + 6x \quad -2x^2 - 6x$

7. $2x^2 - 3x - 4$

8. $5x - x^2 + 9$

9. $5x^2 - 9x - 3$

Use algebra tiles to subtract.

10. $\quad x^2 - 5x + 6$
$(-)\ \underline{x^2 + 2x + 4}$

11. $\quad 3x^2 + 6x - 9$
$(-)\ \underline{2x^2 - \quad x + 4}$

12. $\quad x^2 + 2x - 6$
$(-)\ \underline{2x^2 - 4x + 5}$

13. $\quad -2x^2 + 5x + 4$
$(-)\ \underline{-x^2 + 7x - 5}$

14. $\quad -x^2 + 5x - 6$
$(-)\ \underline{-x^2 + 4x - 8}$

15. $\quad -4x^2 + \quad x - 1$
$(-)\ \underline{x^2 - 3x + 8}$

16. The lengths of the sides of a polygon can be represented as $5x$, $3x - 1$, and $2x + 5$.

   a. Use algebra tiles to represent the perimeter of the polygon in simplest form. $10x + 4$

   b. If $x = 19$ inches, what is the perimeter of the polygon in inches? 194 inches

Using Manipulatives / **559**

Teaching Suggestions
Emphasize that, in building the product rectangle in Example 1, students are building a rectangle with dimensions $(x + 1)$ and $(x + 2)$. Point out that the product of the factors represents an area, and that this area is the area of the rectangle they are building. To find the area of the rectangle, they find the sum of the areas of the tiles that make up the rectangle.

To make this even more concrete for students, have them substitute values for x, such as $x = 2$ and $x = 3$, in the expressions $(x + 1)$, $(x + 2)$, and $(x^2 + 3x + 2)$. By evaluating and comparing the product $(x + 1)(x + 2)$ with $x^2 + 3x + 2$, students can see that the area represented by the trinomial product is the area of a rectangle with the given length and width.

Manipulative 11 *(Use with Lesson 8.5.)*

You can use algebra tiles to find the product of binomials.

EXAMPLE 1   Use algebra tiles to multiply: $(x + 1)(x + 2)$

Solution:

A. Model the factors $(x + 1)$ and $x + 2$.

B. Build a rectangle having dimensions $(x + 1)$ and $(x + 2)$.

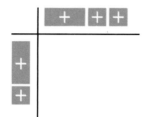
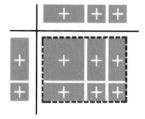

C. Count the tiles in the completed rectangle.

$$(x + 1)(x + 2) = x^2 + 2x + x + 2$$
$$= x^2 + 3x + 2$$

The product $x^2 + 3x + 2$ represents the area of the rectangle. The area is the sum of the areas of the tiles in the rectangle.

EXAMPLE 2   Use algebra tiles to multiply: $(x - 1)(x + 2)$

Solution:

A. Build a rectangle with dimensions $(x - 1)$ and $(x + 2)$. First place the x^2-tile and the x-tiles in the first row. Since the factor at the left and the factors above are positive, the tiles in the first row will be positive.

B. Complete the second row of the rectangle. Since the factor at the left is negative and the factors above are positive, the tiles in the second row will be negative.

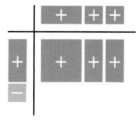
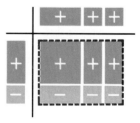

C. Count the tiles in the completed rectangle.

$$(x - 1)(x + 2) = x^2 + 2x - x - 2$$
$$= x^2 + x - 2$$

EXERCISES

Goal: To use algebra tiles to model multiplication of binomials

For each model, name the binomial factors and the product.

1.

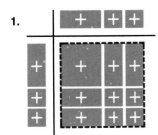

Factors: $(x + 2)$, $(x + 2)$
Product: $x^2 + 4x + 4$

2.

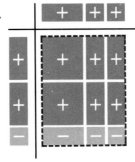

Factors: $(2x - 1)$, $(x + 2)$
Product: $2x^2 + 3x - 2$

Use algebra tiles to find each product.

3. $(x + 3)(x + 2)$
4. $(x + 3)(x + 4)$
5. $(2x + 1)(x + 4)$
6. $(3x + 1)(x + 2)$
7. $(2x - 3)(x + 2)$
8. $(2x + 4)(x - 1)$
9. $(2x + 3)(x - 2)$
10. $(x - 1)(x - 1)$
11. $(3x - 2)(3x - 2)$
12. $(x + 4)(x + 4)$
13. $(x - 4)(x - 4)$
14. $(x + 4)(x - 4)$
15. $(x - 3)(x - 3)$
16. $(x + 3)(x + 3)$
17. $(x - 3)(x + 3)$

18. Explain how the product rectangle in Exercise 2 shows that
$(2x - 1)(x + 2) = x(x + 2) + x(x + 2) - 1(x + 2) = 2x^2 + 3x - 2$.

Solve. Use algebra tiles if you need help.

19. The length of one side of a square piece of land is x feet. If you double the length of one pair of opposite sides and add 3 feet to the other pair of opposite sides, what will be the new dimensions of the piece of land? $2x$ and $x + 3$

20. Write two expressions for the area of the enlarged piece of land in Exercise 19. $2x(x + 3)$ and $2x^2 + 6x$

21. If the length of one side of the original piece of land in Exercise 19 was 12 feet, what is the area of the enlarged piece of land? 360 ft²

Using Manipulatives / 561

Teaching Aid 1
Teaching Aid 3
Teaching Aid 4
Teaching Aid 5
Transparencies 48, 50, 53, 54,
    55, 57

Teaching Suggestions
Focus students' attention on the diagram at the top of page 562. Ask them to name the factors shown in the diagram [$(x + 1)$ and $(x + 1)$] and then to give the product of the factors by referring to the product rectangle ($x^2 + 3x + 2$). Tell them that a polynomial is said to be **factorable** (not a prime polynomial) if it can be expressed as a product of polynomials (other than itself and ±1). Then ask them if the trinomial $x^2 + 3x + 2$ is factorable. (Yes) Why? [It can be expressed as $(x + 2)(x + 3)$]. Now ask the students to suggest how they could use algebra tiles to determine whether a given polynomial is factorable.
(Lead students to the conclusion that if the algebra tiles that model a polynomial can be arranged to form a rectangle, then the polynomial is factorable.)

Now proceed with the Examples and Exercises.

Manipulative 12 *(Use with Lesson 9.3.)*

In the previous lesson, you used algebra tiles to show that

$$(x + 1)(x + 2) = x^2 + 3x + 2.$$

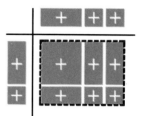

This means that the binomials $(x + 1)$ and $(x + 2)$ are factors of $x^2 + 3x + 2$.

Now consider the trinomial $x^2 + 4x + 3$. Is this trinomial factorable?

EXAMPLE 1   Use tiles to factor $x^2 + 4x + 3$.

Solution: **A.** Model the trinomial.

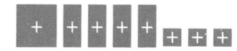

B. The trinomial is **factorable** if the tiles can be arranged to form a rectangle.

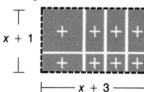

So $x^2 + 4x + 3 = (x + 1)(x + 3)$.

EXAMPLE 2   Use tiles to factor $x^2 + 7x + 10$.

Solution: **A.** Model the trinomial.

B. Build a rectangle. Position the x^2-tile and the x-tiles so that the ten (1)-tiles will fit.

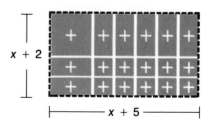

So $x^2 + 7x + 10 = (x + 2)(x + 5)$.

562   /   *Using Manipulatives*

EXERCISES

Goal: To use algebra tiles to factor trinomials

Use algebra tiles to factor each trinomial.

1. $x^2 + 2x + 1$ $(x+1)(x+1)$ 2. $x^2 + 3x + 2$ $(x+2)(x+1)$ 3. $x^2 + 4x + 4$

4. $x^2 + 5x + 6$ $(x+3)(x+2)$ 5. $x^2 + 7x + 12$ $(x+4)(x+3)$ 6. $x^2 + 6x + 9$

7. $x^2 + 4x + 3$ $(x+3)(x+1)$ 8. $x^2 + 5x + 4$ $(x+4)(x+1)$ 9. $x^2 + 6x + 5$

10. $x^2 + 7x + 10$ $(x+5)(x+2)$ 11. $x^2 + 5x + 6$ $(x+3)(x+2)$ 12. $x^2 + 8x + 7$

Use algebra tiles to determine whether the trinomials are factorable.
Answer <u>Yes</u> or <u>No</u>.

13. $x^2 + 3x + 3$ No   14. $x^2 + 8x + 15$ Yes   15. $x^2 + 2x + 16$

16. $x^2 + 2x + 2$ No   17. $x^2 + 7x + 6$ Yes   18. $x^2 + 8x + 16$

Find the missing factor. Use algebra tiles if you need help.

19. $x^2 + 14x + 13 = (x + 13)(\underline{\ ?\ })$ 20. $x^2 + 12x + 11 = (\underline{\ ?\ })(x + 11)$

21. $x^2 + 8x + 12 = (\underline{\ ?\ })(x + 2)$ 22. $x^2 + 10x + 21 = (x + 7)(\underline{\ ?\ })$

23. A car travels $(x^2 + 9x + 14)$ miles in $(x + 2)$ hours.
    Write an expression for the car's speed in miles per hour. $(x+7)$ mi/hr

24. A car travels $(x^2 + 7x + 10)$ miles at an average rate of $(x + 5)$ miles per
    hour. Write an expression for the number of hours the car traveled. $(x+2)$ hours

25. Write an explanation describing how you could use algebra tiles to
    determine whether $x^2 + 11x + 16$ is a factorable or a prime polynomial.

Critical Thinking
See Exercises 23–24.

Additional Answers
3. $(x + 2)(x + 2)$
6. $(x + 3)(x + 3)$
9. $(x + 5)(x + 1)$
12. $(x + 7)(x + 1)$
15. No
18. Yes
19. $(x + 1)$
20. $(x + 1)$
21. $(x + 6)$
22. $(x + 3)$
25. Check students' answers.

Teaching Suggestions
Have students work in pairs to complete Example 1. Then have them use algebra tiles to find these products.

1. $(x + 1)(x - 1)$ $(x^2 - 1)$
2. $(2 - x)(2 + x)$ $(4 - x^2)$

Ask students to look carefully for a pattern in the factors and products they have just modeled, and to suggest a rule for multiplying the sum and difference of the same two numbers. (The product of the sum and difference of the same two numbers equals the square of the first number minus the square of the second number.) Ask: What is the product of $(x + a)(x - a)$? $(x^2 - a^2)$

Then ask students to study carefully the factors and product they modeled once again, and to predict the factors of these binomials.

4. $x^2 - 9$ $[(x + 3)(x - 3)]$
5. $x^2 - 16$ $[(x + 4)(x - 4)]$
6. $x^2 - a^2$ $[(x + a)(x - a)]$

Ask them to suggest a rule for factoring the difference of two squares. (To factor the difference of two squares, multiply the sum of their square roots by the difference of their square roots.)

Now have students proceed with Example 2 and the Exercises.

Manipulative 13 *(Use with Lesson 9.5.)*

You can use algebra tiles to model the product of two binomials of the form $(x + a)(x - a)$.

EXAMPLE 1   Use algebra tiles to multiply: $(x + 3)(x - 3)$

Solution:

A. Model the factors. Then begin to build a rectangle by placing the tiles in the first row. Since the x-tile and the (1)-tiles in the factors differ in sign, the x-tiles in the first row will be negative.

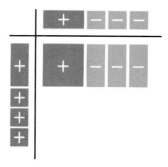

B. Complete the rectangle by placing the tiles in the remaining rows. Since the (1)-tiles in the factors differ in sign, the (1)-tiles in the product will be negative.

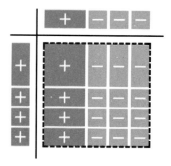

C. Identify neutral pairs. Count the remaining tiles.

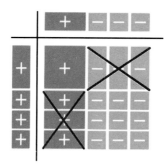

D. Write the product.

$$(x + 3)(x - 3) = x^2 - 9$$

In the expression $x^2 - 9$, x^2 and 9 are perfect squares. That is, $x^2 - 9$ can be written as $(x)^2 - (3)^2$.

This suggests that a binomial that can be expressed as the difference of perfect squares can be factored as the sum and difference of the square roots of the terms of the binomial.

564 / *Using Manipulatives*

EXAMPLE 2   Factor: **a.** $a^2 - 1$ **b.** $r^2 - 16$

Solutions: **a.** $a^2 - 1 = (a)^2 - (1)^2$ **b.** $r^2 - 16 = (r)^2 - (4)^2$
                           $= (a + 1)(a - 1)$ $= (r + 4)(r - 4)$

Critical Thinking
See Exercise 23.

Additional Answers
  3. $4x^2 - 1$
  6. $16x^2 - 1$
15. $(x + 7)(x - 7)$
16. $(x + 12)(x - 12)$
17. $(5 - x)(5 + x)$
18. $(x - 10)(x + 10)$
19. $(xy + 6)(xy - 6)$
20. $(5x + 2)(5x - 2)$
21. $(9x + 1)(9x - 1)$
22. $(x + 3)(x - 3)$
23. Answers will vary, but students should realize that two middle terms which are opposites are formed, and that these terms have a sum of zero. When algebra tiles are used, the middle terms form neutral pairs.

EXERCISES

Goal: To use algebra tiles to find the product of binomials of the form
$(x + a)(x - a)$

Use algebra tiles to find each product.

1. $(x + 2)(x - 2)$ $x^2 - 4$ **2.** $(x + 5)(x - 5)$ $x^2 - 25$ **3.** $(2x - 1)(2x + 1)$
4. $(5 - x)(5 + x)$ $25 - x^2$ **5.** $(3x + 2)(3x - 2)$ $9x^2 - 4$ **6.** $(4x + 1)(4x - 1)$

Draw a model on paper to show each product. Check students' models.

7. $(x + 1)(x - 1)$ **8.** $(x - 7)(x + 7)$ **9.** $(6 - x)(6 + x)$

Match each binomial in Column A with its factors in Column B.

Column A

10. $x^2 - 64$ b
11. $81 - x^2$ c
12. $25x^2 - 64$ f
13. $4x^2 - 49$ g
14. $100 - 9x^2$ h

Column B

a. $(x + 9)(x - 9)$ **f.** $(5x - 8)(5x + 8)$
b. $(x + 8)(x - 8)$ **g.** $(2x + 7)(2x - 7)$
c. $(9 + x)(9 - x)$ **h.** $(10 + 3x)(10 - 3x)$
d. $(5x - 64)(5x + 64)$
e. $(25x + 8)(25x - 8)$

Factor.

15. $x^2 - 49$ **16.** $x^2 - 144$ **17.** $25 - x^2$ **18.** $x^2 - 100$
19. $(xy)^2 - 36$ **20.** $25x^2 - 4$ **21.** $-1 + 81x^2$ **22.** $-9 + x^2$
23. Write an explanation of what happens to the middle term when you multiply
two binomials of the form $(x + a)$ and $(x - a)$.

Using Manipulatives / **565**

Teaching Suggestions
Have students work in pairs to
complete the Examples and Exercises. For Exercises 7–10, be
sure that students realize that
the trinomials $x^2 + 4x + 4$ and
$x^2 - 4x + 4$ differ only in the
sign of the middle term, and that
the signs of their respective binomial factors also differ in the
same way.

Manipulative 14 *(Use with Lesson 9.6.)*

You can use algebra tiles to factor trinomials that have special patterns.

EXAMPLE 1   Use algebra tiles to factor $x^2 + 6x + 9$.

Solution: **A.** Model the polynomial.

B. The polynomial is factorable if the tiles can be arranged to form a rectangle. Arrange the tiles.

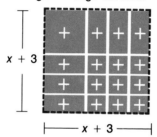

$x + 3$

├──── $x + 3$ ────┤

C. Write the factors.

$$x^2 + 6x + 9 = (x + 3)(x + 3)$$
$$= (x + 3)^2$$

Since the trinomial $x^2 + 6x + 9$ has two equal binomial factors, it is called a **trinomial square**. Notice that the rectangle formed by the tiles is a square.

EXAMPLE 2   Use algebra tiles to factor $4x^2 + 4x + 1$.

Solution: **A.** Model the polynomial.

B. Arrange the tiles to form a rectangle.

C. Write the factors.

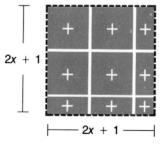

$2x + 1$

├──── $2x + 1$ ────┤

$$4x^2 + 4x + 1 = (2x + 1)(2x + 1)$$
$$= (2x + 1)^2$$

EXERCISES

Goal: To use algebra tiles to factor trinomial squares.

Use algebra tiles to factor each polynomial.

1. $x^2 + 2x + 1$ $(x+1)(x+1)$ **2.** $x^2 + 8x + 16$ $(x+4)(x+4)$ **3.** $x^2 + 4x + 4$

4. $9x^2 + 6x + 1$ $(3x+1)(3x+1)$ **5.** $4x^2 + 8x + 4$ $(2x+2)(2x+2)$ **6.** $25 + 10x + x^2$

7. Use algebra tiles to find this product: $(x - 2)(x - 2)$ $x^2 - 4x + 4$

8. Compare the product you found in Exercise 7 to this product: $(x + 2)(x + 2)$. Which terms are alike? Which are different?

9. What are the factors of $x^2 - 4x + 4$? $(x - 2)(x - 2)$

10. Is $x^2 - 4x + 4$ a trinomial square? Explain.

Use the results of Exercises 7–10 to factor each trinomial.

11. $x^2 - 6x + 9$ $(x-3)(x-3)$ **12.** $x^2 - 8x + 16$ $(x-4)(x-4)$ **13.** $x^2 - 14x + 49$

14. $4x^2 - 4x + 1$ $(2x-1)(2x-1)$ **15.** $9x^2 - 6x + 1$ $(3x-1)(3x-1)$ **16.** $25x^2 - 10x + 1$

17. Write an explanation of how you could use algebra tiles to determine whether a trinomial is a perfect square.

18. Help Yuri solve this problem.
If $x^2 + a^2x + a^2 = (x + a)^2$, what is the value of a?

Critical Thinking
See Exercises 17–18.

Additional Answers

3. $(x + 2)(x + 2)$

6. $(5 + x)(5 + x)$

8. The first and last terms are alike; the middle terms are opposites.

10. Yes; it has two equal binomial factors.

13. $(x - 7)(x - 7)$

16. $(5x - 1)(5x - 1)$

17. If the tiles that model the polynomial can be rearranged to form a square, then the trinomial is a perfect square.

18. $(x + a)^2 = x^2 + 2ax + x^2$
So $x^2 + a^2x + a^2 = x^2 + 2ax + x^2$.
Subtracting x^2 and a^2 from each side, $a^2x = 2ax$.
Dividing each side by ax, where $ax \neq 0$, $a = 2$.

Manipulative 15 *(Use with Lesson 19.5.)*

Theoretical probability is the probability predicted by mathematical reasoning. For example, we say that the probability of a fair coin landing heads up when it is tossed is $\frac{1}{2}$, because a coin has 2 sides and only one of the sides is heads.

If you toss 10 coins yourself, you might get 6 heads and 4 tails. So you would say that the **experimental (empirical) probability** of getting a head is $\frac{6}{10}$.

Since results usually vary somewhat when an experiment is repeated, experimental probabilities are not exact. One way to make an experimental probability more accurate is to repeat the experiment a large number of times. For example, as the number of times you toss a coin increases, the closer the experimental probability will get to $\frac{1}{2}$, the probability predicted by mathematical reasoning.

In the Exercises that follow, you will examine some experimental probabilities.

EXERCISES

Goal: To determine the experimental probability of events

Toss a coin 10 times. Record the outcome of each toss in order in a table like this one.

Toss	1	2	3	4	5	6	7	8	9	10
H or T										

Refer to your table to answer these questions. For Ex. 1–6, answers will vary.

1. How many of the 10 tosses were heads?

2. How many of the 10 tosses were tails?

3. How many times did you get two tails in a row?

4. How many times did you get two heads in a row?

5. What is the greatest number of successive heads or successive tails that you tossed?

6. Suppose that you toss the coin 10 more times. Would you expect that the results will show the same pattern? Explain.

7. Would you expect that the total number of heads in Exercise 6 to be close to the number of heads in Exercise 1? Yes

568 / *Using Manipulatives*

Now copy and complete a table like the one below. Your table will have ten rows. Toss a coin 100 times and record the results, in order, filling the first row, then the second row, then the third row, and so on. Compute the total number of heads and tails for each row.

	1	2	3	4	5	6	7	8	9	10	Total	
											H	T
Row 1												
Row 2												

Use the results in your table to answer these questions.

8. Does each row have the same number of heads?

9. What is the greatest number of heads in any row?

10. Does each row have the same sequence (order) of heads and tails?

11. How many times did you get 4 heads or 5 heads or 6 heads in one of the rows?

12. According to this experiment, what is the probability of getting (4 heads or 5 heads or 6 heads) in one of the rows?

13. According to the experiment, what is the probability of **not** getting (4 heads or 5 heads or 6 heads) in one of the rows?

14. What is the sum of the probabilities in Exercises 12 and 13?

15. Write a summary of the results of your experiment. Include any patterns you observe, which pattern or patterns seem to occur most often, whether the results seem reasonable, and what you predict would happen if you repeated the experiment.

16. Read your summary to the class. Compare your results with those of your classmates and discuss any differences.

GLOSSARY

The following definitions and statements reflect the usage of terms in this textbook.

Absolute Value   The *absolute value* of a positive number equals the number. The *absolute value* of a negative number equals the opposite of the number. The *absolute value* of 0 equals 0. (Page 2)

Algebraic Expression   An *algebraic expression* is an expression with one or more variables. (Page 21)

Axis of a Parabola   A vertical line that contains the vertex of a parabola is called the *axis of the parabola.* (Page 476)

Base   See **Power.**

Binomial   A *binomial* is a polynomial of two terms. (Page 213)

Coefficient   In a monomial of the form ax^n, the number represented by a is the *coefficient* or *numerical coefficient.* (Page 208)

Completing the Square   Writing a quadratic equation in a form that makes the left side a trinomial square is called *completing the square.* (Page 494)

Constant Term   In a polynomial, a term with no variables is called the *constant term.* (Page 220)

Cube Root   A *cube root* of a number is one of its three equal factors. (Page 180)

Degree of a Monomial   The *degree of a monomial* in one variable is determined by the exponent of the variable. (Page 239)

Degree of a Polynomial   The *degree of a polynomial* is the highest degree of the monomials of the polynomial. (Page 239)

Dependent Events   Two events are *dependent* if the occurrence of one event is influenced by the occurrence of the other. (Page 525)

Determinant   A *determinant* is a square array of the form $\begin{vmatrix} a & b \\ c & d \end{vmatrix}$ in which $a, b, c,$ and d are real numbers. (Page 412)

Difference of Two Squares   A binomial such as $x^2 - 16$ is called a *difference of two squares.* The first term and the absolute value of the constant term are perfect squares. (Page 252)

Domain   The set of replacements for a variable is the *domain.* (Page 116)

Domain of a Relation   The *domain of a relation* is the set of x values of its ordered pairs. (Page 333)

Equation   An *equation* is a sentence that contains the equality symbol "=." (Page 58)

Equivalent Equations   *Equivalent equations* have the same solution. (Page 58)

Event   An *event* is a subset of the sample space. It is a set of one or more of the possible outcomes of an experiment. (Page 515)

Exponent   See **Power.**

Experimental Probability   If an experiment is carried out n times and a certain event A occurs a times, then the *experimental probability* for event A is $\frac{a}{n}$. (Page 529)

Expression   An *expression* includes at least one of the operations of addition, subtraction, multiplication, or division. (Page 19)

Fourth Root   A *fourth root* of a number is one of its four equal factors. (Page 180)

Function   A *function* is a relation in which no two ordered pairs have the same *x* value. (Page 333)

Fundamental Counting Principle   If there are *p* choices for the first decision and *q* choices for the second decision, then there are *p* x *q* ways that the two decisions can be made together. (Page 512)

Greatest Common Factor   The greatest common factor of two or more counting numbers is the greatest number that is a factor of all the numbers. (Page 237)

Independent Events   Two events are independent if the occurrence of one event is not influenced by the occurrence of the other. (Page 520)

Integers   The set of integers consists of the set of whole numbers and their opposites. (Page 30)

Irrational Number   An irrational number is a number that cannot be expressed as a quotient of two integers. (Page 31)

Least Common Multiple (LCM)   The least common multiple (LCM) of two or more counting numbers is the smallest counting number that is divisible by the given numbers. (Page 300)

Like Radicals   Radicals such as $6\sqrt{2}$ and $5\sqrt{2}$ are called *like radicals*. They have equal radicands. (Page 187)

Like Terms   *Like terms* have exactly the same variables and the same powers of these variables. (Page 50)

Linear Function   A function that has the points of its graph lying in a straight line is a *linear function*. (Page 351)

Monomial of One Variable   A *monomial of one variable* over $\{$rational numbers$\}$ is an expression

of the form ax^n in which *a* is any rational number and *n* is any nonnegative integer. (Page 208)

Mutually Exclusive Events   Events are *mutually exclusive* if they have no outcomes in common. (Page 521)

Origin   The point where the *x* axis and the *y* axis meet is called the *origin*. (Page 322)

Parallel Lines   *Parallel lines* are lines in the same plane that do not have a common point. (Page 364)

Perfect Square   Positive numbers such as 1, 4, 9, 16, 25, and 36 are *perfect squares*, because each is the square of a counting number. (Page 31)

Perimeter   The *perimeter* of a geometric figure such as a triangle, square, or rectangle is the sum of the lengths of its sides. (Page 95)

Polynomial   A *polynomial* is a monomial or the sum of two or more monomials. (Page 213)

Power   An expression such as $(-2)^3$ is called a *power*. The 3 is the exponent and the −2 is the base. (Page 21)

Prime Factorization   The *prime factorization* of a number is the number expressed as a product of its prime-number factors. (Page 175)

Prime Number   A *prime number* is a counting number greater than 1 that has exactly two counting-number factors, 1 and the number itself. (Page 174)

Probability   The *probability P* of an event *E* is the ratio of the number of favorable outcomes of the event to the total number of possible outcomes in the sample space. (Page 515)

Quadrants   The four regions of the coordinate plane are called *quadrants*. (Page 325)

Quadratic Equation   A *quadratic equation* is an equation of the form $ax^2 + bx + c = 0$, $(a \neq 0)$. (Page 484)

Quadratic Function   A *quadratic function* is a function described by the rule $y = ax^2 + bx + c$, in which *a, b,* and *c* are rational numbers with $a \neq 0$. (Page 456)

Quadratic Polynomials   Second degree polynomials are called *quadratic polynomials*. (Page 239)

Range of a Relation   The *range of a relation* is the set of *y* values of its ordered pairs. (Page 333)

Ratio   A quotient of two numbers is called a *ratio*. (Page 359)

Rational Expression   A *rational expression* is one that can be written as the quotient of two polynomials. (Page 268)

Rational Number   A *rational number* is a number that can be expressed as a quotient of two integers, in which the denominator is not zero. (Page 30)

Real Numbers   The set of *real numbers* contains all the rational numbers and all the irrational numbers. (Page 32)

Reciprocals   Two numbers are *reciprocals* of each other if their product is 1. (Page 16)

Relation   A *relation* is a set of ordered pairs. (Page 328)

Sample Space   The set of all the different possible outcomes of an experiment is called the *sample space*. (Page 515)

Scientific Notation   *Scientific notation* is a numeral of the form $N \times 10^a$, in which *a* is an integer and *N* is a rational number such that $1 \leq N < 10$. (Page 163)

Slope   In the rule for the linear function $y = mx + b$, *m* represents the *slope* of its graph. (Page 356)

Square Root   A *square root* is one of two equal factors of a number. (Page 174)

Solution of an Equation   The *solution* or *root of an equation* is a number that makes the equation true. (Page 58)

Solution Set of a Linear Equation   The *solution set of a linear equation* is the set of ordered paris (x,y) that makes the equation true. (Page 378)

Solution Set of a System of Two Equations
The solution set of a system of two equations is the set of ordered pairs that makes both equations true. (Page 381)

Terms   The monomials that form a polynomial are called its *terms*. (Page 213)

Trinomial   A *trinomial* is a polynomial of three terms. (Page 213)

Trinomial Square   A trinomial that has two equal binomial factors is called a *trinomial square*. (Page 255)

Variable   A *variable* is a letter such as *x* that represents one or more numbers. (Page 21)

Vertex of a Parabola   The *vertex of a parabola* that is a graph of a quadratic function is either the lowest point or the highest point of the graph. (Page 475)

Whole Numbers   The set of *whole numbers* consists of zero and the counting numbers. (Page 30)

x intercept   The *x* value of a point common to the *x* axis and a straight line is called the *x intercept* of the line. (Page 367)

y intercept   The *y* value of the point that a line has in common with the *y* axis is the *y intercept*. (Page 353)

Zeros of a Function   The *zeros of a function* are the values of *x* which make the value of the function 0. (Page 484)

INDEX

ANSWERS TO SELECTED EXERCISES

The answers to the odd-numbered problems in the *Classroom Exercises, Written Exercises, Mid-Chapter Reviews, Chapter Reviews, Cumulative Reviews, Special Topic,* and *Focus on Reasoning* sections are given on the following pages.

The answers are provided for all of the problems in the *Review Capsules* and *Pivotal Exercises* (P-1, P-2, etc.).

CHAPTER 1   OPERATIONS ON NUMBERS

Page 3   Classroom Exercises   1. 25   3. 0.001   5. $\frac{2}{3}$ 7. $-8\frac{7}{8}$ 9. -8.5 11. 9   13. 0   15. $\frac{4}{5}$ 17. 0.053   19. $\frac{1}{4}$ 21. -0.01 23. -2.5 25. $-2\frac{1}{2}$ 27. 12   29. -5

Page 3   Written Exercises   1. -23 3. 0.12   5. 0   7. $-\frac{3}{5}$ 9. $5\frac{6}{7}$ 11. 18   13. -227 15. $\frac{7}{8}$ 17. $5\frac{1}{8}$ 19. 0.083   21. 2.73   23. 23   25. -6.8 27. 12   29. 23   31. -13 33. 13

Pages 5-6   P-1   a. -9 b. -6 c. -4.7 **P-2**   a. 2   b. $1\frac{1}{2}$ c. 2.6   **P-3**   a. -3 b. $-1\frac{1}{3}$ c. -7.4

Page 6   Classroom Exercises   1. b   3. d   5. -40 7. -6 9. -0.7 11. 16   13. 3.4   15. $\frac{1}{5}$ 17. -6 19. -3.3 21. $-7\frac{1}{2}$ 23. 0   25. 0

Page 7   Written Exercises   1. -42 3. -106 5. -0.08 7. 8   9. 4   11. 0.13   13. -25 15. -32 17. -8.9 19. -53 21. 16   23. -45.5 25. 0   27. 0   29. 1   31. 5°C

Page 8   Review Capsule for Section 1.3   1. a. 4   b. 4   Both a and b are the same.   2. a. 4   b. 4   Both a and b are the same.   3. a. 7   b. 7   Both a and b are the same.   4. a. 10   b. 10   Both a and b are the same.   5. a. 21   b. 21   Both a and b are the same.   6. a. 0   b. 0   Both a and b are the same.

Page 8   P-1   10, 0.8, -37, 0   **P-2**   a. 18   b. 20   c. -25 d. 4.2

Page 9   Classroom Exercises   1. $20 + (-15)$ 3. $0 + (-12)$ 5. $(-12) + (-6)$ 7. $-20 + 18$ 9. $5 + 23$ 11. $-3.3 + (-6.2)$ 13. -4 15. -23 17. $-5\frac{1}{3}$ 19. -10 21. 16   23. -14

Pages 9-10   Written Exercises   1. 4   3. 6   5. -5 7. -9 9. -22 11. -10.8 13. -24.7 15. $-\frac{3}{4}$ 17. 31   19. 32   21. 2   23. 42   25. 8.3   27. -4 29. -13 31. -18 33. 8   35. 4.4   37. 1.3   39. $-\frac{1}{7}$ 41. 20°F   43. $38

Page 10   Review Capsule for Section 1.4   1. $\frac{5}{4}$ 2. $\frac{9}{1}$ 3. $\frac{11}{6}$ 4. $\frac{13}{1}$ 5. $\frac{27}{5}$ 6. 30   7. 5.2   8. 9   9. $\frac{1}{3}$ 10. 2.25   11. 0.56   12. 1.44   13. 0   14. 0   15. 1   16. $\frac{2}{3}$ 17. $7\frac{1}{5}$

Pages 11-12   P-1   0   **P-2** $-8, -16, -24$ **P-3**   a. -42 b. -48 c. 0   d. -1 **P-4**   6, 12, 18   **P-5**   a. 45   b. 70   c. 0.26   d. 3

Page 13   Classroom Exercises   1. -36 3. -96 5. -3 7. 0   9. -9.6 11. -2.1 13. 77   15. 150   17. 147   19. 6   21. 1   23. 1

Pages 13-14   Written Exercises   1. −70   3. 0   5. −66   7. −48   9. −10   11. $-\frac{4}{15}$ 13. −20.4   15. −8.36   17. 72   19. 1   21. 87   23. 3200   25. 0.33   27. 99   29. −432   31. 198   33. −3600   35. 0   37. 1.56   39. $-4\frac{2}{3}$ 41. $-3\frac{3}{4}$ 43. −24   45. 8.64   47. Yes, because 2 * (−4) = −16 and −4 * 2 = −16.

Pages 14-15   Mid-Chapter Review   1. −17   3. $\frac{1}{2}$ 5. 0.06   7. 0   9. 2.5   11. −17   13. 10   15. −33   17. 1   19. −26   21. 6°C   23. 11 + (−7)   25. 12.1 + 7.9   27. −7 + (−4)   29. −15   31. 9.25   33. $-6\frac{1}{2}$ 35. 25°F   37. −56   39. −3   41. 1   43. 33.6   45. 22°F

Page 15   Review Capsule for Section 1.5   1. $\frac{3}{2}$ 2. $\frac{12}{1}$ 3. $\frac{8}{5}$ 4. $\frac{7}{3}$ 5. $\frac{22}{7}$ 6. $\frac{22}{5}$ 7. $\frac{17}{3}$ 8. $\frac{13}{6}$ 9. $\frac{7}{1}$ 10. $\frac{15}{4}$ 11. 1   12. 1   13. 1   14. 1   15. 3   16. 3   17. 3   18. $\frac{3}{4}$

Pages 16-17   P-1   a. $\frac{10}{3}$ b. $-\frac{1}{3}$ c. $\frac{3}{4}$ d. 1   e. −1   **P-2**   a. 64   b. $\frac{1}{9}$ c. 0   **P-3**   a. $-1\frac{13}{18}$ b. $-15\frac{3}{7}$ c. −0.3

Page 18   Classroom Exercises   1. −8 × 4   3. 28 × $(-\frac{1}{7})$ 5. 175   7. 3   9. $-\frac{1}{2}$ 11. 0

Page 18   Written Exercises   1. 24   3. 3   5. $10\frac{2}{3}$ 7. $\frac{1}{6}$ 9. $-10\frac{2}{3}$ 11. −5   13. −3   15. −9   17. 0   19. −11   21. −0.9   23. −10   25. −4   27. −5   29. 52   31. 0   33. −3   35. $-\frac{1}{3}$ 37. −10   39. −41   41. −$2350

Page 19   P-1   a. Correct   b. Incorrect   c. Correct   d. Incorrect

Page 20   Classroom Exercises   1. 16   3. 0   5. 5   7. 12   9. 43   11. 12   13. −11   15. −16

Page 20   Written Exercises   1. 9   3. −12   5. −0.5   7. 13   9. 4   11. −60   13. 9   15. −90   17. −14   19. −35   21. 20   23. 5.7   25. −15   27. −22   29. 9   31. −2   33. −5   35. −132   37. 49 years

Pages 21-22   P-1   a. 12   b. 2   c. $-7\frac{1}{2}$ **P-2**   a. 4   b. −4   c. 33

Page 22   Classroom Exercises   1. 36   3. 125   5. −64   7. −4   9. 8   11. −16   13. −110   15. 11   17. −23   19. 18   21. −17   23. 30   25. 25   27. −10   29. 0

Pages 22-23   Written Exercises   1. 121   3. 1000   5. −343   7. 54   9. 37   11. −18   13. 18   15. 13   17. 60   19. 8   21. −2   23. 0   25. −84   27. 30   29. 83   31. 300   33. 190   35. 60   37. −790   39. 256   41. −216   43. 13   45. $-\frac{3}{8}$ 47. 26   49. 234   51. −132   53. −35   55. 72   57. $-12\frac{1}{2}$ 59. 15   61. −1152   63. 49   65. 625   67. 117,649

Pages 24-25   1. a. An estimate will be close enough. Reasons will vary.   b. Answers will vary.   c. His estimate was 17¢ higher than the actual cost.   d. Yes   3. a. An exact answer should be computed. Reasons will vary.   b. Answers will vary.   c. Curry (62.5), Mills (62.4), Jones (52.9), Lang (48.3), Grimm (48.0)   d. Answers will vary.

Pages 26-28   Chapter Review   1. 1.5   3. $\frac{3}{8}$ 5. 0   7. 2.6   9. 35   11. −3   13. 4   15. 11   17. −14   19. −13.5   21. $-3\frac{1}{4}$ points   23. 17 + (−29); −12   25. −27 + 18; 9   27. 19.2 + 4.7; 23.9   29. 19°F   31. −204   33. 450   35. −9   37. −12°F   39. −7   41. 12   43. −52   45. $\frac{2}{3}$ 47. −1.4 inches per month   49. −7   51. −16   53. 22   55. 64   57. 51   59. −28   61. −4   63. −21   65. 21   67. −1   69. −69

CHAPTER 2 REAL NUMBERS

Pages 30-32 **P-1** a. 2 b. −6 c. 0 d. −213 **P-2** a. 0.222 ⋯ b. 0.3636 ⋯ c. 0.1666 ⋯
P-3 a. Irrational b. Irrational c. Rational

Pages 32-33 **Classroom Exercises** 1. $\frac{7}{2}$; $\frac{14}{4}$ 3. $\frac{0}{1}$; $\frac{0}{3}$ 5. $\frac{-2}{8}$; $\frac{-4}{16}$ 7. $\frac{-5}{1}$; $\frac{-10}{2}$ 9. 0.5000 ⋯
11. 0.571428571428 ⋯ 13. −0.375000 ⋯ 15. Integers 17. Counting numbers 19. Irrational
21. Rational 23. 28, −124, 17 25. 28, −1.8, $3\frac{1}{4}$, 0.1010010001 ⋯, −124, 17, $-\frac{22}{7}$, 5.436436436 ⋯
27. 0.1010010001 ⋯

Pages 33-34 **Written Exercises** 1. $\frac{-14}{16}$; $\frac{-28}{32}$ 3. $\frac{35}{8}$; $\frac{70}{16}$ 5. $\frac{-183}{100}$; $\frac{-366}{200}$ 7. $\frac{20}{1}$; $\frac{40}{2}$ 9. $\frac{-3}{1}$; $\frac{-6}{2}$
11. −24.000 ⋯ 13. 0.6363 ⋯ 15. 0.428571428571 ⋯ 17. Integers 19. Whole numbers
21. Rational 23. Irrational 25. Rational 27. Rational 29. Rational 31. Irrational 33. π 35. $\frac{10}{2}$, 19
37. $\frac{10}{2}$, −2, 4.7, 0, π, $-1\frac{3}{4}$, $\frac{7}{8}$, 19, $-\frac{17}{4}$, −6, −0.9 39. b, c, e 41. b, c, e 43. c, e 45. c, e 47. b, c, e
49. b, c, e

Page 34 **Review Capsule for Section 2.2** 1. 6 + (−4) 2. 7 + (−6) 3. 15 + (−8) 4. 4 + (−9)
5. 5 + (−12) 6. 23 + (−19) 7. 14 + (−18) 8. 2 + (−25) 9. 17 10. 17 11. $\frac{1}{16}$ 12. $-\frac{19}{60}$ 13. 1.1
14. 1.2 15. −9 16. −9 17. $-\frac{1}{9}$ 18. $4\frac{1}{6}$ 19. −7.1 20. −5.1 21. 2 22. 3 23. $\frac{1}{12}$ 24. $\frac{37}{60}$
25. −1.8 26. −0.9 27. 1 28. −5 29. $\frac{11}{12}$ 30. $\frac{1}{12}$ 31. −2.6 32. 10.7

Page 36 **P-1** a. 8.2; Commutative Property for Addition b. 6; Associative Property for Addition
P-2 a. 9 b. $-\frac{7}{8}$ c. $\frac{5}{6}$ + y d. b − 0.8

Pages 36-37 **Classroom Exercises** 1. 0 3. $4\frac{1}{3}$ 5. −0.7 7. 0 9. 0 11. $3\frac{1}{2}$ 13. 0.6 15. 3 17. 0
19. 54 − a 21. 2.1 + x 23. x + y − 6 25. 5.1 − a − b

Pages 37-38 **Written Exercises** 1. −8; Addition Property of Opposites 3. 7; Commutative Property
of Addition 5. 0; Addition Property of Opposites 7. 5 9. $1\frac{1}{2}$ 11. 2.5 13. 1.7 15. 2 − u
17. p + 1 19. −a − 0.3 21. f − 4.4 23. −17 + k − b 25. d − $2\frac{1}{8}$ − f 27. a + 9 − b 29. s − t
31. 39 33. $-4\frac{1}{2}$ 35. 10.6 37. −y 39. −82 + q 41. c − d − $11\frac{1}{2}$ 43. 119

Page 38 **Review Capsule for Section 2.3** 1. −30 2. 45 3. $-\frac{1}{4}$ 4. $\frac{6}{25}$ 5. −2.48 6. 7.02 7. −70
8. 88 9. $\frac{5}{8}$ 10. $-4\frac{4}{5}$ 11. 4.94 12. −36.04 13. $\frac{1}{5}$ 14. $\frac{5}{2}$ 15. $-\frac{1}{4}$ 16. $\frac{3}{1}$, or 3 17. $-\frac{8}{3}$ 18. $\frac{1}{10}$
19. 1 20. $-\frac{12}{1}$, or −12 21. $-\frac{1}{13}$ 22. $\frac{1}{11}$ 23. −1 24. $\frac{5}{8}$

Page 40 **P-1** a. 8.1; Commutative Property of Multiplication b. 7; Associative Property of
Multiplication c. $-\frac{1}{9}$; Multiplication Property of Reciprocals **P-2** a. −9.6 b. 48n c. 100pq

Pages 40-41 **Classroom Exercises** 1. 1 3. 62 5. 1 7. 17.8 9. 3.04 11. 16 13. 54 15. $\frac{4}{5}$ 17. $\frac{1}{2}$
19. 570 21. −12 23. 20 25. −0.8yz 27. −120mn 29. −1.08bh

Page 41 Written Exercises 1. $-\frac{5}{7}$; Multiplication Property of 1 3. 8; Associative Property of Multiplication 5. $-\frac{1}{2}$; Commutative Property of Multiplication 7. 40 9. 36 11. 0.42pq 13. 288m 15. $-27t$ 17. $-5.4mp$ 19. 225xy 21. $-1\frac{3}{10}rd$ 23. $-10wz$ 25. p 27. n

Page 42 Mid-Chapter Review 1. $\frac{-10}{12}; \frac{-15}{18}$ 3. $\frac{14}{5}; \frac{28}{10}$ 5. $\frac{35}{10}; \frac{70}{20}$ 7. 0.444 $\cdots$ 9. 0.7272 $\cdots$ 11. 5.25000 $\cdots$ 13. Rational 15. Rational 17. Rational 19. Rational 21. $\sqrt{36}$ 23. $\frac{2}{3}, -\sqrt{5}, -1\frac{2}{7}, \sqrt{36}, -9.3$ 25. $-y$ 27. $y + z - 4$ 29. $r - s - 4\frac{3}{8}$ 31. $-15.5 - a + b$ 33. 4 35. $-3h$ 37. $-1600k$ 39. $-8.82ay$ 41. $-24pf$

Page 42 Review Capsule for Section 2.4 1. $-12xy^2$ 2. $35a^2b^2$ 3. $3p^2qr$ 4. $-2mn^2s$ 5. $x - 8$ 6. $-t - 1$ 7. $6.5f^2gt^2$ 8. $w - 16$ 9. $-22 - p$

Pages 43-44 **P-1** a. $-x - 2$ b. $-4 + t$ c. $-\sqrt{2}$ d. mk **P-2** a. $-\frac{5k}{7}$ b. 1 c. 1 d. -1

Page 45 Classroom Exercises 1. $-3k$ 3. 48y 5. $-x - y$ 7. $x - 2 - y + a$ 9. $-4h$ 11. $-2xy$ 13. $-3x^2y$ 15. $-y^2z^2$ 17. $-2a^2c^2$ 19. $-16m^2n^2$ 21. $-\frac{9}{f}$ 23. 1 25. 1 27. 1

Pages 45-46 Written Exercises 1. ac 3. $2t - r + 1$ 5. $-4d$ 7. $-m - n$ 9. $2a - 11 - b$ 11. 16cd 13. $-4q$ 15. $-9a^2$ 17. $6x^2y^2$ 19. $\frac{4}{3}a^2b^2$ 21. $-1.56r^2s^2$ 23. $-\frac{3}{p}$ 25. 1 27. $\frac{-2 - y}{x}$ 29. 1 31. 1 33. 7fg 35. -1 37. $-0.78r^2$ 39. $\frac{2}{15}x^2y$ 41. $-3 - r + 3s$ 43. $r - 2$ 45. -1 47. 13 triangles

Page 46 Review Capsule for Section 2.5 1. $14a + 16$ 2. $21 + 24x$ 3. $-10 - 30b$ 4. $-8y - 32$ 5. $12pq - 8p$ 6. $-\frac{1}{2}xy - 8x^2y$ 7. Yes 8. No 9. Yes 10. No 11. No 12. Yes

Pages 47-48 **P-1** a. $-15a + 10$ b. $-3m - 2$ c. $6rs - 15r$ **P-2** a. 3x b. 4c c. $-7t$ **P-3** a. $6(y + 4)$ b. $2n(5m - 2)$ c. $3x(5x - 1)$ d. $-3v(3 + w)$

Page 48 Classroom Exercises 1. $10p + 10q$ 3. $2k - 6$ 5. $20x^2 + 28x$ 7. $-3m^2 + m$ 9. $8(m + n)$ 11. $2(3x + 5y)$ 13. $5(x - y)$ 15. $t(4m - 9n)$ 17. $-3(a + b)$ 19. $-r(6r + 1)$

Page 49 Written Exercises 1. $10a + 10b$ 3. $-3x - 6$ 5. $7r - 56$ 7. $24p + 16$ 9. $4r^2 + 12r$ 11. $-2y^2 + 12y$ 13. $24t^2 - 20tu$ 15. $-2t^2 + 3t$ 17. $5(a + b)$ 19. $4(r + 5s)$ 21. $2k(m + 2n)$ 23. $pq(p + q)$ 25. $3(r - s)$ 27. $8(x - 1)$ 29. $8x(kx - 2y^2)$ 31. $\frac{3}{4}r(x - y)$ 33. $-7b(2a + 1)$ 35. $-3x(x - 2)$ 37. $xy(-2 + 5)$, or 3xy 39. $-x(4x + 1$ 41. $5(a - 3b + 2c)$ 43. $3x(2x^2 + x - 6)$ 45. $6(5r^2 - 6x - 23x^2)$ 47. Yes. Let $a = 2$ and $b = 4$. $(2x + 4y) = 2(x + 2y)$. Thus, $c = 2$.

Page 49 Review Capsule for Section 2.6 1. 7 2. -10 3. -6 4. -13 5. -6 6. 5 7. -20 8. 5

Page 50 P-1 a. Like b. Unlike c. Unlike

Page 51 Classroom Exercises 1. 11y 3. 2x 5. 2ab 7. $-3y^2$ 9. $-2t^2$ 11. $8x - 1$ 13. -7 15. $-6x^2 - 3$ 17. $a^2 - a + 5$ 19. $2x^2 + 1$ 21. $\frac{1}{9}a^2 + 2\frac{1}{2}a + \frac{1}{2}$ 23. $2.1m^2 + 5.7m + n$ 25. $-66p^2 - p - 73$

Page 51 **Written Exercises** 1. $20y$ 3. $19t$ 5. $-17x$ 7. $-2t^2$ 9. $7z^2$ 11. $6a - 6$ 13. $-0.2t^2 + 2t$
15. $-3g - 3.4$ 17. $\frac{17}{20}a - 1\frac{1}{2}b$ 19. $-62b - 295bz + 200z$ 21. $3.3t^2 + 2.8t + 0.3$ 23. $\frac{11}{24}r^2 - \frac{8}{15}s^2 + \frac{1}{3}$
25. $-50f^2 - 17f - 45$ 27. $0.5t$ 29. $ab^2 - 2a^2b + ab - a^2b^2$ 31. $8v^2 + 10w^2 - 4w$ 33. $75a^2 + 25$
35. $-4\frac{1}{3}x^2 - \frac{1}{4}x + \frac{1}{5}$

Pages 52-53 1. Median 3. Mean: 28.7 mpg; Median: 27.9 mpg; Mode: 30.4 mpg 5. Mean: 29.2 mpg;
Median: 28.2 mpg; Mode: 28.2 mpg 7. The manufacturer of Car C; Car C has the greatest mean of the
three cars. 9. Median; Car A has the greatest median of the three cars. 11. Median; Reasons will vary.
13. Median; Reasons will vary.

Pages 55-56 **Chapter Review** 1. $\frac{-26}{32}; \frac{-39}{48}$ 3. $\frac{31}{8}; \frac{62}{16}$ 5. $\frac{-37}{1}; \frac{-74}{2}$ 7. $-21.000\cdots$ 9. $-0.333\cdots$
11. $4.5000\cdots$ 13. Positive rational numbers 15. Rational numbers 17. $-\sqrt{5}, \pi$ 19. $-5, -\sqrt{5},$
$-\sqrt{9}, -0.5$ 21. $0, 13$ 23. $5\frac{1}{2}$; Addition Property of Zero 25. -14; Addition Property of Opposites
27. -11 29. -4.5 31. $6 - a$ 33. $10 - r + w$ 35. 400 37. -2000 39. -63 41. -36 43. $-10y$
45. $-10d$ 47. $-12m + 2n + p$ 49. $w - k - 12$ 51. $r + s - p + q$ 53. $12kn$ 55. $7t^2$ 57. $72q^3r^2$
59. -1 61. 1 63. 1 65. $-7r - 70$ 67. $-18a - 9$ 69. $-18m^2 + 30m$ 71. $-1.02ab - 0.78b$
73. $\frac{1}{2}(a - b)$ 75. $5(3r + 2s)$ 77. $-2y(2y - 1)$ 79. $-22xy(y - 2)$ 81. $-27t^2 - 5$ 83. $-1.4q - 12.1$
85. $-2p^2q + pq^2$

CHAPTER 3 EQUATIONS AND PROBLEM SOLVING

Pages 58-59 **P-1** a. True b. True c. True **P-2** a. $y = -1$ b. $x = 29$ c. $t = 0$ d. $k = 22$
P-3 a. $n = -1$ b. $k = 1$ c. $x = 0$

Page 59 **Classroom Exercises** 1. Subtract 12. 3. Subtract 8. 5. Add 13.7. 7. Add $\frac{5}{2}$.
9. Subtract 23. 11. $y = 5$ 13. $q = 26$ 15. $b = 4$ 17. $h = -2\frac{5}{6}$ 19. $y = 17$ 21. $c = -17$

Page 60 **Written Exercises** 1. $x = 33$ 3. $z = 12$ 5. $b = -4.2$ 7. $d = 38$ 9. $f = 24$ 11. $g = 1$
13. $j = 17$ 15. $n = -24$ 17. $x = -1$ 19. $b = 4.8$ 21. $k = 2$ 23. $y = -1\frac{1}{5}$ 25. $x = -33$ 27. $b = 10.1$
29. $d = -2.9$ 31. $f = 8.8$ 33. $x = -3\frac{1}{2}$ 35. $k = 12.8$ 37. $p = 1.3$ 39. $z = -62$ 41. $b = 33.4$

Page 60 **Review Capsule for Section 3.2** 1. 1 2. 1 3. 1 4. 1 5. 1 6. 1 7. 1 8. 1 9. 1 10. 1
11. 1 12. 1 13. 9 14. -7.02 15. $-7\frac{1}{2}$ 16. 117 17. $2\frac{5}{8}$ 18. -988 19. 464 20. -102.5 21. 8
22. $\frac{1}{4}$ 23. 15 24. -52 25. $-25\frac{5}{6}$ 26. 0.4 27. $-2\frac{1}{2}$ 28. $-\frac{1}{20}$

Pages 61-62 **P-1** a. True b. True c. True **P-2** a. $b = -36$ b. $x = -100$ c. $n = 1$ d. $b = 256$
P-3 a. $y = 0.1$ b. $x = -3$ c. $n = 1$ d. $t = -3$ **P-4** a. 2 b. $-\frac{4}{3}$ c. $-\frac{3}{5}$ d. $\frac{5}{4}$

Page 63 **Classroom Exercises** 1. Multiply by 12. 3. Divide by 14. 5. Divide by -8.9.
7. Multiply by $-\frac{3}{2}$. 9. $j = 28$ 11. $z = 140$ 13. $r = -4.8$ 15. $t = 81$ 17. $t = -6$ 19. $a = -0.3$
21. $k = -16$ 23. $k = 36$

Page 63 **Written Exercises** 1. $x = -65$ 3. $y = -96$ 5. $x = -116.2$ 7. $t = 60$ 9. $v = 184$
11. $x = -83.2$ 13. $s = -33$ 15. $r = -0.544$ 17. $a = -7$ 19. $d = -8$ 21. $x = -10.2$ 23. $x = -9$

25. $x = 8\frac{3}{4}$ 27. $r = 0$ 29. $z = -88$ 31. $k = 75$ 33. $w = -30$ 35. $m = 48$ 37. $x = -35$ 39. $p = 56$
41. $x = 0.2$, or $x = -0.2$ 43. $n = 21$, or $n = -21$

Page 64   Review Capsule for Section 3.3   1. $5n$ 2. $3r$ 3. b 4. t 5. x 6. y

Pages 64-65   P-1   a. $t = 7$ b. $k = 8$ c. $r = -4$ **P-2**   a. $b = -7$ b. $s = 25$ c. $d = 60$

Page 65   Classroom Exercises   1. Subtract 5. 3. Add 5.6. 5. Subtract 3. 7. Subtract 5. 9. Add 6.
11. Subtract 6.5. 13. Yes 15. No 17. No 19. Yes 21. No

Page 66   Written Exercises   1. $x = 7$ 3. $b = -13$ 5. $c = -22$ 7. $e = -19$ 9. $y = -8$ 11. $y = -3\frac{3}{4}$
13. $x = -56$ 15. $x = -9.6$ 17. $a = 150$ 19. $x = -5.5$ 21. $h = -300$ 23. $g = 1\frac{1}{2}$ 25. $z = -7\frac{1}{5}$
27. $c = 15$ 29. $x = -1.55$ 31. $j = 33$ 33. $b = 50$ 35. $r = -24$ 37. $x = 25\frac{2}{3}$ 39. $x = 10\frac{1}{2}$ 41. $k = \frac{5}{6}$
43. 16 inches

Page 68   Classroom Exercises   1. $t = 3$ hr 3. $r = 70$ ft per sec. 5. $\ell = 45$ kilometers

Pages 68-69   Written Exercises   1. 300,000 km per sec. 3. 2.3 seconds 5. 0.7 m 7. Correct
9. Correct

Page 69   Mid-Chapter Review   1. $b = -12$ 3. $y = 4.5$ 5. $t = 150$ 7. $a = -3$ 9. $z = -3\frac{1}{3}$ 11. $x = 13$
13. $y = 7$ 15. $t = -15$ 17. $j = 12$ 19. 2.5 seconds

Page 70   1. 3560 pounds 3. decrease 5. 4786.67 pounds

Page 71   Review Capsule for Section 3.5   1. $3n + 3$ 2. $2.7t$ 3. $-2\frac{1}{2}k - \frac{1}{2}$ 4. $14m - 20$ 5. $30 - 5x$
6. $45 + 30n$ 7. $-8k + 48$ 8. $60 + 36x$ 9. $\frac{1}{2}y + \frac{3}{4}$

Pages 71-72   P-1   a. $m = 3$ b. $k = -7$ **P-2**   a. $y = -1$ b. $r = 5$

Pages 72-73   Classroom Exercises   1. $4y + 6 = 13$ 3. $-5r + 8 = 17$ 5. $-9.7 = 3.1m - 4.8$
7. $12x - 3 + 6 = 10$; $12x + 3 = 10$ 9. $\frac{1}{3} - 2q + 10 + \frac{1}{4}q = 6$; $-1\frac{3}{4}q + 10\frac{1}{3} = 6$ 11. $\frac{3}{4}x - \frac{3}{2} + 6x + \frac{1}{2} =$
-11; $6\frac{3}{4}x - 1 = -11$ 13. $-6 = 12 - 8t + 5t - 20$; $-6 = -8 - 3t$ 15. 7 17. -11 19. -3

Page 73   Written Exercises   1. $x = 7$ 3. $z = -0.4$ 5. $x = -8$ 7. $b = 7$ 9. $x = -12$ 11. $f = 14$
13. $k = -5\frac{5}{6}$ 15. $m = -38$ 17. $n = -1$ 19. $x = -\frac{1}{9}$ 21. $h = 11$ 23. $d = -5.1$ 25. $e = 3.8$ 27. $t = 35\frac{1}{2}$
29. $y = 2\frac{1}{2}$ 31. $x = 30$

Pages 75-76   Classroom Exercises   1. $n + 12$ 3. $\frac{n}{8}$ 5. $1800n$ 7. $\frac{1200}{n}$ 9. $26 - n$ (For Ex. 11-20,
variables may vary.) 11. Let n = low temperature; $n + 27 = 50$; $n = 23°$ 13. Let p = total number of
points; $\frac{p}{20} = 18$; $p = 360$ 15. Let n = number of passengers; $104 = n + 11$; $n = 93$ 17. Let r = average
amount of rainfall; $r - 9.4 = 69.3$; $r = 78.7$ cm

Pages 76-77   Written Exercises (For Ex. 1-18, variables may vary.)   1. $2t$ 3. $v + 25,000$ 5. $m - 15.8$
7. $\frac{b}{28}$ 9. $a - 512$ 11. Let c = total cost; $\frac{c}{24} = 86$; $c = \$2064$ 13. Let ℓ = length of rectangle; $14.2 = \ell -$

5.8; ℓ = 20 m   15. Let h = high temperature for the day; −5.3 = h − 7.9; h = 2.6°C   17. Let a = amount of sale; 0.06a = 33.60; a = $560   19. Let s = Tim's weekly salary; s + 283 = 561; s = $278   21. ℓ   23. n   25. o   27. k

Pages 79-80   Classroom Exercises   (For Ex. 1-6, variables may vary.)   1. Let m = number of miles driven on Saturday; m − 12 = 75   3. Let ℓ = loss for August; $\frac{\ell}{-1.2}$ = 5000   5. Let n = number of letters mailed on Monday; 2n − 16 = 132

Pages 80-81   Written Exercises   (For Ex. 1-11, variables may vary.)   1. Let m = number of cars washed in the morning; m + 7 = 18; m = 11   3. Let ℓ = record low temperature; −3.5ℓ = 105; ℓ = −30°F   5. Let t = thermometer reading; 4t + 5 = −23; t = −7°F   7. Let n = number of people attending last year; 2n − 12 = 326; n = 169   9. Let n = number of hours needed to complete a job; $\frac{n}{6}$ = 7.5; n = 45   11. Let ℓ = loss for 1985; −2.5ℓ = 30,000; ℓ = −$12,000

Page 83   1. Alice: drum; Nathan: saxophone; Marie: flute
3.

	Baseball	Soccer	Track	Golf
Alex	X	X	X	✓
Dee	X	✓	X	X
Rod	✓	X	X	X
Sue	X	X	✓	X

Dee is a member of the soccer team.

Pages 84-86   Chapter Review   1. x = −7   3. b = 19   5. f = −42   7. t = 30.3   9. k = −9   11. m = −112   13. t = 36   15. y = −8.3   17. z = $11\frac{1}{4}$ 19. m = −10   21. x = $7\frac{1}{3}$ 23. v = −57.6   25. 19.2 m/sec   27. 83.9 inches   29. x = 11   31. y = 5   33. t = −21.5   35. p = $\frac{2}{7}$ 37. Let d = total distance traveled; d − 25.6 = 278.3; d = 303.9 km   39. Let s = bowler's score; $\frac{s}{5}$ = 168, s = 840   41. Let n = number of rainy days last year; n + 6 = 82; n = 76   43. Let t = thermometer reading; 3t − 7 = −31; t = −8°F

Pages 87-88   Cumulative Review: Chapters 1-3   1. 0.07   3. −12   5. −22   7. −100   9. −12   11. 7   13. −28   15. 170   17. 4   19. −15   21. 2   23. −1   25. 29   27. 85   29. −5, 0, 4, $\frac{18}{3}$ 31. 0   33. k − w + 18   35. −8x   37. −32x²y   39. 30a²b²   41. −48a + 6   43. 8x − 4y   45. 3(4r − 3)   47. −6x(x + 3)   49. 16a − 8   51. x = −4   53. h = 4   55. y = −54   57. x = −$\frac{17}{3}$ 59. $7\frac{1}{2}$ inches   61. n = $\frac{2}{3}$ 63. x = 3   65. Let n = number of apples; $\frac{n}{17}$ = 5; n = 85   67. Let p = programs sold last week; p − 54 = 192; p = 246

CHAPTER 4   PROBLEM SOLVING: ONE VARIABLE

Pages 91-92   Classroom Exercises   1. Let p = number of seats in first class, then p + 186 = number of seats in tourist class. p + p + 186 = 234   3. Let p = cost of labor, then p + 3.58 = cost of parts. p + p + 3.58 = 75.62   5. Let p = city's rainfall in May, then 2p − $4\frac{3}{5}$ = city's rainfall in June. p + 2p − $4\frac{3}{5}$ = $14\frac{9}{10}$ 7. Let p = amount earned first year, then 2p − 5000 = amount earned second year. p + 2p − 5000 = 93,000   9. Let p = number of true/false questions, then 3p − 7 = number of multiple choice questions. p + 3p − 7 = 105

Pages 92-94   Written Exercises   (For Ex. 1-16, variables may vary.)   **1.** Let x = the number of visiting runners, then $x + 448$ = the number of local runners. $x + x + 448 = 3640$; $x = 1596$, $x + 448 = 2044$
3. Let x = number of games in first set, then $x + 4$ = number of games in second set. $x + x + 4 = 20$; $x = 8$, $x + 4 = 12$ **5.** Let x = amount spent on food, then $2x - 150$ = amount spent on housing and transportation. $x + 2x - 150 = 8500$; $x = \$2883.33$ **7.** Let x = fuel capacity of compact car, then $2x - 2.3$ = fuel capacity of large car. $x + 2x - 2.3 = 35.2$; $x = 12.5$ gal., $2x - 2.3 = 22.7$ gal.   **9.** Let x = amount of cargo space, then $5x + 5$ = amount of passenger space. $x + 5x + 5 = 131$; $x = 21$ ft^3, $5x + 5 = 110$ ft^3 **11.** Let x = number of hours worked the first week, then $x + 2\frac{3}{4}$ = number of hours worked in second week. $x + x + 2\frac{3}{4} = 45\frac{3}{4}$; $x = 21\frac{1}{2}$, $x + 2\frac{3}{4} = 24\frac{1}{4}$ **13.** Let x = no. of foreign coins, then $3x - 26$ = no. of U.S. coins. $3x - 26 + x = 998$; $x = 256$, $3x - 26 = 742$ **15.** Let x = number of records entered first day, then $x + 12$ = number of records entered second day. $x + x + 12 = 256$; $x = 122$, $x + 12 = 134$ **17.** 9 coins   **19.** 61 fish

Page 94   Review Capsule for Section 4.2   **1.** $t = 26$ **2.** $w = 61.5$ **3.** $n = 4.2$ **4.** $p = 3.25$ **5.** $x = 6.8$ **6.** $r = 10.3$ **7.** d; $4d$ **8.** $7d$; $2d$ **9.** $5d, 4d, 2d$ **10.** $37d, 9d, 4d$

Page 95   P-1   a. 23 in   b. 24 m   c. 24 cm

Page 97   Classroom Exercises   **1.** 18 cm   **3.** 4s   **5.** Let m = width, $m + 27$ = length, $2m + 2(m + 27) = 174$ **7.** Let m = width, $m + 1.1$ = length. $2m + 2(m + 1.1) = 11.4$ **9.** Let m = length of second leg, $3m$ = length of third leg. $25.3 + m + 3m = 67.3$ **11.** Let $4m$, $5m$, and $6m$ = distances between cities. $4m + 5m + 6m = 184.5$ **13.** Let $20m$ = length, $9m$ = width. $2(20m) + 2(9m) = 232$ **15.** Let $7m$ = length of guy wire, $5m$ = height of pole. $9.8 + 7m + 5m = 33.8$

Pages 98-99   Written Exercises   (For Ex. 1-16, variables may vary.)   **1.** Let x = height, $x + 6$ = length. $2x + 2(x + 6) = 84$; $x = 18$ ft, $x + 6 = 24$ ft   **3.** Let x = shorter side, $x + 1.7$ = longer side. $8.5 + x + x + 1.7 = 51.2$; $x = 20.5$ m, $x + 1.7 = 22.2$ m   **5.** Let sides be $5x$, $12x$, and $13x$. $5x + 12x + 13x = 204$; $x = 6.8$, $5x = 34$ m, $12x = 81.6$ m, $13x = 88.4$ m   **7.** Let $4x$, $7x$, and $9x$ be the lengths of the legs. $4x + 7x + 9x = 684$; $x = 34.2$, $4x = 136.8$ km, $7x = 239.4$ km, $9x = 307.8$ km   **9.** Let x = width, $2x - 4$ = length. $2x + 2(2x - 4) = 148$; $x = 26$ ft, $2x - 4 = 48$ ft   **11.** Let x = height, $x + 2\frac{1}{2}$ = width. $2x + 2(x + 2\frac{1}{2}) = 44$; $x = 9\frac{3}{4}$ in, $x + 2\frac{1}{2} = 12\frac{1}{4}$ in   **13.** Let $18x$ = cable length, $11x$ = tower height. $18x + 11x + 35.6 = 108.1$; $x = 2.5$, $11x = 27.5$ m   **15.** Let $3x$, $4x$, and $5x$ = lengths of sides of triangle, $3x$ = length of each side of square. $3x + 4x + 5x + 4(3x) = 96$; triangle: $3x = 12$ cm, $4x = 16$ cm, $5x = 20$ cm; square: $3x = 12$ cm

Page 99   Review Capsule for Section 4.3   **1.** $\left.\begin{matrix}33\\15\end{matrix}\right\}$ **2.** $\left.\begin{matrix}-4\\-4\end{matrix}\right\}$ **3.** $\left.\begin{matrix}17.5\\17.5\end{matrix}\right\}$ **4.** $\left.\begin{matrix}16\\16\end{matrix}\right\}$ **5.** $\left.\begin{matrix}6\frac{3}{4}\\6\frac{3}{4}\end{matrix}\right\}$ **6.** $\left.\begin{matrix}-18\\-18\end{matrix}\right\}$

Pages 100-101   P-1   a. $t = -2$ b. $j = -4$ **P-2**   a. $p = -14$ b. $f = 1\frac{1}{3}$ **P-3**   a. $y = 1\frac{1}{2}$ b. $g = -4$

Page 101   Classroom Exercises   **1.** $x - 2 = 5$; $-2 = -x + 5$ **3.** $8 + 4a = 6$; $8 = -4a + 6$ **5.** $-3d - 6 = 12$; $-6 = 12 + 3d$ **7.** $5p + 2 = 6$ or $2 = 6 - 5p$ **9.** $-8 = 4z + 5$ or $-4z - 8 = 5$ **11.** $10.8 = 9.2 + 10.5j$ or $10.8 - 10.5j = 9.2$

Page 102   Written Exercises   **1.** $x = 17$ **3.** $a = 13$ **5.** $c = -12$ **7.** $k = -6\frac{2}{3}$ **9.** $x = 7$ **11.** $z = -5$ **13.** $p = -21\frac{1}{4}$ **15.** $t = -0.8$ **17.** $x = -14$ **19.** $y = -11$ **21.** $n = 3.75$ **23.** $y = -\frac{5}{9}$ **25.** $y = -8\frac{1}{2}$ **27.** $q = 2$ **29.** $k = -3$ **31.** $d = -6$ **33.** $d = 6$ **35.** $p = -5$ **37.** $x = b - 2a$; odd

Page 103 **Mid-Chapter Review** (For Ex. 1-10, variables may vary.) **1.** Let x = number of boys, then x + 7 = number of girls. x + x + 7 = 45; x = 19, x + 7 = 26 **3.** Let x = number of cars the second hour, then x − 2900 = number of cars the first hour. x + x − 2900 = 24,500; x = 13,700, x − 2900 = 10,800 **5.** Let x = width, then 2x + 6.2 = length. 2(x) + 2(2x + 6.2) = 106.6; x = 15.7 m, 2x + 6.2 = 37.6 m **7.** Let 2x, 4x, and 5x be the lengths of the sides. 2x + 4x + 5x = 264; x = 24, 2x = 48 in, 4x = 96 in, 5x = 120 in **9.** d = −14 **11.** p = 4 **13.** x = 1

Page 103 **Review Capsule for Section 4.4** **1.** $115.20 **2.** $7 **3.** 139 **4.** $11.98 **5.** 42 **6.** 42.5 **7.** $33.75 **8.** 305

Page 106 **Classroom Exercises** **1.** $128 **3.** $3.80 **5.** $90 **7.** $450; $54 **9.** Monthly Income: x + 240, Amount Spent: 0.30x, 0.25(x + 240); 0.30x = 0.25(x + 240)

Page 107 **Written Exercises** **1.** $575 **3.** $60; $12 (For Ex. 5-7, variables may vary.) **5.** Let x = number of games last season, x + 4 = number of games this season. 0.55x = 0.50(x + 4); x = 40, x + 4 = 44 **7.** Let x = number who tried out for girls' team, x + 4 = number for boys' team. 0.75x = 0.60(x + 4); x = 16, x + 4 = 20

Page 110 **Classroom Exercises** **1.** Rate: x − 6, Distance: 6x, 6.4(x − 6); 6x = 6.4(x − 6) **3.** Fuel Tank Capacity: x − 2, Range: 10x, 12(x − 2); 10x = 12(x − 2) − 112

Page 111 **Written Exercises**

1.

	r	t	d
Going	x + 15	$2\frac{4}{5}$	$2\frac{4}{5}(x + 15)$
Returning	x	4	4x

$2\frac{4}{5}(x + 15) = 4x$, x = 35, x + 15 = 50 mi/hr

3.

	r	t	d
Going	x	1	x
Returning	x − 7	$1\frac{1}{4}$	$1\frac{1}{4}(x − 7)$

$x = 1\frac{1}{4}(x − 7)$, x = 35 mi/hr, x − 7 = 28 mi/hr

5.

	Fuel Capacity	Mileage	Range
Car A	x	5	5x
Car B	x − 35	8.5	8.5(x − 35)

5x = 8.5(x − 35), x = 85 L, x − 35 = 50 L

Page 112 **1.** When r = 3.7, S = 277.3. When r = 3.8, S = 277.0. When r = 3.9, S = 277.1. Thus, to the nearest tenth, 3.8 is the value of r corresponding to the smallest value of S.

Pages 113-114 **Chapter Review** **1.** Let x = number of non-fiction books, x + 82 = number of fiction books. x + x + 82 = 372, x = 145, x + 82 = 227 **3.** Let x = number of miles father drove, then 2x + 100 = number of miles Alice drove. x + 2x + 100 = 1150, x = 350 mi, 2x + 100 = 800 mi **5.** Let x = width, x + 11.8 = length. 2(x) + 2(x + 11.8) = 80, x = 14.1 m, x + 11.8 = 25.9 m **7.** Let length = 7x, width = 3x. 2(7x) + 2(3x) = 166, x = 8.3, 7x = 58.1 mm, 3x = 24.9 mm **9.** h = −7 **11.** $y = 9\frac{1}{2}$ **13.** $t = 1\frac{1}{3}$

15. g = 15 **17.** $457.50 **19.** Let x = Russell's monthly income, x + 200 = Brad's monthly income. 0.32x = 0.28(x + 200); x = $1400, x + 200 = $1600

21.

	r	t	d
Going	x − 5	4.25	4.25(x − 5)
Returning	x	4	4x

4.25(x − 5) = 4x, x = 85 km/hr, x − 5 = 80 km/hr

CHAPTER 5 INEQUALITIES

Pages 116-117 **P-1** To show that all integers less than 2 are solutions. **P-2** 2 is equal to 2. **P-3** 3 is not equal to 2. **P-4** To show that 3 is a solution of $x \leq 3$. **P-5** a. {all real numbers greater than or equal to $3\frac{1}{2}$} b. {all real numbers less than $-2\frac{1}{2}$}; Assume {real numbers} as the domain.

Page 118 **Classroom Exercises** 1. {all integers greater than -2} 3. {all real numbers less than $2\frac{1}{2}$} (For Ex. 5-10, answers will vary.) 5. 0; 4; 9 7. -1.2; 0; 2.1 9. -5; $3\frac{1}{4}$; $\frac{7}{8}$

Pages 118-119 **Written Exercises** 1. {$-2, -1, 0, 1$} 3. {all integers less than 1} 5. {all real numbers greater than -2} 7. {all real numbers less than or equal to -1} 9. {all real numbers greater than or equal to $-1\frac{1}{2}$} 11. On a number line, the points 2, 3, 4, 5, etc. 13. On a number line, the points $-3, -4, -5, -6$, etc. 15. On a number line, all points to the left of, and not including, $-\frac{3}{4}$. 17. On a number line, all points to the right of, and including, 2.7. 19. No 21. Yes 23. Yes 25. Yes

Page 119 **Review Capsule for Section 5.2** 1. 0.3, 0.5; $\frac{5}{10} > \frac{3}{10}$ 2. 0.75, 0.25; $\frac{3}{4} > \frac{1}{4}$ 3. 0.4, -0.6; $\frac{2}{5} > -\frac{3}{5}$ 4. 0.75, 0.875; $\frac{7}{8} > \frac{6}{8}$ 5. $-0.4, -0.5$; $-\frac{4}{10} > -\frac{5}{10}$ 6. $-0.375, -0.625$; $-\frac{3}{8} > -\frac{5}{8}$ 7. $0.666\cdots, 0.8333\cdots$; $\frac{5}{6} > \frac{4}{6}$ 8. $-0.333\cdots, -0.666\cdots$; $-\frac{1}{3} > -\frac{2}{3}$

Pages 120-121 **P-1** $a > b$ **P-2** $a < b$ **P-3** a. $6 > -y$; $-y < 6$ b. $-t < 13$; $13 > -t$ c. $-s < r$; $r > -s$ **P-4** $r < 2$; $2 > r$ **P-5** $2 < t$; $t > 2$ **P-6** $r < t$; $t > r$ **P-7** $x < y$; $y > x$ **P-8** If $a > b$ and $b > c$, then $a > c$.

Page 121 **Classroom Exercises** For Ex. 1-7, only one of two possible inequalities is given for each pair. 1. $-5 < 2$ 3. $-12 < -9$ 5. $-16 < 6$ 7. $1.3 > 1.03$ 9. $<$ 11. $>$

Page 122 **Written Exercises** 1. $-5 < 4$ 3. $-12 < -11$ 5. $1.6 < 1.7$ 7. $-\frac{7}{8} < -\frac{3}{4}$ 9. $-3 > b$ 11. $a > 2$ 13. $a + b > -2$ 15. $b < -a$ 17. True 19. False 21. False 23. True 25. $y < x$ 27. $x < y$ 29. $y < x$ 31. $x < y$

Page 122 **Review Capsule for Section 5.3** 1. $t = 135$ 2. $x = -15$ 3. $y = 13$ 4. $w = 7.9$ 5. $t = -6.1$ 6. $r = 1\frac{5}{8}$ 7. $d = -\frac{5}{6}$ 8. $p = 3.86$ 9. $s = -0.53$

Pages 124-125 **Classroom Exercises** 1. Subtract 5. 3. Subtract 7. 5. Add 0.8. 7. {all real numbers less than 11} 9. {all real numbers less than 8} 11. {all real numbers less than 4} 13. {all real numbers less than 4} 15. {all real numbers greater than -8} 17. {all real numbers less than 8}

Pages 125-126 **Written Exercises** 1. No 3. Yes 5. No 7. Yes 9. No 11. {all real numbers greater than -5} 13. {all real numbers less than -7} 15. {all real numbers less than -1} 17. {all real numbers less than -3.1} 19. {all real numbers greater than 6} 21. {all real numbers greater than 7.9} 23. {all integers greater than -1} 25. {all integers greater than 7} 27. {all integers greater than or equal to -5} 29. $b < a$ 31. Fill the 9-quart pail. Fill the 3-quart pail from the 9-quart pail. There are 6 quarts remaining in the 9-quart pail. Fill the 2-quart pail from the 3-quart pail. Pour out the 1 quart remaining in the 3-quart pail. There are now exactly 8 quarts. 33. First bag: 27; second bag: 21; third bag: 11; fourth bag: 16; fifth bag: 25

Page 126   Mid-Chapter Review   1. On a number line, all points to the right of, and not including, −4.
3. On a number line, the points 4, 3, 2, 1, 0, −1, etc.   **5.** On a number line, the numbers 6, 5, 4, 3, 2, 1, 0.
7. $a < b$ **9.** $b < a$ **11.** $b < a$ **13.** $t < 5$ **15.** $x > -1.2$ **17.** $r < -0.3$

Page 126   Review Capsule for Section 5.4   1. $n = 1\frac{4}{5}$ **2.** $t = 2\frac{2}{3}$ **3.** $w = -7$ **4.** $y = -10$ **5.** $p = 1\frac{2}{3}$
6. $m = 5$

Page 128   Classroom Exercises   1. $-3 < x + 2$ **3.** $a - 2 < -7$ **5.** $-y + \frac{1}{2} > -12\frac{1}{4}$ **7.** $3w - 5 -$
$2w < 2w + 12 - 2w$ **9.** $3\frac{1}{2}x - 2 - 2\frac{1}{2}x > 2\frac{1}{2}x + 10 - 2\frac{1}{2}x$ **11.** $0.6m - 0.7 + 0.4m < -0.4m + 0.2 +$
$0.4m$ **13.** $-3x + 6 < -5 - 2x$ **15.** $5 - x < -2x + 6 - 2$ **17.** $5y - 2 - 3y < 3y - 4$ **19.** $a - 5 <$
$a + 2$ **21.** $5x - 5 < 2x - 6 + 3x$

Page 128   Written Exercises   1. {all real numbers greater than −5}   **3.** {all real numbers less than −5}
5. {all real numbers less than $12\frac{3}{4}$}   **7.** {all real numbers less than 17}   **9.** {all real numbers greater
than 12}   **11.** {all real numbers less than 0.9}   **13.** {all real numbers greater than 11}   **15.** {all real
numbers less than −1}   **17.** {all real numbers greater than 2}   **19.** {real numbers}   **21.** ϕ

Page 128   Review Capsule for Section 5.5   1. $x = 36$ **2.** $n = -84$ **3.** $t = 60$ **4.** $r = -16$ **5.** $w = -5.6$
6. $p = -5\frac{1}{2}$ **7.** $q = 4\frac{1}{5}$ **8.** $s = -26$

Pages 129-130   P-1   a. 2   **b.** −3   **c.** $\frac{5}{4}$ **d.** $-\frac{3}{2}$ **P-2**   Dividing by 3: **a.** $x < 3$ **b.** $-2x > 2$
c. $4x < -1$ **d.** $-5x < \frac{1}{3}$ Dividing by −3: **a.** $-x > -3$ **b.** $2x < -2$ **c.** $-4x > 1$ **d.** $5x > -\frac{1}{3}$

Pages 130-131   Classroom Exercises   1. $5x > 10$ **3.** $10a > 50$ **5.** $-3y < -6$ **7.** $r > -6$ **9.** Divide
by 2.   **11.** Multiply by −2.   **13.** $>$ **15.** $<$

Page 131   Written Exercises   1. {all real numbers less than 5}   **3.** {all real numbers greater than −20}
5. {all real numbers greater than −3}   **7.** {all real numbers greater than −2}   **9.** {all real numbers less
than −6}   **11.** {all real numbers less than 6}   **13.** {all real numbers less than −9}   **15.** {all real
numbers less than $-\frac{11}{3}$}   **17.** {all real numbers less than −10}   **19.** {all real numbers greater than −3}
21. The box marked "GB."

Pages 132-133   P-1   The order of an inequality is unchanged if each side is multiplied by a positive
number.   **P-2**   Each side is multiplied by a negative number.   **P-3**   {all real numbers less than 6}
P-4   {real numbers} ; ϕ

Pages 133-134   Classroom Exercises   1. $2x + 1 - 1 > 3 - 1$ **3.** $\frac{1}{2}x - 3 + 3 < -7 + 3$ **5.** $5 - 2x -$
$4x > 4x - 3 - 4x$ **7.** $10x - 30 < -10$ **9.** $-5x + 3 > -2x - 6 - 6$ **11.** {all real numbers greater
than −2}   **13.** {all real numbers less than 1}   **15.** {all real numbers greater than −9}   **17.** {all real
numbers greater than −1}   **19.** {all real numbers less than −3}   **21.** {real numbers}   **23.** {all real
numbers greater than −4}   **25.** {all real numbers less than 6}

Page 134   Written Exercises   1. {all real numbers less than $\frac{8}{3}$}   **3.** {all real numbers less than $-\frac{13}{2}$}
5. {all real numbers greater than 8}   **7.** {all real numbers greater than $\frac{7}{2}$}   **9.** {all real numbers less
than −2}   **11.** {all real numbers greater than $-\frac{1}{5}$}   **13.** {all real numbers greater than 6}   **15.** {all real
numbers greater than 0}   **17.** {all real numbers less than 2}   **19.** ϕ **21.** {real numbers}

Page 136 **Classroom Exercises** 1. $c < 60$ 3. $2d - 3 > 20$ 5. $\dfrac{32 + 45 + x}{3} < 42$
7. $\dfrac{65 + 70 + 75 + x}{4} > 72$

Page 137 **Written Exercises** 1. Let x = cost of tennis racket, then x − 50 = cost of tennis shoes. $x + x - 50 < 120$; $x < 85$, $84.99 3. Let x = cost of radio, then 4x + 30 = cost of television. $x + 4x + 30 > 525$; $x > 99$; Yes 5. Let x = score on fourth quiz. $\dfrac{85 + 72 + 81 + x}{4} > 80$; $x > 82, 83$ 7. Let x = score in fourth game. $\dfrac{95 + 130 + 125 + x}{4} > 110$; $x > 90, 91$

Pages 138-139 1. 2│5, 7, 7, 8   3. 72   5. a. Answers will vary.
            3│1, 1, 2, 2, 3, 5, 5, 6   b. Reasons will vary.   7. Median: 188;
            4│0, 0, 0   Mode: 196   9. Theater A; Theater A

Pages 140-142 **Chapter Review** 1. $\{-3, -2, -1, 0\}$ 3. $\{$all real numbers less than $1\frac{1}{2}\}$ 5. On a number line, the points 4, 3, 2, 1, 0. 7. On a number line, all points to the right of, and not including, −3. 9. All points to the left of, and not including, 4. 11. $-r > 12$ 13. $-a < 3$ 15. $x < 4.2$ 17. $1 < x$ 19. $c < a$; $a > c$ 21. True 23. No 25. No 27. Yes 29. $\{$all real numbers less than 1$\}$ 31. $\{$all real numbers less than −2$\}$ 33. $\{$all real numbers less than −9.7$\}$ 35. $\{$all real numbers less than −8$\}$ 37. $\{$all real numbers greater than 11$\}$ 39. $\{$all real numbers less than $7\frac{1}{2}\}$ 41. $\{$all real numbers less than −5$\}$ 43. $\{$all real numbers less than 4$\}$ 45. $\{$all real numbers greater than 1.6$\}$ 47. $\{$all real numbers greater than $10\frac{1}{3}\}$ 49. ϕ 51. Let x = cost of pants, then 3x + 20 = cost of coat. $x + 3x + 20 < 105$; $x < 21.25$, $21.24 53. Let x = score on fourth test. $\dfrac{85 + 93 + 78 + x}{4} > 85$; $x > 84, 85$

CHAPTER 6   POWERS

Pages 144-145 **P-1** a. $2 \cdot 2 \cdot 2 \cdot 2$ b. $x \cdot x \cdot x$ c. $y \cdot y \cdot y \cdot y \cdot y$ d. $4 \cdot 4 \cdot 4$ e. $z \cdot z$ **P-2** 5 **P-3** $(-3) \cdot (-3)$; $(-3) \cdot (-3) \cdot (-3) \cdot (-3)$ **P-4** $a + b$ times **P-5** a. $(-1.9)^5$ b. $(\frac{1}{2})^7$ c. t^9 **P-6** a. $t^6 + t^5$ b. $3k^5 - 2k^4 + 3k^3$ **P-7** $x = 4$

Pages 145-146 **Classroom Exercises** 1. 2^5 3. $(-5)^6$ 5. $(\frac{1}{2})^5$ 7. k^{22} 9. 7^6 11. 2^7 13. $x = 5$ 15. $x = 6$ 17. $x = 4$

Page 146 **Written Exercises** 1. 12^{10} 3. 5^4 5. $(3.2)^{12}$ 7. $(-15)^9$ 9. $(-6)^{11}$ 11. $(-\frac{2}{3})^7$ 13. a^7 15. r^9 17. a^{11} 19. 5^9 21. $y^3 + y^4$ 23. $x^9 + x^7$ 25. $2^5 - 2^7$ 27. $n^4 - n^3 + n^2$ 29. $x = 11$ 31. $x = 5$ 33. $x = 5$ 35. $x = 6$

Page 146 **Review Capsule for Section 6.2** 1. True 2. True 3. False 4. False

Pages 147-148 **P-1** $x \cdot x \cdot x \cdot x \cdot x$; $x \cdot x \cdot x$ **P-2** a. 17^3 b. $(1.8)^5$ c. m^7 **P-3** 113^0 **P-4** 1 **P-5** 1 **P-6** a. 2 b. 0 c. 1

Page 149 **Classroom Exercises** 1. 3^3 3. $(-3)^4$ 5. a^2 7. 2^3 9. $(\frac{1}{2})^7$ 11. $(-5)^{t-4}$ 13. False 15. False 17. True 19. 1 21. 1 23. 1 25. 1

Pages 149-150 **Written Exercises** 1. 10^2 3. $(\frac{1}{2})^1$, or $\frac{1}{2}$ 5. x^4 7. 2^3 9. $(0.6)^1$, or 0.6 11. π^3 13. True 15. False 17. False 19. True 21. True 23. $x = 5$ 25. $x = 10$ 27. $x = 5$ 29. $x = 4$ 31. $x = 1$ 33. $y = 1$ 35. $a = 0$ 37. $x = 0$ 39. p = any nonzero real number 41. −1 43. 5 45. −1 47. Sum: 2775; 100 + 330 + 505 + 077 + 099 = 1111

Pages 151-152   **P-1**   a. $\frac{1}{r^3}$ b. $\frac{1}{m^6}$ c. $\frac{1}{3^2}$ or $\frac{1}{9}$ d. $\frac{1}{2^4}$ or $\frac{1}{16}$ e. $\frac{1}{10^3}$ or $\frac{1}{1000}$ **P-2**   Add the exponents. Keep the same base.   **P-3**   a. 17^{-1} b. $(1.2)^4$ c. r^{-3} **P-4**   Subtract the exponents. Keep the same base.   **P-5**   a. 5^2 b. $(3.7)^{-3}$ c. y^2 d. x^5 e. r^{-4}

Page 153   **Classroom Exercises**   1. $\frac{1}{5^{10}}$ 3. $\frac{1}{(5.2)^2}$ 5. $\frac{1}{r^8}$ 7. x^7 9. 2^2

Pages 153-154   **Written Exercises**   1. $\frac{1}{r^4}$ 3. $\frac{2}{y^5}$ 5. $\frac{1}{4^2}$ or $\frac{1}{16}$ 7. $\frac{1}{(-3)^3}$ or $-\frac{1}{27}$ 9. $\frac{a}{b^3}$ 11. 2^5 or 32   13. a^{-3} or $\frac{1}{a^3}$ 15. 2   17. a^2 19. $x^{-5}; \frac{1}{x^5}$ 21. $n^{-1}; \frac{1}{n}$ 23. r^3 25. t^3 27. $p^{-2}; \frac{1}{p^2}$ 29. 10^4 31. $y^{-2}; \frac{1}{y^2}$ 33. $r^0; 1$ 35. 0.0009766   37. 0.0003925   39. 6.1096817   41. 0.0000485   43. positive   45. positive

Page 154   **Mid-Chapter Review**   1. 12^{13} 3. $(-6)^{14}$ 5. $(\frac{3}{4})^{13}$ 7. $(1.4)^7$ 9. 21^2 11. $(\frac{3}{4})^3$ 13. 1   15. 0   17. $(0.6)^{-3}; \frac{1}{(0.6)^3}$ 19. $w^{-6}; \frac{1}{w^6}$ 21. $12^{-2}; \frac{1}{12^2}$ 23. $m^{-3}; \frac{1}{m^3}$

Page 155   1. 490 seconds   3. 4 hours   5. c

Pages 156-157   **P-1**   2   **P-2**   8   **P-3**   a. x^2y^2 b. $9a^2$ c. $8r^3$ d. $9r^2$ **P-4**   a. $\frac{9}{t^2}$ b. $\frac{p^4}{q^4}$ c. $\frac{8r^3}{s^3}$ d. $\frac{1}{r^4}$

Page 158   **Classroom Exercises**   1. $4a^2$ 3. $\frac{1}{4}y^2$ 5. $25n^2$ 7. $4b^2$ 9. $0.25a^2$ 11. $-8r^3$ 13. x^2y^2 15. a^4b^4 17. $\frac{9}{x^2}$ 19. $\frac{m^5}{n^5}$ 21. $\frac{x^2}{y^2}$ 23. $-\frac{8}{x^3}$

Page 158   **Written Exercises**   1. $27r^3$ 3. $4a^2$ 5. $-27p^3$ 7. a^2b^2 9. $-r^3s^3$ 11. $4x^2y^2$ 13. $9p^2q^2$ 15. $-8a^3b^3$ 17. $\frac{8}{x^3}$ 19. $-\frac{a^3}{8}$ 21. $\frac{x^2}{y^2}$ 23. $\frac{y}{2}$ 25. $\frac{4}{pq}$ 27. $-\frac{1}{3t}$ 29. $\frac{27a^3}{b^3}$

Page 158   **Review Capsule for Section 6.5**   1. m^{12} 2. $(-3)^6$ 3. t^{-9} 4. $(2a)^{-6}$ 5. g^{-20} 6. $(-0.5)^6$

Pages 159-161   **P-1** $(x^2)(x^2)(x^2)$ **P-2** $2 \cdot 3 = 6$ **P-3** $(-4)(-3) = 12$ **P-4**   a. m^{15} b. k^8 c. t^{-4} **P-5**   a. 512   b. $\frac{1}{2048}$ c. $\frac{1}{1024}$ d. m^5

Page 161   **Classroom Exercises**   1. 3^6 3. $\frac{1}{a^3}$ 5. y^6 7. $(0.5)^8$ 9. $(-3)^{24}$ 11. $\frac{1}{t^6}$ 13. $\frac{q^4}{9}$ 15. $\frac{1}{x^2}$ 17. $\frac{1}{r^6s^9}$ 19. $\frac{1}{x^6y^9}$ 21. 1   23. $\frac{s^8}{m^4}$

Page 162   **Written Exercises**   1. a^8 3. $\frac{1}{y^{10}}$ 5. m^5 7. $\frac{1}{t^4}$ 9. $9r^6$ 11. $8a^6b^3$ 13. $32x^5y^{10}$ 15. $\frac{x^{15}}{32}$ 17. $\frac{32a^{10}b^{15}}{c^{105}}$ 19. $-8t^9$ 21. $\frac{1}{8x^3y^6}$ 23. $\frac{m^2}{4n^3}$ 25. $\frac{64r^2}{s^6}$ 27. $-\frac{y^5}{32x^{10}z^{15}}$ 29. $0.0000128m^{28}n^{35}$ 31. 120 miles

Page 162   **Review Capsule for Section 6.6**   1. a. 1.5678   b. 156.78   2. a. 826.605027   b. 82,660.5027   3. a. 48,278.3   b. 4,827,830   4. a. 5,280,000   b. 528,000,000   5. a. 67.5928   b. 0.675928   6. a. 502.65   b. 5.0265   7. a. 0.000018   b. 0.00000018   8. a. 16,000   b. 160

Pages 163-164   **P-1**   a. $N = 5.8; a = 3$ b. $N = 3.76; a = -1$ c. $N = 1.0059; a = -6$ **P-2**   The exponent shows how many places to "move" the decimal point.

Page 165 **Classroom Exercises** 1. 300 3. 7030 5. 46.2 7. 50,001 9. 5.3×10^2 11. 4.86×10^1 13. 7.6×10^4 15. 4.83×10^{-4}

Page 165 **Written Exercises** 1. 4200 3. 0.00029 5. 86,200 7. 0.100726 9. 7,280,000 11. 0.00005607 13. 12,800,000 15. 0.000000022 17. 9.2×10^{-3} 19. 2.705×10^3 21. 1.734×10^{-2} 23. 2.7×10^7 25. 2.75×10^{-5} 27. 2×10^6 29. 2.976×10^5 31. 1.5×10^8

Page 166 1. 13, 17, 21, 25 3. 162, 486, $-$1458, $-$4374 5. $-$21.5, $-$26, $-$30.5, $-$35 7. $\frac{1}{13}, \frac{1}{15}, \frac{1}{17}, \frac{1}{19}$ 9. $\frac{1}{49}, \frac{1}{64}, \frac{1}{81}, \frac{1}{100}$ 11. 169, 225, 289, 361 13. $-$30, $-$44, $-$60, $-$78 15. 190, 382, 766, 1534

Page 167 1. 3. 5. 7. 9. 11.

Pages 168-170 **Chapter Review** 1. 27^9 3. r^7 5. t^9 7. 19^5 9. $(2.7)^2$ 11. $(\sqrt{7})^3$ 13. 1 15. 1 17. $-$1 19. $\frac{1}{t^{10}}$ 21. $\frac{5}{x^2}$ 23. $\frac{q}{p^2}$ 25. $k^{-2}; \frac{1}{k^2}$ 27. $m^{-5}; \frac{1}{m^5}$ 29. $p^{-5}; \frac{1}{p^5}$ 31. $16t^2$ 33. $-32r^5$ 35. $\frac{25}{p^2}$ 37. $\frac{1}{16m^2}$ 39. $\frac{1}{y^3}$ 41. $\frac{1}{g^{10}}$ 43. $\frac{27m^6n^9}{p^3}$ 45. $\frac{x^6}{64y^3z^{12}}$ 47. 31,870 49. 0.005279 51. 8000.02 53. 22,240 55. 5.219×10^{-3} 57. 5.12×10^8 59. 3.1536×10^7

Pages 171-172 **Cumulative Review: Chapters 1-6** 1. 8 3. $-$18 5. 40 7. 63 9. $-36x^2y^2$ 11. $10y^2 + 55y$ 13. $4(2c - 3)$ 15. $-4a(a + 1)$ 17. $-4m - 3$ 19. $y = -9$ 21. $k = -9$ 23. $h = 2$ 25. Let x = points scored in second game, then x $-$ 15 = points scored in first game. x + x $-$ 15 = 309; x = 162, x $-$ 15 = 147 27. x = 3 29. $114
31.

	field goals attempted	per cent field goals made	field goals made
this season	x + 6	60%	0.60(x + 6)
last season	x	75%	0.75x

$0.60(x + 6) = 0.75x$; x = 24, x + 6 = 30 33. On a number line, the points to the left of, and not including, 4. 35. On a number line, the points to the right of, and including, $-$3.7. 37. {all real numbers less than $-$6} 39. {all real numbers greater than $-$4} 41. {all real numbers less than $-$25} 43. {all real numbers greater than 4} 45. {all real numbers greater than 5} 47. Let x = score on fifth test. $\frac{73 + 87 + 85 + 82 + x}{5} > 84$; $x > 93,94$ 49. b^{16} 51. x^7 53. r^3 55. m^6 57. $\frac{16}{x^2}$ 59. $\frac{27x^{12}w^3}{y^{15}}$ 61. 0.000412 63. 730,000

CHAPTER 7 ROOTS

Page 174 **P-1** 6; $-$6 **P-2** 0 **P-3** There are no two equal factors of $-$16. **P-4** a. 49 b. $\frac{1}{4}$ c. 0.09 d. 400 e. $\frac{9}{16}$ **P-5** a. $-$7 b. $\frac{1}{2}$ c. $-$0.3 d. 100 **P-6** It has more than two counting-number factors.

Page 176 **Classroom Exercises** 1. 100 3. 144 5. $\frac{1}{9}$ 7. 0.16 9. $\frac{25}{4}$ 11. 3 13. 0 15. 10 17. $-\frac{1}{4}$ 19. $-$0.1 21. $3 \cdot 7$ 23. $2 \cdot 2 \cdot 2$ 25. $2 \cdot 13$ 27. $7 \cdot 7$ 29. $3 \cdot 13$ 31. $3 \cdot 3 \cdot 5$

Page 176 **Written Exercises** 1. 11 3. $-$10 5. $\frac{2}{5}$ 7. $-\frac{1}{3}$ 9. 0.9 11. 10 13. 12 15. 50 17. $5 \cdot 13$ 19. $2 \cdot 3 \cdot 3$ 21. $2 \cdot 3 \cdot 3 \cdot 3$ 23. $2 \cdot 2 \cdot 2 \cdot 3 \cdot 5$ 25. $2 \cdot 3 \cdot 3 \cdot 11$ 27. 22 29. 28 31. 44 33. 56 35. 76 37. 72 39. 95 41. 62

Pages 177-178 **P-1** a. Irrational b. Rational c. Irrational d. Rational **P-2** a. 29 b. $\frac{1}{4}$ c. 1029 **P-3** a. $\sqrt{35}$ b. $\sqrt{66}$ c. 6

Page 179 Classroom Exercises 1. $\sqrt{10}$ 3. 8 5. 10 7. $\sqrt{2}$ 9. $\frac{1}{3}$ 11. 0 13. 20 15. -2

Page 179 Written Exercises 1. 13 3. 12 5. 54 7. 6 9. $\sqrt{51}$ 11. $-\sqrt{10}$ 13. $\sqrt{30}$ 15. $\sqrt{5}$
17. $\frac{1}{2}$ 19. 4 21. 10 23. $\frac{1}{6}$ 25. 6 27. $\sqrt{30}$ 29. 10 31. $x = 6$ 33. $x = \sqrt{14}$ 35. $x = 6$ or $x = -6$
37. $x = \sqrt{14}$ or $x = -\sqrt{14}$ 39. $a = b = c = 1$

Pages 180-181 P-1 a. 7 b. -9 c. 5 d. -13 P-2 a. 2 b. 6 c. 5 d. 4

Pages 181-182 Classroom Exercises 1. 2 3. -5 5. 4 7. 25 9. 11 11. 36 13. 10 15. 14 17. 30
19. $3\sqrt{5}$ 21. 3 23. 6 25. 7 27. $5\sqrt[3]{9}$ 29. $3\sqrt[4]{20}$

Page 182 Written Exercises 1. 6 3. 38 5. $6\sqrt{3}$ 7. $35\sqrt{35}$ 9. $4\sqrt{6}$ 11. $12\sqrt{10}$ 13. $6\sqrt{5}$
15. $10\sqrt{21}$ 17. $2\sqrt{2}$ 19. $3\sqrt{3}$ 21. $2\sqrt{6}$ 23. $2\sqrt{10}$ 25. $-4\sqrt{3}$ 27. $2\sqrt{30}$ 29. $4\sqrt{6}$ 31. $3\sqrt{30}$
33. 2 35. $3\sqrt[3]{3}$ 37. $2\sqrt[4]{9}$ 39. $5\sqrt[3]{10}$ 41. $5\sqrt[4]{2}$ 43. -10 45. $-2\sqrt[3]{7}$ 47. 16

Pages 183-184 P-1 a. 10 b. $\frac{1}{2}$ c. $\sqrt{17}$ d. $\frac{2}{3}$ P-2 A negative number cannot have a real number
as a square root. P-3 a. a^3 b. x^4 c. b^6 d. y^{14} P-4 $a; x^2; y^3$ P-5 a. No b. Yes c. Yes d. No

Pages 184-185 Classroom Exercises 1. a 3. ab 5. xy^2 7. $3y$ 9. $5n$ 11. $3x^3$ 13. $x^2\sqrt{x}$
15. $2a\sqrt{2a}$ 17. $xy\sqrt{2x}$ 19. $x\sqrt{6}$ 21. rt 23. x^4 25. $2k^2$

Pages 185-186 Written Exercises 1. $6y$ 3. yz^3 5. $x^2\sqrt{x}$ 7. $2\sqrt{x}$ 9. $4y\sqrt{y}$ 11. $2x\sqrt{2}$
13. $2a^2\sqrt{2a}$ 15. $2\sqrt{3ab}$ 17. $7ab\sqrt{ab}$ 19. $2ab\sqrt{6b}$ 21. $5r^2s^3\sqrt{3r}$ 23. $2rst^2\sqrt{5st}$ 25. x^4
27. $x^4\sqrt[3]{x}$ 29. $3xy^2$ 31. $2b\sqrt[3]{2a^2}$ 33. $3mn^2\sqrt[3]{2m^2n^2}$ 35. $2r^2s^3\sqrt[3]{4rs}$ 37. x^2 39. $2\sqrt{x}$ 41. $2x^2$
43. $x\sqrt[3]{12}$ 45. $3a^2$ 47. $4b^3$ 49. 8 51. 6.5574385 53. 6

Page 186 Mid-Chapter Review 1. 221 3. 1925 5. 27 7. 36 9. 48 11. 21 13. $-\sqrt{14}$ 15. 5
17. $\sqrt{210}$ 19. $7\sqrt{10}$ 21. $10\sqrt{7}$ 23. $9\sqrt{5}$ 25. $3\sqrt[3]{4}$ 27. $3\sqrt[4]{2}$ 29. $7m^3n$ 31. $3r^2s^2\sqrt{3s}$
33. $6ab^2c^3\sqrt{3abc}$ 35. $3p^2q^2\sqrt[3]{2p}$

Page 186 Review Capsule for Section 7.5 1. $17m$ 2. $2.7t$ 3. $77k$ 4. $-0.3w$

Pages 187-188 P-1 a. $(7 + a)x$ b. $(t - r)y$ c. $(11 + 3)m$ P-2 $11\sqrt{2}$ P-3 a. $5\sqrt{5}$ b. $7\sqrt{3}$
c. $-3\sqrt{10}$ P-4 2, 2, and 3 P-5 a. $\sqrt{3}$ b. $\sqrt{15}$ P-6 $2 \cdot 2 \cdot 2; 2 \cdot 5 \cdot 5$

Pages 188-189 Classroom Exercises 1. $7\sqrt{3}$ 3. $4\sqrt{2}$ 5. $12\sqrt{10}$ 7. $3\sqrt{2}$ 9. $-2\sqrt{11}$ 11. $5\sqrt{x}$
13. $\frac{1}{2}\sqrt{a}$ 15. $10\sqrt{m}$ 17. 5

Page 189 Written Exercises 1. $15\sqrt{3}$ 3. $4\sqrt{2}$ 5. $-3\sqrt{3}$ 7. $4\sqrt{2}$ 9. $2\sqrt{11}$ 11. $\sqrt{5} + \sqrt{15}$
13. $\sqrt{19} - \sqrt{3}$ 15. $12\sqrt{x}$ 17. $-5\sqrt{t}$ 19. $2\sqrt{s}$ 21. $11\sqrt{2}$ 23. 0 25. $5\sqrt{3}$ 27. $7\sqrt{5}$ 29. $11\sqrt{2}$
31. $5\sqrt{7a}$ 33. $4\sqrt{2r}$ 35. $12\sqrt{y}$ 37. $7\sqrt{3}$ 39. $\sqrt{7}$ 41. 0

Page 189 Review Capsule for Section 7.6 1. 6 2. $5x^2$ 3. $13x$ 4. $\frac{1}{3}x^2$

Pages 190-191 P-1 a. $\frac{2}{3}$ b. $\frac{2}{3}$ P-2 a. $\frac{n}{3}$ b. $\frac{\sqrt{2}}{x}$ c. $\frac{r\sqrt{r}}{4}$ d. $\frac{8}{a}$ P-3 a. $\frac{\sqrt{5}}{\sqrt{5}}$ b. $\frac{\sqrt{2n}}{\sqrt{2n}}$ c. $\frac{\sqrt{12s}}{\sqrt{12s}}$
or $\frac{\sqrt{3s}}{\sqrt{3s}}$ d. $\frac{\sqrt{x}}{\sqrt{x}}$

Page 192 Classroom Exercises 1. $\frac{1}{2}$ 3. $\frac{3}{5}$ 5. $\frac{4}{5}$ 7. 2 9. 2 11. 2

Page 192 Written Exercises 1. $\frac{x}{4}$ 3. $\frac{7}{a}$ 5. $\frac{y^2}{10}$ 7. $\frac{\sqrt{2}}{x}$ 9. $\frac{2\sqrt{3}}{a}$ 11. 5 13. 6 15. 7 17. 5 19. x
21. $\frac{2}{y}$ 23. $\frac{b}{a}$ 25. $\sqrt{5}$ 27. $\sqrt{6}$ 29. $\sqrt{11}$ 31. $\frac{\sqrt{15}}{5}$ 33. $\frac{a\sqrt{b}}{b}$ 35. $\frac{\sqrt{15st}}{5t}$

Page 192 Review Capsule for Section 7.7 1. $2 \cdot 2 \cdot 3 \cdot 7$ 2. $2 \cdot 2 \cdot 3 \cdot 3 \cdot 5$ 3. $2 \cdot 2 \cdot 61$
4. $2 \cdot 2 \cdot 2 \cdot 2 \cdot 3 \cdot 3$ 5. $3 \cdot 3 \cdot 23$ 6. $2 \cdot 3 \cdot 3 \cdot 13$

Page 193 **P-1** a. 784 b. 5776 c. 676 **P-2** a. 28 b. −78 c. 27 **P-3** a. 5.099 b. 8.832
c. −5.292

Page 196 Classroom Exercises 1. 841 3. 8649 5. 72 7. −88 9. 6.164 11. 8.832 13. 59
15. −12.247

Page 197 Written Exercises 1. 12.962 3. 15.620 5. 14.388 7. 16.585 9. 21.636 11. 0.866
13. 0.816 15. 0.936 17. 0.958 19. 0.306 21. 69.6 yards 23. 86 miles 25. 239 kilometers

Page 198 **P-1** a

Page 200 Classroom Exercises 1. No 3. No 5. $g^2 = 5^2 + 9^2$ 7. $s^2 = 7^2 + 11^2$

Pages 200-201 Written Exercises 1. Yes 3. No 5. c = 20 7. b = 20 9. a = 24 11. a = 15
13. b = 4.6 15. c = 12.8 17. 12 inches 19. 29.9 feet

Page 203 1. 75 3. 20 miles per hour 5. −16°C

Pages 204-206 Chapter Review 1. −7 3. 0.6 5. 23 7. $-\sqrt{33}$ 9. 6 11. $6\sqrt{3}$ 13. $2\sqrt{13}$
15. $3\sqrt{11}$ 17. $2\sqrt[3]{3}$ 19. r^3 21. $3x^2\sqrt{x}$ 23. $2ab^2\sqrt{7ab}$ 25. $2mn\sqrt[3]{2n}$ 27. $10\sqrt{13}$ 29. $9\sqrt{3}$
31. $-7\sqrt{6}$ 33. $\frac{\sqrt{2}}{4}$ 35. $\frac{2\sqrt{5}}{t}$ 37. $\frac{\sqrt{70}}{10}$ 39. $\frac{x\sqrt{6xy}}{2y}$ 41. 12.369 43. 0.829 45. 0.842
47. 3.5 meters 49. 77 miles 51. Yes 53. No 55. 9.8 inches

CHAPTER 8   POLYNOMIALS

Page 208 **P-1** a. a = 3.7; n = 2 b. a = −1; n = 1 c. a = −15; n = 0 **P-2** n = −2, which is a
negative number. **P-3** a. $\frac{1}{3}$ b. 1 c. −1 d. $-\frac{1}{5}$

Page 209 Classroom Exercises 1. Yes 3. No 5. Yes 7. 2 9. $-7\frac{1}{2}$ 11. $3\frac{2}{3}$

Page 209 Written Exercises 1. Yes 3. Yes 5. No 7. Yes 9. No 11. Yes 13. No 15. No 17. 5
19. $\frac{3}{5}$ 21. 1.8 23. −1 25. $\frac{3}{5}$ 27. $\frac{3}{5}$ 29. −3 31. 24 33. −8.3

Page 209 Review Capsule for Section 8.2 1. m^6 2. t^8 3. r^{10} 4. y^{12} 5. w^4 6. q^3 7. n^4 8. a^5

Pages 210-211 **P-1** a. x^7 b. y^{11} c. r^6 **P-2** 10 **P-3** r^2s^3 **P-4** a. x^3 b. y^3 c. r

Pages 211-212 Classroom Exercises 1. $20y^3$ 3. $-42x^5$ 5. $-8x^3y^2$ 7. $-6x^3y$ 9. $24x^4y^3$ 11. $9x^2$
13. $-2b^4$ 15. $-9s^4t$ 17. $-8p^2qr$

Page 212   Written Exercises   1. $72m^5$ 3. $-96r^3t^5$ 5. $5p^4q^5$ 7. $-24t^4$ 9. $21m^3n^4$ 11. $-27a^4b^6$
13. $-1.52r^2s^3t^4$ 15. $8.4u^3x^3y^3z^3$ 17. $-4k^3$ 19. $4y^2$ 21. $-0.6m^6n^2$ 23. $29abc^2$ 25. $-\frac{22}{7}x^8y^6z^4$
or $-3\frac{1}{7}x^8y^6z^4$ 27. $\frac{4}{15}m^9n^2q^2$

Page 213   **P-1** $2x^2$, $-9x$, and $\frac{1}{4}$ **P-2**   a and d; b and c

Pages 214-215   Classroom Exercises   1. Yes   3. Yes   5. Yes   7. No   9. No   11. Yes   13. Yes
15. Yes   17. $5x^2 - 2x + 3$ 19. $3x^3 + 12$ 21. $5x^2y - 5$ 23. $x^{12} + 5x^8 - 3x^5 - 3x^2$

Pages 215-216   Written Exercises   1. Yes   3. No   5. No   7. Yes   9. Yes   11. $3x^2 - 4x + 7$
13. $x^5 + 4x^3 - 3x^2 + 5x - 1$ 15. $3x^2 - 9x + 8$ 17. $3x^3y + 5x^2y^2 + xy^3 - 5$ 19. $m^3 - 2m^2n + 5mn - 10$
21. $-0.5x^4 - 1.9x^2 + 2.4x$ 23. -23 25. -15 27. 32 29. 49 31. $x^2 - 3x - 5$ 33. $x^2 + x + 12$
35. $67{,}349$ 37. 1472

Page 216   Mid-Chapter Review   1. 32 3. 8 5. $-6r^3t$ 7. $-5s$ 9. $-3x^2 + 6x + 12$ 11. $0.2x^3 - 3.2x^2 +$
$2.8x - 1.4$ 13. 20 15. -180

Page 216   Review Capsule for Section 8.4   1. $-3x^2 + 2x$ 2. $-3x^3 - 4x$ 3. $-6x^2 + 4x + 12$
4. $-5x + 7$ 5. $x^2 + 3x - 5$

Page 217   **P-1**   a. $x^2 - x + 3$ b. $2x^2 + 4x - 5$ **P-2**   a. $2x^2 + 4x + 3$ b. $x^3 - 2x^2 + 5x - 1$
P-3   a. $4x^3$ b. $-2x^2$ c. $-x$ d. 8

Page 218   Classroom Exercises   1. $7x - 7$ 3. $-2x^2 - 5x - 1$ 5. $2x^2 + 4x + 1$ 7. $3x^2 + 3x - 3$
9. $-x^2 - 4x$

Page 219   Written Exercises   1. $11x^2 - 2x - 8$ 3. $1.5x^2 + 2.9x + 0.8$ 5. $-2x^3 - 4x^2 - 3x + 9$
7. $-2x^2 + 6x - 18$ 9. $3x^3 - 5x^2 - 3x + 2$ 11. $4x^2 - 7x + 15$ 13. $-3x^3 + 5x^2 - 3x + 7$
15. $7x^3 - 7x^2 - 4x - 9$ 17. $2.1x^2 + 4.4x + 6.7$ 19. $12p + 3$ 21. $20y + 5$

Pages 220-221   **P-1**   a. $-2x^2$ b. $-15x$ c. 12 **P-2**   a. 5 b. -12 c. 9 **P-3**   a. $2x^2 + 9x + 4$
b. $6x^2 + 5x - 6$ c. $12x^2 - 23x + 5$ d. $2x^2 - 3x - 20$

Page 222   Classroom Exercises   1. $x^2 + 3x + 2$ 3. $x^2 + 2x - 3$ 5. $x^2 - 6x + 8$ 7. $2x^2 + 5x + 2$
9. $4x^2 - 11x - 3$ 11. $2x^2 + 5x$ 13. $x^3 - 4x^2 + 7x - 6$

Pages 222-223   Written Exercises   1. $x^2 + 8x + 15$ 3. $x^2 - 2x - 8$ 5. $x^2 - 8x + 15$ 7. $x^2 + 3x - 40$
9. $2x^2 + 13x + 15$ 11. $3x^2 + 11x - 4$ 13. $5x^2 - 17x + 6$ 15. $8x^2 - 2x - 15$ 17. $x^3 - 6x^2 + 11x - 12$
19. $2x^3 + 13x^2 + 11x - 20$ 21. $x^3 + 6x^2 - 5x - 30$ 23. $6x^3 - 13x^2 + 27x - 14$ 25. $-20x^3 - 17x^2 -$
$6x - 20$ 27. $42x^4 + 57x^3 + 18x^2 + 30x + 15$ 29. $x^2 - 11x + 28$ 31. $3x^2 + 14x - 5$

Page 223   Review Capsule for Section 8.6   1. $-4x + 2$ 2. $-2x - 4$ 3. $8x - 3$ 4. $2x^2 + 8$

Pages 224-225   **P-1** $6x^2$; $4x$; 2 **P-2** $2x^2$; $-x$

Pages 225-226   Classroom Exercises   1. $-5x^2$ 3. $3x^2 - 5$ 5. $x + 5$ 7. $3x + 2$

Page 226   Written Exercises   1. $2x + 3$ 3. $3 - 2xy$ 5. $3x + 4$ 7. $5x - 6$
9. $3x - 4$ 11. $x^2 - 3x + 2$ 13. $2x^2 - 3x + 1$ 15. $x - 5$ 17. $x^2 + 3$ 19. $2x^2 -$
$5x + 10$ 21. $x^2 + 2x + 4$ 23. $x^3 + 3x^2 + 9x + 27$ 25. See diagram at right.

25.

Pages 227-228   **P-1** $3x$ **P-2** $-8x^3 + 10x^2 - 12x - 19$ **P-3** $2x^2$

Page 229   Classroom Exercises   Note: In the answers for Ex. 1-5, Q stands for quotient and R stands for remainder.   1. Q: $x - 5$; R: 3   3. Q: $2x + 2$; R: -1 5. Q: $x + 1$; R: $-2x + 2$ 7. $10x^2 - 15x$; $6x - 9$ 9. $x^3 - 3x$; $2x^2 - 6$; $2x + 11$ 11. $x^3 - 3x^2 + 2x - 6$; $-3x^2 - 6$

Page 230   Written Exercises   1. $2x - 5 + \dfrac{3}{x + 4}$ 3. $2x + 7 - \dfrac{4}{2x - 3}$ 5. $3x + 4 - \dfrac{5}{5x - 6}$ 7. $5x - 6 - \dfrac{10}{2x + 5}$ 9. $6x - 5 + \dfrac{x}{2x^2 - 1}$ 11. $x^2 - 3x + 6 - \dfrac{8x - 8}{x^2 + x - 1}$ 13. $x - 5 - \dfrac{6}{x^2 + 8x + 2}$ 15. $x^2 + 2x - 3 + \dfrac{24}{2x^2 + 9x + 4}$ 17. $x^2 - y^2 - \dfrac{y^3}{x + y}$ 19. $x^2 - 2x + 4 - \dfrac{16}{x + 2}$ 21. $x^2 - 2xy + y^2 + \dfrac{y^3}{2x - 3y}$ 23. $4x^2 + 2x + 1$ 25. $a + 6$ 27. $x^2 - 5x + 29 - \dfrac{148}{x + 5}$ 29. $3a + 4$ 31. $4a^2 + 9a + 3$ 33. $x^2 - xy + y^2$ 35. $x - y$ 37. $x^3 + x^2 + x + 1$

Page 231   1. $212.27   3. $637.01   5. $315.28

Pages 232-234   Chapter Review   1. Yes   3. No   5. -32 7. -12.8 9. $21t^6$ 11. $2a^4b^5$ 13. $-12t^2$ 15. $\frac{1}{6}x^9y^2z^2$ 17. $\frac{13}{4}x^2y^4$ 19. No   21. No   23. $7x^2 + 3x - 16$ 25. $-3x^3 - x^2 + 12x + 5$ 27. $4x^2 - 7x - 17$ 29. $5x^3 + 2x^2 - 11x - 5$ 31. $6x^2 - 7x - 5$ 33. $x^4 - 5x^3 - 2x^2 - 2x + 12$ 35. $6r + 4$ 37. $3n^2 + 8n + 8$ 39. $3x^2 + 4x - 32$ 41. $12x^2 - 64x + 45$ 43. $-16x^4 + 12x^3 - 20x^2 + 27x - 9$ 45. $6x^2 + 22x - 8$ 47. $3xy - 2$ 49. $5x + 8$ 51. $-x^2 + x - 5$ 53. $-3x + 4 - \dfrac{19}{6x - 14}$ 55. $x^2 + 4 - \dfrac{6}{2x^2 - 5}$

CHAPTER 9   FACTORING POLYNOMIALS

Pages 236-237   P-1 $2x(2x - 3)$ P-2 $4x^2$

Page 238   Classroom Exercises   1. Yes   3. No   5. No   7. Yes   9. $2(2x + 1)$ 11. $5(x + 3)$ 13. $2(x^2 + 5)$ 15. $-3(x - 2)$ 17. $5x(x^2 - 2)$ 19. $8(3 - x^2)$

Page 238   Written Exercises   1. $4(x - 1)$ 3. $5(x + 2)$ 5. $x(2x + 3)$ 7. $-3x(x - 3)$ 9. $4x^2(2x - 3)$ 11. $6x(2x^2 - 3)$ 13. $x(x^2 - 2x + 3)$ 15. $2x(x^3 - 2x - 5)$ 17. $-5x(x + 2)$ 19. $4x^2(2x^3 - 3x + 5)$ 21. $5(x^4 - 7x^2 - 2)$ 23. Prime polynomial

Page 238   Review Capsule for Section 9.2   1. $x^2 + 2x + 2x + 2 \cdot 2$ 2. $x^2 + 4x + 2x + 4 \cdot 2$ 3. $x^2 + 7x + 3x + 7 \cdot 3$ 4. $x^2 + 3x + 2x + 3 \cdot 2$ 5. $x^2 + 11x + 1x + 11 \cdot 1$ 6. $x^2 + 5x + 4x + 5 \cdot 4$

Pages 239-240   P-1   a. Second   b. First   c. Zero   d. Eighth   P-2   a. First   b. Third   c. Fourth   P-3   a. $(x + 7)(x + 3)$ b. $(x + 2)(x + 10)$ P-4   5; 2; 10   P-5   9 and 4

Page 240   Classroom Exercises   1. Second   3. First   5. Fifth   7. $(x + 4)(x + 3)$ 9. $(x + 3)(x + 5)$ 11. $(x + 5)(x + 4)$ 13. $(x + 4)(x + 8)$

Page 241   Written Exercises   1. $(x + 3)(x + 4)$ 3. $(x + 5)(x + 3)$ 5. $(x + 4)(x + 5)$ 7. $(x + 8)(x + 4)$ 9. $(x + 3)(x + 11)$ 11. $(x + 5)(x + 2)$ 13. $(x + 3)(x + 2)$ 15. $(x + 7)(x + 3)$ 17. $(x + 13)(x + 2)$ 19. $(x + 10)(x + 3)$ 21. $(x + 7)(x + 4)$ 23. k = 12   25. k = 18

Page 241   Review Capsule for Section 9.3   1. $(x - 2)(x + 5)$ 2. $(x + 2)(x - 3)$ 3. $(x - 4)(x - 6)$ 4. $(x - 2)(x - 10)$ 5. $(x - 5)(x - 5)$ 6. $(x - 9)(x - 9)$

Pages 242-244   P-1   10; 16   P-2 $r + s$ P-3   rs   P-4 -7; 12   P-5 -1; -20 P-6   5; 10

Page 244 **Classroom Exercises** 1. 3, 2 3. 6, 3 5. None 7. 4, −2 9. 2, 4 11. None 13. $x^2 + 4x + 3x + 12$ 15. $x^2 + 5x + 3x + 15$ 17. $x^2 − 3x − 2x + 6$ 19. $x^2 + 5x − x − 5$ 21. $x^2 + 6x − x − 6$ 23. $x^2 + 7x − x − 7$

Page 245 **Written Exercises** 1. 7; 12 3. 14; 33 5. 4; −32 7. −8; −48 9. −17; 72 11. −3; 2 13. $x^2 + 12x + x + 12$ 15. $x^2 + 5x + 6x + 30$ 17. $x^2 − 5x − 3x + 15$ 19. $x^2 − 8x − 3x + 24$ 21. $x^2 − 8x + 5x − 40$ 23. $x^2 + 7x − 3x − 21$ 25. $x^2 − 12x + 3x − 36$ 27. $x^2 − 6x − x + 6$ 29. $x^2 − 14x + 3x − 42$ 31. $(x + 1)(x + 1)$ 33. $(x + 5)(x + 7)$ 35. $(x − 4)(x − 3)$ 37. $(x − 6)(x − 5)$ 39. $(x − 4)(x + 3)$ 41. $(x − 6)(x + 5)$ 43. $(x + 6)(x − 5)$ 45. $(x − 7)(x − 2)$ 47. $(x + 7)(x − 2)$ 49. Prime polynomial 51. $(x − 5)(x + 3)$ 53. $(x + 17)(x + 2)$ 55. $(x − 24)(x + 2)$ 57. $(x + 2)(x + 1)$ 59. $(x − 8)(x + 7)$

Page 246 **P-1** a. 5 b. 1 c. −1

Page 248 **Classroom Exercises** 1. $r + s = 3$; $rs = 2$ 3. $r + s = 10$; $rs = 9$ 5. $r + s = 11$; $rs = 18$ 7. $r + s = 11$; $rs = 18$ 9. $r + s = 17$; $rs = 30$ 11. $r + s = −1$; $rs = −6$ 13. $r + s = −5$; $rs = −50$ 15. $r + s = −7$; $rs = −30$ 17. $r + s = −17$; $rs = 72$

Pages 248-249 **Written Exercises** 1. $2x^2 − 3x + 1$ 3. $3x^2 + x − 2$ 5. $(2x + 1)(x + 1)$ 7. $(3x + 1)(x + 3)$ 9. $(3x + 2)(x + 3)$ 11. $(3x + 1)(2x + 3)$ 13. $(3x + 1)(2x + 5)$ 15. $(3x + 2)(x − 1)$ 17. $(2x + 5)(x − 5)$ 19. $(5x + 1)(3x − 2)$ 21. $(2x − 3)(x − 2)$ 23. $(4x − 1)(2x − 3)$ 25. $(2x − 1)(x + 3)$ 27. $(5x − 2)(3x + 4)$ 29. $(9x + 7)(2x + 1)$ 31. $(12x + 3)(x − 3)$ or $3(4x + 1)(x − 3)$ 33. $(8x + 3)(x + 4)$ 35. 5994 37. 7 39. 3

Page 249 **Mid-Chapter Review** 1. $4x(3x − 1)$ 3. $3x(x^2 + 3x − 5)$ 5. $2xy(x + 3)$ 7. $5x^2y(x − 2 + 5y)$ 9. $(x + 7)(x + 5)$ 11. $(x + 10)(x + 5)$ 13. $(x + 8)(x − 5)$ 15. $(x − 11)(x + 8)$ 17. $(x + 12)(x − 6)$ 19. $(x − 11)(x − 9)$ 21. $(2x − 3)(x + 2)$ 23. $(5x − 2)(4x + 1)$

Page 249 **Review Capsule for Section 9.5** 1. $(x − 12)(x + 12)$ 2. $(x − 2y)(x + 2y)$ 3. $(2x − 3)(2x + 3)$

Page 251 1. 0 pure tall; 64 pure short; 64 hybrid tall 3. 64 pure tall; 0 pure short; 64 hybrid tall 5. 14 pure rough; 28 hybrid rough; 14 pure smooth 7. 28 pure rough; 28 hybrid rough; 0 pure smooth

Pages 252-253 **P-1** a and c **P-2** $(x + a)(x − a)$ **P-3** 11 **P-4** a. x^3 b. r^4 c. t^6

Page 253 **Classroom Exercises** 1. $(a + 1)(a − 1)$ 3. $(p + 3)(p − 3)$ 5. $(x + 5)(x − 5)$ 7. $(t + 7)(t − 7)$ 9. $(b + 9)(b − 9)$ 11. $(p + q)(p − q)$ 13. $(2x + y)(2x − y)$ 15. $(x^2 + 5)(x^2 − 5)$

Page 254 **Written Exercises** 1. $x^2 − 64$ 3. $x^2 − 121$ 5. $4x^2 − y^2$ 7. $x^2y^2 − 9$ 9. $25x^2 − 9$ 11. $x^4 − 4$ 13. $(a + 5)(a − 5)$ 15. $(2x + 1)(2x − 1)$ 17. $(3a + 4)(3a − 4)$ 19. $(x^2 + 2)(x^2 − 2)$ 21. $(2x^2 + 3)(2x^2 − 3)$ 23. $(2x + y)(2x − y)$ 25. $(ab + 1)(ab − 1)$ 27. $(4t + 9)(4t − 9)$ 29. $(xy + r)(xy − r)$ 31. $(4xy^2 + 3)(4xy^2 − 3)$ 33. $(x^3 + 2)(x^3 − 2)$ 35. $(x^5 + y^4)(x^5 − y^4)$ 37. 224 39. 165 41. 396 43. 884 45. 1591 47. 3599

Page 254 **Review Capsule for Section 9.6** 1. $x^2 + 12x + 36$ 2. $x^2 − 6x + 9$ 3. $4x^2 − 20x + 25$ 4. $25x^2 + 10x + 1$ 5. $4x^2 + 28x + 49$ 6. $x^2 − 2xy + y^2$ 7. $9x^2 + 12x + 4$ 8. $25x^2 − 10x + 1$

Pages 255-256 **P-1** 3; 9; 9 **P-2** The value of b is negative. **P-3** a. $4x^2 + 20x + 25$ b. $x^2 − 14x + 49$ c. $9x^2 − 6x + 1$ **P-4** a. $(x + 1)(x + 1)$ b. $(x − 6)(x − 6)$ c. $(3x − 4)(3x − 4)$

Page 256 **Classroom Exercises** 1. Yes 3. Yes 5. Yes 7. No 9. No 11. No 13. $(x + 2)(x + 2)$ 15. $(x − 3)(x − 3)$ 17. $(x + 8)(x + 8)$ 19. $(x − 12)(x − 12)$ 21. $(2x + 1)(2x + 1)$

Page 257 **Written Exercises** 1. No 3. Yes 5. No 7. No 9. No 11. Yes 13. $x^2 − 10x + 25$ 15. $x^2 + 14x + 49$ 17. $4x^2 + 12x + 9$ 19. $25x^2 − 10x + 1$ 21. $16x^2 + 40x + 25$ 23. $4x^2 − 4x + 1$

Chapter 9 / 597

25. (6x − 1)(6x − 1) 27. (5x + 2)(5x + 2) 29. (3x − 4)(3x − 4) 31. (7x + 10)(7x + 10) 33. (7x − 1)
(7x − 1) 35. (x + 2y)(x + 2y) 37. (2x + 3y)(2x + 3y) 39. (xy − 8)(xy − 8) 41. (3xy + 7)(3xy + 7)

Page 257 Review Capsule for Section 9.7 1. B 2. D 3. A 4. C

Page 258 P-1 2 **P-2** 3

Pages 259-260 Classroom Exercises 1. 4; $(x^2 − 3)$ 3. 4; $(x^2 − 5x + 7)$ 5. 5; $(3x^2 − x − 5)$ 7. y;
$(x^2y − 3x + 5)$ 9. −4; $(x^2 + 5)$ 11. 3b; $(a^2b − 3a + 2)$ 13. 3y; $(x^2 − 10x + 25)$

Page 260 Written Exercises 1. 2(y + 3)(y − 3) 3. 2(x + 2)(x + 2) 5. 3(x + 7)(x + 3) 7. 4(x − 6)
(x − 4) 9. 5(x − 5)(x + 4) 11. (2y + 3)(2y − 3) 13. a(x + 2)(x + 2) 15. 2(3x + 1)(x − 2)
17. 4(3x − 1)(x − 3) 19. −3$(x^2 + 4)$ 21. a(2x + 3)(2x − 3) 23. y(x − 5)(x − 5) 25. 3(x + 5)(x − 5)
27. 2(4 − d)(4 − d) 29. 3(x − 4)(x + 2) 31. 2(1 + 8x)(1 − 8x) 33. a(b − 5)(b + 4) 35. 3(2a − 1)
(4a − 3) 37. 6 39. 2

Pages 261-262 1. 16 3. −5 5. 7 7. 3 9. 3 11. 26 13. 8 15. −6 17. 30 19. 7.44 21. $\frac{1}{4}$ 23. 4
25. 2 27. 24 29. −17

Pages 263-264 Chapter Review 1. 2x(2x − 5) 3. 6ab(3a + 5ab − 7) 5. abc$(ac^2 − b + abc)$ 7. (x + 7)
(x + 3) 9. (x + 5)(x + 10) 11. (x + 7)(x + 3) 13. (x + 6)(x − 3) 15. (x − 6)(x − 3) 17. Prime
19. (2x − 3)(x + 1) 21. (4x − 1)(x + 2) 23. (3x − 5)(x + 1) 25. (3a + 4b)(3a − 4b) 27. (rs + xy)
(rs − xy) 29. $(x^2 + 13)(x^2 − 13)$ 31. Yes 33. No 35. No 37. (x − 4)(x − 4) 39. (2x − 3)(2x − 3)
41. (x − 2y)(x − 2y) 43. 5(x + 2)(x − 2) 45. 3(x − 2)(x − 2) 47. 3$(x^2 − 4x + 32)$

Pages 265-266 Cumulative Review: Chapters 1-9 1. −13 3. 28 5. 24x − 56 7. −12n^2 − 6n
9. x = 12 11. Let x = value of lot, then x + 4000 = value of house. x + 4000 = 20,000; x = $16,000
13. Let x = number of multi-family homes, then x + 80 = number of single-family homes. x + x + 80 =
188; x = 54, x + 80 = 134 15. x = −4 17. On a number line, all points to the left of, and including, 2.
19. On a number line, all points to the right of, and not including, −1$\frac{1}{2}$. 21. {all real numbers greater
than −2} 23. {all real numbers less than −2} 25. $\frac{1}{x^4y^4}$ 27. $\frac{25c^2d^4}{x^6}$ 29. −$\frac{3}{4}$ 31. 18 33. 2$\sqrt[3]{3}$
35. 2$x^3y^4\sqrt{2xy}$ 37. −$\sqrt{3}$ 39. 2$\sqrt{2p}$ 41. $\frac{a}{4}$ 43. $\frac{\sqrt{14mp}}{7p}$ 45. 12.247 47. 0.272 49. 16 inches
51. −2 53. −7.2a^5 55. −3c^5 57. 4a^2 − 4a − 12 59. 12w^2 + 17w − 5 61. 4x − 3 63. 5y^2(2y − 3)
65. (x + 2)(x + 7) 67. (g + 2)(g − 12) 69. (3n − 1)(2n + 3) 71. (2y + 5d)(2y − 5d) 73. (3x + 6) ·
(3x − 6) or 9(x − 2)(x + 2)

CHAPTER 10 PRODUCTS AND QUOTIENTS OF RATIONAL EXPRESSIONS

Page 268 P-1 $\frac{x^2 + x − 1}{1}$ **P-2** a. Yes b. Yes c. No d. Yes e. No f. Yes   **P-3** −1

Page 269 Classroom Exercises 1. Yes 3. No 5. Yes 7. Yes 9. 0 11. 3 13. $\frac{1}{2}$ 15. 1, −1

Page 270 Written Exercises 1. Yes 3. Yes 5. Yes 7. No 9. No 11. Yes 13. −$\frac{5}{2}$ 15. $\frac{4}{3}$ 17. $\frac{4}{3}$
19. 0 21. 0 23. 10 25. 0 27. 2, −2 29. $\frac{1}{3}$ 31. 1, −7

Page 270 Review Capsule for Section 10.2 1. 2 · 3 · 3 · x · x · x 2. 2 · 2 · 3 · a · b · b 3. (x + 5)
(x − 5) 4. (c + 9)(c − 9) 5. x(x + 7) 6. (a + b)(a − b) 7. (x − 3)(x − 3) 8. 3b(a + b) 9. 3a(a + 2b)
10. a(5a − b) 11. (x + 2)(x + 2) 12. (x − y)(x + y)

Page 272   **P-1** $x + 2, x - 2; x + 3, x - 2$

Page 273   **Classroom Exercises**   1. $\frac{1}{3}$ 3. $\frac{1}{x}$ 5. $\frac{3}{x}$ 7. $\frac{x+2}{x+3}$ 9. $\frac{1}{x}$

Page 273   **Written Exercises**   1. $\frac{x}{2}$ 3. $\frac{3}{7x}$ 5. $\frac{x}{5}$ 7. $\frac{2}{3}$ 9. $\frac{x+2}{x+5}$ 11. $\frac{4x-4}{3x+9}$ 13. $\frac{3}{5}$ 15. $\frac{x+3}{x+5}$
17. $\frac{1}{x-3}$ 19. $\frac{x+4}{x}$ 21. $\frac{x}{x-2}$ 23. $\frac{x}{x-2}$ 25. $\frac{2x}{5y}$ 27. $\frac{a+b}{c}$ 29. $\frac{a-b}{3}$ 31. $\frac{5a-b}{3a+6b}$

Page 273   **Review Capsule for Section 10.3**   1. $2 \cdot 2 \cdot 3 \cdot x \cdot x \cdot y \cdot y \cdot y$ 2. $2 \cdot 2 \cdot 5 \cdot r \cdot r \cdot r \cdot s \cdot t \cdot t$
3. $2 \cdot 3 \cdot 3 \cdot a \cdot a \cdot b \cdot b \cdot b \cdot c \cdot c \cdot c$ 4. $2 \cdot 2 \cdot 3 \cdot 3 \cdot p \cdot p \cdot p \cdot p \cdot p \cdot q \cdot q \cdot r \cdot r \cdot r$

Pages 274-275   **P-1**   a. 8   b. $\frac{4}{5}$ c. $1\frac{3}{4}$ d. $\frac{1}{20}$ **P-2** x^2 and 20 have no common factors.

Pages 275-276   **Classroom Exercises**   1. $\frac{2}{5}$ 3. 1   5. $\frac{2x}{3y}$ 7. $\frac{5}{6}$ 9. $\frac{3}{5a}$ 11. $\frac{2}{5}$ 13. $\frac{1}{15}$ 15. 6ab
17. $-\frac{x}{3y}$ 19. 1

Page 276   **Written Exercises**   1. $\frac{5}{24}$ 3. $\frac{5y}{3x}$ 5. $\frac{3}{5}$ 7. $\frac{14}{15}$ 9. 2x   11. $\frac{20y}{x}$ 13. $-\frac{21b^2}{5a}$ 15. $\frac{25xy}{4}$
17. $6a^2bc$ 19. 1   21. $\frac{33a^2c}{50}$ 23. $-\frac{3ab}{2}$ 25. 10:53 A.M.

Pages 277-278   **P-1** $-a + b$ **P-2**   a. $2 - x$ b. $t - 5$ c. $m - r$ **P-3** $2 - x$ **P-4** $x^2 + 3x + 2$;
$3x - 9$

Page 279   **Classroom Exercises**   1. $\frac{2}{3}$ 3. $\frac{x^2+4x+3}{x^2+6x+8}$ 5. 1   7. $\frac{1}{x+2}$ 9. $-\frac{5}{6}$ 11. 1   13. $\frac{x+1}{x}$
15. $-\frac{2}{3}$

Pages 279-280   **Written Exercises**   1. $\frac{11}{12}$ 3. $-\frac{14}{15}$ 5. $\frac{x-3}{x-1}$ 7. $\frac{x-3}{2y}$ 9. $x - 3$ 11. $\frac{2}{x+2}$
13. $\frac{x^2-5x+6}{x^2+5x+6}$ 15. $-\frac{5}{x}$ 17. 1   19. $-\frac{1}{3}$ 21. $\frac{x}{x+3}$ 23. 0.6521739   25. 0.06034482

Page 280   **Mid-Chapter Review**   1. $-\frac{7}{5}$ 3. $\frac{9}{5}$ 5. $\frac{x}{2y}$ 7. $\frac{x+6}{2x}$ 9. $\frac{4}{15}$ 11. $-\frac{25rs}{6t^2}$ 13. $\frac{6}{5}$ 15. $2r - 3$

Page 280   **Review Capsule for Section 10.5**   1. $(x + 1)(x - 3)$ 2. $(x - 3)(x - 2)$ 3. $(2x - 1)(x + 3)$
4. $(2x + 3)(2x - 1)$

Pages 281-282   **P-1** $x, x + 2; x + 2, x - 2$ **P-2**   a. $2x - 1, x + 1$ b. $x + 6, x - 6$ c. $2x^2, 3x - 2$
P-3 $x - 5, x + 2; x + 3, x - 3; x + 2, x + 2$

Page 283   **Classroom Exercises**   1. $\frac{x-2}{x+4}$ 3. $\frac{x+5}{x+2}$ 5. $\frac{x-1}{2}$ 7. $\frac{5x^2}{x^3-x^2-9x+9}$ 9. $\frac{x+1}{x}$

Page 283   **Written Exercises**   1. $\frac{x+2}{x}$ 3. $\frac{5x+10}{3}$ 5. $\frac{5x^2+10x}{x^2-25}$ 7. $\frac{x+1}{5}$ 9. $\frac{x+4}{4}$ 11. $\frac{-3x+6}{2}$
13. $\frac{3}{2x^2+9x-5}$ 15. $\frac{2x^3-6x^2+12x-8}{3x-1}$ 17. $x = 2y$ 19. No; a and b can have any values such that
the ratio $\frac{a}{b}$ is equivalent to $\frac{5}{3}$.

Pages 284-285　P-1　a. $\dfrac{x-3}{x+2} \cdot \dfrac{x-2}{x}$　b. $\dfrac{3x}{5} \cdot \dfrac{9x}{4}$　c. $\dfrac{2x^2}{5} \cdot \dfrac{1}{x+1}$　P-2　$\dfrac{x-2}{x} \cdot \dfrac{5}{x-2}$
P-3　$\dfrac{4(r-3)}{9pq}$

Pages 285-286　Classroom Exercises　1. $\dfrac{2}{3} \cdot \dfrac{6}{5}$　3. $\dfrac{2x}{6} \cdot \dfrac{3}{x}$　5. $\dfrac{2}{x-2} \cdot \dfrac{5}{x}$　7. $\dfrac{x-7}{5x} \cdot \dfrac{4}{x+1}$
9. $\dfrac{1}{x+2} \cdot \dfrac{1}{x-3}$　11. $\dfrac{x+1}{x} \cdot \dfrac{2x}{x-10}$　13. $\dfrac{x+8}{x-5} \cdot \dfrac{x-10}{x+2}$

Page 286　Written Exercises　1. $\dfrac{2}{3}$　3. 6　5. $\dfrac{3a}{5}$　7. $\dfrac{1}{2}$　9. $\dfrac{2xy}{3}$　11. $\dfrac{5}{2xy}$　13. $\dfrac{x-5}{5}$　15. $\dfrac{2x^2+4}{x^2}$
17. $\dfrac{1}{2x+10}$　19. $\dfrac{x^4+6x^2+9}{3}$　21. $\dfrac{5x}{2}$　23. $\dfrac{2x^2+4x+2}{3x^2-6x+3}$

Page 287　P-1　$\dfrac{1}{x^2-25}$

Pages 288-289　Classroom Exercises　1. 2, x − 3　3. x, 3x − 1　5. x + 2, x − 2　7. 4, x − 4
9. 3, x − 7　11. x, x − 7　13. 2x, x − 5　15. x + 4, x − 4　17. 3x, x − 5　19. 5x, x + 3

Page 289　Written Exercises　1. $\dfrac{2}{5x^2+15x}$　3. $\dfrac{1}{2}$　5. $\dfrac{x-5}{2}$　7. $\dfrac{x^2-5x+6}{x^2+5x+6}$
9. $\dfrac{x^2+10x+24}{3x^4-21x^3-18x^2+216x}$　11. $\dfrac{xy-y^2}{2}$　13. $\dfrac{2}{7}$　15. $\dfrac{5a+18}{12}$　17. $\dfrac{3a+6b}{10a+5b}$　19. 1　21. $-\dfrac{5}{x+3}$

Page 290　1. 2300 volts　3. 1.625 amperes

Pages 291-292　Chapter Review　1. Yes　3. No　5. −3　7. $\dfrac{1}{2}$　9. $\dfrac{5x}{6}$　11. $\dfrac{1}{3}$　13. $\dfrac{2}{x}$　15. $\dfrac{2xy}{3ab^2}$
17. $\dfrac{1}{ax-a^2}$　19. $\dfrac{x^2-2x}{7}$　21. $-\dfrac{5}{6}$　23. $\dfrac{x+1}{x+2}$　25. 1　27. $-\dfrac{10x}{3a}$　29. $\dfrac{2x-4}{x^2+2x}$　31. $\dfrac{2x^2+8}{3x}$　33. $\dfrac{13a}{25}$
35. $\dfrac{10a-12b}{5a+3b}$

CHAPTER 11　SUMS AND DIFFERENCES OF RATIONAL EXPRESSIONS

Page 294　P-1　a. $\dfrac{5}{7}$　b. $-\dfrac{1}{2}$　P-2　a. $\dfrac{9}{x}$　b. $\dfrac{7}{a}$　c. $\dfrac{x}{y}$　P-3　3

Pages 295-296　Classroom Exercises　1. $\dfrac{3}{5}$　3. $\dfrac{1}{6}$　5. $\dfrac{3}{x}$　7. $\dfrac{7}{3x}$　9. $\dfrac{11}{x+1}$　11. $\dfrac{3x}{x+3}$　13. $-\dfrac{4a}{a+1}$
15. $\dfrac{2x+3y}{x+y}$　17. $\dfrac{3x}{11}$　19. $\dfrac{2x+5}{y+1}$

Page 296　Written Exercises　1. $\dfrac{11}{13}$　3. $-\dfrac{5}{17}$　5. $\dfrac{16}{a}$　7. $-\dfrac{5}{y}$　9. $\dfrac{12}{x+3}$　11. $-\dfrac{1}{a-3}$　13. $\dfrac{7x}{x+2}$
15. $-\dfrac{4a}{a+1}$　17. $\dfrac{a+b}{2a+b}$　19. $\dfrac{3r-5s}{r-2s}$　21. $-\dfrac{4y}{19}$　23. $-\dfrac{2x}{3y}$　25. $-\dfrac{4}{9t}$

Page 296　Review Capsule for Section 11.2　1. x − 3　2. −t + 6　3. 4m − 16　4. r^2-4r-2
5. $3n^2+2n-12$　6. $-2s^2+2s-3$

Page 297　P-1　Their values cannot be the same.　P-2　The numerator and denominator have no common factors.

Page 298　Classroom Exercises　1. $\dfrac{x+5}{x+2}$　3. $\dfrac{3y+3}{y+3}$　5. $\dfrac{4b+1}{b+2}$　7. 1　9. $\dfrac{2x+5}{x-3}$　11. $\dfrac{2x+2}{x+y}$

Page 299 **Written Exercises** 1. $\dfrac{2x+5}{x+y}$ 3. $\dfrac{6r-1}{r+s}$ 5. $\dfrac{4x-7}{x+2}$ 7. $\dfrac{4a-5}{a+b}$ 9. 1 11. 2 13. x 15. 3a

17. $x-4$ 19. $y-4$ 21. 4 23. $-\dfrac{6}{x+5}$

Page 299 **Review Capsule for Section 11.3** 1. 2, 2, 2, 3, x, x, x, y, y 2. 2, 2, 2, 7, r, r, s, s, s, t, t, t
3. 2, 2, (m − 3), (m + 3) 4. (x − 5), (x − 1) 5. 3, 2, (x + 4), (x − 3) 6. 2, 2, (a + 2), (a − 2), (a − 2),
(a + 1)

Page 301 **P-1** $(x-y)(x+y)3xy$

Page 302 **Classroom Exercises** 1. $3\cdot7$ 3. $2\cdot3\cdot x\cdot x\cdot y$ 5. $2\cdot3\cdot5\cdot a\cdot b\cdot c$ 7. $(x-3)(x-5)$
9. $(a+b)(a-b)$ 11. $3(a-b)$ 13. $(2x-y)(x+2y)$ 15. $3\cdot5\cdot x\cdot y$ 17. $3\cdot5(a+b)(a-b)$

Page 302 **Written Exercises** 1. 30 3. $12x^2y$ 5. $60a^2b$ 7. $(x-3)(x+3)$ 9. $3(x-y)(x+y)$
11. $6b(b+5)(b-6)$ 13. $x^2(x-5)(x+5)$ 15. $5(x+2)^2$ 17. $(x-4)^2(x+2)$ 19. $(x-2)(x-3)(x+3)$
21. No. Other possible monomials are 6c, 12c, and 24c.

Pages 303-304 **P-1** Multiplying by these numbers gives the LCM, 6x, as the denominator for both

fractions. **P-2** $40x^2y$ **P-3** There is no common factor. **P-4** $\dfrac{5r}{1}$

Pages 304-305 **Classroom Exercises** 1. $\dfrac{x}{x}$ 3. $\dfrac{2}{2}$ 5. $\dfrac{2x}{2x}$ 7. $\dfrac{3s}{3s}$ 9. $\dfrac{3ab}{3ab}$; 6ab 11. 3a; 10xy

13. $\dfrac{3x}{3x}$; $45x^2$ 15. $\dfrac{2y}{2y}$; $\dfrac{3x}{3x}$; 4y; 4y + 15x

Pages 305-306 **Written Exercises** 1. $\dfrac{29}{35}$ 3. $\dfrac{5b+7a}{ab}$ 5. $-\dfrac{7}{2x}$ 7. $\dfrac{7}{10x}$ 9. $\dfrac{15b-5a}{12ab}$ 11. $\dfrac{2b+3a}{a^2b^2}$

13. $\dfrac{15b^2-a^2}{6a^2b^2}$ 15. $\dfrac{2a^2+3b^2}{abc}$ 17. $\dfrac{2+15a^3b}{3a^2b}$ 19. $\dfrac{40c^2d-3d}{8c}$ 21. c = 2d 23. $\dfrac{2x-2}{2x+3}$ 25. Answers

will vary. One solution is shown. $\boxed{1}\ \boxed{3}\ \boxed{5}$ 27. 400 meters

Page 306 **Mid-Chapter Review** 1. $-\dfrac{4}{t}$ 3. $\dfrac{8m}{m+8}$ 5. $\dfrac{3a-3}{a+1}$ 7. m − 2 9. 6t(t + 3)(t − 3)

11. $\dfrac{2b-15a}{5ab}$ 13. $\dfrac{c^2+2d^2}{6cd}$

Page 306 **Review Capsule for Section 11.5** 1. $\dfrac{x-2}{x-1}$ 2. $\dfrac{5x}{x-3}$ 3. $\dfrac{2(x+1)}{x-4}$ 4. $\dfrac{2x}{3}$

Page 308 **P-1** $\dfrac{x+4}{x+4}$; $\dfrac{x-1}{x-1}$

Pages 308-309 **Classroom Exercises** 1. x(x + 1) 3. (x + 5)(x − 3) 5. r(r + 4) 7. (2x − 1)(x + 3)
9. x(x − 1) 11. $\dfrac{a+3}{a+3}$; 3a + 9 13. $\dfrac{y-2}{y-2}$; 2y − 4 15. $\dfrac{x-2}{x-2}$; x^2-3x+2 17. $\dfrac{x-2}{x-2}$; $x^2+3x-10$

Page 309 **Written Exercises** 1. $\dfrac{3x+2}{x(x+1)}$ 3. $\dfrac{2a+8}{a(a+2)}$ 5. $\dfrac{x^2+2x+25}{(x+5)(x-3)}$ 7. $\dfrac{a^2-2a+10}{a(a-5)}$

9. $\dfrac{a^2+b^2}{(a+b)(a-b)}$ 11. $\dfrac{a^2+6a-7}{(a-3)(a+2)}$ 13. $\dfrac{-7x+14}{(2x-1)(x+3)}$ 15. $\dfrac{2(x^2+2x-5)}{(x-1)(x-2)}$ 17. $\dfrac{-4b+1}{(b-1)(b-2)}$

19. $\dfrac{6x+1}{x(x-1)}$

Page 309 **Review Capsule for Section 11.6** 1. 6(2x − 1) 2. 5(x + 3)(x − 3) 3. x(x − 2)(x + 5)

Pages 310-311   P-1   2, x + 2; 3, x + 2   P-2   There are no common factors.   P-3   x, x + 1
P-4   x + 1, x + 2   P-5   x(x + 1)(x + 2)

Page 312   Classroom Exercises   1. x(x − 1)(x + 1)   3. (x + 3)^2(x − 2)   5. (y + 2)(y − 3)(y + 1)
7. 2(x + 1)(x − 2)^2 9. 2, x − 3   11. 4, x + 3   13. x + 2, x − 1   15. a + 5, a − 2

Pages 312-313   Written Exercises

1. $\dfrac{3}{2x + 2} + \dfrac{1}{x^2 - 1} = \dfrac{3}{2(x + 1)} + \dfrac{1}{(x - 1)(x + 1)}$

$= \dfrac{3}{2(x + 1)}(\dfrac{x - 1}{x - 1}) + \dfrac{1}{(x - 1)(x + 1)}(\dfrac{2}{2})$

$= \dfrac{3(x - 1)}{2(x + 1)(x - 1)} + \dfrac{1(2)}{2(x + 1)(x - 1)} = \dfrac{3x - 1}{2(x + 1)(x - 1)}$

3. $\dfrac{1}{x^2 + x - 6} - \dfrac{2}{x^2 - 2x} = \dfrac{1}{(x - 2)(x + 3)} - \dfrac{2}{x(x - 2)}$

$= \dfrac{1}{(x - 2)(x + 3)}(\dfrac{x}{x}) - \dfrac{2}{x(x - 2)}(\dfrac{x + 3}{x + 3})$

$= \dfrac{1(x)}{x(x + 3)(x - 2)} - \dfrac{2(x + 3)}{x(x + 3)(x - 2)} = \dfrac{-x - 6}{x(x + 3)(x - 2)}$

5. $\dfrac{2x - 3}{3(x - 3)(x - 1)}$ 7. $\dfrac{x - 7}{(x + 2)(x - 1)(x - 3)}$ 9. $\dfrac{7x + 5}{10(x - 5)(x + 5)}$ 11. $\dfrac{x^2 + 4x - 2}{2(x - 2)(x + 2)}$
13. $\dfrac{-2a - 49}{(a - 3)(a + 2)(a - 7)}$ 15. $\dfrac{2(x^2 + x - 3)}{(x + 1)(x - 1)(x + 3)}$ 17. $\dfrac{-a^2 + 5a + 4}{(2a + 1)(a - 3)(a - 1)}$

Page 313   Review Capsule for Section 11.7   1. 9r   2. t(t − 1)   3. 4(q + 2)

Pages 314-315   P-1   3   P-2   3x(x + 2)

Page 316   Classroom Exercises   1. 0, 3   3. 0, −1   5. $-\dfrac{1}{2}$ 7. 4(x − 3)   9. 2x(5x − 1)   11. 2x(x + 3)

Page 316   Written Exercises   1. x = 4   3. x = 0   5. x = 16   7. No solution   9. x = 15   11. x = −8
13. x = $-\dfrac{3}{4}$ 15. x = $\dfrac{2}{5}$ 17. 24

Pages 317-318   1. Tweeter Electronics   3. 600; 700; 900   5. No   7. No; the bars would not be as
easily compared and the reader would be more likely to use the vertical scales.   9. The graph at the right
11. The graph at the right; this graph is a more impressive representation of the company's profits.
13. State: $2000; Out-of-State: $3200; Private: $4800   15. No   17. No   19. No   21. No; since the
scale begins with 0, the lengths of the bars can be compared directly.

Pages 319-320   Chapter Review   1. $\dfrac{8}{3x}$ 3. $\dfrac{a}{a + 5}$ 5. 1   7. $\dfrac{2(a + 2)}{5a - 1}$ 9. 2   11. 6x(x − 2)

13. y^2(y − 2)(y − 3)   15. 2(y + 2)(y − 2)   17. $\dfrac{5}{6a}$ 19. $\dfrac{9y - 10x}{12xy}$ 21. $\dfrac{a^2 + ab + a - b}{(a + b)(a - b)}$ 23. $\dfrac{4a^2 - 23a + 3}{(a + 7)(a - 3)}$

25. $\dfrac{2x^2 - 2x - 7}{(x + 1)(x + 2)}$ 27. $\dfrac{3x - 2}{3(x + 2)}$ 29. $\dfrac{-9x + 16}{(2x + 3)(2x - 3)}$ 31. $\dfrac{5x^2 - 4x + 3}{x(x - 1)^2(x + 1)}$ 33. x = $\dfrac{11}{3}$ 35. x = $-\dfrac{15}{2}$
37. No solution

CHAPTER 12   RELATIONS AND FUNCTIONS

Pages 322-323   P-1   Right; Left   P-2   Up; Down   P-3   3; −2   P-4   They do not have the
same location.   P-5   The first number is x.   P-6   The first coordinate, 5, is positive.   P-7   The
second coordinate, 3, is positive.   P-8   Start at 0, and move left 2 units along the x axis; −2   P-9
Down 4 units parallel to the y axis; −4   P-10   (−2, −4)

Pages 323-324   Classroom Exercises   1. right 3, down 2   3. left 3, up 7   5. right 7, down 1
7. right 8, up 3   9. left 3, down 3   11. (2, 1)   13. (−3, −1)   15. (3, −2)   17. (2, −3)   19. (−4, −3)

Page 324   Written Exercises   For Ex. 1-11, the moves described locate the given points. Always start at origin.   1. Move 1 unit right and 4 units up.   3. left 5, up 3   5. right 6, down 1   7. left 4, down 1
9. left 4, down 4   11. right 3, up 3   13. (−5, −1)   15. (−2, 3)   17. (−4, −2)   19. (4, 3)   21. (−1, 1)
23. (−2, −1)

Page 324   Review Capsule for Section 12.2   1. −4 and −3   2. 5 and 6   3. −1 and 0   4. −4 and −3
5. −4 and −3   6. −2 and −1

Pages 325-326   P-1   x negative, y negative; x positive, y negative   **P-2**   Start at 0 and move $2\frac{1}{2}$ units
left.   **P-3**   Move $3\frac{1}{4}$ units up parallel to the y axis.   **P-4**   (0, 0); (−3, −4)

Page 327   Classroom Exercises   1. II   3. I   5. On the y axis, 5 units down from the origin   7. On the
x axis, $5\frac{1}{2}$ units to the left of the origin

Page 327   Written Exercises   For Ex. 1-19, each point may be located on a graph by the directions following each coordinate pair. Always start at (0, 0), the point of origin. Note in Ex. 5, 9, and 15 there is only one move to locate each point.   1. right $4\frac{1}{2}$ units, up 3 units   3. left 2, down $4\frac{1}{2}$ 5. up 4
7. left 3.5, up 2.2   9. left $3\frac{1}{2}$ 11. left 3, up 2.2   13. left 6, down 2   15. down 2   17. left 1, up 2
19. left 7, up 2   21. parallelogram

Page 327   Review Capsule for Section 12.3   1. $1\frac{1}{4}$ 2. $2\frac{3}{4}$ 3. −30   4. 2.1

Page 328   P-1   (2, 1), (0, 1), (−2, 1), (2, −1), (0, −1), (−2, −1)

Pages 329-330   Classroom Exercises   For Ex. 1 and 3, sample answers are given.   1. (−3, 3), (0, 0),
(3, 3)   3. (−2, 3), (0, 1), (3, −2)   5.

x	−2	−1	0	1	2
y	−2	−1	0	1	2

7.

x	−2	−1	0	1	2
y	4	1	0	1	4

9.

x	−2	−1	0	1	2
y	1	0	1	4	9

11. (1, 1), (−1, −1)   13. (1, 1), (−1, 1)   15. (1, 4), (−1, 0)

Pages 330-331   Written Exercises   1. $\{(1, 0), (2, 1), (3, 2), (4, 3)\}$ 3. $\{(1, 1), (2, 4), (3, 9), (4, 16)\}$
5. $\{(1, 1), (2, 2), (3, 3), (4, 4)\}$ 7. $\{(1, 1), (2, \sqrt{2}), (3, \sqrt{3}), (4, 2)\}$ 9. $\{(1, 1), (2, 1), (3, 1), (4, 1)\}$
11. $\{(1, 1), (2, 3), (3, 7), (4, 13)\}$ For Ex. 13-15, each consists of 4 points. To locate each point, start at origin. Move in the stated direction the stated number of units.   13. (1 right, 2 up), (1 right, 3 up),
(3 left, 5 up), (3 left, $4\frac{1}{2}$ down)   15. (Note: $\pi \approx 3.1$); (origin), (up π), (down π), (right π)   For Ex. 17-
27, each graph consists of 4 points. Ordered pairs are given.   17. (−1, 0), (0, 1), (1, 2), (2, 3)   19. (−1, −6),
(0, −7), (1, −6), (2, −3)   21. (−1, 1), (0, 0), (1, 1), (2, 2)   23. (−1, 1), (0, 1), (1, 1), (2, 1)   25. (−1, −1),
(0, 0), (1, −1), (2, −2)   27. (−1, −1), (0, −1), (1, 1), (2, 5)   29. Loss; the 10% profit is $9 and the 10% loss is $11, resulting in a total loss of $2.   31. 888 + 88 + 8 + 8 + 8

Page 331   Mid-Chapter Review   1. (−4, −2)   3. (−2, 0)   5. (0, 3)   7. (−5, 3)   For Ex. 9-13, each point may be located on a graph by the directions following each coordinate pair. Always start at (0, 0), the point of origin.   9. right 2 units, down 1 unit   11. left 2.5, down 3.5   13. right 4, down $3\frac{1}{4}$
15. $\{(−2, 4), (−1, 3), (0, 2), (1, 3), (2, 4)\}$

Page 331 **Review Capsule for Section 12.4** 1. A(−2, 1), B(−2, −1), C(−4, 2), D(−4, −2) 2. A(−2, 3), B(2, 3), C(−1, 0), D(1, 0) 3. A(−3, −2), B(−1, −3), C(2, 1), D(2, −1)

Page 332 1. 8000 Btu's 3. 6000 Btu's 5. 3500 Btu's

Pages 333-334 **P-1** −1, 0, 4, −3; 3, −2, 0 **P-2** Two x values are equal in A; no two x values are equal in B. **P-3** Show that no two x values of relation T are equal. **P-4** P and S **P-5** Yes; a vertical line will have only one point in common with the graph.

Page 335 **Classroom Exercises** 1. Domain: $\{3, −5, 4, 0\}$; Range: $\{7, 2, −1, 0\}$ 3. Domain: $\left\{3, \sqrt{2}, \pi, \frac{1}{3}\right\}$; Range: $\{8, −1, −2\}$ 5. Domain: $\{2, 3, 4, 5\}$; Range: $\{3, 4, 5, 6\}$ 7. Domain: $\left\{1, −1, 0, \frac{1}{2}, −\frac{1}{2}\right\}$; Range: $\left\{1, 0, \frac{1}{2}\right\}$ 9. Yes 11. Yes 13. Yes 15. Yes

Pages 335-336 **Written Exercises** 1. D = $\{5, −3, −1, 1\}$; R = $\{10, 2, 0, −3\}$ 3. D = $\left\{\frac{1}{2}, −1, −\frac{1}{2}, 2\right\}$ R = $\left\{2, \frac{1}{2}, −1\right\}$ 5. D = $\{−1, −2, \pi, 0\}$; R = $\{\sqrt{2}, \sqrt{3}, −1, −\pi\}$ 7. D = $\{0, 1, 1.5, −1, −\pi\}$; R = $\{0, 1, −1\}$ 9. D = $\left\{0, \frac{1}{2}, 1, 1\frac{1}{2}, −\frac{1}{2}, −1\right\}$; R = $\{0, 1, −1\}$ 11. Yes 13. Yes 15. No 17. No
For Ex. 19-25, apply the vertical line test to each graph. If the line has more than one point in common with a graph, the graph is not a function. 19. Yes 21. No 23. Yes 25. No

Page 336 **Review Capsule for Section 12.5** 1. (−2, 2), (1, 1), (7, −1) 2. (−2, −2), (1, 1), (0, 2) 3. (−3, 3), (−2, 2), (1, 1), (0, 0) 4. (−2, −2), (0, 0) 5. (−2, −2), (0, 0) 6. (0, 0)

Pages 337–338 **P-1** −3 **P-2** {all real numbers greater than or equal to −1} **P-3** $\frac{5}{4}$

Pages 338-339 **Classroom Exercises** 1. 0, −2, 2, −4, 4 3. The vertical line does not cross the graph at more than one point. 5. {real numbers} 7. 0 9. {all real numbers greater than or equal to −1 and less than 3} 11. $1\frac{1}{2}$

Page 339 **Written Exercises** The graph is either a line, two lines, or a curve (parabola). Points on the graph are given. 1. Line; (0, 0), (2, −4) 3. Line; (0, 1), (2, −1) 5. Two lines (inverted V); (0, 0), (−2, −2), (2, −2) 7. Two lines (V-shaped); (−2, 1), (−1, 0), (0, 1), (2, 3) 9. Line parallel to the x axis; (0, −3), (2, −3) 11. Curve; (−2, −2), (−1, 1), (0, 2), (1, 1), (2, −2) 13. Curve; (−2, 4), (−1, 1$\frac{1}{2}$), (0, 0), (1, −$\frac{1}{2}$), (2, 0) 15. Curve; (−2, −6), (−1, −2), (0, 0), (1, 0), (2, −2)

Pages 341-342 **Classroom Exercises** 1. 20 meters 3. 75 meters 5. No 7. 18 kilometers 9. 30 kilometers

Page 342 **Written Exercises** 1. 7 meters 3. 10 meters 5. 38 meters 7. $1.07 9. $1.41 11. $1.41

Page 344 1. $240; Clue 2 only was used. 3. Can't tell. 5. 12 ounces; both Clue 1 and Clue 2 were used. 7. 5 miles per hour; both Clue 1 and Clue 2 were used. 9. 20; Clue 2 only was used.

Pages 345-346 **Chapter Review** Each point in Ex. 1-15 will be located by moving in the stated direction the stated number of units. 1. left 3, up 2 3. left 3, down 4 5. right 3, down 3 7. right 3, up 1 9. right 2, down 3$\frac{1}{2}$ 11. right 2.25, up 1.75 13. up 3 15. left 2$\frac{1}{2}$, down 3$\frac{1}{2}$ For Ex. 17-21, each graph consists of four points, which are listed as a set of ordered pairs. 17. $\{(−1, −8), (0, −5),$

(2, 1), (3, 4)} 19. {(−1, 0), (0, −1), (2, 3), (3, 8)} 21. {(−1, 2), (0, 1), (2, 1), (3, 2)} 23. D = $\left\{-\frac{1}{2}, 2, 4\right\}$; R = {0, 6, −3} 25. D = {−3, 0, −2}; R = $\left\{\frac{1}{2}, -5, 0, -4\right\}$ 27. Yes 29. No 31. The graph is a line through the points (0, −3), (1, −2), (2, −1), (3,0). 33. Two lines (V-shaped) through (−2,0), (0, −2), (2,0). 35. 30 feet 37. 60 feet 39. 83 feet

Pages 347-348   Cumulative Review for Chapters 1-12   1. Let x = cost of labor, then x − 500 = cost of parts. x + x − 500 = 2100; x = 1300, x − 500 = $800   3. y = $\frac{21}{5}$ or $4\frac{1}{5}$ 5. d < 16   7. y > $-\frac{23}{5}$ or y > $-4\frac{3}{5}$ 9. $-\frac{1}{125m^3}$ 11. $\frac{16x^{12}}{y^8}$ 13. 8   15. $2a^3n^4\sqrt{3n}$ 17. $2\sqrt{6}$ 19. $10\sqrt{2g}$ 21. −51   23. $-3c^3d^7e$ 25. 5x − 2   27. 3x(x − 4)   29. (g − 5)(g + 9)   31. (3x − 1)(2x − 5)   33. (2h − 5)(2h − 5)   35. 2(n − 3)(n + 6)   37. $-\frac{1}{2}$ 39. 2, −5   41. $\frac{3}{x-2}$ 43. $\frac{3}{ab}$ 45. $\frac{4x^2y}{3}$ 47. $-\frac{x-1}{x+9}$ 49. $\frac{x-1}{3}$ 51. $\frac{1}{x-2}$ 53. $\frac{17}{x+2}$ 55. $-\frac{1}{10y}$ 57. $\frac{7}{6(x-3)}$ 59. −2   61. 1   63. right 5, down 6   65. left $1\frac{1}{2}$, down 2   67. Yes   69. Two lines (V-shaped); {(−4, 2), (−2, 1), (2, 1), (4, 2)}

CHAPTER 13   LINEAR EQUATIONS

Pages 350-351   P-1   (3, 5), (−4, −9), etc.   **P-2**   Substitute 3 for x and 5 for y in y = 2x − 1.   **P-3**   m: −1, b: $\frac{1}{2}$; m: $\frac{2}{3}$, b: $-\frac{1}{4}$; m: −3, b: 10   **P-4**   Functions a and b

Pages 351-352   Classroom Exercises   1. −6, −3, 0, 6   3. −49, −19, 1, 31   5. m: −3; b: 7   7. m: −1; b: −3   9. m: −7; b: 0   11. m: −0.5; b: 3.8

Page 352   Written Exercises   For Ex. 1-11, four points are given.   1. (3, 0), (0, 3), (1, 2), (−2, 5)   3. (0, −1), (1, 2), (2, 5), (−1, −4)   5. (−1, −2), (2, 4), (0, 0), (3, 6)   7. (−3, −2), (1. −2), (0, −2), (2, −2)   9. (0, 3), (2, −5), (−2, 11), (3, −9)   11. (0, $\frac{1}{2}$), (2, $-\frac{1}{2}$), (−2, $\frac{3}{2}$), (3, −1)   13. Yes   15. Yes   17. Yes   19. Yes   21. No   23. Yes   25. 32 cm²

Pages 353-354   P-1   In each the value of b is 2.   **P-2**   (0, 2)   **P-3**   x value: 0; y values: (A) 1, (B) −3, (C) 3, (D) −1   **P-4**   a. $\frac{5}{2}$ or $2\frac{1}{2}$ b. 4

Pages 354–355   Classroom Exercises   1. 1   3. $\frac{2}{3}$ 5. $1\frac{1}{2}$ 7. $\frac{1}{2}$ 9. $-\frac{1}{3}$ 11. −1   13. 0   15. 6

Page 355   Written Exercises   1. 7   3. $\frac{7}{2}$ 5. $-\frac{5}{3}$ 7. 0.3   9. 6   11. 0   For Ex. 13-21, two points are given.   13. (0, 1), (−1, 0)   15. (0, −3), (−2, 1)   17. (0, 5), (4, 7)   19. (0, 0), (2, −6)   21. (0, $\frac{3}{2}$), (5, 4)

Page 355   Review Capsule for Section 13.3   In each exercises, two points are given.   1. (0,5), (2,6)   2. (0,3), (2,4)   3. (0,1), (2,2)   4. (0,0), (2,1)   5. (0, −3), (2, −2)   6. (0, −5), (2, −4)

Pages 356-357   P-1   2   **P-2**   m   **P-3**   In all four the value of b is −1.   **P-4**   a. Slope: $-\frac{3}{4}$, y intercept: −2   b. Slope: 5, y intercept: $-\frac{1}{2}$ c. Slope: $-\frac{3}{2}$, y intercept: $\frac{5}{2}$ **P-5**   I, II, and IV   **P-6**   I and III; II and IV

Pages 357-358   Classroom Exercises   1. Slope: 3; y intercept: 2   3. Slope: 1; y intercept: 5   5. Slope: $\frac{1}{3}$; y intercept: −6   7. y = 5x + 3   9. y = −1x + 0 or y = −x   11. y = 3x + 10

Page 358 **Written Exercises** 1. 1 3. $-\frac{1}{2}$ 5. 2 7. -3 9. $-\frac{1}{2}$ 11. -4 13. $y = -x + 5$ 15. $y = -\frac{1}{3}x - 3$ 17. $y = -x$ 19. I, III 21. II, IV 23. II, IV 25. I, III 27. II, IV 29. I, III

Page 358 **Review Capsule for Section 13.4** 1. $-\frac{3}{5}$ 2. $-\frac{3}{2}$ 3. 5 4. $\frac{5}{8}$ 5. $\frac{2}{3}$

Page 359 **P-1** $\frac{3}{2}$ **P-2** Vertical motion: 3 units; Horizontal motion: 2 units **P-3** 6 units; 4 units

Page 361 **Classroom Exercises** 1. $\frac{3}{5}$ 3. $\frac{3}{2}$ 5. $-\frac{1}{3}$ 7. -1 9. $\frac{1}{2}$ 11. -1 13. 2 15. $-\frac{1}{2}$ 17. 2

Pages 361-363 **Written Exercises** 1. 1 3. 0 5. $-\frac{3}{4}$ 7. $-\frac{7}{6}$ 9. -1 11. 0 13. $-\frac{3}{4}$ 15. $-\frac{3}{2}$ 17. $y = x + 5$ 19. $y = \frac{9}{4}x - 3$ 21. $y = \frac{8}{5}x$ 23. $y = -6$ 25. $y = -x + 5$ 27. $y = \frac{2}{3}x + 6$ In Ex. 29-33, a second point is given that can be found using the given slope. The line can be drawn using the two points. 29. (1, 3) 31. (6, −6) 33. (0, 3) 35. About 98 inches 37. About 27,000,000 miles

Page 363 **Mid-Chapter Review** For Exercises 1-5, 4 points are given. 1. (0, −3), (1, −2), (2, −1), (3, 0) 3. (−2, 5), (−1, 3), (0, 1), (1, −1) 5. (−2, −3), (0, −2), (2, −1), (4, 0) For Exercises 7 and 9, the y intercept and 2 ordered pairs are given. 7. y intercept: 3; (−1, 4), (1, 2) 9. y intercept: $\frac{5}{2}$; (1, 2), (3, 1) 11. $y = -\frac{1}{4}x - 1$ 13. -1 15. 0

Pages 364-365 **P-1** They appear to be parallel. **P-2** a and c **P-3** 0

Pages 365-366 **Classroom Exercises** 1. Yes 3. Yes 5. $y = 4x + 3$ 7. $y = \frac{3}{2}x + 10$ 9. $y = -5.3x + 0.6$

Page 366 **Written Exercises** In Ex. 1-5, each graph is a pair of parallel lines. Two points are given for each line. 1. (0, 3), (1, 1); (0, −1), (−1, 1) 3. (0, −2), (4, 4); (0, 1), (−2, −2) 5. (0, 3), (4, 8); (0, 0), (4, 5) In Ex. 7-9, the graph is a set of three parallel lines. Two points are given for each line. 7. (0, −3), (2, 1); (0, 3), (−2, −1); (0, 0), (2, 4) 9. (0, 3), (−2, 1); (0, −3), (2, −1); (0, 0), (3, 3) The graphs of Ex. 11 and 13 are horizontal lines. One point is given for each line. 11. (0, 5) 13. $(0, \frac{3}{2})$ 15. Yes; the two lines would have the same slope.

Page 366 **Review Capsule for Section 13.6** 1. $x = -2$ 2. $x = 2$ 3. $x = 12$ 4. $x = \frac{3}{4}$

Pages 367-368 **P-1** (−4, 0) **P-2** a. 5 b. $-\frac{1}{2}$ c. −4 **P-3** $\{-2\}$ **P-4** There is more than one y value for the x value, −2.

Page 369 **Classroom Exercises** 1. −3 3. 1 5. −2 7. $\frac{5}{2}$ 9. −5 11. 1 13. $4\frac{1}{2}$ 15. $x = -5$ 17. $x = -1$ 19. $x = 2\frac{1}{2}$

Page 369 **Written Exercises** 1. −6 3. 3 5. −2 7. 3 9. 4 11. −4 In Ex. 13-19, each graph is a straight line. Two ordered pairs are given for each line, one using the x intercept, the other using the y intercept. 13. (2, 0), (0, −4) 15. (6, 0), (0, −3) 17. (−3, 0), (0, 6) 19. $(\frac{5}{2}, 0)$, (0, −5) In Ex. 21 and 23, the graphs are vertical lines. One point is given for each line. 21. (5, 0) 23. $(2\frac{1}{2}, 0)$

Page 369 **Review Capsule for Section 13.7**   1. slope: 2; y intercept: 3   2. slope: 1.8; y intercept: 32
3. slope: π; y intercept: 0   4. slope: 1; y intercept: 273   5. slope: 2.5; y intercept: 3

Page 371   **P-1** $9

Page 371   **Classroom Exercises**   1. 38°C   3. -7°C   5. -29°C   7. 6 m   9. 7 m

Page 372   **Written Exercises**   1. See Ex. 2 on p. 357 (y = 2x). This graph is similar except that only the
portion in the first quadrant is drawn.   3. The graph is a portion of a line and contains the points
$(-273, 0)$ and $(0, 273)$. The graph is in quadrants I and II only.   5. 68°F   7. 104°F   9. -22°F
11. 18 km/hr   13. 54 km/hr   15. 0.6 cm   17. 1.3 cm

Page 373   1. 1102.5 meters   3. 314 meters

Pages 374-376   **Chapter Review**   1. No   3. Yes   5. The graph is a straight line. The coordinates of
four points are given. $(-1, 3)$, $(0, 1)$, $(1, -1)$, $(2, -3)$ 7. -10 9. 3.8   11. $y = 4x - 1$ 13. $y = -2.8x +$
2.9   15. II, IV   17. I, III   19. $-\frac{1}{2}$ 21. 0.8   23. -2 25. -3 27. $y = 2x - 2$ 29. $y = \frac{3}{4}x - \frac{5}{2}$ 31. The
lines are parallel; two points are given. $(0, -2)$, $(4, 0)$; $(0, 3)$, $(4, 5)$; $(0, 0)$, $(4, 2)$ The graphs of Ex. 33
and 35 are horizontal lines. One point is given.   33. $(0, -3)$ 35. $(0, 3.5)$ 37. $-2\frac{1}{2}$ 39. 8   In Ex. 41
and 43, each graph is a straight line. Two ordered pairs are given for each, one using the x intercept, the
other, the y intercept.   41. $(0, 1)$, $(-1, 0)$ 43. $(0, -6)$, $(2, 0)$ In Ex. 45 and 47, the graphs are vertical
lines. One point is given for each line.   45. $(-1, 0)$ 47. $(\frac{7}{2}, 0)$ 49. 19.6 m   51. 12.6 m

CHAPTER 14   SYSTEMS OF SENTENCES

Pages 378–379   **P-1**   The y intercept; The x intercept   **P-2** $x > 2\frac{1}{2}$; $x \le 2\frac{1}{2}$

Page 380   **Classroom Exercises**

1.
x	0	2	4
y	2	0	-2

3.
x	-2	0	2
y	-1	0	1

5.
x	0	5	2
y	-5	0	-3

7.
x	0	5	-4
y	$-\frac{5}{3}$	0	-3

Page 380   **Written Exercises**   In Ex. 1-7, each graph is a straight line. Three ordered pairs from the
Classroom Exercises are shown for each exercise.   1. $(0, 2)$, $(2, 0)$, $(4, -2)$ 3. $(-2, -1)$, $(0, 0)$, $(2, 1)$
5. $(0, -5)$, $(5, 0)$, $(2, -3)$ 7. $(0, -\frac{5}{3})$, $(5, 0)$, $(-4, -3)$ In Ex. 9-17, each graph is a straight line. Two
points are given for each line.   9. $(1, -1)$, $(3, 2)$ 11. $(-2, -3)$, $(2, -2)$ 13. $(0, 0)$, $(5, 4)$ 15. $(-1, 0)$,
$(0, -1)$ 17. $(0, -6)$, $(2, 0)$

Page 380   **Review Capsule for Section 14.2**   1. $(-1, 2)$ 2. $(-1, 1)$ 3. $(-3, 14)$

Page 381   **P-1** $(-3, -2)$

Page 383   **Classroom Exercises**   1. No   3. Yes   5. No   7. Yes   9. Yes   11. $\{(-2, -1)\}$ 13. $\{(-1\frac{1}{2}, 2)\}$

Page 384   **Written Exercises**   In Ex. 1-17, each graph is a pair of straight lines. The solution set is
given for each exercise.   1. $\{(1, 3)\}$ 3. $\{(-1, 4)\}$ 5. $\{(-3, 4)\}$ 7. ϕ 9. $\{(5, -2)\}$ 11. $\{(0, 0)\}$
13. $\{(0, 0)\}$ 15. $\{(1, -3)\}$ 17. $\{(1, 5)\}$ 19. k = 2   21. k = -1

Page 385   **P-1**   a. $2x = 9$ b. $3y = 12$ c. $5x = 10$ **P-2**   a. $2y = 8$ b. $4x = 12$ c. $2x = -6$

Page 386   **Classroom Exercises**   1. $2x = 6$ 3. $5y = 15$ 5. $3x = 12$ 7. x = 4   9. x = 1

Chapter 14 / 607

Pages 386-387 **Written Exercises** 1. $\{(2, 4)\}$ 3. $\{(-4, 2)\}$ 5. $\{(-2, 1)\}$ 7. $\left\{(4, \frac{3}{2})\right\}$ 9. $\{(6, 0)\}$
11. $\left\{(-1, \frac{3}{2})\right\}$ 13. $\{(-2, 1)\}$ 15. $\{(2, 7)\}$ 17. $\left\{(\frac{1}{8}, -\frac{5}{2})\right\}$

Page 387 **Mid-Chapter Review** For Ex. 1-5, each graph is a line. Two points on each line are given.
1. $(0, 5)$, $(5, 0)$ 3. $(0, 1)$, $(1, -3)$ 5. $(0, 2)$, $(4, 5)$ 7. $\{(-2, -3)\}$ 9. $\{(2, -2)\}$ 11. $\left\{(2, \frac{1}{3})\right\}$

Page 387 **Review Capsule for Section 14.4** 1. 3 2. $\frac{1}{2}$ 3. -4 4. 1 5. -1 6. $2x - 2y$ 7. $3x + 3y$
8. $3x + 6y$ 9. $8x - 12y$ 10. $-9x + 15y$ 11. $-2x + 2y$ 12. $-4x + 2y + 6$ 13. $-15x + 3y + 15$

Pages 388-389 **P-1** Multiply each side of eq. 1 by: (a.) 3; (b.) -1; (c.) -7 **P-2** a. Multiply each
side of equation 1 by 5 and each side of equation 2 by 8. b. Multiply each side of equation 1 by 4 and
each side of equation 2 by 3. c. Multiply each side of equation 1 by 3 and each side of equation 2 by -2.

Pages 389-390 **Classroom Exercises** 1. Multiply equation 1 by 2. 3. Multiply equation 2 by -2.
5. Multiply equation 1 by 3 and equation 2 by 4.

Page 390 **Written Exercises** 1. $\{(4, 2)\}$ 3. $\{(0, 2)\}$ 5. $\left\{(\frac{1}{2}, 4)\right\}$ 7. $\{(2, -3)\}$ 9. $\left\{(\frac{9}{7}, -\frac{10}{7})\right\}$
11. $\{(0, 0)\}$ 13. $\{(0, 0)\}$ 15. $\{(2, -2)\}$ 17. $(1, -\frac{1}{3})$ 19. No 21. Yes 23. No

Page 391 **Review Capsule for Section 14.5** 1. $x + y = -12$ 2. $y - x = 7$ 3. $x = y - 6$ 4. $y = x + 13$

Page 392 **Classroom Exercises** 1. First number: x; second number: y; $x + y = 25$ 3. Jerry's age: x;
Mary's age: y; $x + y = 25$ 5. Width: x; length: y; $x + y = 51$ 7. First number: x; second number: y;
$2x + 3y = 19$

Page 393 **Written Exercises** 1. Larger: x; smaller: y; $x + y = 28$; $x - y = 3$; Answer: $15\frac{1}{2}$, $12\frac{1}{2}$
3. First: x; second: y; $x + y = -12$; $x - y = 42$; Answer: 15, -27 5. Smaller: x; greater: y; $2x + 2 = y$;
$3x + 2y = 39$; Answer: 5, 12 7. Length: x; width: y; $x = 8 + y$; $2x + 2y = 34$; Answer: $4\frac{1}{2}$, $12\frac{1}{2}$
9. Ann's age: x; John's age: y; $y + 7 = x$; $x + y = 45$; Answer: 26, 19 11. Sam's crate: x; Tina's crate:
y; $x = y - 6$; $2y = 3x + 5$; Answer: 13, 7

Page 394 1. 29.6 grams 3. 21.9 grams

Pages 395-396 **Chapter Review** For Ex. 1-5, each graph is a straight line. Two points are given for
each line. 1. $(0, 3)$, $(2, -1)$ 3. $(-1, 0)$, $(4, 1)$ 5. $(0, -2)$, $(3, 0)$ 7. $\{(-3, 2)\}$ 9. $\{(1, 2)\}$
11. $\{(2, -1)\}$ 13. $\{(-4, 9)\}$ 15. $\{(9, -4)\}$ 17. ϕ 19. Greater: x; smaller: y; $x + y = 8$; $4x = 4y + 2$;
Answer: $4\frac{1}{4}$, $3\frac{3}{4}$ 21. Width: x; length: y; $3x + 2y = 40$; $2x + 2y = 34$; Answer: 6, 11

CHAPTER 15 MORE ON SYSTEMS OF SENTENCES

Page 399 **P-1** $2x - 3$ **P-2** $\frac{1}{2}y + 2$

Page 400 **Classroom Exercises** In Ex. 1-9, other answers can also be given. 1. $x = 2y + 3$ 3. $x = 1 - 3y$ 5. $x = 3 - 3y$ 7. $x + (2x - 3) = 5$ 9. $-2(3y + 1) + 5y = 2$

Page 400 **Written Exercises** 1. $\{(6, 2)\}$ 3. $\{(14, 10)\}$ 5. $\{(10, 6)\}$ 7. ϕ 9. $\{(3, -4)\}$
11. $\{(0, -2)\}$

Page 401 **P-1** Because xy represents the product of the digits, not the number.

Page 403 **Classroom Exercises** 1. $x + y = 12$ 3. $y = x - 5$ 5. $5x = 7y$ 7. $x + y = 2x - 1$ 9. $y - x = -8$

Page 404 **Written Exercises** 1. $y = x - 2$; $x + y = 12$; Ans: 75 3. $x = 2y$; $x = y + 4$; Ans: 84
5. $x + y = 15$; $2x = 3y - 5$; Ans: 87 7. $y = x + 1$; $10x + y = 8y + 2$; Ans: 34 9. $x - y = -4$; $10x + y = 2(x + y) + 3$; Ans: 15

Page 404 **Review Capsule for Section 15.3** 1. $x = 48$ 2. $x = 100$ 3. $x = 150$ 4. $x = 450$

Page 405 **P-1** 20 grams

Pages 406-407 **Classroom Exercises** 1. 8 3. 14 5. $0.06y$ 7. 120 9. 15; 485 11. 40; 30

Pages 407-408 **Written Exercises** 1. $0.08y$; $0.05(400)$; $x + y = 400$; $0.04x + 0.08y = 0.05(400)$;
Ans: 4% solution: 300 g, 8% solution: 100 g 3. $0.12x$; $0.20(48)$; 15%; $0.15y$; $x + 48 = y$; $0.12x + 0.20(48) = 0.15y$; Ans: 12% solution: 80 g, 15% solution: 128 g 5. Let x = grams of 15% solution, then y = grams of 30% solution. $x + y = 150$; $0.15x + 0.30y = 0.18(150)$; Ans: 15% solution: 120 g, 30% solution: 30 g 7. Let x = grams of $2\frac{1}{2}$% solution, then y = grams of 4% solution. $x + 250 = y$; $0.025x + 0.10(250) = 0.04y$; Ans: $2\frac{1}{2}$% solution: 1000 g, 4% solution: 1250 g 9. 1806 11. Maria, Carol, Glenn, Luis

Page 409 **Mid-Chapter Review** 1. $\{(-1, -5)\}$ 3. $\{(3, 9)\}$ Let x = ten's digit and y = one's digit. $x = 2y + 2$; $10x + y = 9x + 11$; Ans: 83 7. Let x = grams of 2% solution, then y = grams of 3% solution. $x + y = 800$; $0.02x + 0.03y = 0.025(800)$; Ans: 400 g of 2% solution, 400 g of 3% solution

Page 411 1. 12 hours 3. 440 hours

Pages 412-413 **P-1** First row: π, -2; Second column: -2, $-\sqrt{3}$ **P-2** a. -15 b. 0 c. 2
P-3 a. $2p - 3q$ b. $-5x$ c. $s^2 + t^2$

Page 414 **Classroom Exercises** 1. 2 3. 13 5. -3 7. 17 9. 7 11. -13 13. -48 15. 0 17. 6
19. -3

Page 414 **Written Exercises** 1. -28 3. 0 5. -2 7. 4 9. 0 11. -223 13. $\begin{vmatrix} 2 & -1 \\ 1 & -3 \end{vmatrix}$; -5
15. $\begin{vmatrix} -2 & 5 \\ 1 & -3 \end{vmatrix}$; 1 17. $\begin{vmatrix} 2 & -1 \\ 5 & -1 \end{vmatrix}$; 3 19. $\begin{vmatrix} 5 & 3 \\ -3 & 1 \end{vmatrix}$; 14 21. $\begin{vmatrix} 4 & 5 \\ 5 & 6 \end{vmatrix}$; -1

Page 415 **P-1** 13; -9; 7

Page 417 **Classroom Exercises** 1. $x = \dfrac{\begin{vmatrix} 2 & -1 \\ -1 & 2 \end{vmatrix}}{\begin{vmatrix} 3 & -1 \\ 1 & 2 \end{vmatrix}}$, $y = \dfrac{\begin{vmatrix} 3 & 2 \\ 1 & -1 \end{vmatrix}}{\begin{vmatrix} 3 & -1 \\ 1 & 2 \end{vmatrix}}$ 3. $x = \dfrac{\begin{vmatrix} 1.5 & -0.5 \\ -2.5 & 1.5 \end{vmatrix}}{\begin{vmatrix} 0.5 & -0.5 \\ -1.5 & 1.5 \end{vmatrix}}$, $y = \dfrac{\begin{vmatrix} 0.5 & 1.5 \\ -1.5 & -2.5 \end{vmatrix}}{\begin{vmatrix} 0.5 & -0.5 \\ -1.5 & 1.5 \end{vmatrix}}$

Page 417 **Written Exercises** 1. $\{(5, 1)\}$ 3. $\{(-2, 1)\}$ 5. $\left\{\left(-\dfrac{3}{23}, -\dfrac{42}{23}\right)\right\}$ 7. $\left\{\left(\dfrac{7}{4}, \dfrac{9}{4}\right)\right\}$ 9. $\left\{\left(0, \dfrac{1}{2}\right)\right\}$
11. ϕ

Page 417 **Review Capsule for Section 15.6** 1. $5n$ 2. $100t$ 3. $5w$ 4. $12y$ 5. $850k$ 6. $75n$

Page 420 **Classroom Exercises** 1. $5x$; $10y$; 650; $x + y = 60$; $5x + 10y = 650$ 3. x; $325x$; y; $135y$; 55; $x + y = 55$; $325x + 135y = 11{,}415$ 5. x; $15x$; y; 25¢; $25y$; $x = y + 20$; $15x + 25y = 1500$

Page 421 **Written Exercises** 1. Records: x; tapes: y; $x + y = 9$; $6x + 7y = 58$; Answer: 5 records, 4 tapes 3. Tomatoes: x; beans: y; $x + y = 50$; $35x + 45y = 2030$; Answer: 22 cans of tomatoes, 28 cans of beans 5. Fives: x; tens: y; x + y = 86; 5x + 10y = 675; Answer: 37 fives, 49 tens 7. Students: x; adults: y; x + y = 82; 35x + 50y = 3245; Answer: 57 students, 25 adults

Pages 423-424 Classroom Exercises 1. 20x 3. 10(x + y) 5. 3x + 3y

Pages 424-425 Written Exercises 1. 7; 7y; x + y = 5; 70x + 7y = 224; Ans: 3 hours by bus, 2 hours hiking 3. 3x; 3; 3y; 3720; x = 30y; 3x + 3y = 3720; Ans: speed of plane: 1200 km/hr, speed of carrier: 40 km/hr 5. 4500; 15; 4500; 18(x − y) = 4500; 15(x + y) = 4500; Ans: speed of runner: 275 m/min, speed of wind: 25 m/min 7. 6; 48; 4; 48; 6(x − y) = 48; 4(x + y) = 48; Ans: speed in still water: 10 km/hr, speed of current: 2 km/hr

Pages 426-427 1. B 3. A 5. B 7. B 9. C 11. A 13. B

Pages 428-430 Chapter Review 1. $\{(-3, 3)\}$ 3. $\{(2, -1)\}$ 5. Let x = ten's digit and y = one's digit; x = 3y; x + y = 12; Ans: 93 7. Let x = ten's digit and y = one's digit; 10x + y = 9y; x + y = 9; Ans: 45 9. Let x = grams of 12% solution, then y = grams of 20% solution. x + y = 800; 0.12x + 0.20y = 0.17(800); Ans: 300 g of 12% solution, 500 g of 20% solution 11. 41 13. −1 15. $\{1, -1)\}$ 17. $\{(\frac{53}{11}, -\frac{19}{11})\}$ 19. Adult: x; student: y; x + y = 204; 125x + 75y = 21500; Ans: 124 adult tickets, 80 student tickets 21. 3; 3x; 2; 2y; 1000; x = 6y; 3x + 2y = 1000; Ans: speed of plane: 300 km/hr, speed of jeep: 50 km/hr

Pages 431-432 Cumulative Review: Chapters 1-15 1. $x < -4$ 3. $x < -5$ 5. $\frac{1}{a^5b^5}$ 7. $-\frac{27x^3y^{12}}{125}$ 9. $\sqrt{15}$ 11. $2x^3y^4\sqrt{3y}$ 13. $8c^2d^6e^5$ 15. $(4n + 1)(3n − 5)$ 17. $3(x − 3)(x − 3)$ 19. $\frac{a^2}{a - 5}$ 21. 6 23. $\frac{x^2 − 6x + 13}{x^2 − 4}$ 25. 23 For Ex. 27 and 29, each graph is a line. Two points are given. 27. (0, 3), (1, 1) 29. (0, 0), (2, 8) 31. Slope: −1; y intercept: 7 33. $\frac{2}{5}$ 35. $-\frac{9}{4}$ 37. No 39. $\{(2, 0)\}$ 41. $\{(-2, -3)\}$ 43. $\{(-4, -3)\}$ 45. $\{(12, 4)\}$ 47. Width: x; length: y; x = 2y − 7; 2x + 2y = 19; Ans: width: 4 meters, length: $5\frac{1}{2}$ meters 49. $\{(1, 2)\}$ 51. $\{(\frac{7}{5}, \frac{1}{5})\}$ 53. Adult: x; student: y; x + y = 520; 3x + 2y = 1500; Ans: 460 adult tickets, 60 student tickets

CHAPTER 16 SYSTEMS OF INEQUALITIES

Page 434 P-1 Many answers are possible; (0, −6) is one answer.

Pages 435-436 Classroom Exercises In Ex. 1-7, many other answers are possible. 1. (−2, −3), (0, 0), (2, 2) 3. (−2, 3), (0, 0), (2, −2) 5. (−2, 0), (0, −1), (2, −3) 7. (−2, 4), (0, 4), (2, 6) 9. y < 2x + 1 11. y ≤ 3x + 2 13. y < −x − 10

Page 436 Written Exercises In Ex. 1-11, many other answers are possible. 1. (0, −2), (1, 0), (−1, −4), (2, 0), (−2, −5) 3. (0, 1), (1, 0), (2, 0), (−1, 4), (−2, 9) 5. (0, −4), (2, −3), (4, −2), (−2, −7), (−4, −6) 7. (0, 0), (1, 0), (−1, −1), (2, 0), (−2, 0) 9. (0, 2), (1, 15), (−1, 0), (−2, −18), (2, 25) 11. (0, 2), (1, 3), (−1, 2), (2, 5), (−2, 1) 13. y > 3x + 5 15. y ≤ −4x + 6 17. y > x + 3 19. y > 3x − 15 21. $y ≥ -\frac{2}{3}x + 2$ 23. $y > 2x - \frac{5}{2}$ 25. $y ≥ \frac{1}{2}x + \frac{3}{14}$ 27. $y < -\frac{5}{3}x + \frac{1}{3}$

Page 436 Review Capsule for Section 16.2 In Ex. 1-8, each graph is a straight line. Two points are given for each. 1. (0, 2); (2, 0) 2. (0, −1); (1, 1) 3. (0, −3); (2, 0) 4. (0, 2); (4, 7) 5. (0, $\frac{3}{2}$); (1, 2) 6. (0, −1); (1, −4) 7. (0, −2); (5, −2) 8. (−3, 0); (−3, −2)

Page 437 **P-1** The region where $y > 2x - 1$. The line $y = 2x - 1$. The region where $y < 2x - 1$.

Page 439 **Classroom Exercises** 1. Yes 3. No 5. Yes 7. No 9. $y > -x$ 11. $y < -\frac{1}{3}x - 1$

Page 439 **Written Exercises** In Written Ex. 1-17, each graph is a region that does not include its linear boundary. The location of the region with respect to the boundary is indicated for each exercise.
1. right of the boundary 3. right 5. below 7. right 9. above 11. left 13. left 15. left 17. above

Page 439 **Review Capsule for Section 16.3** 1. $y \leq 2$ 2. $y \leq x - 5$ 3. $y \geq 2x$ 4. $y \geq \frac{1}{2} - 3x$
5. $y \leq -x + 1$ 6. $y \geq 2 - x$

Page 440 **P-1** $-2.9, 0, 5, 17$ **P-2** -3 **P-3** $-3, -2.9, 5, 17$ **P-4** $(2, 0), (7, -5), (0, 2),$
$(1, 1); (-2, 2), (2, -1), (0, 1), (1, 0)$

Page 442 **Classroom Exercises** 1. Yes 3. Yes 5. Yes 7. Yes 9. Yes 11. $y \geq -x$ 13. $y \leq \frac{1}{2}x + 1$

Page 443 **Written Exercises** In Ex. 1-17, each graph is a region that has a straight line as its boundary. The boundary is included in the graphs of Ex. 1-15 but not in the graph of Ex. 17. The location of the region with respect to its boundary is indicated for each exercise. 1. right of the boundary 3. right
5. right 7. below 9. right 11. below 13. above 15. left 17. right 19. $y \leq x + 1$ 21. $y \geq -2x - 2$
23. $y \leq -\frac{1}{3}x + 1$ 25. Yes 27. Yes

Page 444 **Mid-Chapter Review** 1. $y < -5x + 2$ 3. $y < \frac{3}{2}x + \frac{5}{2}$ 5. $y < \frac{4}{3}x - \frac{1}{2}$ In Ex. 7-11, each graph is a region that does not include its linear boundary. The location of the region with respect to the boundary is indicated for each exercise. 7. left of the boundary 9. below 11. left In Ex. 13-17, each graph is a region that has a straight line as its only boundary. The boundary is included in each of the graphs. 13. left 15. left 17. right

Page 444 **Review Capsule for Section 16.4** 1. $(-2, 3), (-4, 3)$ 2. $(0, -1), (3, 5)$ 3. $(2, -4)$
4. $(-4, 3)$

Page 445 **P-1** $3.9, 1, -11$ **P-2** $-1.9, 0, 17$ **P-3** $-1.9, 0, 2, 3.8$

Page 447 **Classroom Exercises** 1. IV 3. II

Page 447 **Written Exercises** In Ex. 1-11, each graph is a region with two straight lines as its boundaries. The boundaries are not included in the graphs. The quadrants that contain the region are indicated, as are the equations of the boundaries. 1. Quadrants: III, IV; Boundaries: Left: $y = x + 1$,
Right: $y = -x - 1$ 3. Quadrants: I, II, and III; Boundaries: Bottom: $y = 2x + 1$, Top: $y = -x + 2$
5. Quadrants: II, III; Boundaries: Top: $x + 2y = 4$, Bottom: $x - 2y = -4$ 7. Quadrants: III and IV;
Boundaries: Left: $x - 3y = 3$, Right: $2x + y = -1$ 9. Quadrants: II and III; Boundaries: Right: $x = -1$,
Top: $y = 2$ 11. Quadrants: I, III, and IV; Boundaries: Left: $x = -1$, Top: $y = 3x$

Page 450 **Classroom Exercises** 1. c 3. d

Page 450 **Written Exercises** In Ex. 1-15, each graph is a region with two straight lines as its boundaries. The quadrants that contain the region are indicated, as are the equations of the boundaries. 1. Quadrants: I and II; Boundaries: Left: $y = -x - 1$, Right: $y = x + 1$; Boundaries included 3. Quadrants: I, II, and III; Boundaries: Top: $y = -x + 1$, Bottom: $y = 2x$; Bottom boundary included 5. Quadrants: II and III; Boundaries: Top: $x + y = 4$, Bottom: $x - y = -4$; Boundaries included 7. Quadrants: III and IV; Boundaries: Left: $x - 3y = 2$, Right: $2x + y = -2$; Right boundary included 9. Quadrants: I and II;

Boundaries: Right: $y = x + 2$, Left: $y = -x - 2$; Right boundary included 11. Quadrant: III;
Boundaries: Top: $y = x$; Right: $y = 3x$; Boundaries included 13. Quadrants: II and III; Boundaries:
Top: $y = 3$, Right: $x = -2$; Right boundary included 15. Quadrants: I and IV; Boundaries: Top:
$y = 3x$; Left: $x = 0$; Top boundary included 17. $y \geq x$; $y \geq -x$ 19. $y \leq x$; $y \geq -x$

Page 452 Table: Plant A: $\frac{x}{60}$, 1.30x; Plant B: $\frac{y}{40}$, 1.35y   Plant A should produce **7200** calculators
and Plant B should produce **3000** calculators. The cost would be **$13,410**.

Pages 453-454 **Chapter Review** In Ex. 1 and 3, many other answers are possible. 1. (0, −2), (1, 0),
(2, −6), (−1, 0), (−2, 2) 3. (0, 1), (2, 5), (−2, −4), (4, 2), (−4, −10) 5. $y \geq 3x - 2$ 7. $y > \frac{1}{3}x + \frac{5}{3}$
In Ex. 9-13, each graph is a region that does not include its linear boundary. The location of the region
with respect to the boundary is indicated for each exercise. 9. above the boundary 11. left 13. below
In Ex. 15-19, each graph is a region that includes its linear boundary. The location of the region with
respect to the boundary is indicated for each exercise. 15. above 17. above 19. above In Ex. 21 and
23, each graph is a region with two straight lines as its boundaries. The boundaries are not included in the
graphs. The quadrants that contain the region are indicated, as are the equations of the boundaries.

21. Quadrants: II, III, and IV; Boundaries: Right: $y = -x - 1$, Top: $y = \frac{1}{2}x + 2$ 23. Quadrant: II;

Boundaries: Top: $2x + y = -3$, Bottom: $x + y = 2$ In Ex. 25, the graph is a region with two straight
lines as its boundaries. The quadrants that contain the region are indicated, as are the equations of the
boundaries. 25. Quadrants: I and IV; Boundaries: Bottom: $x + 2y = 4$, Left: $-3x + y = -1$;
Boundaries included

CHAPTER 17   QUADRATIC FUNCTIONS

Page 456 **P-1** a and c

Page 457 **Classroom Exercises** 1. Yes 3. No 5. No 7. No 9. No 11. Yes 13. Yes 15. No

Page 457 **Written Exercises** 1. No 3. Yes 5. No 7. Yes 9. No 11. Yes 13. (−3, −9), (−1, −1),
(2, −4) 15. (−3, 4), (−1, 0), (2, 9) 17. (−3, 6), (−1, −2), (2, 1) 19. (−3, −9), (−1, 1), (2, −14)

Page 458 **P-1** a. $-1\frac{3}{4}$ b. $2\frac{1}{4}$ c. $-3\frac{3}{4}$ **P-2** {all real numbers greater than or equal to −4}

Pages 459-460 **Classroom Exercises** 1. y: 4, 1, 0, 1, 4 3. y: 6, 3, 2, 3, 6 5. y: 6, 1, 0, 3, 10 7. 2
9. $2\frac{1}{4}$ 11. 0, 2 13. $\frac{1}{2}$, $1\frac{1}{2}$

Page 460 **Written Exercises** In Ex. 1-5, each parabola turns at a high point (maximum), or low point
(minimum). This point is given for each exercise. 1. Turns at (0, 1); low point 3. Turns at (1, 4); high
point 5. Turns at (1, 2); low point In Ex. 7-11, the high or low point and three points on the parabola
are given. 7. Turns at (0, −2); low point; (−1, −1); (0, −2); (1, −1) 9. Turns at $(-\frac{1}{2}, \frac{1}{4})$; high point;
(−1, 0); $(-\frac{1}{2}, \frac{1}{4})$; (0, 0) 11. Turns at (3, 1); low point; (2, 2); (3, 1); (4, 2) 13. 7 + 1 = 8; 9 − 6 = 3;
4 × 5 = 20

Pages 461-462 **P-1** 1. $a = -\frac{5}{2}$; b = 0; c = 0 2. a = 3; b = −4; c = −1 3. a = 1; b = 3; c = 0 **P-2**
$y = \frac{1}{2}x^2$; $y = x^2$; $y = 2x^2$ **P-3** $y = -\frac{1}{2}x^2$; $y = -x^2$; $y = -2x^2$

Pages 462-463   Classroom Exercises   1. The point with coordinates (0, 0)   3. Downward   5. {all real numbers greater than or equal to 0}   7. Downward   9. Downward   11. $y = 3x^2$ 13. $y = 0.09x^2$

Pages 463-464   Written Exercises   In Ex. 1-9, the graphs are all parabolas opening upward or downward.

1.

x	−2	−1	0	1	2
y	4	1	0	1	4

, opens upward at (0, 0);

x	−4	−2	0	2	4
y	4	1	0	1	4

, opens upward at (0, 0)

3.

x	−3	−2	0	2	3
y	$-\frac{9}{2}$	−2	0	−2	$-\frac{9}{2}$

, opens downward at (0, 0);

x	−2	−1	0	1	2
y	−6	$-\frac{3}{2}$	0	$-\frac{3}{2}$	−6

, opens downward at (0, 0)

5.

x	−2	−1	0	1	2
y	−8	−2	0	−2	−8

, opens downward at (0, 0);

x	−2	−1	0	1	2
y	6	$\frac{3}{2}$	0	$\frac{3}{2}$	6

, opens upward at (0, 0)

7.

x	−2	−1	0	1	2
y	2	$\frac{1}{2}$	0	$\frac{1}{2}$	2

, opens upward at (0, 0);

x	−2	−1	0	1	2
y	0	$-\frac{3}{2}$	−2	$-\frac{3}{2}$	0

, opens upward at (0, −2)

9.

x	−4	−2	0	2	4
y	−12	−3	0	−3	−12

, opens downward at (0, 0);

x	−4	−2	0	2	4
y	−11	−2	0	−2	−11

, opens downward at (0, 1)   In Ex. 11 and 13, the parabolas open at the origin to the right or to the left of the y axis.   11.

x	1	1	0	4	4
y	1	−1	0	2	−2

, opens to the right   13.

x	2	2	0	8	8
y	1	−1	0	2	−2

, opens to the right

Page 464   Review Capsule for Section 17.4   1. $x^2 - 4x + 4$ 2. $x^2 + 6x + 9$ 3. $x^2 - 10x + 25$ 4. $x^2 + 12x + 36$ 5. $x^2 + 4.2x + 4.41$ 6. $x^2 - 5.4x + 7.29$

Page 465   P-1 $a = \frac{1}{2}$; $b = -2$; $c = 2$

Page 467   Classroom Exercises   1. 2 units right   3. 1 unit right   5. 2 units left   7. 10 units left   9. 7 units right   11. 6 units left

Pages 467-468   Written Exercises   1. 5 units right   3. 8 units left   5. $3\frac{1}{2}$ units left   7. 2.7 units right   9. Second parabola is 2 units to left of first.   11. Second parabola is 3 units to right of first.   13. Second parabola is $\frac{3}{2}$ units to left of first.   15. Second parabola is 2 units to right of first.   17. Second parabola is 1 unit to left of first.   19. 75   21. −149.36531

Page 468   Mid-Chapter Review   1. $(-2, 4)$, $(-1, \frac{5}{2})$, $(0, 0)$, $(1, -\frac{7}{2})$ 3. $(-2, -1)$, $(-1, -3)$, $(0, -3)$, $(1, -1)$ In Ex. 5, the high point and three points for the graph (parabola) are given.   5. High point: $(0, 2)$; $(-1, 1\frac{1}{2})$, $(0, 2)$, $(1, 1\frac{1}{2})$ In Ex. 7-11, three points are given for each curve (parabola).   7. $y = \frac{3}{4}x^2$: $(0, 0)$, $(2, 3)$, $(-2, 3)$; $y = -\frac{3}{4}x^2$: $(0, 0)$, $(2, -3)$, $(-2, -3)$ 9. $y = 2x^2$: $(0, 0)$, $(1, 2)$, $(-1, 2)$; $y = \frac{1}{2}x^2$: $(0, 0)$, $(2, 2)$, $(-2, 2)$ 11. $y = \frac{1}{2}x^2 + 1$: $(0, 1)$, $(2, 3)$, $(-2, 3)$; $y = -2x^2 - 2$: $(0, -2)$, $(1, -4)$, $(-1, -4)$ 13. 5 units right   15. 4 units left

Page 468   Review Capsule for Section 17.5   1. $(-2, 9)$, $(0, 3)$, $(2, 1)$ 2. $(-2, 7)$, $(0, -3)$, $(2, 3)$ 3. $(-2, -5)$, $(0, -5)$, $(2, -21)$

P-1 Quadratic **P-2** 2 units up **P-3** Move 3 units right and 2 units up. **P-4** 3 units down **P-5** Move 4 units left and 3 units down. **P-6** If h is positive, it is to the right. If h is negative, it is to the left. **P-7** If k is positive, it is above. If k is negative, it is below.

Page 471 Classroom Exercises 1. Yes 3. No 5. Yes 7. a 9. k 11. 5 units right, 2 units down 13. 2 units left, 3 units down 15. 1 unit left, 1 unit down

Page 472 Written Exercises In Ex. 1-7, the ordered pairs for the first parabola and the motions to locate the second parabola are given. 1. $y = \frac{1}{2}x^2$: $(-2, 2)$, $(-1, \frac{1}{2})$, $(0, 0)$, $(1, \frac{1}{2})$, $(2, 2)$; $y = \frac{1}{2}(x - 2)^2 + 3$: move 3 units up, 2 units right 3. $y = -x^2$: $(-2, -4)$, $(-1, -1)$, $(0, 0)$, $(1, -1)$, $(2, -4)$; $y = -(x + 3)^2 + 2$: move 2 units up, 3 units left 5. $y = 2x^2$: $(-2, 8)$, $(-1, 2)$, $(0, 0)$, $(1, 2)$, $(2, 8)$; $y = 2(x + 1)^2 - 3$; move 3 units down, 1 unit left 7. $y = -\frac{3}{2}x^2$: $(-2, -6)$, $(-1, -\frac{3}{2})$, $(0, 0)$, $(1, -\frac{3}{2})$, $(2, -6)$; $y = -\frac{3}{2}(x - 3\frac{1}{2})^2 + 2\frac{1}{2}$: move $2\frac{1}{2}$ units up, $3\frac{1}{2}$ units right 9. $y = 2(x - 3)^2$ 11. $y = \frac{1}{2}(x + 2)^2 + 5$ 13. $y = \frac{1}{3}(x + 1)^2 - 5$ 15. $y = -(x - 3)^2 - 2$ 17. $y = 3x^2 - 5$

Page 472 Review Capsule for Section 17.6 1. 4; $(x - 2)(x - 2)$ 2. 9; $(x + 3)(x + 3)$ 3. $\frac{1}{4}$; $(x + \frac{1}{2})(x + \frac{1}{2})$ 4. $\frac{9}{4}$; $(x - \frac{3}{2})(x - \frac{3}{2})$ 5. 25; $(x + 5)(x + 5)$ 6. 81; $(x + 9)(x + 9)$ 7. $\frac{25}{4}$; $(x - \frac{5}{2})(x - \frac{5}{2})$ 8. $\frac{49}{4}$; $(x - \frac{7}{2})(x - \frac{7}{2})$

Page 473 P-1 a. $a = -1, h = -3, k = -\frac{3}{4}$ b. $a = \frac{3}{2}, h = \frac{1}{4}, k = \frac{5}{4}$

Page 474 Classroom Exercises 1. $(x - (-1))^2$ 3. $(x - 3)^2$ 5. $(x - (-\frac{1}{2}))^2$ 7. $(x - \frac{5}{2})^2$

Page 474 Written Exercises 1. $x^2 - 6x + 11$ 3. $2x^2 - 4x - 5$ 5. $3x^2 + 12x + 15$ 7. $\frac{1}{2}x^2 - 2x + 8$ 9. $(x - (-1))^2 + 2$ 11. $(x - 2)^2 - 3$ 13. $(x - (-3))^2 - 14$ 15. $(x - 5)^2 - 26$ 17. $(x - \frac{1}{2})^2 + \frac{19}{4}$ 19. $(x - (-\frac{3}{2}))^2 - \frac{25}{4}$

Page 474 Review Capsule for Section 17.7 1. $a = 3; h = 2; k = 1$ 2. $a = 4; h = 5; k = -2$ 3. $a = \frac{1}{2}$; $h = -\frac{3}{2}; k = -5$

Pages 475-476 P-1 Lowest point **P-2** $\frac{1}{2}$; 2; 3 **P-3** a. $(\frac{1}{2}, \frac{5}{2})$ b. $(-3, -7)$ **P-4** $x = 2$ **P-5** a. $h = 5$ b. $h = -3$ c. $h = \frac{1}{4}$ d. $h = -\frac{4}{3}$ **P-6** $(-3, -2)$

Pages 476-477 Classroom Exercises 1. $(5, 2)$ 3. $(-2, 6)$ 5. $(0, 0)$ 7. $(\frac{1}{2}, \frac{3}{4})$ 9. $(5, 0)$ 11. $x = -1$ 13. $x = \frac{1}{2}$ 15. $x = 0$ 17. $x = 0$

Page 477 Written Exercises 1. $(1, 10)$ 3. $(7, -2)$ 5. $(-2, 3)$ 7. $(-1, -8)$ 9. $(\frac{1}{2}, -\frac{2}{3})$ 11. $x = 5$ 13. $x = 3$ 15. $x = -1$ 17. $x = \frac{1}{3}$ 19. $x = -\frac{7}{2}$ 21. $x = -\frac{1}{10}$

Page 479 1. b. The median points are $(20, 8)$ and $(110, 58)$. d. About 36 hours 3. b. The median points are $(64, 115)$ and $(72, 180)$. d. About 148 pounds

Pages 480-482   Chapter Review   1. No   **3.** $(-1, -5)$, $(0, -1)$, $(2, -11)$ In Ex. 5-9, the high or low point and three points on the parabola are given.   **5.** Turns at $(0, -1)$; low point; $(-1, 0)$; $(0, -1)$; $(1, 0)$
7. Turns at $(1, \frac{1}{2})$; high point; $(0, 0)$; $(1, \frac{1}{2})$; $(2, 0)$ **9.** Turns at $(-1, -4)$; low point; $(-2, -3)$; $(-1, -4)$; $(0, -3)$ In Ex. 11-17, three points are given for each parabola and the point at which it opens.
11. $(-1, -\frac{1}{2})$; $(0, 0)$; $(1, -\frac{1}{2})$; downward at $(0, 0)$. $(-1, -2)$; $(0, 0)$; $(1, -2)$; downward at $(0, 0)$.
13. $(-1, \frac{5}{2})$; $(0, 0)$; $(1, \frac{5}{2})$; upward at $(0, 0)$. $(-1, -\frac{5}{2})$; $(0, 0)$; $(1, -\frac{5}{2})$; downward at $(0, 0)$.   **15.** $(\frac{1}{2}, -1)$; $(0, 0)$; $(\frac{1}{2}, 1)$; Turns right at $(0, 0)$.   **17.** $(-1, -1)$; $(0, 0)$; $(-1, 1)$; Turns left at $(0, 0)$.   **19.** 3 units left
21. $\frac{5}{2}$ units right   In Ex. 23 and 25, the ordered pairs for the first parabola and the motions to locate the second parabola are given.   **23.** $y = x^2$: $(-2, 4)$, $(-1, 1)$, $(0, 0)$, $(1, 1)$, $(2, 4)$; $y = (x - 3)^2 - 2$; move 2 units down, 3 units right   **25.** $y = -\frac{1}{2}x^2$: $(-2, -2)$, $(-1, -\frac{1}{2})$, $(0, 0)$, $(1, -\frac{1}{2})$, $(2, -2)$; $y = -\frac{1}{2}(x + 1)^2 + 3$: move 3 units up, 1 unit left   **27.** $y = -10(x + 5)^2$ **29.** $y = \frac{5}{2}(x - 2)^2 - 3$ **31.** $x^2 + 2x - 4$ **33.** $-3x^2 + 12x - 11$ **35.** $\frac{1}{2}x^2 + 4x$ **37.** $(x - 3)^2 - 7$ **39.** $(x - (-\frac{3}{2}))^2 - \frac{5}{2}$ **41.** $(x - 1)^2 - \frac{1}{2}$ **43.** $(10, 5)$
45. $(6, 0)$ **47.** $x = 2$ **49.** $x = -\frac{9}{2}$

CHAPTER 18   QUADRATIC EQUATIONS

Page 484   P-1 $-1, 2$

Pages 485-486   Classroom Exercises   1. $x^2 - 2x - 3 = 0$ **3.** $3x^2 - 5x + 6 = 0$ **5.** $x^2 + 2x + 4 = 0$
7. 1, 3   **9.** 0, 4   **11.** 2

Page 486   Written Exercises   1. $x^2 - 5x + 3 = 0$ **3.** $2x^2 - 5x + 10 = 0$ **5.** $x^2 - 9 = 0$ **7.** $-3, 1$
9. $-3.4, 1.4$ **11.** No real solutions   **13.** No real solutions   **15.** 2   **17.** 0.6, 3.4

Page 486   Review Capsule for Section 18.2   1. $(x - 6)(x + 2)$ **2.** $(x - 5)(x - 2)$ **3.** $(x - 6)(x + 4)$
4. $(x + 8)(x - 7)$ **5.** $(2x - 1)(x + 4)$ **6.** $(3x - 2)(x - 1)$ **7.** $(7 + x)(1 - x)$ **8.** $(9x - 2)(x - 7)$

Page 487   P-1 $-5; 2$

Page 489   Classroom Exercises   1. $\{3, -5\}$ **3.** $\{6, -6\}$ **5.** $\{-4, -7\}$ **7.** $\{\frac{1}{2}, -5\}$ **9.** $\{-\frac{3}{2}, -1\}$

Page 489   Written Exercises   1. $\{0, -3\}$ **3.** $\{5, -5\}$ **5.** $\{0, 10\}$ **7.** $\{10, -10\}$ **9.** $\{-5, 1\}$
11. $\{-4, 1\}$ **13.** $\{-4\}$ **15.** $\{2\}$ **17.** $\{2, 6\}$ **19.** $\{3, 8\}$ **21.** $\{-9\}$ **23.** $\{-8, 9\}$ **25.** $\{-3, \frac{1}{3}\}$
27. $\{-\frac{2}{3}, \frac{1}{2}\}$ **29.** $\{-\frac{4}{3}, \frac{1}{4}\}$ **31.** $\{-\frac{5}{4}, \frac{1}{2}\}$ **33.** $\{-\frac{5}{2}, 9\}$ **35.** $\{-1, \frac{3}{4}\}$

Pages 490-491   P-1   a. $\{4, -4\}$ **b.** $\{7, -7\}$ **c.** $\{10, -10\}$ **P-2 -5 P-3   1; 16; 16; 16**

Pages 491-492   Classroom Exercises   1. $\{1, -1\}$ **3.** $\{11, -11\}$ **5.** $x - 6 = 6$ or $x - 6 = -6$
7. $x - \frac{1}{2} = 4$ or $x - \frac{1}{2} = -4$ **9.** 4; 2; No   **11.** 25; 25; Yes   **13.** $\frac{9}{4}$; $\frac{3}{4}$; No   **15.** $x + 10, x + 10$
17. $x - \frac{1}{4}, x - \frac{1}{4}$ **19.** $x - \frac{3}{2}, x - \frac{3}{2}$

Page 492   Written Exercises   1. $\{3, -3\}$ **3.** $\{13, -13\}$ **5.** $\{\sqrt{7}, -\sqrt{7}\}$ **7.** $\{5\sqrt{3}, -5\sqrt{3}\}$
9. $x - 5 = 6$ or $x - 5 = -6$ **11.** $x + 6 = 5$ or $x + 6 = -5$ **13.** $x + \frac{1}{2} = \frac{1}{2}$ or $x + \frac{1}{2} = -\frac{1}{2}$ **15.** $\{-12, 10\}$

17. $\left\{3 + \sqrt{5}, 3 - \sqrt{5}\right\}$ 19. $\{0, 1\}$ 21. $\left\{-\dfrac{2}{3} + \dfrac{\sqrt{5}}{3}, -\dfrac{2}{3} - \dfrac{\sqrt{5}}{3}\right\}$ 23. $\{-4\}$ 25. $\{5\}$ 27. $\left\{\dfrac{1}{4}\right\}$
29. $\left\{\dfrac{2}{5}\right\}$ 31. $\dfrac{9}{4}$ 33. 2.2967226   35. 4.53041   37. 1.4299625   39. 19.453883

Page 493   Mid-Chapter Review   1. $\{-4, 1\}$ 3. $\{1.3, -4.3\}$ 5. $\{0.6, -3.6\}$ 7. $\left\{\dfrac{1}{2}, -5\right\}$
9. $\left\{-\dfrac{1}{3}, 1\right\}$ 11. $\left\{-\dfrac{1}{3}, -\dfrac{1}{2}\right\}$ 13. $\{-9, 15\}$ 15. $\{-1, 2\}$ 17. $\{-2, -8\}$

Page 493   Review Capsule for Section 18.4   1. $x^2 - 6x = -3$ 2. $x^2 + 5x = 4$ 3. $x^2 - 4x = -10$
4. $x^2 - \dfrac{1}{2}x = -\dfrac{3}{2}$ 5. $2x^2 - 4x = -5$ 6. $2x^2 + 3x = 10$

Page 496   Classroom Exercises   1. 4   3. 1   5. $\dfrac{1}{4}$ 7. $\dfrac{1}{36}$ 9. $\left\{\dfrac{\sqrt{3}}{2}, -\dfrac{\sqrt{3}}{2}\right\}$ 11. $\left\{\dfrac{\sqrt{5}}{2}, -\dfrac{\sqrt{5}}{2}\right\}$
13. $\left\{\dfrac{1}{2} + \dfrac{\sqrt{3}}{2}, \dfrac{1}{2} - \dfrac{\sqrt{3}}{2}\right\}$ 15. $x + 9, x + 9$ 17. $x + \dfrac{3}{2}, x + \dfrac{3}{2}$ 19. $x - \dfrac{1}{5}, x - \dfrac{1}{5}$

Page 496   Written Exercises   1. $\{-3, 5\}$ 3. $\{-5, 1\}$ 5. $\{-3, 2\}$ 7. $\left\{-1 + \sqrt{6}, -1 - \sqrt{6}\right\}$ 9. ϕ
11. $\left\{3 + \sqrt{14}, 3 - \sqrt{14}\right\}$ 13. $\left\{-4 + \sqrt{3}, -4 - \sqrt{3}\right\}$ 15. $\left\{\dfrac{3}{2} + \dfrac{\sqrt{13}}{2}, \dfrac{3}{2} - \dfrac{\sqrt{13}}{2}\right\}$ 17. $\left\{-\dfrac{3}{2} + \dfrac{\sqrt{29}}{2},\right.$
$\left.-\dfrac{3}{2} - \dfrac{\sqrt{29}}{2}\right\}$ 19. $\left\{-\dfrac{1}{5} + \dfrac{\sqrt{6}}{5}, -\dfrac{1}{5} - \dfrac{\sqrt{6}}{5}\right\}$ 21. $\left\{-\dfrac{3}{2}, 1\right\}$ 23. $\left\{-\dfrac{1}{3}, 2\right\}$ 25. $\left\{1 + \dfrac{\sqrt{3}}{3}, 1 - \dfrac{\sqrt{3}}{3}\right\}$
27. $\left\{\dfrac{1}{4} + \dfrac{\sqrt{41}}{4}, \dfrac{1}{4} - \dfrac{\sqrt{41}}{4}\right\}$

Page 496   Review Capsule for Section 18.5   1. $\dfrac{5}{2}$; -1 2. 2; -4 3. $\dfrac{2 + \sqrt{6}}{4}$; $\dfrac{2 - \sqrt{6}}{4}$

Page 499   Classroom Exercises   1. $a = 1, b = 3, c = -2$ 3. $a = 1, b = 5, c = 0$ 5. $a = 1, b = 3, c = 5$

Page 499   Written Exercises   1. $\{-6, 1\}$ 3. $\{-7, 1\}$ 5. $\{2, 5\}$ 7. $\{-2, 6\}$ 9. $\{3\}$ 11. $\left\{-\dfrac{2}{3}, \dfrac{1}{2}\right\}$
13. $\left\{\sqrt{3}, -\sqrt{3}\right\}$ 15. $\{0, 6\}$ 17. $\left\{\dfrac{3 + \sqrt{29}}{10}, \dfrac{3 - \sqrt{29}}{10}\right\}$ 19. $\{0.4, -2.4\}$ 21. $\{1.8, -0.3\}$
23. $\{2.4, 0.6\}$

Page 499   Review Capsule for Section 18.6   1. 33   2. -71 3. 44

Page 502   P-1   a. Two rational numbers   b. Two irrational numbers   c. No real numbers   d. One rational number

Page 502   Classroom Exercises   1. Two rationals   3. One rational   5. Two irrationals   7. Two rationals   9. No reals   11. Two rationals   13. 9; Two rationals   15. 1; Two rationals   17. -64; No reals   19. 17; Two irrationals   21. -3; No reals

Page 502   Written Exercises   1. 25; Two rationals   3. 36; Two rationals   5. 17; Two irrationals   7. 0; One rational   9. 28; Two irrationals   11. -100; No reals   13. 49; Two rationals   15. 100; Two rationals   17. 4; Two rationals   19. -24; No reals   21. 145; Two irrationals   23. 120; Two irrationals

Page 503   Review Capsule for Section 18.7   1. $x^2 - 4x - 78 = 0$ 2. $2x^2 + 5x - 102 = 0$ 3. $x^2 - 7x - 1 = 0$ 4. $x^2 + 2x + 4 = 0$ 5. $2x^2 - x - 128 = 0$ 6. $2x^2 + x - 10 = 0$

Page 505   Classroom Exercises   1. Consecutive integers: $n, n + 1$; Equation: $n(n + 1) = 210$
3. Integers: $n, n + 3$; Equation: $n(n + 3) = 108$ 5. Width: w, Length: $w + 5$; Equation: $w(w + 5) = 24$
7. Numbers: $n, 3n$; Equation: $n^2 = (3n)^2 - 18$ or $n^2 = 9n^2 - 18$

Page 506 **Written Exercises** 1. Length: ℓ, Width: $\ell - 2$; Equation: $\ell(\ell - 2) = 80$; Ans: Length: 10, Width: 8 3. Consecutive integers: n, n + 1; Equation: n(n + 1) = 30; Ans: 5 and 6 or −6 and −5 5. Integers: n, n − 13; Equation: n(n − 13) = −40; Ans: −8 and 5 or −5 and 8 7. Equation: (2x + 4) (3x) − (2x)(x) = 40; Ans: Length: 8, Width: 6

Page 508 1. $\frac{3}{2}$; $\frac{5}{4}x$; $x^2 + \frac{5}{2}x$; $\frac{3}{2}$; $\frac{25}{16}$; $\frac{3}{2} + \frac{25}{16} = \frac{49}{16}$; $\frac{7}{4}$; $\frac{2}{4}$, or $\frac{1}{2}$ 3. x = 1 5. x = 2 7. x = $\frac{3}{2}$

Pages 509-510 **Chapter Review** 1. $x^2 + 5x − 2 = 0$ 3. $\{-1.7, 1.7\}$ 5. $\{-1, 10\}$ 7. $\{0, -5\}$ 9. $\{19, -19\}$ 11. $\{-2, 10\}$ 13. $\left\{-\frac{3}{4}, 1\right\}$ 15. $\left\{-\frac{1}{3}, \frac{1}{2}\right\}$ 17. 0; One rational 19. 76; Two irrationals 21. Width: w, Length: w + 4; Equation: w(w + 4) = 60; Ans: Width: 6 cm, Length: 10 cm

CHAPTER 19 PROBABILITY

Page 513 **Classroom Exercises** 1. 8 3. 24 5. 16

Page 514 **Written Exercises** 1. 18 3. 200 5. 676 7. 18

Page 515 **Review Capsule for Section 19.2** 1. $\frac{1}{4}$ 2. $\frac{2}{5}$ 3. $\frac{3}{7}$ 4. $\frac{5}{8}$ 5. $\frac{1}{4}$

Pages 517-518 **Classroom Exercises** 1. $\frac{1}{4}$ 3. $\frac{1}{4}$ 5. $\frac{1}{4}$ 7. $\frac{1}{2}$ 9. $\frac{3}{4}$ 11. 1 13. No; the two events do not include all possible outcomes.

Pages 518-519 **Written Exercises** 1. $\frac{1}{7}$ 3. $\frac{5}{14}$ 5. $\frac{1}{4}$ 7. $\frac{1}{2}$ 9. 1 11. No; the two events do not include all possible outcomes. 13. $\frac{1}{3}$ 15. $\frac{1}{4}$ 17. $\frac{1}{2}$ 19. 0 21. $\frac{3}{8}$ 23. $\frac{1}{2}$ 25. $\frac{1}{2}$ 27. Answers will vary.

Page 521 P-1 a. $\frac{1}{18}$ b. $\frac{1}{4}$

Page 522 **Classroom Exercises** 1. $\frac{9}{16}$ 3. $\frac{1}{8}$ 5. $\frac{4}{7}$ 7. $\frac{3}{7}$

Pages 522-523 **Written Exercises** 1. $\frac{1}{36}$ 3. $\frac{5}{18}$ 5. $\frac{1}{6}$ 7. 0 9. $\frac{13}{20}$ 11. $\frac{9}{20}$ 13. $\frac{1}{2}$

Page 523 **Mid-Chapter Review** 1. 18 3. $\frac{1}{4}$ 5. 1 7. $\frac{1}{6}$ 9. $\frac{3}{5}$

Page 524 1. No; it is possible that the 8 plumbers surveyed are not representative of the larger population. 3. Yes; checking 10% of the pages will give a good indication of the book's quality. 5. No; those viewers of late night television who actually call in to vote are not a random sample of the community population.

Page 527 **Classroom Exercises** 1. $\frac{1}{9}$ 3. $\frac{1}{6}$ 5. $\frac{2}{9}$ 7. $\frac{1}{15}$

Pages 527-528 **Written Exercises** 1. $\frac{8}{45}$ 3. $\frac{1}{45}$ 5. $\frac{2}{15}$ 7. $\frac{7}{15}$ 9. $\frac{1}{45}$ 11. $\frac{7}{24}$ 13. $\frac{2}{45}$ 15. $\frac{1}{180}$

17. They arrive at the same time. The total length of Latasha's paths to the right in the figure is the same as the length of Jerome's path to the right. Likewise, the total length of Latasha's paths down in the figure is the same as the length of Jerome's path down.

Page 528 **Review Capsule for Section 19.5** 1. 192 2. 82 3. 183 4. 195 5. 882 6. 798

Page 530   Classroom Exercises   1. $\frac{1}{50}$ 3. $\frac{33}{50}$ 5. 13   7. 429

Page 531   Written Exercises   1. $\frac{6}{25}$ 3. $\frac{2}{5}$ 5. $\frac{1}{25}$ 7. 276   9. 460   11. 46   13. 420

Pages 532-533   1. 30   3. 50   5. 21   7. 264   9. 12 students

Pages 534-535   Chapter Review   1. 45   3. 252   5. $\frac{3}{10}$ 7. $\frac{1}{5}$ 9. 0   11. $\frac{7}{10}$ 13. $\frac{3}{10}$ 15. $\frac{3}{25}$ 17. $\frac{1}{14}$
19. $\frac{1}{5}$ 21. $\frac{9}{50}$ 23. 270

Pages 536-537   Cumulative Review: Chapters 1-19   1. $a = 2$ 3. $15e^9$ 5. $4x^2 - 49x + 55$ 7. $\frac{x + 3}{2}$
9. 2   11. $\frac{3x + 19}{2(x - 3)(x + 3)}$ 13. $\{(2, -3)\}$ 15. $y > -3x + 1$ 17. $y \leq \frac{7}{2}x + 4$ In Ex. 19, the graph is a
region with two straight lines as its boundaries. The quadrants that contain the region are indicated, as
are the equations of the boundaries.   19. Quadrants: All; Boundaries: Right: $x = 3$, Bottom: $x - y = 3$;
Boundaries included 21. $(-2, -4)$, $(0,0)$, $(3, -9)$ 23. $(-2,9)$, $(0,1)$, $(3,4)$ 25. Turns at $(\frac{1}{4}, -\frac{1}{8})$,
low point   In Ex. 27, three points are given for each curve (parabola).   27. $y = x^2$: $(2,4)$, $(0,0)$,
$(-2, 4)$; $y = 2x^2$: $(1, 2)$, $(0, 0)$, $(-1, 2)$ 29. Move 5 units right.   31. $(x - (-2))^2 - 3$ 33. $(x - (-\frac{3}{2}))^2 - \frac{17}{4}$ 35. $x = 1$ 37. $\{2, 5\}$ 39. $\{-4 + \sqrt{17}, -4 - \sqrt{17}\}$ 41. $\{3, -\frac{1}{2}\}$ 43. Consecutive integers:
n, $n + 1$; Equation: $n(n + 1) = 56$; Ans: 7 and 8 or -8 and -7 45. $\frac{1}{6}$ 47. $\frac{1}{36}$ 49. $\frac{6}{25}$

PICTURE CREDITS

Key: *(t)* top; *(b)* bottom; *(l)* left; *(r)* right; *(c)* center.

CHAPTER ONE: Page 1, *(l)*, IBM Corp., *(tr)*, IBM Corp., *(br)*, HBJ Photo/Jerry White; 24, HBJ Photo/Rodney Jones; 28, Frank Siteman/ Taurus Photos; CHAPTER TWO: 29, *(all)*, E. R. Degginger; 52, Photri; CHAPTER THREE: 57, *(l)*, Jonathon Watts–Science Source Library/Photo Researchers, Inc., *(tr)*, J. Stewart/Bruce Coleman, Inc., *(br)*, John Shaw/Bruce Coleman, Inc.; 70, J. Gordon Miller/Shostal Associates; 79, © Richard Hutchings; 80, E. R. Degginger; 81, Jeff Apoian/Nawrocki Stock Photo; 85, Sullivan and Rogers/Bruce Coleman, Inc.; 86, *(r)*, Brian Brake–Rapho Division/Photo Researchers, Inc., *(bl)*, L. Druskis/Taurus Photos; CHAPTER FOUR: 89, *(l)*, E. R. Degginger, *(tr)*, Paul Silverman/Fundamental Photographs, *(br)*, E. R. Degginger; 92, Lester Sloan/Woodfin Camp & Associates; 93, James Drake for Sports Illustrated © Time, Inc.; 96, Warren Morgan/Focus on Sports; 98, Mark Antman/The Image Works; 99, Michal Heron/Woodfin Camp & Associates; 105, © Richard Hutchings; 107, Jon Feingersh/Tom Stack & Associates; 112, George Hunter/ H. Armstrong Roberts; CHAPTER FIVE: 115, *(l)*, Jeff Hetler/Stock, Boston, *(r)*, Dave Forbert/Shostal Associates; 135, Craig Aurness/ Woodfin Camp & Associates; CHAPTER SIX: 143, *(l)*, Dr. J. Lorre/Photo Researchers, Inc., *(tr)*, Leonard Lessin/Peter Arnold, Inc., *(br)*, M. I. Walker–Science Source Library/Photo Researchers, Inc.; 155, NASA; 170, John Eastcott–Yva Momatuck/The Image Works; CHAPTER SEVEN: 173, *(l)* and *(tr)*, Ronda Bishop–Contact Press Images/Woodfin Camp & Associates; 196, Mark Stevenson/Nawrocki Stock Photo; 203, Mark Segal/Click/Chicago; CHAPTER EIGHT: 207, *(all)*, NASA; 223, Rice Sumner Wagner; 231, HBJ Photo/ Blaise Zito Associates; 234, B. Kulik/Photri; CHAPTER NINE: 235, *(all)*, HBJ Photos; 250, Lowell Georgia/Photo Researchers, Inc.; 251, Grant Heilman Photography; CHAPTER TEN: 267, *(l)*, Rafael Macia/Photo Researchers, Inc., *(tr)*, Steve Solum/Bruce Coleman, Inc., *(br)*, E. R. Degginger; 290, David York/The Stock Shop; CHAPTER ELEVEN: 293, *(l)*, Chuck O'Rear–West Light/Woodfin Camp & Associates, *(tr)*, Chuck O'Rear/Woodfin Camp & Associates, *(br)*, © Nelson Morris/Photo Researchers, Inc.; CHAPTER TWELVE: 321, *(l)*, NASA, *(tr)*, © The Science Museum, London, *(br)*, The Granger Collection; 332, Freda Leinwand/Monkmeyer; 340, Guy Sauvage–Agence Vandiptadt/ Photo Researchers, Inc.; 341, The Granger Collection; 342, Elaine Wicks/Taurus Photos; 344, HBJ Photo/Blaise Zito Associates; CHAPTER THIRTEEN: 349, *(l)*, Donald Miller/Monkmeyer, *(tr)*, Hank Morgan/Photo Researchers, Inc.; *(br)*, Richard Megna/Fundamental Photographs; 373, Jerry Irwin/Shostal Associates; CHAPTER FOURTEEN: 377, *(l)*, Shostal Associates, *(tr)*, HBJ Photo/Blaise Zito Associates, *(br)*, © Mark Greenberg/Visions; 394, Martin Rother/Taurus Photos; CHAPTER FIFTEEN: 397, *(l)*, Jerry Ferrara/Photo Researchers, Inc., *(tr)*, Ellis Herwig/Taurus Photos, *(br)*, © Martin Rogers/Stock, Boston; 410, Mimi Forsyth/Monkmeyer; 411, R. Krubner/H. Armstrong Roberts; 419, Doug Lee/Tom Stack & Associates; 422, Donald Dietz/Stock, Boston; 422, Robert Chase; CHAPTER SIXTEEN: 433, *(l)*, E. R. Degginger, *(tr)*, Sinclair Stammers–Science Source Library/Photo Researchers, Inc., br, NASA; 451, Eric Carle/Shostal Associates; CHAPTER SEVENTEEN: 455, *(t)*, S. L. Craig/Bruce Coleman Inc.,*(c)*, Mickey Palmer/Focus on Sports, *(b)*, Farrell Grehan/FPG; 478, John Lei/Omni Photo Communications; CHAPTER EIGHTEEN: 483, *(l)*, Ben and Miriam Rose/The Image Bank, *(tr)*, Don Katchusky/Taurus Photos; CHAPTER NINETEEN: 511, *(l)*, IBM Corp.; 512, Elaine Wicks/Taurus Photos; 513, *(t)*, © Art Attack, *(b)*, Photo Researchers, Inc.; 514, © Richard Hutchings; 517, Fred Lyons/Photo Researchers, Inc.; 525, Sepp Seitz/Woodfin Camp & Associates; 527, Russ Kinne/Comstock; 531, Bohdan Hrynewych/Stock, Boston; 533, Chris Luneski/Photo Researchers, Inc.